SAFETY SYMBOLS

SAFETY SYMBOLS	HAZARD	EXAMPLES	PRECAUTION	REMEDY
DISPOSAL	Special disposal procedures need to be followed.	certain chemicals, living organisms	Do not dispose of these materials in the sink or trash can.	Dispose of wastes as directed by your teacher.
BIOLOGICAL	Organisms or other biological materials that might be harmful to humans	bacteria, fungi, blood, unpreserved tissues, plant materials	Avoid skin contact with these materials. Wear mask or gloves.	Notify your teacher if you suspect contact with material. Wash hands thoroughly.
EXTREME TEMPERATURE	Objects that can burn skin by being too cold or too hot	boiling liquids, hot plates, dry ice, liquid nitrogen	Use proper protection when handling.	Go to your teacher for first aid.
SHARP OBJECT	Use of tools or glassware that can easily puncture or slice skin	razor blades, pins, scalpels, pointed tools, dissecting probes, broken glass	Practice common-sense behavior and follow guidelines for use of the tool.	Go to your teacher for first aid.
FUME	Possible danger to respiratory tract from fumes	ammonia, acetone, nail polish remover, heated sulfur, moth balls	Make sure there is good ventilation. Never smell fumes directly. Wear a mask.	Leave foul area and notify your teacher immediately.
ELECTRICAL	Possible danger from electrical shock or burn	improper grounding, liquid spills, short circuits, exposed wires	Double-check setup with teacher. Check condition of wires and apparatus.	Do not attempt to fix electrical problems. Notify your teacher immediately.
IRRITANT	Substances that can irritate the skin or mucous membranes of the respiratory tract	pollen, moth balls, steel wool, fiberglass, potassium permanganate	Wear dust mask and gloves. Practice extra care when handling these materials.	Go to your teacher for first aid.
CHEMICAL	Chemicals that can react with and destroy tissue and other materials	bleaches such as hydrogen peroxide; acids such as sulfuric acid, hydrochloric acid; bases such as ammonia, sodium hydroxide	Wear goggles, gloves, and an apron.	Immediately flush the affected area with water and notify your teacher.
TOXIC	Substance may be poisonous if touched, inhaled, or swallowed	mercury, many metal compounds, iodine, poinsettia plant parts	Follow your teacher's instructions.	Always wash hands thoroughly after use. Go to your teacher for first aid.
OPEN FLAME	Open flame may ignite flammable chemicals, loose clothing, or hair	alcohol, kerosene, potassium permanganate, hair, clothing	Tie back hair. Avoid wearing loose clothing. Avoid open flames when using flammable chemicals. Be aware of locations of fire safety equipment.	Notify your teacher immediately. Use fire safety equipment if applicable.

Eye Safety
Proper eye protection should be worn at all times by anyone performing or observing science activities.

Clothing Protection
This symbol appears when substances could stain or burn clothing.

Animal Safety
This symbol appears when safety of animals and students must be ensured.

Radioactivity
This symbol appears when radioactive materials are used.

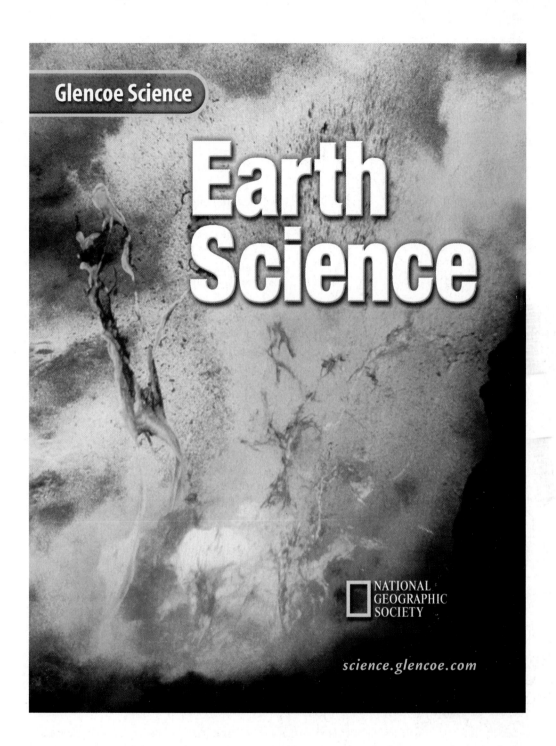

Glencoe Science

Earth Science

NATIONAL GEOGRAPHIC SOCIETY

science.glencoe.com

Glencoe McGraw-Hill

New York, New York   Columbus, Ohio   Woodland Hills, California   Peoria, Illinois

Glencoe Science

Glencoe Earth Science

Student Edition
Teacher Wraparound Edition
Interactive Teacher Edition CD-ROM
Interactive Lesson Planner CD-ROM
Lesson Plans
Content Outline for Teaching
Dinah Zike's Teaching Science with Foldables
Directed Reading for Content Mastery
Foldables: Reading and Study Skills
Assessment
    Chapter Review
    Chapter Tests
    ExamView Pro Test Bank Software
    Assessment Transparencies
    Performance Assessment in the Science Classroom
    The Princeton Review Standardized Test Practice Booklet
Directed Reading for Content Mastery in Spanish
Spanish Resources
English/Spanish Guided Reading Audio Program
Reinforcement

Enrichment
Activity Worksheets
Section Focus Transparencies
Teaching Transparencies
Laboratory Activities
Science Inquiry Labs
Critical Thinking/Problem Solving
Reading and Writing Skill Activities
Mathematics Skill Activities
Cultural Diversity
Laboratory Management and Safety in the Science Classroom
MindJogger Videoquizzes and Teacher Guide
Interactive Explorations and Quizzes CD-ROM with
    Presentation Builder
Vocabulary PuzzleMaker Software
Cooperative Learning in the Science Classroom
Environmental Issues in the Science Classroom
Home and Community Involvement
Using the Internet in the Science Classroom

Glencoe/McGraw-Hill

A Division of The McGraw·Hill Companies

Cover Images: Highly fluid lava erupting from a volcanic vent.

Send all inquiries to:
Glencoe/McGraw-Hill
8787 Orion Place
Columbus, OH 43240

ISBN 0-07-823718-1
Printed in the United States of America.
2 3 4 5 6 7 8 9 10 027/043 06 05 04 03 02 01

Authors

National Geographic Society
Education Division
Washington, D.C.

Ralph M. Feather Jr., PhD
Science Department Chair
Derry Area School District
Derry, Pennsylvania

Susan Leach Snyder
Earth Science Teacher, Consultant
Jones Middle School
Upper Arlington, Ohio

Dinah Zike
Educational Consultant
Dinah-Might Activities, Inc.
San Antonio, Texas

Series Reading Consultants

Elizabeth Babich
Special Education Teacher
Mashpee Public Schools
Mashpee, Massachusetts

Barry Barto
Special Education Teacher
John F. Kennedy Elementary School
Manistee, Michigan

Carol A. Senf, PhD
Associate Professor of English
Georgia Institute of Technology
Atlanta, Georgia

Rachel Swaters
Science Teacher
Rolla Middle Schools
Rolla, Missouri

Nancy Woodson, PhD
Professor of English
Otterbein College
Westerville, Ohio

Series Math Consultants

Michael Hopper, DEng
Manager of Aircraft Certification
Raytheon Company
Greenville, Texas

Teri Willard, EdD
Department of Mathematics
Montana State University
Belgrade, Montana

Series Safety Consultants

Malcolm Cheney, PhD
OSHA Chemical Safety Officer
Hall High School
West Hartford, Connecticut

Aileen Duc, PhD
Science II Teacher
Hendrick Middle School
Plano, Texas

Sandra West, PhD
Associate Professor of Biology
Southwest Texas State University
San Marcos, Texas

Content Consultants

Michelle Anderson
Community Faculty
Marion Technical College
Marion, Ohio

Stephen M. Letro
National Weather Service
Meteorologist In Charge
Jacksonville, Florida

William C. Keel, PhD
Department of Physics and Astronomy
University of Alabama
Tuscaloosa, Alabama

Homer Montgomery, PhD
Department of Geosciences
University of Texas at Dallas
Richardson, Texas

Linda Knight
Associate Director
Rice Model Science Lab
Houston, Texas

Robert Nierste
Science Department Head
Hendrick Middle School
Plano, Texas

Series Activity Testers

José Luis Alvarez, PhD
Math and Science Mentor Teacher
Ysleta ISD
El Paso, Texas

Mary Helen Mariscal-Cholka
Science Teacher
William D. Slider Middle School
Socorro ISD
El Paso, Texas

Nerma Coats Henderson
Teacher
Pickerington Jr. High School
Pickerington, Ohio

José Alberto Marquez
TEKS for Leaders Trainer
Ysleta ISD
El Paso, Texas

Science Kit and Boreal Laboratories
Tonawanda, New York

Reviewers

William Blair
J. Marshall Middle School
Billerica, Massachusetts

Lois Burdette
Green Bank Elementary/Middle School
Green Bank, West Virginia

Marcia Chackan
Pine Crest School
Boca Raton, Florida

Anthony DiSipio
Octorana Middle School
Atglen, Pennsylvania

Mary Ferneau
Westview Middle School
Goose Creek, South Carolina

Connie Cook Fontenot
Bethune Academy
Houston, Texas

Annette Garcia
Kearney Middle School
Commerce City, Colorado

Nerma Coats Henderson
Pickerington Jr. High School
Pickerington, Ohio

Michael Mansour
John Page Middle School
Madison Heights, Michigan

Sharon Mitchell
William D. Slider Middle School
El Paso, Texas

Joanne Stickney
Monticello Middle School
Monticello, New York

CONTENTS IN BRIEF

Contents

CONTENTS

Contents

CONTENTS

UNIT 3 Earth's Internal Processes—272

CONTENTS

CHAPTER 12

Volcanoes—332

UNIT 4

Change and Earth's History—364

CHAPTER 13

Clues to Earth's Past—366

 UNIT 5 Earth's Air and Water—430

CONTENTS

CONTENTS

UNIT 7 Astronomy—636

CHAPTER 22

Exploring Space—638

CONTENTS

xvii

CONTENTS

NATIONAL GEOGRAPHIC Unit Openers

NATIONAL GEOGRAPHIC VISUALIZING

Interdisciplinary Connections

Full Period Labs

Activities

Mini LAB

Feature Contents

Activities

Feature Contents

Math Skills Activities

Skill Builder Activities

Science

Science Connections

SCIENCE *Online*

Collect Data: 76, 222, 622, 652, 677, 681
Data Update: 170, 448, 471, 585, 659
Research: 9, 17, 40, 48, 78, 96, 100, 125, 133, 159, 187, 199, 244, 248, 277, 286, 311, 320, 341, 351, 377, 380, 410, 415, 436, 474, 507, 509, 527, 535, 551, 562, 616, 657, 675, 703, 712, 743, 748

THE PRINCETON REVIEW

31, 59, 87, 117, 149, 150–151, 181, 209, 237, 269, 270–271, 301, 331, 361, 362–363, 395, 427, 428–429, 459, 489, 519, 547, 577, 578–579, 607, 633, 634–635, 669, 699, 733, 765, 766–767

How Are
Rocks &
Fluorescent Lights
Connected?

Around 1600, an Italian cobbler found a rock that contained a mineral that could be made to glow in the dark. The discovery led other people to seek materials with similar properties. Eventually, scientists identified many fluorescent and phosphorescent (fahs fuh RE sunt) substances—substances that react to certain forms of energy by giving off their own light. As seen above, a fluorescent mineral may look one way in ordinary light (front), but may give off a strange glow (back) when exposed to ultraviolet light. In the 1850s, a scientist wondered whether the fluorescent properties of a substance could be harnessed to create a new type of lighting. The scientist put a fluorescent material inside a glass tube and sent an electric charge through the tube, creating the first fluorescent lamp. Today, fluorescent lightbulbs are widely used in office buildings, schools, and factories.

SCIENCE CONNECTION

FLUORESCENT MINERALS Some minerals fluoresce—give off visible light in various colors—when exposed to invisible ultraviolet (UV) light. Using library resources or the Glencoe Science Web site at **science.glencoe.com**, find out more about fluorescence in minerals. Write a paragraph that answers the following questions: Would observing specimens under UV light be a reliable way for a geologist to identify minerals? Why or why not?

The Nature of Science

Inside the chest of a small dinosaur nicknamed Willo is something amazing—what appears to be a heart preserved as stone. An X-ray scan of this 66-million-year-old skeleton shows possible heart structures inside the stone. Scientists still are debating whether this clump of stone is a preserved heart, and more research will be necessary. But that's the nature of science.

What do you think?

Science Journal Look at the photo below. Discuss what you think this might be. Here's a hint: *The one Willo came from may have looked a lot like these.* Write your answer or best guess in your Science Journal.

EXPLORE
ACTIVITY

How big is big? How small is small? Big and small are words people use a lot. But, the meaning of these words depends on your experiences and what you are describing. Your description could be confusing to another person who hasn't had the same experiences you've had. Early in human history, people developed ways to measure things. In the following activity, try some of these measuring devices.

Measure in SI

1. Using only your hands and fingers as measuring devices, measure the length and width of the cover of this book.

2. Compare your measurements with those of other students.

3. Using a metric ruler, repeat the measurement process.

4. Again, compare your measurements with the measurements of other students in the classroom.

Observe

In your Science Journal, infer and describe several advantages of using standardized measuring devices.

Before You Read

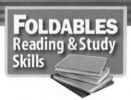

FOLDABLES
Reading & Study
Skills

Making a Vocabulary Study Fold Knowing the definition of vocabulary words is a good way to ensure you understand the content of the chapter.

1. Place a sheet of notebook paper in front of you so the short side is at the top and the holes are on the right side. Fold the paper in half from the left side to the right side.

2. Through the top thickness of paper, cut along every third line from the outside edge to the centerfold, forming tabs.

3. Before you read the chapter, write vocabulary words from each section in this chapter on the front of the tabs. Under each tab, write what you think the word means.

4. As you read the chapter, add to and correct your definitions.

Science All Around

What You'll Learn

- **Describe** scientific methods.
- **Define** science and Earth science.
- **Distinguish** among independent variables, dependent variables, constants, and controls.

Vocabulary

hypothesis
scientific methods
science
Earth science
variable
independent variable

constant
dependent
    variable
control
technology

Why It's Important

Scientific methods are used every day when you solve problems.

Mysteries and Problems

Scientists are often much like detectives trying to solve a mystery. One such mystery occurred in 1996 when Japanese scientists were looking through historical records. They reported finding accounts of a tsunami that had smashed the coast of Honshu Island on January 27, 1700. That led to the question: What had triggered these huge ocean waves?

The Search for Answers The scientists suspected that an earthquake along the coast of North America was to blame. From the coast of British Columbia to northern California is an area called the Cascadia subduction zone, shown in **Figure 1.** A subduction zone is where one section of Earth's outer, rigid layer, called a plate, is sinking beneath another plate. In areas like this, earthquakes are common. However, one problem remained. Based on the size of the tsunami, the earthquake had to have been an extremely powerful one, sending waves rolling all the way across the Pacific Ocean. That would be a much stronger earthquake than any known to have occurred in the area. Could evidence be found for such a large earthquake?

Figure 1
Along the Cascadia subduction zone, the Juan de Fuca Plate is sinking under the North American Plate.

Gathering Evidence Evidence of a large earthquake in the distant past did seem to exist along the coasts of Washington and Oregon. Much of the coast in that area had sunk, submerging coastal forests and killing thousands of trees. However, dating the earthquake to a specific year would be difficult.

A Possible Solution One scientist, whose field of study was tree rings, thought he knew how the earthquake could be dated. He made an educated guess, called a **hypothesis,** that tree rings in the drowned trees could be used to determine when the earthquake occurred.

✔ **Reading Check** *What is a hypothesis?*

The hypothesis was based on what scientists know about tree growth. Each year, a living tree makes a new ring of tissue in its trunk, called an annual growth ring. You can see the annual rings in the cross section of a tree trunk shown in **Figure 2.** Two groups of scientists analyzed the rings in drowned trees, like the remains of cedar trees shown in **Figure 3.** Their data showed that the trees had died or were damaged after August 1699 but before the spring growing season of 1700. That evidence put the date of the earthquake in the same time period as the tsunami on Honshu island.

Importance of Solving the Mystery In addition to solving the mystery of what caused the tsunami, the tree rings also provided a warning for people living in the Pacific Northwest. Earthquakes much stronger than any that have occurred in modern times are possible. Scientists warn that it's only a matter of time until another huge quake occurs.

Figure 2
You can see the growth rings in this tree trunk. *How much time does each ring represent?*

Figure 3
Growth rings from these and other trees linked a huge earthquake along the coast of Washington to a tsunami in Japan that occurred more than 300 years ago.

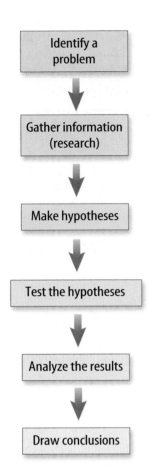

Figure 4
By using scientific methods, you can solve many problems.

Identify a problem

Gather information (research)

Make hypotheses

Test the hypotheses

Analyze the results

Draw conclusions

Scientific Methods

When scientists try to solve a mystery like what caused the tsunami in Japan in 1700, they perform problem-solving procedures called **scientific methods.** As shown in **Figure 4,** some of the scientific methods they use include identifying a problem, gathering information (researching), developing hypotheses, testing the hypotheses, analyzing the results, and drawing conclusions. When you use methods like these, you are solving problems in a scientific way.

Science

Science means "having knowledge." **Science** is a process of observing, studying, and thinking about things in your world to gain knowledge. Many observations can't be explained easily. When people can't explain things, they ask questions. For example, you might observe that the sky appears to be blue during the day but often appears to be red at sunset and sunrise. You might ask yourself why this happens. You might visit or see a picture of Devils Postpile in California, shown in **Figure 5,** and notice that the dark rock is divided into long, thin, six-sided columns. Many fallen columns lie at the base of this mass of rock. You might wonder how and when this strange-looking rock formed. You also might wonder why rocks can be smooth or rough, shiny or dull, and can be so many different colors. Science involves trying to answer questions and solve problems to better understand the world. Every time you attempt to find out how and why things look and behave the way they do, you are performing science.

✓ **Reading Check**   *What is science?*

Figure 5
The columns in Devils Postpile rise between 12 m and 18 m from the valley floor. This unusual formation was created when hot lava cooled and cracked.

Earth Science Science is divided into different areas of study. The kind of science you will learn about this year is Earth science. **Earth science** is the study of Earth and space. Some Earth science topics include rocks, minerals, soil, volcanoes, earthquakes, maps, fossils, mountains, climates, weather, ocean water, and objects in space. Some of these topics are represented in **Figure 6.** Much of the information you'll learn about has been discovered over the ages by people who conducted scientific tests or investigations. However, many unanswered questions remain and much more is waiting to be discovered.

✔ **Reading Check** *What topics do Earth scientists study?*

Working in the Lab

Testing, or experimenting, is an important part of science, and if you really want to learn from an investigation, it must be carefully designed. Suppose that after listening to advertisements for several dishwashing liquids, you want to know which brand of dishwashing liquid cleans dishes the best. To find the answer, you would need to do some library or Internet research on dishwashing liquids. After researching, several thoughts might go through your mind. For example, you might hypothesize that brand X will clean dishes better than any other brand. You also might consider that there would be no difference in how well the different liquids clean.

Next, you would design an experiment that tests the validity of your hypotheses. You would need to think about which dishwashing liquids you would test, the amount of each dishwashing liquid you would use, the temperature of the water, the number of dishes you would wash, the kind and amount of grease you would put on the dishes, and the brand of paper towels you would use. All these factors can affect the outcome.

Research Visit the Glencoe Science Web site at **science.glencoe.com** for information about the different areas of Earth science. Prepare a collage that illustrates what you learn.

Figure 7

Figure 7
Wiping each dish in the same manner with a different paper towel is an important constant. *Why?*

Life Science
INTEGRATION

Suppose you wanted to design an experiment to find out what kind of soil is best for growing cactus plants. What would be your variables and constants in the experiment?

Variables and Constants The different factors that can change in an experiment are **variables.** However, you want to design your experiment so you test only one variable at a time. The variable you want to test is the brand of dishwashing liquid. This is called the **independent variable**—the variable that changes. **Constants** are the variables that do not change in an experiment. Constants in this experiment would be the amount of dishwashing liquid used, the amount of water, the water temperature, the number of dishes, the kind and amount of grease applied to each dish, the brand of paper towels that were used, and the manner in which each dish was wiped. For example, you might use 20 equally greasy dishes that are identical in size, soaked in 20 L of hot water (30°C) to which 10 mL of dishwashing liquid have been added. You might rub each dish with a different dry paper towel of the same brand after it has soaked for 20 min and air dried, as the student in **Figure 7** is doing. If grease does not appear on the towel, you would consider the dish to be clean. The amount of grease on the towel is a measure of how clean each dish is and is called the dependent variable. A **dependent variable** is the variable being measured.

Controls Many experiments also need a control. A **control** is a standard to which your results can be compared. The control in your experiment is the same number of greasy dishes, placed in 20 L of hot water except that no dishwashing liquid is added to the water. These dishes also are allowed to soak for 20 min and air dry. Then they are wiped with paper towels in the same manner as the other dishes were wiped.

✔ **Reading Check**   *Why is a control used in an experiment?*

Repeating Experiments For your results to be valid or reliable, your tests should be repeated many times to see whether you can confirm your original results. For example, you might design your experiment so you repeat the procedures five times for each different dishwashing liquid and control. Also, the number of samples being tested should be large. That is why 20 plates would be chosen for each test of each dishwashing liquid. The control group also would have 20 plates. By repeating an experiment five times, you can be more confident that your conclusions are accurate because your total sample for each dishwashing liquid would be 100 plates. If something in an experiment occurs just once, you can't base a scientific conclusion on it. However, if you can show that brand X cleans best in 100 trials under the same conditions, then you have a conclusion you can feel confident about.

Testing After you have decided how you will conduct an experiment, you can begin testing. During the experiment, you should observe what happens and carefully record your data in a table like the one in **Figure 8.** Your final step is to draw your conclusions. You analyze your results and try to understand what they mean.

When you are making and recording observations, be sure to include any unexpected results. Many discoveries have been made when experiments produced unexpected results.

TRY AT HOME
Mini LAB

Designing an Experiment

Procedure
1. Design an experiment to test the question: *Which flashlight battery lasts the longest?*
2. In your design, be sure to include detailed steps of your experiment.
3. Identify the independent variable, constants, dependent variable, and control.

Analysis
1. List the equipment you would need to do your experiment.
2. Explain why you should repeat the experiment.

Figure 8
Arranging your data in a table makes the information easier to understand and analyze.

Technology

Science doesn't just add to the understanding of your natural surroundings, it also allows people to make discoveries that help others. Science makes the discoveries, and technology puts the discoveries to use. **Technology** is the use of scientific discoveries for practical purposes.

When people first picked up stones to use as tools or weapons, the age of technology had started. The discovery of fire and its ability to change clay into pottery or rocks into metals made the world you live in possible. Think back to the Explore Activity at the beginning of this chapter. Measuring devices like the metric ruler you used are examples of technology.

Everywhere you look, you can see ways that science and technology have shaped your world. Look at **Figure 9** to see how many examples of technology you can identify in each of the pictures. **Figure 10** shows a time line of some important examples of technology used in Earth science. Notice how different cultures have added to discoveries and inventions over the centuries.

Figure 9
Examples of technology are all around you. *What are some ways these examples affect your life?*

A Machines and materials used in construction are examples of technology.

B Cities rely on technology. *How might life in this city be different without it?*

C A modern kitchen has many examples of technology. *How many can you identify?*

Figure 10

For thousands of years, discoveries made by people of many cultures have advanced the study of Earth. This time line shows milestones and inventions that have shaped the development of Earth science technology and led to a greater understanding of the planet and its place in the universe.

10,000 B.C.: First pottery (Japan)

7000 B.C.: Copper metalworking (Turkey)

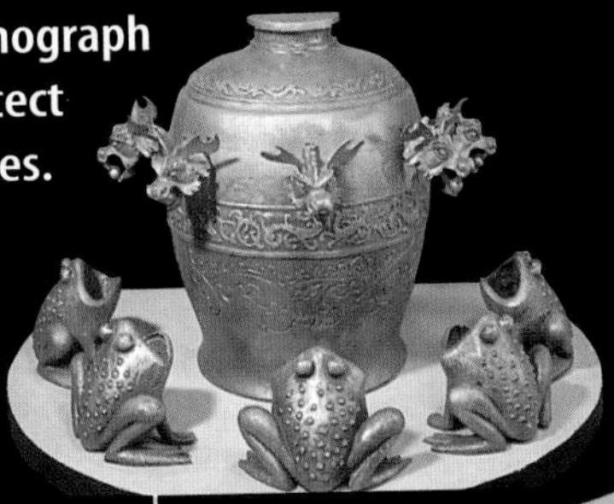

A.D. 132 (CHINA) This early seismograph helped detect earthquakes.

3500 B.C.: Bronze tools and weapons (Mesopotamia)

1000 B.C.

A.D. 100

900: Terraced field for soil conservation (Peru)

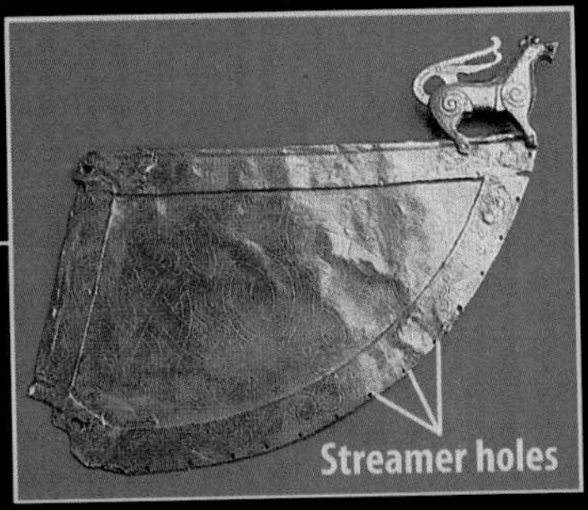

650: Windmill (Persia)

80 B.C.: Astronomical calendar (Greece)

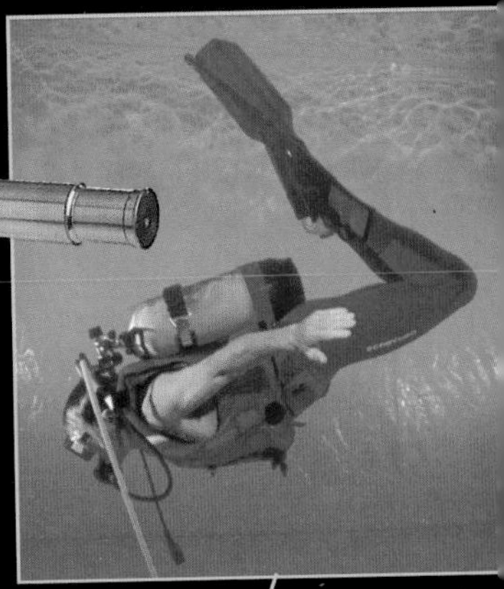

1943 (FRANCE) Breathing from tanks of compressed air lets divers move underwater without being tethered to an air source at the surface.

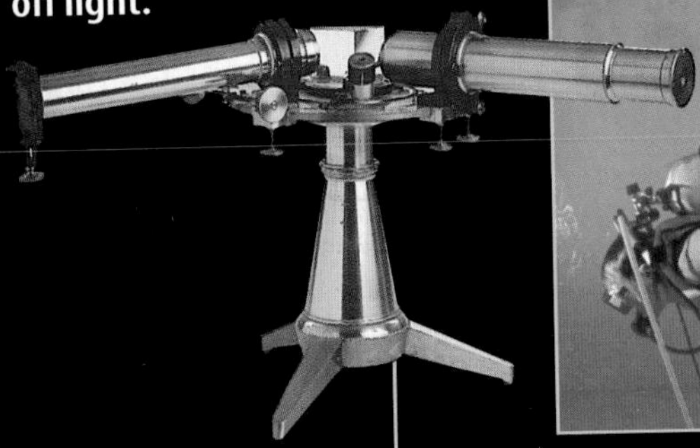

1814 (GERMANY) A spectroscope allowed scientists to determine which elements are present in an object or substance that is giving off light.

1090: Magnetic compass (Arabia, China)

Streamer holes

1000 (NORWAY) Streamers tied through holes in Viking wind vanes indicated wind direction and strength.

1880: Modern seismograph (England)

1538: Diving bell (Spain)

1592: Thermometer (Italy)

1957: Space satellite (former U.S.S.R.)

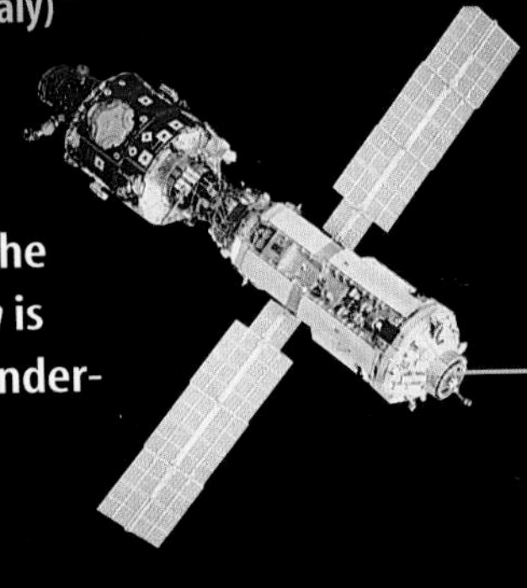

1998–2006: *INTERNATIONAL SPACE STATION* With participants from 16 countries, the *International Space Station* is helping scientists better understand Earth and beyond.

1926: Liquid-fuel rocket (United States)

2000

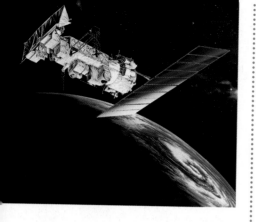

Using Technology Most people immediately think of huge and exotic inventions when the word *technology* is mentioned. However, the use of scientific knowledge has resulted in such common yet important things as paper, can openers, buckets, aspirin, rubber boots, locks and keys, microfiber clothing, ironing boards, bandages, and scissors. It also has resulted in robots that check underwater oil rigs for leaks and others that manufacture cars. Technology also includes calculators and computers that process information.

Transferable Technology Technology is a natural outcome of using scientific knowledge to solve problems and make people's lives easier and better. The wonderful thing about technology is that it is transferable. That means that it can be applied to new situations. For example, many types of technology that are now common were originally developed for use in outer space.

Scientists developed robotic parts, new fibers, and microminiaturized instruments for spacecraft and satellites. After these materials were developed, many were modified for use here on Earth. Technology that once was developed by the military, such as radar and sonar, has applications in the study of space, weather, Earth's structures, and medicine.

Earth scientists rely on information from weather satellites like the one in **Figure 11** to gather weather data. But biologists also use satellites to track animals. A tiny radio transmitter attached to an animal sends signals up to a satellite. The satellite then sends data on the animal's location to a ground station. Some researchers use the data to track bird migration.

Figure 11
Weather satellites help forecasters predict future storms. *How might this same technology be used to protect endangered species?*

Section 1 Assessment

1. What is a hypothesis? Why must hypotheses be testable?

2. What is Earth science?

3. Explain why it is important that scientists perform an experiment more than one time.

4. Explain why it is important to use constants in an experiment.

5. **Think Critically** Name an example of technology you use every day and, as best you can, describe the science behind the technology.

Skill Builder Activities

6. **Comparing and Contrasting** Compare and contrast an independent variable and a dependent variable. **For more help, refer to the** Science Skill Handbook.

7. **Communicating** In your Science Journal, write a paragraph about how you would try to describe a modern device such as a TV, microwave oven, or computer to someone living in 1800. **For more help, refer to the** Science Skill Handbook.

2 Scientific Enterprise

A Work in Progress

Throughout time, people have been frightened by and curious about their surroundings all at the same time. Storms, erupting volcanoes, comets, seasonal changes, and other natural phenomena fascinated people thousands of years ago, and they fascinate people today. As shown in **Figure 12,** early people relied on mythology to explain what they observed. They believed that mythological gods were responsible for creating storms, causing volcanoes to erupt, causing earthquakes, bringing the seasons, and making comets appear in the sky.

Recording Observations Some early civilizations went so far as to record what they saw. They developed calendars that described natural recurring phenomena. Six thousand years ago, Egyptian farmers observed that the Nile River flooded their lands every summer. Their crops had to be planted at the right time in order to make use of this water. The farmers noticed that shortly before flood time, the brightest star in the sky, Sirius, appeared at dawn in the east. The Egyptians developed a calendar based on the appearance of this star, which occurred about every 365 days.

Later, civilizations created instruments to measure with. As you saw in the Explore Activity, instruments allow for precise measurements. As instruments became better, accuracy of observations improved. While observations were being made, people tried to reason why things happened the way they did. They made inferences, or conclusions, to help explain things. Some people developed hypotheses that they tested. Their experimental conclusions allowed them to learn even more.

As You Read

What You'll Learn

■ **Explain** why science is always changing.
■ **Compare and contrast** scientific theories and scientific laws.
■ **Discuss** the limits of science.

Vocabulary

scientific theory   ethics
scientific law   bias

Why It's Important

Science helps you understand the world around you.

Figure 12
Early Scandinavian and Germanic peoples believed that a god named Thor controlled the weather. In this drawing, Thor is creating a storm. Lightning flashed whenever he threw his heavy hammer.

The History of Meteorology

Today, scientists know what they know because of all the knowledge that has been collected over time. The history of meteorology, which is the study of weather, illustrates how an understanding of one area of Earth science has developed over time.

Weather Instruments As you have read, ancient peoples believed that their gods controlled weather. However, even early civilizations observed and recorded some weather information. The rain gauge was probably the first weather instrument. The earliest reference to the use of a rain gauge to record the amount of rainfall appears in a book by the ruler of India from 321 B.C. to 296 B.C.

It wasn't until the 1600s that scientists in Italy began to use instruments extensively to study weather. These instruments included the barometer—to measure air pressure; the thermometer—to measure temperature, shown in **Figure 13;** the hygrometer—to measure water vapor in the air; and the anemometer—to measure wind speed. With these instruments, the scientists set up weather stations across Italy.

 **Reading Check** *What instruments were developed in the 1600s to study weather?*

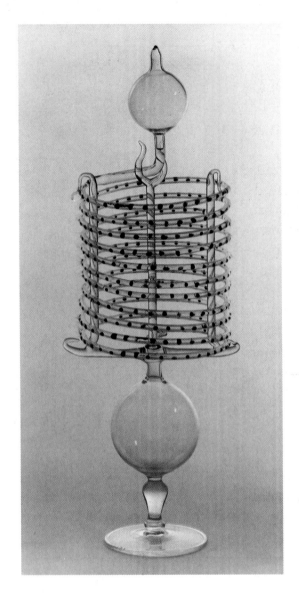

Figure 13
This photo shows a replica of a 1660 Italian alcohol thermometer.

Weather Prediction in the United States Benjamin Franklin was the first American to suggest that weather could be predicted. Franklin read accounts of storms from newspapers across the country. From these articles, Franklin concluded that severe storms generally move across the country from west to east. He also concluded that observers could monitor a storm and notify those ahead of its path that it was coming. Franklin's ideas were finally put to practical use shortly after the telegraph was invented in 1837.

By 1849, an organized system of weather observation sites was set up and weather reports from volunteer weather observers were sent by telegraph to the Smithsonian Institution. In 1850, Joseph Henry, secretary of the Smithsonian Institution in the United States, began drawing maps from the weather data he received. A large weather map was displayed at the Institution and a weather report was sent to the *Washington Evening Post* to be published in the newspaper.

National Weather Service By the late 1800s, the United States Weather Bureau was functioning with more than 350 observing sites across the country. By 1923, weather forecasts were being carried by 140 radio stations across the United States. Finally in 1970, the Bureau name was changed to the National Weather Service and it became part of the National Oceanic and Atmospheric Administration (NOAA).

Today's weather is forecast using orbiting satellites, weather balloons, radar, and other sophisticated technology. Each day about 60,000 reports from weather stations, ships, aircraft, and radar transmitters are gathered and filed. **Figure 14** shows instruments used to gather data at a weather station. All the information gathered is compiled into a report that is distributed to radio stations, television networks, and other news media.

Today, if you want to know about the weather anywhere in the world—at any time of day or night—you could watch a television weather channel, listen to a radio news station, or check an internet site. If you live in an area that has tornadoes, hurricanes, or other severe weather conditions, you know it is important to have weather watches and warnings available to your community.

SCIENCE *Online*

Research Visit the Glencoe Science Web site at **science.glencoe.com** for more information about weather forecasting. Communicate to your class what you learn.

Figure 14
Some weather stations are operated by meteorologists, but many are now automated. Data from automated stations are transmitted to a central

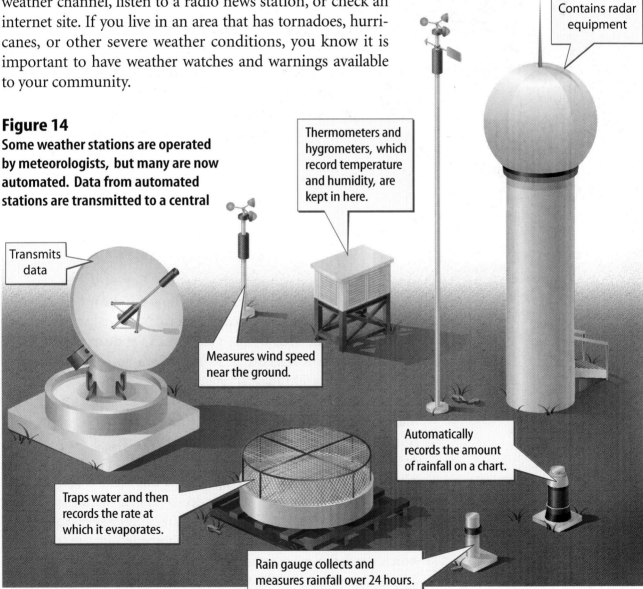

Contains radar equipment

Thermometers and hygrometers, which record temperature and humidity, are kept in here.

Transmits data

Measures wind speed near the ground.

Automatically records the amount of rainfall on a chart.

Traps water and then records the rate at which it evaporates.

Rain gauge collects and measures rainfall over 24 hours.

Continuing Research

Scientific knowledge continues to change as scientists develop better instruments and testing procedures. As it changes, scientists have a greater understanding of nature. As you saw in **Figure 14,** scientists use a variety of technologies to study weather. Scientists have similar technologies to study Earth's interior, the oceans, environmental problems, and space. How could the technology shown in **Figure 15** be used by Earth scientists?

It is impossible to predict the types of instruments scientists will have in the future. But it is easy to predict that as research continues and instruments improve, knowledge will grow. Perhaps one day you will make a scientific breakthrough that changes people's understanding of the world.

Scientific Theories As you learned earlier, scientists test hypotheses. If data gathered over a long period of time support a hypothesis, scientists become convinced that the hypothesis is useful. They use results from many scientists' work to develop a scientific theory. A **scientific theory** is an explanation or model backed by results obtained from many tests or experiments.

Figure 15
The Global Positioning System (GPS) can pinpoint a person's location on Earth. A radio receiver gets signals from several orbiting *Navstar* satellites like this one. By comparing how far the receiver is from each satellite, the receiver's position can be determined and displayed.

✔ Reading Check *How can a scientific hypothesis become a scientific theory?*

Examine how one hypothesis became a theory. Comets once were believed to be the forecasters of disaster. People often were terrified yet fascinated by the ghostly balls appearing in the sky. Slowly over the years, comets lost much of their mystery. However, from the 1800s until 1949, most scientists hypothesized that comets were made of many particles of different kinds of materials swarming in a cluster. Based on this hypothesis, a comet was described as a swirling cloud of dust.

In 1949, American astronomer Fred L. Whipple proposed a hypothesis that a comet was more like a dirty snowball—that the nucleus of a comet contains practically all of a comet's mass and consists of ices and dust. If a comet's orbit brings it close to the Sun, the heat vaporizes some of the ices, releasing dust and gas, which form the comet's tail. Dr. Whipple's hypothesis was published in the March 1950 *Astrophysical Journal.*

Hypothesis Supported Before it became an accepted theory, Dr. Whipple's hypothesis was subjected to many years of tests and observations. Some of the most important were the 1986 observations of Halley's comet, shown in **Figure 16.** A group of astronomers from the University of Arizona, headed by Dr. Susan Wyckoff, studied the composition of the comet. Dr. Wyckoff observed the comet many times, using giant telescopes in Arizona and Chile in South America. At other times, she studied the observations of other astronomers, including those who studied data collected by the *Giotto* and other spacecrafts. All these observations and data supported Dr. Whipple's original hypothesis. With so much support, Dr. Whipple's hypothesis has become an accepted scientific theory.

Scientific Laws A **scientific law** is a rule that describes the behavior of something in nature. Usually, a scientific law describes what will happen in a given situation but doesn't explain why it happens. An example of a scientific law is Newton's first law of motion. According to this law, an object, such as a marble or a spacecraft, will continue in motion or remain at rest until it's acted upon by an outside force. According to Newton's second law of motion, when a force acts on an object, the object will change speed, direction, or both. Finally, according to Newton's third law, for every action, there is an equal and opposite reaction. This law explains how rockets that are used to launch space probes to study Halley's comet and other objects in space work. When a rocket forces burning gases out of its engines, the gases push back on the rocket with a force of equal strength and propel the rocket forward.

Figure 16
The view of Halley's comet from the *Giotto* spacecraft allowed scientists to determine the size of the icy nucleus, and that the nucleus was covered by a black crust of dust. Jets of gas blasted out from holes in the crust to form the comet's tail.

Figure 17
Ethical questions can't be solved
by using scientific methods.

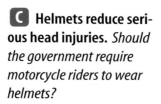

B Disease-carrying
mosquitoes can live
in this swamp. *Should
swamps be drained,
even if other species
lose their habitat?*

A These animals live on
the African plains. *Should
they be hunted as trophies?*

C Helmets reduce serious head injuries. *Should
the government require
motorcycle riders to wear
helmets?*

Health
INTEGRATION

When doing science proj-
ects, you must treat ani-
mals and human subjects
responsibly and safely.
You should not hurt ani-
mals, subject them to any
kind of stress, or embar-
rass or put people at any
kind of physical or emo-
tional risk. Why is the
health of your subjects an
important ethical concern?

Limits of Science

Will science always provide answers to all your questions?
No, science doesn't have answers to all the questions and problems in the universe. Science is limited in what it can explain.
For a question or problem to be scientifically studied, there
must be variables that can be observed, measured, and tested.
Problems that deal with ethics and belief systems cannot be
answered using these methods. **Ethics** deals with moral values
about what is good or bad. Belief systems deal with religious
and/or other beliefs. Examples of ethical and belief-system questions that science cannot answer are: Do humans have more
value on Earth than other life-forms?, Should the federal government regulate car emissions?, and Should animals be used in
medical experiments? Look at **Figure 17.** What's your opinion?

Reading Check *Why can't science be used to answer ethical
questions?*

Doing Science Right

Although ethical questions cannot be answered by science,
there are ethical ways of doing science. The correct approach to
doing science is to perform experiments in a way that honestly
tests hypotheses and draws conclusions in an unbiased way.

Being Objective When you do scientific experiments, be sure that you design your experiments in a way that you objectively test your hypotheses. If you don't, your **bias,** or personal opinion, can affect your observations. For example, in the 1940s, Soviet scientist Trofim Lysenko believed that individuals of the same species would not compete with one another. His ideas were based on the political beliefs held in the Soviet Union at that time. Based on his personal opinion, Lysenko ordered 300,000 tree seedlings planted in groups in a reforestation project. He believed that the trees in each group would aid one another in competing against other plant species. However, the area where the trees were planted was extremely dry and swept by constant winds. Without enough water and nutrients, most of the trees died and Lysenko's personal opinion was not supported. It turned out to be a costly experiment for the Soviet government.

Suppose you wanted to grow as many plants as possible in a single flowerpot. Would you assume that all of the plants in the pot shown in **Figure 18** could survive, or would you set up an experiment to objectively test this hypothesis? Unless you test various numbers of plants in pots under the same conditions, you could not make a valid conclusion.

Figure 18
These seedlings are crowded into a single pot. *How many do you think could survive?*

Problem-Solving Activity

How can bias affect your observations?

Do you think bias can affect a person's observations? With the help of her classmates, Sharon performed an experiment to find out.

Identifying the Problem

Sharon showed ten friends a photograph of an uncut amethyst and asked them to rank the quality of color from 1 to 10. She then wrote the words "prize amethyst" on top of the photo and asked ten more friends to rank the quality of color.

Solving the Problem

1. Examine the tables. Do you think the hint affected the way Sharon's classmates

Rankings Without Hint		Rankings With Hint	
5	7	7	8
4	5	8	9
6	4	9	8
5	6	10	8
5	3	7	9
Average: 5.0		**Average: 8.3**	

rated the amethyst? What effect did the hint have on them?

2. Do you think bias could affect the results of a scientific experiment? Explain. How could this bias be prevented?

Figure 19
Scientists take detailed notes of procedures and observations when they do science experiments. *Why should you do the same thing?*

Being Ethical and Open People who perform science in ethical and unbiased ways keep detailed notes of their procedures like the scientists shown in **Figure 19.** Their conclusions are based on precise measurements and tests. They communicate their discoveries by publishing their research in journals or presenting reports at scientific meetings. This allows other scientists to examine and evaluate their work. Scientific knowledge advances when people work together. Much of the science you know today has come about because of the collaboration of investigations done by many different people over many years.

The opposite of ethical behavior in science is fraud. Scientific fraud involves dishonest acts or statements. Fraud could include such things as making up data, changing the results of experiments, or taking credit for work done by others.

Section Assessment

1. Why is science always changing?
2. How can a hypothesis be supported?
3. What is the difference between a scientific theory and a scientific law?
4. What kinds of questions can't be answered by science?
5. **Think Critically** When reading science articles in newspapers, magazines, or on the Internet, why should you look for the authors' biases?

Skill Builder Activities

6. **Drawing Conclusions** Describe what would have happened if the 1986 observations of Halley's comet had not supported Dr. Whipple's original hypothesis. **For more help, refer to the Science Skill Handbook.**
7. **Using a Word Processor** Use a computer to write a paragraph explaining how fraud could damage science. **For more help, refer to the Technology Skill Handbook.**

Activity

Understanding Science Articles

You know that scientists conduct investigations to learn things about our world. It is important for researchers to share what they learn so other researchers can repeat and expand upon their results. One important way that scientific results are shared is by publishing them in journals and magazines. How can you learn about recent work by scientists? How do you interpret information that you read?

What You'll Investigate

What information about Earth science and scientific methods can you learn by reading an appropriate magazine article?

Materials

magazine articles about Earth science topics

Goals

- **Obtain** a recent magazine article concerning a research topic in Earth science.
- **Identify** aspects of science and scientific methods in the article.

Procedure

1. Locate a recent magazine article about a topic in Earth science research.
2. Read the article, paying attention to details that are related to science, research, and scientific methods.
3. What branch of Earth science does the article discuss?
4. **Describe** what the article is about. Does it describe a particular event or discuss more general research?
5. Are the names of any scientists mentioned? If so, what was the role of each in the research being discussed?

6. Are particular hypotheses being tested? If so, is the research project complete or is it still continuing?
7. **Describe** how the research is conducted. What is being measured? What observations are recorded?

Conclude and Apply

1. Are data available that do or do not support any hypotheses? Explain.
2. What do other scientists think about the research?
3. Are references provided that tell you where you can find more information about this particular research or the more general topic? If not, what are some sources where you might locate more information?

*C*ommunicating
Your Data

Prepare an oral report on the article you read. Present your report to the class. **For more help, refer to the** Science Skill Handbook.

Testing Variables of a Pendulum

A pendulum is an old, but accurate timekeeping device. It works because of two natural phenomena—gravity and inertia—that are important in the study of Earth science. Gravity makes all objects fall toward Earth's surface. Inertia makes matter remain at rest or in motion unless acted upon by an external force. In the following activity, you will test some variables that might affect the swing of a pendulum.

What You'll Investigate

How do the length of a pendulum, the attached mass, and the angle of the release of the mass affect the swing of a pendulum?

Materials

string (60 cm)
metal washers (5)
watch with a second hand
metric ruler
paper clip
protractor

Goals

- **Manipulate** variables of a pendulum.
- **Draw** conclusions from experimentation with pendulums.

Safety Precautions

Table 1 Length of the Pendulum

Length of String (cm)	Swings Per Minute		
	Trial 1	Trial 2	Average
10			
20			
30			
40			
50			

Table 2 Amount of Mass on the Pendulum

Units of Mass	Swings Per Minute		
	Trial 1	Trial 2	Average
1			
2			
3			
4			
5			

Table 3 Angle of the Release of the Mass

Angle of Release	Swings Per Minute		
	Trial 1	Trial 2	Average
90°			
80°			
70°			
60°			
50°			
40°			

Procedure

1. Copy the three data tables into your Science Journal.

2. Bend the paper clip into an S shape and tie it to one end of the string.

3. Hang one washer from the paper clip.

4. **Measure** 10 cm of string from the washer and hold the string at that distance with one hand.

5. Use your other hand to pull back the end of the pendulum with the washer so it is parallel with the ground. Let go of the washer.

6. Count the number of complete swings the pendulum makes in 1 min. Record this number in **Table 1.**

7. Repeat step 5 and record the number of swings in **Table 1** under "Trial 2."

8. Average the results of steps 6 and 7 and record the average swings per minute in **Table 1.**

9. Repeat steps 4 through 8, using string lengths of 20 cm, 30 cm, 40 cm, and 50 cm. Record your data in **Table 1.**

10. Copy the data with the string length of 50 cm in **Table 2.**

11. Repeat steps 5 through 8 using two, three, four, and five washers. Record these data in **Table 2.**

12. Use 50 cm of string and one washer for the third set of tests.

13. Use the protractor to measure a 90° drop of the mass. Repeat this procedure, calculate the average, and record the data in **Table 3.**

14. Repeat procedures 12 and 13, using angles of 80°, 70°, 60°, 50°, and 40°.

Conclude and Apply

1. When you tested the effect of the angle of the drop of the pendulum on the swings per minute, which variables did you keep constant?

2. **Infer** which of the variables you tested affects the swing of a pendulum.

3. Suppose you have a pendulum clock that indicates an earlier time than it really is. (This means it has too few swings per minute.) What could you do to the clock to make it keep better time?

𝒞ommunicating Your Data

Graph the data from your tables. Title and label the graphs. Use different colored pencils for each graph. **Compare** your graphs with the graphs of other members of your class. **For more help, refer to the** Science Skill Handbook.

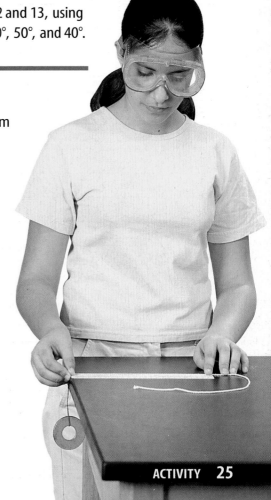

"The Microscope"
by Maxine Kumin

Respond to the Reading

1. Do you think Anton was a scientist by trade? Why or why not?
2. What did Anton find inside a water drop?
3. What are some possible explanations for why some people thought Anton was crazy?

Maxine Kumin

Anton Leeuwenhoek was Dutch.
He sold pincushions, cloth, and such.
The waiting townsfolk fumed and fussed
As Anton's dry goods gathered dust.

He worked, instead of tending store,
At grinding special lenses for
A microscope. Some of the things
He looked at were: mosquitoes' wings,
the hairs of sheep, the legs of lice,
the skin of people, dogs, and mice;
ox eyes, spiders' spinning gear,
fishes' scales, a little smear
of his own blood, and best of all,
the unknown, busy, very small
bugs that swim and bump and hop
inside a simple water drop.

Impossible! Most Dutchmen said.
This Anton's crazy in the head!
We ought to ship him off to Spain.
He says he's seen a housefly's brain.
He says the water that we drink
Is full of bugs. He's mad, we think!

They called him *dumkopf*, which means "dope."
That's how we got the microscope.

Understanding Literature

Rhyming Couplets A couplet is a poetic convention in which every two lines rhyme. The poem you just read rhymes in couplets. Some of the most famous poems that use rhyming couplets describe heroic deeds and often are epic tales. An epic tale is a long story that describes a journey of exploration. You might have heard of Homer, a famous Greek poet who wrote epic tales of adventure such as *The Illiad* and *The Odyssey*. Why do you think the poet used rhyming couplets in the poem you just read?

Science Connection In this chapter, you learned about different scientific instruments that increase scientific knowledge. For example, microscopes have changed the scale at which humans can make observations. Electron microscopes allow observers to obtain images at magnifications ranging from about $10,000\times$ to $1,000,000\times$. In a microscope with a magnification of $10,000\times$, a 0.001-mm object will appear as a 1-cm image.

Microscopes are used in Earth science to observe the arrangement and composition of minerals in rocks. These observations give clues about the conditions that formed the rock.

Linking Science and Writing

Write a heroic poem Write a heroic verse or poem that rhymes. Pick a scientific method or discovery from your textbook. If necessary, research the discovery to find out its origin. Use rhyming words that describe the discovery and the person or people who made it. If possible, try using couplet rhyming, where every two lines of the poem rhyme.

Career Connection

Mineralogist

Julie Sheets Dr. Sheets and other geologists use light and electron microscopes to observe tiny defects in rock such as holes, fractures, and mistakes in the arrangement of atoms, which allow gas and fluid to move through solid rock. These features influence the physical properties of rocks, such as how they deform when Earth's plates collide or how they lose or gain chemical elements. She and other scientists hypothesize that defects in crystal and rock influence how fast minerals weather, or degrade, when exposed to water. This type of research is also important for determining the age of some rocks because elements used to date them may be lost to weathering.

SCIENCE*Online* To learn more about careers in mineralogy, visit the Glencoe Science Web site at **science.glencoe.com.**

Chapter ① Study Guide

Reviewing Main Ideas

Section 1 Science All Around

1. Scientific methods are used to solve problems or answer questions.

2. Scientific methods include identifying a problem or question, gathering information, developing hypotheses, designing an experiment to test the hypotheses, performing the experiment, collecting and analyzing data, and forming conclusions.

3. Science experiments should be repeated to see whether results are consistent. The results of an experiment often allow conclusions to be made and usually lead to additional questions.

4. Science is a process of observing and studying the world. *What do you think the scientist shown here is studying?*

5. In an experiment, the independent variable is the variable being tested. Constants are variables that do not change. The variable being measured is the dependent variable. A control is a standard to which things can be compared.

6. Technology is the use of scientific discoveries. Technology has changed as new scientific discoveries were made.

Section 2 Scientific Enterprise

1. Today, everything known in science results from knowledge that has been collected over time. Science has changed and will continue to change because of continuing research and improvements in instruments and testing procedures. *How have instruments used to study space changed since Galileo used a telescope like the one shown here to study Jupiter's moons?*

2. Scientific theories are explanations or models that are supported by repeated experimentation. If data gathered over a long period of time support a hypothesis, it can become a scientific theory.

3. Scientific laws are rules that describe the behavior of something in nature. Laws are used to predict natural events but usually do not explain them.

4. There are limits to the kinds of questions and problems science can be used to answer and solve. Problems that deal with ethics and belief systems cannot be answered using scientific methods.

FOLDABLES
Reading & Study Skills

After You Read

Exchange your Vocabulary Study Fold with a classmate and quiz each other to see how many vocabulary words you can define without looking under the tabs.

Visualizing Main Ideas

Complete the following concept map about variables and constants.

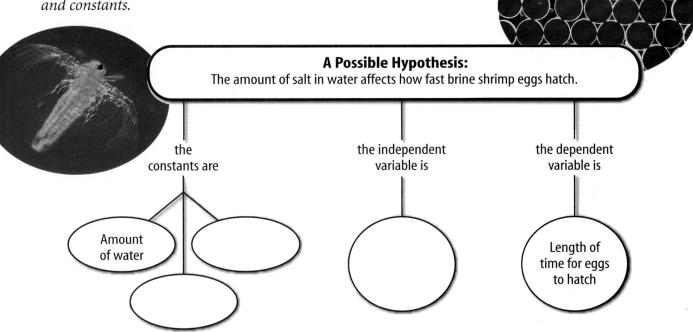

A Possible Hypothesis:
The amount of salt in water affects how fast brine shrimp eggs hatch.

the constants are → Amount of water / ( ) / ( )

the independent variable is → ( )

the dependent variable is → Length of time for eggs to hatch

Vocabulary Review

Vocabulary Words

a. bias
b. constant
c. control
d. dependent variable
e. Earth science
f. ethics
g. hypothesis
h. independent variable
i. science
j. scientific law
k. scientific methods
l. scientific theory
m. technology
n. variable

THE PRINCETON REVIEW Study Tip

Use acronyms to help you remember sequences. For example, the acronym TSMTE helps you remember that the layers of the atmosphere extending from Earth's surface are the troposphere, stratosphere, mesosphere, thermosphere, and exosphere.

Using Vocabulary

Use what you know about the vocabulary words to explain the differences between the words in the following sets. Then explain how the words are related.

1. constant, control
2. dependent variable, independent variable
3. scientific law, scientific theory
4. science, technology
5. hypothesis, scientific theory
6. science, Earth science
7. independent variable, constant
8. variable, control
9. Earth science, technology
10. ethics, bias

Checking Concepts

Choose the word or phrase that best answers the question.

1. Which of these best describes science?
   A) have knowledge   C) study
   B) observe   D) solve

2. Which word is the standard used for comparison in an experiment?
   A) variable   C) control
   B) theory   D) law

3. Which of the following is the first step in using scientific methods?
   A) develop hypotheses
   B) make conclusions
   C) test hypotheses
   D) identify a problem

4. Which word means an educated guess?
   A) theory   C) variable
   B) hypothesis   D) law

5. The idea that a comet is like a dirty snowball is which of the following?
   A) hypothesis   C) law
   B) variable   D) theory

6. The statement that an object at rest will remain at rest unless acted upon by a force is an example of which of the following?
   A) hypothesis   C) law
   B) variable   D) theory

7. Which of the following questions could NOT be answered using scientific methods?
   A) Should lying be illegal?
   B) Does sulfur affect the growth of grass?
   C) How do waves cause erosion?
   D) Does land heat up faster than water?

8. Which of the following describes variables that stay the same in an experiment?
   A) dependent variables   C) constants
   B) independent variables   D) controls

9. Which of the following is a variable that is being tested in a science experiment?
   A) dependent variable   C) constant
   B) independent variable   D) control

10. What should you do if your data are different from what you expected?
    A) Conclude that you made a mistake in the way you collected the data.
    B) Change your data to be consistent with your expectation.
    C) Conclude that you made a mistake when you recorded your data.
    D) Conclude that your expectation might have been wrong.

Thinking Critically

11. Suppose you had two plants— a cactus and a palm. You planted them in soil and watered them daily. After two weeks, the cactus was dead. What scientific methods could you use to find out why the cactus died?

12. How have advances in technology affected society?

13. What is meant by the statement, *Technology is transferable?*

14. Why don't all hypotheses become theories?

15. What are some scientific methods you use every day to answer questions or solve problems?

Developing Skills

16. **Identifying and Manipulating Variables and Controls** How would you set up a simple experiment to test whether salt-crystal growth is affected by temperature?

17. Forming Hypotheses You observe two beakers containing clear liquid and ice cubes. In the first beaker, the ice cubes are floating. In the second, the ice cubes are on the bottom of the beaker. Write a hypothesis to explain the difference in your observations about the two beakers.

18. Recognizing Cause and Effect Explain why scientific methods cannot be used to answer ethical questions.

19. Interpreting Data A friend tells you that dark colors absorb more heat than light colors do. You conduct an experiment to determine which color of fabric absorbs the most heat. Analyze your data below. Was your friend correct? Explain.

Color and Heat Absorption		
Color	**Beginning Temperature (°C)**	**Temperature (°C) after 10 minutes**
Red	24°	26°
Black	24°	28°
Blue	24°	27°
White	24°	25°
Green	24°	27°

Performance Assessment

20. Poster Research an example of Earth science technology that is not shown in **Figure 10.** Create a poster that explains the contribution this technology made to the understanding of Earth science.

TECHNOLOGY

Go to the Glencoe Science Web site at **science.glencoe.com** or use the **Glencoe Science CD-ROM** for additional chapter assessment.

THE PRINCETON REVIEW — Test Practice

Scientists must use different tools to measure amounts and distances during experiments. Some measuring equipment is divided into two groups in the boxes below.

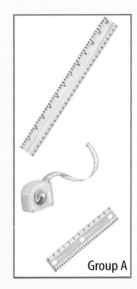

Group A

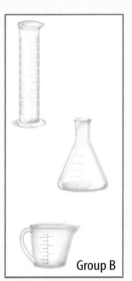

Group B

Study the pictures and answer the following questions.

1. The pieces of equipment in Group A are different from the equipment in Group B because only the equipment in Group A can be used to measure _____ .
  A) distance
  B) volume
  C) weight
  D) mass

2. Which of the following belongs with Group B above?
  F) a bathroom scale
  G) a soda can
  H) a beaker
  J) a stopwatch

2 Matter

What is most striking about this picture—water or ice? What do these things have in common? How are they different? In this chapter, you will find the answers to these questions. You also will learn about the matter that makes up your surroundings and makes up your body.

What do you think?

Science Journal Look at the picture below with a classmate. Discuss what you think these might be. Here's a hint: *They could be part of the landscape of a far-off planet or something in your desk.* Write your answer in your Science Journal.

EXPLORE **A**CTIVITY

On Earth water is unique because it is found as a solid, liquid, or gas. On a cool autumn morning, you might see gaseous water condensing into fog over a lake or a river whose surface soon will be solid ice. The following activity will help you visualize how matter can change states.

Change the state of water

Safety Precautions

1. Pour 500 mL of water into a 1,000-mL glass beaker.
2. Mark the level of water in the beaker with the bottom edge of a piece of tape.
3. Place the beaker on a hot plate.
4. With the help of an adult, heat the water until it boils for 5 min. Let the water cool.
5. With the help of an adult, compare the level of the water to the bottom edge of the tape.

Observe

What came out of the beaker as the water boiled? Did the amount of water that you started with change? In your Science Journal, explain what happened to the water.

Before You Read

FOLDABLES
**Reading & Study
Skills**

Making a Vocabulary Study Fold
Knowing the definition of vocabulary words is a good way to ensure that you understand the content of the chapter.

1. Place a sheet of notebook paper in front of you so the short side is at the top and the holes are on the right side. Fold the paper in half from the left side to the right side.
2. Through the top thickness of paper, cut along every third line from the outside edge to the center fold, forming tabs.
3. Before you read, write vocabulary words from each section in this chapter on the front of the tabs. Under each tab, write what you think the word means.
4. As you read the chapter, add to and correct your definitions.

① Atoms

What You'll Learn

- **Identify** the states of matter.
- **Describe** the internal structure of an atom.
- **Compare** isotopes of an element.

Vocabulary

matter	electron
atom	atomic number
element	mass number
proton	isotope
neutron	

Why It's Important

Nearly everything around you—air, water, food, and clothes—is made of atoms.

The Building Blocks of Matter

What do the objects you see, the air you breathe, and the food you eat have in common? They are matter. **Matter** is anything that has mass and takes up space. Heat and light are not matter, because they have no mass and do not take up space. Glance around the room. If all the objects you see are matter, why do they look so different from one another?

Atoms Matter, in its various forms, surrounds you. You can't see all matter as clearly as you see water, which is a transparent liquid, or rocks, which are colorful solids. You can't see air, for example, because air is colorless gas. The forms or properties of one type of matter differ from the properties of another, because matter is made up of tiny particles called **atoms.** The structures of different types of atoms and how they join together determine all the properties of matter that you can observe. **Figure 1** illustrates how small objects, like atoms, can be put together in different ways.

Figure 1
Like atoms, the same few blocks can combine in many ways. *How could this model help explain the variety of matter?*

The Structure of Matter Matter is joined together much like the blocks shown in **Figure 1.** The building blocks of matter are atoms. The types of atoms in matter and how they attach to each other give matter its properties.

Elements When atoms combine, they form many different types of matter. Your body contains several types of atoms combined in different ways. These atoms form the proteins, DNA, tissues, and other matter that make you the person you are. Most other objects that you see also are made of several different types of atoms. However, some substances are made of only one type of atom. **Elements** are substances that are made of only one type of atom and cannot be broken down into simpler substances by normal chemical or physical means.

Elements are useful for making a variety of items you depend on every day. They also combine to make up the minerals that compose Earth's crust. Some minerals, however, are made up of only one element. These minerals, which include copper and silver, are called native elements. **Table 1** shows some common elements and their uses. A table of the elements, called the periodic table of the elements, is included on the inside back cover of this book.

Table 1 Some Common Uses of Elements

Element	Sulfur	Silver	Copper	Carbon
Native State of the Element	Sulfur	Silver	Copper	Graphite
Uses of the Element	Fertilizer	Tableware	Wire	Ski wax

Figure 2
This model airplane is a small-scale version of a large object.

Modeling the Atom

How can you study things that are too small to be seen with the unaided eye? When something is too large or too small to observe directly, models can be used. The model airplane, shown in **Figure 2,** is a small version of a larger object. A model also can describe tiny objects, such as atoms, that otherwise are difficult or impossible to see.

The History of the Atomic Model The current model of the atom is based on the work of many scientists over hundreds of years. The idea that matter is composed of atoms dates back more than 2,300 years when the Greek philosopher Democritus (dih MAH krih tuhs) proposed that matter is composed of small particles. He called these particles atoms and said that different types of matter were composed of different types of atoms. More than 2,000 years later, John Dalton expanded on these ideas. He theorized that all atoms of an element contain the same type of atom.

Protons and Neutrons In the early 1900s, additional work led to the development of the current model of the atom, shown in **Figure 3.** Three basic particles make up an atom—protons, neutrons (NOO trahnz), and electrons. **Protons** are particles that have a positive electric charge. **Neutrons** have no electric charge. Both particles are located in the nucleus—the center of an atom. With no negative charge to balance the positive charge of the protons, the charge of the nucleus is positive.

Electrons Particles with a negative charge are called **electrons,** and they exist outside of the nucleus. In 1913, Niels Bohr, a Danish scientist, proposed that an atom's electrons travel in orbitlike paths around the nucleus. He also proposed that electrons in an atom have energy that depends on their distance from the nucleus. Electrons in paths that are closer to the nucleus have lower energy, and electrons further from the nucleus have higher energy.

The Current Atomic Model Over the next several decades, research showed that although electrons do have specific amounts of energy, they do not travel in orbitlike paths. Instead, electrons move in an electron cloud surrounding the nucleus. Electrons can be anywhere within the cloud, but evidence suggests that they are located near the nucleus most of the time. To understand how this might work, imagine a beehive. The hive represents the nucleus of an atom. The bees swarming around the hive are like electrons moving around the nucleus. As they swarm, you can't predict their exact location, but they usually stay close to the hive.

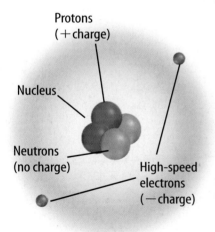

Protons
(+charge)

Nucleus

Neutrons
(no charge)

High-speed
electrons
(—charge)

Figure 3
This model of a helium atom shows two protons and two neutrons in the nucleus, and two electrons in the electron cloud.

Counting Atomic Particles

You now know where protons, neutrons, and electrons are located, but how many of each are in an atom? The number of protons in an atom depends on the element. All atoms of the same element have the same number of protons. For example, all iron atoms—whether in train tracks or breakfast cereal—contain 26 protons, and all atoms with 26 protons are iron atoms. The number of protons in an atom is equal to the **atomic number** of the element. This number can be found above the element symbol on the periodic table. Notice that as you go from left to right on the periodic table, the atomic number of the element increases by one.

 **Reading Check** *How many protons are in an atom of gold, which has an atomic number of 79?*

How many electrons? In a neutral atom, the number of protons is equal to the number of electrons. This makes the overall charge of the atom zero. Therefore, for a neutral atom:

Atomic number = number of protons = number of electrons

Atoms of an element can lose or gain electrons and still be the same element. When this happens, the atom is no longer neutral. Atoms with fewer electrons than protons have a positive charge, and atoms with more electrons than protons have a negative charge.

How many neutrons? Unlike protons, atoms of the same element can have different numbers of neutrons. The number of neutrons in an atom isn't found on the periodic table. Instead, you need to know the atom's mass number. The **mass number** of an atom is equal to the number of protons plus the number of neutrons. The number of neutrons is determined by subtracting the atomic number from the mass number. In **Figure 4,** the number of neutrons can be determined by counting the blue spheres and the number of protons by counting orange spheres. Atoms of the same element that have different numbers of neutrons are called **isotopes.** **Table 2** lists useful isotopes of some elements.

 **Reading Check** *How are isotopes of the same element different?*

 Health INTEGRATION

Some isotopes of elements are radioactive. Physicians can introduce these isotopes into a patient's circulatory system. The low-level radiation they emit allows the isotopes to be tracked as they move throughout the patient's body. Explain how this would be helpful in diagnosing a disease.

Figure 4
This carbon atom is common in organic material. *What is this atom's mass number?*

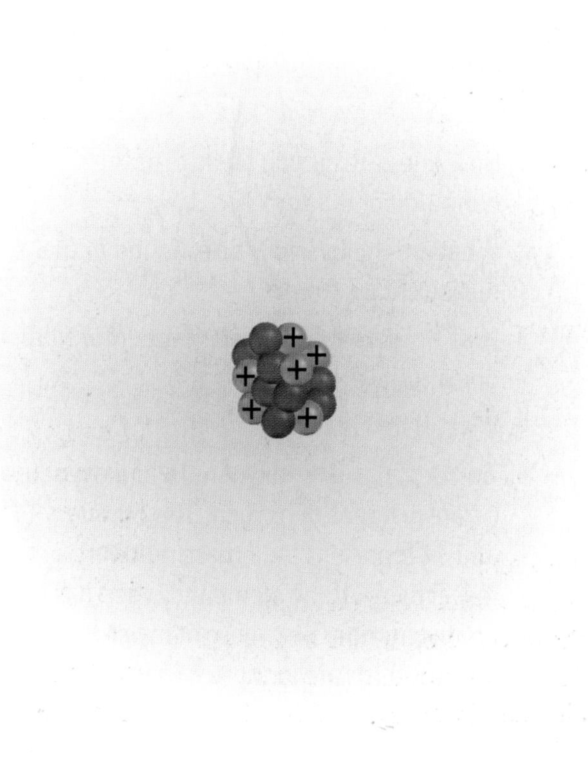

Table 2 Some Useful Isotopes

Isotope	Number of Protons	Number of Neutrons	Number of Electrons	Atomic Number	Mass Number
Hydrogen-1	1	0	1	1	1
Hydrogen-2	1	1	1	1	2
Hydrogen-3	1	2	1	1	3
Carbon-12	6	6	6	6	12
Carbon-14	6	8	6	6	14
Uranium-234	92	142	92	92	234
Uranium-235	92	143	92	92	235
Uranium-238	92	146	92	92	238

Uses of Isotopes Scientists have found uses for isotopes that benefit humans. For example, medical doctors use radioactive isotopes to treat certain types of cancer, such as prostate cancer. Geologists use isotopes to determine the ages of some rocks and fossils.

As you continue to investigate matter in this chapter, you will explore how atoms of different elements combine to form the materials around you.

Section Assessment

1. How does the air you breathe fit the definition of matter?
2. What are the basic particles found in the nucleus of an atom?
3. How do isotopes of an element differ from one another?
4. What is an element?
5. **Think Critically** Oxygen-16 and oxygen-17 are isotopes of oxygen. The numbers 16 and 17 represent their mass numbers, respectively. If the element oxygen has an atomic number of 8, how many protons and neutrons are in these two isotopes?

Skill Builder Activities

6. **Comparing and Contrasting** Review the material in Section 1. How do atoms and elements differ? How are they similar? **For more help, refer to the** Science Skill Handbook.
7. **Solving One-Step Equations** The mass number of a nitrogen atom is 14. Find its atomic number in the periodic table shown on the inside back cover. Then determine the number of neutrons in its nucleus by subtracting the atomic number from the mass number. **For more help, refer to the** Math Skill Handbook.

Combinations of Atoms

Interactions of Atoms

When you take a shower, eat your lunch, or do your homework on the computer, you probably don't think about elements. But everything you touch, eat, or use is made from them. Elements are all around you and in you.

There are about 90 naturally occurring elements on Earth. When you think about the variety of matter in the universe, you might find it difficult to believe that most of it consists of combinations of these same elements. How could so few elements produce so many different things? This happens because elements can combine in countless ways. For example, the same oxygen atoms that you breathe also might be found in many other objects, as shown in **Figure 5.** As you can see, each combination of atoms is unique. How do these combinations form and what holds them together?

As You Read

What **You'll Learn**
- **Describe** ways atoms combine to form compounds.
- **List** differences between compounds and mixtures.

Vocabulary
compound
ion
mixture
heterogeneous mixture
homogeneous mixture
solution

Why **It's Important**
On Earth, most matter exists as compounds or mixtures.

A Solid limestone has oxygen within its structure.

B Oxygen also is present in the juices of these apples.

C This canister contains pure oxygen gas.

Figure 5
Oxygen is a common element found in many different solids, liquids, and gases. *How can the same element, made from the same type of atoms, be found in so many different materials?*

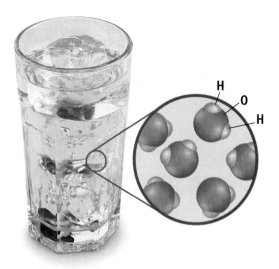

Figure 6
The water you drink is a compound consisting of hydrogen and oxygen atoms.

Compounds When atoms of more than one element combine, they form a compound. A **compound** contains atoms of more than one type of element that are chemically bonded together. Water, shown in **Figure 6,** is a compound consisting of molecules made up of one oxygen atom bonded to two hydrogen atoms. Table salt—sodium chloride—is a compound consisting of sodium atoms bonded to chlorine atoms. Compounds are represented by chemical formulas that show the ratios and types of atoms in the compound. For example, the chemical formula for sodium chloride is NaCl. The formula for water is H_2O.

✔ **Reading Check** *How many atoms does a water molecule have?*

The properties of compounds often are very different from the properties of the elements that combine to form them. Sodium is a soft, silvery metal, and chlorine is a greenish, poisonous gas, but the compound they form is the white, crystalline table salt you use to season food. Under normal conditions on Earth, the hydrogen and oxygen that form water are gases. Water can be solid ice, liquid water, or gaseous vapor. Which state do you think is most common for water at Earth's south pole?

Chemical Properties A property that describes a change that occurs when one substance reacts with another is called a chemical property. For example, one chemical property of water is that it changes to hydrogen and oxygen gas when an electric current passes through it. The chemical properties of a substance depend on what elements are in that substance and how they are arranged. Iron atoms in the mineral biotite will react with water and oxygen to form iron oxide, or rust, but iron mixed with chromium and nickel in stainless steel resists rusting.

Bonding

The forces that hold the atoms in compounds together are called chemical bonds. Some atoms are reactive and form bonds easily. Other atoms are much less reactive. For example, atoms that have eight electrons in the outermost portion of their electron cloud are not likely to combine with other atoms. If an atom has fewer than eight electrons in the outermost portion of its electron cloud, it is unstable and is more likely to combine with other atoms. One exception to this rule is the element helium, with only two electrons in its electron cloud. This atom is stable and does not react easily.

Online

Research Visit the Glencoe Science Web site at **science.glencoe.com** for more information about bonding and the periodic table. Communicate to your class what you learn.

Covalent Bonds Atoms can combine to form compounds in two different ways. One way is by sharing the electrons in the outer portion of their electron clouds. The type of bond that forms by sharing outer electrons is a covalent bond. A group of atoms connected by covalent bonds is called a molecule. For example, two atoms of hydrogen can share outer electrons with one atom of oxygen to form a molecule of water, as shown in **Figure 7.** Each of the hydrogen atoms has one outer electron and the oxygen has six outer electrons. This arrangement causes hydrogen and oxygen atoms to bond together. Each of the hydrogen atoms becomes stable by sharing one electron with the oxygen atom, and the oxygen atom becomes stable by sharing two electrons with the two hydrogen atoms.

Ionic Bonds In addition to sharing electrons, atoms also combine if they become positively or negatively charged. This type of bond is called an ionic bond. Atoms can be neutral, or under certain conditions, atoms can lose or gain electrons. When an atom loses electrons, it has more protons than electrons, so the atom is positively charged. When an atom gains electrons, it has more electrons than protons, so the atom is negatively charged. Electrically charged atoms are called **ions.**

Ions are attracted to each other when they have opposite charges. This is similar to the way magnets behave. If the ends of a pair of magnets have the same type of pole, they repel each other. Conversely, if the ends have opposite poles, they attract one another. Ions form electrically neutral compounds when they join. The mineral halite, commonly used as table salt, forms in this way. A sodium (Na) atom loses an outer electron and becomes a positively charged ion. As shown in **Figure 8,** if the sodium ion comes close to a negatively charged chlorine (Cl) ion, they attract each other and form the salt you use on french fries or popcorn.

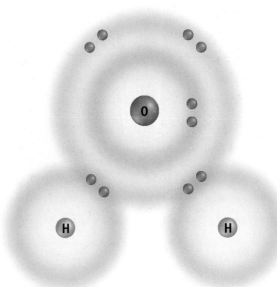

Figure 7
A molecule of water consists of two atoms of hydrogen that share outer electrons with one atom of oxygen.

Figure 8
Table salt forms when a sodium ion and a chlorine ion are attracted to one another. *What kind of bond holds ions together?*

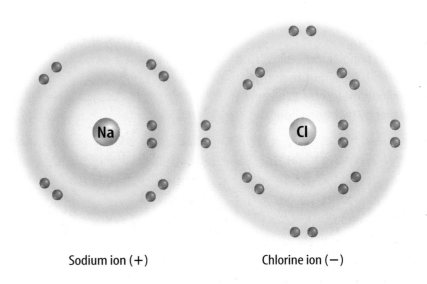

Sodium ion (+)   Chlorine ion (−)

Figure 9
Electrons move freely between the copper atoms in this wire.
What type of bond holds copper atoms together?

Metallic Bonds Metallic bonds are found in metals such as copper, gold, aluminum, and silver. In this type of bond, electrons are free to move from one positively charged ion to another. This free movement of electrons is responsible for key characteristics of metals. The movement of electrons, or conductivity, allows metals like copper, shown in **Figure 9,** to pass an electric current easily.

Hydrogen Bonds Some types of bonds, such as hydrogen bonds, can form without the interactions of electrons. The arrangement of hydrogen and oxygen atoms in water molecules causes them to be polar molecules. A polar molecule has a positive end and a negative end. This happens because the atoms do not share electrons equally. When hydrogen and oxygen atoms form a molecule with covalent bonds, the hydrogen atoms produce an area of partial positive charge and the oxygen atom produces an area of partial negative charge. The positive end of one molecule is attracted to the negative end of another molecule, as shown in **Figure 10,** and a weak hydrogen bond is formed. The different parts of the water molecule are slightly charged, but as a whole, the molecule has no charge. This type of bond is easily broken, indicating that the charges are weak.

Hydrogen bonds are responsible for several properties of water, some of which are unique. Cohesion is the attraction between water molecules that allows them to form raindrops and to form beads on flat surfaces. Hydrogen bonds cause water to exist as a liquid, rather than a gas, at room temperature. As water freezes, hydrogen bonds force water molecules apart, into a structure that is less dense than liquid water.

Figure 10
The ends of polar molecules, such as water, have opposite charges. This allows molecules to be held together by hydrogen bonds.

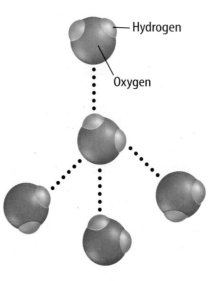

Hydrogen

Oxygen

Figure 11
This rock contains a variety of minerals that together form a mixture.

Mixtures

Sometimes compounds and elements mix together but do not combine chemically. A **mixture** is composed of two or more substances that are not chemically combined. There are two different types of mixtures—heterogeneous and homogeneous. The components of a **heterogeneous mixture** are not mixed evenly and each component retains its own properties. Maybe you've seen a rock like the one in **Figure 11.** Several different minerals are mixed together, but if you were to examine the minerals separately, you would find that they have the same properties and appearance as they have in the rock.

The components of a **homogeneous mixture** are evenly mixed throughout. You can't see the individual components. Another name for a homogeneous mixture is a **solution**. The properties of the components of this type of mixture often are different from the properties of the mixture. Ocean water is an example of a liquid solution that consists of salts mixed with liquid water.

✔ **Reading Check** *What is a solution?*

Separating Mixtures and Compounds

The components of a mixture can be separated by physical means. For example, you can sit at your desk and pick out the separate items in your backpack, or you can let the water evaporate from a saltwater mixture and the salt will remain.

Separating the components of a mixture is a relatively easy task compared to separating those of a compound. The substances in a compound must be separated by chemical means. This means that an existing compound can be changed to one or more new substances by chemically breaking down the original compound. For example, a drop of dilute hydrochloric acid (HCl) can be placed on calcium carbonate ($CaCO_3$) and carbon dioxide (CO_2) is released. To break down most compounds, several steps usually are required.

Mini LAB

Classifying Forms of Matter

Procedure
1. Make a chart with columns titled Mixtures, Compounds, and Elements.
2. Classify each of these items into the proper column on your chart: **air, sand, hydrogen, muddy water, sugar, ice, sugar water, water, salt, oxygen, copper.**
3. Make a solution using two or more of the items listed above.

Analysis
1. How does a solution differ from other types of mixtures?
2. How does an element differ from a compound?

Figure 12
The ocean is a mixture of many different forms of matter. The ocean water itself is a solution.

Chemistry
INTEGRATION

When one substance dissolves in another, some of the properties of the dissolving substance change. When salt dissolves in water, decide whether its chemical or physical properties change.

Exploring Matter

Air, sweetened tea, salt water, and the contents of your backpack are examples of mixtures. The combination of rocks, fish, and coral shown in **Figure 12** also is a mixture. In each case, the materials within the mixture are not chemically combined. The individual components are made of compounds, or elements. The atoms that make up these compounds lost their individual properties when they combined. Even though atoms are known as the building blocks of matter, they are composed of protons, neutrons, and electrons, which are even smaller. As you continue to explore matter, apply what you've learned about atoms, elements, compounds, mixtures, and solutions to your studies.

Section Assessment

1. How do atoms or ions combine to form compounds?

2. Why is sweetened tea considered to be a solution rather than a compound?

3. Describe the chemical property of iron that results in rust.

4. Explain what makes a solution different from a heterogeneous mixture.

5. **Think Critically** How can you determine whether salt water is a solution or a compound?

Skill Builder Activities

6. **Comparing and Contrasting** How are solutions and compounds similar? How are they different? **For more help, refer to the** Science Skill Handbook.

7. **Communicating** Design an investigation that would show whether sugar water is a mixture or a compound. Discuss your design with your teacher. Perform the investigation and write the results in your Science Journal. **For more help, refer to the** Science Skill Handbook.

Activity

Scales of Measurement

How would you describe some of the objects in your classroom? Perhaps your desktop is about one-half the size of a door. Measuring physical properties in a laboratory experiment will help you make better observations.

What You'll Investigate
How are physical properties of objects measured?

Materials
triple beam balance	rock sample
100-mL graduated cylinder	string
metersticks (2)	globe
non-mercury thermometers (3)	water
stick or dowel	

Goals
- **Measure** various physical properties in SI.
- **Determine** sources of error.

Safety Precautions 🥽 🐄 🧤
Never "shake down" lab thermometers.

Procedure

1. Go to every station and determine the measurement requested. Record your observations in a data table and list sources of error.
   a. Use a balance to determine the mass, to the nearest 0.1 g, of the rock sample.
   b. Use a graduated cylinder to measure the water volume to the nearest 0.5 mL.
   c. Use three thermometers to determine the average temperature, to the nearest 0.5°C, at a selected location in the room.
   d. Use a meterstick to measure the length, to the nearest 0.1 cm, of the stick or dowel.
   e. Use a meterstick and string to measure the circumference of the globe. Be accurate to the nearest 0.1 cm.

Measurement and Error		
Sample at Station	**Value of Measurement**	**Causes of Error**
a.	mass = ____ g	
b.	volume = ____ mL	
c. (location)	average temp. = ____ °C	
d.	length = ____ cm	
e.	circumference = ____ cm	

Conclude and Apply

1. **Compare** your results with those of other students who measured the same objects. Review the values provided by your teacher. How do the values you obtained compare with those provided by your teacher and other students?

2. **Calculate** your percentage of error in each case. Use this formula.

$$\% \text{ error} = \frac{\text{your value} - \text{teacher's value}}{\text{teacher's value}} \times 100$$

3. Decide what percentage of error will be acceptable. Generally, being within five percent to seven percent of the correct value is considered good. If your values exceed ten percent error, what could you do to improve your results and reduce error? What was the most common source of error?

Communicating Your Data

Compare your conclusions with those of other students in your class. **For more help, refer to the** Science Skill Handbook.

Properties of Matter

As You Read

What You'll Learn

- **Distinguish** between chemical and physical properties.
- **List** the four states of matter.

Vocabulary
density

Why It's Important
You can recognize many substances by their physical properties.

Physical Properties of Matter

In addition to the chemical properties of matter that you have already investigated in this chapter, matter also has other properties that can be described. You might describe a pair of blue jeans as soft, blue, and about 80 cm long. A sandwich could have two slices of bread, lettuce, tomato, cheese, and turkey. These descriptions can be made without altering the sandwich or the blue jeans in any way. The properties that you can observe without changing a substance into a new substance are physical properties.

One physical property that you will use to describe matter is density. **Density** is a measure of the mass of an object divided by its volume. Generally, this measurement is given in grams per cubic centimeter (g/cm^3). For example, the average density of liquid water is about 1 g/cm^3. So 1 cm^3 of pure water has a mass of about 1 g.

An object that's more dense than water will sink in water. On the other hand, an object that's not as dense as water will float in water. When oil spills occur on the ocean, as shown in **Figure 13,** the oil floats on the surface of the water and washes up on beaches. Because the oil floats, even a small spill can spread out and cover large areas.

Figure 13
Oil spills on the ocean spread across the surface of the water.
How does the density of oil compare to the density of water?

States of Matter

On Earth, matter occurs in four physical states. These four states are solid, liquid, gas, and plasma. You might have had solid toast and liquid milk or juice for breakfast this morning. You breathe air, which is a gas. A lightning bolt during a storm is an example of matter in its plasma state. What are the differences among these four states of matter?

Solids The reason some matter is solid is that its particles are in fixed positions relative to each other. The individual particles vibrate, but they don't switch positions with each other. Solids have a definite shape and take up a definite volume.

Suppose you have a puzzle that is completely assembled. The pieces are connected so one piece cannot switch positions with another piece. However, the pieces can move a little. For example, you can push on one end of the puzzle and move each individual puzzle piece, but the pieces of the puzzle stay attached to one another. The puzzle pieces in this model represent particles of a substance in a solid state. Such particles are strongly attracted to each other and resist being separated.

Math Skills Activity

Calculating Density

You want to find the density of a small cube of an unknown material. It measures 1 cm $\times$ 1 cm $\times$ 2 cm. It has a mass of 8 g.

Solution

1 *This is what you know:*   mass: $m = 8$ g
volume: $v = 1$ cm $\times$ 1 cm $\times$ 2 cm $= 2$ cm^3

2 *This is what you need to find:*   density: d

3 *This is the equation you need to use:* $d = m/v$

4 *Substitute the known values:* $d = 8$ g$/2$ cm^3
$d = 4$ g$/$cm^3

Check your answer by multiplying by the volume. Do you calculate the same mass that was given? Explain.

Practice Problem

Your aunt brings you a souvenir gold bar from her visit to Fort Knox. It measures 10 cm $\times$ 5 cm $\times$ 2 cm. It has a mass of 1,930 g. Find the density of gold.

For more help refer to the Math Skill Handbook.

Liquids Particles in a liquid are attracted to each other, but are not in fixed positions as they are in the solid shown in **Figure 15A.** This is because liquid particles have more energy than solid particles. This energy allows them to move around and change positions with each other.

When you eat breakfast, you might have several liquids at the table such as syrup, juice, and milk. These are substances in the liquid state, even though one flows more freely than the others at room temperature. The particles in a liquid can change positions to fit the shape of the container they are held in. You can pour any liquid into any container, and it will flow until it matches the shape of its new container.

Gases The particles that make up gases have enough energy to overcome any attractions between them. This allows them to move freely and independently. Unlike liquids and solids, gases spread out and fill the container in which they are placed. Air fresheners work in a similar way. If an air freshener is placed in a corner, it isn't long before the particles from the air freshener have spread throughout the room. Look at the hot-air balloon shown in **Figure 15C.** The particles in the balloon are evenly spaced throughout the balloon. The balloon floats in the sky, because the hot air inside the balloon is less dense than the colder air around it.

 **Reading Check** *Why do air fresheners work?*

Figure 14
The Sun is an example of a plasma.

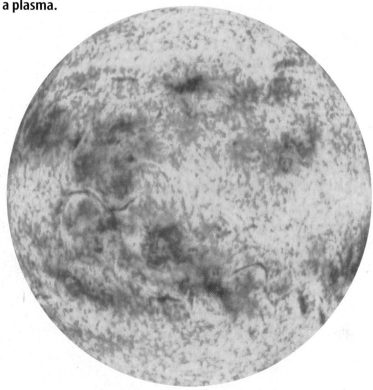

Plasma The most common state of matter in the universe is plasma. This state is associated with high temperatures. Can you name something that is in the plasma state? Stars like the Sun, shown in **Figure 14,** are composed of matter in the plasma state. Plasma also exists in Jupiter's magnetic field. On Earth, plasma is found in lightning bolts, as shown in **Figure 15D.** Plasma is composed of ions and electrons. It forms when high temperatures cause some of the electrons normally found in an atom's electron cloud to escape and move outside of the electron cloud.

Figure 15

Matter on Earth exists naturally in four different states—solid, liquid, gas, and plasma—as shown here. The state of a sample of matter depends upon the amount of energy its atoms or molecules possess. The more energy that matter contains, the more freely its atoms or molecules move, because they are able to overcome the attractive forces that tend to hold them together.

D PLASMA Electrically charged particles in lightning are free moving.

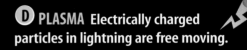

A SOLID In a solid such as galena, the tightly packed atoms or molecules lack the energy to move out of position.

B LIQUID The atoms or molecules in a liquid such as water have enough energy to overcome some attractive forces and move over and around one another.

C GAS In air and other gases, atoms or molecules have sufficient energy to separate from each other completely and move in all directions.

Figure 16
A solid metal can be changed to a liquid by adding thermal energy to its molecules. *What is happening to the molecules during this change?*

Figure 17
If ice were more dense than water, lakes would freeze solid from the bottom up. *What effect might this have on the fish?*

Changing the State of Matter

Matter is changed from a liquid to a solid at its freezing point and from a liquid to a gas at its boiling point. You may know the freezing and boiling points of water. Water changes from a liquid to a solid at its freezing point of 0°C. It boils at 100°C. Water is the only substance that occurs naturally on Earth as a solid, liquid, and gas. Other substances don't naturally occur in these three states on Earth because of the limited temperature range Earth experiences. For example, temperatures on Earth do not get cold enough for solid carbon dioxide to exist naturally. However, it can be produced by humans.

The attraction between particles of a substance and their rate of movement are factors that determine the state of matter. When thermal energy is added to ice, the rate of movement of its molecules increases. This allows the molecules to move more freely and causes the ice to melt. As **Figure 16** shows, even solid metal can be converted into liquid when enough thermal energy is added.

Changes in state also occur because of increases or decreases in pressure. You can demonstrate this with an ice cube. When subjected to pressure, the ice will change to liquid water when no thermal energy is added. This occurs because the melting temperature of the ice is lowered as more pressure is added. This might explain how the base of a glacier can move around some rock obstacles. It is thought that the pressure of the glacier on the rock melts the ice, creating a thin layer of water. The water then flows around the obstacle and refreezes on the other side.

Changes in Physical Properties

Chemical properties of matter don't change when the matter changes state, but some of its physical properties change. For example, the density of water changes as water changes state. Ice floats in liquid water, as seen in **Figure 17,** because it is less dense than liquid water. This is unique, because most materials are denser in their solid state than in their liquid state.

Reading Check   *Why does ice float in water?*

Some physical properties of substances don't change when they change state. For example, water is colorless and transparent in each of its states.

Changing Mars's Matter

Matter in one state often can be changed to another state by adding or removing thermal energy. Changes in thermal energy might explain why Mars appears to have had considerable water on its surface in the past but now has little or no water on its surface. Recent images of Mars reveal that there might still be some groundwater that occasionally reaches the surface, as shown in **Figure 18.** But what could explain the huge water-carved channels that formed long ago? Much of the liquid water on Mars might have changed state as the planet cooled to its current temperature. Scientists believe that some of Mars's liquid water soaked into the ground and froze, forming permafrost. Some of the water might have frozen to form the polar ice caps. Even more of the water might have evaporated into the atmosphere and escaped to space.

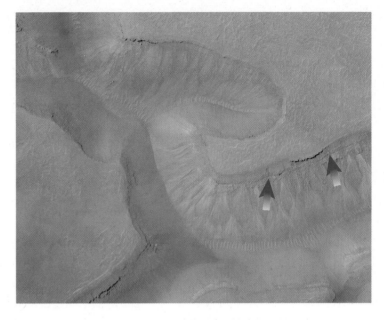

Figure 18
Groundwater might reach the surface of Mars along the edge of this large channel.

Section ③ Assessment

1. List the four states of matter in order from lowest to highest in terms of amount of particle movement.

2. Compare and contrast the movement of water molecules when water is in a solid, liquid, and gaseous state. How is the movement of water molecules dependent upon temperature?

3. As water freezes, what happens to the water molecules that causes ice to float? Why is this unique?

4. What type of properties of a substance can be observed without changing it into a new substance?

5. **Think Critically** Suppose you blow up a balloon and then place it in a freezer. Later, you find that the balloon has shrunk and has drops of frozen liquid in it. Explain what has happened.

Skill Builder Activities

6. **Classifying** Classify the following items into the four types of matter: *groundwater, lightning, lava, snow, textbook, ice cap, notebook, apple juice, eraser, glass, cotton, helium, iron oxide, lake, limestone,* and *water vapor.* Compare and contrast their characteristics. **For more help, refer to the** Science Skill Handbook.

7. **Using Graphics Software** Research the melting and boiling points in degrees Celsius of several compounds, including water. Use your computer to make a line graph showing the temperatures at which these several compounds change state from solid to liquid to gas. Rearrange your data in order of increasing melting and boiling points. **For more help, refer to the** Technology Skill Handbook.

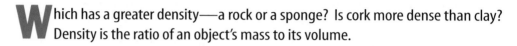

Determining Density

Which has a greater density—a rock or a sponge? Is cork more dense than clay? Density is the ratio of an object's mass to its volume.

Recognize the Problem

How can you determine the densities of several objects in your classroom?

Form a Hypothesis

State a hypothesis about what process you can use to measure and compare the densities of several materials.

Possible Materials
pan
triple-beam balance
100-mL beaker
250-mL graduated cylinder
water
sponge
piece of quartz
piece of clay
small wooden block
small metal block
small cork
rock
ruler

Goals
- **List** some ways that the density of an object can be measured.
- **Design** an experiment that compares the densities of several materials.

Safety Precautions

WARNING: *Be wary of sharp edges on some of the materials and take care not to break the beaker or graduated cylinder. Wash hands thoroughly with soap and water when finished.*

Test Your Hypothesis

Plan

1. As a group, agree upon and write the hypothesis statement.

2. As a group, list the steps that you need to take to test your hypothesis. Be specific, describing exactly what you will do at each step. List your materials.

3. While working as a group, use this equation: density = mass/volume. Devise a method of determining the mass and volume of each material to be tested.

4. **Design** a data table in your Science Journal so that it is ready to use as your group collects data.

Check the Plan

1. Read over your entire experiment to make sure that all steps are in a logical order.

2. Should you run the process more than once for any of the materials?

3. **Identify** any constants, variables, and controls of the experiment.

4. Make sure your teacher approves your plan before you start.

Do

1. Carry out the experiment as planned.

2. While the experiment is going on, write any observations that you make and complete the data table in your Science Journal.

Analyze Your Data

1. Do you observe anything about the way objects with greater density feel compared with objects of lower density?

2. Which of the objects you tested would float in water? Which would sink?

Draw Conclusions

1. Based on your results, would you hypothesize that a cork is more dense, the same density, or less dense than water?

2. Without measuring the density of an object that floats, conclude how you know that it has a density of less than 1.0 g/cm^3.

3. Would the density of the clay be affected if you were to break it into smaller pieces?

*C*ommunicating
Your Data

Write an informational pamphlet on different methods for determining the density of objects. Include equations and a step-by-step procedure.

Amazing Atoms

Did you know . . .

. . . The diameter of an atom is about 100,000 times as great as the diameter of its nucleus. Suppose that when you sit in a chair, you represent the nucleus of an atom. The nearest electron in your atom would be about 120 km away—nearly half the distance across the Florida peninsula.

. . . Uranium has the greatest mass of the abundant natural elements. One atom of uranium has a mass number that is more than 235 times greater than the mass number of one hydrogen atom, the element with the least mass. However, the diameter of a uranium atom is only about three times the size of a hydrogen atom, similar to the difference between a baseball and a volleyball.

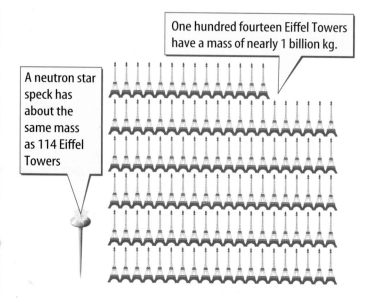

One hundred fourteen Eiffel Towers have a mass of nearly 1 billion kg.

A neutron star speck has about the same mass as 114 Eiffel Towers

. . . The densest material in the universe is found in a neutron star. The core of this type of star is made only of neutrons. Although neutron stars are small, measuring about 10 km to 20 km in diameter, they have a greater mass than the Sun. One tiny pinhead-sized speck of a neutron star would have the same mass as about 114 Eiffel Towers.

. . . The melting point of Cesium is 28.4°C. It would melt in your hand if you held it. You would not want to hold cesium, though, because it would react strongly with your skin. In fact, the metal might even catch fire.

. . . An atomic fountain clock is the world's most accurate timepiece. This timepiece is called an atomic fountain clock because it uses a fountainlike movement of atoms to record time. The atoms are cooled to an extremely low temperature and tossed into a vacuum chamber. The natural vibrations of the atoms are measured. Atomic fountain clocks gain or lose only 1 s in more than 20 million years.

. . . More than ninety elements occur naturally. However, about 98% of Earth's crust consists of only the eight elements shown here.

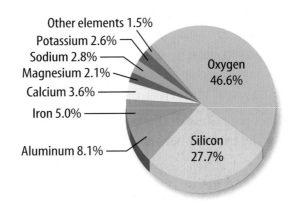

Other elements 1.5%
Potassium 2.6%
Sodium 2.8%
Magnesium 2.1%
Calcium 3.6%
Iron 5.0%
Aluminum 8.1%
Oxygen 46.6%
Silicon 27.7%

Do the Math

1. Looking at the circle graph, which is the third most abundant element in Earth's crust?
2. How much lower is the melting point of cesium than the average human body temperature of 37°C?
3. The diameter of the Sun is 1,392,000 km. How many neutron stars, each measuring 15 km in diameter, would fit along the Sun's diameter when placed side to side?

Go Further

Do research on the Glencoe Science Web site at **science.glencoe.com** to find out more about atoms and isotopes. What is a radioactive isotope of an element? How are isotopes used in science?

Section 1 Atoms

1. Matter is anything that has mass and takes up space. *How is matter similar to the snap-together blocks shown below?*

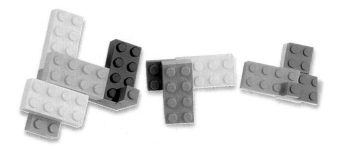

2. Protons and neutrons make up the nucleus of an atom. Protons have a positive charge, and neutrons have no charge. Electrons have a negative charge and surround the nucleus, forming an electron cloud.

3. Isotopes are atoms of the same element that have different numbers of neutrons.

Section 2 Combinations of Atoms

1. Atoms join to form compounds and molecules. A compound is a substance made of two or more elements. The properties of a compound differ from the chemical and physical properties of the elements of which it is composed.

2. A mixture is a substance in which the components are not chemically combined. *Why are the contents of the book bag, shown to the right, considered to be a mixture and not a compound?*

Section 3 Properties of Matter

1. Physical properties can be observed and measured without causing a chemical change in a substance. Chemical properties can be observed only when one substance reacts with another substance.

2. Atoms or molecules in a solid are in fixed positions relative to one another. In a liquid, the atoms or molecules are close together but are freer to change positions. Atoms or molecules in a gas move freely to fill any container. *What makes up plasma, such as the lightning bolts to the right?*

3. Water is the only substance on Earth that occurs naturally as a solid, liquid, and gas. Other substances do not exist in all three states on Earth because of Earth's narrow temperature range.

4. One physical property that is used to describe matter is density. Density is a ratio of the mass of an object to its volume. A material that is less dense will float in a material that is more dense. *Ice floats in liquid water, so which state of water is more dense?*

FOLDABLES
Reading & Study Skills

After You Read

Use each vocabulary word on your Vocabulary Study Fold in a sentence about matter and write it next to the definition of the word.

Visualizing Main Ideas

Complete the following concept map on matter. Use the following terms: liquids, plasma, matter, *and* solids.

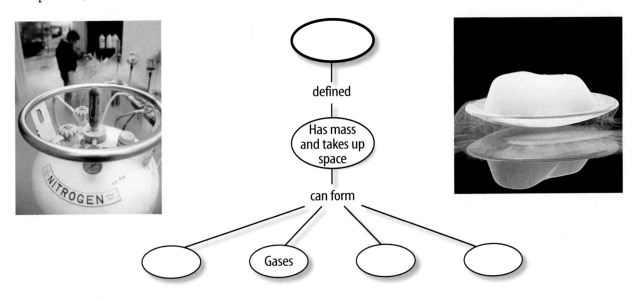

Vocabulary Review

Vocabulary Words

a. atom
b. atomic number
c. compound
d. density
e. electron
f. element
g. ion
h. isotope
i. heterogeneous mixture

j. homogeneous mixture
k. mass number
l. matter
m. mixture
n. neutron
o. proton
p. solution

THE PRINCETON REVIEW — Study Tip

Take good notes, even during lab. Lab experiments reinforce key concepts, and looking back on these notes can help you better understand what happened and why.

Using Vocabulary

Explain the difference between the vocabulary words in each of the following sets.

1. atom, element

2. mass number, atomic number

3. solution, heterogeneous mixture

4. matter, compound, element

5. heterogeneous mixture, homogeneous mixture

6. proton, neutron, electron

7. isotope, atom

8. atom, ion

9. mixture, compound

10. neutron, mass number

Chapter 2 Assessment

Checking Concepts

Choose the word or phrase that best answers the question.

1. Which of the following contains only one type of atom?
   A) compound
   B) mixture
   C) element
   D) solution

2. Which of the following has a positive electric charge?
   A) electron
   B) proton
   C) neutron
   D) atom

3. In an atom, what forms a cloud around the nucleus?
   A) electrons
   B) protons
   C) neutrons
   D) positively charged particles

4. A carbon atom has a mass number of 12. How many protons and how many neutrons does it have?
   A) 6, 6
   B) 12, 12
   C) 6, 12
   D) 12, 6

5. On Earth, oxygen usually exists as which of the following?
   A) solid
   B) gas
   C) liquid
   D) plasma

6. Which of the following isotopes has seven neutrons?
   A) boron-12
   B) nitrogen-12
   C) carbon-14
   D) hydrogen-2

7. Which type of bond occurs because of the polar molecules of water?
   A) ionic
   B) covalent
   C) metallic
   D) hydrogen

8. Which of the following are electrically charged?
   A) molecule
   B) solution
   C) isotope
   D) ion

9. What type of property is the color of your clothes?
   A) chemical property
   B) physical property
   C) isotopic property
   D) molecular property

10. Which of the following is not a physical property of water?
   A) transparent
   B) colorless
   C) higher density than ice
   D) changes to hydrogen and oxygen when electricity passes through it

Thinking Critically

11. If an atom has no electric charge, what can be said about the number of protons and electrons it contains?

12. Carbon has six protons and nitrogen has seven protons. Which has the greatest number of neutrons—carbon-13, carbon-14, or nitrogen-14?

13. Would isotopes of the same element have the same number of electrons? Explain.

14. A chlorine atom comes in contact with a lithium atom. Do they combine to form a compound? Why or why not?

15. You pour cooking oil into a glass of water. You briefly stir the materials in the glass. Does the glass contain a mixture? Does it contain a solution? Explain.

Developing Skills

16. Classifying Use the periodic table of the elements, located on the inside back cover, to classify the following substances as elements or compounds: iron, aluminum, carbon dioxide, gold, water, and sugar.

17. Making and Using Graphs Use the following data to make a line graph. For each isotope, plot the mass number along the *y*-axis and the atomic number along the *x*-axis. What is the relationship between mass number and atomic number?

Atomic Number versus Mass Number		
Element	**Atomic Number**	**Mass Number**
Fluorine	9	19
Lithium	3	7
Carbon	6	12
Nitrogen	7	14
Beryllium	4	9
Boron	5	11
Oxygen	8	16
Neon	10	20

Performance Assessment

18. Song with Lyrics Create a song about how matter changes state by changing the words to a song you know. Include in your song as many states of matter as possible.

TECHNOLOGY

Go to the Glencoe Science Web site at **science.glencoe.com** or use the **Glencoe Science CD-ROM** for additional chapter assessment.

THE PRINCETON REVIEW — Test Practice

Marlena was instructed by her teacher to find examples of elements and their isotopes. The examples she found are presented in the table below.

Element	Number of Protons	Number of Neutrons
Hydrogen-1	1	0
Hydrogen-2	1	1
Hydrogen-3	1	2
Carbon-12	6	6
Carbon-14	6	8
Oxygen-16	8	8
Oxygen-18	8	10

Study the table and answer the following questions.

1. The hydrogen atom listed first is different from the hydrogen-2 and hydrogen-3 isotopes, because the first hydrogen isotope has _____ .
   A) only one neutron
   B) fewer neutrons
   C) more electrons
   D) more neutrons

2. The mass number of an atom is equal to the number of protons and neutrons in its nucleus. According to this definition, which of these has the highest mass number?
   F) hydrogen-3
   G) carbon-14
   H) oxygen-18
   J) oxygen-16

Minerals

These gorgeous diamonds appear so perfect that it seems impossible they formed naturally. Although it's true that these specimens were cut by humans to enhance their beauty, some minerals seem flawless the moment they're found. As you'll soon learn, a requirement for being a mineral is that it must be naturally occurring. In the case of a diamond, this increases its rarity and its price. In this chapter, you'll learn about minerals and gems and how to identify them. You'll also learn how much society depends upon minerals.

What do you think?

Science Journal Look at the picture below with a classmate. Discuss what this might be. Here's a hint: *This "egg" was laid by a volcano.* Write your answer or best guess in your Science Journal.

EXPLORE ACTIVITY

What's the difference between a rock and a mineral? Just as a building is made of many different materials such as concrete, wood, plaster, and steel, rocks are made of many different things. Most rocks are made from crystals of several different minerals. Some rocks are made from many different crystals of the same mineral. When examining a rock, you'll notice that it is made from many different components. However, a mineral looks the same throughout. Can you tell a rock from a mineral?

Distinguish rocks from minerals

1. Use a magnifying glass to examine a quartz crystal, salt grains, and samples of sandstone, granite, calcite, mica, and schist (SHIHST).

2. Determine which samples are made of one type of material and should be classified as minerals.

3. Determine which samples are made of more than one type of material and should be classified as rocks.

Observe

In your Science Journal, compile a list of descriptions for the minerals you examined and a second list of descriptions for the rocks. Compare and contrast your observations of minerals and rocks.

Before You Read

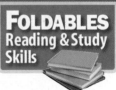

FOLDABLES
Reading & Study Skills

Making a Question Study Fold Asking your-self questions helps you stay focused and better understand minerals when you are reading the chapter.

1. Place a sheet of notebook paper in front of you so the short side is at the top and the holes are on the right side. Fold the paper in half from the left side to the right side.

2. Through the top thickness of paper, cut along every third line from the outside edge to the center fold, forming tabs as shown.

3. Before you read the chapter, write questions you have about minerals on the front of the tabs. As you read the chapter, add more questions and write answers under the tabs.

Minerals

As You Read

What You'll Learn
- **Describe** characteristics that all minerals share.
- **Explain** how minerals form.

Vocabulary
mineral
crystal
magma
silicate

Why It's Important
You use minerals every day.

Figure 1
You probably use minerals or materials made from minerals every day without thinking about it. *How many objects in this picture might be made from minerals?*

What is a mineral?

How important are minerals to you? Very important? You actually own or encounter many things made from minerals every day. Ceramic, metallic, and even some paper items are examples of products that are derived from or include minerals. **Figure 1** shows just a few of these things. Metal bicycle racks, bricks, and the glass in windows would not exist if it weren't for minerals. A **mineral** is a naturally occurring, inorganic solid with a definite chemical composition and an orderly arrangement of atoms. About 4,000 different minerals are found on Earth, but they all share these four characteristics.

Mineral Characteristics First, all minerals are formed by natural processes. These are processes that occur on or inside Earth with no input from humans. For example, salt formed by the natural evaporation of seawater is the mineral halite, but salt formed by evaporation of saltwater solutions in laboratories is not a mineral. Second, minerals are inorganic. This means that they aren't made by life processes. Third, every mineral is an element or compound with a definite chemical composition. For example, halite's composition, NaCl, gives it a distinctive taste that adds flavor to many foods. Fourth, minerals are crystalline solids. All solids have a definite volume and shape. Gases and liquids like air and water have no definite shape, and they aren't crystalline. Only a solid can be a mineral, but not all solids are minerals.

Atom Patterns The word *crystalline* means that atoms are arranged in a pattern that is repeated over and over again. For example, graphite's atoms are arranged in layers. Opal, on the other hand, is not a mineral in the strictest sense because its atoms are not all arranged in a definite, repeating pattern, even though it is a naturally occurring, inorganic solid.

Figure 2
More than 200 years ago, the smooth, flat surfaces on crystals led scientists to infer that minerals had an orderly, internal structure inside.

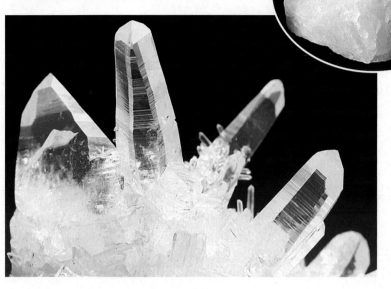

A The well-formed crystal shapes exhibited by these clear quartz crystals suggest an orderly structure. **B** Even though this rose quartz looks uneven on the outside, its atoms have an orderly arrangement on the inside.

The Structure of Minerals

Do you have a favorite mineral sample or gemstone? If so, perhaps it contains well-formed crystals. A **crystal** is a solid in which the atoms are arranged in orderly, repeating patterns. You can see evidence for this orderly arrangement of atoms when you observe the smooth, flat outside surfaces of crystals. A crystal system is a group of crystals that have similar atomic arrangements and therefore similar external crystal shapes.

✓ Reading Check *What is a crystal?*

Crystals Not all mineral crystals have smooth surfaces and regular shapes like the clear quartz crystals in **Figure 2A.** The rose quartz in **Figure 2B** has atoms arranged in repeating patterns, but you can't see the crystal shape on the outside of the mineral. This is because the rose quartz crystals developed in a tight space, while the clear quartz crystals developed freely in an open space. The six-sided, or hexagonal crystal shape of the quartz crystals in **Figure 2A,** and other forms of quartz can be seen in some samples of the mineral. **Figure 3** illustrates the six major crystal systems, which classify minerals according to their crystal structures. The hexagonal system to which quartz belongs is one example of a crystal system.

Crystals form by many processes. Next, you'll learn about two of these processes—crystals that form from magma and crystals that form from solutions of salts.

TRY AT HOME

Mini LAB

Inferring Salt's Crystal System

Procedure
1. Use a **magnifying glass** and a **dissecting probe** to observe grains of common **table salt.** Sketch the shape of a salt grain. **WARNING:** *Do not taste or eat mineral samples. Keep hands away from your face.*
2. Compare the shapes of the salt crystals with the shapes of crystals shown in **Figure 3.**

Analysis
1. Which characteristics do all the grains have in common?
2. Research another mineral with the same crystal system as salt. What is this crystal system called?

Figure 3

A crystal's shape depends on how its atoms are arranged. Crystal shapes can be organized into groups known as crystal systems—shown here in 3-D with geometric models (in blue). Knowing a mineral's crystal system helps researchers understand its atomic structure and physical properties.

▲ HEXAGONAL (hek SA guh nul) In hexagonal crystals, horizontal distances between opposite crystal surfaces are equal. These crystal surfaces intersect to form 60° or 120° angles. The vertical length is longer or shorter than the horizontal lengths.

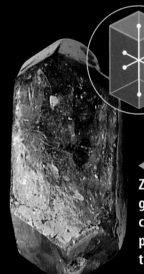

▲ CUBIC Fluorite is an example of a mineral that forms cubic crystals. Minerals in the cubic crystal system are equal in size along all three principal dimensions.

◀ TETRAGONAL (te TRA guh nul) Zircon crystals are tetragonal. Tetragonal crystals are much like cubic crystals, except that one of the principal dimensions is longer or shorter than the other two dimensions.

▲ ORTHORHOMBIC (awr thuh RAHM bihk) Minerals with orthorhombic structure, such as barite, have dimensions that are unequal in length, resulting in crystals with a brick-like shape.

▲ MONOCLINIC (mah nuh KLIH nihk) Minerals in the monoclinic system, such as orthoclase, also exhibit unequal dimensions in their crystal structure. Only one right angle forms where crystal surfaces meet. The other angles are oblique, which means they don't form 90° angles where they intersect.

▲ TRICLINIC (tri KLIH nihk) The triclinic crystal system includes minerals exhibiting the least symmetry. Triclinic crystals, such as rhodonite (ROH dun ite), are unequal in all dimensions, and all angles where crystal surfaces meet are oblique.

Figure 4
Minerals form by many natural processes.

A This rock formed as magma cooled slowly, allowing large mineral grains to form.

Labradorite

B Some minerals form when salt water evaporates, such as these white crystals of halite in Death Valley, California.

Crystals from Magma
Natural processes form minerals in many ways. For example, hot melted rock, called **magma,** cools when it reaches Earth's surface, or even if it's trapped below the surface. As magma cools, its atoms lose heat energy, move closer together, and begin to combine into compounds. Molecules of the different compounds then arrange themselves into orderly, repeating patterns. The type and amount of elements present in a magma partly determine which minerals will form. Also, the size of the crystals that form depends partly on how rapidly the magma cools.

When magma cools slowly, the crystals that form are generally large enough to see with the unaided eye, as shown in **Figure 4A.** This is because the atoms have enough time to move together and form into larger crystals. When magma cools rapidly, the crystals that form will be small. In such cases, you can't easily see individual mineral crystals.

Crystals from Solution
Crystals also can form from minerals dissolved in water. When water evaporates, as in a dry climate, ions that are left behind can come together to form crystals like those in **Figure 4B.** Or, if too much of a substance is dissolved in water, ions can come together and crystals of that substance can begin to form in the solution. Minerals can come out of a solution in this way without the need for evaporation.

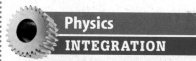

Physics
INTEGRATION

Evaporites commonly form in dry climates. Research the change that takes place when a saline lake or shallow sea evaporates and halite or gypsum forms.

Elements in Earth's Crust

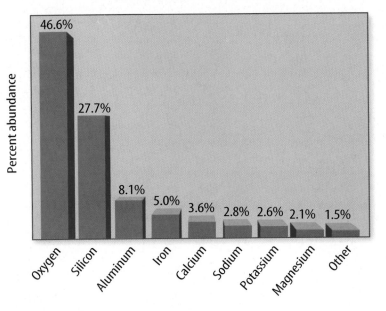

Figure 5
Most of Earth's crust is composed of eight elements.

Mineral Compositions and Groups

Ninety elements occur naturally in Earth's crust. Approximately 98 percent (by weight) of the crust is made of only eight of these elements, as in **Figure 5.** Of the thousands of known minerals, only a few dozen are common, and these are mostly composed of the eight common elements in Earth's crust.

Most of the common rock-forming minerals belong to a group called the silicates. **Silicates** (SIH luh kayts) are minerals that contain silicon (Si) and oxygen (O) and usually one or more other elements. Silicon and oxygen are the two most abundant elements in Earth's crust. These two elements alone combine to form the basic building blocks of most of the minerals in Earth's crust and mantle. Feldspar and quartz, which are silicates, and calcite, which is a carbonate, are examples of common, rock-forming minerals. Other mineral groups also are defined according to their compositions.

Section 1 Assessment

1. What four characteristics must a substance have to be a mineral?

2. Describe two processes involving solutions that form minerals.

3. Diamonds can be made in the laboratory by subjecting starting materials containing carbon to high pressure. Are these diamonds minerals? Explain.

4. How are crystals of minerals classified?

5. **Think Critically** The mineral dolomite, a rock-forming mineral, contains oxygen, carbon, magnesium, and calcium. Is dolomite a silicate? In your Science Journal, compare and contrast rock-forming minerals and silicates.

Skill Builder Activities

6. **Classifying** Discover some items in your home or classroom that are made from minerals. How many items are metals? How many are non-metals? **For more help, refer to the** Science Skill Handbook.

7. **Using an Electronic Spreadsheet** Use spreadsheet software to make a graph of your own design that shows the relative percentages of the eight most common elements in Earth's crust. Using these eight elements, about what percentage of the crust is made up of iron and aluminum? Use the spreadsheet to perform this calculation. **For more help, refer to the** Technology Skill Handbook.

Activity

Crystal Formation

So far in this chapter, you've learned about minerals and how they form. In this activity, you'll have a chance to learn how crystals form from solutions.

What You'll Investigate
How do crystals form from solution?

Materials
250-mL beakers (2)	cotton string
cardboard	hot plate
large paper clip	hand lens
table salt	thermal mitt
flat wooden stick	shallow pan
granulated sugar	spoon

Goals
- **Compare and contrast** the crystals that form from salt and sugar solutions.
- **Observe** crystals and infer how they formed.

Safety Precautions 🥽 👕 🧤 🔪 🔥

WARNING: *Never taste or eat any lab materials.*

Procedure

1. Gently mix separate solutions of salt in water and sugar in water in the two beakers. Keep stirring the solutions as you add salt or sugar to the water. Stop mixing when no more salt or sugar will dissolve in the solutions. Label each beaker as a salt or sugar solution.

2. Place the sugar beaker on a hot plate. Use the hot plate to heat the sugar solution gently. **WARNING:** *The liquid is hot. Do not touch the beaker without protecting your hands.*

3. Tie one end of the thread to the middle of the wooden stick. Tie a large paper clip to the free end of the string for weight. Place the stick across the opening of the sugar beaker so the thread dangles in the sugar solution.

4. Remove the beaker from the hot plate and cover it with cardboard. Place it in a location where it won't be disturbed.

5. Pour a thin layer of the salt solution into the shallow pan.

6. Leave the beaker and the shallow pan undisturbed for at least one week.

7. After one week, examine each solution with a hand lens to see whether crystals have formed.

Conclude and Apply

1. **Compare and contrast** the crystals that formed from the salt and the sugar solutions. How do they compare with samples of table salt and sugar?

2. **Describe** what happened to the saltwater solution in the shallow pan.

3. Did this same process occur in the sugar solution? Explain.

Communicating Your Data

Make a poster that describes your methods of growing salt and sugar crystals. Present your results to your class.

2 Mineral Identification

As You Read

What You'll Learn

- **Describe** physical properties used to identify minerals.
- **Identify** minerals using physical properties such as hardness and streak.

Vocabulary

hardness   streak
luster   cleavage
specific gravity fracture

Why It's Important

Identifying minerals helps you recognize valuable mineral resources.

Physical Properties

Why can you recognize a classmate when you see him or her in a crowd away from school? A person's height or the shape of his or her face helps you tell that person from the rest of your class. Height and facial shape are two properties unique to individuals. Individual minerals also have unique properties that distinguish them.

Mineral Appearance Just like height and facial characteristics help you recognize someone, mineral properties can help you recognize and distinguish minerals. Color and appearance are two obvious clues that can be used to identify minerals.

However, these clues alone aren't enough to recognize most minerals. The minerals pyrite and gold are gold in color and can appear similar, as shown in **Figure 6.** As a matter of fact, pyrite often is called fool's gold. Gold is worth a lot of money, whereas pyrite has little value. You need to look at other properties of minerals to tell them apart. Some other properties to study include how hard a mineral is, how it breaks, and its color when crushed into a powder. Every property you observe in a mineral is a clue to its identity.

Figure 6
The general appearance of a mineral often is not enough to identify it.

Gold

Pyrite

A Using only color, observers can be fooled when trying to distinguish between pyrite and gold.

B The mineral azurite is identified readily by its striking blue color.

Hardness A measure of how easily a mineral can be scratched is its **hardness.** The mineral talc is so soft you can scratch it loose with your fingernail. Talcum powder is made from this soft mineral. Diamonds, on the other hand, are the hardest mineral. Some diamonds are used as cutting tools, as shown in **Figure 7.** A diamond can be scratched only by another diamond. Diamonds can be broken, however.

 *Why is hardness sometimes referred to as scratchability?*

Sometimes the concept of hardness is confused with whether or not a mineral will break. It is important to understand that even though a diamond is extremely hard, it can shatter if given a hard enough blow in the right direction along the crystal.

Mohs Scale In 1824, the Austrian scientist Friedrich Mohs developed a list of common minerals to compare their hardnesses. This list is called Mohs scale of hardness, as seen in **Table 1.** The scale lists the hardness of ten minerals. Talc, the softest mineral, has a hardness value of one, and diamond, the hardest mineral, has a value of ten.

Here's how the scale works. Imagine that you have a clear or whitish-colored mineral that you know is either fluorite or quartz. You try to scratch it with your fingernail and then with an iron nail. You can't scratch it with your fingernail but you can scratch it with the iron nail. Because the hardness of your fingernail is 2.5 and that of the iron nail is 4.5, you can determine the unknown mineral's hardness to be somewhere around 3 or 4. Because it is known that quartz has a hardness of 7 and fluorite has a hardness of 4, the mystery mineral must be fluorite.

Some minerals have a hardness range rather than a single hardness value. This is because atoms are arranged differently in different directions in their crystal structures.

Figure 7
Some saw blades have diamonds embedded in them to help slice through materials, such as this limestone. Blades are kept cool by running water over them.

Table 1 Mineral Hardness

Mohs Scale	Hardness	Hardness of Common Objects	
Talc (softest)	1		
Gypsum	2	fingernail	(2.5)
Calcite	3	piece of copper	(2.5 to 3.0)
Fluorite	4	iron nail	(4.5)
Apatite	5	glass	(5.5)
Feldspar (orthoclase)	6	steel file	(6.5)
Quartz	7	streak plate	(7.0)
Topaz	8		
Corundum	9		
Diamond (hardest)	10		

Graphite

Fluorite

Luster Luster is the way a mineral reflects light. Luster can be metallic or nonmetallic. Minerals with a metallic luster, like the graphite shown in **Figure 8,** shine like metal. Metallic luster can be compared to the shine of a metal belt buckle, the shiny chrome trim on some cars, or the shine of metallic cooking utensils. When a mineral does not shine like metal, its luster is nonmetallic. Examples of terms for non-metallic luster include *dull, pearly, silky,* and *glassy.* Common examples of minerals with glassy luster are quartz, calcite, halite, and fluorite.

Figure 8

Luster is an important physical property that is used to distinguish minerals. Graphite has a metallic luster. Fluorite has a nonmetallic, glassy luster.

Specific Gravity Minerals also can be distinguished by comparing their heft. The **specific gravity** of a mineral is the ratio of its weight compared with the weight of an equal volume of water. Like hardness, specific gravity is expressed as a number. If you were to research the specific gravities of gold and pyrite, you'd find that gold's specific gravity is about 17, and pyrite's is 5. This means that gold is about 17 times heavier than water and pyrite is 5 times heavier than water. You could experience this by comparing equal-sized samples of gold and pyrite in your hands—the pyrite would feel much lighter.

Problem-Solving Activity

How can you identify minerals?

Properties of Minerals

Mineral	Hardness	Streak
Copper	2.5–3	copper-red
Galena	2.5	dark gray
Gold	2.5–3	yellow
Hematite	5.5–6.5	red to brown
Magnetite	6–6.5	black
Silver	2.5–3	silver-white

You have learned that minerals are identified by their physical properties, such as streak, hardness, cleavage, and color. Use your knowledge of mineral properties and your ability to read a table to solve the following problems.

Identifying the Problem

The table includes hardnesses and streak colors for several minerals. How can you use these data to distinguish minerals?

Solving the Problem

1. What test would you perform to distinguish hematite from copper? How would you carry out this test?

2. How could you distinguish copper from galena? What tool would you use?

3. What would you do if two minerals had the same hardness and the same streak color?

Streak When a mineral is rubbed across a piece of unglazed porcelain tile, as in **Figure 9,** a streak of powdered mineral is left behind. **Streak** is the color of a mineral when it is in a powdered form. The streak test works only for minerals that are softer than the streak plate. Gold and pyrite can be distinguished by a streak test. Gold has a yellow streak and pyrite has a greenish-black or brownish-black streak.

Some soft minerals will leave a streak even on paper. The last time you used a pencil to write on paper, you left a streak of the mineral graphite. One reason that graphite is used in pencil lead is because it is soft enough to leave a streak on paper.

 **Reading Check** *Why do gold and pyrite leave a streak, but quartz does not?*

Cleavage and Fracture The way a mineral breaks is another clue to its identity. Minerals that break along smooth, flat surfaces have **cleavage** (KLEE vihj). Cleavage, like hardness, is determined by the arrangement of the mineral's atoms. Mica is a mineral that has one perfect cleavage. You can see in **Figure 10** how it breaks along smooth, flat planes. If you were to take a layer cake and separate its layers, you would show that the cake has cleavage. Not all minerals have cleavage. Minerals that break with uneven, rough, or jagged surfaces have **fracture.** Quartz is a mineral with fracture. If you were to grab a chunk out of the side of that cake, it would be like breaking a mineral that has fracture.

Figure 9
Streak is more useful for mineral identification than is mineral color. Hematite, for example, can be dark red, gray, or silver in color. However, its streak is always dark reddish-brown.

Halite

Figure 10
Weak bonds within the structures of mica and halite allow them to be broken along smooth, flat cleavage planes. *If you broke quartz, would it look the same?*

Mica

Figure 11
Some minerals are natural magnets, such as this lodestone, which is a variety of magnetite.

Mini LAB

Observing Mineral Properties

Procedure

1. Obtain samples of some of the following clear minerals: **gypsum, muscovite mica, halite,** and **calcite.**
2. Place each sample over the print on this page and observe the letters.

Analysis

1. Which mineral can be identified by observing the print's double image?
2. What other special property is used to identify this mineral?

Other Properties Some minerals have unique properties. Magnetite, as you can guess by its name, is attracted to magnets. Lodestone, a form of magnetite, will pick up iron filings like a magnet, as shown in **Figure 11.** Light forms two separate rays when it passes through some calcite specimens, causing you to see a double image. Calcite also can be identified because it fizzes when hydrochloric acid is put on it.

Now you know that you sometimes need more information than color and appearance to identify a mineral. You also might need to test its streak, hardness, luster, and cleavage or fracture. Although the overall appearance of a mineral can be different from sample to sample, its physical properties remain the same.

Section 2 Assessment

1. What is the difference between a mineral that has cleavage and one that has fracture?
2. How can an unglazed porcelain tile be used to identify a mineral?
3. Why is streak often more useful for mineral identification than color?
4. What hardness does a mineral have if it does not scratch glass but it scratches an iron nail?
5. **Think Critically** What does the presence of cleavage planes within a mineral tell you about the chemical bonds that hold the mineral together?

Skill Builder Activities

6. **Drawing Conclusions** A large piece of the mineral halite (salt) is broken repeatedly into several perfect cubes. How can this be explained? **For more help, refer to the** Science Skill Handbook.
7. **Making and Using Tables** In your Science Journal, make a list of three minerals. Next to each mineral, write another mineral that is similar in appearance. In a third column, list one physical property that can be used to distinguish the pair of similar-looking minerals. **For more help, refer to the** Science Skill Handbook.

SECTION 3

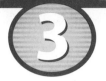

Uses of Minerals

Gems

Walking past the window of a jewelry store, you notice a large selection of beautiful jewelry—a watch sparkling with diamonds, a necklace holding a brilliant red ruby, and a gold ring. For thousands of years, people have worn and prized minerals in their jewelry. What makes some minerals special? What unusual properties do they have that make them so valuable?

Properties of Gems **Gems** or gemstones, like the ones shown in **Figure 12,** are highly prized minerals because they are rare and beautiful. Most gems are special varieties of a particular mineral. They are clearer, brighter, or more colorful than common samples of that mineral. The difference between a gem and the common form of the same mineral can be slight. Amethyst is a gem form of quartz that contains just traces of iron in its structure. This small amount of iron gives amethyst a desirable purple color. Sometimes a gem has a crystal structure that allows it to be cut and polished to a higher quality than that of a non-gem mineral. **Table 2** lists popular gems and some locations where they have been collected.

As You Read

***What* You'll Learn**

- **Describe** characteristics of gems that make them more valuable than other minerals.
- **Identify** useful elements that are contained in minerals.

Vocabulary

gem

***Why* It's Important**

Minerals are necessary materials for decorative items and many manufactured products.

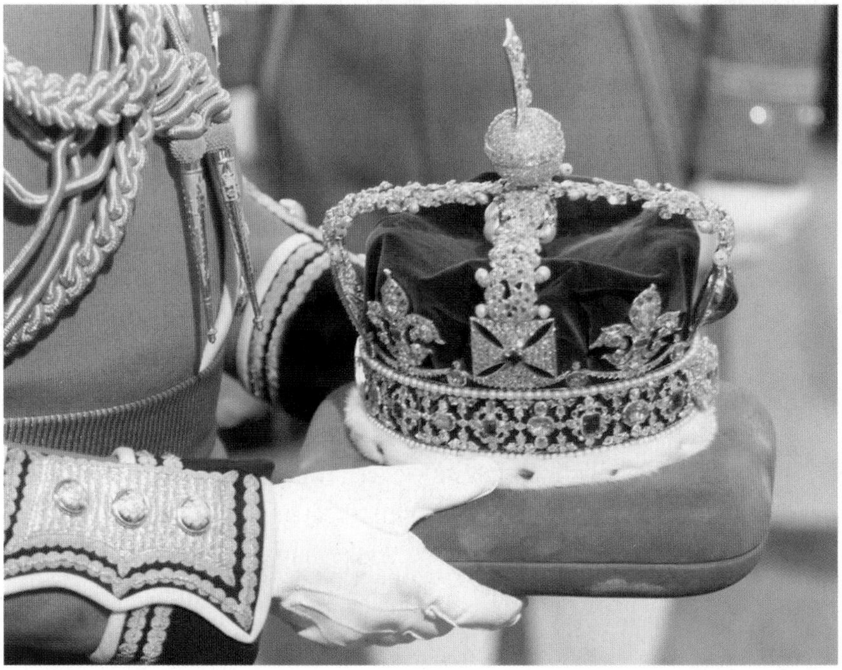

Figure 12
It is easy to see why gems are prized for their beauty and rarity. Shown here is The Imperial State Crown, made for Queen Victoria of England in 1838. It contains thousands of jewels, including diamonds, rubies, sapphires, and emeralds.

Table 2 Minerals and Their Gems

Fun Facts	Mineral	Gem Example	Some Important Locations
Beryl is named for the element beryllium, which it contains. Some crystals reach several meters in length.	**Beryl**	**Emerald**	Colombia, Brazil, South Africa, North Carolina
A red spinel in the British crown jewels has a mass of 352 carats. A carat is 0.2g.	**Spinel**	**Ruby spinel**	Sri Lanka, Thailand, Myanmar (Burma)
Purplish-blue examples of zoisite were discovered in 1967 near Arusha, Tanzania.	**Zoisite**	**Tanzanite**	Tanzania
The most valuable examples are yellow, pink, and blue varieties.	**Topaz** (uncut)	**Topaz** (gem)	Siberia, Germany, Japan, Mexico, Brazil, Colorado, Utah, Texas, California, Maine, Virginia, South Carolina

Fun Facts	Mineral	Gem Example	Some Important Locations
Olivine composes a large part of Earth's upper mantle. It also is present in moon rocks.	Olivine	Peridot	Myanmar (Burma), Zebirget (Saint John's Island, located in the Red Sea), Arizona, New Mexico
Garnet is a common mineral found in a wide variety of rock types. The red color of the variety almandine is caused by iron in its crystal structure.	Garnet	Almandine	Ural Mountains, Italy, Madagascar, Czech Republic, India, Sri Lanka, Brazil, North Carolina, Arizona, New Mexico
Quartz makes up about 30 percent of Earth's continental crust.	Quartz	Amethyst	Colorless varieties in Hot Springs, Arkansas; Amethyst in Brazil, Uruguay, Madagascar, Montana, North Carolina, California, Maine
The blue color of sapphire is caused by iron or titanium in corundum. Chromium in corundum produces the red color of ruby.	Corundum	Blue sapphire	Thailand, Cambodia, Sri Lanka, Kashmir

Important Gems All gems are prized, but some are truly spectacular and have played an important role in history. For example, the Cullinan diamond, found in South Africa in 1905, was the largest uncut diamond ever discovered. Its mass was 3,106.75 carats (about 621 g). The Cullinan diamond was cut into 9 main stones and 96 smaller ones. The largest of these is called the Cullinan 1 or Great Star of Africa. It's mass is 530.20 carats (about 106 g), and it is now part of the British monarchy's crown jewels, shown in **Figure 13A.**

Another well-known diamond is the blue Hope diamond, shown in **Figure 13B.** This is perhaps the most notorious of all diamonds. It was purchased by Henry Philip Hope around 1830, after whom it is named. Because his entire family as well as a later owner suffered misfortune, the Hope diamond has gained a reputation for bringing its owner bad luck. The Hope diamond's mass is 45.52 carats (about 9 g). Currently it is displayed in the Smithsonian Institution in Washington.

Useful Gems In addition to their beauty, some gems serve useful purposes. You learned earlier that diamonds have a hardness of 10 on Mohs scale. They can scratch almost any material—a property that makes them useful as industrial abrasives and cutting tools. Other useful gems include rubies, which are used to produce specific types of laser light. Quartz crystals are used in electronics and as timepieces. When subjected to an electric field, quartz vibrates steadily, which helps control frequencies in electronic devices and allows for accurate timekeeping.

Most industrial diamonds and other gems are synthetic, which means that humans make them. However, the study of natural gems led to their synthesis, allowing the synthetic varieties to be used by humans readily.

Figure 13
These gems are among the most famous examples of precious stones.

A The Great Star of Africa is part of a sceptre in the collection of British crown jewels.

B Beginning in 1668, the Hope diamond was part of the French crown jewels. Then known as the French Blue, it was stolen in 1792 and later surfaced in London, England in 1812.

Figure 14
Bauxite, an ore of aluminum, is processed to make pure aluminum metal for useful products.

Bauxite

Useful Elements in Minerals

Gemstones are perhaps the best-known use of minerals, but they are not the most important. Look around your home. How many things made from minerals can you name? Can you find anything made from iron?

Ores Iron, used in everything from frying pans to ships, is obtained from its ore, hematite. A mineral or rock is an ore if it contains a useful substance that can be mined at a profit. Magnetite is another mineral that contains iron.

Reading Check *When is a mineral also an ore?*

Extracting Elements Aluminum sometimes is refined, or purified, from the ore bauxite, shown in **Figure 14.** In the process of refining aluminum, aluminum oxide powder is separated from unwanted materials that are present in the original bauxite. After this, the aluminum oxide powder is converted to molten aluminum by a process called smelting.

During smelting, a substance is melted to separate it from any unwanted materials that may remain. Aluminum can be made into useful products like bicycles, soft-drink cans, foil, and lightweight parts for airplanes and cars. The plane flown by the Wright brothers during the first flight at Kitty Hawk had an engine made partly of aluminum.

Chemistry
INTEGRATION

In 1886, Charles Martin Hall discovered an important process that led to modern aluminum refinement. Research the Hall process. Describe what you learn in your Science Journal.

Figure 15
The mineral sphalerite (black) is an important source of zinc. Iron often is coated with zinc to prevent rust in a process called galvanization.

SCIENCE *Online*

Research Visit the Glencoe Science Web site at **science.glencoe.com** for recent news or magazine articles on uses of titanium. Communicate to your class what you learn.

Figure 16
Rutile and ilmenite are common ore minerals of the element titanium.

Vein Minerals Under certain conditions, metallic elements can dissolve in fluids. These fluids then travel through weaknesses in rocks and form mineral deposits. Weaknesses in rocks include natural fractures or cracks, faults, and surfaces between layered rock formations. Mineral deposits left behind that fill in the open spaces created by the weaknesses are called vein mineral deposits.

✓ **Reading Check** *How do fluids move through rocks?*

Sometimes, vein mineral deposits fill in the empty spaces after rocks collapse. An example of a mineral that can form in this way is shown in **Figure 15.** This is the shiny mineral sphalerite, a source of the element zinc, which is used in batteries. Sphalerite sometimes fills spaces in collapsed limestone.

Minerals Containing Titanium You might own golf clubs with titanium shafts or a racing bicycle containing titanium. Perhaps you know someone who has a titanium hip or knee replacement. Titanium is a durable, nontoxic, lightweight, metallic element derived from minerals that contain this metal in their atomic structures. Two minerals that are sources of the element titanium are ilmenite (IHL muh nite) and rutile (rew TEEL), shown in **Figure 16.** Ilmenite and rutile are common in rocks that form when magma cools and solidifies. They also occur as vein mineral deposits and in beach sands.

Rutile

Ilmenite

Uses for Titanium Titanium is used in automobile body parts, such as connecting rods, valves, and suspension springs. It is used in the manufacture of aircraft and eyeglass frames. Low density and durability make titanium valuable in the production of these goods and of sports equipment such as tennis rackets and bicycles. Wheelchairs used by people who want to race or play basketball often are made from titanium, as shown in **Figure 17.** Titanium is one of many examples of useful materials that come from minerals and that enrich humans' lives.

Figure 17
Wheelchairs used for racing and playing basketball often have parts made from titanium.

Section Assessment

1. Why is the Cullinan diamond considered to be important?

2. What do rubies and sapphires have in common?

3. Describe how vein minerals form.

4. Why is bauxite considered to be a useful rock?

5. **Think Critically** Titanium is a nontoxic metal. Why is this an important characteristic in the manufacture of artificial body parts?

Skill Builder Activities

6. **Comparing and Contrasting** Compare and contrast gem-quality amethyst with regular quartz. **For more help, refer to the** Science Skill Handbook.

7. **Using Percentages** On average, Earth's continental crust contains 5 percent iron and 0.007 percent zinc. How many times more iron than zinc is present in the average continental crust? **For more help, refer to the** Math Skill Handbook.

Activity
Design Your Own Experiment

Mineral Identification

Some mineral properties can be more useful than others when testing a particular mineral. Although certain minerals can be identified observing only one property, others require testing several properties to identify them.

Recognize the Problem

How can you identify unknown minerals?

Form a Hypothesis

Make a hypothesis about which properties you think will be most useful in identifying particular minerals.

Goals

- **Determine** which properties of a mineral you will use for identification purposes.
- **Design** an experiment that tests these properties to help identify minerals.

Possible Materials

mineral samples
hand lens
pan balance
graduated cylinder
water
piece of copper
copper penny
glass plate
small iron nail
steel file

streak plate
5% hydrochloric
 acid (HCl) with
 dropper
vinegar
Mohs scale
 of hardness
Minerals Appendix
safety goggles
Alternate materials

Safety Precautions

Use care when handling acids. Be careful when handling materials with sharp edges. **WARNING:** *HCl can cause burns. If spillage occurs, notify your teacher and rinse with a generous flow of cool water until you are told to stop. Do not taste, eat, or drink any lab materials. Never hold a glass plate in your hands while testing hardness because the plate could break.*

Test Your Hypothesis

Plan

1. As a group, agree upon and write your hypothesis statement.

2. As a group, list the steps you will need to take to test your hypothesis. Be specific, describing exactly what you will do at each step.

3. **List** the materials you will need to complete your experiment.

4. **List** the various properties of minerals that you will test.

5. **Predict** any special properties you expect to observe or test.

6. Reread your experiment to make sure that steps are in a logical order.

7. Should you test the properties of the minerals more than once?

8. How will you summarize your data?

9. **Identify** all constants, variables, and controls of the experiment.

Do

1. Make sure your teacher approves your plan before you start.

2. Carry out the experiment as planned.

3. While doing the experiment, write any observations that you or other members of your group make. Summarize your data in your Science Journal.

Analyze Your Data

1. Which properties were most useful in identifying your samples? Which properties were least useful?

2. **Compare** the properties that worked best for you with those that worked best for other students.

Draw Conclusions

1. For five minerals, discuss reasons why certain properties are useful and others are not.

2. **Determine** two properties that distinguish clear, transparent quartz from clear, transparent calcite. Explain your choice of properties.

*C*ommunicating
Your Data

For three minerals, **list** physical properties that were important for their identification. **For more help, refer to the** Science Skill Handbook.

Dr. Dorothy

Like X rays, electrons are diffracted by crystalline substances, revealing information about their internal structures and symmetry. This electron diffraction pattern of titanium was obtained with an electron beam focused along a specific direction in the crystal.

What contributions did Dorothy Crowfoot Hodgkin make to science?

Dr. Hodgkin used a method called X-ray crystallography (kris tuh LAH gruh fee) to figure out the structures of crystalline substances, including vitamin B_{12}, vitamin D, penicillin, and insulin. Her studies showed how the atoms in these materials are organized.

What's X-ray crystallography?

Using X rays, scientists are able to figure out the shapes and the atomic structures of crystalline substances.

As X rays travel through a crystal, the crystal diffracts, or scatters, the X rays into a regular pattern. Like an individual's fingerprints, each crystalline substance has a unique diffraction pattern.

Crystallography has applications in the life, Earth, and physical sciences. For example, geologists use X-ray crystallography to study minerals found in Earth's rocks. Minerals like diamond and graphite have the same chemical makeup but different atomic structures. Because the atomic structure of a mineral is unique, X-ray crystallography helps geologists know exactly which substance they are examining.

"She will be remembered as a great chemist, a saintly, gentle, and tolerant lover of people, and a devoted protagonist on peace."

—Max F. Perutz, Recipient of the 1962 Nobel Prize, Chemistry

Crowfoot Hodgkin

Trailblazing scientist and humanitarian

How does Hodgkin's research help people today?

Dr. Hodgkin's discovery of the structure of insulin helped scientists learn how to control diabetes, a disease that affects over 15 million Americans. Diabetics' bodies are unable to process sugar efficiently. Diabetes can be fatal. Fortunately, Dr. Hodgkin's research with insulin has saved many lives.

What were some obstacles Hodgkin overcame?

Dorothy Hodgkin was born in 1910. When she was four, she and her sisters were left in England while their parents traveled to Egypt.

Dr. Hodgkin always said that this helped encourage her independent spirit. During the 1930s, there were few women scientists. It was a man's world. Hodgkin was not even allowed to attend meetings of the chemistry faculty where she taught because she was a woman. Eventually, she won over her stubborn colleagues with her intelligence and tenacity. Hodgkin died in 1994.

What else did she do?

Dr. Hodgkin was involved in world peace, nuclear disarmament, and aiding poorer nations. She also helped make it a bit easier for women scientists to be accepted by men.

CONNECTIONS Research Look in reference books or go to the Glencoe Science Web site for information on how X-ray crystallography is used to study minerals. Write your findings and share them with your class.

Chapter ③ Study Guide

Reviewing Main Ideas

Section 1 Minerals

1. You use minerals every day of your life. The pencil "lead" you write with is the mineral graphite. The pretzels you eat for a snack are sprinkled with halite. Gold, silver, gems, and semiprecious stones are used in jewelry. Much of what you use each day is made at least in some part from minerals.

2. All minerals are formed by natural processes and are inorganic solids with definite chemical compositions and orderly internal atomic structures. *Why do minerals in geodes like the one shown here have nicely formed crystal faces?*

3. Minerals have crystal structures in one of six major crystal systems.

Section 2 Mineral Identification

1. Hardness is a measure of how easily a mineral can be scratched.

2. Luster describes how light reflects from a mineral's surface. *What type of luster does this cube-shaped mineral have?*

3. Streak is the color of the powder left by a mineral on an unglazed porcelain tile.

4. Minerals that break along smooth, flat surfaces have cleavage. When minerals break with rough or jagged surfaces, they are displaying fracture.

5. Some minerals have special properties that aid in identifying them. For example, clear crystals of calcite display a double image when placed over type on a page. The mineral magnetite is identified readily by its attraction to a magnet.

Section 3 Uses of Minerals

1. Gems are minerals that are more rare and beautiful than common minerals.

2. Minerals are useful for their physical properties and for the elements they contain. Titanium is a useful element that is derived from the minerals ilmenite and rutile. Rocks containing ilmenite and rutile are ores of titanium, provided that they can be mined at a profit. *What physical property of diamond makes it useful as an industrial tool as well as a beautiful gem?*

FOLDABLES
Reading & Study Skills

After You Read

Exchange Foldables with a classmate and quiz each other to see if you know the answers to your mineral questions.

Visualizing Main Ideas

Complete the following concept map about minerals. Use the following words and phrases: the way a mineral breaks, the way a mineral reflects light, ore, a rare and beautiful mineral, how easily a mineral is scratched, streak, *and* a useful substance mined for profit.

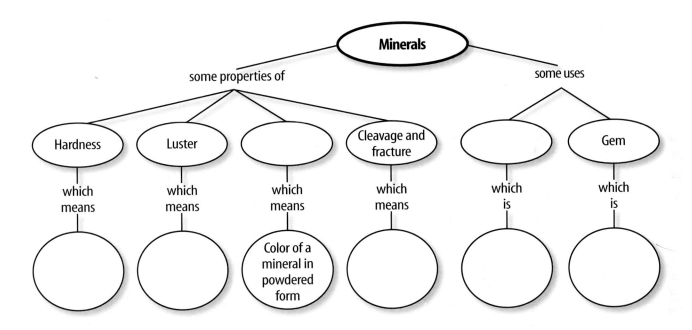

Vocabulary Review

Vocabulary Words

a. cleavage
b. crystal
c. fracture
d. gem
e. hardness
f. luster
g. magma
h. mineral
i. silicate
j. specific gravity
k. streak

Using Vocabulary

Explain the difference between the vocabulary words in each of the following sets.

1. cleavage, fracture
2. crystal, mineral
3. luster, streak
4. magma, crystal
5. hardness, specific gravity
6. magma, mineral
7. crystal, luster
8. mineral, silicate
9. gem, crystal
10. streak, specific gravity

Chapter 3 Assessment

Checking Concepts

Choose the word or phrase that best answers the question.

1. Which is a characteristic of a mineral?
   A) It can be a liquid.
   B) It is organic.
   C) It has no crystal structure.
   D) It is inorganic.

2. What must all silicates contain?
   A) magnesium
   B) silicon and oxygen
   C) silicon and aluminum
   D) oxygen and carbon

3. What is hot, melted rock called?
   A) magma   C) salt water
   B) quartz   D) gypsum

4. Which mineral group does quartz belong to?
   A) oxides   C) silicates
   B) carbonates   D) sulfides

5. What is the measure of how easily a mineral can be scratched?
   A) luster   C) cleavage
   B) hardness   D) fracture

6. What is the color of a powdered mineral formed when rubbing it against an unglazed porcelain tile?
   A) luster   C) hardness
   B) density   D) streak

7. In what way does quartz break?
   A) cleavage   C) luster
   B) fracture   D) flat planes

8. Which of the following must crystalline solids have?
   A) carbonates   C) ordered atoms
   B) cubic structures D) cleavage

9. Which is hardest on Mohs scale?
   A) talc   C) diamond
   B) quartz   D) feldspar

10. Which crystal system do halite crystals belong to?
    A) triclinic   C) monoclinic
    B) hexagonal   D) cubic

Thinking Critically

11. Water is a nonliving substance that is formed by natural processes on Earth. It has a unique composition. Sometimes water is a mineral and other times it is not. Explain.

12. How many sides does a perfect salt crystal have?

13. Suppose you let a sugar solution evaporate, leaving sugar crystals behind. Are these crystals minerals? Explain.

14. Will diamond leave a streak on a streak plate? Explain.

15. Explain how you would use **Table 1** to determine the hardness of any mineral.

Developing Skills

16. **Drawing Conclusions** Suppose you found a white mineral with a glassy, nonmetallic luster that was harder than calcite. You identify the sample as quartz. What are your observations? What is your conclusion?

17. **Interpreting Data** Using **Table 1,** identify a mineral with these properties: pink color, nonmetallic and glassy luster, softer than topaz and quartz, scratches apatite, harder than fluorite, has cleavage, and is scratched by a steel file.

18. **Collecting Data** Make an outline of how at least seven physical properties can be used to identify unknown materials.

19. Measuring in SI The volume of water in a graduated cylinder is 107.5 mL. A specimen of quartz, tied to a piece of string, is immersed in the water. The new water level reads 186 mL. What is the volume of the piece of quartz?

20. Concept Mapping Make a network tree concept map showing two crystal systems and two examples from each group. Use the following words and phrases: *hexagonal, corundum, halite, fluorite,* and *quartz.*

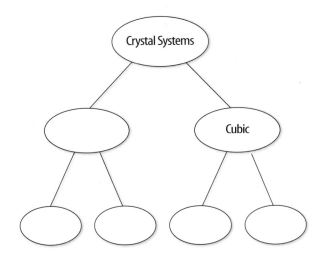

Performance Assessment

21. Poster Make a poster that shows the six crystal systems of minerals. Research the crystal systems of minerals and give three examples for each crystal system. Are any of the minerals found in your state? Do any of the minerals have an important use? Display your poster for the class.

TECHNOLOGY

Go to the Glencoe Science Web site at **science.glencoe.com** or use the **Glencoe Science CD-ROM** for additional chapter assessment.

Test Practice

Stacey's science teacher gave her nine different mineral samples that are on the Mohs Hardness Scale (see chart below). Her task was to identify each sample using hardness tests.

Mohs Hardness Scale	
Talc	1
Gypsum	2
Calcite	3
Fluorite	4
Apatite	5
Feldspar	6
Quartz	7
Topaz	8
Corundum	9
Diamond	10

Approximate Hardness of Common Objects	
Fingernail	(2.5)
Copper penny	(3.0)
Iron nail	(4.5)
Glass	(5.5)
Steel file	(6.5)
Streak plate	(7.0)

Study the diagram and answer the following questions.

1. One of Stacey's mineral samples can scratch glass but is scratched by a steel file. According to the chart, which mineral is it?
A) calcite   **C)** feldspar
B) apatite   **D)** quartz

2. Which of the following is a method that will distinguish samples of topaz and corundum?
F) scratching glass with both samples
G) scratching both samples with a steel file
H) scratching a streak plate with both samples
J) scratching the samples on each other

4 Rocks

Have you ever seen a rock and wondered where it might have come from or how old it is? Whether it is a small pebble found by the side of the road or a mountain of solid rock, every rock is like a history book that tells how and where it was formed. In this chapter, you will learn about how rocks form. You'll learn about the three different groups of rocks—igneous, sedimentary, and metamorphic. You also will learn about the cycle that describes how rocks change from one type to another.

What do you think?

Science Journal Look at the picture below with a classmate. Discuss what you think this might be. Here's a hint: *Your teacher could use this to announce an assignment.* Write your answer or best guess in your Science Journal.

EXPLORE ACTIVITY

El Capitan rises straight up from the valley floor in Yosemite National Park, California. The hard rock that attracts rock climbers and sightseers from around the world is made of small mineral grains that lock together like pieces of a puzzle. Other rocks form from grains of sand tightly held together or from lava flowing from a volcano. If you examine rocks closely, you sometimes can tell what they are made of.

Safety Precautions 🧤 👓 🥽

Determine what rocks are made of

1. Collect three or four different rock fragments near your home or school.

2. Draw a picture of the details you see in each rock.

3. Look for different types of materials within the same rock. If these different materials could be separated by physical means, then the rock would be considered a mixture.

Observe

Describe the characteristics of each rock. Are your rocks mixtures? If so, what might these mixtures contain? Write your observations in your Science Journal.

Before You Read

Making an Organizational Study Fold Make the following Foldable to help you organize your thoughts into clear categories about types of rock.

1. Stack two sheets of paper in front of you so the short side of both sheets is at the top.

2. Slide the top sheet up so about four centimeters of the bottom sheet show.

3. Fold both sheets top to bottom to form four tabs and staple along the topfold as shown.

4. Title the top flap *Three Types of Rocks.* Label the other flaps *Igneous, Metamorphic,* and *Sedimentary.*

5. As you read the chapter, record what you learn about the three types of rocks on the flaps.

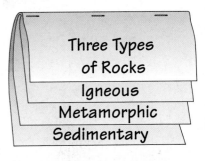

① The Rock Cycle

As You Read

What You'll Learn

■ **Distinguish** between a rock and a mineral.
■ **Describe** the rock cycle and some changes that a rock could undergo.

Vocabulary
rock
rock cycle

Why It's Important
Rocks are everywhere around you—near your school, your home, and even in the sidewalks you walk on.

What is a rock?

Imagine it's the top of the sixth inning. With the sweltering Sun at your back, you scrape your shoe along the infield, waiting for the pitch. Among all the sand and dull-looking, gray stones, your eye catches a glint from a piece of rock that has shiny crystals in it. You quickly snatch it up and put it in your pocket for further examination after the game.

Common Rocks The next time you walk past a large building or monument, stop and take a close look at it. Chances are that it is made out of common rock. In fact, most rock used for building stone contains one or more common minerals, called rock-forming minerals, such as quartz, feldspar, mica, or calcite. When you look closely, the sparkles you see are individual crystals of minerals. A **rock** is a mixture of such minerals, volcanic glass, organic matter, or other materials. **Figure 1** shows minerals mixed together to form the rock granite. You might even find granite on your baseball field.

feldspar

quartz

mica

hornblende

Figure 1
Mount Rushmore, in South Dakota, is made of granite. Granite is a mixture of feldspar, quartz, mica, hornblende, and other minerals.

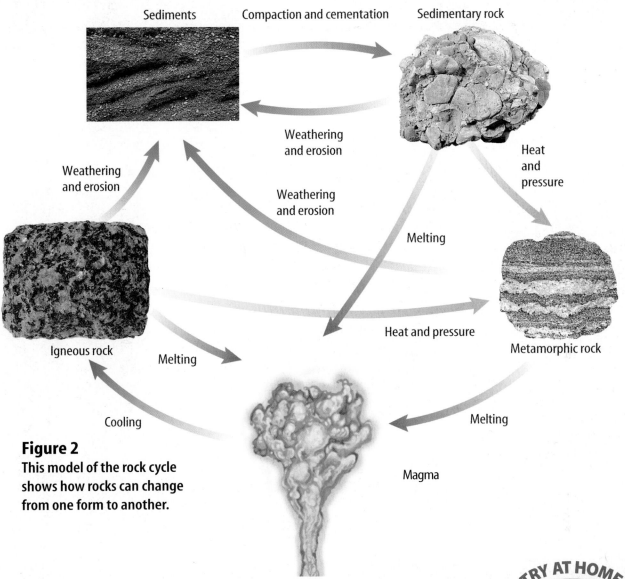

Sediments   Compaction and cementation   Sedimentary rock

Weathering and erosion

Weathering and erosion

Weathering and erosion

Heat and pressure

Melting

Heat and pressure

Igneous rock

Melting

Metamorphic rock

Cooling

Melting

Magma

Figure 2
This model of the rock cycle shows how rocks can change from one form to another.

The Rock Cycle To show how rocks slowly change through time, scientists have created a model called the **rock cycle,** shown in **Figure 2.** It illustrates the processes that create and change rocks. The rock cycle shows the three types of rock—igneous, metamorphic, and sedimentary—and the processes that form them. Look at the rock cycle and notice that rocks change by many processes. For example, a sedimentary rock can change by heat and pressure to form a metamorphic rock. The metamorphic rock then can melt and later cool to form an igneous rock. The igneous rock then could be broken into fragments by weathering and erode away. The fragments might later compact and cement together to form another sedimentary rock. Any given rock can change into any of the three major rock types. A rock even can transform into another rock of the same type. How many paths can you see through the rock cycle?

Reading Check *What is illustrated by the rock cycle?*

TRY AT HOME
Mini LAB

Modeling Rock
Procedure
1. Mix 125 mL of **white glue** with 95 g of **sand** in a **large paper cup.**
2. Stir the mixture and then allow it to harden overnight.
3. Tear away the paper cup carefully from your mixture.

Analysis
1. Which rock type is similar to your hardened mixture?
2. Which part of the rock cycle did you model?

Figure 3

Rocks continuously form and transform in a process that geologists call the rock cycle. For example, molten rock—from volcanoes such as Washington's Mount Rainier, background—cools and solidifies to form igneous rock. It slowly breaks down when exposed to air and water to form sediments. These sediments are compacted or cemented into sedimentary rock. Heat and pressure might transform sedimentary rock into metamorphic rock. When metamorphic rock melts and hardens, igneous rock forms again. There is no distinct beginning, nor is there an end, to the rock cycle.

▲ The black sand beach of this Polynesian island is sediment weathered and eroded from the igneous rock of a volcano nearby.

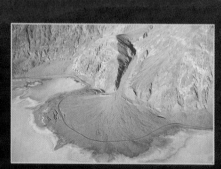

▲ This alluvial fan on the edge of Death Valley, California, was formed when gravel, sand, and finer sediments were deposited by a stream emerging from a

▲ Layers of shale and chalk form Kansas's Monument Rocks. They are remnants of sediments deposited on the floor of the ancient sea

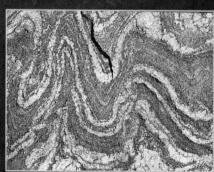

▲ Heat and pressure deep below Earth's surface can change rock into metamorphic rock, like this banded gneiss.

Physics
INTEGRATION

Matter and the Rock Cycle The rock cycle, illustrated in **Figure 3,** shows how rock can be weathered to small rock and mineral grains. This material then can be eroded and carried away by wind, water, or ice. When you think of erosion, it might seem that the material is somehow destroyed and lost from the cycle, but this is not the case. The chemical elements that make up minerals and rocks are not destroyed. This fact illustrates the principle of conservation of matter. The changes that take place in the rock cycle never destroy or create matter. The elements are just redistributed in other forms.

Figure 4
The rock formations at Siccar Point, Scotland show that rocks undergo constant change.

 Reading Check *What is the principle of conservation of matter?*

Discovering the Rock Cycle James Hutton, a Scottish physician and naturalist, first recognized in 1788 that rocks undergo profound changes. Hutton noticed, among other things, that some layers of solid rock in Siccar Point, shown in **Figure 4,** had been altered since they formed. Instead of showing a continuous pattern of horizontal layering, some of the rock layers at Siccar Point are tilted and partly eroded. However, the younger rocks above them are nearly horizontal.

Hutton published these and other observations, which proved that rocks are subject to constant change. Hutton's early recognition of the rock cycle continues to influence geologists—even after over 200 years have passed.

Section ❶ Assessment

1. What materials are rocks made of?
2. What are the three basic types of rock?
3. Describe the major processes of the rock cycle.
4. Describe one way that the rock cycle can illustrate the principle of conservation of matter.
5. **Think Critically** Look at the model of the rock cycle. How would you define magma based on the illustration in **Figure 2?** How would you define sediment and sedimentary rock?

Skill Builder Activities

6. **Concept Mapping** Using a computer, make a concept map explaining how igneous rocks can become sedimentary, then metamorphic, and finally, other igneous rocks. **For more help, refer to the** Science Skill Handbook.

7. **Communicating** Review the model of the rock cycle in **Figure 2.** In your Science Journal, write a story or poem that explains what can happen to a sedimentary rock as it changes through the rock cycle. **For more help, refer to the** Science Skill Handbook.

Igneous Rocks

Formation of Igneous Rocks

Perhaps you've heard of recent volcanic eruptions in the news. When some volcanoes erupt, they eject a flow of molten rock material, as shown in **Figure 5.** Molten rock material, called magma, flows when it is hot and becomes solid when it cools. When hot magma cools and hardens, it forms **igneous** (IHG nee us) **rocks.** Why do volcanoes erupt, and where does the molten material come from?

Magma In certain places within Earth, the temperature and pressure are just right for rocks to melt and form magma. Most magmas come from deep below Earth's surface. Magma is located at depths ranging from near the surface to about 150 km below the surface. Temperatures of magmas range from about 650°C to 1,200°C, depending on their chemical compositions and pressures exerted on them.

The heat that melts rocks comes from sources within Earth's interior. One source is the decay of radioactive elements within Earth. Some heat is left over from the formation of the planet, which originally was molten. Radioactive decay of elements contained in rocks balances some heat loss as Earth continues to cool.

Because magma is less dense than surrounding solid rock, it is forced upward toward the surface, as shown in **Figure 6.** When magma reaches Earth's surface and flows from volcanoes, it is called **lava.**

Figure 5
Some lava is highly fluid and free-flowing, as shown by this spectacular lava fall in Volcano National Park, East Rift, Kilauea, Hawaii.

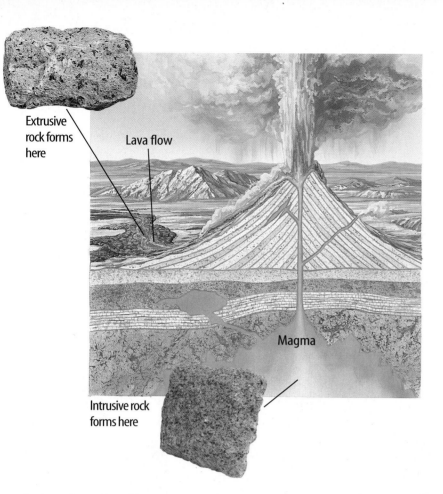

Extrusive rock forms here

Lava flow

Magma

Intrusive rock forms here

Figure 6
Intrusive rocks form from magma trapped below Earth's surface. Extrusive rocks form from lava flowing at the surface.

Field **GUIDE**

Are there any igneous rocks used as building stones in your area? To find out more about rocks used in construction, see the **Building Stones Field Guide** at the back of the book.

Intrusive Rocks Magma is made up of atoms and molecules of melted minerals. As magma cools, the atoms and molecules rearrange themselves into new crystals called mineral grains. Rocks form as these mineral grains grow together. Rocks that form from magma below the surface, as illustrated in **Figure 6,** are called **intrusive** igneous rocks. Intrusive rocks are found at the surface only after the layers of rock and soil that once covered them have been removed by erosion. Erosion occurs when the rocks are pushed up by forces within Earth. Because intrusive rocks form at depth and they are surrounded by other rocks, it takes a long time for them to cool. Slowly cooled magma produces individual mineral grains that are large enough to be observed with the unaided eye.

Extrusive Rocks **Extrusive** igneous rocks are formed as lava cools on the surface of Earth. When lava flows on the surface, as illustrated in **Figure 6,** it is exposed to air and water. Lava, such as the basaltic lava shown in **Figure 5,** cools quickly under these conditions. The quick cooling rate keeps mineral grains from growing large, because the atoms and molecules don't have the time to arrange into large crystals. Therefore, extrusive igneous rocks are fine grained.

✔ **Reading Check**   *What controls the grain size of an igneous rock?*

Table 1 Common Igneous Rocks

Magma Type	Basaltic		Andesitic	Granitic		
Intrusive	Gabbro		Diorite	Granite		
Extrusive	Basalt	Scoria	Andesite	Rhyolite	Pumice	Obsidian

Volcanic Glass Pumice, obsidian, and scoria are examples of volcanic glass. These rocks cooled so quickly that few or no mineral grains formed. Most of the atoms in these rocks are not arranged in orderly patterns, and few crystals are present.

In the case of pumice and scoria, gases become trapped in the gooey molten material as it cools. Some of these gases eventually escape, but holes are left behind where the rock formed around the pockets of gas.

Classifying Igneous Rocks

Igneous rocks are intrusive or extrusive depending on how they are formed. A way to further classify these rocks is by the magma from which they form. As shown in **Table 1,** an igneous rock can form from basaltic, andesitic, or granitic magma. The type of magma that cools to form an igneous rock determines important chemical and physical properties of that rock. These include mineral composition, density, color, and melting temperature.

✔ **Reading Check** *Name two ways igneous rocks are classified.*

Basaltic Rocks **Basaltic** (buh SAWL tihk) igneous rocks are dense, dark-colored rocks. They form from magma that is rich in iron and magnesium and poor in silica, which is the compound SiO_2. The presence of iron and magnesium in minerals in basalt gives basalt its dark color. Basaltic lava is fluid and flows freely from volcanoes in Hawaii, such as Kilauea. How does this explain the black beach sand common in Hawaii?

Granitic Rocks **Granitic** igneous rocks are light-colored rocks of a lower density than basaltic rocks. Granitic magma is thick and stiff and contains lots of silica but lesser amounts of iron and magnesium. Because granitic magma is stiff, it can build up a great deal of gas pressure, which is released explosively during violent volcanic eruptions.

Andesitic Rocks Andesitic igneous rocks have mineral compositions between those of granitic and basaltic rocks. Many volcanoes around the rim of the Pacific Ocean formed from intermediate magmas. Like volcanoes that erupt granitic magma, these volcanoes also can erupt violently.

Take another look at **Table 1.** Basalt forms at the surface of Earth because it is an extrusive rock. Granite forms below Earth's surface from magma with a high concentration of silica. When you identify an igneous rock, you can infer how it formed and the type of magma that it formed from.

Chemistry
INTEGRATION

Inside Earth, materials contained in rocks can melt. In your Science Journal, describe what is happening to the atoms and molecules to cause this change of state.

Section 2 Assessment

1. Why do some types of magma form igneous rocks that are dark colored and dense?

2. How do intrusive and extrusive igneous rocks differ?

3. What property of magma causes it to rise toward Earth's surface?

4. The texture of obsidian is best described as glassy. Why does obsidian contain few or no mineral grains?

5. **Think Critically** Study the photographs of granite and rhyolite in **Table 1.** How are granite and rhyolite similar? How are these rocks different?

Skill Builder Activities

6. **Interpreting Scientific Illustrations** Suppose you are given a photograph of two igneous rocks. You are told one is an intrusive rock and one is extrusive. How could you tell which is which? **For more help, refer to the** Science Skill Handbook.

7. **Making and Using Graphs** Four elements make up most of the rocks in Earth's crust. They are: *oxygen—46.6 percent, aluminum— 8.1 percent, silicon—27.7 percent,* and *iron— 5.0 percent.* Make a graph of these data. What might you infer from the low amount of iron? **For more help, refer to the** Science Skill Handbook.

Activity

Igneous Rock Clues

Y ou've learned how color often is used to esti-mate the composition of an igneous rock. The texture of an igneous rock describes its over-all appearance, including mineral grain sizes and the presence or absence of bubble holes, for example. In most cases, grain size relates to how quickly the magma or lava cooled. Crystals you can see without a magnifying glass indi-cate slower cooling. Smaller, fine-grained crystals indicate quicker cooling, possi-bly due to volcanic activity. Rocks with glassy textures cooled so quickly that there was no time to form mineral grains.

What You'll Investigate
What does an igneous rock's texture and color indicate about its formation history?

Materials
rhyolite	granite
basalt	obsidian
vesicular basalt	gabbro
pumice	magnifying glass

Goals
- **Classify** different samples of igneous rocks by color and infer their composition.
- **Observe** the textures of igneous rocks and infer how they formed.

Safety Precautions 🧤 🥽

Some rock samples might have sharp edges. Always use caution while handling samples.

Procedure

1. **Arrange** rocks according to color (light or dark). Record your observations in your Science Journal.
2. **Arrange** rocks according to similar texture. Consider grain sizes and shapes, presence of holes, etc. Use your magnifying glass to see small features more clearly. Record your observations.

Conclude and Apply

1. Infer which rocks are granitic based on color.
2. Infer which rocks cooled quickly. What obser-vations led you to this inference?
3. Identify any samples that suggest gases were escaping from them as they cooled.
4. Which samples have a glassy appear-ance? How did these rocks form?

5. **Infer** which sam-ples are not vol-canic. Explain your choices.

*C*ommunicating
Your Data

Research the compositions of each of your samples. Did the colors of any samples lead you to infer the wrong compositions? Com-municate to your class what you learned.

Metamorphic Rocks

Formation of Metamorphic Rocks

Have you ever packed your lunch in the morning and not been able to recognize it at lunchtime? You might have packed a sandwich, banana, and a large bottle of water. You know you didn't smash your lunch on the way to school. However, you didn't think about how the heavy water bottle would damage your food if the bottle was allowed to rest on the food all day. The heat in your locker and the pressure from the heavy water bottle have changed the form of your sandwich. Like your lunch, rocks can be affected by temperature changes and pressure.

Metamorphic Rocks Rocks that have changed because of changes in temperature and pressure or the presence of hot, watery fluids are called **metamorphic rocks.** Changes that occur can be in the form of the rock, shown in **Figure 7,** the composition of the rock, or both. Metamorphic rocks can form from igneous, sedimentary, or other metamorphic rocks. What Earth processes can change these rocks?

As You Read

What You'll Learn
- **Describe** the conditions in Earth that cause metamorphic rocks to form.
- **Classify** metamorphic rocks as foliated or nonfoliated.

Vocabulary
metamorphic rock
foliated
nonfoliated

Why It's Important
Metamorphic rocks are useful because of their unique properties.

+ pressure

Figure 7
A The mineral grains in granite are flattened and aligned when heat and pressure are applied to them. **B** As a result, gneiss is formed. *What other conditions can cause metamorphic rocks to form?*

Heat and Pressure Rocks beneath Earth's surface are under great pressure from rock layers above them. Temperature also increases with depth in Earth. In some places, the heat and pressure are just right to cause rocks to melt and magma to form. In other areas where melting doesn't occur, some mineral grains can become flattened like the sandwich in the lunch bag. Sometimes, under these conditions, minerals exchange atoms with surrounding minerals and new, bigger minerals form.

Depending upon the amount of pressure and temperature applied, one type of rock can change into several different metamorphic rocks, and each type of metamorphic rock can come from several kinds of parent rocks. For example, the sedimentary rock shale will change into slate. As increasing pressure and temperature are applied, the slate can change into phyllite, then schist, and eventually gneiss. Schist also can form when basalt is metamorphosed, or changed, and gneiss can come from granite.

✔ **Reading Check** *How can one type of rock change into several different metamorphic rocks?*

Hot Fluids Did you know that fluids can move through solid rock? These fluids, which are mostly water with dissolved elements, can react chemically with a rock and change its composition, especially when the fluids are hot. That's what happens when rock surrounding a hot magma body reacts with hot fluids from the magma, as shown in **Figure 8.** Most fluids that transform rocks during metamorphic processes are hot and mainly are comprised of water and carbon dioxide.

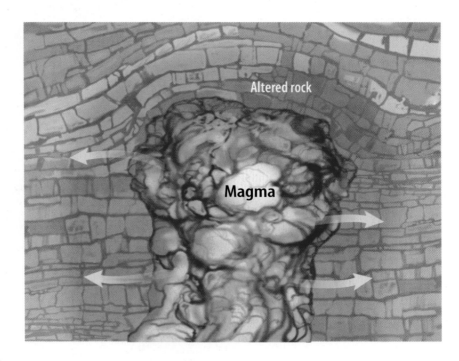

Altered rock

Magma

Figure 8
In the presence of hot, water-rich fluids, solid rock can change in mineral composition without having to melt.

Classifying Metamorphic Rocks

Metamorphic rocks form from igneous, sedimentary, or other metamorphic rocks. Heat, pressure, and hot fluids trigger the changes. Each resulting rock can be classified according to its composition and texture.

Foliated Rocks When mineral grains flatten and line up in parallel layers, the metamorphic rock is said to have a **foliated** or banded texture. Two examples of foliated rocks are slate and gneiss. Slate forms from the sedimentary rock shale. The minerals in shale arrange into layers when they are exposed to heat and pressure. As **Figure 9** shows, slate separates easily along these foliation layers.

The minerals in slate are pressed together so tightly that water can't pass between them easily. Because it's watertight, slate is ideal for paving around pools and patios. The naturally flat nature of slate and the fact that it splits easily make it useful for roofing and tiling many surfaces.

Gneiss (NISE), another foliated rock, forms when granite and other rocks are changed. Quartz, feldspar, mica (MI kuh), and other minerals that make up granite aren't changed much, but they arrange into alternating bands of light and dark minerals.

✔ Reading Check *What type of metamorphic rock is composed of layers of mineral grains?*

Figure 9
Slate often is used as a building or landscaping material. *What property makes slate so useful for these purposes?*

Figure 10
This exhibit in Vermont shows the beauty of carved marble.

Nonfoliated Rocks In some metamorphic rocks, layering does not occur. The mineral grains grow and rearrange, but they don't form layers. This process produces a **nonfoliated** texture.

Sandstone is a sedimentary rock that's often composed mostly of quartz grains. When sandstone is heated under a lot of pressure, the grains of quartz grow in size and become interlocking, like the pieces of a jigsaw puzzle. The resulting rock is called quartzite.

Marble is another nonfoliated metamorphic rock. Marble forms from the sedimentary rock limestone, which is composed of the mineral calcite. Usually, marble contains several other minerals besides calcite. For example, hornblende and serpentine give marble a black or greenish tone, whereas hematite makes it red. As **Figure 10** shows, marble is a popular material for artists to sculpt because it is not as hard as other rocks.

So far, you've investigated only a portion of the rock cycle. You still haven't observed how sedimentary rocks are formed and how igneous and metamorphic rocks evolve from them. The next section will complete your investigation of the rock cycle.

Section Assessment

1. How is the formation of metamorphic rocks different from that of igneous, or sedimentary, rocks?

2. What role do fluids play in rock metamorphism?

3. How are metamorphic rocks classified? What are the characteristics of rocks in each of these classifications?

4. Give an example of a foliated and a nonfoliated metamorphic rock. Name one of their possible parent rocks.

5. **Think Critically** Marble is a common material used to make sculptures, but not just because it's a beautiful stone. What properties of marble make it useful for this purpose?

Skill Builder Activities

6. **Concept Mapping** Put the following events in an events-chain concept map that explains how a metamorphic rock might form from an igneous rock. Here's a hint: *Start with "igneous rock forms."* Use each event just once. **For more help, refer to the** Science Skill Handbook.

   Events: *sedimentary rock forms, weathering occurs, heat and pressure are applied, igneous rock forms, metamorphic rock forms, erosion occurs, sediments are formed, deposition occurs*

7. **Using Graphics Software** Use a graphics program to illustrate how metamorphic rocks form. Be sure to show how directed pressure causes alignment of mineral grains. **For more help, refer to the** Technology Skill Handbook.

Sedimentary Rocks

Formation of Sedimentary Rocks

Igneous rocks are the most common rocks on Earth, but because most of them exist below the surface, you might not have seen too many of them. That's because 75 percent of the rocks exposed at the surface are sedimentary rocks.

Sediments are loose materials such as rock fragments, mineral grains, and bits of shell that have been moved by wind, water, ice, or gravity. If you look at the model of the rock cycle, you will see that sediments come from already-existing rocks that are weathered and eroded. **Sedimentary rocks** form when sediments are pressed and cemented together, or when minerals form from solutions.

Stacked Rocks Sedimentary rocks often form as layers. The older layers are on the bottom because they were deposited first. Sedimentary rock layers are a lot like the books and papers in your locker. Last week's homework is on the bottom, and today's notes will be deposited on top of the stack. However, if you disturb the stack, the order in which the books and papers are stacked will change, as shown in **Figure 11.** Sometimes, forces within Earth overturn layers of rock, and the oldest are no longer on the bottom.

As You Read

What You'll Learn
- **Explain** how sedimentary rocks form from sediments.
- **Classify** sedimentary rocks as detrital, chemical, or organic in origin.

Vocabulary
sediment

sedimentary rock

compaction

cementation

Why It's Important
Some sedimentary rocks, like coal, are important sources of energy.

Figure 11
Like sedimentary rock layers, the oldest paper is at the bottom of the stack. If the stack is disturbed, then it is no longer in order.

Classifying Sediments

Procedure 🖼 🥽 🧤

WARNING: *Use care when handling sharp objects.*

1. Collect different samples of **sediment.**
2. Spread them on a sheet of **paper.**
3. Use **Table 2** to determine the size range of gravel-sized sediment.
4. Use **tweezers or a dissecting probe** and a **magnifying lens** to separate the gravel-sized sediments.
5. Separate the gravel into piles—rounded or angular.

Analysis

1. Describe the grains in both piles.
2. Determine what rock could form from each type of sediment you have.

Figure 12
During compaction, pore space between sediments decreases, causing them to become packed together more tightly.

Classifying Sedimentary Rocks

Sedimentary rocks can be made of just about any material found in nature. Sediments come from weathered and eroded igneous, metamorphic, and sedimentary rocks. Sediments also come from the remains of some organisms. The composition of a sedimentary rock depends upon the composition of the sediments from which it formed.

Like igneous and metamorphic rocks, sedimentary rocks are classified by their composition and by the manner in which they formed. Sedimentary rocks usually are classified as detrital, chemical, or organic.

Detrital Sedimentary Rocks

The word *detrital* (Dih TRY tul) comes from the Latin word *detritus,* which means "to wear away." Detrital sedimentary rocks, such as those shown in **Table 2,** are made from the broken fragments of other rocks. These loose sediments are compacted and cemented together to form solid rock.

Weathering and Erosion When rock is exposed to air, water, or ice, it is unstable and breaks down chemically and mechanically. This process, which breaks rocks into smaller pieces, is called weathering. **Table 2** shows how these pieces are classified by size. The movement of weathered material is called erosion.

Compaction Erosion moves sediments to a new location, where they then are deposited. Here, layer upon layer of sediment builds up. Pressure from the upper layers pushes down on the lower layers. If the sediments are small, they can stick together and form solid rock. This process, shown in **Figure 12,** is called **compaction.**

✓ **Reading Check** *How do rocks form through compaction?*

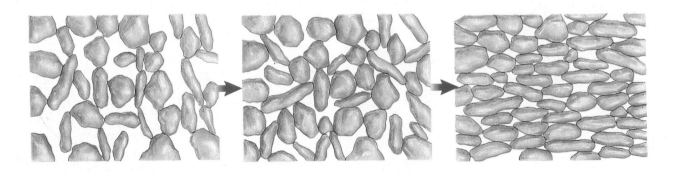

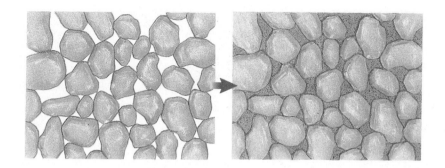

Figure 13
Sediments are cemented together as minerals crystallize between grains.

Cementation If sediments are large, like sand and pebbles, pressure alone can't make them stick together. Large sediments have to be cemented together. **Cementation,** which is shown in **Figure 13,** occurs when water soaks through soil and rock. As water moves through soil and rock, it picks up atoms and molecules released from minerals during weathering. The resulting solution of water and dissolved materials moves through open spaces between sediments. Minerals such as quartz, calcite, hematite, and limonite are deposited between the pieces of sediment. These minerals, acting as natural cements, hold the sediment together like glue, making a detrital sedimentary rock.

Shape and Size of Sediments Detrital rocks have granular textures much like granulated sugar. They are named according to the shapes and sizes of the sediments that form them. For example, conglomerate and breccia both form from large sediments, as shown in **Table 2.** If the sediments are rounded, the rock is called conglomerate. If the sediments have sharp angles, the rock is called breccia. The roundness of sediment particles depends on how far they have been moved by wind or water.

Table 2 Sediment Sizes and Detrital Rocks

Sediment	Clay	Silt	Sand	Gravel
Size Range	<0.004 mm	0.004–0.063 mm	0.063–2 mm	>2 mm
Example	Shale	Siltstone	Sandstone	Conglomerate (shown) or Breccia

Conglomerate

Figure 14
Although concrete strongly resembles conglomerate, concrete is not a rock because it does not occur in nature.

Chemistry

INTEGRATION

You may know of large caves or caverns that have formed underground in limestone deposits. Research the chemical reactions that form caves in Carlsbad Caverns, New Mexico, or Mammoth Cave, Kentucky.

Materials Found in Sedimentary Rocks The gravel-sized sediments in conglomerate and breccia can consist of any type of rock or mineral. Often, they are composed of chunks of the minerals quartz and feldspar. They also can be pieces of rocks such as gneiss, granite, or limestone. The cement that holds the sediments together usually is made of quartz or calcite.

Have you ever looked at the concrete in sidewalks, driveways, and stepping stones? The concrete in **Figure 14** is made of gravel and sand grains that have been cemented together. Although the structure is similar to that of naturally occurring conglomerate, it cannot be considered a rock.

Sandstone is formed from smaller particles than conglomerates and breccias. Its sand-sized sediments can be just about any mineral, but they are usually grains of the minerals quartz and feldspar that are compacted and cemented together. Siltstone is similar to sandstone except it is made of smaller, silt-sized particles. Shale is a detrital sedimentary rock that is made mainly of clay-sized particles. Clay-sized sediments are compacted together by pressure from overlying layers.

Chemical Sedimentary Rocks

Chemical sedimentary rocks form when dissolved minerals come out of solution. You can show that salt is deposited in the bottom of a glass or pan when saltwater solution evaporates. In a similar way, minerals collect when seas or lakes evaporate. The deposits of minerals that come out of solution form sediments and rocks. For example, the sediment making up New Mexico's White Sands desert consists of pieces of a chemical sedimentary rock called rock gypsum. Chemical sedimentary rocks are different. They are not made from pieces of preexisting rocks.

Reading Check *How do chemical sedimentary rocks form?*

Limestone Calcium carbonate is carried in solution in ocean water. When calcium carbonate comes out of solution as calcite and its many crystals grow together, limestone forms. Limestone also can contain other minerals and sediments, but it must be at least 50 percent calcite. Limestone usually is deposited on the bottom of lakes or shallow seas. Large areas of the central United States have limestone bedrock because seas covered much of the country for millions of years. It is hard to imagine Kansas being covered by ocean water, but it has happened several times throughout geological history.

Rock Salt When water that is rich in dissolved salt evaporates, it often deposits the mineral halite. Halite forms rock salt, shown in **Figure 15**. Rock salt deposits can range in thickness from a few meters to more than 400 m. Companies mine these deposits because rock salt is an important resource. It's used in the manufacturing of glass, paper, soap, and dairy products. The halite in rock salt is used as table salt.

Figure 15
Rock salt is extracted from this mine in Germany. The same salt can be processed and used to season your favorite foods.

Organic Sedimentary Rocks

Rocks made of the remains of once-living things are called organic sedimentary rocks. One of the most common organic sedimentary rocks is fossil-rich limestone. Like chemical limestone, fossil-rich limestone is made of the mineral calcite ($CaCO_3$). However, fossil-rich limestone mostly contains remains of once-living ocean organisms instead of only calcite that has come out of ocean water.

Animals such as mussels, clams, corals, and snails make their shells from $CaCO_3$ that eventually becomes calcite. When they die, their shells accumulate on the ocean floor. When these shells are cemented together, fossil-rich limestone forms. If a rock is made completely of shell fragments that you can see, the rock is called coquina (koh KEE nuh).

Chalk Chalk is another organic sedimentary rock that is made of microscopic shells. When you write with naturally occurring chalk, you're crushing and smearing the calcite-shell remains of once-living ocean organisms.

Coal Another useful organic sedimentary rock is coal, shown in **Figure 16.** Coal forms when pieces of dead plants are buried under other sediments in swamps. These plant materials are chemically changed by microorganisms. The resulting sediments are compacted over millions of years to form coal, an important source of energy. Much of the coal in North America and Europe formed during a period of geologic time that is so named because of this important reason. The Carboniferous Period, which spans from approximately 360 to 286 million years ago, was named in Europe. So much coal formed during this interval of time that coal's composition—primarily carbon—was the basis for naming a geologic period.

Math Skills Activity

Calculating Thickness

Example Problem

It took 300 million years for a layer of plant matter about 0.9m thick to produce a bed of bituminous coal 0.3m thick. Estimate the thickness of plant matter that produced a bed of coal 0.15m thick.

Solution

1 *This is what you know:*

original thickness of plant matter = 0.9m
original coal thickness = 0.3m
new coal thickness = 0.15m

2 *This is what you need to know:*

thickness of plant matter needed to form 0.15m of coal

3 *This is the equation you need to use:*

(thickness of plant matter) / (new coal thickness) = (original thickness of plant matter) / original coal thickness

4 *Substitute the known values:*

(?m plant matter) / (0.15m coal) = (0.9m plant matter) / (0.3m coal)

5 *Solve the equation:*

(?m plant matter) = (0.9m plant matter) (0.15m coal) / (0.3m coal) = 0.45m plant matter

Multiply your answer by the original coal thickness. Divide by original plant matter thickness. Do you get the new coal thickness that was given?

> **Practice Problem**
>
> Estimate the thickness of plant matter that produced a bed of coal 0.6m thick.

For more help, refer to the Math Skill Handbook.

Figure 16
This coal layer in Alaska is easily identified by its jet-black color, as compared with other sedimentary layers.

Another Look at the Rock Cycle

You have seen that the rock cycle has no beginning and no end. Rocks change continually from one form to another. Sediments come from rocks and minerals that have been broken apart. Even the magma that forms igneous rocks comes from the melting of rocks that already exist.

All of the rocks that you've learned about in this chapter formed through some process within the rock cycle. All of the rocks around you, including those used to build houses and monuments, are part of the rock cycle. Slowly, they are all changing, because the rock cycle is a continuous, dynamic process.

Section Assessment

1. Where do sediments come from?
2. Explain how compaction and cementation are important in forming sedimentary rocks.
3. Explain the difference between detrital and chemical sedimentary rock.
4. List chemical sedimentary rocks that are essential to your health or that are used to make life more convenient. How is each used?
5. **Think Critically** Use the rock cycle to explain how pieces of granite and slate could be found in the same piece of conglomerate.

Skill Builder Activities

6. **Making and Using Tables** You are told to identify several rocks based on their compositions and other physical properties. What physical properties would you consider? How would you use these properties so the rocks could be identified easily? **For more help, refer to the** Science Skill Handbook.

7. **Calculating Ratios** Sediment sizes of different sedimentary rock types are presented in **Table 2.** Estimate how many times larger the largest grains of silt and sand are compared to clay grains. **For more help, refer to the** Math Skill Handbook.

Activity

Sedimentary Rocks

Sedimentary rocks are formed by compaction and cementation of sediment. Because sediment is found in all shapes and sizes, do you think these characteristics could be used to classify detrital sedimentary rocks? Sedimentary rocks also can be classified as chemical or organic.

What You'll Investigate

How are rock characteristics used to classify sedimentary rocks as detrital, chemical, or organic?

Goals
- **Observe** sedimentary rock characteristics.
- **Compare and contrast** sedimentary rock textures.
- **Classify** sedimentary rocks as detrital, chemical, or organic.

Materials
unknown sedimentary rock samples
marking pen
5 percent hydrochloric acid (HCl)
  vinegar
  dropper
  paper towels
  water
  magnifying lens
  metric ruler
Alternate materials

Safety Precautions

Sedimentary Rock Samples					
Sample	Observations	Minerals or Fossils Present	Sediment Size	Detrital, Chemical, or Organic	Rock Name
A					
B					
C					
D					
E					

Procedure

1. Make a Sedimentary Rock Samples chart similar to the one shown above in your Science Journal.

2. **Determine** the sizes of sediments in each sample, using a magnifying lens and a metric ruler. Using **Table 2,** classify any grains of sediment in the rocks as gravel, sand, silt, or clay. In general, the sediment is silt if it is gritty and just barely visible, and clay if it is smooth and if individual grains are not visible.

3. Place a few drops of HCl or vinegar on each rock sample. Bubbling on a rock indicates the presence of calcite. **WARNING:** *HCl is an acid and can cause burns. Wear goggles and a lab apron. Rinse spills with water and wash hands afterwards.*

4. **Examine** each sample for fossils and describe any that are present.

5. **Determine** whether each sample has a granular or nongranular texture.

6. **Classify** your samples as detrital, chemical, or organic. Identify each rock sample.

Conclude and Apply

1. Explain why you tested the rocks with acid. What minerals react with acid?

2. **Compare and contrast** sedimentary rocks that have a granular texture with sedimentary rocks that have a nongranular texture.

*C*ommunicating
Your Data

Compare your conclusions with those of other students in your class. **For more help, refer to the** Science Skill Handbook.

Australia's controversial rock star

One of the most famous rocks in the world is causing serious problems for Australians

Uluru (yew LEW rew) is one of the most popular tourist destinations in Australia. This sandstone skyscraper is more than 8 km around, over 300 m high, and extends as much as 4.8 km below the surface. One writer describes it as an iceberg in the desert. Geologists hypothesize that the mighty Uluru rock began forming 550 million years ago during Precambrian time. That's when large mountain ranges started to form in Central Australia. About 300 million years later, sediment from these ranges settled on the Amadeus Basin. The sandy material hardened into rock. Faults in Earth weakened the rock. Erosion eventually cut away this weakened surrounding rock to expose the massive sandstone structure called Ayers Rock. Today, erosion continues to slowly cut away at the rock.

For more than 25,000 years, this geological wonder has played an important role in the lives of the Aboriginal peoples, the Anangu (a NA noo). These native Australians are the original owners of the rock and have spiritual explanations for its many caves, holes, and scars.

In 1873, the rock was named in honor of Sir Henry Ayers, an Australian government official. Few people in the early 1900s visited it. But as transportation technology improved, so did the rock's tourist appeal. In 1958, the Australian government made Ayers Rock part of a national park. In the 1980s, 100,000 tourists visited it. In 2000, the rock attracted about 400,000 tourists.

North America

Uluru

Australia

Asia

Europe

Africa

Australia

Tourists Take Over

This increase in tourism is causing serious problems. The Anangu take offense at the climbing of their sacred rock, a holy site. However, if climbing the rock were outlawed, tourism would be seriously hurt. And that would mean less income for Australians. One Australian marketing director sums up the apparent threat to tourism: Going to Uluru and not climbing is like going to the Great Barrier Reef and not diving.

Some steps have been taken to respect the native people's wishes. In 1985, the government returned Ayers rock to the Anangu, and agreed to call it by its traditional name. As part of this arrangement, the Anangu leased back the rock to the Australian government until the year 2084, when management of Uluru and the surrounding park goes back to the Anangu. Until then, the Anangu will collect 25 percent of the money people pay to visit the rock.

The Aboriginal people encourage tourists to respect their beliefs. They offer a walking tour around the rock, and they show videos about Aboriginal traditions. The Anangu sell T-shirts that say "I *didn't* climb Uluru." They hope visitors to Uluru will wear the T-shirt with pride and respect.

Athlete Nova Benis-Kneebone had the honor of receiving the Olympic torch near the sacred Uluru and carried it partway to the Olympic stadium.

CONNECTIONS Write Research a natural landmark or large natural land or water formation in your area. What is the geology behind it? When was it formed? How was it formed? Write a folktale that explains its formation. Share your folktale with the class.

SCIENCE *Online*

For more information, visit science.glencoe.com

Reviewing Main Ideas

Section 1 The Rock Cycle

1. A rock is a mixture of one or more minerals, rock fragments, organic matter, or volcanic glass.

2. The rock cycle includes all processes by which rocks form. *How could an igneous rock change to a metamorphic rock?*

Section 2 Igneous Rocks

1. Magma and lava are molten materials that harden to form igneous rocks.

2. Intrusive igneous rocks form when magma cools slowly below the surface. Extrusive igneous rocks form when lava cools rapidly at the surface.

3. Basalt is dense and dark colored. Granite is light colored and less dense than basalt. Andesite has intermediate density and color, somewhere between basalt and granite. *What kind of rock is shown below?*

Section 3 Metamorphic Rocks

1. Heat, pressure, and fluids can cause metamorphic rocks to form. Metamorphic rocks can have foliated textures or nonfoliated textures.

2. Slate and gneiss are examples of foliated metamorphic rocks. Quartzite and marble are examples of nonfoliated metamorphic rocks. *How are these rocks different?*

Section 4 Sedimentary Rocks

1. Detrital sedimentary rocks form when fragments of rocks and minerals are compacted and cemented together. Detrital rocks always have a granular texture.

2. Chemical sedimentary rocks come out of solution or are left behind by evaporation. Chemical rocks generally have a non-granular texture.

3. Organic sedimentary rocks are made mostly of the remains of once-living organisms. *What kinds of remains make up the rock to the right?*

FOLDABLES
Reading & Study Skills

After You Read

Using the information in your Foldable for help, write examples of each type of rock under the flaps of your Foldable.

Visualizing Main Ideas

Complete the following concept map on rocks. Use the following terms:
organic, metamorphic, foliated, extrusive, igneous, *and* chemical.

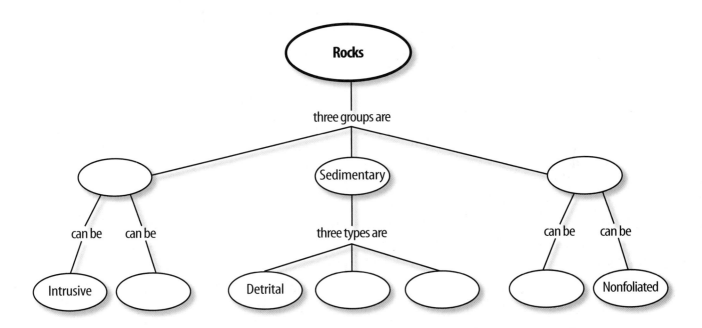

Vocabulary Review

Vocabulary Words

a. basaltic
b. cementation
c. compaction
d. extrusive
e. foliated
f. granitic
g. igneous rock
h. intrusive
i. lava
j. metamorphic rock
k. nonfoliated
l. rock
m. rock cycle
n. sediment
o. sedimentary rock

Using Vocabulary

Explain the difference between the vocabulary words in each of the following sets.

1. foliated, nonfoliated

2. cementation, compaction

3. sediment, lava

4. extrusive, intrusive

5. rock, rock cycle

6. metamorphic rock, igneous rock, sedimentary rock

7. sediment, sedimentary rock

8. lava, igneous rock

9. rock, sediment

10. basaltic, granitic

Checking Concepts

Choose the word or phrase that best answers the question.

1. Why does magma tend to rise toward Earth's surface?
   A) It is more dense than surrounding rocks.
   B) It is more massive than surrounding rocks.
   C) It is cooler than surrounding rocks.
   D) It is less dense than surrounding rocks.

2. During metamorphism of granite into gneiss, what happens to minerals?
   A) They partly melt.
   B) They become new sediments.
   C) They grow smaller.
   D) They align into layers.

3. Which rock has large mineral grains?
   A) granite   C) obsidian
   B) basalt   D) pumice

4. What do igneous rocks form from?
   A) sediments   C) gravel
   B) mud   D) magma

5. What kind of rock is marble?
   A) foliated   C) intrusive
   B) nonfoliated   D) extrusive

6. What sedimentary rock is made of large, angular pieces of sediments?
   A) conglomerate   C) limestone
   B) breccia   D) chalk

7. Which of the following is an example of a detrital sedimentary rock?
   A) limestone   C) breccia
   B) evaporite   D) chalk

8. During what process are sediments pressed together?
   A) cooling   C) melting
   B) weathering   D) compaction

9. What is molten material at Earth's surface called?
   A) limestone   C) breccia
   B) lava   D) granite

10. Which of these is an organic sedimentary rock?
    A) coquina   C) rock salt
    B) sandstone   D) conglomerate

Thinking Critically

11. Granite, pumice, and scoria are igneous rocks. Why doesn't granite have airholes like the other two?

12. Why does marble rarely contain fossils?

13. Why is coquina classified as an organic rock with a granular texture?

14. Would you expect a quartzite or a sandstone to break more easily? Explain your answer.

15. Why are granitic igneous rocks lighter in color than basaltic rocks?

Developing Skills

16. **Comparing and Contrasting** Compare and contrast basaltic and granitic magmas.

17. **Forming Hypotheses** A geologist found a sequence of rocks in which 200-million-year-old shales were on top of 100-million-year-old sandstones. *Hypothesize how this could happen.*

18. **Recognizing Cause and Effect** Explain the effects of pressure and temperature on shale.

19. **Measuring in SI** Assume that the conglomerate shown on the first page of the 2-page Activity is one half of its actual size. Determine the average length of the gravel in the rock.

20. **Concept Mapping** Copy and complete the concept map shown below. Use the following terms and phrases: *magma, sediments, igneous rock, sedimentary rock, metamorphic rock*. Add and label any missing arrows.

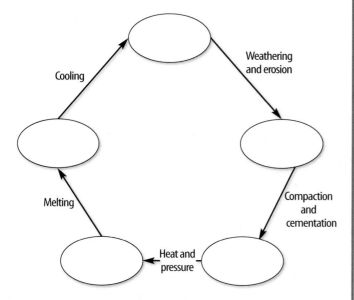

Cooling

Weathering and erosion

Melting

Compaction and cementation

Heat and pressure

Performance Assessment

21. **Poster** Collect a group of small rocks. Make a poster that shows the classifications of rocks, and glue your rocks to the poster under the proper headings. Display your poster to the class. Describe your rocks and explain where you found them.

TECHNOLOGY

Go to the Glencoe Science Web site at **science.glencoe.com** or use the **Glencoe Science CD-ROM** for additional chapter assessment.

Test Practice

Ellen did research on some common rocks. The information she gathered is listed in the table below.

Rock Properties and Uses		
Rocks	**Properties**	**Uses**
Granite	Hard; resistant to attack by air and natural waters	Building stones and monuments
Pumice	Lightweight and porous	Polishing mixtures and cleansers
Limestone	Soft; weathers in presence of natural waters	Gravel, concrete
Slate	Nonporous and splits into thin sheets	Roofing, paving, chalkboards

Study the table and answer the following questions.

1. According to the table, which rock would be a good choice to use as a grave marker in a cemetery?
   A) granite
   B) pumice
   C) sandstone
   D) slate

2. According to the information given, which rock would most likely be used in hand soap?
   F) granite
   G) pumice
   H) sandstone
   J) slate

Earth's Energy and Mineral Resources

Oil fields in Texas, like the one shown here, harbor some of the world's richest reserves of energy resources. In this chapter, you'll learn about many different types of resources that humans extract from Earth. You also will appreciate the importance of conserving these resources—especially those that are extracted faster than nature can produce them.

What do you think?

Science Journal Look at the picture below with a classmate. Discuss what this might be. Here's a hint: *It's on the cutting edge of oil exploration.* Write your answer in your Science Journal.

EXPLORE ACTIVITY

The physical properties of Earth materials determine how easily liquids and gases move through them. Geologists use these properties, in part, to predict where reserves of energy resources like petroleum or natural gas can be found.

Compare rock permeability

1. Obtain a sample of sandstone and a sample of shale from your teacher.

2. Make sure that your samples can be placed on a tabletop so that the sides facing up are reasonably flat and horizontal.

3. Place the two samples side by side in a shallow baking pan.

4. Using a dropper, place three drops of cooking oil on each sample.

5. For ten minutes, observe what happens to the oil on the samples.

Observe

Write your observations in your Science Journal. Infer which rock type might be a good reservoir for petroleum.

Before You Read

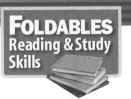

FOLDABLES
Reading & Study Skills

Making a Main Ideas Study Fold Make the following Foldable to help you identify the major topics about energy and mineral resources.

1. Place a sheet of paper in front of you so the long side is at the top. Fold the paper in half from the left side to the right side and then unfold.

2. Fold each side in to the centerfold line to divide the paper into fourths. Fold the paper in half from top to bottom and unfold.

3. Through the top thickness of paper, cut along both of the middle fold lines to form four tabs. Label the tabs *Nonrenewable Energy Resources, Inexhaustible Energy Resources, Renewable Energy Resources,* and *Mineral Resources.*

4. As you read the chapter, list examples on the front of the tabs and write about each type of resource under the tabs.

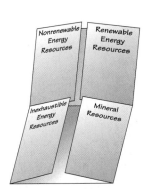

Nonrenewable Energy Resources

As You Read

What You'll Learn

- **Identify** examples of nonrenewable energy resources.
- **Describe** the advantages and disadvantages of using fossil fuels.
- **Explain** the advantages and disadvantages of using nuclear energy.

Vocabulary

fossil fuel	natural gas
coal	reserve
oil	nuclear energy

Why It's Important

Nonrenewable resources should be conserved to ensure their presence for future generations.

Energy

The world's population relies on energy of all kinds. Energy is the ability to cause change. Some energy resources on Earth are being used faster than natural Earth processes can replace them. These resources are referred to as nonrenewable energy resources. Most of the energy resources used to generate electricity are nonrenewable.

Fossil Fuels

Nonrenewable energy resources include fossil fuels. **Fossil fuels** are fuels such as coal, oil, and natural gas that form from the remains of plants and other organisms that were buried and altered over millions of years. Coal is a sedimentary rock formed from the compacted and transformed remains of ancient plant matter. Oil is a liquid hydrocarbon that often is referred to as petroleum. Hydrocarbons are compounds that contain hydrogen and carbon atoms. Other naturally occurring hydrocarbons occur in the gas or semisolid states. Fossil fuels are processed to make gasoline for cars, to heat homes, and for many other uses, as shown in **Table 1.**

Table 1 Uses of Fossil Fuels

Coal	■ To generate electricity	
Oil	■ To produce gasoline and other fuels ■ As lubricants ■ To make plastics, home shingles, and other products	
Natural Gas	■ To heat buildings ■ As a source of sulfur	

Anthracite
Bituminous
Lignite

Coal The most abundant fossil fuel in the world is coal, shown in **Figure 1.** If the consumption of coal continues at the current rate, it is estimated that the coal supply will last for about another 250 years.

Coal is a rock that contains at least 50 percent plant remains. Coal begins to form when plants die in a swampy area. The dead plants are covered by more plants, water, and sediment, preventing atmospheric oxygen from coming into contact with the plant matter. The lack of atmospheric oxygen prevents the plant matter from decaying rapidly. Bacterial growth within the plant material causes a gradual breakdown of molecules in the plant tissue, leaving carbon and some impurities behind. This is the material that eventually will become coal after millions of years. Bacteria also cause the release of methane gas, carbon dioxide, ammonia, and water as the original plant matter breaks down.

✓ Reading Check *What happens to begin the formation of coal in a swampy area?*

Synthetic Fuels Unlike gasoline, which is refined from petroleum, other fuels called synthetic fuels are extracted from solid organic material. Synthetic fuels can be created from coal—a sedimentary rock containing hydrocarbons. The hydrocarbons are extracted from coal to form liquid and gaseous synthetic fuels. Liquid fuels can be processed to produce gasoline for automobiles and fuel oil for home heating. Gaseous synthetic fuels can be used to generate electricity and to heat buildings.

Life Science
INTEGRATION

The coal found in the eastern and midwestern United States formed from plants that lived in great swamps about 300 million years ago during the Pennsylvanian Period of geologic time. Research the Pennsylvanian Period to find out what types of plants lived in these swamps. Describe the plants in your Science Journal.

Stages of Coal Formation
As decaying plant material loses gas and moisture, the concentration of carbon increases. The first step in this process, shown in **Figure 2,** results in the formation of peat. Peat is a layer of organic sediment. When peat burns, it releases large amounts of smoke because it has a high concentration of water and impurities.

As peat is buried under more sediment, it changes into lignite, which is a soft, brown coal with much less moisture. Heat and pressure produced by burial force water out of peat and concentrate carbon in the lignite. Lignite releases more energy and less smoke than peat when it is burned.

As the layers are buried deeper, bituminous coal, or soft coal, forms. Bituminous coal is compact, black, and brittle. It provides lots of heat energy when burned. Bituminous coal contains various levels of sulfur, which can pollute the environment.

If enough heat and pressure are applied to buried layers of bituminous coal, anthracite coal forms. Anthracite coal contains the highest amount of carbon of all forms of coal. Therefore, anthracite coal is the cleanest burning of all coals.

Figure 2
Coal is formed in four basic stages.

A Plant material accumulates in swamps and eventually forms a layer of peat.

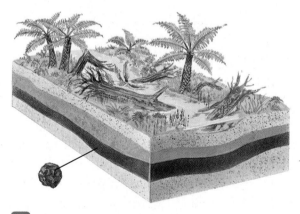

B Over time, heat and pressure cause the peat to change into lignite coal.

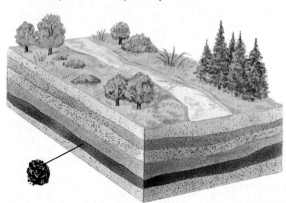

C As the lignite coal becomes buried by more sediments, heat and pressure change it into bituminous coal.

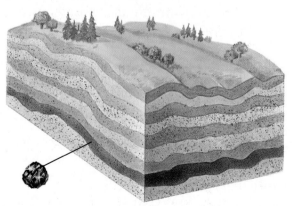

D When bituminous coal is heated and squeezed during metamorphism, anthracite coal forms.

Oil and Natural Gas Coal isn't the only fossil fuel used to obtain energy. Two other fossil fuels that provide large quantities of the energy used today are oil and natural gas. **Oil** is a thick, black liquid formed from the buried remains of microscopic marine organisms. **Natural gas** forms under similar conditions and often with oil, but it forms in a gaseous state. Oil and natural gas are hydrocarbons. However, natural gas is composed of hydrocarbon molecules that are lighter than those in oil.

Residents of the United States burn vast quantities of oil and natural gas for daily energy requirements. As shown in **Figure 3,** Americans obtain most of their energy from these sources. Natural gas is used mostly for heating and cooking. Oil is used in many ways, including as heating oil, gasoline, lubricants, and in the manufacture of plastics and other important compounds.

Formation of Oil and Natural Gas Most geologists agree that petroleum forms over millions of years from the remains of tiny marine organisms in ocean sediment. The process begins when marine organisms called plankton die and fall to the seafloor. Similar to the way that coal is buried, sediment is deposited over them. The temperature rises with depth in Earth, and increased heat eventually causes the dead plankton to change to oil and gas after they have been buried deeply by sediment.

Oil and natural gas often are found in layers of rock that have become tilted or folded. Because they are less dense than water, oil and natural gas are forced upward. Rock layers that are impermeable, such as shale, stop this upward movement. When this happens, a folded shale layer can trap the oil and natural gas below it. Such a trap for oil and gas is shown in **Figure 4.** The rock layer beneath the shale in which the petroleum and natural gas accumulate is called a reservoir rock.

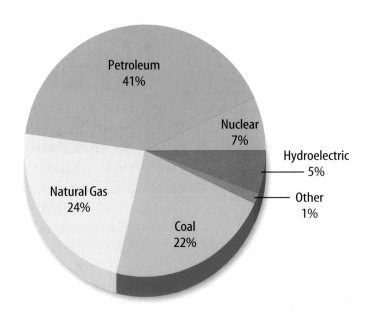

Figure 3
This graph shows the percentages of energy that the United States derives from various energy resources.

Figure 4
Oil and natural gas are fossil fuels formed by the burial of marine organisms. These fuels can be trapped and accumulate beneath Earth's surface.

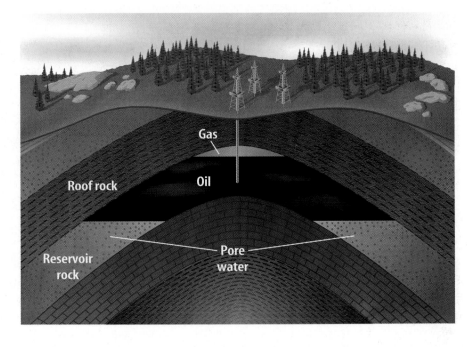

Removing Fossil Fuels from the Ground

Coal is removed from the ground using one of several methods of excavation. The two most common methods are strip mining, also called open-pit mining, and underground mining, shown in **Figure 5.** Oil and natural gas are removed by pumping them out of the ground.

Coal Mining During strip mining, shown in **Figure 5A,** layers of soil and rock above coal are removed and piled to one side. The exposed coal then is removed and loaded into trucks or trains and transported elsewhere. After the coal has been removed, mining companies often return the soil and rock to the open pit and cover it with topsoil. Trees and grass are planted in a process called land reclamation. If possible, animals native to the area are reintroduced. Strip mining is used only when the coal deposits are close to the surface.

In one method of underground coal mining, tunnels are dug and pillars of rock are left to support the rocks surrounding the tunnels. Two types of coal mines are drift mines and slope mines. Drift mining, shown in **Figure 5B,** is the removal of coal that is not close to Earth's surface through a horizontal opening in the side of a hill or mountain. In slope mining, an angled opening and air shaft are made in the side of a mountain to remove coal.

Figure 5
Coal is a fossil fuel that can be removed from Earth in many different ways.

A During strip mining, coal is accessed by removing the soil and rock above it.

B During underground coal mining, tunnels are made into the earth.
How do you think the coal is removed from these tunnels?

Drilling for Oil and Gas Oil and natural gas are fossil fuels that can be pumped from underground deposits. Geologists and engineers drill wells through rocks where these resources might be trapped, as shown in **Figure 6.** As the well is being drilled, it is lined with pipe to prevent it from caving in. When the drill bit reaches the rock layer containing oil, drilling is stopped. Equipment is installed to control the flow of oil. The surrounding rock then is fractured to allow oil and gas to flow into the well. The oil and gas are pumped to the surface.

 **Reading Check** *How are oil and natural gas brought to Earth's surface?*

Fossil Fuel Reserves

The amount of a fossil fuel that can be extracted at a profit using current technology is known as a **reserve.** This is not the same as a fossil fuel resource. A fossil fuel resource has fossil fuels that are concentrated enough that they can be extracted from Earth in useful amounts. However, a resource is not classified as a reserve unless the fuel can be extracted economically. What might cause a known fossil fuel resource to become classified as a reserve?

Methane Hydrates You have learned that current reserves of coal will last about 250 years. Enough natural gas is located in the United States to last about 60 more years. However, recent studies indicate that a new source of methane, which is the main component of natural gas, might be located beneath the seafloor. Icelike substances known as methane hydrates could provide tremendous reserves of methane.

Methane hydrates are stable molecules found hundreds of meters below sea level in ocean floor sediment. They form under conditions of relatively low temperatures and high pressures. The hydrocarbons are trapped within the cagelike structure of ice, as described in **Figure 7.** Scientists estimate that more carbon is contained in methane hydrates than in all current fossil fuel deposits combined. Large accumulations of methane hydrates are estimated to exist off the eastern coast of the United States. Can you imagine what it would mean to the world's energy supply if relatively clean-burning methane could be extracted economically from methane hydrates?

Figure 6
Oil and natural gas are recovered from Earth by drilling deep wells.

Research Visit the Glencoe Science Web site at **science.glencoe.com** for more information about methane hydrates. Communicate to your class what you learn.

Figure 7

Reserves of fossil fuels—such as oil, coal, and natural gas—are limited and will one day be used up. Methane hydrates could be an alternative energy source. This icelike substance, background, has been discovered in ocean floor sediments and in permafrost regions worldwide. If scientists can harness this energy, the world's gas supply could be met for years to come.

Methane hydrates are highly flammable compounds made up of methane—the main component of natural gas—trapped in a cage of frozen water. Methane hydrates represent an enormous source of potential energy. However, they contain a greenhouse gas that might intensify global warming. More research is needed to determine how to safely extract them from the seafloor.

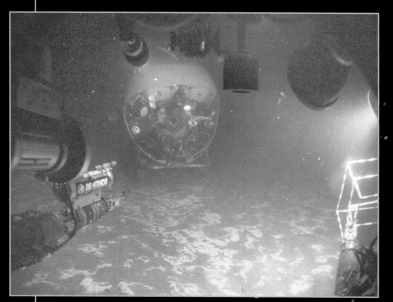

In the photo above, a Russian submersible explores a site in the North Atlantic that contains methane hydrate deposits.

Conserving Fossil Fuels Do you sometimes forget to turn off the lights when you walk out of a room? Wasteful habits might mean that electricity to run homes and industries will not always be as plentiful and cheap as it is today. Fossil fuels take millions of years to form and are used much faster than Earth processes can replenish them.

Today, coal provides about 25 percent of the energy that is used worldwide and 22 percent of the energy used in the United States. Oil and natural gas provide almost 61 percent of the world's energy and about 65 percent of the U.S. energy supply. At the rate these fuels are being used, they could run out some-day. How can this be avoided?

By remembering to turn off lights and appliances, you can avoid wasting fossil fuels. Another way to conserve fossil fuels is to make sure doors and windows are shut tightly during cold weather so heat doesn't leak out of your home. If you have air-conditioning, run it as little as possible. Ask the adults you live with if more insulation could be added to your home or if an insulated jacket could be put on the water heater.

Energy from Atoms

Most electricity in the United States is generated in power plants that use fossil fuels. However, alternate sources of energy exist. **Nuclear energy** is an alternate energy source produced from atomic reactions. When the nucleus of a heavy element is split, lighter elements form and energy is released. This energy can be used to light a home or power the submarines shown in **Figure 8.**

The splitting of heavy elements to produce energy is called nuclear fission. Nuclear fission is carried out in nuclear power plants using a type of uranium as fuel.

TRY AT HOME
Mini LAB

Practicing Energy Conservation

Procedure
1. Have an adult help you find the **electric meter** for your home and record the reading in your **Science Journal.**
2. Do this for several days, taking your meter readings at about the same time each day.
3. List things you and your family can do to reduce your electricity use.
4. Encourage your family to try some of the listed ideas for several days.

Analysis
1. Keep taking meter readings and infer whether the changes make any difference.
2. Have you and your family helped conserve energy?

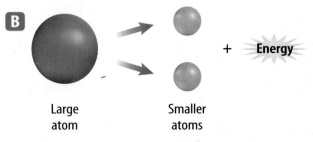

A

Figure 8
Atoms can be a source of energy.
A These submarines are powered by nuclear fission. **B** Energy is given off when a larger atom, like uranium, splits into smaller atoms.

B

Large atom

Smaller atoms

+ Energy

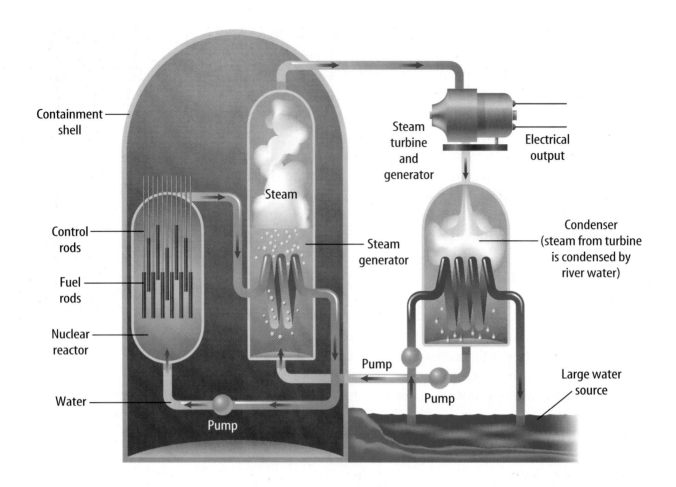

Figure 9
Heat released in nuclear reactors produces steam, which in turn is used to produce electricity. This is an example of transforming nuclear energy into electrical energy.

In the figure, the following labels appear:

- Containment shell
- Control rods
- Fuel rods
- Nuclear reactor
- Water
- Pump
- Steam
- Steam generator
- Steam turbine and generator
- Electrical output
- Condenser (steam from turbine is condensed by river water)
- Pump
- Pump
- Large water source

Electricity from Nuclear Energy A nuclear power plant, shown in **Figure 9,** has a large chamber called a nuclear reactor. Within the nuclear reactor, uranium fuel rods sit in a pool of cooling water. Neutrons are fired into the fuel rods. When the uranium-235 atoms are hit, they break apart and fire out neutrons that hit other atoms, beginning a chain reaction. As each atom splits, it not only fires neutrons but also releases heat that is used to boil water to make steam. The steam drives a turbine, which turns a generator that produces electricity.

✔ Reading Check *How is nuclear energy used to produce electricity?*

Nuclear energy from fission is considered to be a nonrenewable energy resource because it uses uranium-235 as fuel. A limited amount of uranium-235 is available for use. Another problem with nuclear energy is the waste material that it produces. Nuclear waste from power plants consists of highly radioactive elements formed by the fission process. Some of this waste will remain radioactive for thousands of years. The Environmental Protection Agency (EPA) has determined that nuclear waste must be stored safely and contained for at least 10,000 years before reentering the environment.

Fusion Environmental problems related to nuclear power could be eliminated if usable energy could be obtained from fusion. The Sun is a natural fusion power plant that provides energy for Earth and the solar system. Someday fusion also might provide energy for your home.

During fusion, materials of low mass are fused together to form a substance of higher mass. No fuel problem exists if the low-mass material is a commonly occurring substance. Also, if the end product is not radioactive, storing nuclear

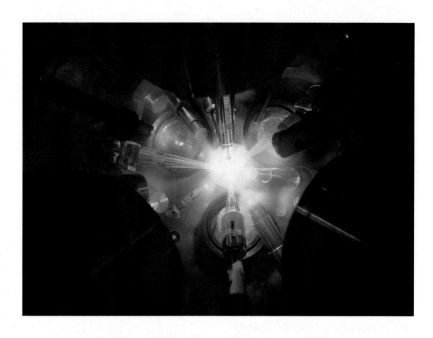

waste is not a problem. In fact, fusion of hydrogen into helium would satisfy both of these conditions. However, technologies do not currently exist to enable humans to fuse hydrogen into helium at reasonably low temperatures in a controlled manner. But research is being conducted, as shown in **Figure 10.** If this is accomplished, nuclear energy could be considered an inexhaustible fuel resource. You will learn the importance of inexhaustible and renewable energy resources in the next section.

Figure 10
Lasers are used in research facilities to help people understand and control fusion.

Section 1 Assessment

1. Why are coal, oil, and natural gas considered to be fossil fuels?

2. Why are fossil fuels considered to be non-renewable energy resources?

3. Describe similarities and differences among peat, lignite, bituminous coal, and anthracite coal.

4. How is nuclear energy obtained? What are two disadvantages of nuclear energy?

5. **Think Critically** Fossil fuels form in specific geologic environments. Why are you likely to find natural gas and oil deposits in the same location, but less likely to find coal and petroleum deposits at the same location?

Skill Builder Activities

6. **Making and Using Tables** Current energy consumption by source in the United States is as follows: *petroleum, 41 percent; natural gas, 24 percent; coal, 22 percent; nuclear energy, seven percent;* and *inexhaustible and renewable resources, six percent.* Make a bar graph of these data. **For more help, refer to the** Science Skill Handbook.

7. **Using Percentages** Using the data from question 6, what total percent of energy in the United States comes from petroleum and natural gas? How many times more energy is obtained from these sources than from coal? **For more help, refer to the** Math Skill Handbook.

2 Inexhaustible and Renewable Energy Resources

Inexhaustible Energy Resources

How soon the world runs out of fossil fuels depends on how they are used and conserved. Fortunately, there are inexhaustible energy resources. These sources of energy are constant and will not run out in the future as fossil fuels will. Inexhaustible energy resources include the Sun, wind, water, and geothermal energy.

Energy from the Sun When you sit in the Sun, walk into the wind, or sail against an ocean current, you are experiencing the power of solar energy. **Solar energy** is energy from the Sun. You already know that the Sun's energy heats Earth, and it causes circulation in Earth's atmosphere and oceans. Ocean currents and global winds are examples of nature's use of solar energy. Thus, solar energy is used indirectly when the wind and some types of moving water are used to do work.

People can use solar energy in a passive way or in an active way. South-facing windows on buildings act as passive solar collectors, warming exposed rooms. Solar cells actively collect energy from the Sun and transform it into electricity. Solar cells were invented to generate electricity for satellites. Now they also are used to power calculators, streetlights, and experimental cars. Some people have installed solar energy cells on their roofs, as shown in **Figure 11.**

Figure 11
Solar panels, such as on this home in Laguna Niguel, California, can be used to collect inexhaustible solar energy to power appliances and heat water.

Figure 12
Wind farms are used to produce electricity.

Disadvantages of Solar Energy Solar energy is clean and inexhaustible, but it does have some disadvantages. Solar cells work less efficiently on cloudy days and cannot work at all at night. Some systems use batteries to store solar energy for use at night or on cloudy days, but it is difficult to store large amounts of energy in batteries. Worn out batteries also must be discarded. This can pollute the environment if not done properly.

Energy from Wind What is better to do on a warm, windy day than fly a kite? A strong wind can lift a kite high in the sky and whip it around. The pull of the wind is so great that you wonder if it will whip the kite right out of your hands. Wind is a source of energy. It was and still is used to power sailing ships. Windmills have used wind energy to grind corn and pump water. Today, windmills can be used to generate electricity. When a large number of windmills are placed in one area for the purpose of generating electricity, the area is called a **wind farm,** as shown in **Figure 12.**

Wind energy has advantages and disadvantages. Wind is non-polluting and free. It does little harm to the environment and produces no waste. However, only a few regions of the world have winds strong enough to generate electricity. Also, wind isn't steady. Sometimes it blows too hard and at other times it is too weak or stops entirely. For an area to use wind energy consistently, the area must have a persistent wind that blows at an appropriate speed.

Physics
INTEGRATION

Wind is an inexhaustible energy resource that is used by windmills to produce energy. As the blades rotate, turbines are turned to produce electricity. Find out which areas utilize wind farms and report in your Science Journal how much electricity is produced and what it is used for.

 **Reading Check** *Why are some regions better suited for wind farms than others?*

Energy from Water For a long time, waterwheels steadily spun next to streams and rivers. The energy in the flowing water powered the wheels that ground grain or cut lumber. More than a pretty picture, using a waterwheel in this way is an example of microhydropower. Microhydropower has been used throughout the world to do work.

Today energy from running water is used to generate electricity. Electricity produced by waterpower is called **hydroelectric energy.** To generate electricity from water running in a river, a large concrete dam is built to retain water, as illustrated in **Figure 13.** A lake forms behind the dam. As water is released, its force turns turbines at the base of the dam. The turbines then turn generators that make electricity.

At first it might appear that hydroelectric energy doesn't create any environmental problems and that the water is used with little additional cost. However, when dams are built, upstream lakes fill with sediment and downstream erosion increases. Land above the dam is flooded, and wildlife habitats are damaged.

Energy from Earth Erupting volcanoes and geysers like Old Faithful are examples of geothermal energy in action. The energy that causes volcanoes to erupt or water to shoot up as a geyser also can be used to generate electricity. Energy obtained by using hot magma or hot, dry rocks inside Earth is called **geothermal energy.**

Bodies of magma can heat large reservoirs of groundwater. Geothermal power plants use steam from the reservoirs to produce electricity, as shown in **Figure 14.** In a developing method, water becomes steam when it is pumped through broken, hot, dry rocks. The steam then is used to turn turbines that run generators to make electricity. The advantage of using hot, dry rocks is that they are found just about everywhere. Geothermal energy presently is being used in Hawaii and in parts of the western United States.

Figure 13
Hydroelectric power is important in many regions of the United States. **A** Hoover Dam was built on the Colorado River to supply electricity for a large area.
B The power of running water is converted to usable energy in a hydroelectric power plant.

A

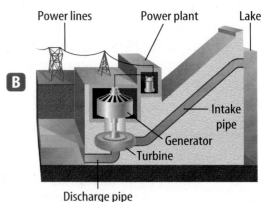

Power lines   Power plant   Lake

B

Intake pipe

Generator

Turbine

Discharge pipe

Figure 14
Geothermal energy is used to supply
electricity to industries and homes.

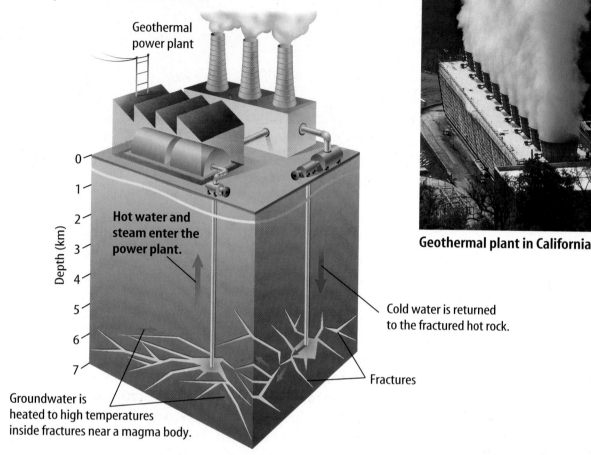

Geothermal
power plant

Geothermal power plant

Depth (km)

0
1
2
3
4
5
6
7

Hot water and
steam enter the
power plant.

Cold water is returned
to the fractured hot rock.

Fractures

Groundwater is
heated to high temperatures
inside fractures near a magma body.

Geothermal plant in California

Renewable Energy Resources

Energy resources that can be replaced in nature or by
humans within a relatively short period of time are referred to
as renewable energy resources. This short period of time is
defined generally as being within a human life span. For example, trees can be considered a renewable energy resource. As one
tree is cut down, another can be planted in its place. The new
tree might be left untouched or harvested.

You have learned that most energy used in the United
States—about 90 percent—comes from fossil fuels, which are
nonrenewable energy resources. Next, you'll look at some
renewable energy resources and how they might fit into the
world's total energy needs now and in the future.

Biomass Energy

A major renewable energy resource is biomass materials.
Biomass energy is energy derived from burning organic material such as wood, alcohol, and garbage. The term *biomass* is
derived from the words *biological* and *mass.*

SCIENCE
Online

Research Visit the
Glencoe Science Web site at
science.glencoe.com for
more information about biomass energy. Communicate
to your class what you learn.

Figure 15
These campers are using wood, a renewable energy resource, to produce heat and light.

Energy from Wood If you've ever sat around a campfire, like the campers shown in **Figure 15,** or close to a wood-burning fireplace to keep warm, you have used energy from wood. The burning wood is releasing stored solar energy as heat energy. Humans have long used wood as an energy resource. Much of the world still cooks with wood. In fact, firewood is used more widely today than any other type of biomass fuel.

Using wood as a biomass fuel has its problems. Gases and small particles are released when wood is burned. These materials can pollute the air. When trees are cut down for firewood, natural habitats are destroyed. However, if proper conservation methods are employed or if tree farms are maintained specifically for use as fuel, energy from wood can be a part of future energy resources.

Energy from Alcohol Biomass fuel can be burned directly, such as when wood or peat is used for heating and cooking. However, it also can be transformed into other materials that might provide cleaner, more efficient fuels.

For example, during distillation, biomass fuel, such as corn, is changed to an alcohol such as ethanol. Ethanol then can be mixed with another fuel. When the other fuel is gasoline, the mixture is called gasohol. Gasohol can be used in the same way as gasoline, as shown in **Figure 16,** but it cuts down on the amount of fossil fuel needed to produce gasoline. Fluid biomass fuels are more efficient and have more uses than solid biomass fuels do.

The problem with this process is that presently, growing the corn and distilling the ethanol often uses more energy from burning fossil fuels than the amount of energy that is derived from burning ethanol. At present, biomass fuel is best used locally.

Figure 16
Gasohol sometimes is used to reduce dependence on fossil fuels.

✓ **Reading Check** *What are the drawbacks of biomass fuels?*

Energy from Garbage Every day humans throw away a tremendous amount of burnable garbage. As much as two thirds of what is thrown away could be burned. If more garbage were used for fuel, as shown in **Figure 17,** human dependence on fossil fuels would decrease. Burning garbage is a cheap source of energy and also helps reduce the amount of material that must be dumped into landfills.

Compared to other nations, the United States lags in the use of municipal waste as a renewable energy resource. For example, in some countries in Western Europe, as much as half of the waste generated is used for biomass fuel. When the garbage is burned, heat is produced, which turns water to steam. The steam turns turbines that run generators to produce electricity.

Unfortunately, some problems can be associated with using energy from garbage. Burning municipal waste can produce toxic ash residue and air pollution. Substances such as heavy metals could find their way into the smoke from garbage and thus into the atmosphere.

Figure 17
Garbage can be burned to produce electricity at trash-burning power plants such as this one in Virginia.

Section Assessment

1. What are some advantages and disadvantages of using solar energy, wind energy, and hydroelectric energy?

2. What is the difference between inexhaustible and renewable energy resources? Give specific examples of each.

3. How is geothermal energy used to create electricity?

4. Why are nonrenewable resources used more than inexhaustible and renewable resources?

5. **Think Critically** How could nuclear energy, which normally is classified as a nonrenewable energy resource, be reclassified as an inexhaustible energy resource?

Skill Builder Activities

6. **Classifying** Classify the following energy resources as inexhaustible or renewable: *wood, solar energy, hydroelectric energy, geothermal energy, ethanol, garbage,* and *wind energy.* **For more help, refer to the** Science Skill Handbook.

7. **Using an Electronic Spreadsheet** Using the spreadsheet capabilities of a computer, make a table of energy resources. Include an example of how each resource could be used. Then write a description in your Science Journal about how you could reduce the use of energy resources at home. **For more help, refer to the** Technology Skill Handbook.

Soaking up Solar Energy

W inter clothing tends to be darker in color than summer clothing. The color of the material used in the clothing affects its ability to absorb energy. In this activity, you will use different colors of soil to study this effect.

Time and Temperature			
Time (min)	Temperature Dish A (°C)	Temperature Dish B (°C)	Temperature Dish C (°C)
0.0			
0.5			
1.0			
1.5			

What You'll Investigate
How does color affect the absorption of energy?

Materials
dry, black soil
dry, brown soil
dry, sandy, white soil
thermometers (3)
ring stand
graph paper
colored pencils (3)
metric ruler

clear glass or plastic dishes (3)
200-watt gooseneck lamp
*200-watt lamp with reflector and clamp
watch or clock with second hand
*stopwatch
*Alternate materials

Safety Precautions

WARNING: *Handle glass with care so as not to break it. Wear thermal mitts when handling the light source.*

Goals
■ **Determine** whether color has an effect on the absorption of solar energy.
■ **Relate** the concept of whether color affects absorption to other applications.

Procedure
1. Fill each plastic dish with a different color of soil to a depth of 2.5 cm.

2. Arrange the dishes close together on your desk and place a thermometer in each dish. Be sure to cover the thermometer bulb in each dish completely with the soil.

3. Position the lamp over all three dishes.

4. **Design** a data table for your observations similar to the sample table above. You will need to read the temperature of each dish every 30 s for 20 min after the light is turned on.

5. Turn on the light and begin your experiment.

6. Use the data to construct a graph. Time should be plotted on the horizontal axis and temperature on the vertical axis. Use a different colored pencil to plot the data for each type of soil, or use a computer to design a graph that illustrates your data.

Conclude and Apply
1. Which soil had the greatest temperature change? The least?

2. Why do the curves on the graph flatten?

3. Why do flat-plate solar collectors have black plates behind the water pipes?

4. How does the color of a material affect its ability to absorb energy?

5. Why is most winter clothing darker in color than summer clothing?

Mineral Resources

Metallic Mineral Resources

If your room at home is anything like the one shown in **Figure 18,** you will find many metal items. Metals are obtained from Earth materials called metallic mineral resources. A **mineral resource** is a deposit of useful minerals. See how many metals you can find. Is there anything in your room that contains iron? What about the metal in the frame of your bed? Is it made of iron? If so, the iron might have come from the mineral hematite. What about the framing around the windows in your room? Is it aluminum? Aluminum, like that in a soft-drink can, comes from a mixture of minerals known as bauxite. Many minerals contain these and other useful elements. Which minerals are mined as sources for the materials you use every day?

Ores Deposits in which a mineral or minerals exist in large enough amounts to be mined at a profit are called **ores.** Generally, the term *ore* is used for metallic deposits, but this is not always the case. The hematite that was mentioned earlier as an iron ore and the bauxite that was mentioned earlier as an aluminum ore are metallic ores.

✔ **Reading Check** *What is an ore?*

As You Read

What You'll Learn

- **Explain** the conditions needed for a mineral to be classified as an ore.
- **Describe** how market conditions can cause a mineral to lose its value as an ore.
- **Compare and contrast** metallic and nonmetallic mineral resources.

Vocabulary
mineral resource
ore
recycling

Why It's Important
Many products you use are made from mineral resources.

Many bed frames contain iron, which is extracted from minerals such as hematite.

Copper in wires found in electrical equipment comes from the mineral chalcopyrite.

Aluminum comes from a mixture of minerals called bauxite.

Stainless steel contains chromium, which comes from the mineral chromite.

Figure 18
Many items in your home are made from metals obtained from metallic mineral resources.

Figure 19
Iron ores are smelted to produce nearly pure iron. *What could this iron be used for?*

Economic Effects When is a mineral deposit considered an ore? The mineral in question must be in demand. Enough of it must be present in the deposit to make it worth removing. Some mining operations are profitable only if a large amount of the mineral is needed. It also must be fairly easy to separate the mineral from the material in which it is found. If any one of these conditions isn't met, the deposit might not be considered an ore.

Supply and demand is an important part of life. You might have noticed that when the supply of fresh fruit is down, the price you pay for it at the store goes up. Economic factors largely determine what an ore is.

Refining Ore The process of extracting a useful substance from an ore involves two operations—concentrating and refining. After a metallic ore is mined from Earth's crust, it is crushed and the waste rock is removed. The waste rock that must be removed before a mineral can be used is called gangue (GANG).

Refining produces a pure or nearly pure substance from ore. For example, iron can be concentrated from the ore hematite, which is composed of iron oxide. The concentrated ore then is refined to be as close to pure iron as possible. One method of refining is smelting, illustrated in **Figure 19.** Smelting is a chemical process that removes unwanted elements from the metal that is being processed. During one smelting process, a concentrated ore of iron is heated with a specific chemical. The chemical combines with oxygen in the iron oxide, resulting in pure iron. Note that one resource, fossil fuel, is burned to produce the heat that is needed to obtain the finished product of another resource, in this case iron.

Nonmetallic Mineral Resources

Any mineral resources not used as fuels or as sources of metals are nonmetallic mineral resources. These resources are mined for the nonmetallic elements contained in them and for the specific physical and chemical properties they have. Generally, nonmetallic mineral resources can be divided into two different groups—industrial minerals and building materials. Some materials, such as limestone, belong to both groups of nonmetallic mineral resources, and others are specific to one group or the other.

Industrial Minerals Many useful chemicals are obtained from industrial minerals. Sandstone is a source of silica (SiO_2), which is a compound that is used to make glass. Some industrial minerals are processed to make fertilizers for farms and gardens. For example, sylvite, a mineral that forms when seawater evaporates, is used to make potassium fertilizer.

Many people enjoy a little sprinkle of salt on french fries and pretzels. Table salt is a product derived from halite, a nonmetallic mineral resource. Halite also is used to help melt ice on roads and sidewalks during winter and to help soften water.

Other industrial minerals are useful because of their characteristic physical properties. For example, abrasives are made from deposits of corundum and garnet. Both of these minerals are hard and able to scratch most other materials they come into contact with. Small particles of garnet can be glued onto a sheet of heavy paper to make abrasive sandpaper. **Figure 20** illustrates just a few ways in which nonmetallic mineral resources help make your life more convenient.

Figure 20
You benefit from the use of industrial minerals every day.

A Road salt melts ice on streets.

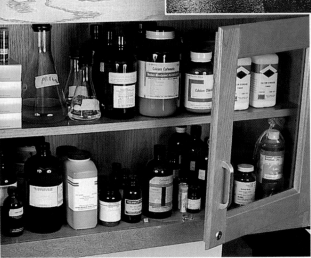

B Many important chemicals are made from industrial minerals.

C An industrial mineral called trona is important for making glass.

What stones are used in buildings where you live? To find out more about building stones, see the **Building Stones Field Guide** at the back of the book.

Building Materials One of the most important nonmetallic mineral resources is aggregate. Aggregate is composed of crushed stone or a mixture of gravel and sand and has many uses in the building industry. For example, aggregates can be mixed with cement and water to form concrete. Quality concrete is vital to the building industry. Limestone also has industrial uses. It is used as paving stone and as part of concrete mixtures. Have you ever seen the crushed rock in a walking path or driveway? The individual pieces might be crushed limestone. Gypsum, a mineral that forms when seawater evaporates, is soft and lightweight and is used in the production of plaster and wallboard. If you handle a piece of broken plaster or wallboard, note its appearance, which is similar to the mineral gypsum.

Rock also is used as building stone. You might know of buildings in your region that are made from granite, limestone, or sandstone. These rocks and others are quarried and cut into blocks and sheets. The pieces then can be used to construct buildings. Some rock also is used to sculpt statues and other pieces of art.

Reading Check *What are some important nonmetallic mineral resources?*

Problem-Solving Activity

Why should you recycle?

Recycling in the United States has become a way of life. In 2000, 88 percent of Americans participated in recycling. Recycling is important because it saves precious raw materials and energy. Recycling aluminum saves 95 percent of the energy required to obtain it from its ore. Recycling steel saves up to 74 percent in energy costs, and recycling glass saves up to 22 percent.

Recycling Rates in the United States			
Material	**1995 (%)**	**1997 (%)**	**2000 (%) (estimated)**
Glass	24.5	24.3	29–33
Steel	36.5	38.4	41–46
Aluminum	34.6	31.2	37–39
Plastics	5.3	5.2	6–7

Identifying the Problem

The following table includes materials that currently are being recycled and rates of recycling for the years 1995, 1997, and 2000. Examine the table to determine materials for which recycling increased or decreased between 1995 and 2000.

Solving the problem

1. Has the recycling of materials increased or decreased over time? Which materials are recycled most? Which materials are recycled least? Suggest reasons why some materials might be recycled more than others are.
2. How can recycling benefit society? Explain your answer.

Recycling Mineral Resources

Mineral resources are nonrenewable. You've learned that nonrenewable resources are those that Earth processes cannot replace within an average human's lifetime. Most mineral resources take millions of years to form. Have you ever thrown away an empty soft-drink can? Many people do. These cans become solid waste. Wouldn't it be better if these cans and other items made from mineral resources were recycled into new items?

Recycling is using old materials to make new ones. Recycling has many advantages. It reduces the demand for new mineral resources. The recycling process often uses less energy than it takes to obtain new material. Because supplies of some minerals might become limited in the future, recycling could be required to meet needs for certain materials, as shown in **Figure 21.**

Recycling also can be a profitable experience. Some companies purchase scrap metal and empty soft-drink cans for the aluminum and tin content. The seller receives a small amount of money for turning in the material. Schools and other groups earn money by recycling soft-drink cans.

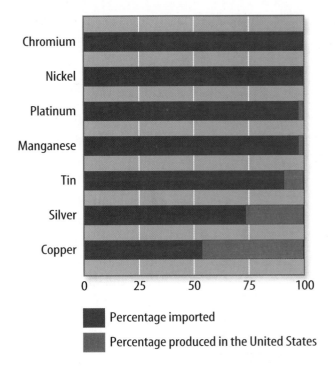

- Percentage imported
- Percentage produced in the United States

Figure 21
The United States produces only a small percentage of the metallic resources it consumes.

Section Assessment

1. How are metals obtained from metallic mineral resources used in your home and school? Which of these products could be recycled easily?

2. What characteristics define a mineral deposit as an ore?

3. List two industrial uses for nonmetallic mineral resources.

4. How can supply and demand of a material cause a mineral to become an ore?

5. **Think Critically** Gangue is waste rock remaining after a mineral ore is removed. Why is gangue sometimes reprocessed?

Skill Builder Activities

6. **Classifying** Classify the following mineral resources as metallic or nonmetallic: *hematite, limestone, bauxite, sandstone, garnet,* and *chalcopyrite.* Explain why you classified each one as you did. **For more help, refer to the** Science Skill Handbook.

7. **Communicating** Research the element titanium and its uses. Write a description in your Science Journal explaining the importance of this element and the mineral resources it comes from. Explain why this element is used in the ways that you found. **For more help, refer to the** Science Skill Handbook.

Home Sweet Home

As fossil fuel supplies continue to be depleted, an increasing U.S. population has recognized the need for alternative energy sources. United States residents might be forced to consider using renewable and inexhaustible energy resources to meet some of their energy needs. The need for energy-efficient housing is more relevant now than ever before. A designer of energy-efficient homes considers proper design and structure, a well chosen building site with wise material selection, and selection of efficient energy generation systems to power the home. Energy-efficient housing uses less energy and produces fewer pollutants.

Recognize the Problem

What does the floor plan, building plan, or a model of an energy efficient home look like?

Thinking Critically

How and where should your house be designed and built to use the alternative energy resources you've chosen efficiently?

Goals
- **Research** various renewable and inexhaustible energy resources available to use in the home.
- **Design** blueprints for an energy-efficient home and/or design and build a model of an energy-efficient home.

Possible Materials
paper
ruler
pencils
cardboard
glue
aluminum foil

Data Source
SCIENCE*Online* Go to the Glencoe Science Web site at **science.glencoe.com** for more information about designing an energy-efficient home.

FOR SALE
AWARD WINNING
PASSIVE SOLAR HOUSE

Planning the Model

Plan

1. **Research** current information about energy-efficient homes.

2. **Research** inexhaustible energy resources such as wind, hydroelectric power, or solar power, as well as energy conservation. Decide which energy resources are most efficient for your home design.

3. Decide where your house should be built to use energy efficiently.

4. Decide how your house will be laid out and draw mock blueprints for your home. Highlight energy issues such as where solar panels can be placed.

5. Build a model of your energy-efficient home.

Do

1. Ask your peers for input on your home. As you research, become an expert in one area of alternative energy generation and share your information with your classmates.

2. **Compare** your home's design to energy-efficient homes you learn about through your research.

Making the Model

1. Think about how most of the energy in a home is used. Remember as you plan your home that energy-efficient homes not only generate energy—they also use it more efficiently.

2. Carefully consider where your home should be built. For instance, if you plan to use wind power, will your house be built in an area that receives adequate wind?

3. Be sure to plan for backup energy generation. For instance, if you plan to use mostly solar energy, what will you do if it's a cloudy day?

Analyzing and Applying Results

1. Devise a budget for building your home. Could your energy-efficient home be built at a reasonable price? Could anyone afford to build it?

2. Create a list of pro and con statements about the use of energy-efficient homes. Why aren't inexhaustible and renewable energy sources widely used in homes today?

*C*ommunicating

Your Data

Present your model to the class. **Explain** which energy resources you chose to use in your home and why. Have an open house. Take prospective home owners/classmates on a tour of your home and sell it.

Oil's Well That

What if you went out to your backyard, started digging a hole, and all of a sudden oil spurted out of the ground? Dollar signs might flash before your eyes.

It wasn't quite that exciting for Charles Tripp. Tripp, a Canadian, is credited with being the first person to strike oil. And he wasn't even looking for what has become known as "black gold."

In 1851, Tripp built a factory in Ontario, Canada, not far from Lake Erie. He used a natural, black, thick, sticky substance that could be found nearby to make asphalt for paving roads and to construct buildings. He also sold the product to waterproof boats.

The Titusville, Pennsylvania, oil well drilled by Edwin Drake. This photo was taken in 1864.

144

In 1855, Tripp dug a well looking for fresh-water for his factory. After digging just 2 m or so, he unexpectedly came upon liquid. It wasn't clear, clean, and delicious; it was smelly, thick, and black. You guessed it—oil! Tripp didn't understand the importance of his find and how it might help his business. Nor did he realize that it might make him rich someday. So, two years after his accidental discovery, Tripp sold his company to James Williams.

In 1858, Williams continued to search for water for the factory, but, as luck would have it, diggers kept finding oil. So workers barreled up the oil and sold it as liquid grease for the train industry. By 1861, 1.5 million L of crude oil had been barreled and shipped. The finds also led to a short boom in oil production in the area.

Some people argue that the first oil well in North America was in Titusville, Pennsylvania, when Edwin Drake hit oil in 1859. However, most historians agree that Williams was first in 1858. But they also agree that it was Edwin Drake's discovery that led to the growth of the oil industry. So, Drake and Williams can share the credit!

Ends

Some people used TNT to search for oil. This photo was taken in 1943.

EXPLOSIVES

The accidental discovery of a gloppy black substance helped change the planet

Well

Today, many oil companies are drilling beneath the sea for oil.

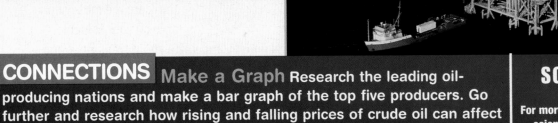

CONNECTIONS Make a Graph Research the leading oil-producing nations and make a bar graph of the top five producers. Go further and research how rising and falling prices of crude oil can affect the U.S. and world economies. Share your findings with your class.

SCIENCE *Online*

For more information, visit science.glencoe.com.

Chapter 5 Study Guide

Reviewing Main Ideas

Section 1 Nonrenewable Energy Resources

1. Fossil fuels are considered to be non-renewable energy resources.

2. The higher the concentration of carbon in coal is, the cleaner it burns. *What might the coal in this truck be used for?*

3. Oil and natural gas form from altered and buried marine organisms and often are found near one another.

4. Nuclear energy is obtained from the fission of heavy isotopes.

Section 2 Inexhaustible and Renewable Energy Resources

1. Inexhaustible energy resources—solar energy, wind energy, water energy, and geothermal energy—are constant and will not run out. *What are some advantages of using solar panels to produce electricity?*

2. Renewable energy resources are replaced within a relatively short period of time.

3. Biomass energy is derived from organic material such as wood and corn.

Section 3 Mineral Resources

1. Metallic mineral resources provide metals when they are processed.

2. Ores are mineral resources that contain a usable substance that can be mined at a profit.

3. Smelting is a chemical process that removes unwanted elements from a metal that is being processed.

4. Nonmetallic mineral resources are classified as industrial minerals or building materials. *Is the aggregate that is being produced here an industrial mineral or a building material?*

FOLDABLES
Reading & Study Skills

After You Read

Circle the resources you think are the most valuable on your Foldable. Explain why and write about similarities and differences among the resources.

Visualizing Main Ideas

Fill in the following table that lists advantages and disadvantages of different energy resources.

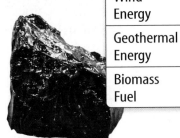

Energy Resources		
Resource	**Advantages**	**Disadvantages**
Fossil Fuels		
Nuclear Energy		
Solar Energy		
Wind Energy		
Geothermal Energy		
Biomass Fuel		

Vocabulary Review

Vocabulary Words

a. biomass energy
b. coal
c. fossil fuel
d. geothermal energy
e. hydroelectric energy
f. mineral resource
g. natural gas
h. nuclear energy
i. oil
j. ore
k. recycling
l. reserve
m. solar energy
n. wind farm

Using Vocabulary

Each phrase below describes a vocabulary word from the list. Write the word that matches the phrase describing it.

1. mineral resource that can be mined at a profit

2. fuel that is composed mainly of the remains of dead plants

3. method of conservation in which items are processed to be used again

4. inexhaustible energy resource that is used to power the *Hubble Space Telescope*

5. energy resource that is based on the fission of atoms

6. liquid from remains of marine organisms

Chapter 5 Assessment

Checking Concepts

Choose the word or phrase that best answers the question.

1. Which of the following has the highest content of carbon?
   - **A)** peat
   - **B)** lignite
   - **C)** bituminous coal
   - **D)** anthracite coal

2. Which of the following is the first step in the evolution of coal?
   - **A)** formation of peat
   - **B)** formation of lignite
   - **C)** formation of bituminous coal
   - **D)** formation of anthracite coal

3. Which of the following is an example of a fossil fuel?
   - **A)** wind
   - **B)** water
   - **C)** natural gas
   - **D)** uranium-235

4. What is the waste material that must be separated from an ore?
   - **A)** smelter
   - **B)** mineral resource
   - **C)** gangue
   - **D)** petroleum

5. What common rock structure can trap oil and natural gas under it?
   - **A)** folded rock
   - **B)** sandstone rock
   - **C)** porous rock
   - **D)** unconsolidated rock

6. Which type of energy resource uses large dams in a river?
   - **A)** wind
   - **B)** nuclear
   - **C)** hydroelectric
   - **D)** solar

7. What is a region where many windmills are located in order to generate electricity from wind called?
   - **A)** wind farm
   - **B)** hydroelectric dam
   - **C)** oil well
   - **D)** steam-driven turbine

8. Which of the following is a deposit of hematite that can be mined at a profit?
   - **A)** ore
   - **B)** anthracite
   - **C)** gangue
   - **D)** energy resource

9. What is an important use of petroleum?
   - **A)** making plaster
   - **B)** making glass
   - **C)** as abrasives
   - **D)** making gasoline

10. Which of the following is a nonrenewable energy resource?
    - **A)** water
    - **B)** wind
    - **C)** geothermal
    - **D)** petroleum

Thinking Critically

11. Explain how solar energy becomes stored in plants and other organisms and is released later when fossil fuels are burned.

12. Describe the major problems associated with generating electricity using nuclear power plants.

13. Why is wind considered to be an inexhaustible energy resource?

14. Which type of energy resources are considered to be biomass fuels?

15. What conditions could occur to cause gangue to be reclassified as an ore?

Developing Skills

16. **Predicting** If a well were drilled into a rock layer containing petroleum, natural gas, and water, which substance would be encountered first? Explain.

17. **Comparing and Contrasting** Create a table comparing and contrasting solar energy and wind energy.

18. Concept Mapping Complete the following concept map about mineral resources.

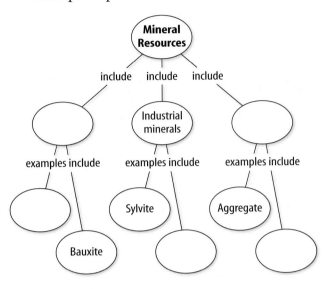

19. Making Models Make a model of a house that has been built to use passive solar energy.

Performance Assessment

20. Oral Presentation Research the latest information on inexhaustible energy resources. Develop a presentation that tries to persuade consumers to reduce their use of fossil fuels and to use inexhaustible energy resources instead.

21. Letter Write a letter to the Department of Energy asking how usable energy might be obtained from methane hydrates in the future. Also inquire about methods to extract methane hydrates.

TECHNOLOGY

Go to the Glencoe Science Web site at **science.glencoe.com** or use the **Glencoe Science CD-ROM** for additional chapter assessment.

Test Practice

The graph below shows the average abundance of some important metals in Earth's crust.

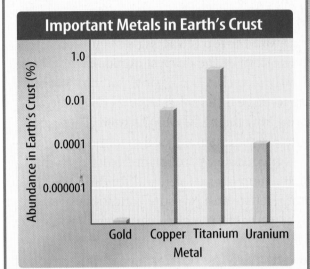

Study the bar graph and answer the following questions.

1. After examining this graph, you should conclude that these important metals _____ .
   A) are concentrated in Earth's crust
   B) are valuable for jewelry
   C) are mined out
   D) are generally rare in Earth's crust

2. Which of the following best explains the occurrence of rich ores of these metals?
   F) The metals are concentrated in some places by Earth processes.
   G) Metal-rich meteorites frequently strike Earth.
   H) Some metals are more abundant in Earth's core.
   J) Percentages often are misleading.

Read the passage. Then read each question that follows the passage. Decide which is the best answer to each question.

Obsidian Uses

From the top of the highest mountain to the bottom of the deepest ocean, Earth is made mostly of rock. Geologists classify rocks according to three different categories depending upon how the rocks were formed. These are igneous, sedimentary, and metamorphic.

Igneous rocks are formed from rock that melted and later cooled and solidified. When temperature and pressure conditions are just right, often deep within Earth, rocks will melt. This <u>molten</u> rock, called magma, moves up toward the surface of Earth over time or might even reach the surface quickly in a volcanic eruption. When magma erupts at Earth's surface, it is called lava. When lava cools quickly, small mineral crystals will form. If it cools even more quickly, volcanic glass can form.

Obsidian is a type of volcanic rock. It is often referred to as volcanic glass because it has a smooth surface. It was a prized material among prehistoric cultures because it fractures with sharp edges and can be used as a weapon or tool. Knives, arrowheads, and spear points were made from obsidian. Prehistoric people also used obsidian as mirrors. In modern times, obsidian has been used for surgical scalpel blades.

The beauty and mystery of igneous rocks have inspired many artists. Some artists care-

This is obsidian, a type of volcanic glass.

fully sculpt already cooled volcanic rock into beautiful, one-of-a-kind pieces of art.

1. In this passage, the word <u>molten</u> means
   A) deep
   B) melted
   C) igneous
   D) cooled

2. According to the passage, the three categories, or groups, of rocks are _____.
   F) igneous, metamorphic, and sedimentary
   G) volcanic, glassy, and irregular
   H) chemical, organic, and detrital
   J) residual, original, and primitive

3. Which conclusion is best supported by information given in the passage?
   A) When magma cools rapidly, it produces rocks with many large crystals.
   B) Igneous rocks can be made into tools and pieces of art.
   C) Obsidian was used by prehistoric cultures to build stone houses.
   D) Igneous rock forms when rocks weather and erode.

Test-Taking Tip As you read the passage, write a one-sentence summary for each paragraph.

Reasoning and Skills

Read each question and choose the best answer.

Igneous Rocks

Formed	Light-colored	Dark-colored
Below Earth's Surface	granite	gabbro
At Earth's Surface	rhyolite	basalt

1. According to the chart, lava that flows onto the surface from a volcano should cool to form the dark-colored rock _____.

A) granite
C) rhyolite
B) gabbro
D) basalt

Test-Taking Tip Reread the question and think about the color of the rock and where the rock was formed.

2. Earth's crust is estimated to be composed of about 46% oxygen, 28% silicon, 8% aluminum, and 18% other elements. Which area of the graph represents aluminum?

F) Q
G) R
H) S
J) T

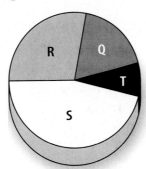

Test-Taking Tip Think about the quantities of the element the question refers to compared to the sizes of each portion of the graph.

Selected Rock Data

Volume of Sample (cm^3)	Mass of Sample (g)
1	3
2	6
3	9
4	12
5	?

3. These data were collected after determining the mass and volume of three different samples of the same rock. If everything remains the same, what will be the mass of a fifth sample whose volume is 5 cm^3?

A) 9 g
C) 15 g
B) 10 g
D) 20 g

Test-Taking Tip If you know the mass of a 1 cm^3 sample of the rock, you can use math to calculate the mass of a 5-cm^3 sample.

Consider this question carefully before writing your answer on a separate sheet of paper.

4. Many igneous rocks contain crystals. Geologists have observed that igneous rocks containing large crystals formed slowly, while those containing small crystals formed rapidly. Design an experiment, using sugar and water, to show that the rate at which crystals form affects their size.

Test-Taking Tip You can use a supersaturated solution and seed crystals to speed up the rate at which the sugar crystals form.

How Are Rivers & Writing Connected?

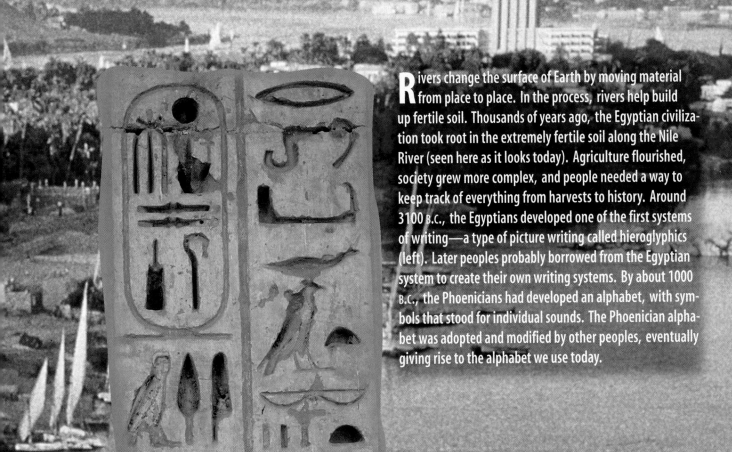

Rivers change the surface of Earth by moving material from place to place. In the process, rivers help build up fertile soil. Thousands of years ago, the Egyptian civilization took root in the extremely fertile soil along the Nile River (seen here as it looks today). Agriculture flourished, society grew more complex, and people needed a way to keep track of everything from harvests to history. Around 3100 B.C., the Egyptians developed one of the first systems of writing—a type of picture writing called hieroglyphics (left). Later peoples probably borrowed from the Egyptian system to create their own writing systems. By about 1000 B.C., the Phoenicians had developed an alphabet, with symbols that stood for individual sounds. The Phoenician alphabet was adopted and modified by other peoples, eventually giving rise to the alphabet we use today.

SCIENCE CONNECTION

SEDIMENTATION The Blue Nile, which joins the White Nile to form the Nile River, picks up sediment as it courses through the volcanic highlands of Ethiopia. Each year when the Nile flooded its banks, ancient Egyptian farmers trapped the floodwaters in their fields to capture the rich sediment. Working in small groups, research the details of this agricultural process. Make drawings showing how the process worked.

Views of Earth

Viewing Earth from satellites, often called remote sensing, is a powerful way to learn about Earth's landforms, weather, and vegetation. This colorful image shows the metropolitan area of New York City and surrounding regions. Vegetation shows up as green, uncovered land is red, water is blue, and human-made structures appear gray. In this chapter, you will learn about studying Earth from space. You'll learn about Earth's major landforms, and you'll learn how to locate places on Earth's surface.

What do you think?

Science Journal Look at the picture below with a classmate. Discuss what you think this might be. Here's a hint: *It can keep you from getting lost on land or at sea.* Write your answer or best guess in your Science Journal.

EXPLORE ACTIVITY

Pictures of Earth from space are acquired by instruments attached to satellites. Scientists use these images to make maps because they show features of Earth's surface, such as mountains and rivers. In the activity below, use a map or a globe to explore Earth's surface.

Describe landforms

Using a globe, atlas, or a world map, locate the following features and describe their positions on Earth relative to other major features. Provide any other details that would help someone else find them.

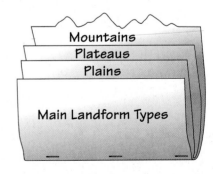

1. Andes mountains
2. Amazon, Ganges, and Mississippi Rivers
3. Indian Ocean, the Sea of Japan, and the Baltic Sea
4. Australia, South America, and North America

Observe

Choose one country on the globe or map and describe its major physical features in your Science Journal.

Before You Read

FOLDABLES
Reading & Study Skills

Making a Main Ideas Study Fold Make the following Foldable to help you identify the major topics about landforms.

1. Stack two sheets of paper in front of you so the short side of both sheets is at the top.
2. Slide the top sheet up so about 4 cm of the bottom sheet show.
3. Fold both sheets top to bottom to form four tabs and staple along the fold. Turn the Foldable so the staples are at the bottom. Cut mountain shapes on the top tab.
4. Label the tabs *Main Landform Types, Plains, Plateaus,* and *Mountains.* Before you read the chapter, write what you know about each landform under the tabs.
5. As you read the chapter, add to and correct what you have written.

Landforms

As You Read

What You'll Learn

- **Discuss** differences between plains and plateaus.
- **Describe** folded, upwarped, fault-block, and volcanic mountains.

Vocabulary

plain
plateau
folded mountain
upwarped mountain
fault-block mountain
volcanic mountain

Why It's Important

Landforms influence how people can use land.

Plains

Earth offers abundant variety—from tropics to tundras, deserts to rain forests, and freshwater mountain streams to salt-water tidal marshes. Some of Earth's most stunning features are its landforms, which can provide beautiful vistas, such as vast, flat, fertile plains; deep gorges that cut through steep walls of rock; and towering, snowcapped peaks. **Figure 1** shows the three basic types of landforms—plains, plateaus, and mountains.

Even if you haven't ever visited mountains, you might have seen hundreds of pictures of them in your lifetime. Plains are more common than mountains, but they are more difficult to visualize. **Plains** are large, flat areas, often found in the interior regions of continents. The flat land of plains is ideal for agriculture. Plains often have thick, fertile soils and abundant, grassy meadows suitable for grazing animals. Plains also are home to a variety of wildlife, including foxes, ground squirrels, and snakes. When plains are found near the ocean, they're called coastal plains. Together, interior plains and coastal plains make up half of all the land in the United States.

Figure 1
Three basic types of landforms are plains, plateaus, and mountains.

Mountains

Plateau

Plain

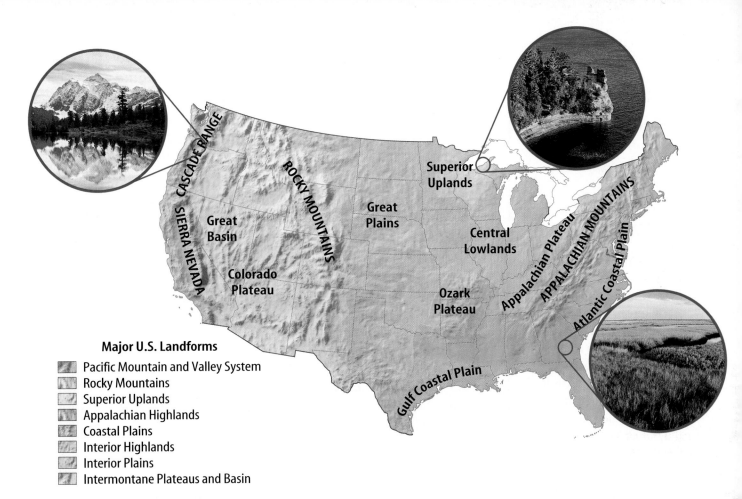

Major U.S. Landforms

- Pacific Mountain and Valley System
- Rocky Mountains
- Superior Uplands
- Appalachian Highlands
- Coastal Plains
- Interior Highlands
- Interior Plains
- Intermontane Plateaus and Basin

Coastal Plains A coastal plain often is called a lowland because it is lower in elevation, or distance above sea level, than the land around it. You can think of the coastal plains as being the exposed portion of a continental shelf. The continental shelf is the part of a continent that extends into the ocean. The Atlantic Coastal Plain is a good example of this type of landform. It stretches along the east coast of the United States from New Jersey to Florida. This area has low rolling hills, swamps, and marshes. A marsh is a grassy wetland that usually is flooded with water.

The Atlantic Coastal Plain, shown in **Figure 2,** began forming about 70 million years ago as sediment began accumulating on the ocean floor. Sea level eventually dropped, and the seafloor was exposed. As a result, the coastal plain was born. The size of the coastal plain varies over time. That's because sea level rises and falls. During the last ice age, the coastal plain was larger than it is now because so much of Earth's water was contained in glaciers.

The Gulf Coastal Plain includes the lowlands in the southern United States that surround the Gulf of Mexico. Much of this plain was formed from sediment deposited in deltas by the many rivers that enter the Gulf of Mexico.

✓ Reading Check *How are coastal plains formed?*

Figure 2
The United States has eight major landform regions, which include plains, mountains, and plateaus. After looking at this map, describe the region that you live in.

Interior Plains The central portion of the United States is comprised largely of interior plains. Shown in **Figure 3,** you'll find them between the Rocky Mountains, the Appalachian Mountains, and the Gulf Coastal Plain. They include the Central Lowlands around the Missouri and Mississippi Rivers and the rolling hills of the Great Lakes area.

A large part of the interior plains is known as the Great Plains. This area lies between the Mississippi River and the Rocky Mountains. It is a flat, grassy, dry area with few trees. The Great Plains also are referred to as the high plains because of their elevation, which ranges from 350 m above sea level at the eastern border to 1,500 m in the west. The Great Plains consist of nearly horizontal layers of sedimentary rocks.

Plateaus

At somewhat higher elevations, you will find plateaus (pla TOHZ). **Plateaus** are flat, raised areas of land made up of nearly horizontal rocks that have been uplifted by forces within Earth. They are different from plains in that their edges rise steeply from the land around them. Because of this uplifting, it is common for plateaus, such as the Colorado Plateau, to be cut through by deep river valleys and canyons. The Colorado River, as shown in **Figure 3,** has cut deeply into the rock layers of the plateau, forming the Grand Canyon. Because the Colorado Plateau is located mostly in what is now a dry region, only a few rivers have developed on its surface. If you hiked around on this plateau, you would encounter a high, rugged environment.

Figure 3
Plains and plateaus are fairly flat, but plateaus have higher elevation. **A** This short-grass prairie in Kansas is part of an interior plain. **B** The Colorado River has carved the Grand Canyon into the Colorado Plateau.

Mountains

Mountains with snowcapped peaks often are shrouded in clouds and tower high above the surrounding land. If you climb them, the views are spectacular. The world's highest mountain peak is Mount Everest in the Himalaya—more than 8,800 m above sea level. By contrast, the highest mountain peaks in the United States reach just over 6,000 m. Mountains also vary in how they are formed. The four main types of mountains are folded, upwarped, fault-block, and volcanic.

Reading Check *What is the highest mountain peak on Earth?*

Folded Mountains The Appalachian Mountains and the Rocky Mountains in Canada, shown in **Figure 4,** are comprised of folded rock layers. In **folded mountains,** the rock layers are folded like a rug that has been pushed up against a wall.

To form folded mountains, tremendous forces inside Earth squeeze horizontal rock layers, causing them to fold. The Appalachian Mountains formed between 480 million and 250 million years ago and are among the oldest and longest mountain ranges in North America. The Appalachians once were higher than the Rocky Mountains, but weathering and erosion have worn them down. They now are less than 2,000 m above sea level. The Ouachita (WAH shuh tah) Mountains of Arkansas are extensions of the same mountain range.

SCIENCE *Online*

Research Visit the Glencoe Science Web site at **science.glencoe.com** to learn how landforms can affect economic development.

Figure 4
Folded mountains form when rock layers are squeezed from opposite sides. These mountains in Banff National Park, Canada, consist of folded rock layers.

Figure 5
The southern Rocky Mountains are upwarped mountains that formed when crust was pushed up by forces inside Earth.

Upwarped Mountains The Adirondack Mountains in New York, the southern Rocky Mountains in Colorado and New Mexico, and the Black Hills in South Dakota are upwarped mountains. **Figure 5** shows a mountain range in Colorado. Notice the high peaks and sharp ridges that are common to this type of mountain. **Upwarped mountains** form when blocks of Earth's crust are pushed up by forces inside Earth. Over time, the soil and sedimentary rocks at the top of Earth's crust erode, exposing the hard, crystalline rock underneath. As these rocks erode, they form the peaks and ridges.

Fault-Block Mountains **Fault-block mountains** are made of huge, tilted blocks of rock that are separated from surrounding rock by faults. These faults are large fractures in rock along which mostly vertical movement has occurred. The Grand Tetons of Wyoming, shown in **Figure 6,** and the Sierra Nevada in California, are examples of fault-block mountains. As **Figure 6** shows, when these mountains formed, one block was pushed up, while the adjacent block dropped down. This mountain-building process produces majestic peaks and steep slopes.

Figure 6
Fault-block mountains such as the Grand Tetons are formed when faults occur. Some rock blocks move up, and others move down. *How are fault-block mountains different from upwarped mountains?*

Figure 7
Mount Shasta is a volcanic mountain made up of layers of lava flows and ash.

Volcanic Mountains **Volcanic mountains,** like the one shown in **Figure 7,** begin to form when molten material reaches the surface through a weak area of the crust. The deposited materials pile up, layer upon layer, until a cone-shaped structure forms. Two volcanic mountains in the United States are Mount St. Helens in Washington and Mount Shasta in California. The Hawaiian Islands are the peaks of huge volcanoes that sit on the ocean floor. Measured from the base, Mauna Loa in Hawaii would be higher than Mount Everest.

Plains, plateaus, and mountains offer different kinds of landforms to explore. They range from low, coastal plains and high, desert plateaus to mountain ranges thousands of meters high.

Section Assessment

1. Describe the eight major landform regions in the United States that are mentioned in this chapter.
2. How do plains and plateaus differ?
3. Why are some mountains folded and others upwarped?
4. How are volcanic mountains different from other mountains?
5. **Think Critically** If you wanted to know whether a particular mountain was formed by movement along a fault, what would you look for?

Skill Builder Activities

6. **Concept Mapping** Make an events-chain concept map to explain how upwarped mountains form. **For more help, refer to the** Science Skill Handbook.
7. **Using an Electronic Spreadsheet** Design a spreadsheet that compares the origin and features of the following: *folded, upwarped, fault-block,* and *volcanic mountains.* Then, explain an advantage of using a spreadsheet to compare different types of mountains. **For more help, refer to the** Technology Skill Handbook.

2 Viewpoints

What You'll Learn

- **Define** latitude and longitude.
- **Explain** how latitude and longitude are used to identify locations on Earth.
- **Determine** the time and date in different time zones.

Vocabulary

equator   prime meridian
latitude   longitude

Why It's Important

Latitude and longitude allow you to locate places on Earth.

Figure 8
Latitude and longitude are measurements that are used to indicate locations on Earth's surface.

Latitude and Longitude

During hurricane season, meteorologists track storms as they form in the Atlantic Ocean. To identify the exact location of a storm, latitude and longitude lines are used. These lines form an imaginary grid system that allows people to locate any place on Earth accurately.

Latitude Look at **Figure 8.** The **equator** is an imaginary line around Earth exactly halfway between the north and south poles. It separates Earth into two equal halves called the northern hemisphere and the southern hemisphere. Lines running parallel to the equator are called lines of **latitude,** or parallels. Latitude is the distance, measured in degrees, either north or south of the equator. Because they are parallel, lines of latitude do not intersect, or cross, one another.

The equator is at 0° latitude, and the poles are each at 90° latitude. Locations north and south of the equator are referred to by degrees north latitude and degrees south latitude, respectively. Each degree is further divided into segments called minutes and seconds. There are 60 minutes in one degree and 60 seconds in one minute.

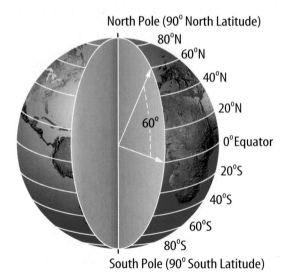

A Latitude is the measurement of the imaginary angle created by the equator, the center of Earth, and a location on Earth.

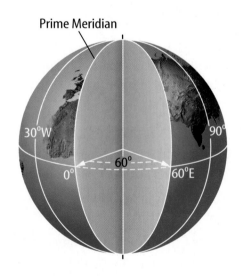

B Longitude is the measurement of the angle along the equator, between the prime meridian, the center of Earth, and a meridian on Earth.

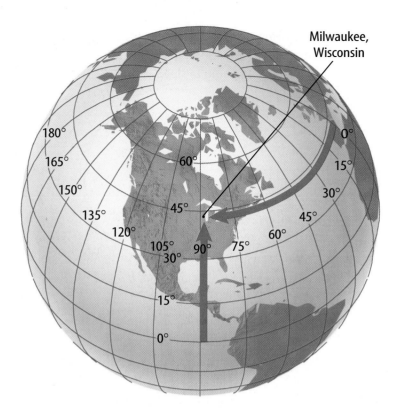

Milwaukee, Wisconsin

Figure 9
The city of Milwaukee, Wisconsin is located at about 43°N, 88°W. *How is latitude different from longitude?*

Longitude

The vertical lines, seen in **Figure 8B,** have two names—meridians and lines of longitude. Longitude lines are different from latitude lines in many important ways. Just as the equator is used as a reference point for lines of latitude, there's a reference point for lines of longitude—the **prime meridian.** This imaginary line represents 0° longitude. In 1884, astronomers decided the prime meridian should go through the Greenwich (GREN ihtch) Observatory near London, England. The prime meridian had to be agreed upon, because no natural point of reference exists.

Longitude refers to distances in degrees east or west of the prime meridian. Points west of the prime meridian have west longitude measured from 0° to 180°, and points east of the prime meridian have east longitude, measured similarly.

Prime Meridian

The prime meridian does not circle Earth as the equator does. Rather, it runs from the north pole through Greenwich, England, to the south pole. The line of longitude on the opposite side of Earth from the prime meridian is the 180° meridian. East lines of longitude meet west lines of longitude at the 180° meridian. You can locate places accurately using latitude and longitude as shown in **Figure 9.** Note that latitude position always comes first when a location is given.

 *What line of longitude is found opposite the prime meridian?*

Mini LAB

Interpreting Latitude and Longitude

Procedure
1. Find the equator and prime meridian on a **world map.**
2. Move your finger to latitudes north of the equator, then south of the equator. Move your finger to longitudes west of the prime meridian, then east of the prime meridian.

Analysis
1. Identify the cities that have the following coordinates:
   a. 56°N, 38°E
   b. 34°S, 18°E
   c. 23°N, 82°W
2. Determine the latitude and longitude of the following cities:
   a. London, England
   b. Melbourne, Australia
   c. Buenos Aires, Argentina

Figure 10
The United States has six time zones.

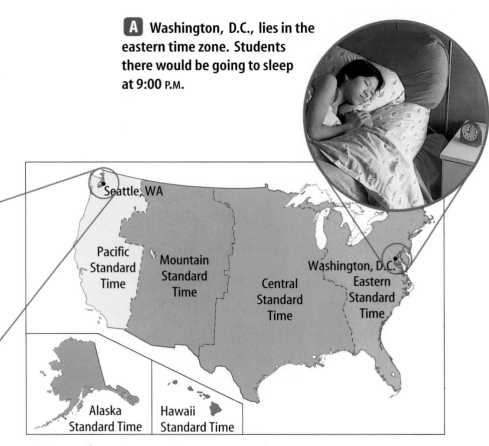

A Washington, D.C., lies in the eastern time zone. Students there would be going to sleep at 9:00 P.M.

B But students in Seattle, Washington, which lies in the Pacific time zone, are eating dinner. *What time would it be in Seattle when the students in Washington, D.C., are sleeping at 9:00 P.M.?*

Seattle, WA

Pacific Standard Time

Mountain Standard Time

Central Standard Time

Washington, D.C.
Eastern Standard Time

Alaska Standard Time

Hawaii Standard Time

Life Science
INTEGRATION

If you travel east or west across three or more time zones, you could suffer from jet lag. Jet lag occurs when your internal time clock does not match the new time zone. Jet lag can disrupt the daily rhythms of sleeping and eating. Have you or any of your classmates ever suffered from jet lag?

Time Zones

What time it is depends on where you are on Earth. Time is measured by tracking Earth's movement in relation to the Sun. Each day has 24 h, so Earth is divided into 24 time zones. Each time zone is about 15° of longitude wide and is 1 h different from the zones on each side of it. The United States has six different time zones. As you can see in **Figure 10,** people in different parts of the country don't experience dusk simultaneously. Because Earth rotates, the eastern states end a day while the western states are still in sunlight.

✓ Reading Check *What is the basis for dividing Earth into 24 time zones?*

Time zones do not follow lines of longitude strictly. Time zone boundaries are adjusted in local areas. For example, if a city were split by a time zone boundary, the results would be confusing. In such a situation, the time zone boundary is moved outside of the city.

Calendar Dates

In each time zone, one day ends and the next day begins at midnight. If it is 11:59 P.M. Tuesday, then 2 min later it will be 12:01 A.M. Wednesday in that particular time zone.

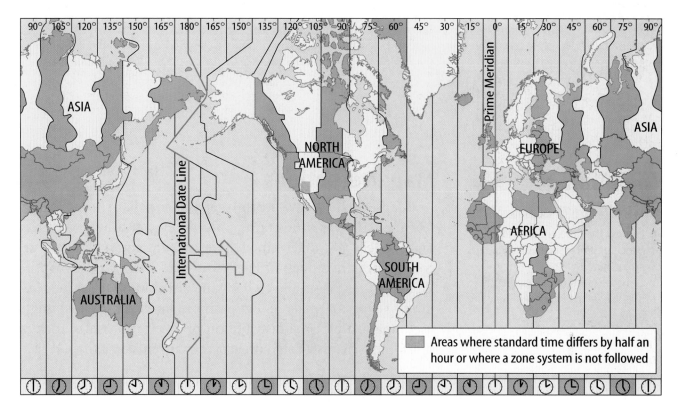

Areas where standard time differs by half an hour or where a zone system is not followed

International Date Line You gain or lose time when you enter a new time zone. If you travel far enough, you can gain or lose a whole day. The International Date Line, shown on **Figure 11**, is the transition line for calendar days. If you were traveling west across the International Date Line, located near the 180° meridian, you would move your calendar forward one day. If you were traveling east when you crossed it, you would move your calendar back one day.

Figure 11
Lines of longitude roughly determine the locations of time zone boundaries. These boundaries are adjusted locally to avoid splitting cities and other political subdivisions, such as counties, into different time zones.

Section 2 Assessment

1. What are latitude and longitude?

2. How do lines of latitude and longitude help people find locations on Earth?

3. What are the latitude and longitude of New Orleans, Louisiana?

4. If it were 7:00 P.M. in New York City, what time would it be in Los Angeles?

5. **Think Critically** How could you leave home on Monday to go sailing on the ocean, sail for 1 h on Sunday, and return home on Monday?

Skill Builder Activities

6. **Interpreting Scientific Illustrations** Use a world map to find the latitude and longitude of the following locations: Sri Lanka; Tokyo, Japan; and the Falkland Islands. **For more help, refer to the** Science Skill Handbook.

7. **Using Fractions** If you started at the prime meridian and traveled east one fourth of the way around Earth, what line of longitude would you reach? **For more help, refer to the** Math Skill Handbook.

3 Maps

As You Read

What You'll Learn

- **Explain** the differences among Mercator, Robinson, and conic projections.
- **Describe** features of topographic maps, geologic maps, and satellite maps.

Vocabulary

conic projection map scale
topographic map map legend
contour line

Why It's Important

Maps help people navigate and understand Earth.

Figure 12

Lines of longitude are drawn parallel to one another in Mercator projections. *What happens near the poles in Mercator projections?*

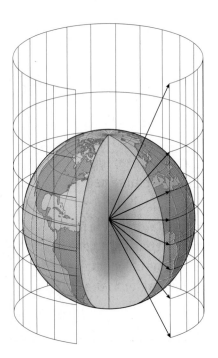

Map Projections

Maps—road maps, world maps, maps that show physical features such as mountains and valleys, and even treasure maps—help you determine where you are and where you are going. They are models of Earth's surface. Scientists use maps to locate various places and to show the distribution of various features or types of material. For example, an Earth scientist might use a map to plot the distribution of a certain type of rock or soil. Other scientists could draw ocean currents on a map.

✔ Reading Check *What are possible uses a scientist would have for maps?*

Many maps are made as projections. A map projection is made when points and lines on a globe's surface are transferred onto paper, as shown in **Figure 12.** Map projections can be made in several different ways, but all types of projections distort the shapes of landmasses or their areas. Antarctica, for instance, might look smaller or larger than it is as a result of the projection that is used for a particular map.

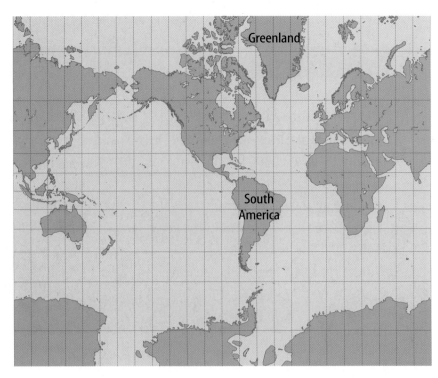

Figure 13
Robinson projections show little distortion in continent shapes and sizes.

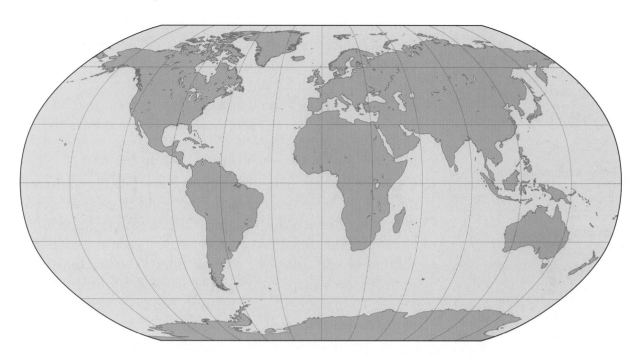

Mercator Projection Mercator (mer KAY ter) projections are used mainly on ships. They project correct shapes of continents, but the areas are distorted. Lines of longitude are projected onto the map parallel to each other. As you learned earlier, only latitude lines are parallel. Longitude lines meet at the poles. When longitude lines are projected as parallel, areas near the poles appear bigger than they are. Greenland, in the Mercator projection in **Figure 12,** appears to be larger than South America, but Greenland is actually smaller.

Robinson Projection A Robinson projection shows accurate continent shapes and more accurate land areas. As shown in **Figure 13,** lines of latitude remain parallel, and lines of longitude are curved as they are on a globe. This results in less distortion near the poles.

Conic Projection When you look at a road map or a weather map, you are using a conic (KAH nihk) projection. Conic projections, like the one shown in **Figure 14,** often are used to produce maps of small areas. These maps are well suited for middle latitude regions but are not as useful for mapping polar or equatorial regions. **Conic projections** are made by projecting points and lines from a globe onto a cone.

Reading Check *How are conic projections made?*

Figure 14
Small areas are mapped accurately using conic projections.

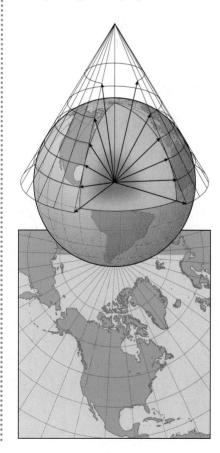

Topographic Maps

For nature hiking, a conic map projection can be helpful by directing you to the location where you will start your hike. On your hike, however, you would need a detailed map identifying the hills and valleys of that specific area. A **topographic map,** shown in **Figure 15,** models the changes in elevation of Earth's surface. With such a map, you can determine your location relative to identifiable natural features. Topographic maps also indicate cultural features such as roads, cities, dams, and other structures built by people.

Contour Lines Before your hike, you study the contour lines on your topographic map to see the trail's changes in elevation. A **contour line** is a line on a map that connects points of equal elevation. The difference in elevation between two side-by-side contour lines is called the contour interval, which remains constant for each map. For example, if the contour interval on a map is 10 m and you walk between two lines anywhere on that map, you will have walked up or down 10 m.

In mountainous areas, the contour lines are close together. This situation models a steep slope. However, if the change in elevation is slight, the contour lines will be far apart. Often large contour intervals are used for mountainous terrain, and small contour intervals are used for fairly flat areas. Why? **Table 1** gives additional tips for examining contour lines.

Index Contours Some contour lines, called index contours, are marked with their elevation. If the contour interval is 5 m, you can determine the elevation of other lines around the index contour by adding or subtracting 5 m from the elevation shown on the index contour.

Physics
INTEGRATION

Topographic maps of Venus and Mars have been made by space probes. The probes send a radar beam or laser pulses to the surface and measure how long it takes for the beam or pulses to return to the probe.

Table 1 Contour Rules

1. **Contour lines close around hills and basins.** To decide whether you're looking at a hill or basin, you can read the elevation numbers or look for hachures (ha SHOORZ). These are short lines drawn at right angles to the contour line. They show depressions by pointing toward lower elevations.

2. **Contour lines never cross.** If they did, it would mean that the spot where they cross would have two different elevations.

3. **Contour lines form Vs that point upstream when they cross streams.** This is because streams flow in depressions that are beneath the elevation of the surrounding land surface. When the contour lines cross the depression, they appear as Vs pointing upstream on the map.

Figure 15

Planning a hike? A topographic map will show you changes in elevation. With such a map, you can see at a glance how steep a mountain trail is, as well as its location relative to rivers, lakes, roads, and cities nearby. The steps in creating a topographic map are shown here.

A To create a topographic map of Old Rag Mountain in Shenandoah National Park, Virginia, mapmakers first measure the elevation of the mountain at various points.

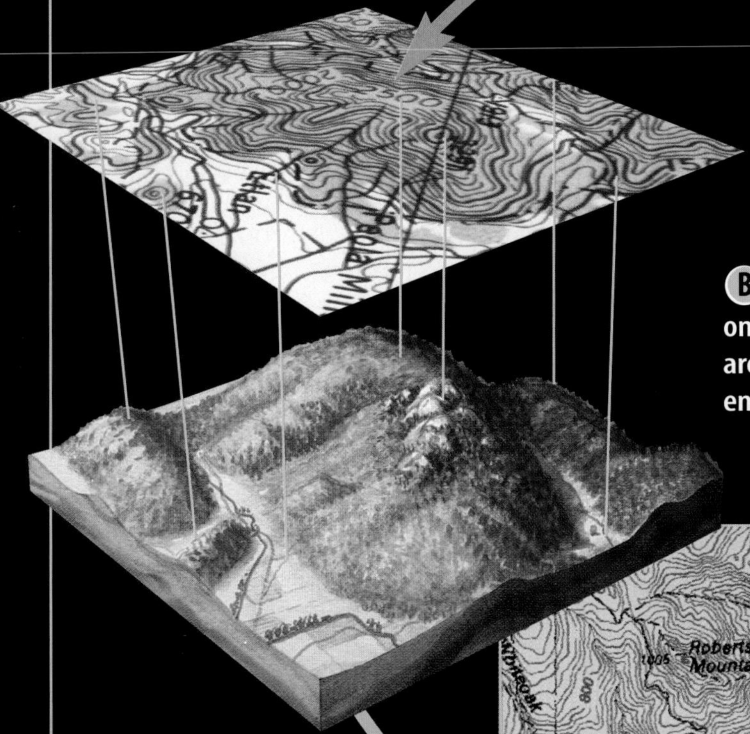

B These points are then projected onto paper. Points at the same elevation are connected, forming contour lines that encircle the mountain.

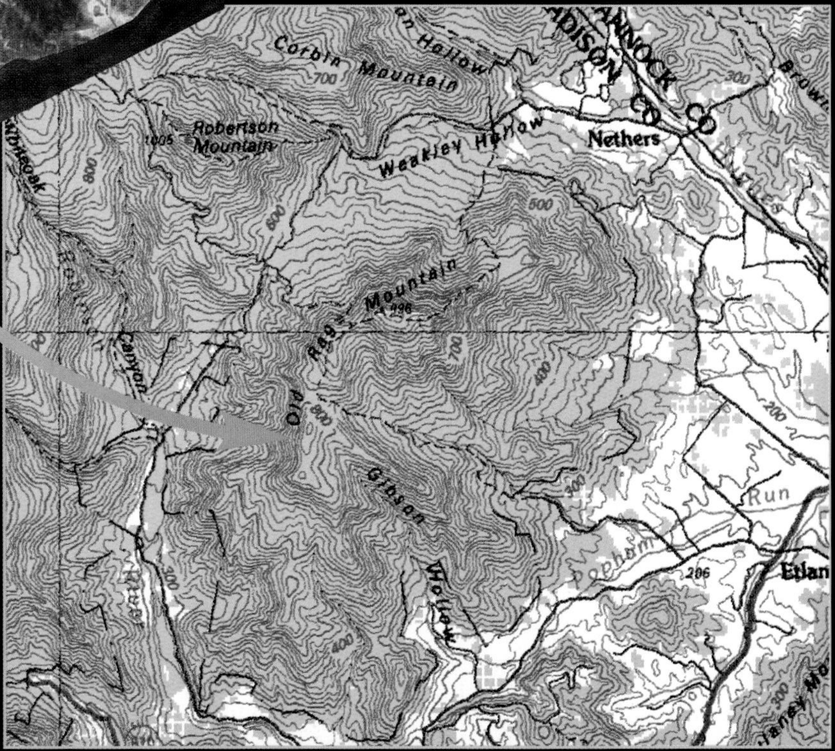

C Where contour lines on a topographic map are close together, elevation is changing rapidly—and the trail is very steep!

Map Scale When planning your hike, you'll want to determine the distance to your destination before you leave. Because maps are small models of Earth's surface, distances and sizes of things shown on a map are proportional to the real thing on Earth. Therefore, real distances can be found by using a scale.

The **map scale** is the relationship between the distances on the map and distances on Earth's surface. Scale often is represented as a ratio. For example, a topographic map of the Grand Canyon might have a scale that reads 1:80,000. This means that one unit on the map represents 80,000 units on land. If the unit you wanted to use was a centimeter, then 1 cm on the map would equal 80,000 cm on land. The unit of distance could be feet or millimeters or any other measure of distance. However, the units of measure on each side of the ratio must always be the same. A map scale also can be shown in the form of a small bar that is divided into sections and scaled down to match real distances on Earth.

Map Legend Topographic maps and most other maps have a legend. A **map legend** explains what the symbols used on the map mean. Some frequently used symbols for topographic maps are shown in the appendix at the back of the book.

Map Series Topographic maps are made to cover different amounts of Earth's surface. A map series includes maps that have the same dimensions of latitude and longitude. For example, one map series includes maps that are 7.5 minutes of latitude by 7.5 minutes of longitude. Other map series include maps covering larger areas of Earth's surface.

Geologic Maps

One of the more important tools to Earth scientists is the geologic map. Geologic maps show the arrangement and types of rocks at Earth's surface. Using geologic maps and data collected from rock exposures, a geologist can infer how rock layers might look below Earth's surface. The block diagram in **Figure 16** is a 3-D model that illustrates a solid section of Earth. The top surface of the block is the geologic map. Side views of the block are called cross sections, which are derived from the surface map. Developing geologic maps and cross sections is extremely important for the exploration and extraction of natural resources. What can a scientist do to determine whether a cross section accurately represents the underground features?

Figure 16
Geologists use block diagrams to understand Earth's subsurface. The different colors represent different rock layers.

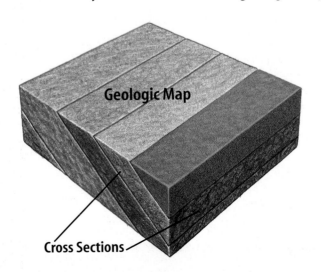

Geologic Map

Cross Sections

Three-Dimensional Maps Topographic maps and geologic maps are two-dimensional models that are used to study features of Earth's surface. To visualize Earth three dimensionally, scientists often rely on computers. Using computers, information is digitized to create a three-dimensional view of features such as rock layers or river systems. Digitizing is a process by which points are plotted on a coordinate grid.

Map Uses As you have learned, Earth can be viewed in many different ways. Maps are chosen depending upon the situation. If you wanted to determine New Zealand's location relative to Canada and you didn't have a globe, you probably would examine a Mercator projection. In your search, you would use lines of latitude and longitude, and a map scale. If you wanted to travel across the country, you would rely on a road map, or conic projection. You also would use a map legend to help locate features along the way. To climb the highest peak in your region, you would take along a topographic map.

Problem-Solving Activity

How can you create a cross section from a geologic map?

Earth scientists are interested in knowing the types of rocks and their configurations underground. To help them visualize this, they use geologic maps. Geologic maps offer a two-dimensional view of the three-dimensional situation found under Earth's surface. You don't have to be a professional geologist to understand a geologic map. Use your ability to create graphs to interpret this geologic map.

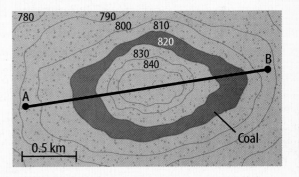

Identifying the Problem

Above is a simple geologic map showing where a coal seam is found on Earth's surface. Place a straight edge of paper along the line marked A–B and mark the points where it meets a contour. Make a different color mark where it meets the exposure of coal. Make a graph on which the various elevations (in meters) are marked on the y-axis. Lay your marked edge of paper along the x-axis and transfer the points directly above onto the proper elevation line. Now connect the dots to draw in the land's surface and connect the marks you made for the coal seam separately.

Solving the Problem

1. What type of topography does the map represent?
2. At what elevation is the coal seam?
3. Does this seam tilt, or is it horizontal? Explain how you know.

Figure 17
Sensors on *Landsat 7* detect light reflected off landforms on Earth.

Remote Sensing

Scientists use remote-sensing techniques to collect much of the data used for making maps. Remote sensing is a way of collecting information about Earth from a distance, often using satellites.

Landsat One way that Earth's surface has been studied is with data collected from Landsat satellites, as shown in **Figure 17.** These satellites take pictures of Earth's surface using different wavelengths of light. The images can be used to make maps of snow cover over the United States or to evaluate the impact of forest fires, such as those that occurred in the western United States during the summer of 2000. The newest Landsat satellite is *Landsat 7,* which was launched in April of 1999. It can acquire the most detailed Landsat images yet.

Global Positioning System The Global Positioning System, or GPS, is a satellite-based, radio-navigation system that allows users to determine their exact position anywhere on Earth. Twenty-four satellites orbit 20,200 km above the planet. Each satellite sends a position signal and a time signal. The satellites are arranged in their orbits so that signals from at least six can be picked up at any given moment by someone using a GPS receiver. By processing the signals, the receiver calculates the user's exact location. GPS technology is used to navigate, to create detailed maps, and to track wildlife.

Section Assessment

1. How do Mercator, Robinson, and conic projections differ?
2. Why does Greenland appear to be larger on a Mercator projection than it does on a Robinson projection?
3. Why can't contour lines ever cross?
4. What is a geologic map?
5. **Think Critically** Would a map that covers a large area have the same map scale as a map that covers a small region? How would the scales differ?

Skill Builder Activities

6. **Making Models** Architects make detailed maps called scale drawings to help them plan their work. Make a scale drawing of your classroom. **For more help, refer to the** Science Skill Handbook.
7. **Communicating** Draw a map in your Science Journal that your friends could use to get from school to your home. Include a map legend and a map scale. **For more help, refer to the** Science Skill Handbook.

Activity

Making a Topographic Map

Have you ever wondered how topographic maps are made? Today, radar and remote-sensing devices aboard satellites collect data, and computers and graphic systems make the maps. In the past, surveyors and aerial photographers collected data. Then, maps were hand drawn by cartographers, or mapmakers. In this activity, you can practice cartography.

Materials

plastic model of a landform
water tinted with food coloring
transparency
clear, plastic storage box with lid
beaker
metric ruler
tape
transparency marker

What You'll Investigate

How is a topographic map made?

Goals

- **Draw** a topographic map.
- **Compare and contrast** contour intervals.

Procedure

1. Using the ruler and the transparency marker, make marks up the side of the storage box that are 2 cm apart.

2. Secure the transparency to the outside of the box lid with tape.

3. Place the plastic model in the box. The bottom of the box will be zero elevation.

4. Using the beaker, pour water into the box to a height of 2 cm. Place the lid on the box.

5. Use the transparency marker to trace the top of the water line on the transparency.

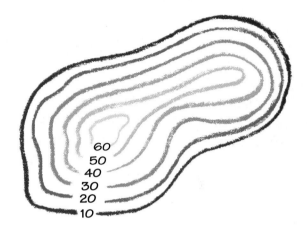

6. Using the scale 2 cm = 10 m, mark the elevation on the line.

7. Remove the lid and add water until a depth of 4 cm is reached.

8. Map this level on the storage box lid and record the elevation.

9. Repeat the process of adding 2 cm of water and tracing until the landform is mapped.

10. Transfer the tracing of the landform onto a sheet of white paper.

Conclude and Apply

1. What is the contour interval of this topographic map?

2. How does the distance between contour lines on the map show the steepness of the slope on the landform model?

3. **Determine** the total elevation of the landform you have selected.

4. How was elevation represented on your map?

5. How are elevations shown on topographic maps?

6. Must all topographic maps have a contour line that represents 0 m of elevation? Explain.

Activity
Model and Invent

Constructing Landforms

Most maps perform well in helping you get from place to place. A road map, for example, will allow you to choose the shortest route from one place to another. If you are hiking, though, distance might not be so important. You might want to choose a route that avoids steep terrain. In this case you need a map that shows the highs and lows of Earth's surface, called relief. Topographic maps use contour lines to show the landscape in three dimensions. Among their many uses, such maps allow hikers to choose routes that maximize the scenery and minimize the physical exertion.

Recognize the Problem

What does a landscape depicted on a two-dimensional topographic map look like in three dimensions?

Thinking Critically

How can you model a landscape?

Goals

- **Research** how contour lines show relief on a topographic map.
- **Determine** what scale you can best use to model a landscape of your choice.

- Working cooperatively with your classmates, model a landscape in three dimensions from the information given on a topographic map.

Possible Materials
U.S. Geological Survey 7.5 minute
    quadrangle maps
sandbox sand
rolls of brown paper towels
spray bottle filled with water
ruler

Data Source
SCIENCE *Online* Go to the Glencoe Science Web site at **science.glencoe.com** for more information about topographic maps.

Planning the Model

1. **Choose** a topographic map showing a landscape easily modeled using sand. Check to see what contour interval is used on the map. Use the index contours to find the difference between the lowest and the highest elevations shown on the landscape. Check the distance scale to determine how much area the landscape covers.

2. **Determine** the scale you will use to convert the elevations shown on your map to heights on your model. Make sure the scale is proportional to the distances on your map.

3. **Plan** a model of the landscape in sand by sketching the main features and their scaled heights onto paper. Note the degree of steepness found on all sides of the features.

Check the Model Plans

1. **Prepare** a document that shows the scale you plan to use for your model and the calculations you used to derive that scale. Remember to use the same scale for distance as you use for height. If your landscape is fairly flat, you can exaggerate the vertical scale by a factor of two or three. Be sure your paper is neat, is easy to follow, and includes all units. Present the document to your teacher for approval.

2. **Research** how the U.S. Geological Survey creates topographic maps and find out how it decides upon a contour interval for each map. This information can be obtained from the Glencoe Science Web site.

Making the Model

1. Using the sand, spray bottle, and ruler, create a scale model of your landscape on the brown paper towels.

2. **Check** your topographic map to be sure your model includes the landscape features at their proper heights and proper degrees of steepness.

Analyzing and Applying Results

1. Did your model accurately represent the landscape depicted on your topographic map? Discuss the strengths and weaknesses of your model.

2. Why was it important to use the same scale for height and distance? If you exaggerated the height, why was it important to indicate the exaggeration on your model?

3. Why did the mapmakers choose the contour interval used on your topographic map?

4. **Predict** the contour intervals mapmakers might choose for topographic maps of the world's tallest mountains—the Himalaya—and for topographic maps of Kansas, which is fairly flat.

LOCATION,

New York Harbor in 1849

Rich Midwest farmland

Georgia peaches

Why is New York City at the mouth of the Hudson River and not 300 km inland? Why are there more farms in Iowa than in Alaska? What's the reason for growing lots of peaches in Georgia but not in California's Death Valley? It's all about location. The landforms, climate, soil, and resources in an area determine where cities and farms grow and what people connected with them do.

Landforms Are Key

When many American cities were founded hundreds of years ago, waterways were the best means of transportation. Old cities such as New York City and Boston are located on deep harbors where ships could land with people and goods. Rivers also were major highways centuries ago. They still are. A city such as New Orleans, located at the mouth of the Mississippi River, receives goods from the entire river valley.

It then ships the goods from its port to places far away.

Topography and soil also play a role in where activities such as farming take root. States such as Iowa and Illinois have many farms because they have lots of flat land and fertile soil. Growing crops is more difficult in mountainous areas or where soil is stony and poor.

Climate and Soil

Climate limits the locations of cities and farms, as well. The fertile soil and warm, moist climate of Georgia make it a perfect place to grow peaches. California's Death Valley can't support such crops because it's a hot desert. Deserts are too dry to grow much of anything without irrigation. Deserts also don't have large population centers unless water is brought in from far away. Los Angeles and Las Vegas are both desert cities that are huge only because they pipe in water from hundreds of miles away.

Resources Rule

The location of an important natural resource can change the rules. A gold deposit or an oil field can cause a town to grow in a place where the topography, soil, and climate are not favorable. For example, thousands of people now live in parts of Alaska only because of the great supply of oil there. People settled in rugged areas of the Rocky Mountains to mine gold and silver. Maine has a harsh climate and poor soil. But people settled along its coast because they could catch lobsters and fish in the nearby North Atlantic.

LOCATION

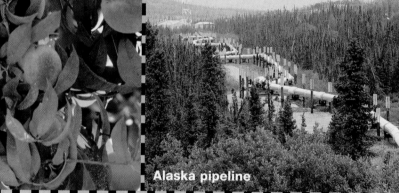

Alaska pipeline

Maine fishing and lobster industry

The rules that govern where towns grow and where people live are a bit different now than they used to be. Often information, not goods, moves from place to place on computers that can be anywhere. But as long as people farm, use minerals, and transport goods from place to place, the natural environment and natural resources will always help determine where people are and what they do.

Cities, farms, and industries grow in logical places

CONNECTIONS Research Why was your community built where it is? Research its history. What types of economic activity were important when it was founded? Did topography, climate, or resources determine its location? Are they important today? Report to the class.

SCIENCE Online

For more information, visit science.glencoe.com

Chapter 6 Study Guide

Reviewing Main Ideas

Section 1 Landforms

1. The three main types of landforms are plains, plateaus, and mountains.

2. Plains are large, flat areas. Plateaus are relatively flat, raised areas of land made up of nearly horizontal rocks that have been uplifted. Mountains rise high above the surrounding land. *Which type of landform is shown in the photograph below?*

Section 2 Viewpoints

1. Latitude and longitude form an imaginary grid system that enables points on Earth to be located exactly.

2. Latitude is the distance in degrees north or south of the equator. Longitude is the distance in degrees east or west of the prime meridian.

3. Reference lines have been established for measuring latitude and longitude. Latitude is measured from Earth's equator, an imaginary line halfway between Earth's poles. Longitude is measured from the prime meridian. The prime meridian runs from pole to pole through Greenwich, England.

4. Earth is divided into 24 time zones. Each time zone represents a 1-h difference. The International Date Line separates different calendar days. *How many time zones are in the United States?*

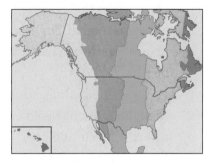

Section 3 Maps

1. Mercator, Robinson, and conic projections are made by transferring points and lines on a globe's surface onto paper.

2. Topographic maps show the elevation of Earth's surface. Geologic maps show the types of rocks that make up Earth's surface. *What type of map is shown here?*

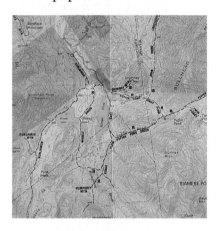

3. Remote sensing is a way of collecting information about Earth from a distance. Satellites are important remote-sensing devices.

FOLDABLES
Reading & Study Skills

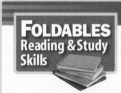

After You Read

To help you review the three main landform types, use the Foldable you made at the beginning of this chapter.

Visualizing Main Ideas

Complete the following concept map on landforms.

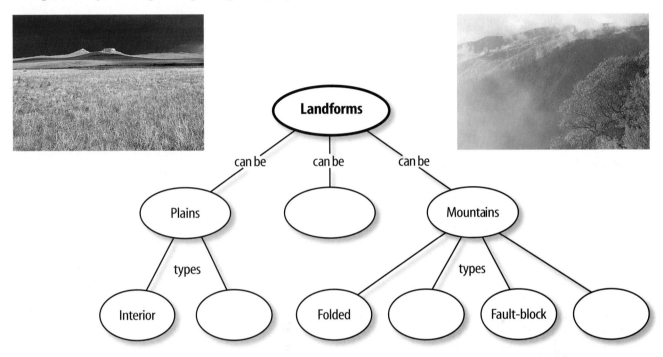

Vocabulary Review

Vocabulary Words

a. conic projection
b. contour line
c. equator
d. fault-block mountain
e. folded mountain
f. latitude
g. longitude
h. map legend
i. map scale
j. plain
k. plateau
l. prime meridian
m. topographic map
n. upwarped mountain
o. volcanic mountain

THE PRINCETON REVIEW **Study Tip**

Make a plan! Before you start your homework, write a checklist of what you need to do for each subject. As you finish each item, check it off.

Using Vocabulary

For each set of terms below, choose the one term that does not belong and explain why it does not belong.

1. upwarped mountain, equator, volcanic mountain

2. plain, plateau, prime meridian

3. topographic map, contour line, volcanic mountain

4. prime meridian, equator, folded mountain

5. fault-block mountain, upwarped mountain, plateau

6. prime meridian, map scale, contour line

Checking Concepts

Choose the word or phrase that best answers the question.

1. What makes up about 50 percent of all land areas in the United States?
   - **A)** plateaus
   - **B)** plains
   - **C)** mountains
   - **D)** volcanoes

2. Where is the north pole located?
   - **A)** 0°N
   - **B)** 180°N
   - **C)** 50°N
   - **D)** 90°N

3. What kind of mountains are the Hawaiian Islands?
   - **A)** fault-block
   - **B)** volcanic
   - **C)** upwarped
   - **D)** folded

4. What are lines that are parallel to the equator called?
   - **A)** lines of latitude
   - **B)** prime meridians
   - **C)** lines of longitude
   - **D)** contour lines

5. How many degrees apart are the 24 time zones?
   - **A)** 10
   - **B)** 34
   - **C)** 15
   - **D)** 25

6. Which type of map is most distorted at the poles?
   - **A)** conic
   - **B)** topographic
   - **C)** Robinson
   - **D)** Mercator

7. Which type of map shows changes in elevation at Earth's surface?
   - **A)** conic
   - **B)** topographic
   - **C)** Robinson
   - **D)** Mercator

8. What is measured with respect to sea level?
   - **A)** contour interval
   - **B)** elevation
   - **C)** conic projection
   - **D)** sonar

9. What kind of map shows rock types making up Earth's surface?
   - **A)** topographic
   - **B)** Robinson
   - **C)** geologic
   - **D)** Mercator

10. Which major U.S. landform includes the Grand Canyon?
    - **A)** Great Plains
    - **B)** Colorado Plateau
    - **C)** Gulf Coastal Plain
    - **D)** Appalachian Mountains

Thinking Critically

11. How would a topographic map of the Atlantic Coastal Plain differ from a topographic map of the Rocky Mountains?

12. If you left Korea early Wednesday morning and flew to Hawaii, on what day of the week would you arrive?

13. If you were flying directly south from the north pole and reached 70° north latitude, how many more degrees of latitude would you pass over before reaching the south pole?

14. Using the map below, arrange these cities in order from the city with the earliest time to the one with the latest time on a given day: Anchorage, Alaska; San Francisco, California; Bangor, Maine; Denver, Colorado; Houston, Texas.

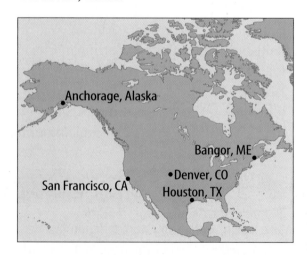

15. How is a map with a scale of 1:50,000 different from a map with a scale of 1:24,000?

Developing Skills

16. Comparing and Contrasting Compare and contrast Mercator, Robinson, and conic map projections.

17. Forming Hypotheses You are visiting a mountain in the northwest part of the United States. The mountain has steep sides and is not part of a mountain range. A crater can be seen at the top of the mountain. Hypothesize about what type of mountain you are visiting.

18. Concept Mapping Complete the following concept map about parts of a topographic map.

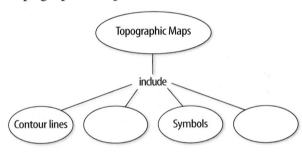

Performance Assessment

19. Poem Create a poem about the different types of landforms. Include characteristics of each landform in your poem. Display your poem with those of your classmates.

20. Poster Create a poster showing how satellites can be used for remote sensing.

TECHNOLOGY

Go to the Glencoe Science Web site at **science.glencoe.com** or use the **Glencoe Science CD-ROM** for additional chapter assessment.

THE PRINCETON REVIEW   Test Practice

Alicia was looking at a map of the United States because her science teacher suggested that she learn about the landform regions in the United States.

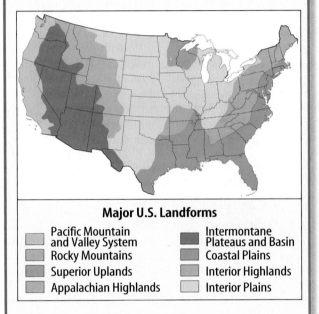

Major U.S. Landforms

Pacific Mountain and Valley System	Intermontane Plateaus and Basin
Rocky Mountains	Coastal Plains
Superior Uplands	Interior Highlands
Appalachian Highlands	Interior Plains

Study the diagram and answer the following questions.

1. Which technological development would have had the greatest impact on the accuracy of Alicia's map?
   A) radio communications
   B) measurement with lasers
   C) computer-assisted design
   D) satellite imaging

2. Which of the following landform regions would contain high, rugged mountains?
   F) Coastal Plains
   G) Interior Plains
   H) Appalachian Highlands
   J) Rocky Mountains

Weathering and Soil

Can you imagine how these balanced rocks were formed? For millions of years, nature has been working on them, wearing away softer materials and leaving behind more resistant rock. In this chapter, you'll read about how rocks are weathered into small fragments such as sand and clay. You also will learn how soil forms, how it erodes, and how to prevent soil erosion.

What do you think?

Science Journal Examine the picture below with a classmate. Discuss what you think this might be or what is happening. Here's a hint: *It's a strange place to grow.* Write your answer or best guess in your Science Journal.

EXPLORE ACTIVITY

Weathering breaks apart rock by exposing it to natural elements such as water and ice. Do you think you can simulate weathering on Earth's surface?

Model weathering

1. Form groups of four or five students as instructed by your teacher.

2. Fill a coffee can one third of the way full with small cobbles and pebbles obtained from your school yard.

3. Observe the cobbles and pebbles that you collected. Sketch their shapes in your Science Journal.

4. Put the lid on the coffee can and take turns shaking the can vigorously from side to side for several minutes.

5. Remove the lid and examine the rocks.

Observe

Describe in your Science Journal what happened to the rocks. How is this similar to weathering?

Before You Read

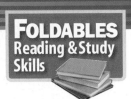

FOLDABLES
Reading & Study Skills

Making a Vocabulary Study Fold To help you study this chapter, make the following vocabulary Foldable. Knowing the definition of vocabulary words in a chapter is a good way to ensure you have understood the content.

1. Place a sheet of notebook paper in front of you so the short side is at the top and the holes are on the right side. Fold the paper in half from the left side to the right side.

2. Through the top thickness of paper, cut along every third line from the outside edge to the center fold, forming tabs as shown.

3. On the front of each tab, write a vocabulary word listed on the first page of each section in this chapter. On the back of each tab, define the word.

4. As you read the chapter, write a sentence using the vocabulary words.

Weathering

As You Read

What **You'll Learn**

- **Describe** the difference between mechanical weathering and chemical weathering.
- **Explain** the effects of climate on weathering.

Vocabulary

weathering
mechanical weathering
ice wedging
chemical weathering
oxidation
climate

Why **It's Important**

Weathering causes rocks to crumble and shapes many landforms.

Weathering and Its Effects

Can you believe that tiny moss plants, a burrowing vole shrew, and even oxygen in the air can affect solid rock? These things and many more weaken and break apart rock at Earth's surface. Together, surface processes that work to break down rock are called **weathering.**

Weathering breaks rock into smaller and smaller pieces, such as sand, silt, and clay. These smaller, loose pieces are called sediment. The terms *sand, silt,* and *clay* are used to describe specific sizes of sediment. Sediment then changes gradually into soil. The formation of soil depends upon the amount of weathering that occurs in a specific place.

Over millions of years, weathering has changed Earth's surface. The process continues today. Weathering wears mountains down to hills, as shown in **Figure 1.** Rocks at the top of mountains are broken down by weathering and then carried downhill by gravity, water, and ice. Weathering also helps produce strange rock formations like those shown at the beginning of this chapter. Two different types of weathering—mechanical weathering and chemical weathering—work together to shape Earth's surface.

Figure 1
Over long periods of time, weathering wears mountains down to rolling hills.

Figure 2
Growing tree roots can be agents of mechanical weathering.

A Tree roots can grow beneath a sidewalk, cracking the concrete and pushing it up.

B Tree roots also can grow into cracks in rock, breaking it apart.

Mechanical Weathering

Mechanical weathering occurs when rocks are broken apart by physical processes. This means that the overall chemical makeup of the rock stays the same. Each fragment has characteristics similar to the original rock. Growing plants, burrowing animals, and expanding ice are some of the things that can mechanically weather rock. These physical processes produce enough force to break rocks into smaller pieces.

 **Reading Check** *What can cause mechanical weathering?*

Figure 3
Small animals mechanically weather rock when they burrow by breaking apart sediment and moving it to the surface.

Plants and Animals Water and nutrients that collect in the cracks of rocks result in conditions in which plants can grow. As the roots grow, they enlarge the cracks. You've seen this kind of mechanical weathering if you've ever tripped on a crack in a sidewalk near a tree, as shown in **Figure 2A.** Sometimes the roots will wedge rock apart, as shown in **Figure 2B.**

Burrowing animals also cause mechanical weathering, as shown in **Figure 3.** As these animals burrow, they loosen sediments and push them to the surface. Once the sediments are brought to the surface, other weathering processes can act on them.

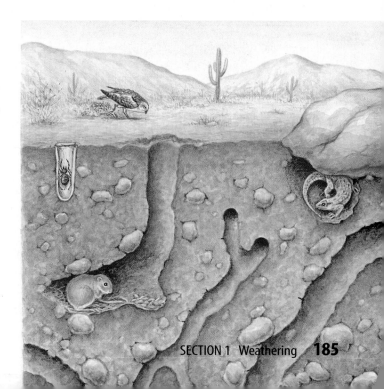

Figure 4
When water enters cracks in rock and freezes, it expands, causing the cracks to enlarge and the rock to break apart.

Figure 5
As rock is broken apart by mechanical weathering, the amount of rock surface exposed to air and water increases. The background squares show the total number of surfaces exposed.

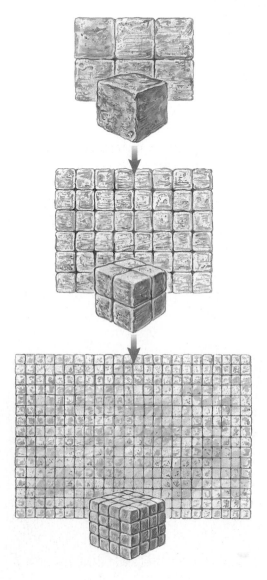

Ice Wedging The mechanical weathering process known as ice wedging is illustrated in **Figure 4. Ice wedging** occurs in temperate and cold climates where water enters cracks in rocks and freezes. Because water expands when it turns to ice, pressure builds up in the cracks. This pressure can extend the cracks and break apart rock. The ice then melts, allowing more water to enter the crack, where it freezes and breaks the rock even more. Ice wedging is most noticeable in the mountains, where warm days and cold nights are common. It is one process that wears down mountain peaks. This cycle of freezing and thawing not only breaks up rocks, but also can break up roads and highways. When water enters cracks in road pavement and freezes, it forces the pavement apart. This causes potholes to form in roads.

Surface Area Mechanical weathering by plants, animals, and ice wedging reduces rocks to smaller pieces. These small pieces have more surface area than the original rock body, as shown in **Figure 5.** As the amount of surface area increases, more rock is exposed to water and oxygen. This results in a different type of weathering called chemical weathering.

Chemical Weathering

Chemistry INTEGRATION

The second type of weathering, **chemical weathering,** occurs when chemical reactions dissolve the minerals in rocks or change them into different minerals. This type of weathering changes the chemical composition of the rock, which can weaken the rock. Next, you will see how chemical weathering happens.

Natural Acids Naturally formed acids can weather rocks chemically. When water mixes with carbon dioxide gas in the air or soil, a weak acid, called carbonic acid, forms. This is the same weak acid that makes soft drinks fizzy, especially when shaken. Carbonic acid reacts with minerals such as calcite, which is the main mineral that makes up limestone. This reaction causes the calcite to dissolve. Over many thousands of years, carbonic acid has weathered so much limestone that caves, such as the one shown in **Figure 6,** have formed.

Chemical weathering also occurs when naturally formed acids come in contact with other rocks. Over a long period of time, the mineral feldspar, which is found in granite, some types of sandstone, and other rocks, is broken down into a clay mineral called kaolinite (KAY oh luh nite). Kaolinite clay is common in some soils. Clay is an end product of weathering.

✔ Reading Check *How does kaolinite clay form?*

SCIENCE *Online*

Research Visit the Glencoe Science Web site at **science.glencoe.com** for more information about chemical weathering. Communicate to your class what you learn.

Figure 6
Caves form when slightly acidic groundwater dissolves large amounts of limestone.

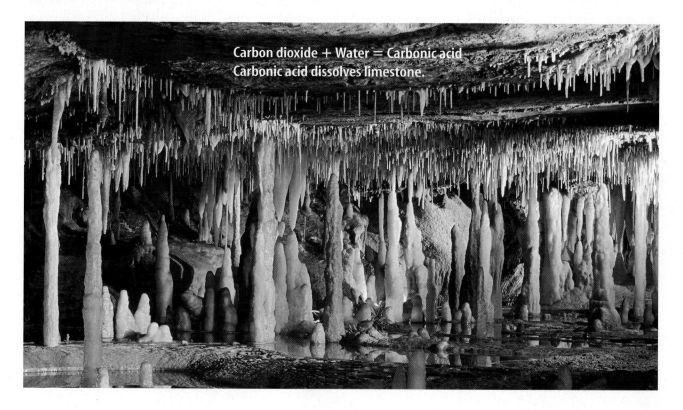

Carbon dioxide + Water = Carbonic acid
Carbonic acid dissolves limestone.

Observing the Formation of Rust

Procedure

1. Place some **steel wool** in a **glass petri dish** with 1 cm of **water**.
2. Observe for several days.

Analysis

1. What changes occurred?
2. What caused the changes?
3. How are these changes related to weathering?

Plant Acids Some roots and decaying plants give off acids that also can dissolve minerals in rock. When these minerals are dissolved, the rock is weakened. Eventually, the rock will break into smaller pieces. Do you know how a plant can benefit by being able to dissolve rock?

Oxygen Oxygen helps cause chemical weathering. You've seen rusty swing sets, nails, and cars. Rust is caused by oxidation. **Oxidation** (ahk sih DAY shun) occurs when metallic materials are exposed to oxygen and water over prolonged periods of time. For example, when minerals containing iron are exposed to water and the oxygen in air, the iron in the mineral can form a new material that resembles rust. One common example of this type of weathering is the alteration of the iron-bearing mineral magnetite to a rustlike material called limonite, as shown in **Figure 7.** What factors do you think would affect how fast minerals are oxidized?

Effects of Climate

Mechanical and chemical weathering occur everywhere. However, climate can affect the rate of weathering in different parts of the world. **Climate** is the pattern of weather that occurs in a particular area over many years. In cold climates, where freezing and thawing are frequent, mechanical weathering rapidly breaks down rocks through the process of ice wedging. Chemical weathering is more rapid in warm, wet climates. Thus, chemical weathering occurs quickly in tropical areas such as the Amazon River region of South America. Lack of moisture in deserts and low temperatures in polar regions slow down chemical weathering. Which type of weathering do you think is most rapid in the area where you live?

Figure 7

Iron-containing minerals like the magnetite shown here can weather to form a rustlike material called limonite. *How is this similar to rust forming on your bicycle chain?*

A

B

Figure 8
Different types of rock weather at different rates. **A** In humid climates, marble statues weather rapidly and become discolored. **B** Granite statues weather more slowly.

Effects of Rock Type Rock type also can affect the rate of weathering in a particular climate. In wet climates, for example, marble weathers more rapidly than granite, as shown in **Figure 8.**

Now you can understand how weathering affects rocks, caves, mountains, and even buildings and streets. Weathering is an important part of the rock cycle. When weathering breaks down rocks, it produces sediment that can form sedimentary rocks. Weathering also begins the process of forming soil from rock and sediment. This is discussed in the next section.

Section Assessment

1. What is the difference between mechanical and chemical weathering?

2. Explain how tree roots can weather rock. How can prairie dogs weather rock?

3. What effect does carbonic acid have on limestone?

4. How does climate affect the rate of chemical weathering?

5. **Think Critically** Why does limestone often form cliffs in dry climates but not in wet climates?

Skill Builder Activities

6. **Concept Mapping** Make a network tree concept map about mechanical weathering. **For more help, refer to the** Science Skill Handbook.

7. **Using an Electronic Spreadsheet** Create a spreadsheet that identifies examples of weathering that you see around your neighborhood and school. Classify each example as the result of mechanical weathering, chemical weathering, or both. **For more help, refer to the** Technology Skill Handbook.

The Nature of Soil

As You Read

What You'll Learn

- **Explain** how soil develops from rock.
- **Describe** soil by comparing soil horizons.
- **Describe** factors that affect the development of soils.

Vocabulary

soil
humus
horizon
soil profile
litter
leaching

Why It's Important

Much of the food you eat is grown in soil. The quality of food you eat reflects the quality of the soil it grows in.

Formation of Soil

How often have you been told "Take off those dirty shoes before you come into this house"? Ever since you were a child, you've had experience with soil. Soil is found in many places—backyards, empty city lots, farm fields, gardens, and forests.

What is soil and where does it come from? A layer of rock and mineral fragments produced by weathering covers the surface of Earth. As you learned in Section 1, weathering gradually breaks rocks into smaller and smaller fragments. However, these fragments do not become high-quality soil until plants and animals live in them. Plants and animals add organic matter, the remains of once-living organisms, to the rock fragments. Organic matter can include leaves, twigs, roots, and dead worms and insects. After organic matter has been added, soil as you know it begins to develop. **Soil** is a mixture of weathered rock, decayed organic matter, mineral fragments, water, and air.

Soil can take thousands of years to form and ranges from 60 m thick in some areas to just a few centimeters thick in others. Climate, slope, types of rock, types of vegetation, and length of time that rock has been weathering all affect the formation of soil, as shown in **Figure 9.** For example, different kinds of soils develop in tropical regions than in polar regions. Soils that develop on steep slopes are different from soils that develop on flat land. **Figure 10** illustrates how soil develops from rock.

Figure 9
Five different factors affect soil formation. *How does time influence the development of soils?*

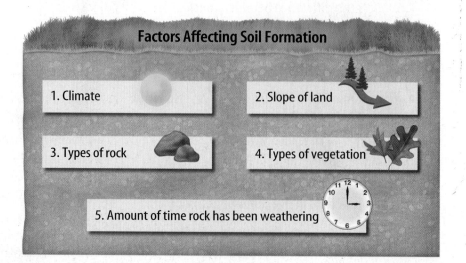

Factors Affecting Soil Formation

1. Climate
2. Slope of land
3. Types of rock
4. Types of vegetation
5. Amount of time rock has been weathering

Figure 10

It may take hundreds of years to form, but soil is constantly evolving from solid rock, as this series of illustrations shows. Soil is a mixture of weathered rock, mineral fragments, and organic material—the remains of dead plants and animals—along with water and air.

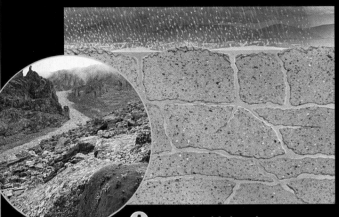

A Natural acids in rainwater weather the surface of exposed bedrock. Water can also freeze in cracks, causing rocks to fracture and break apart. The inset photo shows weathered rock in the Tien Shan Mountains of Central Asia.

B Plants take root in the cracks and among bits of weathered rock—shown in the inset photo above. As they grow, plants, along with other natural forces, continue the process of breaking down rocks, and a thin layer of soil begins to form.

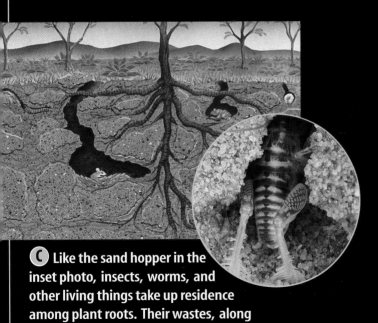

C Like the sand hopper in the inset photo, insects, worms, and other living things take up residence among plant roots. Their wastes, along with dead plant material, add organic matter to the soil.

D As organic matter increases and underlying bedrock continues to break down, the soil layer thickens. Rich topsoil supports trees and other plants with large root systems.

Comparing Components of Soil

Procedure

1. Collect a sample of **soil.**
2. Observe it closely with a **magnifying glass.**

Analysis

1. Describe the different particles found in your sample. Did you find any remains of once-living organisms?
2. Compare and contrast your sample with those other students have collected.

Composition of Soil

As you have seen already, soil is made up of rock and mineral fragments, organic matter, air, and water. The rock and mineral fragments found in soils come from rocks that have been weathered. Most of these fragments are small particles of sediment such as clay, silt, and sand. However, some larger pieces of rock also might be present.

Most organic matter in soil comes from plants. Plant leaves, stems, and roots all contribute organic matter to soil. Animals and microorganisms provide additional organic matter when they die. After plant and animal material gets into soil, fungi and bacteria cause it to decay. The decayed organic matter turns into a dark-colored material called **humus** (HYEW mus). Humus serves as a source of nutrients for plants. As worms, insects, and rodents burrow throughout soil, they mix the humus with the fragments of rock. Good-quality surface soil has about as much humus as weathered rock material.

✓ **Reading Check** *What does humus do for soil?*

Soil has many small spaces between individual soil particles that are filled with air or water. In swampy areas, water may fill the spaces year-round. In other areas, the spaces in soil are filled mostly with air.

Soil Profile

You have seen layers of soil if you've ever dug a deep hole or driven along a road that has been cut into a hillside. You probably observed that most plant roots grow in the top layer of soil. The top layer typically is darker than the soil layers below it. These different layers of soil are called **horizons.** All the horizons of a soil form a **soil profile.** Most soils have three horizons—labeled A, B, and C, as shown in **Figure 11.**

A Horizon

B Horizon

C Horizon

Figure 11
This soil, which developed beneath a grassy prairie, has three main horizons. *How is the A horizon different from the other two horizons?*

A Horizon The A horizon is the top layer of soil. In a forest or unplowed area, the A horizon might be covered with litter. **Litter** consists of leaves, twigs, and other organic material that eventually can be changed to humus by decomposing organisms. Litter helps prevent erosion and holds water. The A horizon also is known as topsoil. Topsoil has more humus and fewer rock and mineral particles than the other layers in a soil profile. The A horizon generally is dark and fertile. The dark color of soil is caused by organic material, which provides nutrients for plant growth and development.

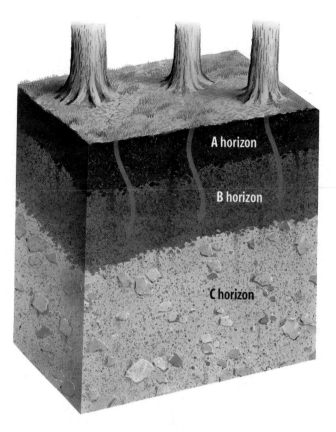

B Horizon The layer below the A horizon is the B horizon. Because litter does not add organic matter to this horizon, it is lighter in color than the A horizon and contains less humus. As a result, the B horizon is less fertile. The B horizon contains material moved down from the A horizon by the process of leaching.

Figure 12
Leaching removes material from the upper layer of soil. Much of this material then is deposited in the B horizon.

Leaching is the removal of minerals that have been dissolved in water. The process of leaching resembles making coffee in a drip coffeemaker. In a coffeemaker, water drips through ground coffee. In soil, water seeps through the A horizon. In a coffeemaker, water absorbs the flavor and color from the coffee and flows into a coffeepot below. In soil, water reacts with humus and carbon dioxide to form acid. The acid dissolves some of the minerals in the A horizon and carries the material into the B horizon, as shown in **Figure 12.**

Reading Check *How does leaching transport material from the A horizon to the B horizon?*

C Horizon The C horizon consists of partially weathered rock and is the bottom horizon in a soil profile. It is often the thickest soil horizon. This horizon does not contain much organic matter and is not strongly affected by leaching. It usually is composed of coarser sediment than the soil horizons above it. What would you find if you dug to the bottom of the C horizon? As you might have guessed, you would find rock—the rock that gave rise to the soil horizons above it. This rock is called the parent material of the soil. The C horizon is the soil layer that is most like the parent material.

Life Science
INTEGRATION

Earthworms in the A horizon swallow sediment to obtain food. These worms then excrete waste into the soil. This material is fertile and helps produce high-quality soil. Do research to find out how other animals affect soil development.

Glacial Deposits At many places on Earth, the land is covered by material that was deposited by glaciers. This unsorted mass of clay, silt, sand, and boulders covers much of the United States, creating, for example, the flat landscapes of the Midwest. The soils that developed on this glacial material are extremely fertile and are an important part of the Midwest's agricultural industry. How does this soil profile differ from the one described earlier? If you were to dig through the C horizon, you would find bedrock as before, but it would not be the rock the soil formed from. What material did this soil develop from?

Math Skills Activity

Calculating Percentages of Soil Particles

The properties of soil, such as the ability to hold water, are determined by the abundances of different types of particles. Therefore, it is important to determine particle percentages in soils.

Sample Problem

The circle graph represents a soil sample containing particles of clay, silt, and sand. Determine what percentage of the sample is clay.

20 g
Sand particles

15 g
Clay particles

15 g
Silt particles

1 *This is what you know:*

sand weight: 20 g
clay weight: 15 g
silt weight: 15 g

2 *This is what you need to find:*

total weight of the sample
percentage of clay particles

3 *These are the equations you need to use:*

sand weight + clay weight + silt weight = total weight
decimal equivalent of percentage = clay weight/total weight
decimal equivalent of percentage × 100 = percentage of clay

4 *Solve for the total weight:* 20 g + 15 g + 15 g = 50 g

5 *Solve for the decimal equivalent of percentage:* 15 g/50 g = 0.3

6 *Solve for the percentage of particle type:* 0.3 × 100 = 30% clay particles

Practice Problem

A second soil sample is taken that contains 30 g of sand, 30 g of clay, and 15 g of silt. What percentage of the entire soil sample is silt? Draw a circle graph for this sample.

For more help, refer to the Math Skill Handbook.

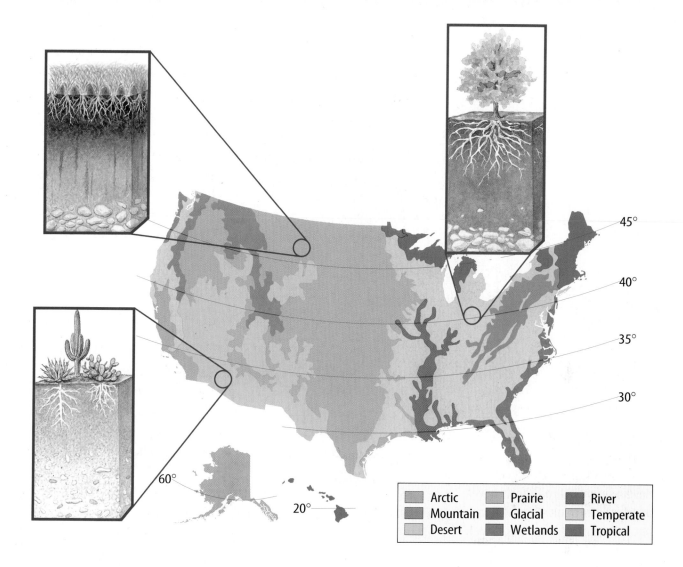

Figure 13
The United States has nine different soil types. They vary in color, depth, texture, and fertility.

Legend:
- Arctic
- Mountain
- Desert
- Prairie
- Glacial
- Wetlands
- River
- Temperate
- Tropical

Map latitude labels: 45°, 40°, 35°, 30°, 60°, 20°

Soil Types

If you travel across the country and look at soils, you will notice that they are not all the same. Some soils are thick and red, some are brown with hard, rounded rock nodules, and some have thick, black A horizons. Many different types of soils exist, as shown in **Figure 13.**

Soil Types Reflect Climate Different regions on Earth have different climates. Deserts are dry, prairies are semidry, and temperate forests are mild and moist. These places also have different types of soils. Soils in deserts contain little organic material and also are thinner than soil horizons in wetter climates. Prairie soils have thick, dark A horizons because the grasses that grow there contribute lots of organic matter. Temperate forest soils have thinner A horizons and B horizons that have been enriched in many elements because of leaching. The abundant rainfall in forests promotes leaching. Other regions such as tundra and tropical areas also have distinct soils.

Figure 14
The slope of the land affects soil development. Thin, poorly developed soils form on steep slopes, but valleys often have thick, well-developed soils. *Why is this so?*

Other Factors Parent rock has a strong effect on the soils that develop from it. Clay soils often develop on rocks like basalt, because minerals in the rock weather to clay. What type of soil do you think might develop on sandstone? Rock type also can affect the type of vegetation that grows in a region, because different rocks provide different nutrients for plant growth. Type of vegetation then affects soil formation.

Time also affects soil development. If weathering has been occurring for only a short time, the parent rock determines the soil characteristics. As weathering continues, the soil resembles the parent rock less and less.

Slope also is a factor affecting soil profiles, as shown in **Figure 14.** On steep slopes, soil horizons often are poorly developed, because material moves downhill before it can be weathered much. In bottomlands, sediments and water are plentiful. Bottomland soils are often thick, dark, and full of organic material.

Section Assessment

1. List five factors that affect soil development.
2. Why do soil profiles contain layers or horizons?
3. How do organisms help soils develop dark A horizons?
4. How does the B horizon differ from the A horizon and the C horizon?
5. **Think Critically** Why is the soil profile in a tropical rain forest different from one in a desert?

Skill Builder Activities

6. **Concept Mapping** Make an events chain concept map that explains how soil develops. **For more help, refer to the** Science Skill Handbook.
7. **Using Statistics** A farmer collected five soil samples from a field and tested their acidity or pH. His data were the following: 7.5, 8.2, 7.7, 8.1, and 8.0. Calculate the mean of these data. Also, determine the range and median. **For more help, refer to the** Math Skill Handbook.

Activity

Soil Characteristics

Y ou've probably noticed a number of different types of soils. Collect samples of soil to compare from around your neighborhood and from designated areas of your school grounds.

What You'll Investigate
What are the characteristics of soils?

Materials

soil sample	water
cheesecloth squares (3)	250-mL beakers (3)
sand	watch
100-mL graduated cylinder	large polystyrene or plastic cups (3)
gravel	pie pans
plastic coffee-can lids (3)	hand lens
clay	scissors
rubber bands (3)	thumbtack

Goals
■ **Analyze** permeability of different soils.

Safety Precautions

WARNING: *Use care when punching holes in the bottom of the cups with the thumbtack.*

Procedure

1. Spread the soil sample in a pie pan.
2. **Describe** the color of the soil and examine the soil with a hand lens. Describe the different particles.
3. Rub a small amount of soil between your fingers. Describe how it feels. Also, press the soil sample together. Does it stick together? Wet the sample and try this again. Record all your observations.

4. **Test** sediment to see how water moves through it. Label the three cups A, B, and C. Using a thumbtack, punch ten holes in and around the bottom of each cup.
5. Cover the area of holes with a square of cheesecloth and secure with a rubber band.
6. To hold the cups over the beakers, cut the three coffee-can lids so that the cups will just fit inside the hole (see photo). Place a cup and lid over each beaker.
7. Fill cup A halfway with dry sand and cup B halfway with clay. Fill cup C halfway with a mixture of equal parts of sand, gravel, and clay.
8. Use the graduated cylinder to pour 100 mL of water into each cup. Record the time when the water is first poured into each cup and when the water first drips from each cup.
9. Allow the water to drip for 25 min, then measure and record the amount of water in each beaker.

Conclude and Apply

1. How does the addition of gravel and sand affect the permeability of clay?
2. **Describe** three characteristics of soil. Which characteristics affect permeability?

3 Soil Erosion

What You'll Learn

- **Explain** why soil is important.
- **Identify** human activities that lead to soil loss.
- **Describe** ways to reduce soil loss.

Vocabulary
terracing

Why It's Important
If topsoil is eroded, soil becomes much less fertile.

Figure 15
Removing vegetation can lead to severe soil erosion. **A** Trees protect the soil from erosion in this forested region. **B** When forest is removed, soil erodes rapidly.

Soil—An Important Resource

While picnicking at a local park, a flash of lightning and a clap of thunder tell you that a storm is upon you. Watching the pounding rain from the park shelter, you notice that the water flowing off of the ball diamond is muddy, not clear. The flowing water is carrying away some of the soil that used to be on the field. This process is called soil erosion. Soil erosion is harmful because plants do not grow as well when topsoil has been lost.

Causes and Effects of Soil Erosion

Soil is eroded when it is moved from the place where it formed. Erosion occurs when water flows along Earth's surface or when wind picks up and transports sediments. Generally, erosion is more common on steep slopes than on gentle slopes. It's also more common in areas where there is little vegetation. Under normal conditions, a balance between soil production and soil erosion often is maintained. This means that soil forms at about the same rate as it is eroded. However, when vegetation is removed or slopes are steepened, soil erodes much faster than it can be produced. Many human activities result in the removal of vegetation and the steepening of slopes, as shown in **Figure 15**. Some of these activities are described in this section.

A

B

Figure 16
Tropical rain forests often are cleared by burning. *How can this increase soil erosion?*

Physics
INTEGRATION

Rain falling on farm fields can be an important agent of erosion. The erosive force of rain depends on how much rain falls and on how hard the rain falls. What type of rains do you think would have the most energy to erode soil?

Agricultural Cultivation Did you know that the population of Earth increases by nearly 95 million people every year? More people means that more food is needed. This has led to increased use of farmable land and to rapid soil erosion in many places. Plowing mechanically turns and loosens the soil, improving it for crops. However, this practice also removes the plant cover that holds soil particles in place, leaving soils vulnerable to wind and water erosion. Over time, this practice can reduce soil quality.

Forest Harvesting When forests are removed, soil is exposed and erosion increases. This creates severe problems in many parts of the world, but tropical regions are especially at risk. Each year, thousands of square kilometers of tropical rain forest are cleared for lumber, farming, and grazing, as shown in **Figure 16.** Soils in tropical rain forests appear rich in nutrients but are almost infertile below the first few centimeters. The soil is useful to farmers for only a few years before the nutrients are gone. Farmers then clear new land, repeating the process and increasing the damage to the soil.

Overgrazing In most places, land can be grazed with little damage to soil. However, overgrazing can increase soil erosion. In some arid regions of the world, sheep and cattle raised for food are grazed on grasses until almost no ground cover remains to protect the soil. When natural vegetation is removed from land that receives little rain, plants are slow to grow back. Without protection, soil is carried away by wind, and the moisture in the soil evaporates.

Research Visit the Glencoe Science Web site at **science.glencoe.com** for information on soil erosion research and how this research can benefit the environment. Communicate to your class what you learn.

✔ **Reading Check** *How does overgrazing affect soil?*

Urban Construction Each year in the United States, about 6,100 km^2 of land are developed for roadways and other structures. You've probably noticed that when construction takes place, land is cleared of vegetation and soil is moved. During the construction process, water and wind erode soils. Areas that have been strip mined also are susceptible to soil erosion. Eroded soil can enter streams, causing them to fill up with sediment, as shown in **Figure 17.**

Figure 17
Erosion from exposed land can cause streams to fill with excessive amounts of sediment. *How could this damage streams?*

Figure 18
No-till farming helps prevent soil erosion because fields are not plowed before planting.

Preventing Soil Erosion

Each year more than 4 billion metric tons of soil are eroded in the United States. Soil is a resource that must be managed and protected. People can do several things to conserve soil.

Manage Crops All over the world, farmers work to slow down soil erosion. They plant shelter belts of trees to break the force of the wind and cover bare soils with decaying plants to hold soil particles in place. In dry areas, instead of plowing under vegetation, many farmers graze animals on the vegetation. Proper grazing management can maintain vegetation and reduce soil erosion.

In recent years, many farmers have begun to practice no-till farming. Normally, farmers till or plow their fields one or more times each year. In no-till farming, seen in **Figure 18,** plant stalks are left in the field over the winter months. At the next planting, farmers seed crops without destroying these stalks and without plowing the soil. No-till farming provides cover for the soil year-round, which reduces water runoff and slows soil erosion. The leftover stalks also keep weeds from growing in the fields.

Reading Check *How can farmers reduce soil erosion?*

Reduce Erosion on Slopes On gentle slopes, planting along the natural contours of the land, as in **Figure 19,** helps reduce soil erosion. This practice, called contour farming, slows the flow of water down the slope and helps prevent the formation of gullies.

Where slopes are steep, terracing often is used. **Terracing** (TER uh sing) is a method in which steep-sided, level topped areas are built onto the sides of steep hills and mountains so that crops can be grown. These terraces reduce runoff by creating flat areas and shorter sections of slope. In the Philippines, Japan, China, and Peru, terraces have been used for centuries.

Reduce Erosion at Construction Sites A variety of methods are used at construction sites to help reduce erosion. During the construction process, exposed ground sometimes is covered with mulch, mats, or plastic coverings. Water is sprayed onto bare soil to prevent erosion by wind. When construction is complete, topsoil is added in areas where it was removed and trees are planted. Some areas are sodded, but if an area is to be seeded, soil is reinforced using netting and straw. Netting and straw make the surface layer more stable, which allows plant seeds to germinate quickly and vegetation to uniformly cover any slopes. Along steeper slopes, retaining walls made of concrete, stones, or wood keep soil and rocks from sliding downhill.

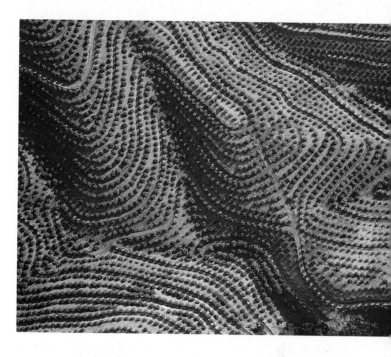

Figure 19
This orchard was planted along the natural contours of the land.
Why was this done?

Section 3 Assessment

1. Why is soil important?
2. Why does soil erosion increase when soil is plowed?
3. How can urban construction cause soil erosion?
4. What are two ways erosion can be reduced at construction sites?
5. **Think Critically** How can contour farming help water soak into the ground?

Skill Builder Activities

6. **Recognizing Cause and Effect** Explain how soil erosion contributes to pollution of streams and lakes. **For more help,** refer to the Science Skill Handbook.
7. **Communicating** In your Science Journal, write a poem about the causes and effects of soil erosion. **For more help,** refer to the Science Skill Handbook.

Weathering Chalk

Chalk is a type of limestone made of the shells of microscopic organisms. The famous White Cliffs of Dover, England, are made up of chalk. This experiment will help you understand how chalk can be chemically weathered.

Recognize the Problem

How can you simulate chemical weathering of chalk?

Form a Hypothesis

How do you think acidity, surface area, and temperature affect the rate of chemical weathering of chalk? What happens to chalk in water or acid (vinegar)? How will the size of the chalk pieces affect the rate of weathering? What will happen if you heat the acid? Make hypotheses to support your ideas.

Possible Materials
equal-sized pieces of chalk (6)
small beakers or clear plastic cups (2)
metric ruler
water
white vinegar (100 mL)
hot plate
250-mL graduated cylinder
computer probe for temperature
*thermometer
*Alternate materials

Goals
- **Design** experiments to compare the effects of acidity, surface area, and temperature on the rate of chemical weathering of chalk.
- **Describe** factors that affect chemical weathering.
- **Explain** how the chemical weathering of chalk is similar to the chemical weathering of rocks.
- **Describe** how the factors used in this experiment apply to different parts of the world.

Safety Precautions

Wear safety goggles when pouring acids. Be careful when using a hot plate and heated solutions.
WARNING: *If mixing liquids, always add acid to water.*

Test Your Hypothesis

Plan

1. **Develop** hypotheses about the effects of acidity, surface area, and temperature on the rate of chemical weathering.

2. **Decide** how to test your first hypothesis. List the steps needed to test the hypothesis.

3. Repeat step 2 for your other two hypotheses.

4. **Design** data tables in your Science Journal. Make one for acidity, one for surface area, and one for temperature.

5. **Identify** what remains constant in your experiment and what varies. Each test should have only one variable being tested. Determine the control for each experiment.

6. **Summarize** your data in a graph. Decide from reading the **Science Skill Handbook** which type of graph to use.

Do

1. Make sure your teacher approves your plan before you start.

2. Carry out the three experiments as planned.

3. While you are conducting the experiments, record your observations and complete the data tables in your Science Journal.

4. Graph your data to show how each variable affected the rate of weathering.

Analyze Your Data

1. **Analyze** your graph to find out which substance—water or acid—weathered the chalk more quickly. Was your hypothesis supported by your data?

2. **Infer** from your data whether the amount of surface area makes a difference in the rate of chemical weathering. Explain.

Draw Conclusions

1. **Explain** how the chalk was chemically weathered.

2. How does heat affect the rate of chemical weathering?

3. What does this imply about weathering in the tropics and in polar regions?

*C*ommunicating Your Data

Compare your results with those of your classmates. How were your data similar? How were they different? **For more help, refer to the** Science Skill Handbook.

Landscape, History, and the Pueblo Imagination
by Leslie Marmon Silko

In this excerpt, Leslie Marmon Silko, a woman of Pueblo, Hispanic, and American heritage, explains what ancient Pueblo people believed about the circle of life on Earth.

You see that after a thing is dead, it dries up. It might take weeks or years, but eventually if you touch the thing, it crumbles under your fingers. It goes back to dust. The soul of the thing has long since departed. With the plants and wild game the soul may have already been borne back into bones and blood or thick green stalk and leaves. Nothing is wasted. What cannot be eaten by people or in some way used must then be left where other living creatures may benefit. What domestic animals or wild scavengers can't eat will be fed to the plants. The plants feed on the dust of these few remains.

. . . Corn cobs and husks, the rinds and stalks and animal bones were not regarded by the ancient people as filth or garbage. The remains were merely resting at a midpoint in their journey back to dust. . . .

The dead become dust . . . . The ancient Pueblo people called the earth the Mother Creator of all things in this world. Her sister, the Corn mother, occasionally merges with her because all . . . green life rises out of the depths of the earth.

Rocks and clay . . . become what they once were. Dust.

A rock shares this fate with us and with animals and plants as well.

Understanding Literature

Repetition Many authors use repetition as a literary tool. Repetition is the recurrence of sounds, words, or phrases in a piece of writing. Poets often use repetition to give their poems a particular rhythm. In Silko's passage, she repeatedly uses the word *dust*. This repetition reminds the reader of the common link between rocks, clay, plants, and animals. Also, it gives this passage a storytelling quality that makes the reader feel like he or she is reading a myth or legend instead of a work of nonfiction.

Science Connection In this chapter, you learned how weathered rocks and mineral fragments combine with organic matter to make soil. Silko's writing explains how the ancient Pueblo people understood that all living matter returns to the earth, or becomes dust. Lines such as "green life rises out of the depths of the earth," show that the Pueblo people understood that the earth, or rocks and mineral fragments, must combine with living matter in order to make soil and support plant life. Are there other ways in which your scientific understanding of the formation of soil is similar to the ancient Pueblo beliefs outlined in Silko's writing?

Linking Science and Writing

Using Repetition Write a one-page, how-to paper for your classmates on a type of soil conservation practice. For instance, the subject can be no-till agriculture, strip-cropping, or contour plowing. Use repetition as a tool to remind your readers of important steps in the process.

Career Connection

Soil Scientist

Elvia Niebla works for the U.S. Environmental Protection Agency (EPA) studying soil and soil pollution. Soil scientists explain how people can use soil effectively. They develop plans that keep people from contaminating soil and that keep soil from eroding. Niebla's research has even helped keep hamburger safe to eat. How? In a report for the EPA, she explained how meat could be contaminated when cows graze on contaminated soil. She is now the National Coordinator of the Global Change Research Program.

SCIENCE*Online* To learn more about careers in soil science, visit the Glencoe Science Web site at **science.glencoe.com**.

Reviewing Main Ideas

Section 1 Weathering

1. Mechanical weathering breaks apart rock without changing its chemical composition. Plants, animals, and ice wedging are important agents of mechanical weathering. *How can this prairie dog weather rock?*

2. Chemical weathering changes the composition of rocks. Acidic water can dissolve rock or certain minerals within a rock. Some plants cause chemical weathering by secreting acids. Exposure to oxygen causes some rocks to weather, forming rustlike minerals.

Section 2 The Nature of Soil

1. Soil is a mixture of rock and mineral fragments, organic matter, air, and water.

2. Soil develops as rock is weathered and organic matter is added by organisms. Soil has horizons that differ in their color and composition. *How could the plants in the photo help soil develop?*

3. Climate, parent rock, slope of the land, type of vegetation, and the time that rock has been weathering affect the development of soil and cause different soils to have different characteristics.

Section 3 Soil Erosion

1. Soil is eroded when it is transported by wind and water. Erosion is more common on steep slopes and in areas where there is no ground cover.

2. Human activities like plowing, harvesting forests, overgrazing animals, and construction can increase soil erosion. *How does this netting help reduce soil erosion?*

3. Windbreaks, no-till farming, contour planting, and terracing help reduce soil erosion in farm fields. The best way to reduce soil erosion at construction sites is to replace vegetation quickly.

FOLDABLES
Reading & Study Skills

After You Read

To help you review the vocabulary words, use the Foldable you made at the beginning of the chapter.

Visualizing Main Ideas

Complete the following concept map on weathering.

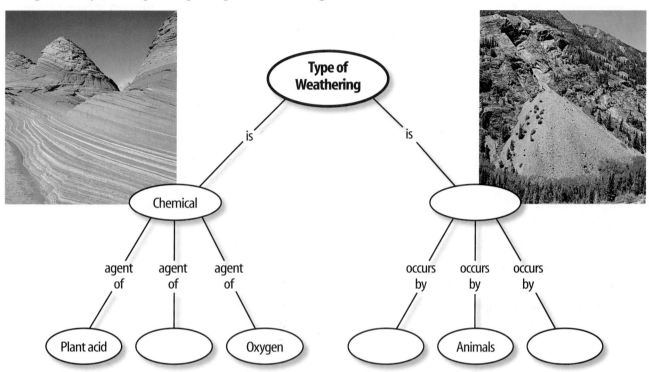

Vocabulary Review

Vocabulary Words

a. chemical weathering
b. climate
c. horizon
d. humus
e. ice wedging
f. leaching
g. litter
h. mechanical weathering
i. oxidation
j. soil
k. soil profile
l. terracing
m. weathering

Study Tip

Make flash cards for new vocabulary words. Put the word on one side and the definition on the other. Use them to quiz yourself.

Using Vocabulary

The sentences below include vocabulary words that have been used incorrectly. Change the incorrect vocabulary words so that the sentence reads correctly. Underline your change.

1. Mechanical weathering causes a change in a rock's composition.

2. Leaching forms from organic matter such as leaves and roots.

3. The A, B, and C horizons of a soil make up the climate.

4. Ice wedging transports materials to the B horizon.

5. Litter occurs when many materials containing iron are exposed to oxygen and water.

Checking Concepts

Choose the word or phrase that best answers the question.

1. Which of the following is caused by acids produced by plants?
   A) soil erosion
   B) overgrazing
   C) mechanical weathering
   D) chemical weathering

2. What occurs when roots force rocks apart?
   A) mechanical weathering
   B) leaching
   C) ice wedging
   D) chemical weathering

3. What reacts with iron to form rust?
   A) oxygen   C) feldspar
   B) carbon dioxide   D) paint

4. Which of the following is an agent of mechanical weathering?
   A) ice wedging   C) leaching
   B) oxidation   D) desert formation

5. In which region is chemical weathering most rapid?
   A) cold, dry   C) warm, moist
   B) cold, moist   D) warm, dry

6. What is a mixture of weathered rock, organic matter, air, and water called?
   A) soil   C) carbon dioxide
   B) limestone   D) clay

7. What is decayed organic matter called?
   A) leaching   C) soil
   B) humus   D) sediment

8. Where is most humus found?
   A) A horizon   C) C horizon
   B) B horizon   D) D horizon

9. What does no-till farming help prevent?
   A) leaching   C) overgrazing
   B) crop rotation D) soil erosion

10. What is done to reduce erosion on steep slopes?
    A) weathering   C) terracing
    B) overgrazing   D) forest harvesting

Thinking Critically

11. Which type of weathering, mechanical or chemical, would you expect to have more effect in a polar region? Explain.

12. Explain how off-road vehicles affect soil erosion.

13. How does increasing human population affect soil erosion?

14. Why is it difficult to replace lost topsoil?

15. Describe how chemical weathering can form a cave.

Developing Skills

16. **Using Variables, Constants, and Controls** Juan wanted to know if planting grass on a slope would prevent soil from being washed away. To find out, he put the same amount and kind of soil in two identical pans. In one of the pans, he planted sod. To create equal slopes for his test, he placed identical wooden wedges under one end of each pan. He carefully poured the same amount of water at the same rate over the soil in the two pans. What is Juan's control? What factors in his activity are constants? What variable is he testing?

17. Classifying Classify the following as examples of either chemical or mechanical weathering: rocks oxidize to form rustlike minerals, freezing and thawing of water causes a cliff face to break apart, acids from mosses discolor rocks, tree roots break rocks apart, and water seeping through cracks in limestone dissolves away some of the rock.

18. Concept Mapping Complete the events chain concept map that shows two ways in which acids can cause chemical weathering.

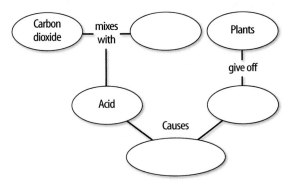

19. Forming a Hypothesis Over time, crop yields on a farm field diminish in spite of repeated applications of fertilizer. What hypothesis could be tested to explain this? How could an explanation benefit the farm owner?

Performance Assessment

20. Design a Landscape Find a slope in your area that might benefit from erosion maintenance. Design a landscape map showing the types of plants you would introduce and where you would put them.

TECHNOLOGY

Go to the Glencoe Science Web site at **science.glencoe.com** or use the **Glencoe Science CD-ROM** for additional chapter assessment.

THE PRINCETON REVIEW Test Practice

Scientists in North Dakota have recently analyzed a sample of soil for sand, silt, and clay. They placed their results in the following table.

Horizon	Percent		
	Sand	Silt	Clay
A	16.2	54.4	29.4
B	10.5	50.2	39.3
C	31.4	48.4	20.2
R (Bedrock)	31.7	50.1	18.2

Study the table and answer the following questions.

1. According to this information, which soil horizon has the lowest percentage of sand?
A) A horizon
B) B horizon
C) C horizon
D) R horizon

2. According to the table, R horizon refers to _____ .
F) topsoil
G) bedrock
H) humus
J) gravel

3. According to the information in the table, the best description of the soil in the table above would be _____ .
A) sandy
B) silty
C) clayey
D) organic

8

Erosional Forces

Loosened by years of wind, water, and ice, these displaced rocks came crashing down on the Karakoram Highway in Pakistan. Blocked by the boulders, travelers observe the aftermath of a powerful erosional force. As you can see, transport of rocks and sediment downslope can impact the lives of humans. In this chapter, you'll learn about some of the forces that cause erosion. You'll see how gravity, glaciers, and wind erode and deposit Earth materials—constantly changing landscapes.

What do you think?

Science Journal Look at the picture below with a classmate. Discuss what you think this might be or what is happening. Here's a hint: *Someone used to live here.* Write your answer or best guess in your Science Journal.

EXPLORE ACTIVITY

Can you think of ways to move something without touching it? In nature, sediment is moved from one location to another by a variety of forces. What are some of these forces? In this activity, you will investigate to find out.

Demonstrate sediment movement

WARNING: *Do not pour sand or gravel down the drain.*

1. Place a small pile of a sand-and-gravel mixture in one end of a large shoe-box lid.

2. Move the sediment pile to the other end of the lid without touching the particles with your hands. You can touch and manipulate the box lid.

3. Try to move the mixture in a number of different ways.

Observe

In your Science Journal, describe the methods you used to move the sediment. Which method was most effective? Explain how your methods compare with forces of nature that move sediment.

Before You Read

FOLDABLES
Reading & Study Skills

Making an Organizational Study Fold
Make the following Foldable to help you organize your thoughts about erosion and deposition.

1. Place a sheet of paper in front of you so the short side is at the top.

2. Fold the paper in half from the top to the bottom. Then fold it in half again. Unfold all the folds.

3. Trace over all the fold lines. Label the columns *Erosion* and *Deposition*. Label the rows *Gravity, Glaciers,* and *Wind* as shown.

4. As you read the chapter, record information including specific examples of erosion and deposition in the rows of your Foldable.

Erosion by Gravity

As You Read

What You'll Learn

- **Explain** the differences between erosion and deposition.
- **Compare and contrast** slumps, creep, rockfalls, rock slides, and mudflows.
- **Explain** why building on steep slopes might not be wise.

Vocabulary

erosion   slump
deposition   creep
mass movement

Why It's Important

Many natural features throughout the world were shaped by erosion.

Erosion and Deposition

Do you live in an area where landslides occur? As **Figure 1** shows, large piles of sediment and rock can move downhill with devastating results. Such events often are triggered by heavy rainfall. The muddy debris at the lower end of the slide comes from material that once was further up the hillside. The displaced soil and rock debris is a product of erosion (ih ROH zhun). **Erosion** is a process that wears away surface materials and moves them from one place to another.

What wears away sediments? How were you able to move the pile of sediments in the Explore Activity? If you happened to tilt the box lid, you took advantage of an important erosional force—gravity. Gravity is the force of attraction that pulls all objects toward Earth's center. Other causes of erosion, also called agents of erosion, are water, wind, and glaciers.

Water and wind erode materials only when they have enough energy of motion to do work. For example, air can't move much sediment on a calm day, but a strong wind can move dust and even larger particles. Glacial erosion works differently by slowly moving sediment that is trapped in solid ice. As the ice melts, sediment is deposited, or dropped. Sometimes sediment is carried farther by moving meltwater.

Figure 1
The jumbled sediment at the base of a landslide is material that once was located farther uphill. *What force moves materials toward the center of Earth?*

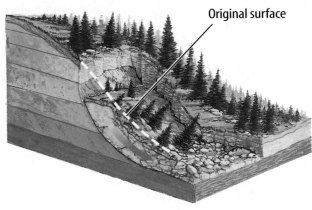

Original surface

Figure 2
Slump occurs when material slips downslope as one large mass.
What might have caused this slump to happen?

Dropping Sediments Agents of erosion drop the sediments they are carrying as they lose energy. This is called **deposition.** When sediments are eroded, they are not lost from Earth—they are just relocated.

Mass Movement

The greater an object's mass is, the greater its gravitational force is. Earth has such a great mass that gravity is a major force of erosion and deposition. Rocks and other materials, especially on steep slopes, are pulled toward the center of Earth by gravity.

A **mass movement** is any type of erosion that happens as gravity moves materials downslope. Some mass movements are so slow that you hardly notice they're happening. Others happen quickly—possibly causing catastrophes. Common types of mass movement include slump, creep, rockfalls, rock slides, and mudflows. Landslides are mass movements that can be one of these types or a combination of these types of mass movement.

✔ **Reading Check** *What is a mass movement?*

Slump When a mass of material slips down along a curved surface, the mass movement is called **slump.** Often, when a slope becomes too steep, the base material no longer can support the rock and sediment above it. The soil and rock slip downslope as one large mass or break into several sections.

Sometimes a slump happens when water moves to the base of a slipping mass of sediment. This water weakens the slipping mass and can cause movement of material downhill. Or, if a strong rock layer lies on top of a weaker layer—commonly clay—the clay can weaken further under the weight of the rock. The clay no longer can support the strong rock on the hillside. As shown in **Figure 2,** a curved scar is left where the slumped materials originally rested.

Mini LAB

Modeling Slump

Procedure 🥽 👕 🚫

WARNING: *Do not pour lab materials down the drain.*

1. Place one end of a **baking pan** on **two bricks** and position the other end over a sink with a sealed drain.
2. Fill the bottom half of the pan with **gelatin powder** and the top half of the pan with **aquarium gravel.** Place a large, **flat rock** on the gravel.
3. Using a **watering can,** sprinkle water on the materials in the pan for several minutes. Record your observations in your **Science Journal.**

Analysis

1. What happened to the different sediments in the pan?
2. Explain how your experiment models slump.

Figure 3
Over time, creep has caused these tree trunks to lean downhill. The trees then curved back toward the Sun.

Figure 4
Rockfalls and rockslides move solid materials rapidly downslope.

A Rockfalls, such as this one, occur as material free falls through the air.

Creep The next time you travel, look along the roadway or trail for slopes where trees and fence posts lean downhill. Leaning trees and human-built structures show another mass movement called creep. **Creep** occurs when sediments slowly shift their positions downhill, as **Figure 3** illustrates. Creep is common in areas of freezing and thawing.

Rockfalls and Rock Slides Signs along mountainous roadways warn of another type of mass movement called rockfalls. Rockfalls happen when blocks of rock break loose from a steep slope and tumble through the air. As they fall, these rocks crash into other rocks and knock them loose. More and more rocks break loose and tumble to the bottom. The fall of a single, large rock down a steep slope can cause serious damage to structures at the bottom. During the winter, when ice freezes in the cracks of rocks, the cracks expand and extend. In the spring, the pieces of rock break loose and fall down the mountainside, as shown in **Figure 4A.**

Rock slides occur when layers of rock—usually steep layers—slip downslope suddenly. Rock slides, like rockfalls, are fast and can be destructive in populated areas. They commonly occur in mountainous areas or in areas with steep cliffs, as shown in **Figure 4B.** Rock slides happen most often after heavy rains or during earthquakes, but they can happen on any rocky slope at any time without warning.

B Rock slides are common in regions where layers of rock are steep.

Mudflows What would happen if you took a long trip and forgot to turn off the sprinkler in your hillside garden before you left? If the soil is usually dry, the sprinkler water could change your yard into a muddy mass of material much like chocolate pudding. Part of your garden might slide downhill. You would have made a mudflow, a thick mixture of sediments and water flowing down a slope. The mudflow in **Figure 5** caused a lot of destruction.

Mudflows usually occur in areas that have thick layers of loose sediments. They often happen after vegetation has been removed by fire. When heavy rains fall on these areas, water mixes with sediment, causing it to become thick and pasty. Gravity causes this mass to flow downhill. When a mudflow finally reaches the bottom of a slope, it loses its energy of motion and deposits all the sediment and everything else it has been carrying. These deposits often form a mass that spreads out in a fan shape. Why might mudflows cause more damage than floodwaters?

Figure 5
Mudflows, such as these in the town of Sarno, Italy, have enough energy to move almost anything in their paths. *How do mudflows differ from slumps, creep, and rock slides?*

✔ Reading Check *What conditions are favorable for triggering mudflows?*

Mudflows, rock slides, rockfalls, creep, and slump are similar in some ways. They all are most likely to occur on steep slopes, and they all depend on gravity to make them happen. Also, all types of mass movement occur more often after a heavy rain. The water adds mass and creates fluid pressure between grains and layers of sediment. This makes the sediment expand—possibly weakening it.

Consequences of Erosion

People like to have a great view and live in scenic areas away from noise and traffic. To live this way, they might build or move into houses and apartments on the sides of hills and mountains. When you consider gravity as an agent of erosion, do you think steep slopes are safe places to live?

Building on Steep Slopes When people build homes on steep slopes, they constantly must battle naturally occurring erosion. Sometimes builders or residents make a slope steeper or remove vegetation. This speeds up the erosion process and creates additional problems. Some steep slopes are prone to slumps because of weak sediment layers underneath.

Making Steep Slopes Safe Plants can be beautiful or weedlike—but they all have root structures that hold soil in place. One of the best ways to reduce erosion is to plant vegetation. Deep tree roots and fibrous grass roots bind soil together, reducing the risk of mass movement. Plants also absorb large amounts of water. Drainage pipes or tiles inserted into slopes can prevent water from building up, too. These materials help increase the stability of a slope by allowing excess water to flow out of a hillside more easily.

Walls made of concrete or boulders also can reduce erosion by holding soil in place, as shown in **Figure 6.** However, preventing mass movements on a slope is difficult because rain or earthquakes can weaken all types of Earth materials, eventually causing them to move downhill.

 Reading Check *What can be done to slow erosion on steep slopes?*

Figure 6
Some slopes are stabilized by building walls made from concrete or stone.

People who live in areas with erosion problems spend a lot of time and money trying to preserve their land. Sometimes they're successful in slowing down erosion, but they never can eliminate erosion and the danger of mass movement. Eventually, gravity wins. Sediment moves from place to place, constantly reducing elevation and changing the shape of the land.

Section ① Assessment

1. Define the term *erosion* and name the forces that cause it.
2. Explain how deposition changes the surface of Earth.
3. What characteristics do all types of mass movements have in common?
4. Describe ways to help slow erosion on steep slopes.
5. **Think Critically** When people build houses and roads, they often pile up dirt or cut into the sides of hills. Predict how this might affect sediment on a slope. Explain how to control the effects of such activities.

Skill Builder Activities

6. **Comparing and Contrasting** Compare and contrast rockfalls and rock slides. **For more help, refer to the** Science Skill Handbook.
7. **Using an Electronic Spreadsheet** Pretend that you live along a shore where the water is 500 m from your front door. Each year, slumping causes about 1.5 m of land to cave into the water. Design a spreadsheet that will predict how much property will be left each year for ten years. Type a formula that will compute the amount of land left the second year. **For more help, refer to the** Technology Skill Handbook.

Glaciers

How Glaciers Form and Move

If you've ever gone sledding, snowboarding, or skiing, you might have noticed that after awhile, the snow starts to pack down under your weight. A snowy hillside can become icy if it is well traveled. In much the same way, glaciers form in regions where snow accumulates. Some areas of the world, as shown in **Figure 7,** are so cold that snow remains on the ground year-round. When snow doesn't melt, it piles up. As it accumulates slowly, the increasing weight of the snow becomes great enough to compress the lower layers into ice. Eventually, there can be enough pressure on the ice so that it becomes plasticlike. The mass slowly begins to flow in a thick, plasticlike lower layer, and ice moves away from its source. A large mass of ice and snow moving on land under its own weight is a **glacier.**

Ice Eroding Rock

Glaciers are agents of erosion. As glaciers pass over land, they erode it, changing features on the surface. Glaciers then carry eroded material along and deposit it somewhere else. Glacial erosion and deposition change large areas of Earth's surface. How is it possible that something as fragile as snow or ice can push aside trees, drag rocks along, and change the surface of Earth?

Figure 7
The white regions on this map show areas that are glaciated today.

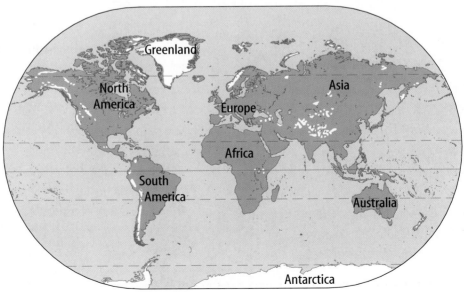

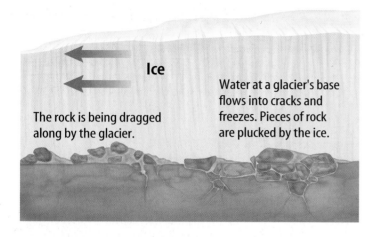

Ice

The rock is being dragged along by the glacier.

Water at a glacier's base flows into cracks and freezes. Pieces of rock are plucked by the ice.

Plucking Glaciers weather and erode solid rock. When glacial ice melts, water flows into cracks in rocks. Later, the water refreezes in these cracks, expands, and fractures the rock. Pieces of rock then are lifted out by the ice, as shown in **Figure 8.** This process, called **plucking,** results in boulders, gravel, and sand being added to the bottom and sides of a glacier.

☑ **Reading Check**   *What is plucking?*

Figure 8
Plucking is a process that occurs when a moving glacier picks up loosened rock particles.

Transporting and Scouring As it moves forward over land, a glacier can transport huge volumes of sediment and rock. Plucked rock fragments and sand at its base scour and scrape the soil and bedrock like sandpaper against wood, eroding the ground below even more. When bedrock is gouged deeply by rock fragments being dragged along, marks such as those in **Figure 9** are left behind. These marks, called grooves, are deep, long, parallel scars on rocks. Shallower marks are called striations (stri AY shuns). Grooves and striations indicate the direction in which the glacier moved.

Ice Depositing Sediment

When glaciers begin to melt, they are unable to carry much sediment. The sediment drops, or is deposited, on the land. When a glacier melts and begins to shrink back, it is said to retreat. As it retreats, a jumble of boulders, sand, clay, and silt is left behind. This mixture of different-sized sediments is called

Figure 9
When glaciers melt, striations or grooves can be found on the rocks beneath. These glacial grooves on Kelley's Island, Ohio, give evidence of past glacial erosion and movement.

till. Till deposits can cover huge areas of land. Thousands of years ago, huge ice sheets in the northern United States left enough till behind to fill valleys completely and make these areas appear flat. Till areas include the wide swath of what are now wheat farms running northwestward from Iowa to northern Montana. Some farmland in parts of Ohio, Indiana, and Illinois and the rocky pastures of New England are also regions that contain till deposits.

Moraine Deposits

Till also is deposited at the end of a glacier when it is not moving forward. Unlike the till that is left behind as a sheet of sediment over the land, this type of deposit doesn't cover such a wide area. Rocks and soil are moved to the end of the glacier, much like items on a grocery store conveyor belt. Because of this, a big ridge of material piles up that looks as though it has been pushed along by a bulldozer. Such a ridge is called a **moraine.** Moraines also are deposited along the sides of a glacier, as shown in **Figure 10.**

Outwash Deposits

When glacial ice starts to melt, the meltwater can deposit sediment that is different from till. Material deposited by the meltwater from a glacier, most often beyond the end of the glacier, is called **outwash.** Meltwater carries sediments and deposits them in layers. Heavier sediments drop first, so bigger pieces of rock are deposited closer to the glacier. The outwash from a glacier also can form into a fan-shaped deposit when the stream of meltwater deposits sand and gravel in front of the glacier.

✔ Reading Check *What is outwash?*

Eskers

Another type of outwash deposit looks like a long, winding ridge. This deposit forms in a melting glacier when meltwater forms a river within the ice, as shown in **Figure 11A.** This river carries sand and gravel and deposits them within its channel. When the glacier melts, a winding ridge of sand and gravel, called an esker (ES kur), is left behind. An esker is shown in **Figure 11B.**

Figure 10
Moraines are forming along the sides of this glacier. Unlike the moraines that form at the ends of glaciers, these moraines form as rock and sediment fall from nearby slopes.

Figure 11
Eskers are glacial deposits formed by meltwater.

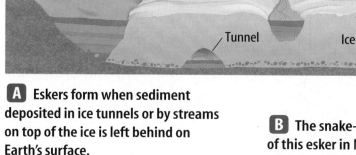

Meltwater stream

Ice

Tunnel

Ice

A Eskers form when sediment deposited in ice tunnels or by streams on top of the ice is left behind on Earth's surface.

B The snake-like shape of this esker in North Dakota is characteristic of this type of glacial deposit.

Continental Glaciers

The two types of glaciers are continental glaciers and valley glaciers. Today, continental glaciers like the one in **Figure 12** cover only ten percent of Earth, mostly near the poles in Antarctica and Greenland. These continental glaciers are huge masses of ice and snow. Continental glaciers are thicker than some mountain ranges. Glaciers make it impossible to see most of the land features in Antarctica and Greenland.

✓ Reading Check *In what regions on Earth would you expect to find continental glaciers?*

Figure 12
Continental glaciers and valley glaciers are agents of erosion and deposition. This continental glacier covers a large area in Antarctica.

Climate Changes In the past, continental glaciers covered as much as 28 percent of Earth. **Figure 13** shows how much of North America was covered by glaciers during the most recent ice advance. These periods of widespread glaciation are known as ice ages. Extensive glaciers have covered large portions of Earth many times over the last 2 million to 3 million years. During this time, glaciers advanced and retreated many times over much of North America. The average air temperature on Earth was about 5°C lower during these ice ages than it is today. The last major advance of ice reached its maximum extent about 18,000 years ago. After this last advance of glaciers, the ends of the ice sheets began to recede, or move back, by melting.

Figure 13
This map shows how much of North America was covered by continental glaciers about 18,000 years ago. *Was your location covered? If so, what evidence of glaciers does your area show?*

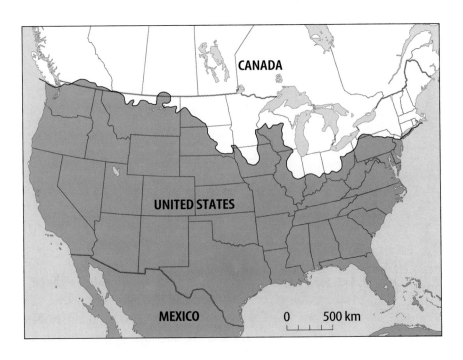

Valley Glaciers

Valley glaciers occur even in today's warmer global climate. In the high mountains where the average temperature is low enough to prevent snow from melting during the summer, valley glaciers grow and creep along. **Figure 14** shows valley glaciers in Africa.

Evidence of Valley Glaciers If you visit the mountains, you can tell whether valley glaciers ever existed there. You might look for striations, then search for evidence of plucking. Glacial plucking often occurs near the top of a mountain where a glacier is mainly in contact with solid rock. Valley glaciers erode bowl-shaped basins, called cirques (SURKS), into the sides of mountains. If two valley glaciers side by side erode a mountain, a long ridge called an arête (ah RAYT) forms between them. If valley glaciers erode a mountain from several directions, a sharpened peak called a horn might form. **Figure 15A** shows some features formed by valley glaciers.

Valley glaciers flow down mountain slopes and along valleys, eroding as they go. Valleys that have been eroded by glaciers have a different shape from those eroded by streams. Stream-eroded valleys are normally V-shaped. Glacially eroded valleys are U-shaped because a glacier plucks and scrapes soil and rock from the sides as well as from the bottom. A U-shaped valley is illustrated in **Figure 15B**.

Figure 14
Valley glaciers, like these on Mount Kilimanjaro in north Tanzania, Africa, form between mountain peaks that lie above the snow line, where snow lasts all year.

Figure 15
Valley glaciers transform the mountains over which they pass.

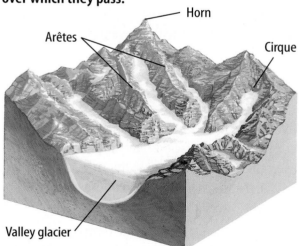

A Bowl-shaped basins called cirques form by erosion at the start of a valley glacier. Arêtes form where two adjacent valley glaciers meet and erode a long, sharp ridge. Horns are sharpened peaks formed by glacial action in three or more cirques.

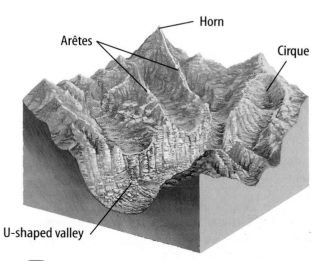

B U-shaped valleys result when valley glaciers move through regions once occupied by streams.

Figure 16
Sand and gravel deposits left by glaciers are important starting materials for the construction of roadways and buildings.

SCIENCE *Online*

Collect Data Visit the Glencoe Science Web site at **science.glencoe.com** for data about uses of glacial deposits. Communicate to your class what you learn.

Importance of Glaciers

Glaciers have had a profound effect on Earth's surface. They have eroded mountaintops and transformed valleys. Vast areas of the continents have sediments that were deposited by great ice sheets. Today, glaciers in polar regions and in mountains continue to change the surface features of Earth.

In addition to changing the appearance of Earth's surface, glaciers leave behind sediments that are economically important, as illustrated in **Figure 16.** The sand and gravel deposits from glacial outwash and eskers are important resources. These deposits are excellent starting materials for the construction of roads and buildings.

Section ② Assessment

1. How do glaciers move?
2. Describe two ways in which a glacier can cause erosion.
3. Till and outwash are glacial deposits. Explain how till and outwash are different.
4. How do moraines form? What are moraines made of?
5. **Think Critically** Rivers and lakes that receive water from glacial meltwater often appear milky blue in color. What do you think might cause the milky appearance of these waters?

Skill Builder Activities

6. **Recognizing Cause and Effect** Since 1900, the Alps have lost 50 percent of their ice caps, and New Zealand's glaciers have shrunk by 26 percent. Describe what you think the effects of glacial melting have been. **For more help, refer to the** Science Skill Handbook.

7. **Researching Information** The term *erratic* comes from the Latin word *errare*, meaning "to wander." Research how glaciers move erratics. Write a poem about the "life" of an erratic. **For more help, refer to the** Science Skill Handbook.

Activity

Glacial Grooving

Throughout the world's mountainous regions, 200,000 valley glaciers are moving in response to gravity.

What You'll Investigate
How is the land affected when a valley glacier moves downslope?

Materials
sand
large plastic or metal tray
stream table
ice block
books (2 or 3)

wood block
metric ruler
overhead light source with reflector
Alternate materials

Safety Precautions

WARNINGS: *Do not pour sand down the drain. Make sure source is plugged into a GFI electrical outlet. Do not touch light source—it may be hot.*

Goals
■ **Compare** stream and glacial valleys.

Procedure

1. Set up the large tray of sand as shown. Place books under one end of the tray to make a slope.
2. Cut a narrow channel, like a river, through the sand. Measure and record its width and depth. Draw a sketch that includes these measurements.
3. Position the overhead light source to shine on the channel as shown.
4. Force the ice block into the channel at the upper end of the tray.

5. Gently push the ice along the channel until it's halfway between the top and bottom of the tray, and directly under the light.
6. Turn on the light and allow the ice to melt. Record what happens.
7. **Record** the width and depth of the ice channel. Make a scale drawing.

Conclude and Apply

1. **Explain** how you can determine the direction that a glacier traveled from the location of deposits.
2. **Explain** how you can determine the direction of glacial movement from sediments deposited by meltwater.
3. How do valley glaciers affect the surface over which they move?

Glacier Data			
Sample Data	**Width (cm)**	**Depth (cm)**	**Observations**
Original Channel	1–2	3	Stream channel looked V-shaped
Glacier Channel			
Meltwater Channel			

③ Wind

As You Read

What **You'll Learn**
- **Explain** how wind causes deflation and abrasion.
- **Recognize** how loess and dunes form.

Vocabulary
deflation   loess
abrasion   dune

Why **It's Important**
Wind erosion and deposition change landscapes, especially in dry climates.

Figure 17
The odd shape of this boulder was produced by wind abrasion.

Wind Erosion

When air moves, it picks up loose material and transports it to other places. Air differs from other erosional forces because it usually cannot pick up heavy sediments. Unlike rivers that move in confined places like channels and valleys, wind carries and deposits sediments over large areas. For example, wind is capable of picking up and carrying dust particles from fields or volcanic ash high into the atmosphere and depositing them thousands of kilometers away.

Deflation Wind erodes Earth's surface by deflation (dih FLAY shun) and abrasion (uh BRAY zhun). When wind erodes by **deflation,** it blows across loose sediment, removing small particles such as silt and sand. The heavier, coarser material is left behind.

Abrasion When windblown sediment strikes rock, the surface of the rock gets scraped and worn away by a process called abrasion. **Abrasion,** shown in **Figure 17,** is similar to sandblasting. Workers use machines that spray a mixture of sand and water under high pressure against a building. The friction wears away dirt from stone, concrete, or brick walls. It also polishes the walls of buildings by breaking away small pieces and leaving an even, smooth finish. Wind acts like a sandblasting machine, bouncing and blowing sand grains along. These sand grains strike against rocks and break off small fragments. The rocks become pitted and are worn down gradually.

✓ **Reading Check** *How is wind abrasion similar to sandblasting?*

Deflation and abrasion happen to all land surfaces but occur mostly in deserts, beaches, and plowed fields. These areas have fewer plants to hold the sediments in place. When winds blow over them, they can be eroded rapidly.

Sandstorms Even when the wind blows strongly, it seldom carries sand grains higher than 0.5 m from the ground. However, sandstorms do occur. When the wind blows forcefully in the sandy parts of deserts, sand grains bounce along and hit other sand grains, causing more and more grains to rise into the air. These windblown sand grains form a low cloud just above the ground. Most sandstorms occur in deserts, but they can occur in other arid regions, as shown in **Figure 18.**

Dust Storms When soil is moist, it stays packed on the ground, but when it dries out, it can be eroded by wind. Soil is composed largely of silt- and clay-sized particles. Because these small particles weigh less than sand-sized particles of the same material, wind can move them high into the air.

Silt and clay particles are small and stick together. A faster wind is needed to lift these fine particles of soil than is needed to lift grains of sand. However, after they are airborne, the wind can carry them long distances. Where the land is dry, dust storms can cover hundreds of kilometers. These storms blow topsoil from open fields, overgrazed areas, and places where vegetation has disappeared. In the 1930s, silt and dust picked up in Kansas fell in New England and in the North Atlantic Ocean. Dust blown from the Sahara has been traced as far away as the West Indies—a distance of at least 6,000 km.

Figure 18
Sandstorms can obscure visibility.

Problem-Solving Activity

What factors affect wind erosion?

Many factors compound the effects of wind erosion. But can anything be done to minimize erosion?

Identifying the Problem
Wind velocity and duration, the size of sediment particles, the size of the area subjected to the wind, and the amount of vegetation present all affect how much soil is eroded by wind. The table shows different combinations of these factors. It also includes an erosion rating that depends upon what factors pertain to an area.

Factors That Affect Wind Erosion					
Factor	**Descriptions**				
Wind velocity	high	high	low	low	low
Duration of wind	long	long	short	long	long
Particle size	coarse	medium	coarse	coarse	medium
Surface area	large	large	small	small	large
Amount of vegetation	high	low	high	high	high
Erosion rating	some	a lot	a little	some	?

Solving the Problem
1. Looking at the table, can you figure out which factors increase and which factors decrease the amount of erosion?
2. From what you've discovered, can you estimate the missing erosion rating?

Observing How Soil Is Held in Place

Procedure
1. Obtain a piece of **sod** (a chunk of soil about 5 cm thick with grass growing from it).
2. Carefully remove soil from the sod roots by hand. Examine the roots with a **magnifying glass or hand lens.**
3. Wash hands thoroughly with soap and water.

Analysis
1. Draw several of these roots in your **Science Journal.**
2. What characteristics of grass roots help hold soil in place and thus reduce erosion?

Reducing Wind Erosion

Life Science INTEGRATION

As you've learned, wind erosion is most common where there are no plants to protect the soil. Therefore, one of the best ways to slow or stop wind erosion is to plant vegetation. This practice helps conserve soil and protect valuable farmland.

Windbreaks People in many countries plant vegetation to reduce wind erosion. For centuries, farmers have planted trees along their fields to act as windbreaks that prevent soil erosion. As the wind hits the trees, its energy of motion is reduced. It no longer is able to lift particles.

In one study, a thin belt of cottonwood trees reduced the effect of a 25-km/h wind to about 66 percent of its normal speed, or to about 16.5 km/h. Tree belts also trap snow and hold it on land. This increases the moisture level of the soil, which helps prevent further erosion.

Roots Along many seacoasts and deserts, vegetation is planted to reduce erosion. Plants with fibrous root systems, such as grasses, work best at stopping wind erosion. Grass roots are shallow and slender with many fibers. They twist and turn between particles in the soil and hold it in place.

Planting vegetation is a good way to reduce the effects of deflation and abrasion. Even so, if the wind is strong and the soil is dry, nothing can stop erosion completely. **Figure 19** shows a project designed to decrease wind erosion.

Figure 19
Rows of grasses and rocks were installed on these dunes in Qinghai, China to reduce wind erosion.

Deposition by Wind

Sediments blown away by wind eventually are deposited. Over time, these windblown deposits develop into landforms, such as accumulations of loess and dunes.

Loess Some examples of large deposits of windblown sediments are found near the Mississippi and Missouri Rivers. These wind deposits of fine-grained sediments known as **loess** (LOOS) are shown in **Figure 20.** Strong winds that blew across glacial outwash areas carried the sediments and deposited them. The sediments settled on hilltops and in valleys. Once there, the particles packed together, creating a thick, unlayered, yellowish-brown-colored deposit. Loess is as fine as talcum powder. Many farmlands of the midwestern United States have fertile soils that developed from loess deposits.

Dunes Do you notice what happens when wind blows sediments against an obstacle such as a rock or a clump of vegetation? The wind sweeps around or over the obstacle. Like a river, air drops sediment when its energy decreases. Sediment starts to build up behind the obstacle. The sediment itself then becomes an obstacle, trapping even more material. If the wind blows long enough, the mound will become a dune, as shown in **Figure 21.** A **dune** (DOON) is a mound of sediments drifted by the wind.

Reading Check *What is a dune?*

Dunes are common in desert regions. You also can see sand dunes along the shores of oceans, seas, or lakes. If dry sediments exist in an area where prevailing winds or sea breezes blow daily, dunes build up. Sand or other sediment will continue to build up and form a dune until the sand runs out or the obstruction is removed. Some desert sand dunes can grow to 100 m high, but most are much shorter.

Moving Dunes A sand dune has two sides. The side facing the wind has a gentler slope. The side away from the wind is steeper. Examining the shape of a dune tells you the direction from which the wind usually blows.

Unless sand dunes are planted with grasses, most dunes move, or migrate away from the direction of the wind. This process is shown in **Figure 22.** Some dunes are known as traveling dunes because they move rapidly across desert areas. As they lose sand on one side, they build it up on the other.

Figure 20
This sediment deposit is composed partially of wind-blown loess.

Figure 21
Loose sediment of any type can form a dune if enough of it is present and an obstacle lies in the path of the wind.

Figure 22

Sand blown loose from dry desert soil often builds up into dunes. A dune may begin to form when windblown sand is deposited in the sheltered area behind an obstacle, such as a rock outcrop. The sand pile grows as more grains accumulate. As shown in the diagram at right, dunes are mobile, gradually moved along by wind.

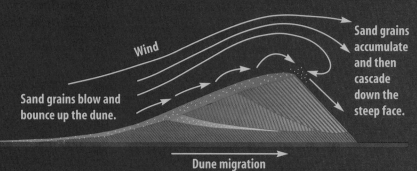

Wind

Sand grains accumulate and then cascade down the steep face.

Sand grains blow and bounce up the dune.

Dune migration

▲ A dune migrates as sand blows up its sloping side and then cascades down the steeper side. Gradually, a dune moves forward—in the same direction that the wind is blowing—as sand, lost from one side, piles up on the other side.

▲ Dunes are made of sediments eroded from local materials. Although many dunes are composed of quartz and feldspar, the brilliant white dunes in White Sands National Park, New Mexico, are made of gypsum.

▲ Deserts may expand when humans move into the transition zone between habitable land and desert. Here, villagers in Mauritania in northwestern Africa shovel the sand that encroaches on their schoolhouse daily.

◀ The dunes at left are coastal dunes from the Laguna Madre region of South Texas on the Gulf of Mexico. Note the vegetation in the photo, which has served as an obstacle to trap sand.

Dune Shape The shape of a dune depends on the amount of sand or other sediment available, the wind speed and direction, and the amount of vegetation present. One common dune shape is a crescent-shaped dune known as a barchan (BAR kun) dune. The open side of a barchan dune faces the direction that the wind is blowing. When viewed from above, the points of the crescent are directed downwind. This type of dune forms on hard surfaces where the sand supply is limited.

Another common type of dune, called a transverse dune, forms where sand is abundant. Transverse dunes are so named because the long directions of these dunes are perpendicular to the general wind direction. In regions where the wind direction changes, star dunes, shown in **Figure 23,** form pointed structures. Other dune forms also exist, some of which show a combination of features.

Shifting Sediments When dunes and loess form, the landscape changes. Wind, like gravity, running water, and glaciers, shapes the land. New landforms created by these agents of erosion are themselves being eroded. Erosion and deposition are part of a cycle of change that constantly shapes and reshapes the land.

Figure 23
Star dunes form in areas where the wind blows from several different directions.

Section 3 Assessment

1. Compare and contrast abrasion and deflation. Describe how they affect the surface of Earth.

2. Explain the differences between dust storms and sandstorms. Describe how the energy of motion affects the deposition of sand and dust by these storms.

3. Explain what loess is made of and how a loess deposit forms.

4. Why do farmers plant trees along the edges of their fields?

5. **Think Critically** You notice that sand is piling up behind a fence outside your apartment building. Explain why this occurs.

Skill Builder Activities

6. **Predicting** Predict the sequence of the following events about dune formation. **For more help,** refer to the Science Skill Handbook.
   a. Grains collect to form a mound.
   b. Wind blows sand grains around an obstacle.
   c. Wind blows over an area and causes deflation.
   d. Vegetation grows on the dune.

7. **Solving One-Step Equations** Between 1972 and 1992, the Sahara Desert increased by nearly 700 km^2 in Mali and the Sudan. Calculate the average number of square kilometers the desert increased each year between 1972 and 1992. **For more help,** refer to the Math Skill Handbook.

Activity

Design Your Own Experiment

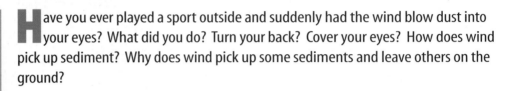

Blowing in the Wind

Have you ever played a sport outside and suddenly had the wind blow dust into your eyes? What did you do? Turn your back? Cover your eyes? How does wind pick up sediment? Why does wind pick up some sediments and leave others on the ground?

Recognize the Problem

What factors affect wind erosion?

Form a Hypothesis

How does moisture in sediment affect the ability of wind to erode sediments? Does the speed of the wind limit the size of sediments it can transport? Form a hypothesis about how sediment moisture affects wind erosion. Form another hypothesis about how wind speed affects the size of the sediment the wind can transport.

Goals

- **Observe** the effects of soil moisture and wind speed on wind erosion.
- **Design** and carry out experiments that test the effects of soil moisture and wind speed on wind erosion.

Possible Materials

flat pans (4)
fine sand (400 mL)
gravel (400 mL)
hair dryer
sprinkling can
water
28-cm × 35-cm cardboard sheets (4)
tape
mixing bowl
metric ruler
wind speed indicator

Safety Precautions

Wear your safety goggles at all times when using the hair dryer on sediments. Make sure the dryer is plugged into a GFI electrical outlet.

Test Your Hypothesis

Plan

1. As a group, agree upon and write your hypothesis statements.

2. **List** the steps needed to test your first hypothesis. Plan specific steps and vary only one factor at a time. Then, list the steps needed to test your second hypothesis. Test only one factor at a time.

3. Mix the sediments in the pans. Plan how you will fold cardboard sheets and attach them to the pans to keep sediments contained.

4. **Design** data tables in your Science Journal. Use them as your group collects data.

5. **Identify** all constants, variables, and controls of the experiment. One example of a control is a pan of sediment not subjected to any wind.

Do

1. Make sure your teacher approves your plan before you start.

2. Carry out the experiments as planned.

3. While doing the experiments, write any observations that you or other members of your group make. Summarize your data in the data tables you designed in your Science Journal.

Sediment Movement		
Sediment	**Wind Speed**	**Sediment Moved**
Fine Sand (dry)	low	
	high	
Fine Sand (wet)	low	
	high	
Gravel (dry)	low	
	high	
Gravel (wet)	low	
	high	
Fine Sand and Gravel (dry)	low	
	high	
Fine Sand and Gravel (wet)	low	
	high	

Analyze Your Data

1. **Compare** your results with those of other groups. Explain what might have caused any differences among the groups.

2. **Explain** the relationship that exists between the speed of the wind and the size of the sediments it transports.

Draw Conclusions

1. How does energy of motion of the wind influence sediment transport? What is the general relationship between wind speed and erosion?

2. **Explain** the relationship between the sediment moisture and the amount of sediment moved by the wind.

*C*ommunicating Your Data

Design a table on poster board or construction paper that summarizes the results of your experiment. Use your table to explain your interpretations to others in the class. **For more help, refer to the** Science Skill Handbook.

Losing Against Erosion

Did you know...

...Some sand dunes migrate as much as 30 m per year. In a coastal region of France, traveling dunes have buried farms and villages. The dunes were halted by anti-erosion practices, such as planting grass in the sand and growing a barrier of trees between the dunes and farmland.

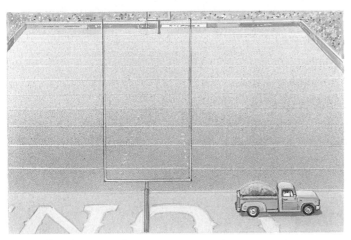

...Wind erosion never stops. Wind erosion robs rural areas in the United States of as much as 5,590 kg of soil per hectare per year. One hectare is equivalent to 10,000 m^2. This amounts to removing one small truckload of soil from an area about the size of a football field each year.

...Erosion can move mountains. Between 1982 and 1983, El Niño caused rockfalls along Highway 1 in California. The falling rock from the mountain alongside the road forced a section of road to close. The rocks stranded the residents of Half Moon Bay until the road was cleared.

...In 1959, an earthquake triggered a mass movement in Madison River Canyon, Montana. About 21 million km³ of rock and soil slid down the canyon at an estimated 160 km/h. This type of mass movement of earth is called a rock slide.

Homes in Danger from Shoreline Erosion by the Year 2060

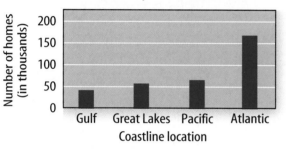

...Glaciers, one of nature's most powerful erosional forces, can move more than 30 m per day. In one week, a fast-moving glacier can travel the length of almost two football fields. Glaciers such as these are unusual—most move less than 10 cm per day.

Do the Math

1. If you could devise a method to slow soil loss due to wind erosion by 25 percent, how much soil would be saved per hectare each year in the United States?
2. If erosion destroys 300 m of shoreline every 100 years, how long would it take to destroy 1 km of shoreline?
3. If a sand dune is traveling at 30 m per year, how many meters does it travel in one month?

Go Further

Go to **science.glencoe.com** to learn about landslides. When is a landslide called a mudflow? In which U.S. states are mudflows most likely to occur?

Reviewing Main Ideas

Section 1 Erosion by Gravity

1. Erosion is the process that picks up and transports sediment.

2. Deposition occurs when an agent of erosion loses its energy and can no longer carry its load of sediment.

3. Slump, creep, rock slides, and mudflows are all mass movements caused by gravity. *Explain the differences between rock slides and mudflows.*

Section 2 Glaciers

1. Glaciers are powerful agents of erosion. As water freezes and thaws in cracks, it breaks off pieces of surrounding rock. These pieces then are incorporated into glacial ice by plucking.

2. As sediment embedded in the base of a glacier moves across the land, grooves and striations form. Glaciers deposit two kinds of material—till and outwash. *Which of these two sediments describes a jumbled pile of rocks and smaller sediment?*

Section 3 Wind

1. Deflation occurs when wind erodes only fine-grained sediments, leaving coarse sediments behind. *Why doesn't wind normally pick up large-grained sediments?*

2. The pitting and polishing of rocks and grains of sediment by windblown sediment is called abrasion.

3. Wind deposits include loess and dunes. Loess consists of fine-grained particles such as silt and clay. Dunes form when windblown sediments accumulate behind an obstacle.

4. Dunes are common landforms in desert regions. The shapes and orientations of dunes can provide clues about the prevailing wind directions in an area.

FOLDABLES
Reading & Study Skills

After You Read

Use the Foldable you made at the beginning of the chapter to compare and contrast the forces causing erosion and deposition.

Visualizing Main Ideas

Complete the following concept map on erosional forces. Use the following terms and phrases: striations, leaning trees and structures, curved scar on slope, deflation, *and* mudflows.

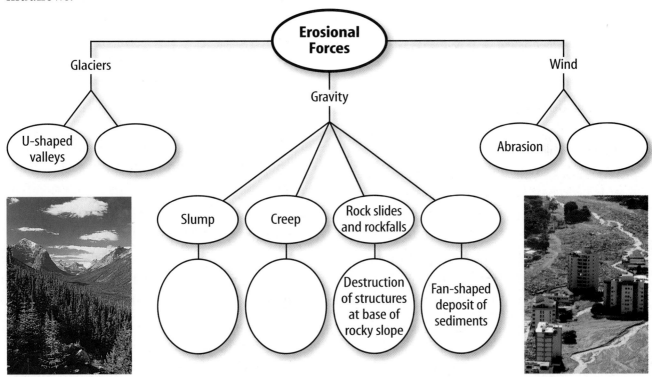

Vocabulary Review

Vocabulary Words

a. abrasion
b. creep
c. deflation
d. deposition
e. dune
f. erosion
g. glacier
h. loess
i. mass movement
j. moraine
k. outwash
l. plucking
m. slump
n. till

THE PRINCETON REVIEW

Study Tip

Make sure to read your class notes after each lesson. Reading them will help you better understand what you've learned, as well as prepare you for the next day's lesson.

Using Vocabulary

Each phrase below describes a vocabulary word from the list. In your Science Journal, write the term that matches each description.

1. loess, dunes, and moraines are examples
2. slowest mass movement
3. ice picking up pieces of rock
4. much like sandblasting
5. gravity transport of material downslope
6. sand and gravel deposited by meltwater
7. glacial deposit composed of sediment with many sizes and shapes

Chapter 8 Assessment

Checking Concepts

Choose the word or phrase that best answers the question.

1. Which of the following is suggested by leaning trees on a hillside?
   A) abrasion
   B) creep
   C) slump
   D) mudflow

2. The best plants for reducing wind erosion have what type of root system?
   A) taproot
   B) striated
   C) fibrous
   D) sheet

3. What does a valley glacier create at the point where it starts?
   A) esker
   B) moraine
   C) till
   D) cirque

4. Which is caused by glacial erosion?
   A) eskers
   B) arêtes
   C) moraines
   D) warmer climate

5. What term describes a mass of snow and ice in motion?
   A) loess deposit
   B) glacier
   C) outwash
   D) abrasion

6. What shape do glacier-created valleys have?
   A) V-shaped
   B) L-shaped
   C) U-shaped
   D) S-shaped

7. Which term is an example of a feature created by deposition?
   A) cirque
   B) abrasion
   C) striation
   D) dune

8. Which characteristic is common to all agents of erosion?
   A) They carry sediments when they have enough energy of motion.
   B) They are most likely to erode when sediments are moist.
   C) They create deposits called dunes.
   D) They erode large sediments before they erode small ones.

9. What type of wind erosion leaves pebbles and boulders behind?
   A) deflation
   B) loess
   C) abrasion
   D) sandblasting

10. What is a ridge formed by deposition of till called?
    A) striation
    B) esker
    C) cirque
    D) moraine

Thinking Critically

11. How can striations give information about the direction that a glacier moved?

12. How effective would a retaining wall made of fine wire mesh be against erosion?

13. Sand dunes often migrate. What can be done to prevent the migration of beach dunes?

14. A researcher finds evidence of movement of ice within a glacier. Explain how this could occur.

15. The end of a valley glacier is at a lower elevation than its point of origin is. How does this help explain melting at its end while snow and ice still are accumulating where it originated?

Developing Skills

16. **Making and Using Tables** Make a table to contrast continental and valley glaciers.

17. **Testing a Hypothesis** Explain how to test the effect of glacial thickness on a glacier's ability to erode.

18. **Forming Hypotheses** Hypothesize why silt in loess deposits is transported farther than sand in dune deposits.

19. Concept Mapping Copy and complete the events chain concept map below to show how a sand dune forms. Use the terms and phrases: *sand accumulates, dune, dry sand,* and *obstruction traps.*

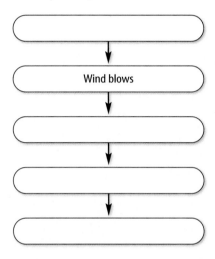

Wind blows

20. Classifying Classify the following as erosional or depositional features: loess, cirque, U-shaped valley, sand dune, abraded rock, striation, and moraine.

Performance Assessment

21. Poster Make a poster with magazine photos showing glacial features in North America. Add a map to locate each feature.

22. Design an Experiment Design an experiment to see how the amount of moisture added to sediments affects mass movement. Keep all variables constant except the amount of moisture in the sediment. Try your experiment.

TECHNOLOGY

Go to the Glencoe Science Web site at **science.glencoe.com** or use the **Glencoe Science CD-ROM** for additional chapter assessment.

THE PRINCETON REVIEW
Test Practice

Ms. Lee was reviewing examples of erosion caused by gravity for her science class.

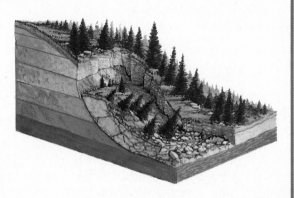

Examine the diagram above and answer the following questions.

1. This diagram shows the shape, in cross section, of one example of erosion by gravity. Which of the following choices best matches this type of erosion?
A) slump   **C)** rock slide
B) creep   **D)** mudflow

2. Select the most likely cause of this type of mass wasting from the choices below.
F) Pieces of rock became loosened and fell freely through the air.
G) Heavy rains caused a fluid mass of mud to form and flow rapidly downhill.
H) After a heavy rain, the underlying sediment was weakened and could no longer support the sediment and rock above it.
J) Sediments slowly shifted their positions downhill.

Water Erosion and Deposition

Bryce Canyon National Park in Utah is home to the Hoodoo Formations. What carved these canyons and helped create these magnificent sculptures? They were made by one of the most powerful forces on Earth—water. Water, in the form of light rain or a great river, causes erosion. In this chapter, you will learn how rainwater runoff forms streams. You also will learn how water carries sediment and deposits it far from its source.

What do you think?

Science Journal Look at the picture below with a classmate. Discuss what you think this might be. Here's a hint: *It's something out of this world.* Write your answer in your Science Journal.

EXPLORE ACTIVITY

Moving water has great energy. Sometimes rainwater falls softly and soaks slowly into soil. Other times it rushes down a slope with tremendous force and carries away valuable topsoil. What determines whether rain soaks into the ground or runs off and wears away the surface?

Model how erosion works

1. Place an aluminum pie pan on your desktop.
2. Put a pile of dry soil about 7 cm high into the pan.
3. Slowly drip water from a dropper onto the pile and observe what happens next.
4. Drip the water faster and continue to observe what happens.
5. Repeat steps 1 through 4, but this time change the slope of the hill. Start again with dry soil.

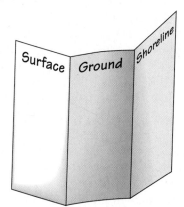

Observe
Record in your Science Journal what effect the water had on the different slopes.

Before You Read

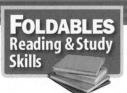
FOLDABLES
Reading & Study Skills

Making a Main Ideas Study Fold A main idea consists of the major concepts or topics talked about in a chapter. Before you read the chapter, make the following Foldable to help you identify the main ideas of this chapter.

1. Place a sheet of paper in front of you so the long side is at the top. Fold the left side and right side of the paper in to divide the paper into equal thirds.
2. Open the paper and label the three columns *Surface, Ground,* and *Shoreline,* as shown.
3. As you read the chapter, list characteristics of each type of water. Then, across the bottom of each fold, draw a picture of surface water, groundwater, and a shoreline.

Surface Water

As You Read

What You'll Learn

- **Identify** the causes of runoff.
- **Compare** rill, gully, sheet, and stream erosion.
- **Identify** three different stages of stream development.
- **Explain** how alluvial fans and deltas form.

Vocabulary

runoff
channel
sheet erosion

drainage basin
meander

Why It's Important
Runoff and streams shape Earth's surface.

Runoff

Picture this. You pour a glass of milk, and it overflows, spilling onto the table. You grab a towel to clean up the mess, but the milk is already running through a crack in the table, over the edge, and onto the floor. This is similar to what happens to rainwater when it falls to Earth. Some rainwater soaks into the ground and some evaporates, turning into a gas. The rainwater that doesn't soak into the ground or evaporate runs over the ground. Eventually, it enters streams, lakes, or the ocean. Water that doesn't soak into the ground or evaporate but instead flows across Earth's surface is called **runoff.** If you've ever spilled milk while pouring it, you've experienced something similar to runoff.

Factors Affecting Runoff What determines whether rain soaks into the ground or runs off? The amount of rain and the length of time it falls are two factors that affect runoff. Light rain falling over several hours probably will have time to soak into the ground. Heavy rain falling in less than an hour or so will run off because it cannot soak in fast enough, or it can't soak in because the ground cannot hold any more water.

Figure 1
A In areas with gentle slopes and vegetation, little runoff and erosion take place. **B** Lack of vegetation has led to severe soil erosion in some areas.

Other Factors Another factor that affects the amount of runoff is the steepness, or slope, of the land. Gravity, the attractive force between all objects, causes water to move down slopes. Water moves rapidly down steep slopes so it has little chance to soak into the ground. Water moves more slowly down gentle slopes and across flat areas. Slower movement allows water more time to soak into the ground.

Vegetation, such as grass and trees, also affects the amount of runoff. Just like milk running off the table, water will run off smooth surfaces that have little or no vegetation. Imagine a tablecloth on the table. What would happen to the milk then? Runoff slows down when it flows around plants. Slower-moving water has a greater chance to sink into the ground. By slowing down runoff, plants and their roots help prevent soil from being carried away. Large amounts of soil may be carried away in areas that lack vegetation, as shown in **Figure 1.**

Effects of Gravity When you lie on the ground and feel as if you are being held in place, you are experiencing the effects of gravity. Gravity is the attracting force all objects have for one another. The greater the mass of an object is, the greater its force of gravity is. Because Earth has a much greater mass than any of the objects on it, Earth's gravitational force pulls objects toward its center. Water runs downhill because of Earth's gravitational pull. When water begins to run down a slope, it picks up speed. As its speed increases, so does its energy. Fast-moving water, shown in **Figure 2,** carries more soil than slow-moving water does.

The force that drives most types of erosion is gravity. Gravity gives water its potential, or stored, energy. When this energy is changed into kinetic energy, or energy of motion, water becomes a powerful force strong enough to move mountains. Find out how water has shaped the region in which you live.

Figure 2
During floods, the high volume of fast-moving water erodes large amounts of soil.

Figure 3
Heavy rains can remove large amounts of sediment, forming deep gullies in the side of a slope.

Figure 4
When water accumulates, it can flow in sheets like the water seen flowing over the hood of this car.

Water Erosion

Suppose you and several friends walk the same way to school each day through a field or an empty lot. You always walk in the same footsteps as you did the day before. After a few weeks, you've worn a path through the field. When water travels down the same slope time after time, it also wears a path. The wearing away of soil and rock is called erosion.

Rill and Gully Erosion You may have noticed a groove or small ditch on the side of a slope that was left behind by running water. This is evidence of rill erosion. Rill erosion begins when a small stream forms during a heavy rain. As this stream flows along, it has enough energy to erode and carry away soil. Water moving down the same path creates a groove, called a **channel**, on the slope where the water eroded the soil. If water frequently flows in the same channel, rill erosion may change over time into another type of erosion called gully erosion.

During gully erosion, a rill channel becomes broader and deeper. **Figure 3** shows gullies that were formed when water carried away large amounts of soil.

Sheet Erosion Water often erodes without being in a channel. Rainwater that begins to run off during a rainstorm often flows as thin, broad sheets before forming rills and streams. For example, when it rains over an area, the rainwater accumulates until it eventually begins moving down a slope as a sheet, like the water flowing off the hood of the car in **Figure 4.** Water also can flow as sheets if it breaks out of its channel.

Floodwaters spilling out of a river can flow as sheets over the surrounding flatlands. Streams flowing out of mountains fan out and may flow as sheets away from the foot of the mountain. **Sheet erosion** occurs when water that is flowing as sheets picks up and carries away sediments.

Stream Erosion Sometimes water continues to flow along a low place it has formed. As the water in a stream moves along, it picks up sediments from the bottom and sides of its channel. By this process, a stream channel becomes deeper and wider.

The sediment that a stream carries is called its load. Water picks up and carries some of the lightweight sediments, called the suspended load. Larger, heavy particles called the bed load just roll along the bottom of the stream channel, as shown in **Figure 5.** Water can even dissolve some rocks and carry them away in solution. The different-sized sediments scrape against the bottom and sides of the channel like a piece of sandpaper. Gradually, these sediments can wear away the rock by a process called abrasion.

Figure 5
This cross section of a stream channel shows the location of the suspended load and the bed load. *How does the stream carry dissolved material?*

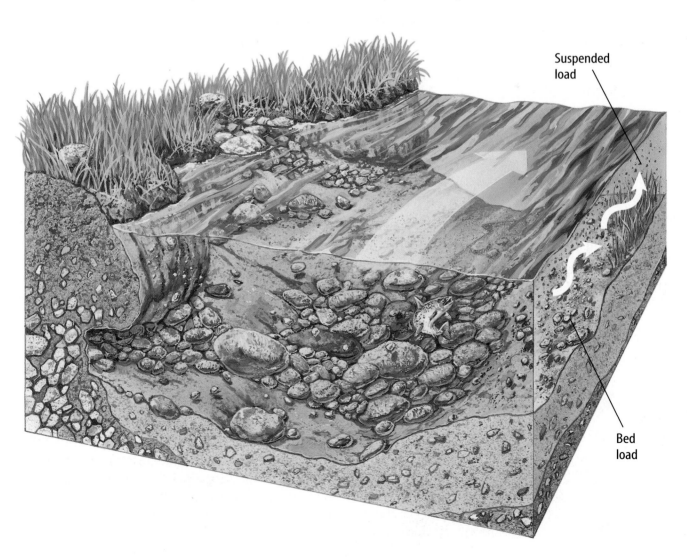

Suspended load

Bed load

River System Development

Have you spent time near a river or stream in your community? Each day, probably millions of liters of water flow through that stream. Where does all the water come from? Where is it flowing to?

River Systems Streams are parts of river systems. The water comes from rills, gullies, and smaller streams located upstream. Just as the tree in **Figure 6A** is a system containing twigs, branches, and a trunk, a river system also has many parts. Runoff enters small streams, which join together to form larger streams. Larger streams come together to form rivers. Rivers grow and carry more water as more streams join.

Drainage Basins A **drainage basin** is the area of land from which a stream or river collects runoff. Compare a drainage basin to a bathtub. Water that collects in a bathtub flows toward one location—the drain. Likewise, all of the water in a river system eventually flows to one location—the main river, or trunk. The largest drainage basin in the United States is the Mississippi River drainage basin shown in **Figure 6B.**

✔ **Reading Check** *What is a drainage basin?*

SCIENCE *Online*

Research Visit the Glencoe Science Web site at **science.glencoe.com** to learn more about drainage basins in your region. Make a poster that summarizes your findings, and share it with your class.

Figure 6
River systems can be compared with the structure of a tree.

A The system of twigs, branches, and the trunk that make up a tree is similar to the system of streams and rivers that make up a river system.

Columbia River Drainage Basin

CANADA

Mississippi River Drainage Basin

Colorado River Drainage Basin

MEXICO

B A large number of the streams and rivers in the United States are part of the Mississippi River drainage basin, or watershed. *What river represents the trunk of this system?*

Stages of Stream Development

Streams come in a variety of forms. Some are narrow and swift moving, and others are wide and slow moving. Streams differ because they are in different stages of development. These stages depend on the slope of the ground over which the stream flows. Streams are classified as young, mature, or old. **Figure 8** shows how the stages come together to form a river system.

The names of the stages of development aren't always related to the actual age of a river. The New River in West Virginia is one of the oldest rivers in North America. However, it has a steep valley and flows swiftly. As a result, it is classified as a young stream.

Young Streams A stream that flows swiftly through a steep valley is a young stream. A young stream may have white-water rapids and waterfalls. Water flowing through a steep channel with a rough bottom has a high level of energy and erodes the stream bottom faster than its sides.

Figure 7
A meander is a broad bend in a river or stream. As time passes, erosion of the outer bank increases the bend.

Mature Streams The next stage in the development of a stream is the mature stage. A mature stream flows more smoothly through its valley. Over time, most of the rocks in the streambed that cause waterfalls and rapids are eroded by running water and the sediments it carries.

Erosion is no longer concentrated on the bottom in a mature stream. A mature stream starts to erode more along its sides, and curves develop. These curves form because the speed of the water changes throughout the width of the channel.

Water in a shallow area of a stream moves slower because it drags along the bottom. In the deeper part of the channel, the water flows faster. If the deep part of the channel is next to one side of the river, water will erode that side and form a slight curve. Over time, the curve grows to become a broad arc called a **meander** (mee AN dur), as shown in **Figure 7.**

The broad, flat valley floor formed by a meandering stream is called a floodplain. When a stream floods, it often will cover part or all of the floodplain.

Figure 8

Although no two streams are exactly alike, all go through three main stages—young, mature, and old—as they flow from higher to lower ground. A young stream, below, surging over steep terrain, moves rapidly. In a less steep landscape, right, a mature stream flows more smoothly. On nearly level ground, the stream—considered old—winds leisurely through its valley. The various stages of a stream's development are illustrated here.

Waterfall

Rapids

A A young stream begins at a source—here, a melting mountain glacier. From its source, the stream flows swiftly downhill, cutting a narrow valley.

B A mature stream flows smoothly through its valley. Mature streams often develop broad curves called meanders.

C Old streams flow through broad, flat flood-plains. Near its mouth, the stream gradually drops its load of silt. This sediment forms a delta, an area of flat, fertile land extending into the ocean.

Oxbow lake

SCIENCE
Online

Research Visit the Glencoe Science Web site at **science.glencoe.com** to find out about major rivers in the United States. Classify two of these streams as young, mature, or old.

Old Streams The last stage in the development of a stream is the old stage. An old stream flows smoothly through a broad, flat floodplain that it has deposited. South of St. Louis, Missouri, the lower Mississippi River is in the old stage.

Major river systems, such as the Mississippi River, usually contain streams in all stages of development. In the upstream portion of a river system, you find whitewater streams moving swiftly down mountains and hills. At the bottom of mountains and hills, you find streams that start to meander and are in the mature stage of development. These streams meet at the trunk of the drainage basin and form a major river.

✔ Reading Check *How do old streams differ from young streams?*

Too Much Water

Sometimes heavy rains or a sudden melting of snow can cause large amounts of water to enter a river system. What happens when a river system has too much water in it? The water needs to go somewhere, and out and over the banks is the only choice. A river that overflows its banks can bring disaster by flooding homes or washing away bridges or crops.

Dams and levees are built in an attempt to prevent this type of flooding. A dam is built to control the water flow downstream. It may be built of soil, sand, or steel and concrete. Levees are mounds of earth that are built along the sides of a river. Dams and levees are built to prevent rivers from overflowing their banks. Unfortunately, they do not stop the water when flooding is great. This was the case in 1993 when heavy rains caused the Mississippi River to flood parts of nine midwestern states. Flooding resulted in billions of dollars in property damage. **Figure 9** shows some of the damage caused by this flood.

As you have seen, floods can cause great amounts of damage. But at certain times in Earth's past, great floods have completely changed the surface of Earth in a large region. Such floods are called catastrophic floods.

Figure 9
Flooding causes problems for people who live along major rivers. Floodwater broke through a levee during the Mississippi River flooding in 1993.

Figure 10

A The Channeled Scablands formed when Lake Missoula drained catastrophically. **B** These channels were formed by the floodwaters.

Catastrophic Floods During Earth's long history, many catastrophic floods have dramatically changed the face of the surrounding area. One catastrophic flood formed the Channeled Scablands in eastern Washington State, shown here in **Figure 10.** A vast lake named Lake Missoula covered much of western Montana. A natural dam of ice formed this lake. As the dam melted or was eroded away, tremendous amounts of water suddenly escaped through what is now the state of Idaho into Washington. In a short period of time, the floodwater removed overlying soil and carved channels into the underlying rock, some as deep as 50 m. Flooding occurred several more times as the lake refilled with water and the dam broke loose again. Scientists say the last such flood occurred about 13,000 years ago.

Deposition by Surface Water

You know how hard it is to carry a heavy object for a long time without putting it down. As water moves throughout a river system, it loses some of its energy of motion. The water can no longer carry some of its sediment. As a result, it drops, or is deposited, to the bottom of the stream.

Some stream sediment is carried only a short distance. In fact, sediment often is deposited within the stream channel itself. Other stream sediment is carried great distances before being deposited. Sediment picked up when rill and gully erosion occur are examples of this. Water usually has a lot of energy as it moves down a steep slope. When water begins flowing on a level surface, it slows, loses energy, and deposits its sediment. Water also loses energy and deposits sediment when it empties into an ocean or lake.

Figure 11

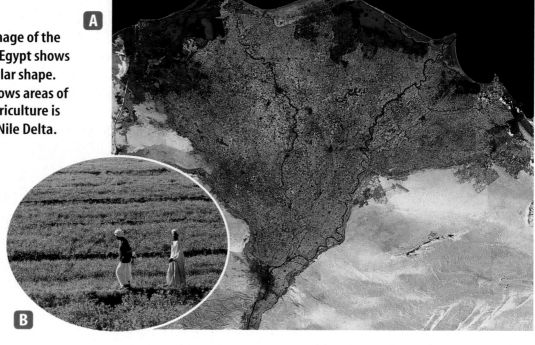

A This satellite image of the Nile River Delta in Egypt shows the typical triangular shape. The green color shows areas of vegetation. **B** Agriculture is important on the Nile Delta.

Deltas and Fans Sediment that is deposited as water empties into an ocean or lake forms a triangular, or fan-shaped, deposit called a delta, shown in **Figure 11.** When the river waters empty from a mountain valley onto an open plain, the deposit is called an alluvial (uh LEW vee ul) fan. The Mississippi River is an example of the topics presented in this section. Runoff causes rill and gully erosion. Sediment is picked up and carried into the larger streams that flow into the Mississippi River. As the Mississippi River flows, it cuts into its banks and picks up more sediment. Where the land is flat, the river deposits some of its sediment in its own channel. As the Mississippi enters the Gulf of Mexico, it slows, dropping much of its sediment and forming the Mississippi River delta.

Section 1 Assessment

1. How does the slope of an area affect runoff?
2. Compare rill and gully erosion.
3. Describe the three stages of stream development.
4. What is a delta?
5. **Think Critically** How is a stream's rate of flow related to the amount of erosion it causes? How is it related to the size of the sediments it deposits?

Skill Builder Activities

6. **Comparing and Contrasting** Compare and contrast the characteristics and formation of deltas and alluvial fans. **For more help, refer to the** Science Skill Handbook.

7. **Using an Electronic Spreadsheet** Design a table to compare and contrast sheet, rill, gully, and stream erosion. **For more help, refer to the** Technology Skill Handbook.

Groundwater

Groundwater Systems

What would have happened if the spilled milk in Section 1 ran off the table onto a carpeted floor? It probably would have quickly soaked into the carpet. Water that falls on Earth can soak into the ground just like the milk into the carpet.

Water that soaks into the ground becomes part of a system, just as water that stays above ground becomes part of a river system. Soil is made up of many small rock and mineral fragments. These fragments are all touching one another, as shown in **Figure 12,** but some empty space remains between them. Holes, cracks, and crevices exist in the rock underlying the soil. Water that soaks into the ground collects in these pores and empty spaces and becomes part of what is called **groundwater.**

How much of Earth's water do you think is held in the small openings in rock? Scientists estimate that 14 percent of all fresh-water on Earth exists as groundwater. This is almost 30 times more water than is contained in all of Earth's lakes and rivers.

Figure 12
Soil has many small, connected pores that are filled with water when soil is wet.

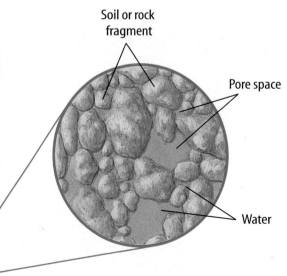

Soil or rock fragment

Pore space

Water

Measuring Pore Space

Procedure

1. Use two identical, **clear-plastic containers.**
2. Put 3 cm of **sand** in one container and 3 cm of **gravel** in the other.
3. Pour **water** slowly into the containers and stop when the water just covers the top of the sediment.
4. Record the volume of water used in each.

Analysis

Which substance has more pore space—sand or gravel?

Permeability A groundwater system is similar to a river system. However, instead of having channels that connect different parts of the drainage basin, the groundwater system has connecting pores. Soil and rock are **permeable** (PUR mee uh bul) if the pore spaces are connected and water can pass through them. Sandstone is an example of a permeable rock.

Soil or rock that has many large, connected pores is permeable. Water can pass through it easily. However, if a rock or sediment has few pore spaces or they are not well connected, then the flow of groundwater is blocked. These materials are **impermeable,** which means that water cannot pass through them. Granite has few or no pore spaces at all. Clay has many small pore spaces, but the spaces are not well connected.

✔ **Reading Check** *How does water move through permeable rock?*

Groundwater Movement How deep into Earth's crust does groundwater go? **Figure 13** shows a model of a groundwater system. Groundwater keeps going deeper until it reaches a layer of impermeable rock. When this happens, the water stops moving down. As a result, water begins filling up the pores in the rocks above. A layer of permeable rock that lets water move freely is an **aquifer** (AK wuh fur). The area where all of the pores in the rock are filled with water is the zone of saturation. The upper surface of this zone is the **water table.**

Figure 13
A stream's surface level is the water table. Below that is the zone of saturation.

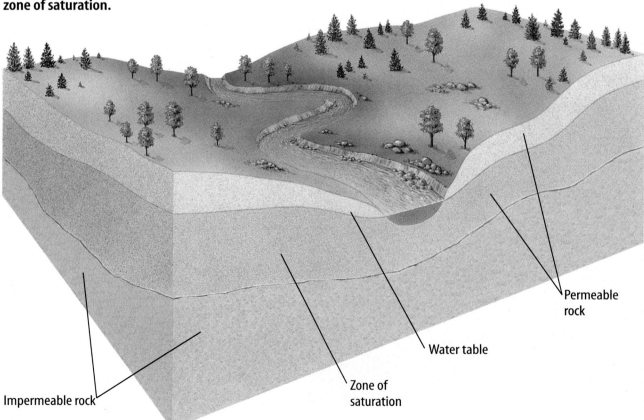

Permeable rock

Water table

Zone of saturation

Impermeable rock

Water Table

Why are the zone of saturation and the water table so important? An average United States resident uses about 626 L of water per day. That's enough to fill nearly two thousand soft drink cans. Many people get their water from groundwater through wells that have been drilled into the zone of saturation. However, the supply of groundwater is limited. During a drought, the water table drops. This is why you should conserve water.

Math Skills Activity

Calculating the Rate of Groundwater Flow

You and your family are hiking and the temperature is hot. You feel as if you can't walk one step farther. Luckily, relief is in sight. On the side of a nearby hill you see a stream, and you rush to splash some water on your face. Although you probably feel that it's taking you forever to reach the stream, your pace is quick when compared to how long it takes groundwater to flow through the aquifer that feeds the stream. The following problem will give you some idea of just how slowly groundwater flows through an aquifer.

Example Problem

You've run 200 m to get some water from a stream. How long does it take the groundwater in the aquifer to travel the same distance? The groundwater flows at a rate of 0.6 m/day.

Solution

1 *This is what you know:*
the distance that the
groundwater has to travel: $d = 200$ m
the rate that groundwater
flows through the aquifer: $r = 0.6$ m/day

2 *This is what you want to find:* time $= t$

3 *This is the equation you use:* $r \times t = d$ (rate $\times$ time $=$ distance)

4 *Solve the equation for* t *and
then substitute known values:* $t = \dfrac{d}{r} = \dfrac{(200 \text{ m})}{(0.6 \text{ m/day})} = 333.33$ days

Practice Problem

The groundwater in an aquifer flows at a rate of 0.5 m/day. How far does the groundwater move in a year?

For more help, refer to the Math Skill Handbook.

Figure 14
The years on the pole show how much the ground level dropped in the San Joaquin Valley, California, between 1925 and 1977.

Figure 15
The pressure of water in a sloping aquifer keeps an artesian well flowing. *What limits how high water can flow in an artesian well?*

Wells A good well extends deep into the zone of saturation, past the top of the water table. Groundwater flows into the well, and a pump brings it to the surface. Because the water table sometimes drops during very dry seasons, even a good well can go dry. Then time is needed for the water table to rise, either from rainfall or through groundwater flowing from other areas of the aquifer.

Where groundwater is the main source of drinking water, the number of wells and how much water is pumped out are important. If a large factory were built in such a town, the demand on the groundwater supply would be even greater. Even in times of normal rainfall, the wells could go dry if water were taken out at a rate greater than the rate at which it can be replaced.

In areas where too much water is pumped out, the land level can sink from the weight of the sediments above the now-empty pore spaces. **Figure 14** shows what occurred when too much groundwater was removed in a region of California.

One type of well doesn't need a pump to bring water to the surface. An artesian well is a well in which water rises to the surface under pressure. Artesian wells are less common than other types of wells because of the special conditions they require.

As shown in **Figure 15,** the aquifer for an artesian well needs to be located between two impermeable layers that are sloping. Water enters at the high part of the sloping aquifer. The weight of the water in the higher part of the aquifer puts pressure on the water in the lower part. If a well is drilled into the lower part of the aquifer, the pressurized water will flow to the surface. Sometimes, the pressure is great enough to force the water into the air, forming a fountain.

✔ **Reading Check** *Why does water flow from an artesian well?*

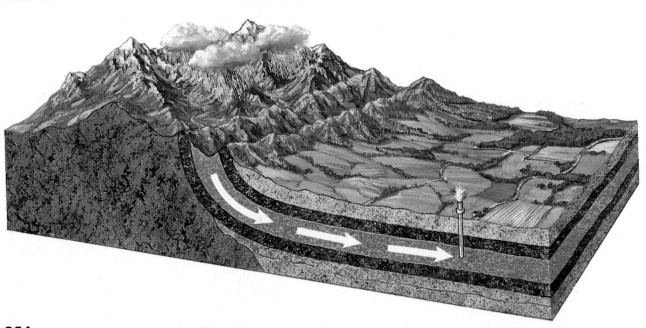

Springs In some places, the water table is so close to Earth's surface that water flows out and forms a **spring.** Springs are found on hillsides or other places where the water table meets a sloping surface. Springs often are used as a source of freshwater.

The water from most springs is a constant, cool temperature because soil and rock are good insulators and protect the groundwater from changes in temperature on Earth's surface. However, in some places, magma rises to within a few kilometers of Earth's surface and heats the surrounding rock. Groundwater that comes in contact with these hot rocks is heated and can come to the surface as a hot spring.

Geysers When water is put into a teakettle to boil, it heats slowly at first. Then some steam starts to come out of the cap on the spout, and suddenly the water starts boiling. The teakettle starts whistling as steam is forced through the cap. A similar process can occur with groundwater. One of the places where groundwater is heated is in Yellowstone National Park in Wyoming. Yellowstone has hot springs and geysers. A **geyser** is a hot spring that erupts periodically, shooting water and steam into the air. Groundwater is heated to high temperatures, causing it to expand underground. This expansion forces some of the water out of the ground, taking the pressure off of the remaining water. The remaining water boils quickly, with much of it turning to steam. The steam shoots out of the opening like steam out of a teakettle, forcing the remaining water out with it. Yellowstone's famous geyser, Old Faithful, pictured in **Figure 16,** shoots between 14,000 and 32,000 L of water and steam into the air about once every 80 min.

The Work of Groundwater

Although water is the most powerful agent of erosion on Earth's surface, it also can have a great effect underground. Water mixes with carbon dioxide gas to form a weak acid called carbonic acid. Some of this carbon dioxide is absorbed from the air by rainwater or surface water. Most carbon dioxide is absorbed by groundwater moving through soil. One type of rock that is dissolved easily by this acid is limestone. Acidic groundwater moves through natural cracks and pores in limestone, dissolving the rock. Gradually, the cracks in the limestone enlarge until an underground opening called a **cave** is formed.

Chemistry
INTEGRATION

Acid rain occurs when gases released by burning oil and coal mix with water in the air. Infer what effect acid rain can have on a statue made of limestone.

Figure 16
Yellowstone's famous geyser, Old Faithful, used to erupt once about every 76 min. An earthquake on January 9, 1998, slowed Old Faithful's "clock" by 4 min to an average of one eruption about every 80 min. The average height of the geyser's water is 40.5 m.

Cave Formation

You've probably seen a picture of the inside of a cave, like the one shown in **Figure 17,** or perhaps you've visited one. Groundwater not only dissolves limestone to make caves, but it also can make deposits on the insides of caves.

Water often drips slowly from cracks in the cave walls and ceilings. This water contains calcium ions dissolved from the limestone. If the water evaporates while hanging from the ceiling of a cave, a deposit of calcium carbonate is left behind. Stalactites form when this happens over and over. Where drops of water fall to the floor of the cave, a stalagmite forms. The words *stalactite* and *stalagmite* come from Greek words that mean "to drip."

Sinkholes

If underground rock is dissolved near the surface, a sinkhole may form. A sinkhole is a depression on the surface of the ground that forms when the roof of a cave collapses or when material near the surface dissolves. Sinkholes are common features in places like Florida and Kentucky that have lots of limestone and enough rainfall to keep the groundwater system supplied with water. Sinkholes can cause property damage if they form in a populated area.

In summary, when rain falls and becomes groundwater, it might dissolve limestone and form a cave, erupt from a geyser, or be pumped from a well to be used at your home.

Figure 17
Water dissolves rock to form caves and also deposits material to form spectacular formations, such as these in Carlsbad Caverns in New Mexico.

Section ② Assessment

1. Describe how water enters the groundwater system.
2. How does the permeability of soil and rocks affect the flow of groundwater?
3. Describe why a well might go dry.
4. Explain how caves form.
5. **Think Critically** Would you expect water in wells, geysers, and hot springs to contain dissolved materials? Why or why not?

Skill Builder Activities

6. **Comparing and Contrasting** Compare and contrast wells, geysers, and hot springs. **For more help, refer to the** Science Skill Handbook.

7. **Drawing Conclusions** Read an article about geothermal energy and draw a diagram in your Science Journal explaining how it works. Also, list the limitations of this energy source. **For more help, refer to the** Science Skill Handbook.

Ocean Shoreline

The Shore

Picture yourself sitting on a beautiful, sandy beach like the one shown in **Figure 18A.** Nearby, palm trees sway in the breeze. Children play in the quiet waves lapping at the water's edge. It's hard to imagine a place more peaceful. Now, picture yourself sitting along another shore. You're on a high cliff watching waves crash onto boulders far below. Both of these places are shorelines. An ocean shoreline is where land meets the ocean.

The two shorelines just described are different even though both experience surface waves, tides, and currents. These actions cause shorelines to change constantly. Sometimes you can see these changes from hour to hour. Why are shorelines so different? You'll understand why they look different when you learn about the forces that shape shorelines.

As You Read

What You'll Learn

- **Identify** the different causes of shoreline erosion.
- **Compare and contrast** different types of shorelines.
- **Describe** some origins of sand.

Vocabulary

longshore current
beach

Why It's Important

Constantly changing shorelines impact the people who live and work by them.

Figure 18

A Waves, tides, and currents cause shorelines to change constantly. **B** Waves approaching the shoreline at an angle create a longshore current.

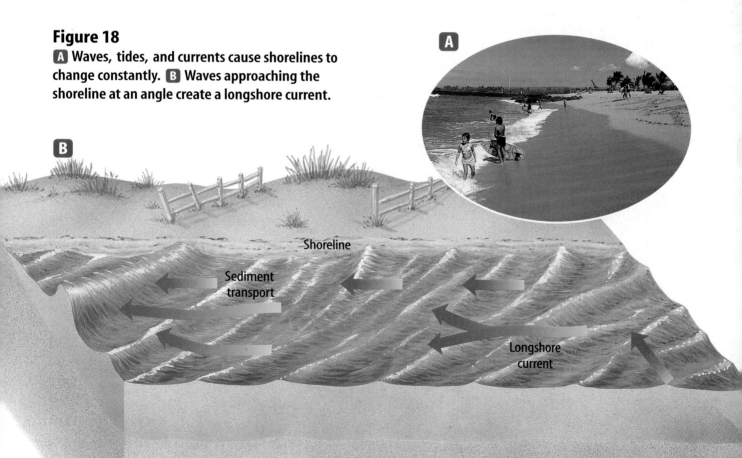

What features would you find on the shoreline? To learn more about shoreline features, see the **Shoreline Field Guide** at the back of the book.

Figure 19
Along a rocky shoreline, the force of pounding waves breaks rock fragments loose, then grinds them into smaller and smaller pieces.

Shoreline Forces When waves constantly pound against the shore, they break rocks into ever-smaller pieces. Currents move many metric tons of sediment along the shoreline. The sediment grains grind against each other like sandpaper. The tide goes out carrying sediment to deeper water. When the tide returns, it brings new sediment with it. These forces are always at work, slowly changing the shape of the shoreline. Water is always in motion along the shore.

The three major forces at work on the shoreline are waves, currents, and tides. Winds blowing across the water make waves. Waves, crashing against a shoreline, are a powerful force. They can erode and move large amounts of material in a short time. Waves usually collide with a shore at slight angles. This creates a **longshore current** of water that runs parallel to the shoreline. Longshore currents, shown in **Figure 18B,** carry many metric tons of loose sediments and act like rivers of sand in the ocean.

✔ **Reading Check** *How does a longshore current form?*

Tides create currents that move at right angles to the shore. These are called tidal currents. Outgoing tides carry sediments away from the shore, and incoming tides bring new sediments toward the shore. Tides work with waves to shape shorelines. You've seen the forces that affect all shorelines. Now you will see the differences that make one shore a flat, sandy beach and another shore a steep, rocky cliff.

Rocky Shorelines

Rocks and cliffs are the most common features along rocky shorelines like the one in **Figure 19.** Waves crash against the rocks and cliffs. Sediments in the water grind against the cliffs, slowly wearing the rock away. Then rock fragments broken from the cliffs are ground up by the endless motion of waves. They are transported as sediment by longshore currents.

Softer rocks become eroded before harder rocks do, leaving islands of harder rocks. This takes thousands of years, but remember that the ocean never stops. In a single day, about 14,000 waves crash onto shore.

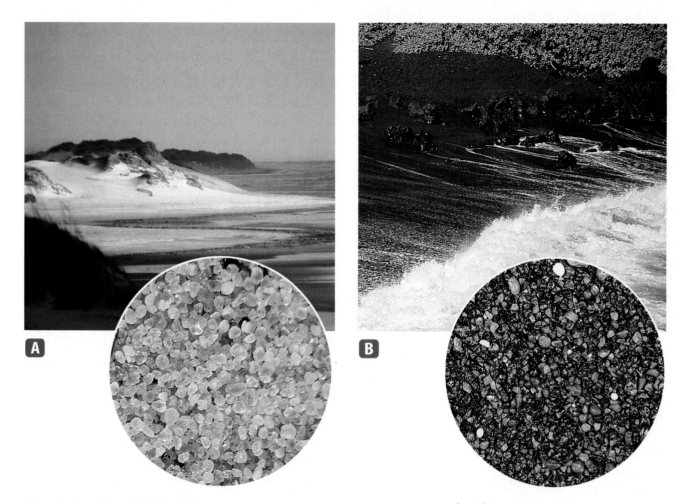

Sandy Beaches

Smooth, gently sloping shorelines are different from steep, rocky shorelines. Beaches are the main feature here. **Beaches** are deposits of sediment that are parallel to the shore.

Beaches are made up of different materials. Some are made of rock fragments from the shoreline. Many sands are made of grains of quartz, and others are made of seashell fragments. These fragments range in size from stones larger than your hand to fine sand. Sand grains range from 0.06 mm to 2 mm in diameter. Why do many beaches have particles of this size? Waves break rocks and seashells down to sand-sized particles like those shown in **Figure 20.** The constant wave motion bumps sand grains together. This bumping not only breaks particles into smaller pieces but also smooths off their jagged corners, making them more rounded.

Reading Check *How do waves affect beach particles?*

Sand in some places is made of other things. For example, Hawaii's black sands are made of basalt, and its green sands are made of the mineral olivine. Jamaica's white sands are made of coral and shell fragments.

Figure 20
Beach sand varies in size, color, and composition. **A** This quartz sand from a Texas beach is clear and glassy. **B** Some Hawaiian beaches are composed of black basalt sand.

Sand Erosion and Deposition

Longshore currents carry sand along beaches to form features such as barrier islands, spits, and sandbars. Storms and wind also move sand. Thus, beaches are fragile, short-term land features that are damaged easily by storms and human activities such as some types of construction. Communities in widely separated places such as Long Island, New York; Malibu, California; and Padre Island, Texas, have problems because of beach erosion.

Barrier Islands Barrier islands are sand deposits that lay parallel to the shore but are separated from the mainland. These islands start as underwater sand ridges formed by breaking waves. Hurricanes and storms add sediment to them, raising some to sea level. When a barrier island becomes large enough, the wind blows the loose sand into dunes, keeping the new island above sea level. As with all seashore features, barrier islands are short term, lasting from a few years to a few centuries.

The forces that build barrier islands also can erode them. Storms and waves carry sediments away. Beachfront development, as in **Figure 21,** can be affected by shoreline erosion.

Figure 21
Shorelines change constantly. Human development is often at risk from shoreline erosion.

Section 3 Assessment

1. What major forces cause shoreline erosion?
2. Contrast the features you would find along a steep, rocky shoreline with the features you would find along a gently sloping, sandy shoreline.
3. How could the type of shoreline affect the types of sediments you might find there?
4. List several materials that beach sand might be composed of. Where do these materials come from?
5. **Think Critically** How would erosion and deposition of sediment along a shoreline be affected if the longshore current was blocked by a wall built out into the water?

Skill Builder Activities

6. **Concept Mapping** Make an events chain concept map that shows how the sand from a barrier island can become a new barrier island 100 years from now. Use these terms: *barrier island, breaking waves, wind, longshore currents,* and *new barrier island.* **For more help, refer to the** Science Skill Handbook.

7. **Solving One-Step Equations** If 14,000 waves crash onto a shore daily, how many waves crash onto it in a year? Calculate how many have crashed onto it since you were born. Explain how you found your answer. **For more help, refer to the** Math Skill Handbook.

Activity

Classifying Types of Sand

You know that sand is made of many different kinds of grains, but did you realize that the slope of a beach is related to the size of its grains? The coarser the grain size is, the steeper the beach is. The composition of sand also is important. Many sands are mined because they have economic value.

What You'll Investigate
What characteristics can be used to classify different types of beach sand?

Materials
samples of different sands (3)
magnifying glass
*stereomicroscope
magnet
*Alternate materials

Goals
- **Observe** differences in sand.
- **Identify** characteristics of beach sand.
- **Infer** sediment sources.

Procedure

1. **Design** a data table in which to record your data when you compare the three sand samples. You will need five columns in your table. One column will be for the samples and the others for the characteristics you will be examining.

| Angular | Sub-Angular | Sub-Rounded | Rounded |

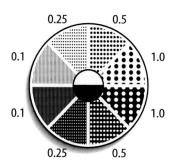

Sand gauge
(measurements in mm)

2. Use the diagram to determine the average roundness of each sample.

3. **Identify** the grain size of your samples by using the sand gauge above. To determine the grain size, place sand grains in the middle of the circle of the sand gauge. Use the upper half of the circle for dark-colored particles and the bottom half for light-colored particles.

4. Decide on two other characteristics to examine that will help you classify your samples.

Conclude and Apply

1. **Compare and contrast** some characteristics of beach sand.

2. Why are there variations in the characteristics of different sand samples?

3. What do your observations tell you about the sources of the three samples?

Communicating Your Data

Compare your results with those of other students. **For more help, refer to the Science Skill Handbook.**

Water Speed and Erosion

What would it be like to make a raft and use it to float on a river? Would it be easy? Would you feel like Tom Sawyer? Probably not. You'd be at the mercy of the current. Strong currents create fast rivers. But does fast moving water affect more than just floating rafts and other objects? How does the speed of a stream or river affect its ability to erode?

What You'll Investigate

How does the speed of water affect its ability to erode?

Materials

paint roller pan
disposable wallpaper trays
sand
1-L beaker
rubber tubing (20 cm)
metric ruler
water
stopwatch
fine-mesh screen
wood block
Alternate materials

Goals

- **Assemble** an apparatus for measuring the effect of water speed on erosion.
- **Observe and measure** the ability of water traveling at different speeds to erode sand.

Safety Precautions

Wash your hands after you handle the sand. Immediately clean up any water that spills on the floor.

Procedure

1. Copy the data table below.

2. Place the screen in the sink. Pour moist sand into your pan and smooth out the sand. Set one end of the pan on the wood block and hang the other end over the screen in the sink. Excess water will flow onto the screen in the sink.

3. Attach one end of the hose to the faucet and place the other end in the beaker. Turn on the water so that it trickles into the beaker. Time how long it takes for the trickle of water to fill the beaker to the 1-L mark. Divide 1 L by your time in seconds to calculate the water speed. Record the speed in your data table.

4. Without altering the water speed, hold the hose over the end of the pan that is resting on the wood block. Allow the water to flow into the sand for 2 min. At the end of 2 min, turn off the water.

5. **Measure** the depth and length of the eroded channel formed by the water. Count the number of branches formed on the channel. Record your measurements and observations in your data table.

6. Empty the excess water from the tray and smooth out the sand. Repeat steps 3 through 5 two more times, increasing your water speed each time.

Water Speed and Erosion			
Water Speed (Liters per Second)	Depth of Channel	Length of Channel	Number of Channel Branches

Conclude and Apply

1. **Identify** the constants and variables in your experiment.

2. Which water speed created the deepest and longest channel?

3. Which water speed created the greatest number of branches?

4. **Infer** the effect that water speed has on erosion.

5. **Predict** how your results would have differed if one end of the pan had been raised higher.

6. **Infer** how streams and rivers can shape Earth's surface.

𝒞ommunicating
Your Data

Write a pamphlet for people buying homes near rivers or streams that outlines the different effects that water erosion could have on their property.

Is there hope for America's coastlines or is beach erosion a "shore" thing?

Sands

Water levels are rising along the coastline of the United States. Serious storms and the building of homes and businesses along the shore are leading to the eroding of anywhere from 70 percent to 90 percent of the U.S. coastline. A report from the Federal Emergency Management Agency (FEMA) confirms this. The report says that one meter of United States beaches will be eaten away each year for the next 60 years. Since 1965, the federal government has spent millions of dollars replenishing more than 1,300 eroding sandy shores around the country. And still beaches continue to disappear.

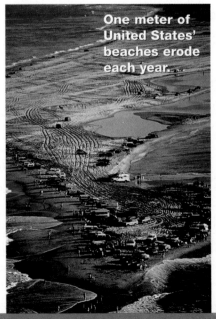

One meter of United States' beaches erode each year.

The slowly eroding beaches are upsetting to residents and officials of many communities, who depend on their shore to earn money from visitors. Some city and state governments are turning to beach nourishment—a process in which sand is taken from the seafloor and dumped on beaches. The process is expensive, however. The state of Delaware, for example, is spending 7,000,000 dollars to bring in sand for its beaches.

Some geologists believe that beach nourishment does not work. They explain that the grains of the new sand are often smaller and finer than the original beach sand, so it erodes faster, carried back out to sea by waves.

Dredged sand is pumped onto an eroded beach in Long Beach, North Carolina.

This beach house will collapse as its underpinnings are eroded.

in Time

Despite the odds, some people think beach nourishment is still the way to go even if it isn't a permanent solution. One person in favor of nourishing beaches answers those against beach nourishment this way: "That's like saying it's a losing battle to pave streets because some-day grass is going to poke through."

Other methods of saving eroding beaches are being tried. In places along the Great Lakes shores and coastal shores, one company has installed fabrics underwater to slow currents. By slowing currents, sand is naturally deposited and kept in place.

Another shore-saving device is a synthetic barrier that is shaped like a plastic snowflake.

A string of these barriers is secured just off-shore. They absorb the energy of incoming waves. Reducing wave energy can prevent sand from being eroded from the beach. New sand also might accumulate because the barriers slow down the currents that flow along the shore.

Many people believe that communities along the shore must restrict the beachfront building of homes, hotels, and stores. Since some estimates claim that by the year 2025, nearly 75 percent of the U.S. population will live in coastal areas, it's a tough solution. Says one geologist, "We can retreat now and save our beaches or we can retreat later and probably ruin the beaches in the process."

Chapter 9 Study Guide

Reviewing Main Ideas

Section 1 Surface Water

1. Rainwater that does not soak into the ground is pulled down the slope by gravity. This water is called runoff.

2. Runoff has the ability to erode and carry sediment. Factors such as steepness of slope and number and type of plants affect the amount of erosion. Rill, gully, and sheet erosion are types of surface water erosion caused by runoff. *What type of erosion is occurring in this photograph?*

3. Runoff generally flows into streams that merge with larger and larger rivers until emptying into a lake or ocean. Major river systems usually contain several different types of streams.

4. In the mountains, young streams flow through steep valleys and have rapids and waterfalls. Mature streams flow through gentler terrain and have less energy. Old streams are often wide and snake back and forth across their floodplains.

Section 2 Groundwater

1. When water soaks into the ground, it becomes part of a vast groundwater system.

2. Although rock may seem solid, many types are filled with connected spaces called pores. Such rocks are permeable and can contain large amounts of groundwater. *What would happen to the permeability of this material if the pores filled with clay or some other material?*

Section 3 Ocean Shoreline

1. Ocean shorelines are always changing. *Would you expect the rocky coast, shown to the right, to erode slower or faster than a sandy coast? Why?*

2. Waves and currents have tremendous amounts of energy. Pounding waves break up rocks into tiny fragments called sediment. Over time, the waves and currents move this sediment and deposit it elsewhere constantly changing beaches, sandbars, and barrier islands.

FOLDABLES
Reading & Study Skills

After You Read

Use the information in your Main Ideas Study Fold to review the different characteristics of surface water, groundwater, and ocean shorelines.

Visualizing Main Ideas

Complete the following concept map on caves.

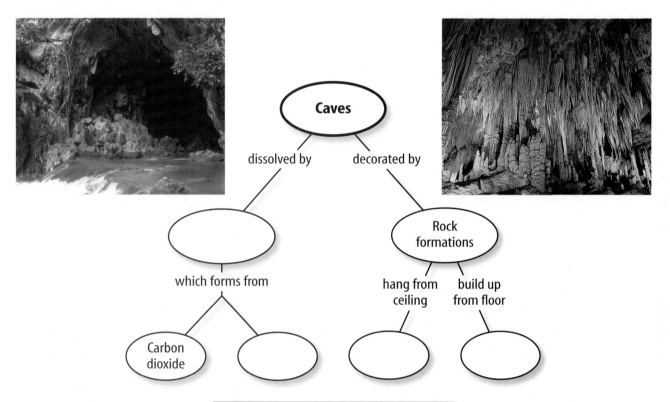

Caves

dissolved by   decorated by

which forms from

Rock formations

hang from ceiling   build up from floor

Carbon dioxide

Vocabulary Review

Vocabulary Words

a. aquifer
b. beach
c. cave
d. channel
e. drainage basin
f. geyser
g. groundwater
h. impermeable

i. longshore current
j. meander
k. permeable
l. runoff
m. sheet erosion
n. spring
o. water table

Study Tip

Don't just memorize definitions. Write complete sentences using new vocabulary words to be certain you understand what they mean.

Using Vocabulary

Explain the difference between the vocabulary words in each of the following sets.

1. runoff, sheet erosion

2. channel, drainage basin

3. aquifer, cave

4. spring, geyser

5. permeable, impermeable

6. sheet erosion, meander

7. groundwater, water table

8. permeable, aquifer

9. longshore current, beach

10. meander, channel

Checking Concepts

Choose the word or phrase that best answers the question.

1. Identify an example of a structure created by deposition.
   A) beach
   B) rill
   C) cave
   D) geyser

2. Name the deposit that forms when a mountain river runs onto a plain.
   A) subsidence
   B) an alluvial fan
   C) infiltration
   D) water diversion

3. What is a layer of permeable rock that water flows through?
   A) an aquifer
   B) a pore
   C) a water table
   D) impermeable

4. What is the network formed by a river and all the smaller streams that contribute to it?
   A) groundwater system
   B) zone of saturation
   C) river system
   D) water table

5. Which term describes rock through which fluids can flow easily?
   A) impermeable
   B) meanders
   C) saturated
   D) permeable

6. Which stage of development are mountain streams in?
   A) young
   B) mature
   C) old
   D) meandering

7. What forms as a result of the water table meeting Earth's surface?
   A) meander
   B) spring
   C) aquifer
   D) stalactite

8. What contains heated groundwater that reaches Earth's surface?
   A) water table
   B) cave
   C) aquifer
   D) hot spring

9. Where are beaches most common?
   A) rocky shorelines
   B) flat shorelines
   C) aquifers
   D) young streams

10. Why does water rise in an artesian well?
    A) a pump
    B) erosion
    C) heat
    D) pressure

Thinking Critically

11. Explain why the Mississippi River has meanders along its course.

12. What determines whether a stream erodes its bottom or its sides?

13. Why might you be concerned if developers of a new housing project started drilling wells near your well?

14. Along what kind of shoreline would you find barrier islands? Explain.

15. Explain why beach sands collected from different locations will differ.

Developing Skills

16. **Interpreting Data** The rate of water flowing out of the Brahmaputra River in India, the La Plata River in South America, and the Mississippi River in North America are given in the table below. Infer which river carries the most sediment.

River Flow Rates	
River	Flow, m^3/s
Brahmaputra River, India	19,800
La Plata River, South America	79,300
Mississippi River, North America	175,000

17. **Forming Hypotheses** Hypothesize why most of the silt in the Mississippi delta is found farther out to sea than the sand-sized particles are.

18. **Communicating** Make an outline that explains the three stages of stream development. Share it with the class.

19. **Concept Mapping** Complete the concept map below using the following terms: *developed meanders, gentle curves, gentle gradient, old, rapids, steep gradient, wide floodplain,* and *young.*

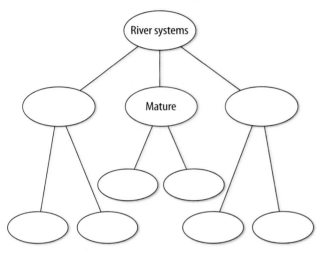

20. **Using Variables, Constants, and Controls** Explain how you could test the effect of slope on the amount of runoff produced.

Performance Assessment

21. **Poster** Research a beach that you might be interested in. Make a poster that shows different features you would find at a beach. Display your poster for the class.

22. **Poem** Write a brief poem about groundwater. Include things such as the source of the groundwater and how it is used.

TECHNOLOGY

Go to the Glencoe Science Web site at **science.glencoe.com** or use the **Glencoe Science CD-ROM** for additional chapter assessment.

Test Practice

A scientist is gathering data and looking at trends in water erosion. The scientist's data are shown in the graph below.

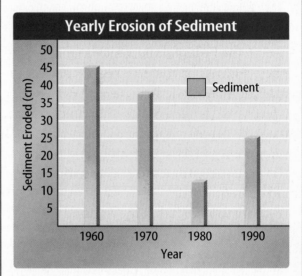

Study the graph and answer the following questions.

1. According to this information, which year had the most erosion?
   A) 1960   C) 1980
   B) 1970   D) 1990

2. According to the graph, between which years was erosion increasing?
   F) 1960–1970   H) 1980–1990
   G) 1970–1980   J) 1960–1980

3. According to the graph, which year had the least erosion?
   A) 1960   C) 1980
   B) 1970   D) 1990

4. According to the graph, how many centimeters of erosion were there in 1990?
   F) 5   H) 15
   G) 25   J) 20

Reading Comprehension

Read the passage carefully. Then read the questions that follow the passage. Decide which is the best answer to each question.

Test-Taking Tip   After reading a passage, quickly glance over the questions and then skim the passage again, looking for important information.

Shorelines

A recent study on shoreline erosion completed by the Federal Emergency Management Agency (FEMA) has determined that about one quarter of man-made structures that are within 500 feet of the United States' shorelines will be damaged by erosion during the next 60 years. If development of coasts continues and the level of the oceans continues to rise, the situation could be even worse.

Scientists predict that the Atlantic and Gulf of Mexico coastlines will experience the most erosion. The cost to homeowners is predicted to be about a half billion dollars per year. With yearly average erosion rates of 1m on the Atlantic Coast and 2m on the Gulf of Mexico coastlines, many homes will be lost forever. In fact, these areas will account for approximately 60 percent of nationwide losses.

The grain size of sediment is one factor that accounts for some of the difference in erosion rates between coasts. The sediment on the shores of the Gulf of Mexico generally is fine grained and easily picked up and carried away by wave action. On the Atlantic coast, the sediment generally is more coarse grained and therefore erodes at a slower rate. Along the Pacific Coast, rates of erosion vary from one area to another. Erosion rates of 2.4 m per year have been recorded along parts of the Alaskan coast, while the state of Washington's shoreline actually is growing in some places.

1. Based on the passage, which of the following is most likely true?
   A) Building a summer home along the Gulf of Mexico would be more secure than building one along the Atlantic Coast.
   B) Erosion is a major threat to shorelines in many parts of the country.
   C) All man-made structures within 500 feet of the Atlantic Coast will disappear.
   D) Ocean water levels will decrease during the next 60 years.

2. The main purpose of the Federal Emergency Management Agency's study was most likely to _____.
   F) discourage people from building homes near the coastline
   G) explain how coastline erosion of various types of sediment occurs
   H) help to limit erosion and protect property
   I) figure out the costs of coastline erosion in the United States

Reasoning and Skills

Read each question and choose the best answer.

Group R

Group S

1. The weathering examples in Group R are different from weathering examples in Group S because only the weathering examples in Group R have been caused by _____.
   A) chemical weathering
   B) ice wedging
   C) mechanical weathering
   D) oxidation

Test-Taking Tip   Think about the different kinds of weathering and how they affect Earth.

2. The rock slide pictured in the left column is most likely caused by the forces of _____.
   F) water
   G) gravity
   H) wind
   J) glacier movement

Test-Taking Tip   Think about the forces that help cause erosion.

3. Which of the following is most responsible for the development of these landscapes?
   A) ice wedging
   B) the movement of wind
   C) the effects of acid rain
   D) the movement of water

Test-Taking Tip   Consider what you know about erosion and specific examples of erosion, such as the Grand Canyon.

Consider this question carefully before writing your answer on a separate sheet of paper.

4. Earth's surface has been shaped by erosion. Discuss some of the different agents of erosion that affect Earth's surface. How do these agents of erosion reshape the landscape?

How Are Volcanoes & Fish Connected?

It's hard to know exactly what happened four and a half billion years ago, when Earth was very young. But it's likely that Earth was much more volcanically active than it is today. Along with lava and ash, volcanoes emit gases—including water vapor. Some scientists think that ancient volcanoes spewed tremendous amounts of water vapor into the early atmosphere. When the water vapor cooled, it would have condensed to form liquid water. Then the water would have fallen to the surface and collected in low areas, creating the oceans. Scientists hypothesize that, roughly three and a half billion to four billion years ago, the first living things developed in the oceans. According to this hypothesis, these early life forms gradually gave rise to more and more complex organisms—including the multitudes of fish that swim through the world's waters.

SCIENCE CONNECTION

VOLCANOES AND COMETS Not all scientists agree with the hypothesis that Earth's oceans were formed primarily by emissions from ancient volcanoes. For example, some researchers suggest that the water may have come largely from comets. Divide the class into two teams. Have one team investigate the volcano hypothesis, while the other team researches the comet hypothesis. Then hold a class debate, with each team presenting evidence in support of its hypothesis.

10 Plate Tectonics

Characterized by volcanoes and scenic vistas, the East African Rift Valley marks a place where Earth's crust is being pulled apart. If the pulling continues over millions of years, Africa will separate into two landmasses. In this chapter, you'll learn about Rift Valleys and other features explained by the theory of plate tectonics. You'll also learn about the fossil, climate, and rock clues that indicate that Earth's continents have drifted over time.

What do you think?

Science Journal Look at the picture below with a classmate. Discuss what you think this might be or what is happening. Here's a hint: *A river runs through this dog leg.* Write your answer or best guess in your Science Journal.

EXPLORE ACTIVITY

Can you imagine a giant landmass that broke into many separate continents and Earth scientists working to reconstruct Earth's past? Do this activity to learn about clues that can be used to reassemble a supercontinent.

Reassemble an image

1. Collect interesting photographs from an old magazine.

2. You and a partner each select one photo, but don't show them to each other. Then each of you cut your photos into pieces no smaller than about 5 cm or 6 cm.

3. Trade your cut-up photo for your partner's.

4. Observe the pieces, and reassemble the photograph your partner has cut up.

Observe

In your Science Journal, describe the characteristics of the cut-up photograph that helped you put the image back together. Think of other examples in which characteristics of objects are used to match them up with other objects.

Before You Read

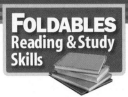
FOLDABLES
Reading & Study Skills

Making a Know-Want-Learn Study Fold It would be helpful to identify what you already know and what you want to know. Make the following Foldable to help you focus on reading about plate tectonics.

1. Place a sheet of paper in front of you so the long side is at the top. Fold the paper in half from top to bottom.

2. Fold both sides in to divide the paper into thirds. Unfold the paper so three sections show.

3. Through the top thickness of paper, cut along each of the fold lines to the topfold, forming three tabs. Label the tabs *Know, Want,* and *Learn,* as shown.

4. Before you read the chapter, write what you know about plate tectonics under the left tab and what you want to know under the middle tab.

5. As you read the chapter, write what you learn about plate tectonics under the right tab.

1 Continental Drift

As You Read

What You'll Learn

- **Describe** the hypothesis of continental drift.
- **Identify** evidence supporting continental drift.

Vocabulary

continental drift
Pangaea

Why It's Important

The hypothesis of continental drift led to plate tectonics—a theory that explains many processes in Earth.

Figure 1
This illustration represents how the continents once were joined to form Pangaea. This fitting together of continents according to shape is not the only evidence supporting the past existence of Pangaea.

Evidence for Continental Drift

If you look at a map of Earth's surface, you can see that the edges of some continents look as though they could fit together like a puzzle. Other people also have noticed this fact. For example, Dutch mapmaker Abraham Ortelius noted the fit between the coastlines of South America and Africa more than 400 years ago.

Pangaea German meteorologist Alfred Wegener (VEG nur) thought that the fit of the continents wasn't just a coincidence. He suggested that all the continents were joined together at some time in the past. In a 1912 lecture, he proposed the hypothesis of continental drift. According to the hypothesis of **continental drift,** continents have moved slowly to their current locations. Wegener suggested that all continents once were connected as one large landmass, shown in **Figure 1,** that broke apart about 200 million years ago. He called this large landmass **Pangaea** (pan JEE uh), which means "all land."

✔ **Reading Check** *Who proposed continental drift?*

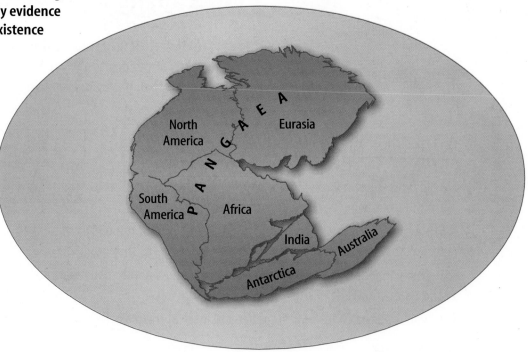

A Controversial Idea Wegener's ideas about continental drift were controversial. It wasn't until long after Wegener's death in 1930 that his basic hypothesis was accepted. The evidence Wegener presented hadn't been enough to convince many people during his lifetime. He was unable to explain exactly how the continents drifted apart. He proposed that the continents plowed through the ocean floor, driven by the spin of Earth. Physicists and geologists of the time strongly disagreed with Wegener's explanation. They pointed out that continental drift would not be necessary to explain many of Wegener's observations. Other important observations that came later eventually supported Wegener's earlier evidence.

Fossil Clues Besides the puzzlelike fit of the continents, fossils provided support for continental drift. Fossils of the reptile *Mesosaurus* have been found in South America and Africa, as shown in **Figure 2.** This swimming reptile lived in freshwater and on land. How could fossils of *Mesosaurus* be found on land areas separated by a large ocean of salt water? It probably couldn't swim between the continents. Wegener hypothesized that this reptile lived on both continents when they were joined.

☑ **Reading Check** *How do* Mesosaurus *fossils support the past existence of Pangaea?*

SCIENCE Online

Research Visit the Glencoe Science Web site at **science.glencoe.com** for more information about the continental drift hypothesis. Communicate to your class what you learn.

Figure 2
Fossil remains of plants and animals that lived in Pangaea have been found on more than one continent. *How do the locations of* Glossopteris, Mesosaurus, Kannemeyerid, Labyrinthodont, *and other fossils support Wegener's hypothesis of continental drift?*

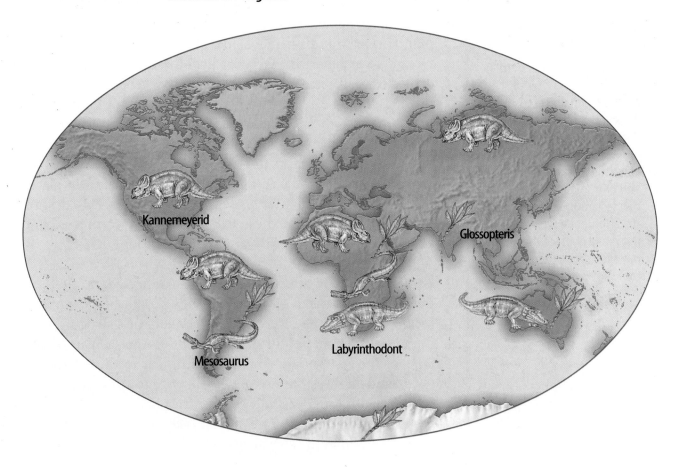

Figure 3
This fossil plant, *Glossopteris*, grew in a temperate climate.

Interpreting Fossil Data

Procedure 🥽 👕

1. Build a three-layer landmass using **clay or modeling dough.**
2. Mold the clay into mountain ranges.
3. Place similar **"fossils"** into the clay at various locations around the landmass.
4. Form five continents from the one landmass. Also, form two smaller landmasses out of different clay with different mountain ranges and fossils.
5. Place the five continents and two smaller landmasses around the room.
6. Have someone who did not make or place the landmasses make a model that shows how they once were positioned.
7. Return the clay to its container so it can be used again.

Analysis
What clues were useful in reconstructing the original landmass?

A Widespread Plant Another fossil that supports the hypothesis of continental drift is *Glossopteris* (glahs AHP tur us). **Figure 3** shows this fossil plant, which has been found in Africa, Australia, India, South America, and Antarctica. The presence of *Glossopteris* in so many areas also supported Wegener's idea that all of these regions once were connected and had similar climates.

Climate Clues Wegener used continental drift to explain evidence of changing climates. For example, fossils of warm-weather plants were found on the island of Spitsbergen in the Arctic Ocean. To explain this, Wegener hypothesized that Spitsbergen drifted from tropical regions to the arctic. Wegener also used continental drift to explain evidence of glaciers found in temperate and tropical areas. Glacial deposits and rock surfaces scoured and polished by glaciers are found in South America, Africa, India, and Australia. This shows that parts of these continents were covered with glaciers in the past. How could you explain why glacial deposits are found in areas where no glaciers exist today? Wegener thought that these continents were connected and partly covered with ice near Earth's south pole long ago.

Rock Clues If the continents were connected at one time, then rocks that make up the continents should be the same in locations where they were joined. Similar rock structures are found on different continents. Parts of the Appalachian Mountains of the eastern United States are similar to those found in Greenland and western Europe. If you were to study rocks from eastern South America and western Africa, you would find other rock structures that also are similar. Rock clues like these support the idea that the continents were connected in the past.

250 million years ago

135 million years ago

Present day

How could continents drift?

Although Wegener provided evidence to support his hypothesis of continental drift, he couldn't explain how, when, or why these changes, shown in **Figure 4,** took place. The idea suggested that lower-density, continental material somehow had to plow through higher-density, ocean-floor material. The force behind this plowing was thought to be the spin of Earth on its axis—a notion that was quickly rejected by physicists. Because other scientists could not provide explanations either, Wegener's idea of continental drift was initially rejected. The idea was so radically different at that time that most people closed their minds to it.

Rock, fossil, and climate clues were the main types of evidence for continental drift. After Wegener's death, more clues were found, largely because of advances in technology, and new ideas that related to continental drift were developed. You'll learn about one of these new ideas, seafloor spreading, in the next section. Seafloor spreading helped provide an explanation of how the continents could move.

Figure 4
These computer models show the probable course the continents have taken. On the far left is their position 250 million years ago. In the middle is their position 135 million years ago. At right is their current position.

Section 1 Assessment

1. Why were Wegener's ideas about continental drift initially rejected?

2. How did Wegener use climate clues to support his hypothesis of continental drift?

3. What rock clues were used to support the hypothesis of continental drift?

4. In what ways do fossils help support the hypothesis of continental drift?

5. **Think Critically** Why would you expect to see similar rocks and rock structures on two landmasses that were connected at one time?

Skill Builder Activities

6. **Comparing and Contrasting** Compare and contrast the locations of fossils of the temperate plant *Glossopteris,* as shown in **Figure 2,** with the climate that exists at each location today. **For more help, refer to the** Science Skill Handbook.

7. **Communicating** Imagine that you are Alfred Wegener in the year 1912. In your Science Journal, write a letter to another scientist explaining your idea about continental drift. Try to convince this scientist that your hypothesis is correct. **For more help, refer to the** Science Skill Handbook.

2 Seafloor Spreading

As You Read

***What* You'll Learn**
- **Explain** seafloor spreading.
- **Recognize** how age and magnetic clues support seafloor spreading.

Vocabulary
seafloor spreading

***Why* It's Important**
Seafloor spreading helps explain how continents moved apart.

Figure 5
As the seafloor spreads apart at a mid-ocean ridge, new seafloor is created. The older seafloor moves away from the ridge in opposite directions.

Mapping the Ocean Floor

If you were to lower a rope from a boat until it reached the seafloor, you could record the depth of the ocean at that particular point. In how many different locations would you have to do this to create an accurate map of the seafloor? This is exactly how it was done until World War I, when the use of sound waves was introduced to detect submarines. During the 1940s and 1950s, scientists began using sound waves on moving ships to map large areas of the ocean floor in detail. Sound waves echo off the ocean bottom—the longer the sound waves take to return to the ship, the deeper the water is.

Using sound waves, researchers discovered an underwater system of ridges, or mountains, and valleys like those found on the continents. In the Atlantic, the Pacific, and in other oceans around the world, a system of ridges, called the mid-ocean ridges, is present. These underwater mountain ranges, shown in **Figure 5,** stretch along the center of much of Earth's ocean floor. This discovery raised the curiosity of many scientists. What formed these mid-ocean ridges?

✔ **Reading Check** *How were mid-ocean ridges discovered?*

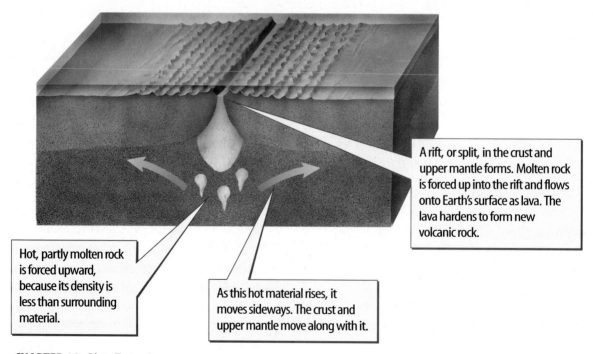

A rift, or split, in the crust and upper mantle forms. Molten rock is forced up into the rift and flows onto Earth's surface as lava. The lava hardens to form new volcanic rock.

Hot, partly molten rock is forced upward, because its density is less than surrounding material.

As this hot material rises, it moves sideways. The crust and upper mantle move along with it.

The Seafloor Moves In the early 1960s, Princeton University scientist Harry Hess suggested an explanation. His now-famous theory is known as **seafloor spreading.** Hess proposed that hot, less dense material below Earth's crust rises toward the surface at the mid-ocean ridges. Then, it flows sideways, carrying the seafloor away from the ridge in both directions, as seen in **Figure 5.**

As the seafloor spreads apart, magma moves upward and flows from the cracks. It becomes solid as it cools and forms new seafloor. As new seafloor moves away from the mid-ocean ridge, it cools, contracts, and becomes denser. This denser, colder seafloor sinks, helping to form the ridge. The theory of seafloor spreading was later supported by the following observations.

 **Reading Check** *How does new seafloor form at mid-ocean ridges?*

Figure 6
Many new discoveries have been made on the seafloor. These giant tube worms inhabit areas near hot water vents along mid-ocean ridges.

Evidence for Spreading In 1968, scientists aboard the research ship *Glomar Challenger* began gathering information about the rocks on the seafloor. *Glomar Challenger* was equipped with a drilling rig that allowed scientists to drill into the seafloor to obtain rock samples. They made a remarkable discovery as they studied the ages of the rocks. Scientists found that the youngest rocks are located at the mid-ocean ridges. The ages of the rocks become increasingly older in samples obtained farther from the ridges, adding to the evidence for seafloor spreading.

Using submersibles along mid-ocean ridges, new seafloor features and life-forms also were discovered there, as shown in **Figure 6.** As molten material rises along the ridges, it brings heat and chemicals that support exotic life-forms in deep, ocean water. Among these are giant clams, mussels, and tube worms.

 **Chemistry**
INTEGRATION

Find out what the Curie point is and describe in your Science Journal what happens to iron-bearing minerals when they are heated to the Curie point. Explain how this is important to studies of seafloor spreading.

 **Physics**
INTEGRATION

Magnetic Clues Earth's magnetic field has a north and a south pole. Magnetic lines, or directions, of force leave Earth near the south pole and enter Earth near the north pole. During a magnetic reversal, the lines of magnetic force run the opposite way. Scientists have determined that Earth's magnetic field has reversed itself many times in the past. These reversals occur over intervals of thousands or even millions of years. The reversals are recorded in rocks forming along mid-ocean ridges.

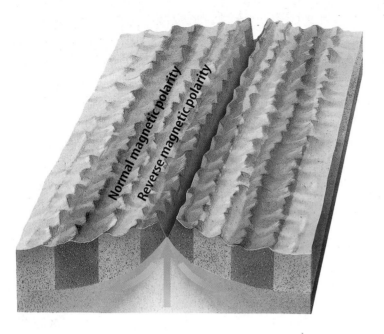

Figure 7
Changes in Earth's magnetic field are preserved in rock that forms on both sides of mid-ocean ridges. *Why is this considered to be evidence of seafloor spreading?*

Magnetic Time Scale Iron-bearing minerals, such as magnetite, that are found in the rocks of the seafloor can record Earth's magnetic field direction when they form. Whenever Earth's magnetic field reverses, newly forming iron minerals will record the magnetic reversal.

Using a sensing device called a magnetometer (mag nuh TAH muh tur) to detect magnetic fields, scientists found that rocks on the ocean floor show many periods of magnetic reversal. The magnetic alignment in the rocks reverses back and forth over time in strips parallel to the mid-ocean ridges, as shown in **Figure 7.** A strong magnetic reading is recorded when the polarity of a rock is the same as the polarity of Earth's magnetic field today. Because of this, normal polarities in rocks show up as large peaks. This discovery provided strong support that seafloor spreading was indeed occurring. The magnetic reversals showed that new rock was being formed at the mid-ocean ridges. This helped explain how the crust could move—something that the continental drift hypothesis could not do.

Section 2 Assessment

1. What properties of iron-bearing minerals on the seafloor support the theory of seafloor spreading?

2. How do the ages of the rocks on the ocean floor support the theory of seafloor spreading?

3. How did Harry Hess's hypothesis explain seafloor movement?

4. Why does some partly molten material rise toward Earth's surface?

5. **Think Critically** The ideas of Hess, Wegener, and others emphasize that Earth is a dynamic planet. How is seafloor spreading different from continental drift?

Skill Builder Activities

6. **Concept Mapping** Make a concept map that includes evidence for seafloor spreading using the following phrases: *ages increase away from ridge, pattern of magnetic field reversals, mid-ocean ridge, pattern of ages,* and *reverses back and forth.* **For more help,** refer to the Science Skill Handbook.

7. **Solving Simple Equations** North America is moving about 1.25 cm per year away from a ridge in the middle of the Atlantic Ocean. Using this rate, how much farther apart will North America and the ridge be in 200 million years? **For more help,** refer to the Math Skill Handbook.

Activity

Seafloor Spreading Rates

How did scientists use their knowledge of seafloor spreading and magnetic field reversals to reconstruct Pangaea? Try this activity to see how you can determine where a continent may have been located in the past.

What You'll Investigate

Can you use clues, such as magnetic field reversals on Earth, to help reconstruct Pangaea?

Materials

metric ruler
pencil

Goals

■ **Interpret** data about magnetic field reversals. Use these magnetic clues to reconstruct Pangaea.

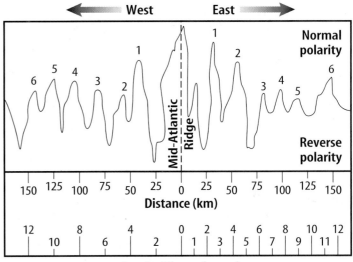

Procedure

1. Study the magnetic field graph above. You will be working only with normal polarity readings, which are the peaks above the baseline in the top half of the graph.

2. Place the long edge of a ruler vertically on the graph. Slide the ruler so that it lines up with the center of peak 1 west of the Mid-Atlantic Ridge.

3. **Determine** and record the distance and age that line up with the center of peak 1 west. Repeat this process for peak 1 east of the ridge.

4. **Calculate** the average age and distance for this pair of peaks.

5. Repeat steps 2 through 4 for the remaining pairs of normal-polarity peaks.

6. **Calculate** the rate of movement in cm per year for the six pairs of peaks. Use the formula rate = distance/time. Convert kilometers to centimeters. For example, to calculate a rate using normal-polarity peak 5, west of the ridge:

$$\text{rate} = \frac{125 \text{ km}}{10 \text{ million years}} = \frac{12.5 \text{ km}}{\text{million years}} =$$

$$\frac{1,250,000 \text{ cm}}{1,000,000 \text{ years}} = 1.25 \text{ cm/year}$$

Conclude and Apply

1. **Compare** the age of igneous rock found near the mid-ocean ridge with that of igneous rock found farther away from the ridge.

2. If the distance from a point on the coast of Africa to the Mid-Atlantic Ridge is approximately 2,400 km, calculate how long ago that point in Africa was at or near the Mid-Atlantic Ridge.

3. How could you use this method to reconstruct Pangaea?

Theory of Plate Tectonics

As You Read

What You'll Learn

- **Compare and contrast** different types of plate boundaries.
- **Explain** how heat inside Earth causes plate tectonics.
- **Recognize** features caused by plate tectonics.

Vocabulary

plate tectonics
plate
lithosphere
asthenosphere
convection current

Why It's Important

Plate tectonics explains how many of Earth's features form.

Plate Tectonics

The idea of seafloor spreading showed that more than just continents were moving, as Wegener had thought. It was now clear to scientists that sections of the seafloor and continents move in relation to one another.

Plate Movements In the 1960s, scientists developed a new theory that combined continental drift and seafloor spreading. According to the theory of **plate tectonics,** Earth's crust and part of the upper mantle are broken into sections. These sections, called **plates,** move on a plasticlike layer of the mantle. The plates can be thought of as rafts that float and move on this layer.

Composition of Earth's Plates Plates are made of the crust and a part of the upper mantle, as shown in **Figure 8.** These two parts combined are the **lithosphere** (LIH thuh sfihr). This rigid layer is about 100 km thick and generally is less dense than material underneath. The plasticlike layer below the lithosphere is called the **asthenosphere** (as THE nuh sfihr). The rigid plates of the lithosphere float and move around on the asthenosphere.

Figure 8
Plates of the lithosphere are composed of oceanic crust, continental crust, and rigid upper mantle.

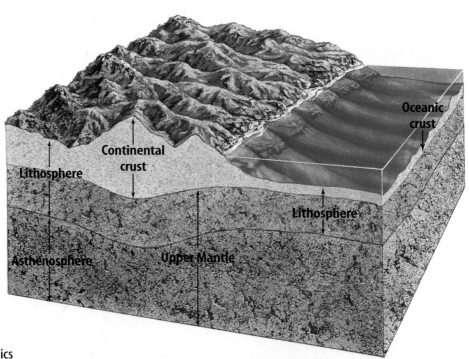

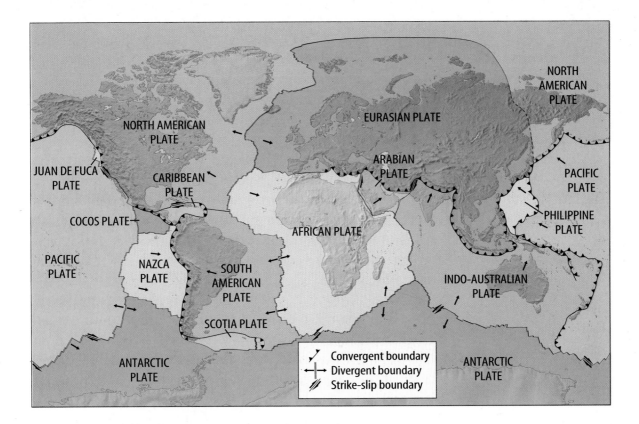

⟋	Convergent boundary
⟵—⟶	Divergent boundary
⟋⟋	Strike-slip boundary

Plate Boundaries

When plates move, they can interact in several ways. They can move toward each other and converge, or collide. They also can pull apart or slide alongside one another. When the plates interact, the result of their movement is seen at the plate boundaries, as in **Figure 9.**

Reading Check *What are the general ways that plates interact?*

Movement along any plate boundary means that changes must happen at other boundaries. What is happening to the Atlantic Ocean floor between the North American and African Plates? Compare this with what is happening along the western margin of South America.

Plates Moving Apart The boundary between two plates that are moving apart is called a divergent boundary. You learned about divergent boundaries when you read about seafloor spreading. In the Atlantic Ocean, the North American Plate is moving away from the Eurasian and the African Plates, as shown in **Figure 9.** That divergent boundary is called the Mid-Atlantic Ridge. The Great Rift Valley in eastern Africa might become a divergent plate boundary. There, a valley has formed where a continental plate is being pulled apart. **Figure 10** shows a side view of what a rift valley might look like and illustrates how the hot material rises up where plates separate.

Figure 9
This diagram shows the major plates of the lithosphere, their direction of movement, and the type of boundary between them. *Based on what is shown in this figure, what is happening where the Nazca Plate meets the Pacific Plate?*

Plates Moving Together If new crust is being added at one location, why doesn't Earth's surface keep expanding? As new crust is added in one place, it disappears below the surface at another. The disappearance of crust can occur when seafloor cools, becomes denser, and sinks. This occurs where two plates move together at a convergent boundary.

When an oceanic plate converges with a less dense continental plate, the denser oceanic plate sinks under the continental plate. The area where an oceanic plate subducts, or goes down, into the mantle is called a subduction zone. Some volcanoes form above subduction zones. **Figure 10** shows how this type of convergent boundary creates a deep-sea trench where one plate bends and sinks beneath the other. High temperatures cause rock to melt around the subducting slab as it goes under the other plate. The newly formed magma is forced upward along these plate boundaries, forming volcanoes. The Andes mountain range of South America contains many volcanoes. They were formed at the convergent boundary of the Nazca and the South American Plates.

Problem-Solving Activity

How well do the continents fit together?

Recall the Explore Activity you performed at the beginning of this chapter. While you were trying to fit pieces of a cut-up photograph together, what clues did you use?

Identifying the Problem

Take a copy of a map of the world and cut out each continent. Lay them on a tabletop and try to fit them together, using techniques you used in the Explore Activity. You will find that the pieces of your Earth puzzle—the continents—do not fit together well. Yet, several of the areas on some continents fit together extremely well.

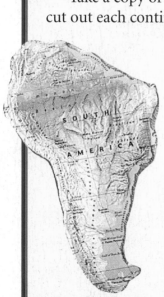

Take out another world map—one that shows the continental shelves as well as the continents. Copy it and cut out the continents, this time including the continental shelves.

Solving the Problem

1. Does including the continental shelves solve the problem of fitting the continents together?
2. Why should continental shelves be included with maps of the continents?

Figure 10

B y diverging at some boundaries and converging at others, Earth's plates are continually—but gradually—reshaping the landscape around you. The Mid-Atlantic Ridge, for example, was formed when the North and South American Plates pulled apart from the Eurasian and African Plates (see globe). Some features that occur along plate boundaries— rift valleys, volcanoes, and mountain ranges—are shown on the right and below.

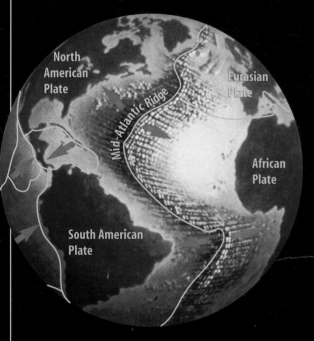

North American Plate

Eurasian Plate

Mid-Atlantic Ridge

African Plate

South American Plate

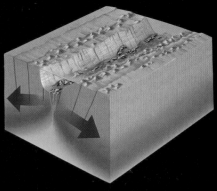

A RIFT VALLEY When continental plates pull apart, they can form rift valleys. The African continent is separating now along the East African Rift Valley.

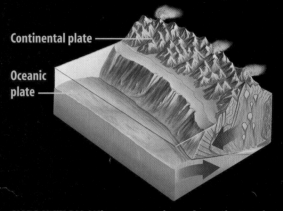

Continental plate

Oceanic plate

SUBDUCTION Where oceanic and continental plates collide, the oceanic plate plunges beneath the less dense continental plate. As the plate descends, molten rock (yellow) forms and rises toward the surface, creating volcanoes.

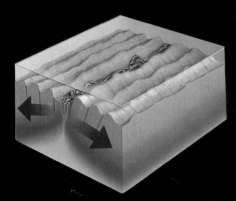

SEA-FLOOR SPREADING A mid-ocean ridge, like the Mid-Atlantic Ridge, forms where oceanic plates continue to separate. As rising magma (yellow) cools, it forms new oceanic crust.

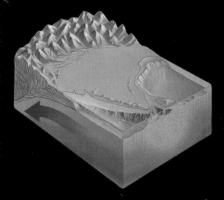

CONTINENTAL COLLISION Where two continental plates collide, they push up the crust to form mountain ranges such as the Himalaya.

287

Where Plates Collide A subduction zone also can form where two oceanic plates converge. In this case, the colder, older, denser oceanic plate bends and sinks down into the mantle. The Mariana Islands in the western Pacific are a chain of volcanic islands formed where two oceanic plates collide.

Usually, no subduction occurs when two continental plates collide, as shown in **Figure 10.** Because both of these plates are less dense than the material in the asthenosphere, the two plates collide and crumple up, forming mountain ranges. Earthquakes are common at these convergent boundaries. However, volcanoes do not form because there is no, or little, subduction. The Himalaya in Asia are forming where the Indo-Australian Plate collides with the Eurasian Plate.

Where Plates Slide Past Each Other The third type of plate boundary is called a transform boundary. Transform boundaries occur where two plates slide past one another. They move in opposite directions or in the same direction at different rates. When one plate slips past another suddenly, earthquakes occur. The Pacific Plate is sliding past the North American Plate, forming the famous San Andreas Fault in California, as seen in **Figure 11.** The San Andreas Fault is part of a transform plate boundary. It has been the site of many earthquakes.

Figure 11
The San Andreas Fault in California occurs along the transform plate boundary where the Pacific Plate is sliding past the North American Plate.

A Overall, the two plates are moving in roughly the same direction. *Why, then, do the red arrows show movement in opposite directions?*

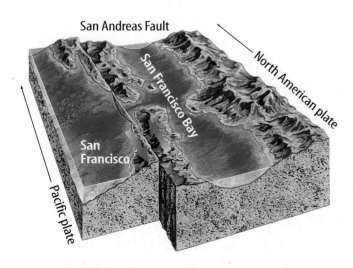

San Andreas Fault

San Francisco Bay

North American plate

San Francisco

Pacific plate

B This photograph shows an aerial view of the San Andreas Fault.

Causes of Plate Tectonics

Many new discoveries have been made about Earth's crust since Wegener's day, but one question still remains. What causes the plates to move? Scientists now think they have a good idea. They think that plates move by the same basic process that occurs when you heat soup.

Convection Inside Earth Soup that is cooking in a pan on the stove contains currents caused by an unequal distribution of heat in the pan. Hot, less dense soup is forced upward by the surrounding, cooler soup. As the hot soup reaches the surface, it cools and sinks back down into the pan. This entire cycle of heating, rising, cooling, and sinking is called a **convection current.** A version of this same process, occurring in the mantle, is thought to be the force behind plate tectonics. Scientists suggest that differences in density cause hot, plasticlike rock to be forced upward toward the surface.

Moving Mantle Material Wegener wasn't able to come up with an explanation for why plates move. Today, researchers who study the movement of heat in Earth's interior have proposed several possible explanations. All of the hypotheses use convection in one way or another. It is, therefore, the transfer of heat inside Earth that provides the energy to move plates and causes many of Earth's surface features. One hypothesis is shown in **Figure 12.** It relates plate motion directly to the movement of convection currents. According to this hypothesis, convection currents cause the movements of plates.

Mini LAB

Modeling Convection Currents

Procedure
1. Pour **water** into a **clear, colorless casserole dish** until it is 5 cm from the top.
2. Center the dish on a **hot plate** and heat it. WARNING: *Wear **thermal mitts** to protect your hands.*
3. Add a few drops of **food coloring** to the water above the center of the hot plate.
4. Looking from the side of the dish, observe what happens in the water.
5. Illustrate your observations in your **Science Journal.**

Analysis
1. Determine whether any currents form in the water.
2. Infer what causes the currents to form.

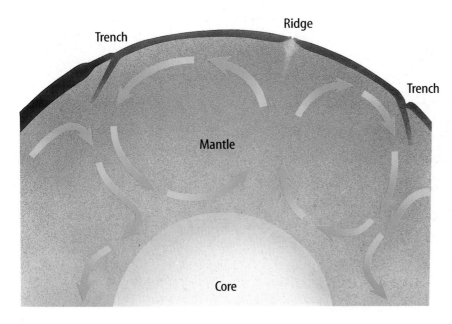

Figure 12
In one hypothesis, convection currents occur throughout the mantle. Such convection currents (see arrows) are the driving force of plate tectonics.

Features Caused by Plate Tectonics

Earth is a dynamic planet with a hot interior. This heat leads to convection, which powers the movement of plates. As the plates move, they interact. The interaction of plates produces forces that build mountains, create ocean basins, and cause volcanoes. When rocks in Earth's crust break and move, energy is released in the form of seismic waves. Humans feel this release as earthquakes. You can see some of the effects of plate tectonics in mountainous regions, where volcanoes erupt, or where landscapes have changed from past earthquake or volcanic activity.

Reading Check *What happens when seismic energy is released as rocks in Earth's crust break and move?*

Normal Faults and Rift Valleys Tension forces, which are forces that pull apart, can stretch Earth's crust. This causes large blocks of crust to break and tilt or slide down the broken surfaces of crust. When rocks break and move along surfaces, a fault forms. Faults interrupt rock layers by moving them out of place. Entire mountain ranges can form in the process, called fault-block mountains, as shown in **Figure 13.** Generally, the faults that form from pull-apart forces are normal faults—faults in which the rock layers above the fault move down when compared with rock layers below the fault.

Rift valleys and mid-ocean ridges can form where Earth's crust separates. Examples of rift valleys are the Great Rift Valley in Africa, and the valleys that occur in the middle of mid-ocean ridges. Examples of mid-ocean ridges include the Mid-Atlantic Ridge and the East Pacific Rise.

Figure 13
Fault-block mountains can form when Earth's crust is stretched by tectonic forces. The arrows indicate the directions of moving blocks. *What type of force occurs when Earth's crust is pulled in opposite directions?*

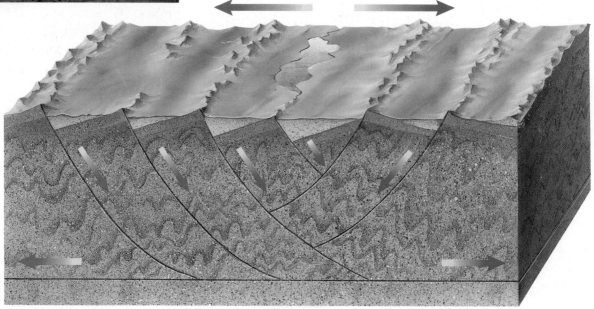

Mountains and Volcanoes Compression forces squeeze objects together. Where plates come together, compression forces produce several effects. As continental plates collide, the forces that are generated cause massive folding and faulting of rock layers into mountain ranges such as the Himalaya, shown in **Figure 14,** or the Appalachian Mountains. The type of faulting produced is generally reverse faulting. Along a reverse fault, the rock layers above the fault surface move up relative to the rock layers below the fault.

☑ **Reading Check** *What features occur where plates converge?*

As you learned earlier, when two oceanic plates converge, the denser plate is forced beneath the other plate. Curved chains of volcanic islands called island arcs form above the sinking plate. If an oceanic plate converges with a continental plate, the denser oceanic plate slides under the continental plate. Folding and faulting at the continental plate margin can thicken the continental crust to produce mountain ranges. Volcanoes also typically are formed at this type of convergent boundary.

Are there any features caused by plate tectonics in your area? To find out more about these features, see the **Field Guide to Faults and Folds** at the back of the book.

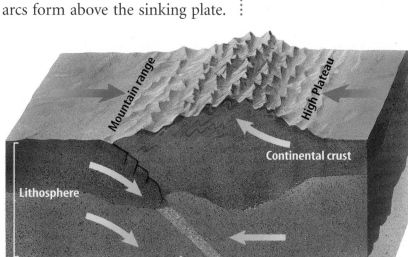

Figure 14
The Himalaya still are forming today as the Indo-Australian Plate collides with the Eurasian Plate.

Figure 15
Most of the movement along a strike-slip fault is parallel to Earth's surface. When movement occurs, human-built structures along a strike-slip fault are off-set, as shown here in this road.

Strike-Slip Faults At transform boundaries, two plates slide past one another without converging or diverging. The plates stick and then slide, mostly in a horizontal direction, along large strike-slip faults. In a strike-slip fault, rocks on opposite sides of the fault move in opposite directions, or in the same direction at different rates. This type of fault movement is shown in **Figure 15.** One such example is the San Andreas Fault. When plates move suddenly, vibrations are generated inside Earth that are felt as an earthquake.

Earthquakes, volcanoes, and mountain ranges are evidence of plate motion. Plate tectonics explains how activity inside Earth can affect Earth's crust differently in different locations. You've seen how plates have moved since Pangaea separated. Is it possible to measure how far plates move each year?

Testing for Plate Tectonics

Until recently, the only tests scientists could use to check for plate movement were indirect. They could study the magnetic characteristics of rocks on the seafloor. They could study volcanoes and earthquakes. These methods supported the theory that the plates have moved and still are moving. However, they did not provide proof—only support—of the idea.

New methods had to be discovered to be able to measure the small amounts of movement of Earth's plates. One method, shown in **Figure 16,** uses lasers and a satellite. Now, scientists can measure exact movements of Earth's plates of as little as 1 cm per year.

Physics
INTEGRATION

In which directions do forces act at convergent, divergent, and transform boundaries? Demonstrate these forces using wooden blocks or your hands.

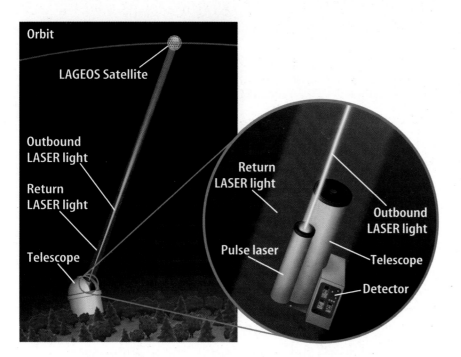

Current Data Satellite data show that Hawaii is moving toward Japan at a rate of about 8.3 cm per year. Maryland is moving away from England at a rate of 1.7 cm per year. Using such methods, scientists have observed that the plates move at rates ranging from about 1 cm to 12 cm per year.

Section Assessment

1. What happens to plates at a transform plate boundary?

2. What occurs at plate boundaries that are associated with seafloor spreading?

3. Describe three types of plate boundaries where volcanic eruptions can occur.

4. How are convection currents related to plate tectonics?

5. **Think Critically** Using **Figure 9** and a world map, determine what natural disasters might occur in Iceland. Also determine what disasters might occur in Tibet. Explain why some Icelandic disasters are not expected to occur in Tibet.

Skill Builder Activities

6. **Predicting** Plate tectonic activity causes many events that can be dangerous to humans. One of these events is a seismic sea wave, or tsunami. Learn how scientists predict the arrival time of a tsunami in a coastal area. **For more help, refer to the** Science Skill Handbook.

7. **Using a Word Processor** Write three separate descriptions of the three basic types of plate boundaries—divergent boundaries, convergent boundaries, and transform boundaries. Then draw a sketch of an example of each boundary next to your description. **For more help, refer to the** Technology Skill Handbook.

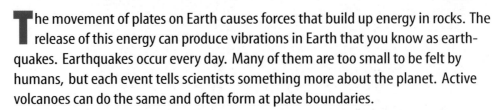

Predicting Tectonic Activity

The movement of plates on Earth causes forces that build up energy in rocks. The release of this energy can produce vibrations in Earth that you know as earthquakes. Earthquakes occur every day. Many of them are too small to be felt by humans, but each event tells scientists something more about the planet. Active volcanoes can do the same and often form at plate boundaries.

Recognize the Problem

Can you predict tectonically active areas by plotting locations of earthquake epicenters and volcanic eruptions?

Form a Hypothesis

Think about where earthquakes and volcanoes have occurred in the past. Make a hypothesis about whether the locations of earthquake epicenters and active volcanoes can be used to predict tectonically active areas.

Goals
- **Research** the locations of earthquakes and volcanic eruptions around the world.
- **Plot** earthquake epicenters and the locations of volcanic eruptions obtained from the Glencoe Science Web site.
- **Predict** locations that are tectonically active based on a plot of the locations of earthquake epicenters and active volcanoes.

Data Sources
SCIENCE*Online* Go to the Glencoe Science Web site at **science.glencoe.com** for more information about earthquake and volcano sites, hints about earthquake and volcano sites, and data from other students.

Test Your Hypothesis

Plan

1. Make a data table in your Science Journal like the one shown.

2. Collect data for earthquake epicenters and volcanic eruptions for at least the past two weeks. Your data should include the longitude and latitude for each location. For help, refer to the data sources given on the opposite page.

Locations of Epicenters and Eruptions

Earthquake Epicenter/ Volcanic Eruption	Longitude	Latitude

Do

1. Make sure your teacher approves your plan before you start.

2. **Plot** the locations of earthquake epicenters and volcanic eruptions on a map of the world. Use an overlay of tissue paper or plastic.

3. After you have collected the necessary data, predict where the tectonically active areas on Earth are.

4. **Compare and contrast** the areas that you predicted to be tectonically active with the plate boundary map shown in **Figure 9.**

Analyze Your Data

1. What areas on Earth do you predict to be the locations of tectonic activity?

2. How close did your prediction come to the actual location of tectonically active areas?

Draw Conclusions

1. How could you make your predictions closer to the locations of actual tectonic activity?

2. Would data from a longer period of time help? Explain.

3. What types of plate boundaries were close to your locations of earthquake epicenters? Volcanic eruptions?

4. **Explain** which types of plate boundaries produce volcanic eruptions. Be specific.

*C*ommunicating
Your Data

SCIENCE *Online* Find this Internet activity on the Glencoe Science Web site at **science.glencoe .com. Post** your data in the table provided. **Compare** your data to those of other students. Combine your data with those of other students and **plot** these combined data on a map to **recognize** the relationship between plate boundaries, volcanic eruptions, and earthquake epicenters.

Listening In
by Gordon Judge

Respond to the Reading

1. Who is narrating the poem?

2. Why might the narrator think he or she hasn't "moved for ages"?

3. Who or what is the narrator's "best mate"?

I'm just a bit of seafloor on this mighty solid sphere.
With no mind to be broadened, I'm quite content down here.
The mantle churns below me, and the sea's in turmoil, too;
But nothing much disturbs me, I'm rock solid through and
    through.

I do pick up occasional low-frequency vibrations –
(I think, although I can't be sure, they're sperm whales'
    conversations).
I know I shouldn't listen in, but what else can I do?
It seems they are all studying for degrees from the OU.

They've mentioned me in passing, as their minds begin improving:
I think I've heard them say "The theory says the sea-floor's
    moving…".
Well, that shook me, I can tell you; yes, it gave me quite a fright.
Yet I've not moved for ages, so I *know* it can't be right.

They call it "Plate Tectonics", this new theory in their noddle.
If they would only ask me, I could tell them it's all twaddle.
Apparently, I "oozed out from a mid-Atlantic split,
Solidified and cooled right down, then moved out bit by bit".

But, how can I be moving, when I know full well myself
That I'm quite firmly anchored to a continental shelf?
"Well, the continent is moving, too; you're *pushing* it, you see,"
I hear those OU whales intone, hydro-acoustically.

Now, my best mate's a sea floor in the mighty East Pacific.
He reckons life is balmy there: the summers are terrific!
He's heard the whale-talk, too, and found it pretty scary.
"Subduction" was the word he heard, which sounded rather hairy.

It was to be his fate, they claimed with undisguised great relish:
A hot and fiery end to things – it really would be hellish.
In fact, he'd end up underneath *my* continent, lengthwise,
So I would be the one to blame for my poor mate's demise.

Well, thank you very much, OU. You've upset my composure.
Next time you send your student whales to look at my exposure
I'll tell them it's a load of tosh: it's *they* who move, not me,
Those arty-smarty blobs of blubber, clogging up the sea!

Understanding Literature

Point of View Point of view refers to the perspective from which an author writes. This poem begins, "I'm just a bit of sea floor…." Right away, you know that the poem, or story, is being told from the point of view of the speaker, or the "first person." Not all first-person stories are told from the point of view of a person. The narrator in this poem is a geological feature, not a person. This point of view helps give the poem a fantastic or outlandish quality. It also gives a playful tone to the poem. What other effects does the first-person narration have on the story?

Science Connection Volcanoes can occur where two plates move toward each other. In the poem, the author gives several clues that a volcano will form. First, the narrator's "best mate" is a seafloor in the Pacific Ocean. When an oceanic plate and a continental plate collide, a volcano will form. The narrator also hears the word *subduction* spoken. Subduction zones occur when one plate sinks under another plate. Rocks melt in the zones where these plates converge, causing magma to move upward and form volcanic mountains. What other clues does the author give that a volcano will form?

Linking Science and Writing

Using Point of View Using the first-person point of view, write an account from the point of view of a living or nonliving thing. You could write an account of an object, such as a pencil, that you use or encounter every day. You also could write from the point of view of a living thing, such as a family pet. Be sure to use the personal pronoun "I" in your account.

Career Connection

Volcanologist

Ed Klimasauskas is a volcanologist at the Cascades Volcano Observatory in Washington State. His job is to study volcanoes in order to predict eruptions. Volcanologists' predictions can save lives: people can be evacuated from danger areas before an eruption occurs. Klimasauskas also educates the public about the hazards of volcanic eruptions and tells people who live near active volcanoes what they can do to be safe in case a volcano erupts. Volcanologists travel all over the world to study new sites.

SCIENCE*Online* To learn more about careers in volcanology, visit the Glencoe Science Web site at **science.glencoe.com.**

Reviewing Main Ideas

Section 1 Continental Drift

1. Alfred Wegener suggested that the continents were joined together at some point in the past in a large landmass he called Pangaea. Wegener proposed that continents have moved slowly, over millions of years, to their current locations.

2. The puzzlelike fit of the continents, fossils, climatic evidence, and similar rock structures support Wegener's idea of continental drift. However, Wegener could not explain what process could cause the movement of the landmasses. *How do fossils support the hypothesis of continental drift?*

Section 2 Seafloor Spreading

1. Detailed mapping of the ocean floor in the 1950s showed underwater mountains and rift valleys.

2. In the 1960s, Harry Hess suggested seafloor spreading as an explanation for the formation of mid-ocean ridges. *How is magnetic evidence preserved in rocks forming along a mid-ocean ridge?*

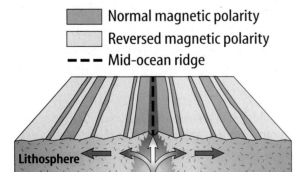

■ Normal magnetic polarity
□ Reversed magnetic polarity
--- Mid-ocean ridge

Lithosphere

3. The theory of seafloor spreading is supported by magnetic evidence in rocks and by the ages of rocks on the ocean floor.

Section 3 Theory of Plate Tectonics

1. In the 1960s, scientists combined the ideas of continental drift and seafloor spreading to develop the theory of plate tectonics. The theory states that the surface of Earth is broken into sections called plates that move around on the asthenosphere.

2. Currents in Earth's mantle called convection currents transfer heat in Earth's interior. It is thought that this transfer of heat energy moves plates.

3. Earth is a dynamic planet. As the plates move, they interact, resulting in many of the features of Earth's surface. *How do converging plates form mountains?*

After You Read

FOLDABLES
Reading & Study Skills

To help you review what you learned about plate tectonics, use the Foldable you made at the beginning of the chapter.

Visualizing Main Ideas

Complete the concept map below about continental drift, seafloor spreading, and plate tectonics.

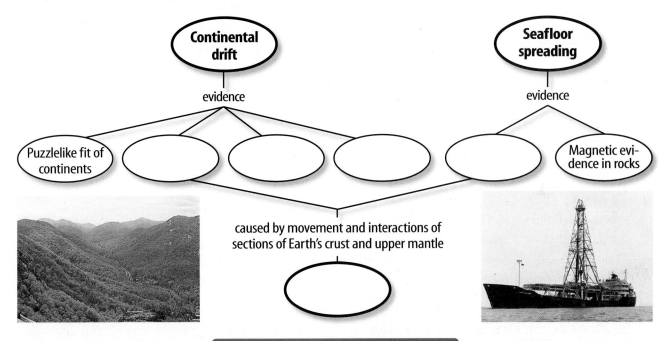

Vocabulary Review

Vocabulary Words

a. asthenosphere
b. continental drift
c. convection current
d. lithosphere
e. Pangaea
f. plate
g. plate tectonics
h. seafloor spreading

Using Vocabulary

Each phrase below describes a vocabulary term from the list. Write the term that matches the phrase describing it.

1. plasticlike layer below the lithosphere

Study Tip

Make a note of anything you don't understand so that you'll remember to ask your teacher about it.

2. idea that continents move slowly across Earth's surface

3. large, ancient landmass that consisted of all the continents on Earth

4. process that forms new seafloor as hot material is forced upward

5. driving force for plate movement

6. composed of oceanic or continental crust and upper mantle

7. explains locations of mountains, trenches, and volcanoes

8. piece of the lithosphere that moves over a plasticlike layer

9. theory proposed by Harry Hess that includes processes along mid-ocean ridges

10. forms as warm material rises and cold material sinks

Checking Concepts

Choose the word or phrase that best answers the question.

1. Which layer of Earth contains the asthenosphere?
   A) crust   C) outer core
   B) mantle   D) inner core

2. What type of plate boundary is the San Andreas Fault part of?
   A) divergent   C) convergent
   B) subduction   D) transform

3. What hypothesis states that continents slowly moved to their present positions on Earth?
   A) subduction   C) continental drift
   B) seafloor spreading D) erosion

4. Which plate is subducting beneath the South American Plate to form the Andes mountain range?
   A) North American   C) Indo-Australian
   B) African   D) Nazca

5. Which of the following features indicates that many continents were once near Earth's south pole?
   A) glacial deposits   C) volcanoes
   B) mid-ocean ridges   D) earthquakes

6. What evidence in rocks supports the theory of seafloor spreading?
   A) plate movement   C) subduction
   B) magnetic reversals D) convergence

7. Which type of plate boundary is the Mid-Atlantic Ridge a part of?
   A) convergent   C) transform
   B) divergent   D) lithosphere

8. What theory states that plates move around on the asthenosphere?
   A) continental drift   C) subduction
   B) seafloor spreading   D) plate tectonics

9. What forms when one plate slides past another plate?
   A) transform boundary
   B) divergent boundary
   C) subduction zone
   D) mid-ocean ridge

10. When oceanic plates collide, what volcanic landforms are made?
    A) folded mountains
    B) island arcs
    C) strike-slip faults
    D) mid-ocean ridges

Thinking Critically

11. Why do many earthquakes but few volcanic eruptions occur in the Himalaya?

12. Glacial deposits often form at high latitudes near the poles. Explain why glacial deposits have been found in Africa.

13. How is magnetism used to support the theory of seafloor spreading?

14. Explain why volcanoes do not form along the San Andreas Fault.

15. Explain why the fossil of an ocean fish found on two different continents would not be good evidence of continental drift.

Developing Skills

16. **Forming Hypotheses** Mount St. Helens in the Cascade Range is a volcano. Use **Figure 9** and a U.S. map to hypothesize how it might have formed.

17. **Measuring in SI** Movement along the African Rift Valley is about 2.1 cm per year. If plates continue to move apart at this rate, how much larger will the rift be (in meters) in 1,000 years? In 15,500 years?

18. Concept Mapping Make an events chain concept map that describes seafloor spreading along a divergent plate boundary. Choose from the following phrases: *magma cools to form new seafloor, convection currents circulate hot material along divergent boundary,* and *older seafloor is forced apart.*

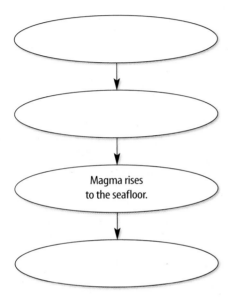

Magma rises to the seafloor.

Performance Assessment

19. Observe and Infer In the MiniLab Modeling Convection Currents, you observed convection currents produced in water as it was heated. Repeat the experiment, placing sequins, pieces of wood, or pieces of rubber bands into the water. How do their movements support your observations and inferences from the MiniLab?

THE PRINCETON REVIEW Test Practice

Ms. Fernandez was leading a class discussion on plate tectonics and Earth's interior.

Study the diagram below and then answer the following questions.

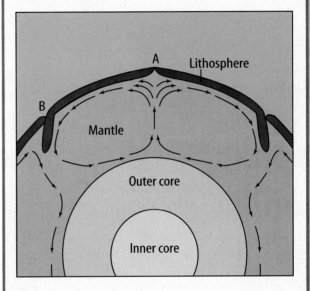

1. Suppose that the arrows in the diagram represent patterns of convection in Earth's mantle. Which type of plate boundary is most likely to form along the region labeled "A"?

A) transform
B) reverse
C) convergent
D) divergent

2. Which statement is true of the region marked "B" on the diagram?

F) Plates separate and slip past one another sideways.
G) Plates diverge and volcanoes form.
H) Plates converge and volcanoes form.
J) Plates collapse and form a strike-slip boundary.

11 Earthquakes

More than 20,000 deaths and at least 166,000 injuries resulted from a powerful earthquake in India on January 26, 2001. Collapse of structures, such as this building in Ahmedabad, India, is among the greatest dangers associated with earthquakes. What causes earthquakes? Why do some areas experience repeated earthquakes while other areas rarely do? In this chapter you'll learn the answers to these and other questions. You'll also learn how earthquake damage can be reduced.

What do you think?

Science Journal Look at the picture below with a classmate. Discuss what you think this might be. Here's a hint: *Something must have been shaking to make these waves.* Write your answer or best guess in your Science Journal.

EXPLORE ACTIVITY

Why do earthquakes occur? The bedrock beneath the soil can break and form cracks known as faults. When blocks of rock move past each other along a fault, they cause the ground to shake. Why don't rocks move all the time, causing constant earthquakes? You'll find out during this activity.

Model stress buildup along faults

1. Tape a sheet of medium-grain sandpaper to the tabletop.

2. Tape a second sheet of sandpaper to the book cover on a textbook.

3. Place the book on the table so that both sheets of sandpaper meet.

4. Tie two large, thick rubber bands together and loop one of the rubber bands around the edge of the book so that it is not touching the sandpaper.

5. Pull on the free rubber band until the book moves and observe this movement.

Observe

Write a paragraph in your Science Journal describing how the book moved and explaining how this activity modeled the buildup of stress along a fault.

Before You Read

FOLDABLES
Reading & Study Skills

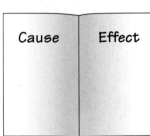

Making a Cause and Effect Study Fold Make the following Foldable to help you understand the cause and effect relationship of earthquakes and Earth's crust.

1. Place a sheet of paper in front of you so the long side is at the top. Fold the paper in half from the left side to the right side and then unfold.

2. Label the left side of the paper *Cause* and the right side *Effect*. Refold the paper.

3. Before you read the chapter, draw a cross section of Earth's crust showing what you think happens during an earthquake on the outside of your Foldable.

4. As you read the chapter, change your drawing and list causes of earthquakes on the inside of your Foldable.

Cause	Effect

Forces Inside Earth

As You Read

***What* You'll Learn**

- **Explain** how earthquakes result from the buildup of energy in rocks.
- **Describe** how compression, tension, and shear forces make rocks move along faults.
- **Distinguish** among normal, reverse, and strike-slip faults.

Vocabulary

fault   reverse fault
earthquake   strike-slip fault
normal fault

***Why* It's Important**

Earthquakes are among the most dramatic of all natural disasters on Earth.

Earthquake Causes

Recall the last time you used a rubber band. Rubber bands stretch when you pull them. Because they are elastic, they return to their original shape once the force is released. However, if you stretch a rubber band too far, it will break. A wooden craft stick behaves in a similar way. When a force is first applied to the stick, it will bend and change shape, as shown in **Figure 1A.** The energy needed to bend the stick is stored inside the stick as potential energy. If the force keeping the stick bent is removed, the stick will return to its original shape, and the stored energy will be released as energy of motion.

Fault Formation   There is a limit to how far a wooden craft stick can bend. This is called its elastic limit. Once its elastic limit is passed, the stick breaks, as shown in **Figure 1B.** Rocks behave in a similar way. Up to a point, applied forces cause rocks to bend and stretch, undergoing what is called elastic deformation. Once the elastic limit is passed, the rocks may break. When rocks break, they move along surfaces called **faults.** A tremendous amount of force is required to overcome the strength of rocks and to cause movement along a fault. Rock along one side of a fault can move up, down, or sideways in relation to rock along the other side of the fault.

A   **B**

Figure 1
The **A** bending and
B breaking of
wooden craft sticks
are similar to how
rocks bend and break.

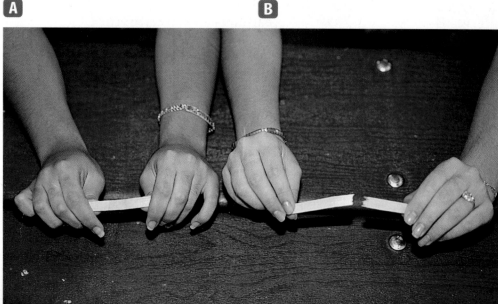

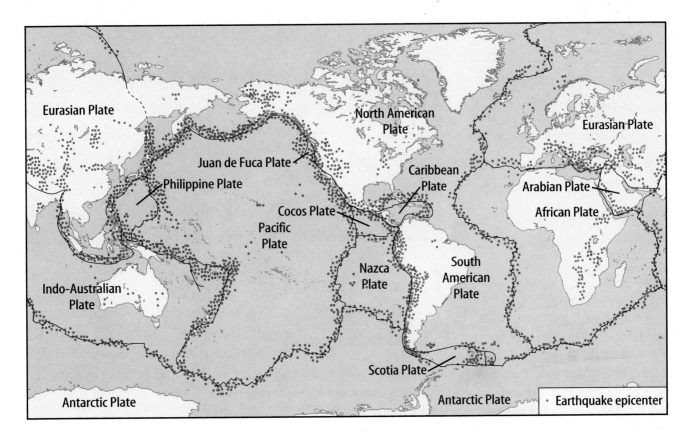

Eurasian Plate

North American Plate

Eurasian Plate

Juan de Fuca Plate

Philippine Plate

Caribbean Plate

Arabian Plate

Cocos Plate

African Plate

Pacific Plate

Nazca Plate

South American Plate

Indo-Australian Plate

Scotia Plate

Antarctic Plate

Antarctic Plate

· Earthquake epicenter

What causes faults? What produces the forces that cause rocks to break and faults to form? The surface of Earth is in constant motion because of forces inside the planet. These forces cause sections of Earth's surface, called plates, to move. This movement puts stress on the rocks near the plate edges. To relieve this stress, the rocks tend to bend, compress, or stretch. If the force is great enough, the rocks will break. An **earthquake** is the vibrations produced by the breaking of rock. **Figure 2** shows how the locations of earthquakes outline the plates that make up Earth's surface.

Reading Check *Why do most earthquakes occur near plate boundaries?*

How Earthquakes Occur As rocks move past each other along a fault, their rough surfaces catch, temporarily halting movement along the fault. However, forces keep driving the rocks to move. This action builds up stress at the points where the rocks are stuck. The stress causes the rocks to bend and change shape. When the rocks are stressed beyond their elastic limit, they break, move along the fault, and return to their original shapes. An earthquake results. Earthquakes range from unnoticeable vibrations to devastating waves of energy. Regardless of their intensity, most earthquakes result from rocks moving over, under, or past each other along fault surfaces.

Figure 2
The dots represent the epicenters of major earthquakes over a ten-year period. Note that most earthquakes occur near plate boundaries. *Why do earthquakes rarely occur in the middle of plates?*

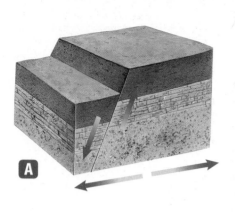

B

Figure 3

A When rock moves along a fracture caused by tension forces, the break is called a normal fault. Rock above the normal fault moves downward in relation to rock below the fault surface.
B This normal fault formed near Kanab, Utah.

Figure 4

A Compression forces in rocks form reverse faults. The rock above the reverse fault surface moves upward in relation to the rock below the fault surface.
B Rock layers have been offset along this reverse fault.

Types of Faults

Physics INTEGRATION

Three types of forces—tension, compression, and shear—act on rocks. Tension is the force that pulls rocks apart, and compression is the force that squeezes rocks together. Shear is the force that causes rocks on either side of a fault to slide past each other.

Normal Faults Tensional forces inside Earth cause rocks to be pulled apart. When rocks are stretched by these forces, a normal fault can form. Along a **normal fault,** rock above the fault surface moves downward in relation to rock below the fault surface. The motion along a normal fault is shown in **Figure 3A.** Notice the normal fault shown in the photograph in **Figure 3B.**

Reverse Faults Reverse faults result from compression forces that squeeze rock. **Figure 4A** shows the motion along a reverse fault. If rock breaks from forces pushing from opposite directions, rock above a **reverse fault** surface is forced up and over the rock below the fault surface. **Figure 4B** shows a large reverse fault in California.

B

A

Figure 5

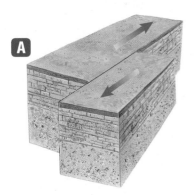 Shear forces push on rock in opposite—but not directly opposite—horizontal directions. When they are strong enough, these forces split rock and create strike-slip faults. Little vertical movement occurs along a strike-slip fault.  The North American Plate and the Pacific Plate slide past each other along the San Andreas Fault, a strike-slip fault, in California.

Strike-Slip Faults At a **strike-slip fault,** shown in **Figure 5A,** rocks on either side of the fault are moving past each other without much upward or downward movement. **Figure 5B** shows the largest fault in California—the San Andreas Fault—which stretches more than 1,100 km through the state. The San Andreas Fault is the boundary between two of Earth's plates that are moving sideways past each other.

 Reading Check *What is a strike-slip fault?*

Section 1 Assessment

1. What is an earthquake?
2. The Himalaya in Tibet formed when two of Earth's plates collided. What type of faults would you expect to find in these mountains? Why?
3. In what direction do rocks above a normal fault surface move?
4. Why is California's San Andreas Fault a strike-slip fault?
5. **Think Critically** Why is it easier to predict where an earthquake will occur than it is to predict when it will occur?

Skill Builder Activities

6. **Forming Hypotheses** Hypothesize why the chances of an earthquake occurring along a fault increase rather than decrease as time since the last earthquake passes. **For more help, refer to the** Science Skill Handbook.
7. **Using Graphics Software** Use a graphics program to make models of the three types of faults—normal, reverse, and strike-slip. Add arrows to show the directions of movement along both sides of each type. **For more help, refer to the** Technology Skill Handbook.

Features of Earthquakes

As You Read

What You'll Learn

- **Explain** how earthquake energy travels in seismic waves.
- **Distinguish** among primary, secondary, and surface waves.
- **Describe** the structure of Earth's interior.

Vocabulary

seismic wave   surface wave
focus   epicenter
primary wave   seismograph
secondary wave

Why It's Important

Seismic waves are responsible for most damage caused by earthquakes.

Seismic Waves

When two people hold opposite ends of a rope and shake one end, as shown in **Figure 6,** they send energy through the rope in the form of waves. Like the waves that travel through the rope, **seismic** (SIZE mihk) **waves** generated by an earthquake travel through Earth. During a strong earthquake, the ground moves forward and backward, heaves up and down, and shifts from side to side. The surface of the ground can ripple like waves do in water. Imagine trying to stand on ground that had waves traveling through it. This is what you might experience during a strong earthquake.

Origin of Seismic Waves You learned earlier that rocks move past each other along faults, creating stress at points where the rocks' irregular surfaces catch each other. The stress continues to build up until the elastic limit is exceeded and energy is released in the form of seismic waves. The point where this energy release first occurs is the **focus** (plural, *foci*) of the earthquake. The foci of most earthquakes are within 65 km of Earth's surface. A few have been recorded as deep as 700 km. Seismic waves are produced and travel outward from the earthquake focus.

Figure 6
Some seismic waves are similar to the wave that is traveling through the rope. Note that the rope moves perpendicular to the wave direction.

Primary Waves When earthquakes occur, three different types of seismic waves are produced. All of the waves are generated at the same time, but each behaves differently within Earth. **Primary waves** (P-waves) cause particles in rocks to move back and forth in the same direction that the wave is traveling. If you squeeze one end of a coiled spring and then release it, you cause it to compress and then stretch as the wave travels through the spring, as shown in **Figure 7.** Particles in rocks also compress and then stretch apart, transmitting primary waves through the rock.

Secondary and Surface Waves Secondary waves (S-waves) move through Earth by causing particles in rocks to move at right angles to the direction of wave travel. The wave traveling through the rope shown in **Figure 6** is an example of a secondary wave.

Surface waves cause most of the destruction resulting from earthquakes. **Surface waves** move rock particles in a backward, rolling motion and a side-to-side, swaying motion, as shown in **Figure 8.** Many buildings are unable to withstand intense shaking because they are made with stiff materials. The buildings fall apart when surface waves cause different parts of the building to move in different directions.

✓ Reading Check *Why do surface waves damage buildings?*

Surface waves are produced when earthquake energy reaches the surface of Earth. Surface waves travel outward from the epicenter. The earthquake **epicenter** (EH pi sen tur) is the point on Earth's surface directly above the earthquake focus. Find the focus and epicenter in **Figure 9.**

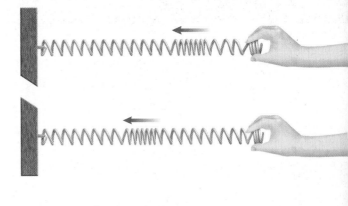

Figure 7
Primary waves move through Earth the same way that a wave travels through a coiled spring.

Physics
INTEGRATION

When sound is produced, waves move through air or some other material. Research sound waves to find out which type of seismic wave they are similar to.

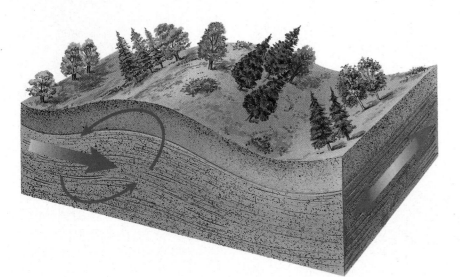

Figure 8
Surface waves move rock particles in a backward, rolling motion and a side-to-side, swaying motion. *How does this movement differ from rock movement caused by secondary waves?*

Figure 9

As the plates that form Earth's lithosphere move, great stress is placed on rocks. They bend, stretch, and compress. Occasionally, rocks break, producing earthquakes that generate seismic waves. As shown here, different kinds of seismic waves—each with distinctive characteristics—move outward from the focus of the earthquake.

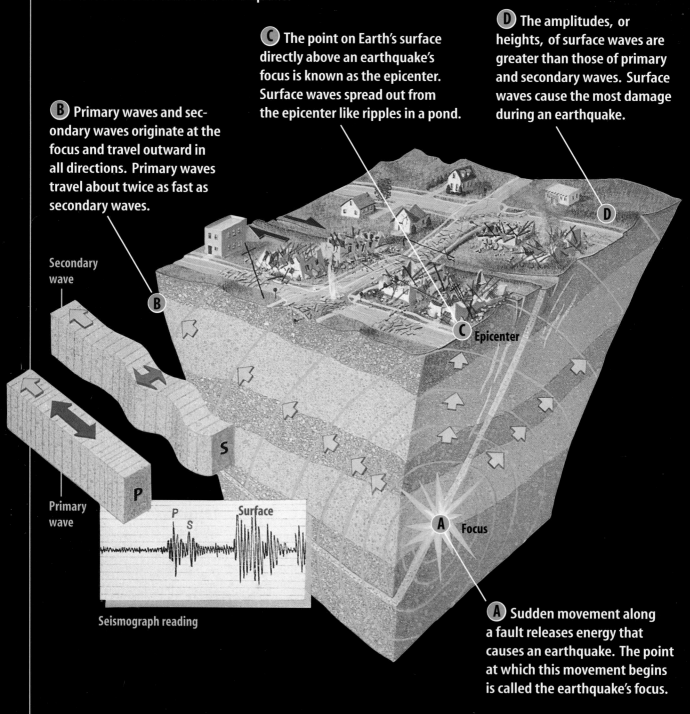

C The point on Earth's surface directly above an earthquake's focus is known as the epicenter. Surface waves spread out from the epicenter like ripples in a pond.

D The amplitudes, or heights, of surface waves are greater than those of primary and secondary waves. Surface waves cause the most damage during an earthquake.

B Primary waves and secondary waves originate at the focus and travel outward in all directions. Primary waves travel about twice as fast as secondary waves.

Secondary wave

Primary wave

Seismograph reading

Epicenter

Focus

A Sudden movement along a fault releases energy that causes an earthquake. The point at which this movement begins is called the earthquake's focus.

Locating an Epicenter

Different seismic waves travel through Earth at different speeds. Primary waves are the fastest, secondary waves are slower, and surface waves are the slowest. Can you think of a way this information could be used to determine how far away an earthquake epicenter is? Think of the last time you saw two people running in a race. You probably noticed that the faster person got further ahead as the race continued. Like runners in a race, seismic waves travel at different speeds.

Scientists have learned how to use the different speeds of seismic waves to determine the distance to an earthquake epicenter. When an epicenter is far from a location, the primary wave has more time to put distance between it and the secondary and surface waves, just like the fastest runner in a race.

Measuring Seismic Waves Seismic waves from earthquakes are measured with an instrument known as a **seismograph.** Seismographs register the waves and record the time that each arrived. Seismographs consist of a rotating drum of paper and a pendulum with an attached pen. When seismic waves reach the seismograph, the drum vibrates but the pendulum remains at rest. The stationary pen traces a record of the vibrations on the moving drum of paper. The paper record of the seismic event is called a seismogram. **Figure 10** shows two types of seismographs that measure either vertical or horizontal ground movement, depending on the orientation of the drum.

SCIENCE *Online*

Research Visit the Glencoe Science Web site at **science.glencoe.com** to learn about the National Earthquake Information Center and the World Data Center for Seismology. Share what you learn with your class.

Figure 10
Seismographs differ according to whether they are intended to measure horizontal or vertical seismic motions. *Why can't one seismograph measure both horizontal and vertical motions?*

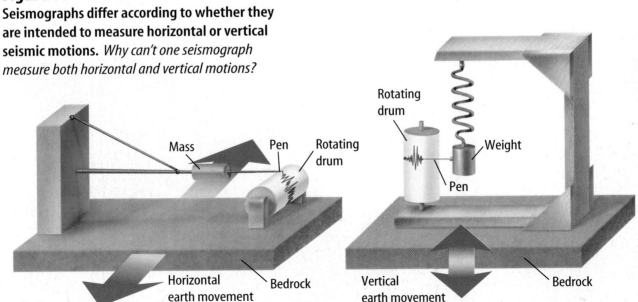

Mass   Pen   Rotating drum

Horizontal earth movement   Bedrock

Rotating drum   Weight   Pen

Vertical earth movement   Bedrock

Figure 11
Primary waves arrive at a seismograph station before secondary waves do.

A This graph shows the distance that primary and secondary waves travel over time. By measuring the difference in arrival times, a seismologist can determine the distance to the epicenter.

Seismograph Stations Each type of seismic wave reaches a seismograph station at a different time based on its speed. Primary waves arrive first at seismograph stations, and secondary waves, which travel slower, arrive second, as shown in the graph in **Figure 11A.** Because surface waves travel slowest, they arrive at seismograph stations last.

If seismic waves reach three or more seismograph stations, the location of the epicenter can be determined. To locate an epicenter, scientists draw circles around each station on a map. The radius of each circle equals that station's distance from the earthquake epicenter. The point where all three circles intersect, shown in **Figure 11B,** is the location of the earthquake epicenter.

Seismologists usually describe earthquakes based on their distances from the seismograph. Local events occur less than 100 km away. Regional events occur 100 km to 1,400 km away. Teleseismic events are those that occur at distances greater than 1,400 km.

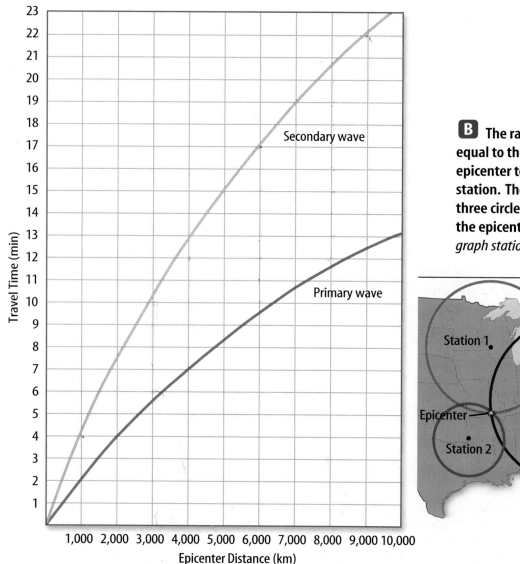

B The radius of each circle is equal to the distance from the epicenter to each seismograph station. The intersection of the three circles is the location of the epicenter. *Why is one seismograph station not enough?*

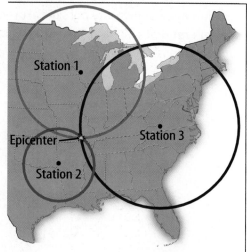

Basic Structure of Earth

Figure 12 shows Earth's internal structure. At the very center of Earth is a solid, dense inner core made mostly of iron with smaller amounts of nickel, oxygen, silicon, and sulfur. Pressure from the layers above causes the inner core to be solid. Above the solid inner core lies the liquid outer core, which also is made mainly of iron.

☑ **Reading Check** *How do the inner and outer cores differ?*

Earth's mantle is the largest layer, lying directly above the outer core. It is made mostly of silicon, oxygen, magnesium, and iron. The mantle often is divided into an upper part and a lower part based on changing seismic wave speeds. A portion of the upper mantle, called the asthenosphere (as THE nuh sfihr), consists of weak rock that can flow slowly.

Earth's Crust The outermost layer of Earth is the crust. Together, the crust and a part of the mantle just beneath it make up Earth's lithosphere (LIH thuh sfihr). The lithosphere is broken into a number of plates that move over the asthenosphere beneath it.

The thickness of Earth's crust varies. It is more than 60 km thick in some mountainous regions and less than 5 km thick under some parts of the oceans. Compared to the mantle, the crust contains more silicon and aluminum and less magnesium and iron. Earth's crust generally is less dense than the mantle beneath it.

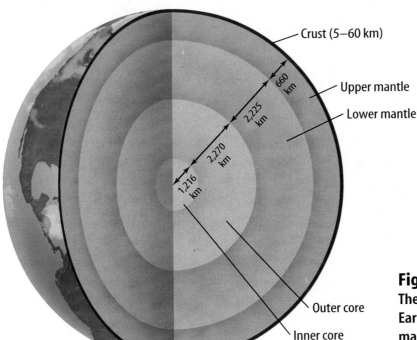

Figure 12
The internal structure of Earth shows that it is made of different layers.

Mini LAB

Interpreting Seismic Wave Data

Procedure

1. Use the **graph** in **Figure 11** to determine the difference in arrival times for primary and secondary waves at the distances listed in the data table below. Two examples are provided for you.

Wave Data	
Distance (km)	**Difference in Arrival Time**
1,500	2 min, 50 s
2,250	
2,750	
3,000	
4,000	5 min, 55 s
7,000	
9,000	

2. Use the graph to determine the differences in arrival times for the other distances in the table.

Analysis

1. What happens to the difference in arrival times as the distance from the earthquake increases?
2. If the difference in arrival times at a seismograph station is 6 min, 30 s, how far away is the epicenter?

Mapping Earth's Internal Structure As shown in **Figure 13,** the speeds and paths of seismic waves change as they travel through materials with different densities. By studying seismic waves that have traveled through Earth, scientists have identified different layers with different densities. In general, the densities increase with depth as pressures increase. Studying seismic waves has allowed scientists to map Earth's internal structure without being there.

Early in the twentieth century, scientists discovered that large areas of Earth don't receive seismic waves from an earthquake. In the area on Earth between 105° and 140° from the earthquake focus, no waves are detected. This area, called the shadow zone, is shown in **Figure 13.** Secondary waves are not transmitted through a liquid, so they stop when they hit the liquid outer core. Primary waves are slowed and bent but not stopped by the liquid outer core. Because of this, scientists concluded that the outer core and mantle are made of different materials. Primary waves speed up again as they travel through the solid inner core. The bending of primary waves and the stopping of secondary waves create the shadow zone.

✔ Reading Check *Why do seismic waves change speed as they travel through Earth?*

Figure 13
Seismic waves bend and change speed as the density of rock changes. Primary waves bend when they contact the outer core, and secondary waves are stopped completely. This creates a shadow zone where no seismic waves are received.

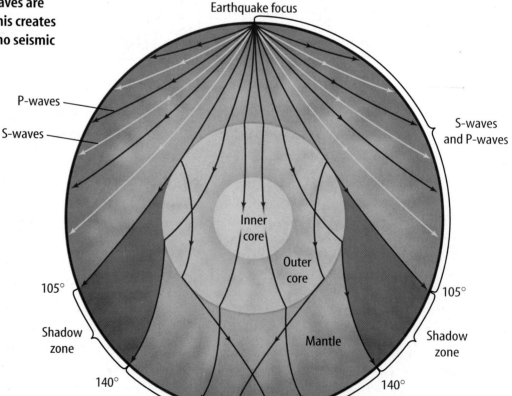

Earthquake focus

P-waves

S-waves

S-waves and P-waves

Inner core

Outer core

105°

Shadow zone

140°

Mantle

105°

Shadow zone

140°

P-waves only

Layer Boundaries **Figure 14** shows how seismic waves change speed as they pass through layers of Earth. Seismic waves speed up when they pass through the bottom of the crust and enter the upper mantle, shown on the far left of the graph. This boundary between the crust and upper mantle is called the Mohorovicic discontinuity (moh huh ROH vee chihch • dis kahn tuh NEW uh tee), or Moho.

The mantle is divided into layers based on changes in seismic wave speeds. For example, primary and secondary waves slow down again when they reach the asthenosphere. Then, they generally speed up as they move through a more solid region of the mantle below the asthenosphere.

The core is divided into two layers based on how seismic waves travel through it. Secondary waves do not travel through the liquid outer core, as you can see in the graph. Primary waves slow down when they reach the outer core, but they speed up again upon reaching the solid inner core.

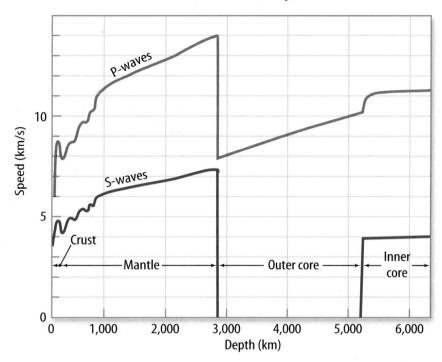

Seismic Wave Speeds

Figure 14
Changes in the speeds of seismic waves allowed scientists to detect boundaries between Earth's layers. S waves in the inner core form when P waves strike its surface.

Section 2 Assessment

1. How many seismograph stations are needed to determine the location of an epicenter? Explain.

2. Name the layers of Earth's interior.

3. What makes up most of Earth's inner core?

4. What are the three types of seismic waves? Which one does the most damage to property?

5. **Think Critically** Why do some seismograph stations receive both primary and secondary waves from an earthquake but other stations don't?

Skill Builder Activities

6. **Predicting** What will happen to the distance between two opposite walls of a room as primary waves move through the room? **For more help,** refer to the Science Skill Handbook.

7. **Solving One-Step Equations** Primary waves travel about 6 km/s through Earth's crust. The distance from Los Angeles, California, to Phoenix, Arizona, is about 600 km. How long would it take primary waves to travel between the two cities? **For more help,** refer to the Math Skill Handbook.

Activity

Epicenter Location

In this activity you can plot the distance of seismograph stations from the epicenters of earthquakes and determine the earthquake epicenters.

What You'll Investigate

Can plotting the distance of several seismograph stations from two earthquake epicenters allow you to determine the locations of the two epicenters?

Materials

string   globe
metric ruler   chalk

Earthquake Data			
Location of Seismograph	**Wave**	**Wave Arrival Times**	
		Earthquake A	**Earthquake B**
New York, New York	P	2:24:05 P.M.	1:19:42 P.M.
	S	2:29:15 P.M.	1:25:27 P.M.
Seattle, Washington	P	2:24:40 P.M.	1:14:37 P.M.
	S	2:30:10 P.M.	1:16:57 P.M.
Rio de Janeiro, Brazil	P	2:29:10 P.M.	—
	S	2:37:50 P.M.	—
Paris, France	P	2:30:30 P.M.	1:24:57 P.M.
	S	2:40:10 P.M.	1:34:27 P.M.
Tokyo, Japan	P	—	1:24:27 P.M.
	S	—	1:33:27 P.M.

Goals

- **Plot** the distances from several seismograph stations based on primary and secondary wave arrival times.
- **Interpret** the location of earthquake epicenters from these plots.

Procedure

1. Determine the difference in arrival time between the primary and secondary waves at each station for each earthquake listed in the table.

2. After you determine the arrival time differences for each seismograph station, use the graph in **Figure 11** to determine the distance in kilometers of each seismograph from the epicenter of each earthquake. Record these data in a data table. For example, the difference in arrival times in Paris for earthquake B is 9 min, 30 s. On the graph, the primary and secondary waves are separated along the vertical axis by 9 min, 30 s at a distance of 8,975 km.

3. Using the string, measure the circumference of the globe. Determine a scale of centimeters of string to kilometers on Earth's surface. (Earth's circumference is 40,000 km.)

4. For each earthquake, place one end of the string at each seismic station location on the globe. Use the chalk to draw a circle with a radius equal to the distance to the earthquake's epicenter.

5. **Identify** the epicenter for each earthquake.

Conclude and Apply

1. How is the distance of a seismograph from the earthquake related to the arrival times of the waves?

2. What is the location of the epicenter for each earthquake?

3. How many stations were needed to locate each epicenter accurately?

4. **Explain** why some seismographs didn't receive secondary waves from some quakes.

3 People and Earthquakes

Earthquake Activity

Imagine awakening in the middle of the night with your bed shaking, windows shattering, and furniture crashing together. That's what many people in Northridge, California, experienced at 4:30 A.M. on January 17, 1994. The ground beneath Northridge shook violently—it was an earthquake.

Although the earthquake lasted only 15 s, it killed 51 people, injured more than 9,000 people, and caused $44 billion in damage. More than 22,000 people were left homeless. **Figure 15A** shows some of the damage caused by the Northridge earthquake. **Figure 15B** shows the record of the Northridge earthquake on a seismogram.

Earthquakes are natural geological events that provide information about Earth. Unfortunately, they also cause billions of dollars in property damage and kill an average of 10,000 people every year. With so many lives lost and such destruction, it is important for scientists to learn as much as possible about earthquakes to try to reduce their impact on society.

As You Read

What **You'll Learn**

■ **Explain** where most earthquakes in the United States occur.
■ **Describe** how scientists measure earthquakes.
■ **List** ways to make your classroom and home more earthquake-safe.

Vocabulary
magnitude
liquefaction
tsunami

Why **It's Important**
Earthquake preparation can save lives and reduce damage.

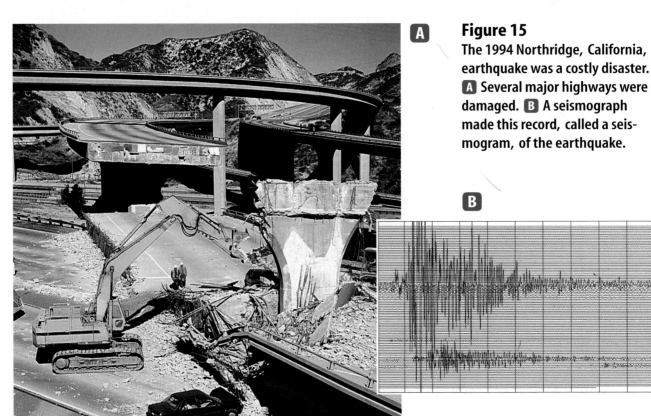

Figure 15
The 1994 Northridge, California, earthquake was a costly disaster.
Ⓐ Several major highways were damaged. Ⓑ A seismograph made this record, called a seismogram, of the earthquake.

Figure 16
The 1999 earthquake in Turkey released about 32 times more energy than the 1994 Northridge earthquake did.

Studying Earthquakes Scientists who study earthquakes and seismic waves are seismologists. As you learned earlier, the instrument that is used to record primary, secondary, and surface waves from earthquakes all over the world is called a seismograph. Seismologists can use records from seismographs, called seismograms, to learn more than just where the epicenter of an earthquake is located.

Measuring Earthquake Magnitude The height of the lines traced on the paper record of a seismograph is a measure of the energy that is released, or the **magnitude,** of the earthquake. The Richter magnitude scale is used to describe the strength of an earthquake and is based on the height of the lines on the seismogram. The Richter scale has no upper limit. However, scientists think that a value of about 9.5 would be the maximum strength an earthquake could register. For each increase of 1.0 on the Richter scale, the height of the line on a seismogram is ten times greater. However, about 32 times as much energy is released for every increase of 1.0 on the scale. For example, an earthquake with a magnitude of 8.5 releases about 32 times more energy than an earthquake with a magnitude of 7.5. **Figure 16** shows damage from the 7.8-magnitude earthquake in Turkey in 1999. **Table 1** is a list of some large-magnitude earthquakes that have occurred around the world and the damage they have caused.

Most of the earthquakes you hear about are large ones that cause great damage. However, of all the earthquakes detected throughout the world each year, most have magnitudes too low to be felt by humans. Scientists record thousands of earthquakes every day with magnitudes of less than 3.0. Each year, about 55,000 earthquakes are felt but cause little or no damage. These minor earthquakes have magnitudes that range from approximately 3.0 to 4.9 on the Richter scale.

Table 1	Large-Magnitude Earthquakes		
Year	**Location**	**Magnitude**	**Deaths**
1556	Shensi, China	?	830,000
1755	Lisbon, Portugal	8.8 (est.)	70,000
1811–12	New Madrid, MO	8.3 (est.)	few
1886	Charleston, SC	?	60
1906	San Francisco, CA	8.3	700 to 800
1923	Tokyo, Japan	9.2	143,000
1960	Chile	9.5	490 to 2,290
1964	Prince William Sound, AK	8.5	131
1976	Tangshan, China	8.2	242,000
1990	Iran	7.7	50,000
1995	Kobe, Japan	6.9	5,378
2000	Indonesia	7.9	90
2001	India	7.7	>20,000

Describing Earthquake Intensity Earthquakes also can be described by the amount of damage they cause. The modified Mercalli intensity scale describes the intensity of an earthquake using the amount of structural and geologic damage in a specific location. The amount of damage done depends on the strength of the earthquake, the nature of surface material, the design of structures, and the distance from the epicenter. Under ideal conditions, only a few people would feel an intensity-I earthquake, and it would cause no damage. An intensity-IV earthquake would be felt by everyone indoors during the day but would be felt by only a few people outdoors. Pictures might fall off walls and books might fall from shelves. However, an intensity-IX earthquake would cause considerable damage to buildings and would cause cracks in the ground. An intensity-XII earthquake would cause total destruction of buildings, and objects such as cars would be thrown upward into the air. The 1994 6.8-magnitude earthquake in Northridge, California, was listed at an intensity of IX because of the damage it caused.

Liquefaction Have you ever tried to drink a thick milkshake from a cup? Sometimes the milkshake is so thick that it won't flow. How do you make the milkshake flow? You shake it. Something similar can happen to very wet soil during an earthquake. Wet soil can be strong most of the time, but the shaking from an earthquake can cause it to act more like a liquid. This is called **liquefaction.** When liquefaction occurs in soil under buildings, the buildings can sink into the soil and collapse, as shown in **Figure 17.** People living in earthquake regions should avoid building on loose soils.

Physics
INTEGRATION

In 1975, Chinese scientists successfully predicted an earthquake by measuring a slow tilt of Earth's surface and small changes in Earth's magnetism. Many lives were saved as a result of this prediction. Do research to find out why most earthquakes have not been predicted.

Figure 17
San Francisco's Marina district suffered extensive damage from liquefaction in a 1989 earthquake because it is built on a landfilled marsh.

Tsunamis Most earthquake damage occurs when surface waves cause buildings, bridges, and roads to collapse. People living near the seashore, however, have another problem. An earthquake under the ocean causes a sudden movement of the ocean floor. The movement pushes against the water, causing a powerful wave that can travel thousands of kilometers in all directions. Far from shore, a wave caused by an earthquake is so long that a large ship might ride over it without anyone noticing. But when one of these waves breaks on a shore, as shown in **Figure 18,** it forms a towering crest that can reach 30 m in height.

Ocean waves caused by earthquakes are called seismic sea waves, or **tsunamis** (soo NAH meez). Just before a tsunami crashes onto shore, the water along a shoreline might move rapidly toward the sea, exposing a large portion of land that normally is underwater. This should be taken as a warning sign that a tsunami could strike soon, and you should head for higher ground immediately.

Because of the number of earthquakes that occur around the Pacific Ocean, the threat of tsunamis is constant. To protect lives and property, a warning system has been set up in coastal areas and for the Pacific islands to alert people if a tsunami is likely to occur. The Pacific Tsunami Warning Center, located near Hilo, Hawaii, provides warning information including predicted tsunami arrival times at coastal areas.

However, even tsunami warnings can't prevent all loss of life. In the 1960 tsunami that struck Hawaii, 61 people died when they ignored the warning to move away from coastal areas.

SCIENCE *Online*

Research Visit the Glencoe Science Web site at **science.glencoe.com** for more information about tsunamis. Make a poster to illustrate what you learn.

Figure 18
A tsunami begins over the earthquake focus. *What might happen to towns located near the shore?*

Earthquake Safety

You have learned that earthquakes can be destructive, but the damage and loss of life can be minimized. Although earthquakes cannot be predicted reliably, **Figure 19** shows where earthquakes are most likely to occur in the United States.

Knowing where large earthquakes are likely to occur helps in long-term planning. Cities in such regions can take action to prevent damage to buildings and loss of life. Many buildings withstood the 1989 San Francisco earthquake because they were built with the expectation that such an earthquake would occur someday.

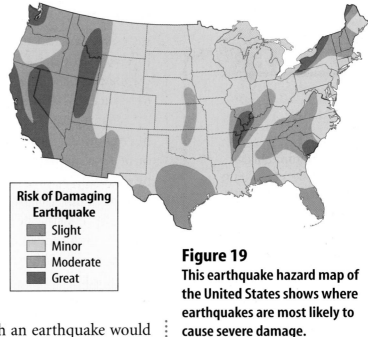

Risk of Damaging Earthquake
- Slight
- Minor
- Moderate
- Great

Figure 19
This earthquake hazard map of the United States shows where earthquakes are most likely to cause severe damage.

Math Skills Activity

Using Multiplication to Compare Earthquake Energy

Example Problem

The Richter scale is used to measure the magnitude of earthquakes. For each number increase on the Richter scale, 32 times more energy is released. How much more energy is released by a magnitude 6 earthquake than by a magnitude 3 earthquake?

Solution

1. *This is what you know:* magnitude 6 earthquake, magnitude 3 earthquake, energy increases 32 times per magnitude number

2. *This is what you need to find out:* amount of additional energy released

3. *This is the procedure you need to use:* Find the difference in magnitude numbers, then use that number of <u>multiples</u> of 32 to find the amount of additional energy released.

4. *Solve the equation:* difference in magnitude = 6 − 3 = 3
   multiply 32 times itself 3 times: $32 \times 32 \times 32 = 32{,}768$
   32,768 times more energy is released

Practice Problem

Calculate the difference in the amount of energy released between a magnitude 7 earthquake and a magnitude 2 earthquake.

For more help, refer to the Math Skill Handbook.

Mini LAB

Modeling Seismic-Safe Structures

Procedure 🥽

1. On a **tabletop,** build a structure out of **building blocks** by simply placing one block on top of another.
2. Build a second structure by wrapping sections of three blocks together with **rubber bands.** Then, wrap larger rubber bands around the entire completed structure.
3. Set the second structure on the tabletop next to the first one and pound on the side of the table with a slow, steady rhythm.

Analysis

1. Which of your two structures was better able to withstand the "earthquake" caused by pounding on the table?
2. How might the idea of wrapping the blocks with rubber bands be used in construction of supports for elevated highways?

Quake-Resistant Structures During earthquakes, buildings, bridges, and highways can be damaged or destroyed. Most loss of life during an earthquake occurs when people are trapped in or on these crumbling structures. What can be done to reduce loss of life?

Seismic-safe structures stand up to vibrations that occur during an earthquake. **Figure 20** shows how buildings can be built to resist earthquake damage. Today in California, some new buildings are supported by flexible, circular moorings placed under the buildings. The moorings are made of steel plates filled with alternating layers of rubber and steel. The rubber acts like a cushion to absorb earthquake waves. Tests have shown that buildings supported in this way should be able to withstand an earthquake measuring up to 8.3 on the Richter scale without major damage.

In older buildings, workers often install steel rods to reinforce building walls. Such measures protect buildings in areas that are likely to experience earthquakes.

✔️ **Reading Check** *What are seismic-safe structures?*

Figure 20

The rubber portions of this building's moorings absorb most of the wave motion of an earthquake. The building itself only sways gently. *What purpose does the rubber serve?*

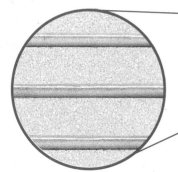

Before an Earthquake To make your home as earthquake-safe as possible, certain steps can be taken. To reduce the danger of injuries from falling objects, move heavy objects from high shelves and place them on lower shelves. Learn how to turn off the gas, water, and electricity in your home. To reduce the chance of fire from broken gas lines, make sure that hot-water heaters and other gas appliances are held securely in place, as shown in **Figure 21.** A newer method that is being used to minimize the danger of fire involves placing sensors on gas lines. The sensors automatically shut off the gas when earthquake vibrations are detected.

During an Earthquake If you're indoors, move away from windows and any objects that could fall on you. Seek shelter in a doorway or under a sturdy table or desk. If you're outdoors, stay in the open—away from power lines or anything that might fall. Stay away from buildings—chimneys or other parts of buildings could fall on you.

After an Earthquake Check water and gas lines for damage. If any are damaged, shut off the valves. If you smell gas, leave the building immediately and call authorities from a phone away from the leak area. Stay out of and away from damaged buildings. Be careful around broken glass and rubble that could contain sharp edges and wear boots or sturdy shoes to keep from cutting your feet. Finally, stay away from beaches. Tsunamis sometimes hit after the ground has stopped shaking.

Figure 21
Securing gas water heaters to walls with sturdy straps helps reduce the danger of fires from broken gas lines.

Section 3 Assessment

1. How can you determine whether or not you live in an area where an earthquake is likely to occur?
2. What can you do to make your home more safe during an earthquake?
3. How is earthquake magnitude measured?
4. Name three ways that an earthquake can cause damage.
5. **Think Critically** How are shock absorbers on a car similar to the circular moorings used in modern earthquake-safe buildings? How do they absorb shock?

Skill Builder Activities

6. **Forming Hypotheses** Hypothesize why some earthquakes with smaller magnitudes result in more deaths than earthquakes with larger magnitudes. **For more help, refer to the** Science Skill Handbook.

7. **Solving One-Step Equations** What is the difference in energy released between an earthquake of Richter magnitude 8.5 and one of magnitude 4.5? Between one of magnitude 3.5 and one of magnitude 5.5? **For more help, refer to the** Math Skill Handbook.

Activity

Earthquake Depths

Y ou learned earlier in this chapter that Earth's crust is broken into sections called plates. Stresses caused by movement of these plates generate energy within rocks that must be released. When this release of energy is sudden and rocks break, an earthquake occurs.

What You'll Investigate

Can a study of the foci of earthquakes tell you anything about plate movement in a particular region?

Goals
- **Observe** any connection between earthquake-focus depth and epicenter location using the data provided on the next page.
- **Describe** any observed relationship between earthquake-focus depth and the movement of plates at Earth's surface.

Materials
graph paper
pencil

Procedure

1. Use graph paper and the data table on the right to make a graph plotting the depths of earthquake foci and the distances from the coast of a continent for each earthquake epicenter.

2. Use the graph on the previous page as a reference to draw your own graph. Place *Distance from the Coast* on the horizontal axis. Begin labeling at the far left with 100 km west. To the right of it should be 0 km, then 100 km east, 200 km east, 300 km east, and so on through 700 km east. What point on your graph represents the coast?

3. Label the vertical axis *Depth Below Earth's Surface.* Label the top of the graph 0 km to represent Earth's surface. Label the bottom of the vertical axis −800 km.

4. **Plot** the focus depths against the distance and direction from the coast for each earthquake in the table to the right.

Conclude and Apply

1. **Describe** any observed relationship between the location of earthquake epicenters and the depth of earthquake foci.

2. Based on the graph you have completed, hypothesize what is happening to the plates at Earth's surface in the vicinity of the plotted earthquake foci.

3. **Infer** what process is causing the earthquakes you plotted on your graph paper.

4. Hypothesize why none of the plotted earthquakes occurred below 700 km.

5. Based on what you have plotted, infer what continent these data could apply to. Explain what you based your answer on.

Focus and Epicenter Data		
Earthquake	Focus Depth (km)	Distance of Epicenter from Coast (km)
A	−55	0
B	−295	100 east
C	−390	455 east
D	−60	75 east
E	−130	255 east
F	−195	65 east
G	−695	400 east
H	−20	40 west
I	−505	695 east
J	−520	390 east
K	−385	335 east
L	−45	95 east
M	−305	495 east
N	−480	285 east
O	−665	545 east
P	−85	90 west
Q	−525	205 east
R	−85	25 west
S	−445	595 east
T	−635	665 east
U	−55	95 west
V	−70	100 west

Communicating Your Data

Compare your graph with those of other members of your class. **For more help, refer to the** Science Skill Handbook.

Science Stats

Moving Earth!

Did you know...

... The most powerful earthquake

to hit the United States in recorded history shook Alaska in 1964. At 8.5 on the Richter scale, the quake shook all of Alaska for nearly 5 min, which is a long time for an earthquake. Nearly 320 km of roads near Anchorage suffered damage, and almost half of the 204 bridges had to be rebuilt.

... Snakes can sense the vibrations

made by a small rodent up to 23 m away. Does this mean that they can detect vibrations prior to major earthquakes? Unusual animal behavior was observed just before a 1969 earthquake in China—an event that was successfully predicted.

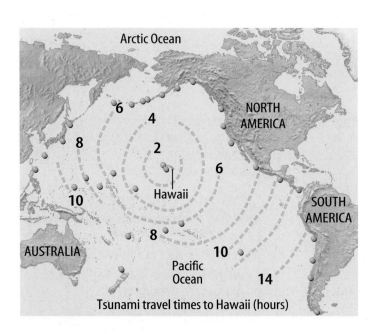

Arctic Ocean

NORTH AMERICA

6
4
8
2
6
10
Hawaii
SOUTH AMERICA
8
10
AUSTRALIA
Pacific Ocean
14

Tsunami travel times to Hawaii (hours)

... Earthquakes beneath the ocean floor can cause

seismic sea waves, or tsunamis. Traveling at speeds of up to 950 km/h—as fast as a commercial jet—a tsunami can strike with little warning. Since 1945, more people have been killed by tsunamis than by the ground shaking from earthquakes.

... Tsunamis can reach heights of 30 m. A wave that tall would knock over this lighthouse.

... On December 16, 1811, a strong earthquake occurred near New Madrid, Missouri. This earthquake was so strong that it changed the course of the Mississippi River. The earthquake also was reported to have rung the bell of St. Phillip's Steeple in Charleston, South Carolina.

Do the Math

1. On the Richter scale, a whole number increase means that the height of the largest recorded seismic wave increases by ten. How much higher is the largest wave from an 8.5 earthquake than the largest wave from a 3.5 earthquake?
2. Look at the tsunami warning system map on the previous page. About how long would a tsunami triggered near the Aleutian Islands take to reach Hawaii?

Go Further

Visit the Glencoe Science Web site at **science.glencoe.com.** Research the history and effects of earthquakes in the United States. Describe how the San Francisco earthquake of 1906 stimulated earthquake research.

Reviewing Main Ideas

Section 1 Forces Inside Earth

1. Plate movements put stress on rocks. To a certain point, the rocks bend and stretch. If the force is beyond the elastic limit, the rocks might break.

2. Earthquakes are vibrations that are created when rocks break along a fault.

3. Normal faults form when rocks undergo tension. Compression produces reverse faults. Strike-slip faults result from shearing forces. *What type of fault is shown here?*

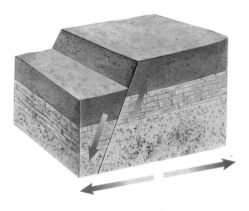

Section 2 Features of Earthquakes

1. Primary waves compress and stretch rock particles as the waves move. Secondary waves move by causing particles in rocks to move at right angles to the direction of the

waves. Surface waves move in a backward rolling motion and a side-to-side swaying motion. *Which kind of earthquake wave caused the damage shown here?*

2. Scientists can locate earthquake epicenters by recording seismic waves. *Where is the epicenter of the earthquake shown here?*

3. By observing the speeds and paths of seismic waves, scientists are able to determine the boundaries between Earth's internal layers.

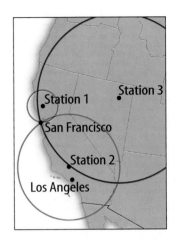

Section 3 People and Earthquakes

1. A seismograph is the instrument used to measure earthquake magnitude. *What is this record of an earthquake produced by a seismograph called?*

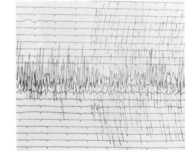

2. The magnitude of an earthquake is a measure of the energy released by the quake. The Richter scale describes how much energy an earthquake releases. The scale has no upper limit.

3. Earthquakes can cause liquefaction of wet soil and tsunamis, both of which increase the amount of structural damage produced by an earthquake.

FOLDABLES
Reading & Study Skills

After You Read

Using what you learned in this chapter, list and explain the effects of earthquakes on the inside of your Foldable.

Complete the following concept map on earthquake damage.

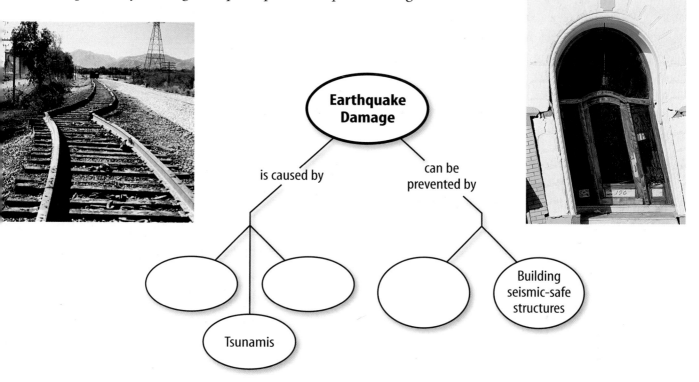

Vocabulary Review

Vocabulary Words

a. earthquake
b. epicenter
c. fault
d. focus
e. liquefaction
f. magnitude
g. normal fault
h. primary wave
i. reverse fault
j. secondary wave
k. seismic wave
l. seismograph
m. strike-slip fault
n. surface wave
o. tsunami

THE PRINCETON REVIEW **Study Tip**

Be a teacher! Gather a group of friends and assign each one a section of the chapter to teach. Teaching helps you remember and understand information.

Using Vocabulary

Replace the underlined words with the correct vocabulary words.

1. Most earthquake damage results from <u>primary waves</u>.

2. At a <u>normal fault</u>, rocks move past each other without much upward or downward movement.

3. The point on Earth's surface directly above the earthquake focus is the <u>fault</u>.

4. The measure of the energy released during an earthquake is its <u>seismograph</u>.

5. An earthquake under the ocean can cause a <u>surface wave</u> that travels thousands of kilometers.

Chapter 11 Assessment

Checking Concepts

Choose the word or phrase that best answers the question.

1. Earthquakes can occur when which of the following is passed?
   A) tension limit
   B) seismic unit
   C) elastic limit
   D) shear limit

2. When the rock above the fault surface moves down relative to the rock below the fault surface, what kind of fault forms?
   A) normal   C) reverse
   B) strike-slip D) shearing

3. From which of the following do primary and secondary waves move outward?
   A) epicenter   C) Moho
   B) focus   D) tsunami

4. What kind of earthquake waves stretch and compress rocks?
   A) surface   C) secondary
   B) primary   D) shear

5. What are the slowest seismic waves?
   A) surface   C) secondary
   B) primary   D) pressure

6. What is the fewest number of seismograph stations that are needed to locate the epicenter of an earthquake?
   A) two   C) four
   B) three   D) five

7. What happens to primary waves when they pass from liquids into solids?
   A) slow down   C) stay the same
   B) speed up   D) stop

8. What part of a seismograph does not move during an earthquake?
   A) sheet of paper C) drum
   B) fixed frame   D) pendulum

9. How much more energy does an earthquake of magnitude 7.5 release than an earthquake of magnitude 6.5?
   A) 32 times more C) twice as much
   B) 32 times less D) about half as much

10. What are the recorded lines from an earthquake called?
   A) seismograph   C) seismogram
   B) Mercalli scale D) Richter scale

Thinking Critically

11. The 1960 earthquake in the Pacific Ocean off the coast of Chile caused damage and loss of life in Chile and also in Hawaii, Japan, and other areas along the Pacific Ocean border. How could this earthquake do so much damage to areas thousands of kilometers from its epicenter?

12. Why is a person who is standing outside in an open field relatively safe during a strong earthquake?

13. Explain why the pendulum of a seismograph remains at rest.

14. Tsunamis often are called tidal waves. Explain why this is incorrect.

15. Which probably would be more stable during an earthquake—a single-story wood-frame house or a brick building? Explain.

Developing Skills

16. **Communicating** Imagine you are a science reporter assigned to interview the mayor about the earthquake safety of buildings in your city. What buildings would you be most concerned about? Make a list of questions about earthquake safety that you would ask the mayor.

17. Measuring in SI Use an atlas and a metric ruler to answer the following question. Primary waves travel at about 6 km/s in continental crust. How long would it take a primary wave to travel from San Francisco, California, to Reno, Nevada?

18. Interpreting Data Use the data table below and a map of the United States to determine the location of the earthquake epicenter.

Seismograph Station Data			
Station	Latitude	Longitude	Distance from Earthquake
1	45° N	120° W	1,300 km
2	35° N	105° W	1,200 km
3	40° N	115° W	790 km

19. Forming Hypotheses Hypothesize how seismologists could assign magnitudes to earthquakes that occurred before modern seismographs and the Richter scale were developed.

Performance Assessment

20. Model Use layers of different colors of clay to illustrate the three different kinds of faults. Label each model, explaining the forces involved and the rock movement.

21. Display Make a display showing why data from two seismograph stations are not enough to determine the location of an earthquake epicenter.

TECHNOLOGY

Go to the Glencoe Science Web site at **science.glencoe.com** or use the **Glencoe Science CD-ROM** for additional chapter assessment.

THE PRINCETON REVIEW **Test Practice**

Seismologists used the modified Mercalli intensity scale to determine the intensity of the same earthquake from four different cities. They recorded their data in the following table.

Earthquake Intensity	
City	Intensity
A	VII
B	X
C	V
D	IX

Study the table and answer the following questions.

1. According to the table, which city probably was the closest to the epicenter of the earthquake?
   A) city A
   B) city B
   C) city C
   D) city D

2. Which of the following would be an accurate conclusion based on the intensity in city B?
   F) The earthquake was not felt by very many people.
   G) The earthquake destroyed well-built wooden and stone structures.
   H) Destruction was minimal. Dishes rattled in cabinets, and pictures fell off of walls.
   J) The earthquake was only felt indoors.

Volcanoes

Every few months, villagers in Bronte, Italy, watch as Mount Etna roars to life and oozes rivers of molten lava. The village, eleven kilometers from the mountain, was out of harm's way, and no one was injured in this October 29, 1999, eruption. In this chapter, you will learn about types of volcanoes and how they form. You will learn how volcanoes affect humans and the surrounding environment and you will see the rock features they leave behind.

What do you think?

Science Journal Look at the picture below with a classmate. Discuss what you think this might be or what is happening. Here's a hint: *Not all volcanoes occur where you can see them.* Write your answer or best guess in your Science Journal.

EXPLORE ACTIVITY

You've seen pictures of volcanoes from the ground, but what would a volcano look like on a map? Volcanoes can be represented on maps that show the elevation of the land, as well as other important features. These maps are called topographic maps.

Map a volcano

1. Obtain half of a foam ball from your teacher and place it on the top of a table with the flat side down.

2. Using a metric ruler and a permanent marker, mark 1-cm intervals on the foam ball. Start at the base of the ball and mark up at several places around the ball.

3. Connect the marks of equal elevation by drawing a line around the ball at the 1-cm mark, at the 2-cm mark, etc.

4. Look directly down on the top of the ball and make a drawing of what you see in your Science Journal.

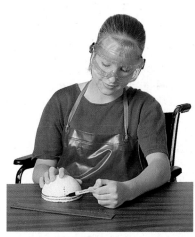

Observe

In your Science Journal, write a paragraph that explains how your drawing shows the general shape of a volcano. What might the lines drawn around the foam ball represent?

Before You Read

FOLDABLES
Reading & Study Skills

Making a Venn Diagram Study Fold As you prepare to read this chapter, make the following Foldable.

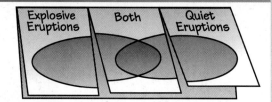

1. Place a sheet of paper in front of you so the short side is at the top. Fold the paper in half.

2. Label "Explosive Eruptions," "Quiet Eruptions," and "Both" across the front of the paper.

3. Fold both sides in to divide the paper into equal thirds. Unfold the paper so three columns show.

4. Through the top thickness of paper, cut along each of the fold lines to the top fold, forming three tabs.

5. As you read the chapter, collect information about each type of eruption under the left and right tabs. Under the middle tab, write what both types of eruptions have in common.

Volcanoes and Earth's Moving Plates

What are volcanoes?

A **volcano** is an opening in Earth that erupts gases, ash, and lava. Volcanic mountains form when layers of lava, ash, and other material build up around these openings. Can you name any volcanoes? Did you know that Earth has more than 600 active volcanoes?

Most Active Volcanoes Kilauea (kee low AY ah), located in Hawaii, is the world's most active volcano. For centuries, this volcano has been erupting, but not explosively. In May of 1990, most of the town of Kalapana Gardens was destroyed, but no one was hurt because the lava moved slowly and people could escape. The most recent series of eruptions from Kilauea began in January 1983 and still continues.

The island country of Iceland is also famous for its active volcanoes. It sits on an area where Earth's plates move apart and is known as the land of fire and ice. The February 26, 2000, eruption of Hekla, in Iceland, is shown in **Figure 1.**

Figure 1
This photo of the February 26, 2000 eruption of Hekla shows why Iceland is known as the land of fire and ice.

Figure 2
This town on Montserrat was devastated by the eruption of Soufrière Hills volcano.

Effects of Eruptions

When volcanoes erupt, they often have direct, dramatic effects on the lives of people and their property. Lava flows destroy everything in their path. Falling volcanic ash can collapse buildings, block roads, and in some cases cause lung disease in people and animals. Sometimes, volcanic ash and debris rush down the side of the volcano. This is called a pyroclastic flow. The temperatures inside the flow can be high enough to ignite wood. When big eruptions occur, people often are forced to abandon their land and homes. People who live farther away from volcanoes are more likely to survive, but cities, towns, crops, and buildings in the area can be damaged by falling debris.

Human and Environmental Impacts The eruption of Soufrière (sew free ER) Hills volcano in Montserrat, which began in July of 1995, was one of the largest recent volcanic eruptions near North America. Geologists knew it was about to erupt, and the people who lived near it were evacuated. On June 25, 1997, large pyroclastic flows swept down the volcano. As shown in **Figure 2,** they buried cities and towns that were in their path. The eruption killed 20 people who ignored the evacuation order.

When sulfurous gases from volcanoes mix with water vapor in the atmosphere, acid rain forms. The vegetation, lakes, and streams around Soufrière Hills volcano were impacted significantly by acid rain. As the vegetation died, shown in **Figure 3,** the organisms that lived in the forest were forced to leave or also died.

Figure 3
The vegetation near the volcano on Chances Peak, on the island of Montserrat in the West Indies, was destroyed by acid rain, heat, and ash.

How do volcanoes form?

What happens inside Earth to create volcanoes? Why are some areas of Earth more likely to have volcanoes than others? Deep inside Earth, heat and pressure cause rock to melt, forming liquid rock or magma. Some deep rocks already are melted. Others are hot enough that a small rise in temperature or drop in pressure can cause them to melt and form magma. What makes magma come to the surface?

Magma Forced Upward Magma is less dense than the rock around it, so it is forced slowly toward Earth's surface. You can see this process if you turn a bottle of cold syrup upside down. Watch the dense syrup force the less dense air bubbles slowly toward the top.

✔ **Reading Check** *Why is magma forced toward Earth's surface?*

After many thousands or even millions of years, magma reaches Earth's surface and flows out through an opening called a **vent.** As lava flows out, it cools quickly and becomes solid, forming layers of igneous rock around the vent. The steep-walled depression around a volcano's vent is the **crater. Figure 4** shows magma being forced out of a volcano.

Figure 4
This cutaway diagram shows how a volcano is formed and how magma from the mantle is forced to the surface.

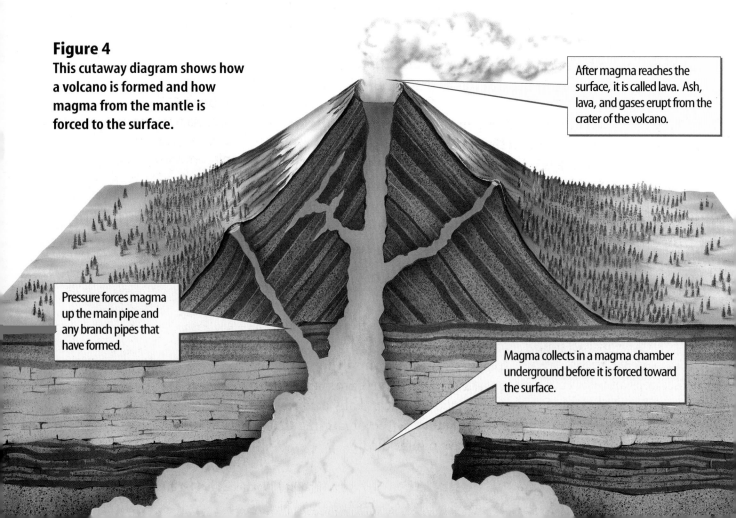

After magma reaches the surface, it is called lava. Ash, lava, and gases erupt from the crater of the volcano.

Pressure forces magma up the main pipe and any branch pipes that have formed.

Magma collects in a magma chamber underground before it is forced toward the surface.

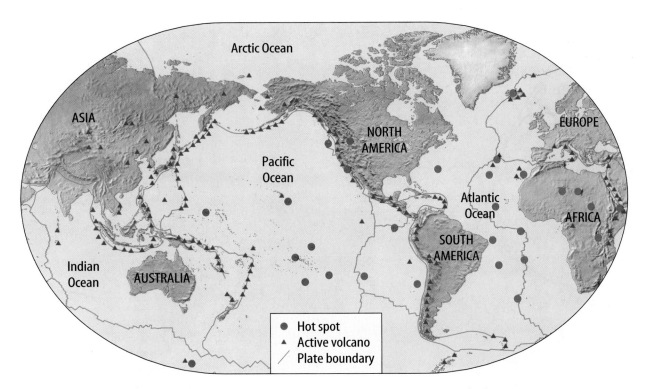

Where do volcanoes occur?

Volcanoes often form in places where plates are moving apart, where plates are moving together, and at locations called hot spots. You can find locations of active volcanoes at plate boundaries and at hot spots on the map in **Figure 5.** Many examples can be found of volcanoes around the world that form at these three different kinds of areas. You'll explore volcanoes in Iceland, on the island of Montserrat, and in Hawaii.

Divergent Plate Boundaries Iceland is a large island in the North Atlantic Ocean. It is near the Arctic Circle and therefore has some glaciers. Iceland has volcanic activity because it sits on top of the Mid-Atlantic Ridge.

The Mid-Atlantic Ridge is a divergent plate boundary, which is an area where Earth's plates are moving apart. When plates separate, they form long, deep cracks called rifts. Lava flows from these rifts and is cooled quickly by seawater. **Figure 6** shows how magma rises at rifts to form new volcanic rock. As more lava flows and hardens, it builds up on the seafloor. Sometimes, the volcanoes and rift eruptions rise above sea level, forming islands such as Iceland. In 1963, the new island Surtsey was formed during a volcanic eruption.

Figure 5
This map shows the locations of volcanoes, hot spots, and plate boundaries around the world. The Ring of Fire is a belt of active volcanoes that circles the Pacific Ocean.

Figure 6
This diagram shows how volcanic activity occurs where Earth's plates move apart.

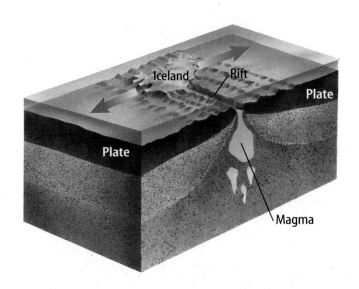

Figure 7
Volcanoes can form where plates collide and one plate slides below the other.

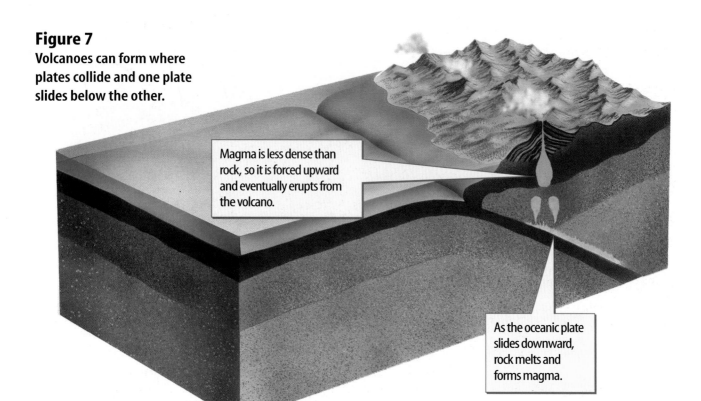

Magma is less dense than rock, so it is forced upward and eventually erupts from the volcano.

As the oceanic plate slides downward, rock melts and forms magma.

TRY AT HOME
Mini LAB

Modeling Magma Movement

Procedure
1. Pour **water** into a **transparent, plastic cup.**
2. Pour a small amount of **olive oil** into a separate plastic cup.
3. Extract a small amount of oil with a **dropper.**
4. Submerge the dropper tip into the water cup and slowly squeeze oil drops into the water.

Analysis
1. Describe what happened to the oil.
2. How do your observations compare with the movement of magma within Earth's crust?

Convergent Plate Boundaries Places where Earth's plates move together are called convergent plate boundaries. They include areas where an oceanic plate slides below a continental plate as in **Figure 7,** and where one oceanic plate slides below another oceanic plate. The Andes in South America began forming when an oceanic plate started sliding below a continental plate. Volcanoes that form on convergent plate boundaries tend to erupt more violently than other volcanoes do.

Magma forms when the plate sliding below another plate and the overlying mantle melt partially. The magma then is forced upward to the surface, forming volcanoes like Soufrière Hills on the island of Montserrat.

Hot Spots The Hawaiian Islands are forming as a result of volcanic activity. However, unlike Iceland, they haven't formed at a plate boundary. The Hawaiian Islands are in the middle of the Pacific Plate, far from its edges. What process could be forming them?

It is thought that some areas at the boundary between Earth's mantle and core are unusually hot. Hot rock at these areas is forced toward the crust where it melts partially to form a **hot spot.** The Hawaiian Islands sit on top of a hot spot under the Pacific Plate. Magma has broken through the crust to form several volcanoes. The volcanoes that rise above the water form the Hawaiian Islands, shown in **Figure 8A.**

Figure 8
The Hawaiian Islands are actually volcanoes.

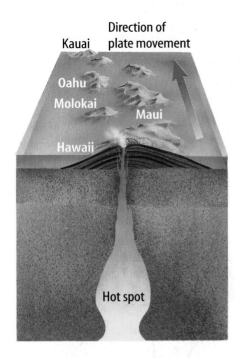

Direction of plate movement

Kauai
Oahu
Molokai
Maui
Hawaii

Hot spot

A This satellite photo shows the Hawaiian Islands. *Why are they in a relatively straight line?*

The Hawaiian Islands As you can see in **Figure 8,** the Hawaiian Islands are all in a line. This is because the Pacific Plate is moving over a stationary hot spot. Kauai, the oldest Hawaiian island, was once located where the big island, Hawaii, is situated today. As the plate moved, Kauai moved away from the hot spot and became dormant. As the Pacific Plate continued to move, the islands of Oahu, Molokai, Maui, and Hawaii were formed. The Hawaiian Islands formed over a period of about 5 million years.

B This illustration shows how the Hawaiian Islands were formed over a hot spot.

Section Assessment

1. How are volcanoes related to Earth's moving plates?

2. Hot lava flows are not the only danger associated with active volcanoes. What effects can pyroclastic flows have on people?

3. Why does lava cool rapidly along a mid-ocean ridge?

4. Describe processes that are occurring to cause Soufrière Hills volcano to erupt.

5. **Think Critically** If the Pacific Plate stopped moving, what might happen to the island of Hawaii?

Skill Builder Activities

6. **Concept Mapping** Make a concept map that shows how the Hawaiian Islands formed. Use the following phrases: *volcano forms, plate moves, volcano becomes dormant,* and *new volcano forms.* **For more help, refer to the** Science Skill Handbook.

7. **Communicating** Scientists were able to predict approximately when Mount Pinatubo in the Philippines would erupt in 1991. In your Science Journal, write a report on equipment used to predict volcanic eruptions. **For more help, refer to the** Science Skill Handbook.

Types of Volcanoes

As You Read

What You'll Learn

■ **Explain** how the explosiveness of a volcanic eruption is related to the silica and water vapor content of its magma.
■ **List** three forms of volcanoes.

Vocabulary

shield volcano   cinder cone volcano
tephra   composite volcano

Why It's Important

If you know the type of volcano, you can predict how it will erupt.

Figure 9
A calm day in Washington state was shattered suddenly when Mount St. Helens erupted on May 18, 1980, as shown in this sequence of photographs.

What controls eruptions?

Some volcanic eruptions are explosive, like those from Soufrière Hills volcano, Mount Pinatubo, and Mount St. Helens. In others, the lava quietly flows from a vent, as in the Kilauea eruptions. What causes these differences?

Two important factors control whether an eruption will be explosive or quiet. One factor is the amount of water vapor and other gases that are trapped in the magma. The second factor is how much silica is present in the magma. Silica is a compound composed of the elements silicon and oxygen.

Trapped Gases When you shake a soft-drink container and then quickly open it, the pressure from the gas in the drink is released suddenly, spraying the drink all over. In the same way, gases such as water vapor and carbon dioxide are trapped in magma by the pressure of the surrounding magma and rock. As magma nears the surface, it is under less pressure. This allows the gas to escape from the magma. Gas escapes easily from some magma during quiet eruptions. However, gas that builds up to high pressures eventually causes explosive eruptions such as the one shown in **Figure 9.**

A 8:32 A.M.

B 38 seconds later

Water Vapor The magma at some convergent plate boundaries contains a lot of water vapor. This is because oceanic plate material and some of its water slide under other plate material at some convergent plate boundaries. The trapped water vapor in the magma can cause explosive eruptions.

Composition of Magma

The second major factor that affects the nature of the eruption is the composition of the magma. Magma can be divided into two major types—silica poor and silica rich.

Quiet Eruptions Magma that is relatively low in silica is called basaltic magma. It is fluid and produces quiet, non-explosive eruptions such as those at Kilauea. This type of lava pours from volcanic vents and runs down the sides of a volcano. As this *pahoehoe* (pa-HOY-hoy) lava cools, it forms a ropelike structure. If the same lava flows at a lower temperature, a stiff, slowly moving *aa* (AH-ah) lava forms. In fact, you can walk right up to some aa lava flows on Kilauea.

Figure 10 shows some different types of lava. These quiet eruptions form volcanoes over hot spots such as the Hawaiian volcanoes. Basaltic magmas also flow from rift zones, which are long, deep cracks in Earth's surface. Many lava flows in Iceland are of this type. Because basaltic magma is fluid when it is forced upward in a vent, trapped gases can escape easily in a nonexplosive manner, sometimes forming lava fountains. Lavas that flow underwater form pillow lava formations. They consist of rock structures shaped like tubes, balloons, or pillows.

SCIENCE *Online*

Research Visit the Glencoe Science Web site at **science.glencoe.com** to learn more about Kilauea volcano in Hawaii. Draw a map of Hawaii that shows the location of Kilauea.

C 42 seconds later

D 53 seconds later

341

Figure 10

Lava rarely travels faster than a few kilometers an hour. Therefore, it poses little danger to people. However, homes and property can be damaged. On land, there are two main types of lava flows—aa (AH ah) and pahoehoe (pa HOY hoy). When lava comes out of cracks in the ocean floor, it is called pillow lava. The lava cooling here came from a volcanic eruption on the island of Hawaii.

Aa flows, like this one on Mount Etna in Italy, carry sharp angular chunks of rock called scoria. Aa flows move slowly and are intensely hot.

Pillow lava occurs where lava oozes out of cracks in the ocean floor. It forms pillow-shaped lumps as it cools. Pillow lava is the most common type of lava on Earth.

Pahoehoe flows, like this one near Kilauea's Mauna Ulu Crater in Hawaii, are more fluid than aa flows. They develop a smooth skin and form rope-like patterns when they cool.

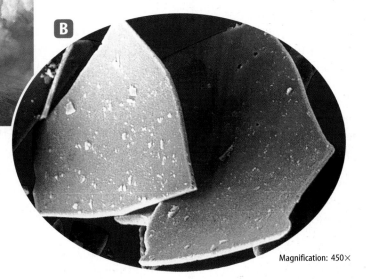

Figure 11

A Magmas that are rich in silica produce violent eruptions, such as this one in Alaska.
B This color enhanced view of volcanic ash, from a 10 million year old volcano in Nebraska, shows the glass particles that make up ash.

Magnification: 450×

Explosive Magma Silica-rich, or granitic, magma on the other hand produces explosive eruptions such as those at Soufrière Hills volcano. This magma sometimes forms where Earth's plates are moving together and one plate slides under another. As the plate that is sliding under the other goes deeper, some rock is melted. The magma is forced upward by denser surrounding rock, comes in contact with the crust, and becomes enriched in silica. Silica-rich granitic magma is thick, and gas gets trapped inside, causing pressure to build up. When an explosive eruption occurs, as shown in **Figure 11,** the gases expand rapidly, often carrying pieces of lava in the explosion.

✓ Reading Check *What type of magmas produce violent eruptions?*

Some magmas have an andesitic composition. Andesitic magma is more silica rich than basaltic magma is, but it is less silica rich than granitic magma. It often forms at convergent plate boundaries where one plate slides under the other. Because of their higher silica content, they also erupt more violently than basaltic magmas. One of the biggest eruptions in recorded history, Krakatau, was primarily andesitic in composition. The word *andesitic* comes from the Andes, which are mountains located along the western edge of South America, where andesite rock is common. Many of the volcanoes encircling the Pacific Ocean also are made of andesite.

Health
INTEGRATION

When volcanoes erupt, ash often is spread over a great distance. People who live near volcanoes must be careful not to inhale too much of the ash particles because the particles can cause respiratory problems. In your Science Journal, describe what people can do to prevent exposure to volcanic ash.

Forms of Volcanoes

A volcano's form depends on whether it is the result of a quiet or an explosive eruption and the type of lava it is made of—basaltic, granitic, or andesitic (intermediate). The three basic types of volcanoes are shield volcanoes, cinder cone volcanoes, and composite volcanoes.

Shield Volcano Quiet eruptions of basaltic lava spread out in flat layers. The buildup of these layers forms a broad volcano with gently sloping sides called a **shield volcano,** as seen in **Figure 12.** The Hawaiian Islands are examples of shield volcanoes. Basaltic lava also can flow onto Earth's surface through large cracks called fissures. This type of eruption forms flood basalts, not volcanoes, and accounts for the greatest volume of erupted volcanic material. The basaltic lava flows over Earth's surface, covering large areas with thick deposits of basaltic igneous rock when it cools. The Columbia Plateau located in the northwestern United States was formed in this way. Much of the new seafloor that originates at mid-ocean ridges forms as underwater flood basalts.

Cinder Cone Volcano Explosive eruptions throw lava and rock high into the air. Bits of rock or solidified lava dropped from the air are called **tephra** (TEH fruh). Tephra varies in size from volcanic ash, to cinders, to larger rocks called bombs and blocks. When tephra falls to the ground, it forms a steep-sided, loosely packed **cinder cone volcano,** as seen in **Figure 13.**

Figure 12
A shield volcano like Mauna Loa, shown here, is formed when lava flows from one or more vents without erupting violently.

Vent

Magma

Paricutín On February 20, 1943, a Mexican farmer learned about cinder cones when he went to his cornfield. He noticed that a hole in his cornfield that had been there for as long as he could remember was giving off smoke. Throughout the night, hot glowing cinders were thrown high into the air. In just a few days, a cinder cone several hundred meters high covered his cornfield. This is the volcano named Paricutín.

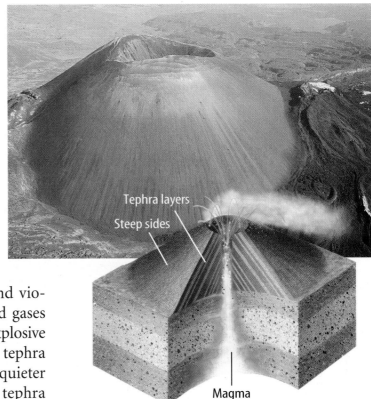

Composite Volcano Some volcanic eruptions can vary between quiet and violent, depending on the amount of trapped gases and how rich in silica the magma is. An explosive period can release gas and ash, forming a tephra layer. Then, the eruption can switch to a quieter period, erupting lava over the top of the tephra layer. When this cycle of lava and tephra is repeated over and over in alternating layers, a **composite volcano** is formed. Composite volcanoes, shown in **Figure 14,** are found mostly where Earth's plates come together and one plate slides below the other. Soufrière Hills volcano is an example. As you can see in **Table 1** on the next page, many things affect eruptions and the form of a volcano.

Figure 13
Paricutín is a large, cinder cone volcano located in Mexico.

Figure 14
Mount Rainier in the state of Washington is an example of a composite volcano.

Violent Eruptions Soufrière Hills volcano formed as ocean floor of the North American Plate and the South American Plate slid beneath the Caribbean Plate, causing magma to form. Successive eruptions of lava and tephra produced the majestic composite volcanoes that tower above the surrounding landscape on Montserrat and other islands in the Lesser Antilles. Before the 1995 eruption, silica-rich magma rose and was trapped beneath the surface. As the magma was forced toward Earth's surface, the pressure on the underlying magma was released. This started a series of eruptions that were still continuing in the year 2001.

Table 1 Thirteen Selected Eruptions

Volcano and Location	Year	Type	Eruptive Force	Magma Content		Ability of Magma to Flow	Products of Eruption
				Silica	H$_2$O		
Mount Etna, Sicily	1669	composite	moderate	high	low	medium	lava, ash
Tambora, Indonesia	1815	cinder cone	high	high	high	low	cinders, ash
Krakatau, Indonesia	1883	composite	high	high	high	low	cinders, ash
Mount Pelée, Martinique	1902	cinder cone	high	high	high	low	gas, ash
Vesuvius, Italy	1906	composite	moderate	high	low	medium	lava, ash
Mount Katmai, Alaska	1912	composite	high	high	high	low	lava, ash, gas
Paricutìn, Mexico	1943	cinder cone	moderate	high	low	medium	ash, cinders
Surtsey, Iceland	1963	shield	moderate	low	low	high	lava, ash
Mount St. Helens, Washington	1980	composite	high	high	high	low	gas, ash
Kilauea, Hawaii	1983	shield	low	low	low	high	lava
Mount Pinatubo, Philippines	1991	composite	high	high	high	low	gas, ash
Soufrière Hills, Montserrat	1995	composite	high	high	high	low	gas, ash, rocks
Popocatépetl, Mexico	2000	composite	moderate	high	low	medium	gas, ash

Figure 15
Not much was left after Krakatau erupted in 1883.

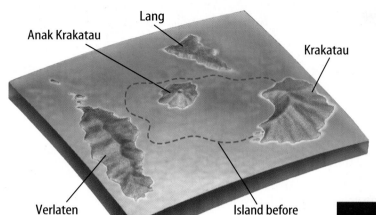

Lang
Anak Krakatau
Krakatau
Verlaten
Island before August 26, 1883

A The dotted lines on this illustration show what Krakatau looked like before the eruption.

Krakatau One of the most violent eruptions in recent times occurred on an island in the Sunda Straits near Indonesia in August of 1883. Krakatau, a volcano on the island, erupted with such force that the island disappeared as shown in **Figure 15A.** Most of the island collapsed into the emptied magma chamber. The noise of the eruption was so loud that it woke people in Australia and was heard as far away as 4,653 km from the island. Ash from the eruption fell in Singapore, which is 840 km to the north, and the area around the volcano was in complete darkness for 24 h. More than 36,000 people were killed, most by the giant tsunami waves created by the eruption. Global temperatures were lowered as much as 1.2°C by particles blown into the atmosphere and didn't return to normal until 1888.

B Anak Krakatau formed in the early 1900s. The name means "Child of Krakatau."

Section 2 Assessment

1. Some eruptions are quiet and others are violent. What causes this difference?
2. Compare and contrast the different types of lava.
3. How is a composite volcano like a shield and a cinder cone volcano?
4. Describe how the Hawaiian Islands formed in the Pacific Ocean.
5. **Think Critically** In 1883, Krakatau in Indonesia erupted. Infer which kind of lava Krakatau erupted—lava rich in silica or lava low in silica. Support your inference using data in **Table 1.**

Skill Builder Activities

6. **Comparing and Contrasting** Use **Table 1** to compare and contrast Kilauea in Hawaii and Mount Pinatubo in the Philippines. **For more help, refer to the** Science Skill Handbook.
7. **Calculating Ratios** When Mount St. Helens erupted in 1980, about 1.3 km³ of material were ejected from the volcano. Tambora in Indonesia gave off 131 km³ of material in 1815. How many times larger was the volume of material given off by Tambora? Based on your information, which volcano did the most damage? **For more help, refer to the** Math Skill Handbook.

Activity

Identifying Types of Volcanoes

You have learned that certain properties of magma are related to the type of eruption and the form of the volcano that will develop. Do this activity to see how to make and use a table that relates the properties of magma to the form of volcano that develops.

What You'll Investigate
Are the silica and water content of a magma related to the form of volcano that develops?

Materials
Table 1 of thirteen selected eruptions
paper
pencil

Goals
- **Determine** any relationship between the ability of magma to flow and eruptive force.
- **Determine** any relationship between magma composition and eruptive force.

Procedure

1. Copy the graph shown above.
2. Using the information from **Table 1,** plot the magma content for each of the volcanoes listed by writing the name of the basic type of volcano in the correct spot on the graph.
3. After you plot all 13 volcanoes, analyze the patterns of volcanic types on the diagram to answer the questions.

Conclude and Apply

1. What relationship appears to exist between the ability of the magma to flow and the eruptive force of the volcano?

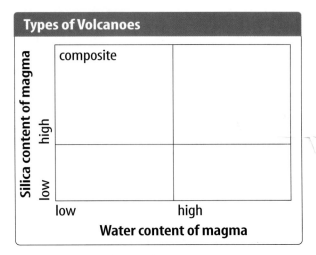

Types of Volcanoes

composite

Silica content of magma — high / low

Water content of magma — low / high

2. Which would be more liquidlike: magma that flows easily or magma that flows with difficulty?

3. What relationship appears to exist between the silica or water content of the magma and the nature of the material ejected from the volcano?

4. How is the ability of a magma to flow related to its silica content?

5. **Infer** which of the two variables, silica or water content, appears to have the greater effect on the eruptive force of the volcano.

6. **Describe** the relationship that appears to exist between the silica and water content of the magma and the type of volcano that is produced.

Communicating Your Data

Create a poster that shows the relationship between magma composition and the type of volcano formed. **For more help, refer to the** Science Skill Handbook.

Igneous Rock Features

Intrusive Features

You can observe volcanic eruptions because they occur at Earth's surface. However, far more activity occurs underground. In fact, most magma never reaches Earth's surface to form volcanoes or to flow as flood basalts. This magma cools slowly underground and produces underground rock bodies that could become exposed later at Earth's surface by erosion. These rock bodies are called intrusive igneous rock features. There are several different types of intrusive features. Some of the most common are batholiths, sills, dikes, and volcanic necks. What do intrusive igneous rock bodies look like? You can see illustrations of these features in **Figure 16.**

As You Read

***What* You'll Learn**
- **Describe** intrusive igneous rock features and how they form.
- **Explain** how a volcanic neck and a caldera form.

Vocabulary

batholith	volcanic neck
dike	caldera
sill	

***Why* It's Important**
Many features formed underground by igneous activity are exposed at Earth's surface by erosion.

Figure 16
This diagram shows intrusive and other features associated with volcanic activity. *Which features shown are formed above ground? Which are formed by intrusive activities?*

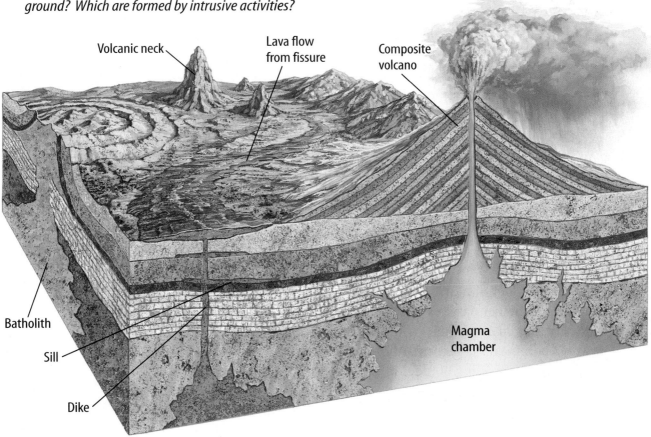

Volcanic neck

Lava flow from fissure

Composite volcano

Batholith

Sill

Dike

Magma chamber

Batholiths The largest intrusive igneous rock bodies are **batholiths.** They can be many hundreds of kilometers in width and length and several kilometers thick. Batholiths form when magma bodies that are being forced upward from inside Earth cool slowly and solidify before reaching the surface. However, not all of them remain hidden inside Earth. Some batholiths have been exposed at Earth's surface by many years of erosion. The granite domes of Yosemite National Park are the remains of a huge batholith that stretches across much of the length of California.

Math Skills Activity

Classifying Igneous Rocks

Igneous rocks are classified into three types depending on the amount of silica they contain. Basaltic rocks contain approximately 45 percent to 52 percent silica. Andesitic, or intermediate, rocks contain about 52 percent to 66 percent silica, and granitic rocks have more than 66 percent silica. The lighter the color is, the higher the silica content is.

Example Problem

A 900-kg block of igneous rock contains 630 kg of silica. Calculate the percent of silica in the rock to classify it.

Solution

1 *This is what you know:*   rock = 900 kg
silica = 630 kg

2 *This is what you need to find:*   The percentage of silica: x

3 *This is the equation you need to use:*   Mass of silica / mass of rock = x / 100

4 *Solve the equation for x:* $x = (630 \text{ kg}/900 \text{ kg}) \times 100$
$x = 70$ percent, therefore, the rock is granitic.

Check your answer by dividing it by 100, then multiplying by 900. Did you get the given amount of silica?

Practice Problems

1. A 250-kg boulder of basalt contains 125 kg of silica. Use the classification system to determine whether basalt is light or dark.
2. Andesite is an intermediate, medium-colored rock with a silica content ranging from 52 percent to 66 percent. About how many kilograms of silica would you predict to be in a 68-kg boulder of andesite?

For help with solving equations, refer to the Math Skill Handbook.

Dikes and Sills Magma sometimes squeezes into cracks in rock below the surface. This is like squeezing toothpaste into the spaces between your teeth. Magma that is forced into a crack that cuts across rock layers and hardens is called a **dike.** Magma that is forced into a crack parallel to rock layers and hardens is called a **sill.** These features are shown in **Figures 17A** and **17B.** Most dikes and sills run from a few meters to hundreds of meters long.

Other Features

When a volcano stops erupting, the magma hardens inside the vent. Erosion, usually by water and wind, begins to wear away the volcano. The cone is much softer than the solid igneous rock in the vent. Thus, the cone erodes first, leaving behind the solid igneous core as a **volcanic neck.** Devil's Tower in Wyoming, shown in **Figure 17C,** might be an example of a volcanic neck.

SCIENCE *Online*

Research Visit the Glencoe Science Web site at **science.glencoe.com** to learn more about igneous rock features. Share your research with your class.

A A sill is formed when magma is forced between rock layers.

Figure 17
Igneous features can form in many different sizes and shapes.

B The vertical dikes shown here near Shiprock, New Mexico, were formed when magma squeezed into vertical cracks in the surrounding rock layers.

C Some people have suggested that Devil's Tower is a volcanic neck.

Figure 18
Calderas are formed when the top of a volcano collapses.

Calderas Sometimes after an eruption, the top of a volcano can collapse, as seen in **Figure 18.** This produces a large depression called a **caldera.** Crater Lake in Oregon, shown in **Figure 19,** is a caldera that filled with water and is now a lake. Crater Lake formed after the violent eruption and destruction of Mount Mazama about 7,000 years ago.

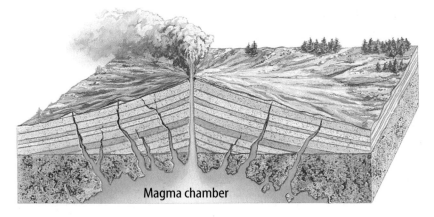

Magma chamber

A Magma rises, causing volcanic activity to occur.

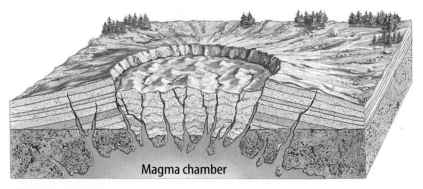

Magma chamber

B The magma chamber partially empties, causing rock to collapse into the emptied chamber below the surface. This forms a circular-shaped caldera.

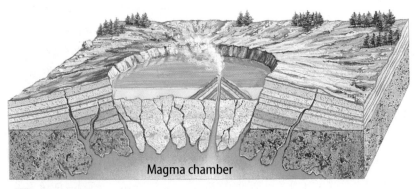

Magma chamber

C Crater Lake in Oregon formed when water collected in the circular space left when surface material collapsed.

Figure 19
Wizard Island in Crater Lake is a cinder cone volcano that erupted after the formation of the caldera. *What causes a caldera to form?*

Igneous Features Exposed You have learned in this chapter that Earth's surface is built up and worn down continually. The surface of Earth is built up by volcanoes. Also, igneous rock is formed when magma hardens below ground. Eventually, the processes of weathering and erosion wear down rock at the surface, exposing features like batholiths, dikes, and sills.

Reading Check *What exposes igneous features that formed below the surface?*

Section 3 Assessment

1. What's the difference between a caldera and a crater?
2. Describe how a sill forms. How is it different from a dike?
3. What is a volcanic neck and how does it form?
4. Explain how a batholith forms.
5. **Think Critically** Why are the large, granite dome features of Yosemite National Park in California considered to be intrusive volcanic features when they are exposed at the surface?

Skill Builder Activities

6. **Comparing and Contrasting** Compare and contrast dikes, sills, batholiths, and volcanic necks. **For more help, refer to the** Science Skill Handbook.
7. **Using Graphics Software** Use the graphics software available on your computer to produce an illustration of igneous rock features based on **Figure 16.** Be sure to include intrusive features and features that form above ground. **For more help, refer to the** Technology Skill Handbook.

Activity
Design Your Own Experiment

How do calderas form?

A caldera is a depression formed when the top of a volcano collapses after an eruption. What might cause the top of a volcano to collapse? What would happen if the magma inside the magma chamber suddenly were removed?

Recognize the Problem

How does the removal of magma from the magma chamber affect a volcano?

Form a Hypothesis

Based on your reading about volcanoes, state a hypothesis about what would happen if the magma inside the magma chamber of a volcano were suddenly removed.

Goals
- **Design** a volcano setup that will demonstrate how a caldera could form.
- **Observe** what happens during trials with your volcano setup.
- **Describe** what you observe.

Possible Materials
small box
small balloon
paper
newspaper
flour
plastic tubing
clamp for tubing
tape
scissors

Safety Precautions

Test Your Hypothesis

Plan

1. As a group, agree upon the hypothesis and identify which results will support the hypothesis.

2. **Design** a volcano that allows you to test your hypothesis. What materials will you use to build your volcano?

3. What will you remove from inside your volcano to represent the loss of magma? How will you remove it?

4. Where will you place your volcano? What will you do to minimize messes?

5. **Identify** all constants, variables, and controls of the experiment.

Do

1. Make sure your teacher approves your plan before you start.

2. **Construct** your volcano with any features that will be required to test your hypothesis.

3. **Conduct** one or more appropriate trials to test your hypothesis. Record any observations that you make and any other data that are appropriate to test your hypothesis.

Analyze Your Data

1. **Describe** in words or with a drawing what your volcano looked like before you began.

2. What happened to your volcano during the experiment that you conducted? Did its appearance change?

3. **Describe** in words or with a drawing what your volcano looked like after the trial.

4. What other observations did you make?

5. **Describe** any other data that you recorded.

Draw Conclusions

1. Did your observations support your hypothesis? Explain.

2. How was your demonstration similar to what might happen to a real volcano? How was it different?

Communicating
Your Data

Make a poster with diagrams and descriptions of how a caldera forms. Use your visual aid to **describe** caldera formation to students in another class.

Buried in Ash

A long-forgotten city is accidentally found after 2,000 years

In the heat of the Italian Sun, a tired farmer wipes his brow. The farmer has spent the morning digging a new well for water. The hole is deep and the ground is dusty. Heaving a sigh, the farmer thrusts the shovel into the ground one more time.

But instead of hitting water, the shovel strikes something hard. It is a slab of smooth white marble.

This richly decorated public bath was unearthed at Herculaneum (large photo).

The small photo shows excavated ruins with Mount Vesuvius in the background.

The farmer didn't know it at the time (the early 1700s), but that marble was the first clue that something very big and very important lay beneath the farm fields. Under the ground people walked on every day, lay the ancient city of Herculaneum (her kew LAY nee um). The city, and its neighbor Pompeii (pom PAY) had been buried for more than 1,600 years. Why? Because on another summer day, August 24, 79 A.D., to be exact, Mount Vesuvius, a nearby volcano, erupted and buried both cities with pumice, rocks, mud, and ash.

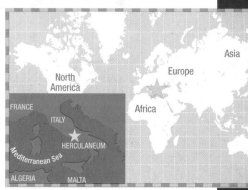
An archaeologist excavates a skeleton in Herculaneum.

Back in Time

The sun shone over the town of Herculaneum on that August morning almost 2,000 years ago. Nestled at the foot of the mountain, overlooking the Gulf of Naples, it was a peaceful place. But at about 1 P.M., that peace was shattered forever.

With massive force, the peak of Vesuvius exploded, sending six cubic kilometers of ash and pumice into the sky. Hours later, a fiery surge made its way from the volcano to the city. These pyroclastic flows continued as gray pumice fell from the sky. Buildings were crushed and buried by falling ash and pumice. Within six hours, much of Herculaneum was totally buried under the flows. After six surges from Vesuvius, the deadly eruption ceased. But the city had disappeared under approximately 21 m of ash, rock, and mud.

A City Vanishes

More than 3,600 people were killed in the natural disaster. Scientists believe that most were killed by the pyroclastic surges. Many died trying to protect their faces from the air that was filled with hot ash. Those lucky enough to escape returned to find no trace of their city. Over hundreds of years, grass and fields covered Herculaneum, erasing it from human memory. Eventually, a town called Resina was built on the site.

In the last couple of hundred years, archaeologists have unearthed colorful and perfectly preserved mosaics and an amazing library with ancient scrolls in excellent condition. Archaeologists found skeletons and voids that were filled with plaster to form casts of people who died when Vesuvius erupted. Visitors to the site can see a Roman woman, a teen-aged girl, and a soldier with his sword still in his hand.

Much of Herculaneum still lies buried beneath thick layers of volcanic ash, and archaeologists still are digging to expose more of the ruins. Their work is helping scientists better understand everyday life in an ancient Italian town. But if it weren't for a farmer's search for water, Herculaneum might not have been discovered at all!

CONNECTIONS Research the history of your town. Ask your local librarian to help "unearth" maps, drawings, or photos that let you travel back in time! Share your finds with your class.

SCIENCE *Online*

For more information, visit science.glencoe.com

Reviewing Main Ideas

Section 1 Volcanoes and Earth's Moving Plates

1. Volcanoes can be dangerous to people because they can cause deaths and destroy property.

2. Rocks in the crust and mantle melt to form magma, which is forced toward Earth's surface. When the magma flows through vents, it's called lava and forms volcanoes. *What is happening to this lava?*

3. Volcanoes can form over hot spots when magma flows onto Earth's surface. Sometimes the lava builds up from the seafloor to form an island. Volcanoes also form when Earth's plates pull apart or come together.

Section 2 Types of Volcanoes

1. The three types of volcanoes are shield volcanoes, cinder cone volcanoes, and composite volcanoes. *Which type of volcano is pictured below?*

2. Shield volcanoes produce quiet eruptions. Cinder cone and composite volcanoes can produce explosive eruptions.

3. Some lavas are thin and flow easily, producing quiet eruptions. Other lavas are thick and stiff, producing violent eruptions.

4. Water vapor and silica in magma add to its explosiveness.

Section 3 Igneous Rock Features

1. Intrusive igneous rock bodies such as batholiths, dikes, and sills form when magma solidifies underground.

2. Batholiths are the most massive igneous rock bodies. Dikes form when magma squeezes into cracks, cutting across rock layers. Sills form when magma squeezes in between rock layers. *Which rock feature is shown in this photo?*

3. A caldera forms when the top of a volcano collapses, forming a large depression. Crater Lake is a caldera in Oregon.

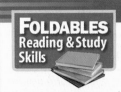

FOLDABLES
Reading & Study Skills

After You Read

Use your Foldable to help review the similarities and differences between quiet and explosive volcanic eruptions. Decide what type of volcano produces each style of eruption.

Visualizing Main Ideas

Complete the following concept map on types of volcanic eruptions.

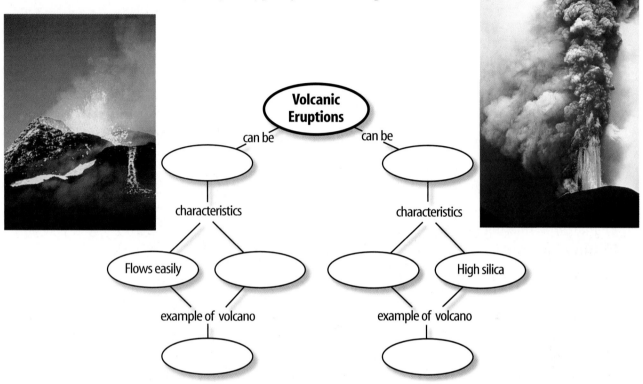

Vocabulary Review

Vocabulary Words

a. batholith
b. caldera
c. cinder cone volcano
d. composite volcano
e. crater
f. dike
g. hot spot

h. shield volcano
i. sill
j. tephra
k. vent
l. volcanic neck
m. volcano

Study Tip

When you encounter new vocabulary, write it down in a sentence. This will help you understand, remember, and use new vocabulary words.

Using Vocabulary

Each of the following sentences is false. Make the sentence true by replacing each underlined word(s) with the correct vocabulary word(s).

1. A broad volcano with gently sloping sides is called a <u>composite volcano</u>.

2. <u>Sills</u> are bits of rock or solidified lava dropped from the air after a volcanic eruption.

3. Magma squeezed into a horizontal crack between rock layers is called a <u>caldera</u>.

4. The steep-walled depression around a volcano's vent is called a <u>tephra</u>.

5. Magma squeezed into a vertical crack across rock layers is called a <u>crater</u>.

Chapter 12 Assessment

Checking Concepts

Choose the word or phrase that best answers the question.

1. What type of boundary is associated with composite volcanoes?
   A) plates moving apart
   B) plates sticking and slipping
   C) plates moving together
   D) plates sliding past each other

2. Why is Hawaii made of volcanoes?
   A) Plates are moving apart.
   B) A hot spot exists.
   C) Plates are moving together.
   D) Rift zones exist.

3. What kind of magmas produce violent volcanic eruptions?
   A) those rich in silica
   B) those that are fluid
   C) those forming shield volcanoes
   D) those rich in iron

4. Magma that is low in silica generally produces what kind of eruptions?
   A) thick
   C) quiet
   B) caldera
   D) explosive

5. What is made entirely of tephra?
   A) shield volcano
   C) cinder cone volcano
   B) caldera
   D) composite volcano

6. What kind of volcano is Kilauea?
   A) shield volcano
   C) cinder cone volcano
   B) composite
   D) caldera cone volcano

7. What is magma that hardens in a crack cutting across rock layers called?
   A) sill
   C) volcanic neck
   B) dike
   D) batholith

8. What is the largest intrusive igneous rock body?
   A) dike
   C) sill
   B) volcanic neck
   D) batholith

9. Which describes bits of material that fall to Earth after an eruption?
   A) dike
   C) tephra
   B) sand
   D) sill

10. What is the process that formed Soufrière Hills volcano on Montserrat?
    A) plates sticking and slipping
    B) caldera formation
    C) plates sliding sideways
    D) plates moving together

Thinking Critically

11. Explain how glaciers and volcanoes can exist on Iceland.

12. What kind of eruption is produced when basaltic lava that is low in silica flows from a volcano? Explain.

13. How are volcanoes related to earthquakes?

14. Misti is a volcano in Peru. Peru is on the western edge of South America. How might this volcano have formed?

15. Describe the layers of a composite volcano. Which layers represent violent eruptions?

Developing Skills

16. **Classifying** Classify Fuji, which has steep sides and is made of layers of silica-rich lava and ash.

17. **Measuring in SI** The base of the volcano Mauna Loa is about 5,000 m below sea level. The total height of the volcano is 9,170 m. What percentage of the volcano is above sea level? Below sea level?

18. **Comparing and Contrasting** Compare and contrast shield volcanoes, cinder cone volcanoes, and composite volcanoes.

19. Interpreting Scientific Illustrations Look at the map below. The Hawaiian Islands and Emperor Seamounts were formed when the Pacific Plate moved over a fixed hot spot. If the Emperor chain trends in a direction different from the Hawaiian Islands, what can you infer about the Pacific Plate?

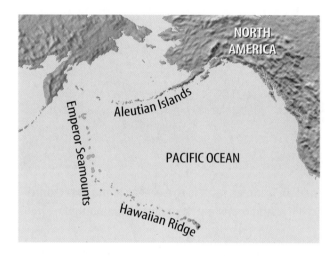

20. Concept Mapping Make a network tree concept map about where volcanoes can occur. Include the following words and phrases: *hot spots, divergent plate boundaries, convergent plate boundaries, volcanoes, can occur, examples, Iceland, Soufrière Hills,* and *Hawaiian Islands.*

Performance Assessment

21. Poster Make a poster of the three basic types of volcanoes. Label them and indicate what type of eruption occurs from each one.

TECHNOLOGY

Go to the Glencoe Science Web site at **science.glencoe.com** or use the **Glencoe Science CD-ROM** for additional chapter assessment.

Test Practice

A scientist who studies volcanoes brought the following information to a lecture he did at a middle school for the entire sixth grade.

Study the table and answer the following questions.

Data for Various Volcanoes		
Volcano	Silica Content in Magma	Trapped Gases in Magma
1	Medium to High	High
2	Low	Medium
3	Low to Medium	Medium
4	High	High
5	Low	Medium

1. According to this information, which volcano probably had the most violent eruption?
 A) Volcano 1
 B) Volcano 2
 C) Volcano 4
 D) Volcano 5

2. Based on this information, choose the most reasonable hypothesis.
 F) The higher the silica content is, the stronger the eruption is.
 G) The lower the trapped gas content is, the stronger the eruption is.
 H) The lower the silica content is, the stronger the eruption is.
 J) The higher the trapped gas content is, the weaker the eruption is.

Reading Comprehension

Read the passage carefully. Then read the questions that follow the passage. Decide which is the best answer to each question.

Earthquakes and Volcanoes

Earthquakes are destructive and potentially fatal natural disasters. Geologists have been working to learn what they can about earthquakes in order to better protect property and to save human lives.

Scientists know that many earthquakes occur because tectonic plates interact with one another at plate boundaries. They also know that many earthquakes occur every day all over the world. They forecast that major earthquakes will occur sometime in the future in certain regions, such as along the San Andreas Fault near Parkfield, California. But scientists cannot predict exactly when and where an earthquake will occur.

One approach to forecasting earthquakes is known as paleoseismology. Paleoseismology involves the study of past movement of rock and sediment along faults. Motion along a fault results in an earthquake, therefore, studying this movement is one way to study ancient earthquake occurrences.

This movement can be measured in the field by observing shifted rock and sediment along a fault. If this displaced sediment can be dated, the time at which the earthquake occurred also can be estimated. With information on several past earthquakes, scientists can estimate how long, on average, time intervals are between the earthquakes. This is one way to estimate how many earthquakes might occur along a fault over a period of time.

Although scientists are not yet able to predict earthquakes, the information gained from their research advances the field.

The San Andreas fault is the center of much tectonic activity, and many earthquakes occur along it.

1. Based on the information in this passage, what can the reader conclude?
   A) Earthquakes always occur during heavy rainstorms.
   B) Earthquakes can not be predicted reliably, but they can be forecasted over estimated periods of time.
   C) Earthquakes are extremely rare natural occurrences.
   D) Earthquakes do not affect property and human lives very much.

2. What does the information in this passage suggest?
   F) Studies of displacement along faults can provide information for forecasting earthquakes.
   G) Studying faults allows scientists to determine the time and date of the next earthquake.
   H) Paleoseismology is the study of fossils along the San Andreas fault.
   J) Scientists do not know what causes earthquakes.

Reasoning and Skills

Read each question and decide which is the best answer.

1. All of these statements are true about earthquakes EXCEPT _____.
   A) Earthquakes can range from unnoticeable vibrations to devastating waves of energy.
   B) Earthquakes occur along fault surfaces.
   C) When earthquakes occur, only primary waves and surface waves are produced.
   D) Surface waves are responsible for most of the damage caused by earthquakes.

Test-Taking Tip Think about all that you know about how earthquakes form and what they are.

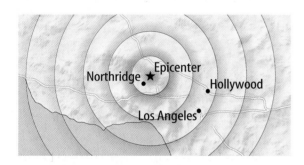

2. Refer to the diagram above. Which statement best describes the movement of surface waves generated by an earthquake?
   F) Surface waves move towards the epicenter just before an earthquake strikes.
   G) Surface waves travel throughout all layers of Earth.
   H) Surface waves travel faster than primary waves do.
   J) Surface waves travel outward from the epicenter of an earthquake.

Test-Taking Tip Review what you know about wave movement and speed.

3. All of these statements are true about volcanoes EXCEPT ____.
   A) Three basic types of volcanoes are shield, cinder cone, and composite.
   B) The amount of water vapor, other gases, and silica in magma determine what kind of eruption will take place.
   C) The lower the silica content in magma, the more explosive an eruption will be.
   D) A volcano's form depends upon the type of eruption and the composition of erupted material.

Test-Taking Tip Think about the different kinds of volcanoes, how they form, and what factors influence the type of eruption.

Consider this question carefully before writing your answer on a separate sheet of paper.

4. Volcanic eruptions transform the environment around them. Discuss some of the ways in which volcanoes change Earth's surface.

Test-Taking Tip Consider the details of specific eruptions that you have learned about.

How Are Canals & the Paleozoic Era Connected?

Before the invention of the locomotive, canals, such as the one at upper right, were an important means of transportation. In the 1790s, an engineer traveled around England to study new canals. The engineer noticed something odd: All across the country, certain types of rocks seemed to lie in predictable layers, or strata. And the same strata always had the same kinds of fossils in them. Since each layer of sedimentary rock typically forms on top of the previous one, scientists realized that the strata recorded the history of life on Earth. By the mid-1800s, the known rock strata had been organized into a system that we now know as the geologic time scale. In this system, Earth's history is divided into units called eras, which in turn are divided into periods. Many of the rock layers in the Grand Canyon (background) date from the Paleozoic, or "ancient life," Era.

SCIENCE CONNECTION

FOSSIL RECORD Choose a period in the geologic time scale and find out what the fossil record tells us about the organisms that were alive during that time. Create a drawing or painting that shows what you think might have been a typical scene during that period in Earth's history, complete with appropriate plants and animals. What aspects of the scene are difficult or impossible to determine based on the fossil record?

13

Clues to Earth's Past

Studying the history of Earth is not as easy as studying the history of a country. This is because no people were around to record the events in Earth's past. Also, the long stretches of time between some events in geologic history are mind-boggling and difficult for humans to grasp. In a study of Earth's past, clues found in rocks and fossils of long-dead organisms take the place of words and pictures. In this chapter, you will encounter some of those long stretches of geologic time and learn how fossils form so that you can interpret some of Earth's past for yourself.

What do you think?

Science Journal Study the picture below. Discuss what you think this might be with a classmate. Here's a hint: *It lived in the ocean millions of years ago.* Write your answer or best guess in your Science Journal.

The process of fossil formation begins when dead plants or animals are buried in sediment. In time, if conditions are right, the sediment hardens into sedimentary rock. Parts of the organism are preserved along with the impressions of parts that don't survive. Any evidence of once-living things contained in the rock record is a fossil.

Make a model of a fossil

1. Fill a small jar (about 500 mL) one-third full of plaster of paris. Add water until the jar is half full.

2. Drop in a few small shells.

3. To model a swift, muddy stream, cover the jar and shake it.

4. Now model the stream flowing into a lake by uncovering the jar and pouring the contents into a paper or plastic bowl. Let the mixture sit for an hour.

5. Crack open the hardened plaster to locate the model fossils.

Observe

Pick the shells out of the plaster and focus your attention on the impressions left by them. In your Science Journal, list what the impressions could tell you.

Before You Read

FOLDABLES
Reading & Study Skills

Making a Question Study Fold Asking yourself questions helps you stay focused and better understand Earth's past when you are reading the chapter.

1. Place a sheet of paper in front of you so the long side is at the top. Fold the paper in half from the left side to the right side and then unfold.

2. Fold each side in to the center fold line to divide the paper into fourths. Fold the paper in half from top to bottom and unfold.

3. Through the top thickness of paper, cut along both of the middle fold lines to form four tabs. Write these questions on the tabs: *What are fossils? How do fossils form? How is the relative age of rocks determined? How is the absolute age of rocks determined?*

4. As you read the chapter, write answers to the questions under the tabs.

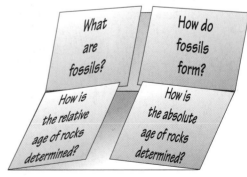

1 Fossils

As You Read

What You'll Learn

- **List** the conditions necessary for fossils to form.
- **Describe** several processes of fossil formation.
- **Explain** how fossil correlation is used to determine rock ages.

Vocabulary

fossil
permineralized remains
carbon film

mold
cast
index fossil

Why It's Important

Fossils help scientists find oil and other sources of energy necessary for society.

Traces of the Distant Past

A giant crocodile lurks in the shallow water of a river. A herd of *Triceratops* emerges from the edge of the forest and cautiously moves toward the river. The dinosaurs are thirsty, but they know danger waits for them in the water. A large bull *Triceratops* moves into the river. The others follow.

Does this scene sound familiar to you? It's likely that you've read about dinosaurs and other past inhabitants of Earth. But how do you know that they really existed or what they were like? What evidence do humans have of past life on Earth? The answer is fossils. Paleontologists, scientists who study fossils, can learn about extinct animals from their fossil remains, as shown in **Figure 1.**

Figure 1
Scientists can learn how dinosaurs looked and moved using fossil remains.

A A paleontologist carefully examines a fossil skeleton.

B The skeleton can be re-assembled and displayed in a museum.

Formation of Fossils

Fossils are the remains, imprints, or traces of prehistoric organisms. Fossils have helped scientists determine approximately when life first appeared, when plants and animals first lived on land, and when organisms became extinct. Fossils are evidence of not only when and where organisms once lived, but also how they lived.

For the most part, the remains of dead plants and animals disappear quickly. Scavengers eat and scatter the remains of dead organisms. Fungi and bacteria invade, causing the remains to rot and disappear. If you've ever left a banana on the counter too long, you've seen this process begin. In time, compounds within the banana cause it to break down chemically and soften. Microorganisms, such as bacteria, cause it to decay. What keeps some plants and animals from disappearing before they become fossils? Which organisms are more likely to become fossils?

Figure 2
These fossil shark teeth are hard parts. Soft parts of animals do not become fossilized as easily.

Conditions Needed for Fossil Formation Whether or not a dead organism becomes a fossil depends upon how well it is protected from scavengers and agents of physical destruction, such as waves and currents. One way a dead organism can be protected is for sediment to bury the body quickly. If a fish dies and sinks to the bottom of a lake, sediment carried into the lake by a stream can cover the fish rapidly. As a result, no waves or scavengers can get to it and tear it apart. The body parts then might be fossilized and included in a sedimentary rock like shale. However, quick burial alone isn't always enough to make a fossil.

Organisms have a better chance of becoming fossils if they have hard parts such as bones, shells, or teeth. One reason is that scavengers are less likely to eat these hard parts. Hard parts also decay more slowly than soft parts do. Most fossils are the hard parts of organisms, such as the fossil teeth in **Figure 2**.

Types of Preservation

Perhaps you've seen skeletal remains of *Tyrannosaurus rex* towering above you in a museum. You also have some idea of what this dinosaur looked like because you've seen illustrations. Artists who draw *Tyrannosaurus rex* and other dinosaurs base their illustrations on fossil bones. What preserves fossil bones?

TRY AT HOME

Mini LAB

Predicting Fossil Preservation

Procedure
1. Take a brief walk outside and observe your neighborhood.
2. Look around and notice what kinds of plants and animals live nearby.

Analysis
1. Predict what remains from your time might be preserved far into the future.
2. Explain what conditions would need to exist for these remains to be fossilized.

Figure 3
Opal and various minerals have replaced original materials and filled the hollow spaces in this permineralized dinosaur bone.
Why has this fossil retained the shape of the original bone?

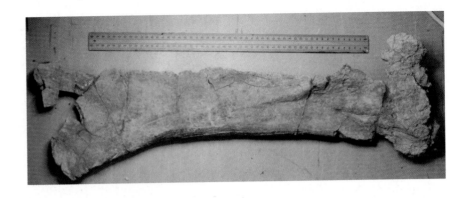

Mineral Replacement Most hard parts of organisms such as bones, teeth, and shells have tiny spaces within them. In life, these spaces can be filled with cells, blood vessels, nerves, or air. When the organism dies and the soft materials inside the hard parts decay, the tiny spaces become empty. If the hard part is buried, groundwater can seep in and deposit minerals in the spaces. **Permineralized remains** are fossils in which the spaces inside are filled with minerals from groundwater. In permineralized remains, some original material from the fossil organism's body might be preserved—encased within the minerals from groundwater. It is from these original materials that DNA, the chemical that contains an organism's genetic code, can sometimes be recovered.

Sometimes minerals replace the hard parts of fossil organisms. For example, a solution of water and dissolved silica (the compound SiO_2) might flow into and through the shell of a dead organism. If the water dissolves the shell and leaves silica in its place, the original shell is replaced.

Often people learn about past forms of life from bones, wood, and other remains that became permineralized or replaced with minerals from groundwater, as shown in **Figure 3,** but many other types of fossils can be found.

Figure 4
Graptolites lived hundreds of millions of years ago and drifted on currents in the oceans. These organisms often are preserved as carbon films.

Carbon Films The tissues of most organisms are made of compounds that contain carbon. Sometimes fossils contain only carbon. Fossils usually form when sediments bury a dead organism. As sediment piles up, the organism's remains are subjected to pressure and heat. These conditions force gases and liquids from the body. A thin film of carbon residue is left, forming a silhouette of the original organism called a **carbon film. Figure 4** shows the carbonized remains of graptolites, which were small marine animals. Graptolites have been found in rocks as old as 500 million years.

Coal In swampy regions, large volumes of plant matter accumulate. Over millions of years, these deposits become completely carbonized, forming coal. Coal is more important as a source of fuel than as a fossil because the structure of the original plant often is lost when coal forms.

✔ **Reading Check** *In what sort of environment does coal form?*

Molds and Casts In nature, impressions form when seashells or other hard parts of organisms fall into a soft sediment such as mud. The object and sediment then are buried by more sediment. Compaction and cementation, the deposition of minerals from water into the pore spaces between sediment particles, turn the sediment into rock. Other open pores in the rock then let water and air reach the shell or hard part. The hard part might decay or dissolve, leaving behind a cavity in the rock called a **mold.** Later, mineral-rich water or other sediment might enter the cavity, form new rock, and produce a copy or **cast** of the original object, as shown in **Figure 5.**

Chemistry
INTEGRATION

The bones of vertebrates contain calcium phosphate, and the shells of many invertebrates contain calcium carbonate. Calcium carbonate dissolves more easily in acid than calcium phosphate does. The groundwater in some swampy environments is acidic. In your Science Journal, design an experiment to find out whether bones or shells would preserve best as fossils in a swamp.

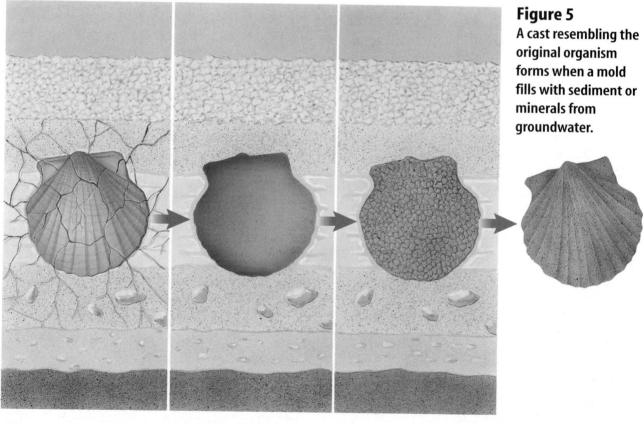

Figure 5
A cast resembling the original organism forms when a mold fills with sediment or minerals from groundwater.

A The fossil begins to dissolve as water moves through spaces in the rock layers.

B The fossil has been dissolved away. The harder rock once surrounding it forms a mold.

C Sediment washes into the mold and is deposited, or mineral crystals form.

D A cast results.

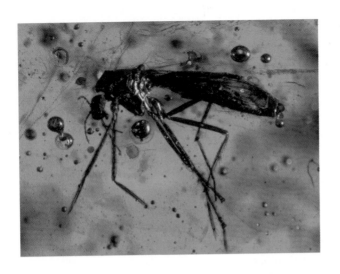

Original Remains Sometimes conditions allow original soft parts of organisms to be preserved for thousands or millions of years. For example, insects can be trapped in amber, a hardened form of sticky tree resin. The amber surrounds and protects the original material of the insect's exoskeleton from destruction, as shown in **Figure 6.** Some organisms, such as the mammoth, have been found preserved in frozen ground in Siberia. Original remains also have been found in natural tar deposits at Earth's surface, such as the La Brea tar pits in California.

Figure 6
The original soft parts of this mosquito have been preserved in amber for millions of years.

Trace Fossils Do you have a handprint in plaster that you made when you were in kindergarten? If so, it's a record that tells something about you. From it, others can guess your size and maybe your weight at that age. Animals walking on Earth long ago left similar tracks, such as those in **Figure 7.** Trace fossils are fossilized tracks and other evidence of the activity of organisms. In some cases, tracks can tell you more about how an organism lived than any other type of fossil. For example, from a set of tracks at Davenport Ranch, Texas, you might be able to learn something about the social life of sauropods, which were large, plant-eating dinosaurs. The largest tracks of the herd are on the outer edges and the smallest are on the inside. These tracks cause some scientists to hypothesize that adult sauropods surrounded their young as they traveled—probably to protect them from predators. A nearby set of tracks might mean that another type of dinosaur, an allosaur, was stalking the herd.

Figure 7
Tracks made in soft mud, and now preserved in solid rock, can provide information about animal size, speed, and behavior.

A This dinosaur track is from the Glen Rose Formation in north-central Texas.

B These tracks are located on a Navajo Reservation in Arizona.

Trails and Burrows Other trace fossils include trails and burrows made by worms and other animals. These, too, tell something about how these animals lived. For example, by examining fossil burrows you can sometimes tell how firm the sediment the animals lived in was. As you can see, fossils can tell a great deal about the organisms that have inhabited Earth.

Reading Check *How are trace fossils different from fossils that are the remains of an organism's body?*

Index Fossils

One thing you can learn by studying fossils is that species of organisms have changed over time. Some species of organisms inhabited Earth for long periods of time without changing. Other species changed a lot in comparatively short amounts of time. It is these organisms that became index fossils.

Index fossils are the remains of species that existed on Earth for relatively short periods of time, were abundant, and were widespread geographically. Because the organisms that became index fossils lived only during specific intervals of geologic time, geologists can estimate the ages of rock layers based on the particular index fossils they contain. However, not all rocks contain index fossils. Another way to approximate the age of a rock layer is to compare the spans of time, or ranges, over which more than one fossil lived. The estimated age is the time interval where fossil ranges overlap, as shown in **Figure 8.**

Figure 8
A The fossils in a sequence of sedimentary rock can be used to estimate the ages of each layer. **B** The chart shows when each organism inhabited Earth. *Why is it possible to say that the middle layer of rock was deposited between 440 million and 410 million years ago?*

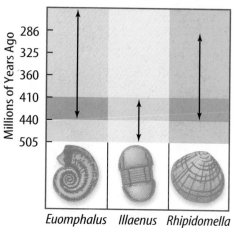

A

Fossil Range Chart

B

Millions of Years Ago

286
325
360
410
440
505

Euomphalus Illaenus Rhipidomella

Ecology is the study of how organisms interact with each other and with their environment. Some paleontologists study the ecology of ancient organisms. Hypothesize about the kinds of information you could use to determine how ancient organisms interacted with their environment.

Fossils and Ancient Environments

Scientists can use fossils to determine what the environment of an area was like long ago. Using fossils, you might be able to find out whether an area was land or whether it was covered by an ocean at a particular time. If the region was covered by ocean, it might even be possible to learn the depth of the water. What clues about the depth of water do you think fosssils could provide?

Fossils also are used to determine the past climate of a region. For example, rocks in parts of the eastern United States contain fossils of tropical plants. The environment of this part of the United States today isn't tropical. However, because of the fossils, scientists know that it was tropical when these fossilized plants were living. **Figure 9** shows that North America was located near the equator when these fossils formed.

Figure 9
Because the continents on the surface of Earth slowly move, the climate of a given region changes through geologic time.

A The equator passed through North America 310 million years ago. At this time, warm shallow seas and coal swamps covered much of the continent.

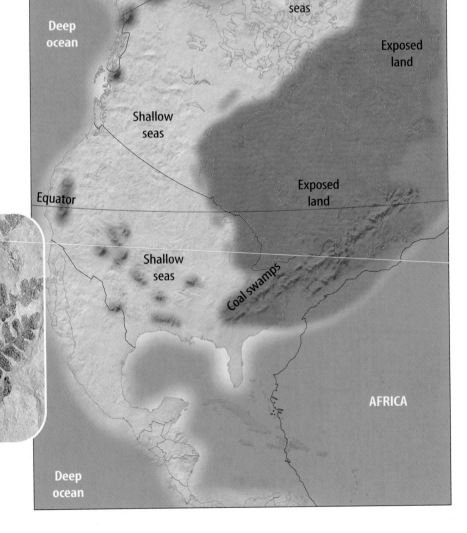

B *Neuropteris* was a common fern that grew in swamps during this time.

Shallow Seas How would you explain the presence of fossilized crinoids—animals that lived in shallow seas—in rocks just west of those containing the fern fossils? **Figure 10** shows a fossil crinoid and a living crinoid. When the fossil crinoids were alive, a shallow sea covered much of western and central North America. The crinoid hard parts were included in rocks that formed from the sediments at the bottom of this sea. Fossils provide information about past life on Earth and also about the history of the rock layers that contain them. Fossils can provide information about the ages of rocks and the climate and type of environment that existed when the rocks formed.

Figure 10
A This fossil crinoid lived in the warm, shallow seas that once covered part of North America.
B This crinoid lives in shallow water near Papua New Guinea. *How do the habitats of these crinoids compare?*

Section 1 Assessment

1. What conditions must exist for most fossils to form?

2. Describe how a fossil mold could form. Explain how a fossil mold is different from a fossil cast.

3. What characteristics do the organisms that become index fossils have? How are these characteristics useful to geologists?

4. How do carbon films form?

5. **Think Critically** What can you say about the ages of two widely separated layers of rock that contain the same type of fossil? What can you say about the environments of these widely separated layers?

Skill Builder Activities

6. **Concept Mapping** Make a concept map that compares and contrasts permineralized remains and original remains. Use the following phrases: *types of fossils, original remains, evidence of former life, permineralized remains, replaced by minerals,* and *parts of organisms.* **For more help, refer to the** Science Skill Handbook.

7. **Communicating** Visit a museum that has fossils on display. In your Science Journal, make an illustration of each fossil. Write a brief description, noting key facts about each. Also, write about how each fossil might have formed. **For more help, refer to the** Science Skill Handbook.

2 Relative Ages of Rocks

As You Read

What You'll Learn

- **Describe** methods used to assign relative ages to rock layers.
- **Interpret** gaps in the rock record.
- **Give** an example of how rock layers can be correlated with other rock layers.

Vocabulary

principle of superposition
relative age
unconformity

Why It's Important

Being able to determine the age of rock layers is important in trying to understand a history of Earth.

Superposition

Imagine that you are walking to your favorite store and you happen to notice an interesting car go by. You're not sure what kind it is, but you remember that you read an article about it. You decide to look it up. At home you have a stack of magazines from the past year, as seen in **Figure 11.**

You know that the article you're thinking of came out in the January edition, so it must be near the bottom of the pile. As you dig downward, you find magazines from March, then February. January must be next. How did you know that the January issue of the magazine would be on the bottom? To find the older edition under newer ones, you applied the principle of superposition.

Oldest Rocks on the Bottom According to the **principle of superposition,** in undisturbed layers of rock, the oldest rocks are on the bottom and the rocks become progressively younger toward the top. Why is this the case?

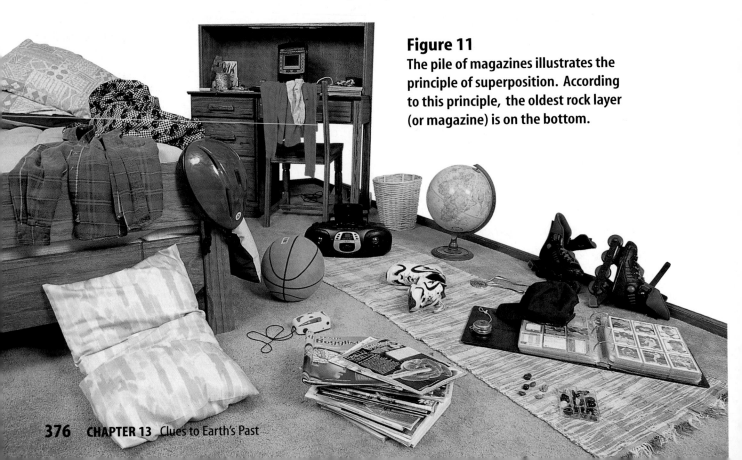

Figure 11
The pile of magazines illustrates the principle of superposition. According to this principle, the oldest rock layer (or magazine) is on the bottom.

Rock Layers Sediment accumulates in horizontal beds, forming layers of sedimentary rock. The first layer to form is on the bottom. The next layer forms on top of the previous one. Because of this, the oldest rocks are at the bottom. However, forces generated by mountain formation sometimes can turn layers over. When layers have been turned upside down, it's necessary to use other clues in the rock layers to determine their original positions and relative ages.

Relative Ages

Now you want to look for another magazine. You're not sure how old it is, but you know it arrived after the January issue. You can find it in the stack by using the principle of relative age.

The **relative age** of something is its age in comparison to other things. Geologists determine the relative ages of rocks and other structures by examining their places in a sequence. For example, if layers of sedimentary rock are offset by a fault, which is a break in Earth's surface, you know that the layers had to be there before a fault could cut through them. The relative age of the rocks is older than the relative age of the fault. Relative age determination doesn't tell you anything about the age of rock layers in actual years. You don't know if a layer is 100 million or 10,000 years old. You only know that it's younger than the layers below it and older than the fault cutting through it.

Other Clues Help Determination of relative age is easy if the rocks haven't been faulted or turned upside down. For example, look at **Figure 12A.** Which layer is the oldest? In cases where rock layers have been disturbed as in **Figure 12B,** you might have to look for fossils and other clues to date the rocks. If you find a fossil in the top layer that's older than a fossil in a lower layer, you can hypothesize that layers have been turned upside down by folding during mountain building.

Figure 12
A In a stack of undisturbed sedimentary rocks, the oldest rocks are at the bottom. **B** This similar stack of rocks has been folded by forces within Earth. *How can you tell if an older rock is above a younger one?*

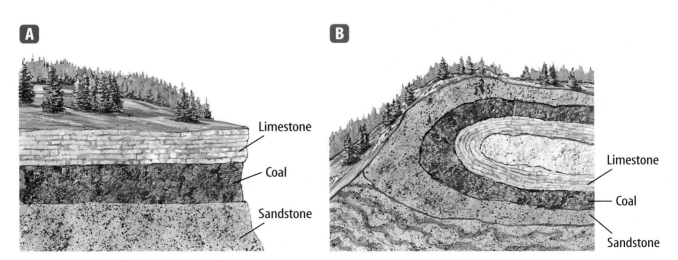

A

Limestone

Coal

Sandstone

B

Limestone

Coal

Sandstone

Unconformities

A sequence of rock is a record of past events. But most rock sequences are incomplete—layers are missing. These gaps in rock sequences are called **unconformities** (un kun FOR mih teez). Unconformities develop when agents of erosion such as running water or glaciers remove rock layers by washing or scraping them away. They also form when a period of time passes without any new deposition occurring to form new layers of rock.

Reading Check *How do unconformities form?*

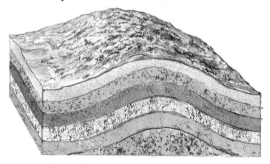

A Sedimentary rocks are deposited originally as horizontal layers.

Angular Unconformities Horizontal layers of sedimentary rock often are tilted and uplifted. Erosion and weathering then wear down these tilted rock layers. Eventually, younger sediment layers are deposited horizontally on top of the eroded and tilted layers. Geologists call such an unconformity an angular unconformity. **Figure 13** shows how angular unconformities develop.

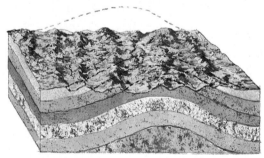

B The horizontal rock layers are tilted as forces within Earth deform them.

Disconformity Suppose you're looking at a stack of sedimentary rock layers. They look complete, but layers are missing. If you look closely, you might find an old surface of erosion. This records a time when the rocks were exposed and eroded. Later, younger rocks formed above the erosion surface when deposition of sediment began again. Even though all the layers are parallel, the rock record still has a gap. This type of unconformity is called a disconformity.

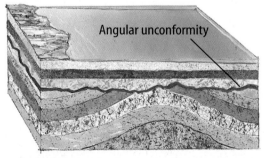

C The tilted layers erode.

Nonconformity Another type of unconformity, called a nonconformity, occurs when metamorphic or igneous rocks are uplifted and eroded. Sedimentary rocks are then deposited on top of this erosion surface. The surface between the two rock types is a nonconformity. Sometimes rock fragments from below are incorporated into sediments deposited above the nonconformity. All types of unconformities are shown in **Figure 14.**

Angular unconformity

D An angular unconformity results when new layers form on the tilted layers as deposition resumes.

Figure 14

An unconformity is a gap in the rock record caused by erosion or a pause in deposition. There are three major kinds of unconformities—nonconformity, angular unconformity, and disconformity.

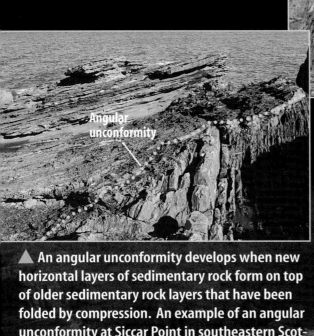

Angular unconformity

Nonconformity

▲ In a nonconformity, horizontal layers of sedimentary rock overlie older igneous or metamorphic rocks. A nonconformity in Big Bend National Park, Texas, is shown above.

▲ An angular unconformity develops when new horizontal layers of sedimentary rock form on top of older sedimentary rock layers that have been folded by compression. An example of an angular unconformity at Siccar Point in southeastern Scotland is shown above.

▼ A disconformity develops when horizontal rock layers are exposed and eroded, and new horizontal layers of rock are deposited on the eroded surface. The disconformity shown below is in the Grand Canyon.

Disconformity

Matching Up Rock Layers

Suppose you're studying a layer of sandstone in Bryce Canyon in Utah. Later, when you visit Canyonlands National Park, Utah, you notice that a layer of sandstone there looks just like the sandstone in Bryce Canyon, 250 km away. Above the sandstone in the Canyonlands is a layer of limestone and then another sandstone layer. You return to Bryce Canyon and find the same sequence—sandstone, limestone, and sandstone. What do you infer? It's likely that you're looking at the same layers of rocks in two different locations. **Figure 15** shows that these rocks are parts of huge deposits that covered this whole area of the western United States. Geologists often can match up, or correlate, layers of rocks over great distances.

Evidence Used for Correlation It's not always easy to say that a rock layer exposed in one area is the same as a rock layer exposed in another area. Sometimes it's possible to walk along the layer for kilometers and prove that it's continuous. In other cases, such as at the Canyonlands area and Bryce Canyon as seen in **Figure 16,** the rock layers are exposed only where rivers have cut through overlying layers of rock and sediment. How can you show that the limestone sandwiched between the two layers of sandstone in Canyonlands is likely the same limestone as at Bryce Canyon? One way is to use fossil evidence. If the same types of fossils were found in the limestone layer in both places, it's a good indication that the limestone at each location is the same age, and, therefore, one continuous deposit.

 **Reading Check** *How do fossils help show that rocks at different locations belong to the same rock layer?*

Research Visit the Glencoe Science Web site at **science.glencoe.com** for a link to more information about the process of correlation.

Figure 15
These rock layers, exposed at Hopi Point in Grand Canyon National Park, Arizona, can be correlated, or matched up, with rocks from across large areas of the western United States.

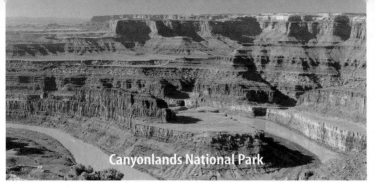
Canyonlands National Park

Bryce Canyon National Park

Date deposited (millions of years ago)

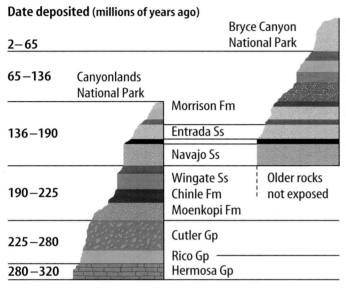

2–65	Bryce Canyon National Park — Wasatch Fm
65–136	Canyonlands National Park — Kaiparowits Fm, Straight Cliffs Ss
	Morrison Fm — Dakota Ss, Winsor Fm
136–190	Entrada Ss — Entrada Ss, Carmel Fm
	Navajo Ss — Navajo Ss
190–225	Wingate Ss, Chinle Fm, Moenkopi Fm — Older rocks not exposed
225–280	Cutler Gp
	Rico Gp
280–320	Hermosa Gp

Figure 16

Geologists have named the many rock layers, or formations, in Canyonlands and in Bryce Canyon, Utah. They also have correlated some formations between the two canyons. *Which of the labeled layers are present at both canyons?*

Can layers of rock be correlated in other ways? Sometimes determining relative ages isn't enough, and other dating methods must be used. In Section 3, you'll see how the numerical ages of rocks can be determined, and how geologists have used this information to estimate the age of Earth.

Section 2 Assessment

1. Suppose you haven't cleaned out your locker all year. Where would you expect to find papers from the beginning of the year? What principle of geology would you use to find these old papers?

2. Explain the concept of relative age.

3. What is a disconformity?

4. What is one way to correlate similar rock layers in two different areas?

5. **Think Critically** Explain the relationship between the concept of relative age and the principle of superposition.

Skill Builder Activities

6. **Interpreting Data** A sandstone contains a 400-million-year-old fossil. A shale contains fossils that are over 500 million years old. A limestone, underlying a sandstone, contains fossils that are between 400 million and 500 million years old. Which rock bed is oldest? Explain. **For more help, refer to the Science Skill Handbook.**

7. **Using an Electronic Spreadsheet** Use this section to prepare a table comparing and contrasting types of unconformities. **For more help, refer to the Technology Skill Handbook.**

Activity

Relative Ages

Which of your two friends is older? To answer this question, you'd need to know their relative ages. You wouldn't need to know the exact age of either of your friends—just who was born first. The same is sometimes true for rock layers.

What You'll Investigate
Can you determine the relative ages of rock layers?

Materials
paper   pencil

Goals
■ **Interpret** illustrations of rock layers and other geological structures and determine the relative order of events.

Procedure

1. **Analyze Figures A** and **B.**
2. Make a sketch of **Figure A.** On it, identify the relative age of each rock layer, igneous intrusion, fault, and unconformity. For example, the shale layer is the oldest, so mark it with a 1. Mark the next-oldest feature with a 2, and so on.
3. Repeat step 2 for **Figure B.**

Conclude and Apply

Figure A

1. **Identify** the type of unconformity shown. Is it possible that there were originally more layers of rock than are shown?
2. **Describe** how the rocks above the fault moved in relation to rocks below the fault.
3. **Hypothesize** how the hill on the left side of the figure formed.

A

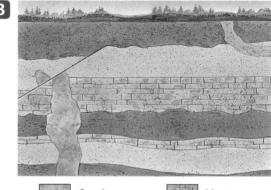

B

	Granite		Limestone
	Sandstone		Shale

Figure B

4. Is it possible to conclude if the igneous intrusion on the left is older or younger than the unconformity nearest the surface?
5. **Describe** the relative ages of the two igneous intrusions. How did you know?
6. **Hypothesize** which two layers of rock might have been much thicker in the past.

Communicating Your Data

Compare your results with other students' results. **For more help, refer to the** Science Skill Handbook.

SECTION 3

Absolute Ages of Rocks

Absolute Ages

As you sort through your stack of magazines looking for that article about the car you saw, you decide that you need to restack them into a neat pile. By now, they're in a jumble and no longer in order of their relative age, as shown in **Figure 17.** How can you stack them so the oldest are on the bottom and the newest are on top? Fortunately, magazine dates are printed on the cover. Thus, stacking magazines in order is a simple process. Unfortunately, rocks don't have their ages stamped on them. Or do they? **Absolute age** is the age, in years, of a rock or other object. Geologists determine absolute ages by using properties of the atoms that make up materials.

Radioactive Decay

Atoms consist of a dense central region called the nucleus, which is surrounded by a cloud of negatively charged particles called electrons. The nucleus is made up of protons, which have a positive charge, and neutrons, which have no electric charge. The number of protons determines the identity of the element, and the number of neutrons determines the form of the element, or isotope. For example, every atom with a single proton is a hydrogen atom. Hydrogen atoms can have no neutrons, a single neutron, or two neutrons. This means that there are three isotopes of hydrogen.

✓ Reading Check *What particles make up an atom's nucleus?*

Some isotopes are unstable and break down into other isotopes and particles. Sometimes a lot of energy is given off during this process. The process of breaking down is called **radioactive decay.** In the case of hydrogen, atoms with one proton and two neutrons are unstable and tend to break down. Many other elements have stable and unstable isotopes.

As You Read

***What* You'll Learn**

- **Identify** how absolute age differs from relative age.
- **Describe** how the half-lives of isotopes are used to determine a rock's age.

Vocabulary
absolute age
radioactive decay
half-life
radiometric dating
uniformitarianism

***Why* It's Important**
Events in Earth's history can be better understood if their absolute ages are known.

Figure 17
The magazines that have been shuffled through no longer illustrate the principle of superposition.

Modeling Carbon-14 Dating

Procedure:
1. Count out 80 **red jelly beans.**
2. Remove half the red jelly beans and replace them with **green jelly beans.**
3. Continue replacing half the red jelly beans with green jelly beans until only 5 red jelly beans remain. Count the number of times you replace half the red jelly beans.

Analysis
1. How did this activity model the decay of carbon-14 atoms?
2. How many half lives of carbon-14 did you model during this activity?
3. If the atoms in a bone experienced the same number of half lives as your jelly beans, how old would the bone be?

Alpha and Beta Decay In some isotopes, a neutron breaks down into a proton and an electron. This type of radioactive decay is called beta decay because the electron leaves the atom as a beta particle. The nucleus loses a neutron but gains a proton. When the number of protons in an atom is changed, a new element forms. Other isotopes give off two protons and two neutrons in the form of an alpha particle. Alpha and beta decay are shown in **Figure 18.**

Half-Life In radioactive decay reactions, the parent isotope undergoes radioactive decay. The daughter product is produced by radioactive decay. Each radioactive parent isotope decays to its daughter product at a certain rate. Based on this decay rate, it takes a certain period of time for one half of the parent isotope to decay to its daughter product. The **half-life** of an isotope is the time it takes for half of the atoms in the isotope to decay. For example, the half-life of carbon-14 is 5,730 years. So it will take 5,730 years for half of the carbon-14 atoms in an object to change into nitrogen-14 atoms. You might guess that in another 5,730 years, all of the remaining carbon-14 atoms will decay to nitrogen-14. However, this is not the case. Only half of the atoms of carbon-14 remaining after the first 5,730 years will decay during the second 5,730 years. So, after two half-lives, one fourth of the original carbon-14 atoms still remain. Half of them will decay during another 5,730 years. After three half-lives, one eighth of the original carbon-14 atoms still remain. After many half-lives, such a small amount of the parent isotope remains that it might not be measurable.

Figure 18

A In beta decay, a neutron changes into a proton by giving off an electron. This electron has a lot of energy and is called a beta particle.

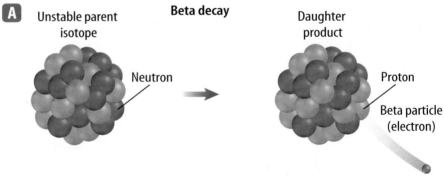

A Unstable parent isotope — Beta decay — Daughter product

Neutron

Proton
Beta particle (electron)

B In the process of alpha decay, an unstable parent isotope nucleus gives off an alpha particle and changes into a new daughter product. Alpha particles contain two neutrons and two protons.

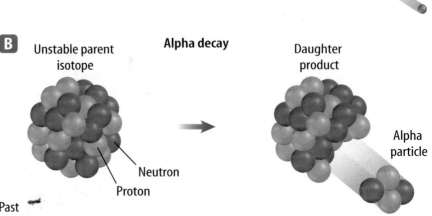

B Unstable parent isotope — Alpha decay — Daughter product

Alpha particle

Neutron
Proton

Radiometric Ages

Decay of radioactive isotopes is like a clock keeping track of time that has passed since rocks have formed. As time passes, the amount of parent isotope in a rock decreases as the amount of daughter product increases, as in **Figure 19.** By measuring the ratio of parent isotope to daughter product in a mineral and by knowing the half-life of the parent, in many cases you can calculate the absolute age of a rock. This process is called **radiometric dating.**

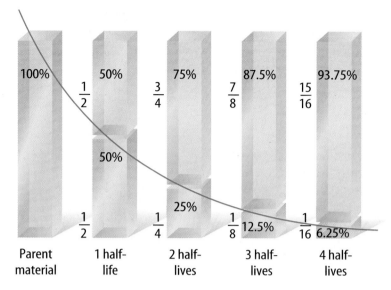

Figure 19
During each half-life, one half of the parent material decays to the daughter product.

A scientist must decide which parent isotope to use when measuring the age of a rock. If the object to be dated seems old, then the geologist will use an isotope with a long half-life. The half-life for the decay of potassium-40 to argon-40 is 1.25 billion years. As a result, this isotope can be used to date rocks that are many millions of years old. To avoid error, conditions must be met for the ratios to give a correct indication of age. For example, the rock being studied must still retain all of the argon-40 that was produced by the decay of potassium-40. Also, it cannot contain any contamination of daughter product from other sources. Potassium-argon dating is good for rocks containing potassium, but what about other things?

Radiocarbon Dating Carbon-14 is useful for dating bones, wood, and charcoal up to 75,000 years old. Living things take in carbon from the environment to build their bodies. After the organism dies, the carbon-14 slowly decays and escapes as nitrogen-14 gas. If scientists can determine the amount of carbon-14 remaining in a sample, they can determine its age. For example, during much of human history, people built campfires. The wood from these fires often is preserved as charcoal. Scientists can determine the amount of carbon-14 remaining in a sample of charcoal by measuring the amount of radiation emitted by the carbon-14 isotope in labs like the one in **Figure 20.** Once they know the amount of carbon-14 in a charcoal sample, scientists can determine the age of the wood used to make the fire.

Figure 20
Radiometric ages are determined in labs like this one.

Figure 21
Meteorites like this one are thought to be as old as Earth.

Age Determinations Aside from carbon-14 dating, rocks that can be radiometrically dated are mostly igneous and metamorphic rocks. Most sedimentary rocks cannot be dated by this method. This is because many sedimentary rocks are made up of particles eroded from older rocks. Dating these pieces only gives the age of the preexisting rock from which it came.

The Oldest Known Rocks Radiometric dating has been used to date the oldest rocks on Earth. These rocks are about 3.96 billion years old. By dating meteorites like the one shown in **Figure 21** and using other evidence, scientists have estimated the age of Earth to be about 4.6 billion years. Earth rocks greater than 3.96 billion years old probably were eroded or changed by heat and pressure.

 **Reading Check** *Why can't most sedimentary rocks be dated radiometrically?*

Problem-Solving Activity

When did the Iceman die?

Carbon-14 dating has been used to date charcoal, wood, bones, mummies from Egypt and Peru, the Dead Sea Scrolls, and the Italian Iceman. The Iceman was found in 1991 in the Italian Alps, near the Austrian border. Based on carbon-14 analysis, scientists determined that the Iceman is 5,300 years old. Determine approximately in what year the Iceman died.

Half-Life of Carbon-14	
Percent Carbon-14	Years Passed
100	0
50	5,730
25	11,460
12.5	17,190
6.25	22,920
3.125	

Reconstruction of Iceman

Identifying the Problem

The half-life chart shows the decay of carbon-14 over time. Half-life is the time it takes for half of a sample to decay. Fill in the years passed when only 3.125 percent of carbon-14 remain. Is there a point at which no carbon-14 would be present? Explain.

Solving the Problem

1. Estimate, using the data table, how much carbon-14 still was present in the Iceman's body that allowed scientists to determine his age.
2. If you had an artifact that contained 10.0 g of carbon-14, how many grams would remain after 17,190 years?

Uniformitarianism

Can you imagine trying to determine the age of Earth without some of the information you know today? Before the discovery of radiometric dating, many people estimated that Earth is only a few thousand years old. But in the 1700s, Scottish scientist James Hutton estimated that Earth is much older. He used the principle of **uniformitarianism.** This principle states that Earth processes occurring today are similar to those that occurred in the past. Hutton's principle is often paraphrased as "the present is the key to the past."

Hutton observed that the processes that changed the landscape around him were slow, and he inferred that they were just as slow throughout Earth's history. Hutton hypothesized that it took much longer than a few thousand years to form the layers of rock around him and to erode mountains that once stood kilometers high. **Figure 22** shows Hutton's native Scotland, a region shaped by millions of years of geologic processes.

Today, scientists recognize that Earth has been shaped by two types of change: slow, everyday processes that take place over millions of years, and violent, unusual events such as the collision of a comet or asteroid about 65 million years ago that might have caused the extinction of the dinosaurs.

Figure 22
The rugged highlands of Scotland were shaped by erosion and uplift.

Section Assessment

1. You discover three rock layers that have not been turned upside down. The absolute age of the middle layer is 120 million years. What can you say about the ages of the layers above and below it?

2. How old would a fossil be if it had only one eighth of its original carbon-14 content remaining?

3. Explain the concept of uniformitarianism.

4. How do radioactive isotopes decay?

5. **Think Critically** Why can't scientists use carbon-14 to determine the age of an igneous rock?

Skill Builder Activities

6. **Making and Using Tables** Make a table that shows the amount of parent material of a radioactive element that is left after four half-lives if the original parent material had a mass of 100 g. **For more help, refer to the** Science Skill Handbook.

7. **Using Fractions** You know that after every half-life, one half of the parent isotope has changed to daughter product. In general, how would you determine how much parent isotope remains after any specified number of half-lives? **For more help, refer to the** Math Skill Handbook.

Activity

Model and Invent

Trace Fossils

Trace fossils can tell you a lot about the activities of organisms that left them. They can tell you how an organism fed or what kind of home it had. What else can you learn from trace fossils?

Recognize the Problem

How can you model trace fossils that can provide information about the behavior of organisms?

Thinking Critically

What can you use to model trace fossils? What types of behavior could you show with your trace fossil model?

Goals
- **Construct** a model of trace fossils.
- **Describe** the information that you can learn from looking at your model.

Possible Materials
construction paper	wire
plastic (a fairly rigid type)	scissors
plaster of paris	toothpicks
sturdy cardboard	clay
pipe cleaners	glue

Safety Precautions

Data Source
SCIENCE *Online* Go to the Glencoe Science Web site at **science.glencoe.com** for more information about trace fossils and what can be learned from them.

Planning the Model

1. **Decide** how you are going to make your model. What materials will you need?

2. **Decide** what types of activities you will demonstrate with your model. Were the organisms feeding? Resting? Traveling? Were they predators? Prey? How will your model indicate the activities you chose?

3. What is the setting of your model? Are you modeling the organism's home? Feeding areas? Is your model on land or water? How can the setting affect the way you build your model?

4. Will you only show trace fossils from a single species or multiple species? If you include more than one species, how will you provide evidence of any interaction between the species?

Check the Model Plans

1. Compare your plans with those of others in your class. Did other groups mention details that you had forgotten to think about? Are there any changes you would like to make to your plan before you continue?

2. Make sure your teacher approves your plan before you continue.

Making the Model

1. Following your plan, **construct** your model of trace fossils.

2. Have you included evidence of all the behaviors you intended to model?

Analyzing and Applying Results

1. Now that your model is complete, do you think that it adequately shows the behaviors you planned to demonstrate? Is there anything that you think you might want to do differently if you were going to make the model again?

2. Describe how using different kinds of materials might have affected your model. Can you think of other materials that would have allowed you to show more detail than you did?

3. Compare and contrast your model of trace fossils with trace fossils left by real organisms. Is one more easily interpreted than the other? Explain.

*C*ommunicating
Your Data

Ask other students in your class or another class to look at your model and describe what information they can learn from the trace fossils. Did their interpretations agree with what you intended to show?

The World's Oldest Fish Story

A catch-of-the-day set science on its ears

Second Dorsal Fin

Anal Fin

Pelvic Fin

On a December day in 1938, just before Christmas, Marjorie Courtenay-Latimer went to say hello to her friends on board a fishing boat that had just returned to port in South Africa. Courtenay-Latimer, who worked at a museum, often went aboard her friends' ship to check out the catch. On this visit, she received a surprise Christmas present—an odd-looking fish. As soon as the woman spotted its strange blue fins among the piles of sharks and rays, she knew it was special.

Courtenay-Latimer took the fish back to her museum to study it. "It was the most beautiful fish I had ever seen, five feet long, and a pale mauve blue with iridescent silver markings," she later wrote. Courtenay-Latimer sketched it and sent the drawing to a friend of hers, J. L. B. Smith.

Smith was a chemistry teacher who was passionate about fish. After a time, he realized it was a coelacanth (SEE luh kanth). Fish experts knew that coelacanths had first appeared on Earth 400 million years ago. But the experts thought the fish were extinct. People had found fossils of coelacanths, but no one had seen one alive. It was assumed that the last coelacanth species had died out 65 million years ago. They were wrong. The ship's crew had caught one by accident.

Smith figured there might be more living coelacanths. So he decided to offer a reward for anyone who could find a living specimen.

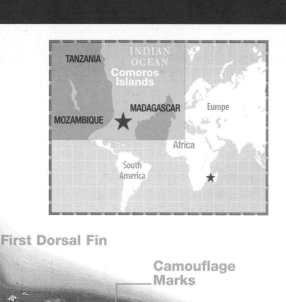

Marjorie Courtenay-Latimer poses with her fish (above).

Courtenay-Latimer's original sketch of the fish

First Dorsal Fin

Camouflage Marks

Pectoral Fin

Some scientists call the coelacanth "Old Four Legs." It got its nickname because the fish has paired fins that look something like legs.

After 14 years of silence, a report came in that a coelacanth had been caught off the east coast of Africa.

Today, scientists know that there are at least several hundred coelacanths living in the Indian Ocean, just east of central Africa. Many of these fish live near the Comoros Islands. The coelacanths live in underwater caves during the day but move out at night to feed. The rare fish are now a protected species. With any luck, they will survive for another hundred million years.

CONNECTIONS **Write** Find out how people responded to the discovery of living coelacanths. Why do you think people were so fascinated by this fish? Use the Glencoe Science Web site to find out more about the coelacanth.

SCIENCE *Online*

For more information, visit
science.glencoe.com

Chapter **13** Study Guide

Reviewing Main Ideas

Section 1 Fossils

1. Fossils are more likely to form if hard parts of the dead organisms are buried quickly.

2. Some fossils form when original materials that made up the organisms are replaced with minerals. Other fossils form when remains are subjected to heat and pressure, leaving only a carbon film behind. Some fossils are the tracks or traces left by ancient organisms. *What type of fossil is shown to the right?*

3. A rock layer can be no older than the age of the fossils embedded in it.

Section 2 Relative Ages of Rocks

1. The principle of superposition states that older rocks lie underneath younger rocks in areas where the rocks haven't been disturbed. Faults and igneous intrusions are always younger than the rocks they cut across. *Which of the rock layers shown below is youngest?*

2. Unconformities, or gaps in the rock record, are due to erosion or periods of time during which no deposition occurred.

3. Rock layers can be correlated using rock types and fossils.

Section 3 Absolute Ages of Rocks

1. Absolute dating of rocks and minerals, unlike relative age determination, provides an age in years for the rocks.

2. The half-life of a radioactive isotope is the time it takes for half of the atoms of the isotope to decay into another isotope. *Which isotope could be used to date the buried tree shown below?*

3. Because half-lives don't change, scientists can determine absolute ages of rocks and minerals containing radioactive elements.

FOLDABLES
Reading & Study Skills

After You Read

To help review fossils and how to determine the age of rocks, use the Foldable you made at the beginning of the chapter.

Visualizing Main Ideas

Complete the following concept map on fossils.

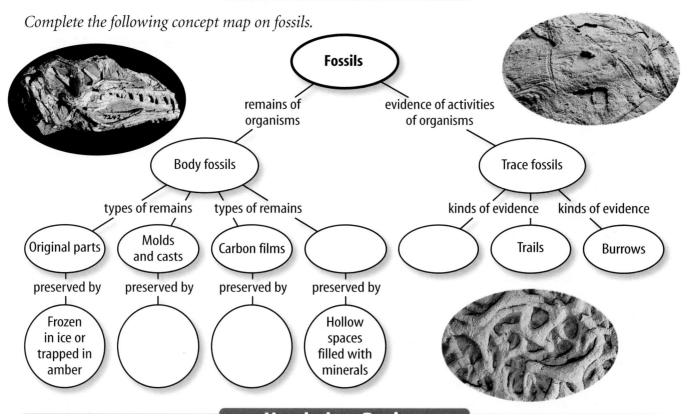

Vocabulary Review

Vocabulary Words

a. absolute age
b. carbon film
c. cast
d. fossil
e. half-life
f. index fossil
g. mold
h. permineralized remains

i. principle of superposition
j. radioactive decay
k. radiometric dating
l. relative age
m. unconformity
n. uniformitarianism

Study Tip

Reading material before your teacher explains it gives you a better understanding and provides you with an opportunity to ask questions.

Using Vocabulary

Write an original sentence using the vocabulary word to which each phrase refers.

1. thin film of carbon preserved as a fossil
2. older rocks lie under younger rocks
3. processes occur today as they did in the past
4. gap in the rock record
5. time needed for half the atoms to decay
6. fossil organism that lived for a short time
7. gives the age of rocks in years
8. minerals fill spaces inside fossil
9. a copy of a fossil produced by filling a mold with sediment or crystals
10. evidence of ancient life

Chapter (13) Assessment

Checking Concepts

Choose the word or phrase that best answers the question.

1. What is any evidence of ancient life called?
   - **A)** half-life
   - **B)** a fossil
   - **C)** unconformity
   - **D)** disconformity

2. Which of the following conditions makes fossil formation more likely?
   - **A)** buried slowly
   - **B)** attacked by scavengers
   - **C)** made of hard parts
   - **D)** composed of soft parts

3. What are cavities left in rocks when a shell or bone dissolves called?
   - **A)** casts
   - **B)** original remains
   - **C)** molds
   - **D)** carbon films

4. When the tiny spaces inside a fossil are filled in with minerals, you have what kind of preservation?
   - **A)** original remains
   - **B)** permineralized remains
   - **C)** molds and casts
   - **D)** carbon films

5. To say "the present is the key to the past" is a way to describe which of the following principles?
   - **A)** superposition
   - **B)** succession
   - **C)** radioactivity
   - **D)** uniformitarianism

6. A fault can be useful in determining which of the following for a group of rocks?
   - **A)** absolute age
   - **B)** radiometric age
   - **C)** index age
   - **D)** relative age

7. Which of the following is an unconformity between parallel rock layers?
   - **A)** angular unconformity
   - **B)** fault
   - **C)** disconformity
   - **D)** nonconformity

8. Which process forms new elements?
   - **A)** superposition
   - **B)** uniformitarianism
   - **C)** permineralization
   - **D)** radioactive decay

9. In one type of radioactive decay, which of the following breaks down, releasing an electron?
   - **A)** alpha particle
   - **B)** proton
   - **C)** beta particle
   - **D)** neutron

10. About how old is Earth, based on radiometric dating?
    - **A)** 2,000 years
    - **B)** 5,000 years
    - **C)** 3.5 billion years
    - **D)** 4.6 billion years

Thinking Critically

11. How do relative and absolute ages differ?

12. How many half-lives have passed in a rock containing one eighth of the original radioactive material and seven eighths of the daughter product?

13. The fossil record of life on Earth is incomplete. Give some reasons why.

14. Suppose a lava flow were found between two sedimentary rock layers. How could you use the lava flow to learn about the ages of the sedimentary rock layers? (Hint: Most lava contains radioactive isotopes.)

15. Suppose you're correlating rock layers in the western United States. You find a layer of volcanic ash deposits. How can this layer help you in your correlation over a large area?

Developing Skills

16. **Recognizing Cause and Effect** Explain how some woolly mammoths could have been preserved intact in frozen ground. What conditions must have persisted since the deaths of these animals?

17. Classifying Copy the table below and place each of the following fossils in the correct category: *dinosaur footprint, worm burrow, dinosaur skull, insect in amber, fossil woodpecker hole,* and *fish tooth.*

Types of Fossils	
Trace Fossils	**Body Fossils**

18. Concept Mapping Make a concept map using the following possible steps in the process of making a cast of a fossil: *organism dies, burial, protection from scavengers and bacteria, replacement by minerals, fossil erodes away,* and *mineral crystals form from solution.*

19. Communicating Make an outline of Section 1 that discusses the ways in which fossils form.

Performance Assessment

20. Write a Poem Write a poem explaining the principle of uniformitarianism.

21. Use a Classification System Start your own fossil collection. Label each find as to type, approximate age, and the place where it was found. Most state geological surveys can provide you with reference materials on local fossils.

TECHNOLOGY

Go to the Glencoe Science Web site at **science.glencoe.com** or use the **Glencoe Science CD-ROM** for additional chapter assessment.

THE PRINCETON REVIEW **Test Practice**

A group of scientists found several different fossils. The scientists listed the different fossils in the table below.

Type of Fossils	
Fossils A	**Fossils B**
Large footprint	Frozen mammoth
Trackway with small footprints	Dinosaur bones
Animal burrow	?

Study the table and answer the following questions.

1. Fossils A are different from Fossils B because only Fossils A are _____ .
A) index fossils
B) trace fossils
C) permineralized fossils
D) original remains

2. Which of these belongs with Fossils B?
F) an imprint of a leaf in mud
G) a footprint in cement
H) an insect in amber
J) a fossilized nest

3. All of the fossils in the table above are _____ .
A) evidence of past life
B) types of rocks
C) soft parts
D) hard parts

Geologic Time

The discovery of a new fossil often changes scientists' ideas about how ancient organisms looked and lived. Pictured here is Sue, the most complete *Tyrannosaurus rex* fossil ever found. Discovered in 1990, Sue has provided evidence indicating that this type of dinosaur relied heavily on a good sense of smell. In this chapter, you will learn how dinosaurs and other types of organisms changed over time. You also will come to appreciate the long span of geologic time.

What do you think?

Science Journal Look at the picture below with a classmate. Discuss what you think this might be or what is happening. Here's a hint: *Sue couldn't have survived without them.* Write your answer or best guess in your Science Journal.

EXPLORE ACTIVITY

nvironments include the living and nonliving things that surround and affect organisms. Whether or not an organism survives in its environment depends upon the characteristics that it has. Only if an organism survives until adulthood can it reproduce and pass on its characteristics to its offspring. In this activity, you will use a model to find out how one characteristic can determine whether individuals can compete and survive in an environment.

Make a model environment

1. Cut 15 pieces each of green, orange, and blue yarn into 3-cm lengths.
2. Scatter them on a sheet of green construction paper.
3. Have your partner use a pair of tweezers to pick up as many pieces as possible in 15 s.

Observe

In your Science Journal, discuss which colors your partner selected. Which color was least selected? Suppose that the construction paper represents grass, the yarn pieces represent insects, and the tweezers represent an insect-eating bird. Which color of insect do you predict would survive to adulthood?

Before You Read

FOLDABLES
Reading & Study Skills

Making an Organizational Study Fold Make the following Foldable to help you organize your thoughts into clear categories about geological time.

1. Place a sheet of paper in front of you so the short side is at the top. Fold the paper in half from top to bottom.
2. Fold both sides in. Unfold the paper so three sections show.
3. Through the top thickness of paper, cut along each of the fold lines to the top fold, forming three tabs. Label the tabs *Paleozoic Era, Mesozoic Era,* and *Cenozoic Era.*
4. Before you read the chapter, find an example of an animal from each era and write it on the front of the tabs. As you read the chapter, write information about each era under the tabs.

Life and Geologic Time

As You Read

What You'll Learn

- **Explain** how geologic time can be divided into units.
- **Relate** changes of Earth's organisms to divisions on the geologic time scale.
- **Describe** how plate tectonics affects species.

Vocabulary

trilobite
geologic
  time scale
eon
era
period

epoch
organic evolution
species
natural selection
Pangaea

Why It's Important

The life and landscape around you are the products of change through geologic time.

Geologic Time

A group of students is searching for fossils. By looking in rocks that are hundreds of millions of years old, they hope to find many examples of **trilobites** (TRI loh bites) so that they can help piece together a puzzle. That puzzle is to find out what caused the extinction of these organisms. **Figure 1** shows some examples of what they are finding. The fossils are small, and their bodies are divided into segments. Some of them seem to have eyes. Could these be trilobites?

Trilobites are interesting fossils to search for. These small, hard-shelled organisms crawled on the seafloor and sometimes swam through the water. Most ranged in size from 2 cm to 7 cm in length and from 1 cm to 3 cm in width. They are considered to be index fossils because they lived over vast regions of the world during specific periods of geologic time.

The Geologic Time Scale The appearance or disappearance of types of organisms throughout Earth's history marks important occurrences in geologic time. Paleontologists have been able to divide Earth's history into time units based on the life-forms that lived only during certain periods. This division of Earth's history makes up the **geologic time scale.** However, sometimes fossils are not present, so certain divisions of the geologic time scale are based on other criteria.

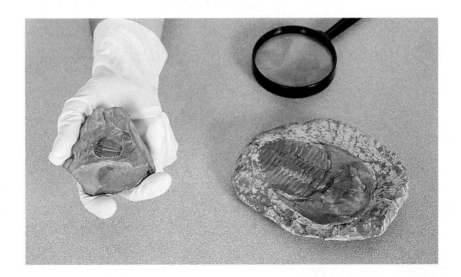

Figure 1
Many sedimentary rocks in the United States are rich in invertebrate fossils such as these trilobites.

Major Subdivisions of Geologic Time The oldest rocks on Earth contain no fossils. Then, for many millions of years after the first appearance of fossils, the fossil record remained sparse. Later in Earth's history came an explosion in the abundance and diversity of organisms. These organisms created a rich fossil record. As shown in **Figure 2,** four major subdivisions of geologic time are used—eons, eras, periods, and epochs. The longest subdivisions—**eons**—are based upon the abundance of fossils.

✔ **Reading Check**   *What are the major subdivisions of geologic time?*

Next to eons, the longest subdivisions are the **eras,** which are marked by major, striking, and worldwide changes in the types of fossils present. For example, at the end of the Mesozoic Era, many kinds of invertebrates, birds, mammals, and reptiles became extinct.

Eras are subdivided into periods. **Periods** are units of geologic time characterized by the types of life existing worldwide at the time. Periods can be divided into smaller units of time called **epochs.** Epochs also are characterized by differences in life-forms, but some of these differences can vary regionally such as from continent to continent. Epochs of periods in the Cenozoic Era have been given specific names. Epochs of other periods usually are referred to simply as early, middle, or late. Epochs are further subdivided into units of shorter duration.

Dividing Geologic Time There is a limit to how finely geologic time can be subdivided. It depends upon the kind of rock record that is being studied. Sometimes it is possible to distinguish layers of rock that formed during a single year or season. In other cases, thick stacks of rock that have no fossils provide little information that could help in subdividing geologic time.

Figure 2
Scientists have divided the geologic time scale into subunits based upon the appearance and disappearance of types of organisms.

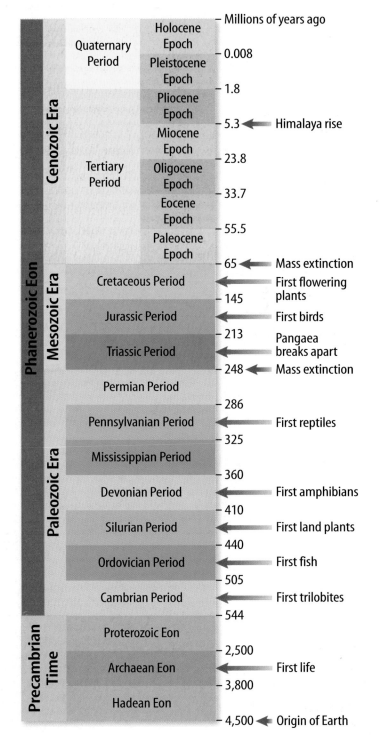

Organic Evolution

The fossil record shows that species have changed over geologic time. This change through time is known as **organic evolution.** According to most theories about organic evolution, environmental changes can affect an organism's survival. Those organisms that are not adapted to changes are less likely to survive or reproduce. Over time, the elimination of individuals that are not adapted can cause changes to species of organisms.

Species Many ways of defining the term **species** (SPEE sheez) have been proposed. Life scientists often define a **species** as a group of organisms that normally reproduces only with other members of their group. For example, dogs are a species because dogs mate and reproduce only with other dogs. In some rare cases, members of two different species, such as lions and tigers, can mate and produce offspring. These offspring, however, are usually sterile and cannot produce offspring of their own. Even though two organisms look nearly alike, if the populations they each come from do not interbreed naturally and produce offspring that can reproduce, the two individuals do not belong to the same species. **Figure 3** shows an example of two species that look similar to each other but live in different areas and do not mate naturally with each other.

Figure 3
Just because two organisms look alike does not mean that they belong to the same species.

A The coast horned lizard lives along the coast of central and southern California.

B The desert horned lizard lives in arid regions of the southwestern United States.

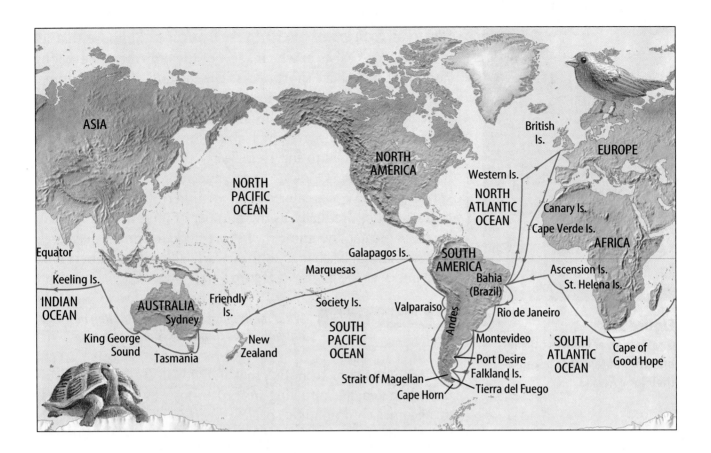

Natural Selection Charles Darwin was a naturalist who sailed around the world from 1831 to 1836 to study biology and geology. **Figure 4** shows a map of his journey. With some of the information about the plants and animals he observed on this trip in mind, he later published a book about the theory of evolution by natural selection.

In his book, he proposed that **natural selection** is a process by which organisms with characteristics that are suited to a certain environment have a better chance of surviving and reproducing than organisms that do not have these characteristics. Darwin knew that many organisms are capable of producing more offspring than can survive. This means that organisms compete with each other for resources necessary for life, such as food and living space. He also knew that individual organisms within the same species could be different, or show variations, and that these differences could help or hurt the individual organism's chance of surviving.

Some organisms that were well suited to their environment lived longer and had a better chance of producing offspring. Organisms that were poorly adapted to their environment produced few or no offspring. Because many characteristics are inherited, the characteristics of organisms that are better adapted to the environment get passed on to offspring more often. According to Darwin, this can cause a species to change over time.

Figure 4
Charles Darwin sailed around the world between 1831 and 1836 aboard the HMS *Beagle* as a naturalist. On his journey he saw an abundance of evidence for natural selection, especially on the Galápagos Islands off the western coast of South America.

Figure 5
Giraffes can eat leaves off the branches of tall trees because of their long necks.

Natural Selection Within a Species Suppose that an animal species exists in which a few of the individuals have long necks, but most have short necks. The main food for the animal is the leafy foliage on trees in the area. What happens if the climate changes and the area becomes dry? The lower branches of the trees might not have any leaves. Now which of the animals will be better suited to survive? Clearly, the long-necked animals have a better chance of surviving and reproducing. Their offspring will have a greater chance of inheriting the important characteristic. Gradually, as the number of long-necked animals becomes greater, the number of short-necked animals decreases. The species might change so that nearly all of its members have long necks, as the giraffe in **Figure 5** has.

✓ Reading Check *What might happen to the population of animals if the climate became wet again?*

It is important to notice that individual, short-necked animals didn't change into long-necked animals. A new characteristic becomes common in a species only if some members already possess that characteristic and if the trait increases the animal's chance of survival. If no animal in the species possessed a long neck in the first place, a long-necked species could not have evolved by means of natural selection.

Artificial Selection Humans have long used the principle of selection when breeding domestic animals. By carefully choosing individuals with desired characteristics, animal breeders have created many breeds of cats, dogs, cattle, and chickens. **Figure 6** shows the great variety of cats produced by artificial selection.

The Evolution of New Species Natural selection explains how characteristics change and how new species arise. For example, if the short-necked animals migrated to a different location, they might have survived. They could have continued to reproduce in the new location, eventually developing enough different characteristics from the long-necked animals that they might not be able to breed with each other. At this point, at least one new species would have evolved.

Figure 6
Cat breeders have succeeded in producing a great variety of cats by using the principle of artificial selection.

Trilobites

Remember the trilobites? The term *trilobite* comes from the structure of the animal's hard outer skeleton or exoskeleton. The exoskeleton of a trilobite consists of three lobes that run the length of the body. As shown in **Figure 7,** the trilobite's body also has a head (cephalon), a segmented middle section (thorax), and a tail (pygidium).

Changing Characteristics of Trilobites Trilobites inhabited Earth's oceans for more than 200 million years. Throughout the Paleozoic Era, some species of trilobites became extinct and other new species evolved. Species of trilobites that lived during one period of the Paleozoic Era showed different characteristics than species from other periods of this era. As **Figure 8** shows, paleontologists can use these different characteristics to demonstrate changes in trilobites through geologic time. These changes can tell you about how different trilobites from different periods lived and responded to changes in their environments.

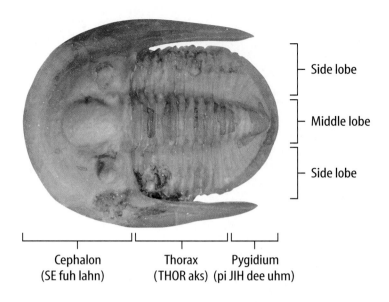

Cephalon (SE fuh lahn)   Thorax (THOR aks)   Pygidium (pi JIH dee uhm)

Figure 7
The trilobite's body was divided into three lobes that run the length of the body—two side lobes and one middle lobe.

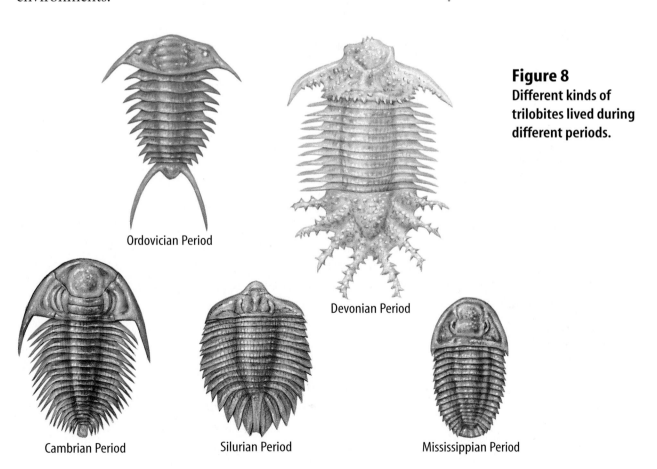

Ordovician Period

Devonian Period

Cambrian Period   Silurian Period   Mississippian Period

Figure 8
Different kinds of trilobites lived during different periods.

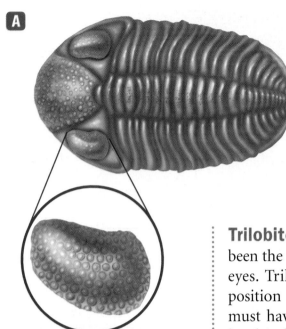

Figure 9
Trilobites had many different types of eyes. **A** Some had eyes that contained hundreds of small circular lenses, somewhat like an insect. **B** This blind trilobite had no eyes.

Figure 10
Olenellus is one of the most primitive trilobite species.

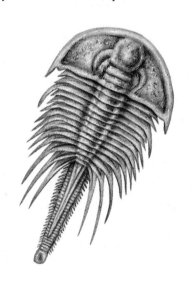

Trilobite Eyes Trilobites, shown in **Figure 9A,** might have been the first organisms that could view the world with complex eyes. Trilobite eyes show the result of natural selection. The position of the eyes on an organism gives clues about where it must have lived. Eyes that are located toward the front of the head indicate an organism that was adapted for active swimming. If the eyes are located toward the back of the head, the organism could have been a bottom dweller. In most species of trilobites, the eyes were located midway on the head—a compromise for an organism that was adapted for crawling on the seafloor and swimming in the water.

Over time, the eyes in trilobites changed. In many trilobite species, the eyes became progressively smaller until they completely disappeared. Blind trilobites, such as the one shown in **Figure 9B,** might have burrowed into sediments on the seafloor or lived deeper than light could penetrate. In other species, however, the eyes became more complex. One kind of trilobite, *Aeglina,* developed large compound eyes that had numerous individual lenses. Some trilobites developed stalks that held the eyes upward. Where would this be useful?

Trilobite Bodies The trilobite body and tail also underwent significant changes in form through time, as you can see in **Figure 8.** A special case is *Olenellus,* shown in **Figure 10.** This trilobite, which lived during the Early Cambrian Period, had an extremely segmented body—perhaps more so than any other known species of trilobite. It is thought that *Olenellus,* and other species that have so many body segments, are primitive trilobites.

Fossils Show Changes Trilobite exoskeletons changed as trilobites adapted to changing environments. Species that could not adapt became extinct. What processes on Earth caused environments to change so drastically that species adapted or became extinct?

Plate Tectonics and Earth History

Plate tectonics is one possible answer to the riddle of trilobite extinction. Earth's moving plates caused continents to collide and separate many times. Continental collisions formed mountains and closed seas. Continental separations created wider, deeper seas between continents. By the end of the Paleozoic Era, sea levels had dropped and the continents had come together to form one giant landmass, the supercontinent **Pangaea** (pan JEE uh). Because trilobites lived in the oceans, their environment was changed or destroyed. **Figure 11** shows the arrangement of continents at the end of the Paleozoic Era. What effect might these changes have had on the trilobite populations?

Not all scientists accept the above explanation for the extinctions at the end of the Paleozoic Era, and other possibilities—such as climate change—have been proposed. As in all scientific debates, you must consider the evidence carefully and come to conclusions based on the evidence.

Figure 11
The amount of shallow water environment was reduced when Pangaea formed. *How do you think this change affected organisms that lived along the coasts of continents?*

Section 1 Assessment

1. Discuss how fossils relate to the geologic time scale.
2. How might plate tectonics affect species of organisms?
3. Relate trilobite eye type to lifestyle type.
4. Why can paleontologists use some trilobite fossils as index fossils for the Cambrian Period and other trilobite fossils as index fossils for other geologic time periods?
5. **Think Critically** Aside from moving continents, what other factors could cause an organism's environment to change? What effects could changing environments have on species?

Skill Builder Activities

6. **Recognizing Cause and Effect** Answer the questions below. **For more help, refer to the** Science Skill Handbook.
   a. How does natural selection cause evolutionary change to take place?
   b. How could the evolution of a characteristic within one species affect the evolution of a characteristic within another species? Give an example.
7. **Communicating** Write a short poem in your Science Journal that describes a day in the life of a trilobite. **For more help, refer to the** Science Skill Handbook.

Early Earth History

***What* You'll Learn**

- **Identify** characteristic Precambrian and Paleozoic life-forms.
- **Draw** conclusions about how species adapted to changing environments in Precambrian time and the Paleozoic Era.
- **Describe** changes in Earth and its life-forms at the end of the Paleozoic Era.

Vocabulary

Precambrian time   Paleozoic Era
cyanobacteria

***Why* It's Important**

The Precambrian includes most of Earth's history.

Figure 12
During the early Precambrian, Earth was a lifeless planet with many volcanoes.

Precambrian Time

Can you imagine a barren Earth with only rock and sea, as shown in **Figure 12?** This seems strange, but it's probably an accurate picture of Earth's first billion years. Over the next 3 billion years, simple life-forms began to colonize the oceans.

Look again at the geologic time scale shown in **Figure 2. Precambrian** (pree KAM bree un) **time** is the longest part of Earth's history and includes the Hadean, Archaean, and Proterozoic Eons. Precambrian time lasted from about 4.5 billion years ago to about 544 million years ago. The oldest rocks that have been found on Earth are about 4 billion years old. However, rocks older than about 3.5 billion years are rare. This probably is due to remelting and erosion. Although the Precambrian was the longest interval of geologic time, relatively little is known about the organisms that lived during this time.

Why is the fossil record from Precambrian time so sparse? One reason is that many Precambrian rocks have been so deeply buried that they have been changed by heat and pressure. Many fossils can't withstand these conditions. In addition, most Precambrian organisms didn't have hard parts that otherwise would have increased their chances to be preserved as fossils.

Lava flow   Ash   Lava flow

Ash deposits

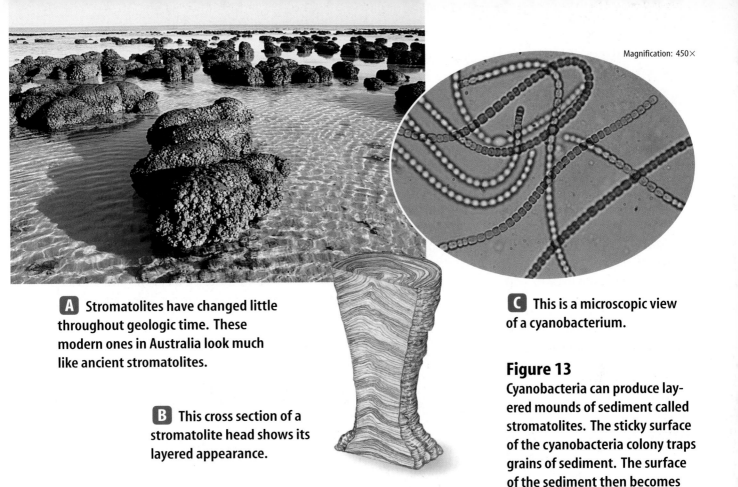

Magnification: 450×

A Stromatolites have changed little throughout geologic time. These modern ones in Australia look much like ancient stromatolites.

B This cross section of a stromatolite head shows its layered appearance.

C This is a microscopic view of a cyanobacterium.

Figure 13
Cyanobacteria can produce layered mounds of sediment called stromatolites. The sticky surface of the cyanobacteria colony traps grains of sediment. The surface of the sediment then becomes colonized with cyanobacteria again, and the cycle repeats, producing the layers inside the stromatolite.

Early Life Many studies of the early history of life involve ancient stromatolites (stroh MAT oh lytes). **Figure 13** shows stromatolites, which are layered mats formed by cyanobacteria colonies. **Cyanobacteria** are blue-green algae thought to be one of the earliest forms of life on Earth. Cyanobacteria first appeared about 3.5 billion years ago. They contained chlorophyll and photosynthesized. This is important because during photosynthesis, they produced oxygen, which helped change Earth's atmosphere. For the few billion years following the appearance of cyanobacteria, oxygen became a major atmospheric gas. Also of importance was that the ozone layer in the atmosphere began to develop, shielding Earth from ultraviolet rays. It is hypothesized that these changes allowed species of single-celled organisms to evolve into more complex organisms.

✓ Reading Check *What atmospheric gas is produced by photosynthesis?*

Animals without backbones, called invertebrates (in VURT uh brayts), appeared toward the end of Precambrian time. Imprints of invertebrates have been found in late Precambrian rocks, but because these early invertebrates were soft bodied, they weren't often preserved as fossils. Because of this, many Precambrian fossils are trace fossils.

Chemistry
INTEGRATION

Cyanobacteria are thought to have been one of the mechanisms by which Earth's early atmosphere became richer in oxygen. Research the composition of Earth's early atmosphere and where these gases probably came from. Record your findings in your Science Journal.

Figure 14
This giant predatory fish lived in North America during the Devonian Period. It grew to about 6 m in length.

Unusual Life-Forms Living late in Precambrian time was a group of animals with shapes similar to modern jellyfish, worms, and soft corals. Fossils of these organisms were first found in the Ediacara Hills in southern Australia. This group of organisms has become known as the Ediacaran (eed ee uh KAR un) fauna. **Figure 15** shows some of these fossils.

Reading Check *What modern organisms do some Ediacaran organisms resemble?*

Ediacaran animals were bottom dwellers and might have had tough outer coverings like air mattresses. Trilobites and other invertebrates might have outcompeted the Ediacarans and caused them to become extinct, but nobody knows for sure why these creatures disappeared.

The Paleozoic Era

As you have learned, fossils are unlikely to form if organisms have only soft parts. An abundance of organisms with hard parts, such as shells, marks the beginning of the Paleozoic (pay lee uh ZOH ihk) Era. The **Paleozoic Era,** or era of ancient life, began about 544 million years ago and ended about 248 million years ago. Traces of life are much easier to find in Paleozoic rocks than in Precambrian rocks.

Paleozoic Life Warm, shallow seas covered large parts of the continents during much of the Paleozoic Era, so many of the life-forms scientists know about were marine, meaning they lived in the ocean. Trilobites were common, especially early in the Paleozoic. Other organisms developed shells that were easily preserved as fossils. Therefore, the fossil record of this era contains abundant shells. However, invertebrates were not the only animals to live in the shallow, Paleozoic seas.

Vertebrates, or animals with backbones, also evolved during this era. The first vertebrates were fishlike creatures without jaws. Armoured fish with jaws such as the one shown in **Figure 14** lived during the Devonian Period. Some of these fish were so huge that they could eat large sharks with their powerful jaws. By the Devonian Period, forests had appeared and vertebrates began to adapt to land environments, as well.

Figure 15

A variety of 600-million-year-old fossils—
known as Ediacaran (eed ee uh KAR un)
fauna—have been found on every continent
except Antarctica. These unusual organisms were originally thought to be descendants of early animals such as
jellyfish, worms, and coral. Today, paleontologists debate
whether these organisms were part of the animal kingdom or belonged to an entirely new kingdom whose
members became extinct about 545 million years ago.

DICKENSONIA (dihk un suh NEE uh) Impressions
of *Dickensonia,* a bottom-dwelling wormlike
creature, have been discovered. Some are
nearly one meter long.

RANGEA (rayn JEE uh) As it lay rooted in
sea-bottom sediments, *Rangea* may have
snagged tiny bits of food by filtering water
through its body.

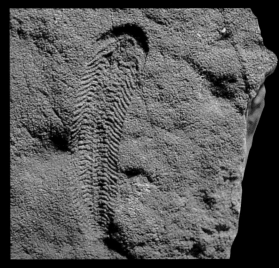

SPRIGGINA (sprih GIHN uh) Some scientists hypothesize that the four-centimeter-long *Spriggina* was a
type of crawling, segmented organism. Others suggest that it sat upright while attached to the seafloor.

CYCLOMEDUSA (si kloh muh DEW suh) Although it
looks a lot like a jellyfish, *Cyclomedusa* may have had
more in common with modern sea anemones. Some
paleontologists, however, hypothesize that it is unrelated to any living organism.

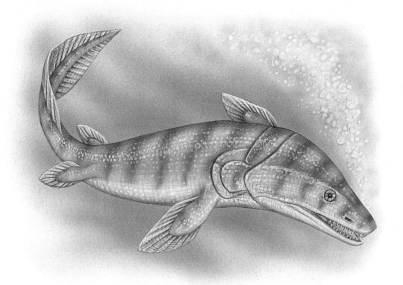

Figure 16
Amphibians probably evolved from fish like *Panderichthys* (pan dur IHK theez), which had leglike fins and lungs.

Research Visit the Glencoe Science Web site at **science.glencoe.com** for more information about Paleozoic life. Communicate to your class what you learn.

Life on Land Based on their structure, paleontologists know that many ancient fish had lungs as well as gills. Lungs enabled these fish to live in water with low oxygen levels— when needed they could swim to the surface and breathe air. Today's lungfish also can get oxygen from the water through gills and from the air through lungs.

One kind of ancient fish had lungs and leglike fins, which were used to swim and crawl around on the bottom. Paleontologists hypothesize that amphibians might have evolved from this kind of fish, shown in **Figure 16.** The characteristics that helped animals survive in oxygen-poor waters also made living on land possible. Today, amphibians live in a variety of habitats in water and on land. They all have at least one thing in common, though. They must lay their eggs in water or moist places.

✔ Reading Check *What are some characteristics of the fish from which amphibians might have evolved?*

By the Pennsylvanian Period, some amphibians evolved an egg with a membrane that protected it from drying out. Because of this, these animals, called reptiles, no longer needed to lay eggs in water. Reptiles also have skin with hard scales that prevent loss of body fluids. This adaptation enables them to survive farther from water and in relatively dry climates, as shown in **Figure 17,** where many amphibians cannot live.

Figure 17
Reptiles have scaly skins that allow them to live in dry places.

Figure 18
The Appalachian Mountains formed in several steps.

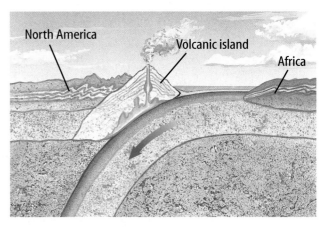

A More than 375 million years ago, volcanic island chains formed in the ocean and were pushed against the coast as Africa moved toward North America.

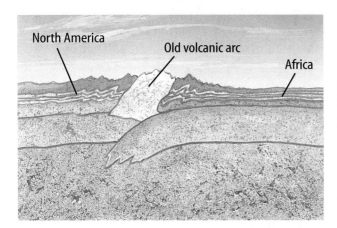

B About 375 million years ago, the African plate collided with the North American plate, forming mountains on both continents.

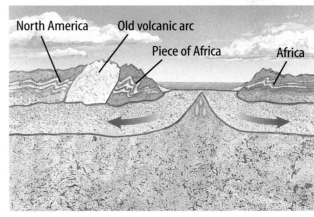

C About 200 million years ago, the Atlantic Ocean opened up, separating the two continents.

Mountain Building Several mountain-building episodes occurred during the Paleozoic Era. The Appalachian Mountains, for example, formed during this time. This happened in several stages, as shown in **Figure 18.** The first mountain-building episode occurred as the ocean separating North America from Europe and Africa closed. Several volcanic island chains that had formed in the ocean collided with the North American Plate, as shown in **Figure 18A.** The collision of the island chains generated high mountains.

The next mountain-building episode was a result of the African Plate colliding with the North American Plate, as shown in **Figure 18B.** When Africa and North America collided, rock layers were folded and faulted. Some rocks originally deposited near the eastern coast of the North American Plate were pushed along faults as much as 65 km westward by the collision. Sediments were uplifted to form an immense mountain belt, part of which still remains today.

Figure 19
Hyoliths were organisms that became extinct at the end of the Paleozoic Era.

End of an Era At the end of the Paleozoic Era, as many as 85 percent of all marine species and 70 percent of all land species died off. **Figure 19** shows one such animal. The cause of these extinctions might have been changes in climate and a lowering of sea level.

Near the end of the Permian Period, all of the continental plates came together and formed Pangaea, and glaciers formed over most of its southern part. The slow, gradual collision of continental plates caused mountain building. Mountain-building processes caused seas to close and deserts to spread over North America and Europe. Many species, especially marine organisms, couldn't adapt to these changes, so they became extinct.

Other Hypotheses Other explanations also have been proposed for this mass extinction. During the late Paleozoic Era, volcanoes were extremely active. If the volcanic activity was great enough, it could have affected the entire globe. Another recent theory is similar to the one proposed to explain the extinction of dinosaurs. Perhaps a large asteroid or comet collided with Earth some 248 million years ago. This event could have caused widespread extinctions just as many paleontologists suggest happened at the end of the Mesozoic Era, 65 million years ago. Perhaps the extinction at the end of the Paleozoic Era was caused by several or all of these events happening at about the same time.

Section Assessment

1. What geologic events occurred at the end of the Paleozoic Era?

2. How might geologic events at the end of the Paleozoic Era have caused extinctions?

3. What advance allowed reptiles to reproduce away from water? Why was this an advantage?

4. What major change in life-forms occurred at the end of Precambrian time?

5. **Think Critically** How, and through what process, did cyanobacteria contribute to the evolution of complex life on land? Do you think cyanobacteria are as significant to this process today as they were during Precambrian time?

Skill Builder Activities

6. **Making and Using Tables** Use **Figure 2** to answer these questions. **For more help, refer to the** Science Skill Handbook.
   a. When did the Paleozoic Era begin?
   b. How long did the Silurian Period last?
   c. When did vertebrates invade dry land?

7. **Using a Database** Research trilobites in a geology book or computer database. Write a paragraph in your Science Journal describing these organisms and their habitat. Include hand-drawn illustrations and compare them with the illustrations in the computer database on geology. **For more help, refer to the** Technology Skill Handbook.

Activity

Changing Species

In this activity, you will observe how adaptation within a species might cause the evolution of a particular trait, leading to the development of a new species.

What You'll Investigate

How might adaptation within a species cause the evolution of a particular trait?

Materials

Deck of playing cards

Goals

- **Model** adaptation within a species.

Procedure

1. **Remove** all of the kings, queens, jacks, and aces from a deck of playing cards.

2. Each remaining card represents an individual in a population of animals called "varimals." The number on each card represents the height of the individual. For example, the 5 of diamonds is a varimal that's 5 units tall.

3. **Calculate** the average height of the population of varimals represented by your cards.

4. Suppose varimals eat grass, shrubs, and leaves from trees. A drought causes many of these plants to die. All that's left are a few tall trees. Only varimals at least 6 units tall can reach the leaves on these trees.

5. All the varimals under 6 units leave the area to seek food elsewhere or die from starvation. Discard all of the cards with a number value less than 6. Calculate the new average height of the population of varimals.

6. **Shuffle** the deck of remaining cards.

7. **Draw** two cards at a time. Each pair represents a pair of varimals that will mate.

8. The offspring of each pair reaches a height equal to the average height of his or her parents. Calculate and record the height of each offspring.

9. Repeat by discarding all parents and offspring under 8 units tall. Now calculate the new average height of varimals. Include both the parents and offspring in your calculation.

Conclude and Apply

1. How did the height of the population change?

2. If you hadn't discarded the shortest varimals, would the average height of the population have changed as much? **Explain.**

3. Why didn't every member of the original population reproduce?

4. If there had been no varimals over 6 units tall in step 5, what would have happened to the population?

5. If there had been no variation in height in the population before the droughts occurred, would the species have been able to evolve into a taller species? **Explain.**

6. How does this activity demonstrate that traits evolve in species?

Middle and Recent Earth History

What You'll Learn

- **Compare and contrast** characteristic life-forms in the Mesozoic and Cenozoic Eras.
- **Explain** how changes caused by plate tectonics affected organisms during the Mesozoic Era.
- **Identify** when humans first appeared on Earth.

Vocabulary

Mesozoic Era
Cenozoic Era

Why It's Important

Many important groups of animals, like birds and mammals, appeared during the Mesozoic Era.

The Mesozoic Era

Dinosaurs have captured people's imaginations since their bones first were unearthed more than 150 years ago. Dinosaurs and other interesting animals lived during the Mesozoic Era, which was between 248 and 65 million years ago. The Mesozoic Era also was marked by rapid movement of Earth's continents.

The Breakup of Pangaea The **Mesozoic** (meh zuh ZOH ihk) **Era,** or era of middle life, was a time of many changes on Earth. At the beginning of the Mesozoic Era, all continents were joined as a single landmass called Pangaea, as shown in **Figure 11.**

Pangaea separated into two large landmasses during the Triassic Period, as shown in **Figure 20.** The northern mass was Laurasia (law RAY zhuh), and Gondwanaland (gahn DWAH nuh land) was the southern landmass. As the Mesozoic Era continued, Laurasia and Gondwanaland broke apart and eventually formed the present-day continents.

Species adapted to the new environments survived the mass extinction at the end of the Paleozoic Era. Recall that a reptile's skin helps it retain bodily fluids. This characteristic, along with their shelled eggs, enabled reptiles to adapt readily to the drier climate of the Mesozoic Era. Reptiles became the most conspicuous animals on land by the Triassic Period.

Figure 20
The supercontinent Pangaea formed at the end of the Paleozoic Era. At the end of the Triassic Period, it began to break up into the northern supercontinent, Laurasia, and the southern supercontinent, Gondwanaland.

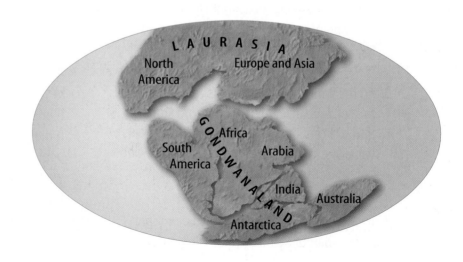

Dinosaurs What were the dinosaurs like? Dinosaurs ranged in height from less than 1 m to enormous creatures like *Apatosaurus* and *Tyrannosaurus.* The first small dinosaurs appeared during the Triassic Period. Larger species appeared during the Jurassic and Cretaceous Periods. Throughout the Mesozoic Era, new species of dinosaurs evolved and other species became extinct.

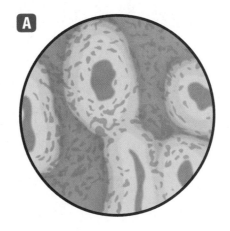

Dinosaurs Were Active Studying fossil footprints sometimes allows paleontologists to calculate how fast animals walked or ran. Some dinosaur tracks that have been found indicate that these animals were much faster runners than you might think. *Gallimimus* was 4 m long and could reach speeds of 65 km/h—as fast as a modern race horse.

Some studies also indicate that dinosaurs might have been warm blooded, not cold blooded like present-day reptiles. The evidence that leads to this conclusion has to do with their bone structure. Slices through some cold-blooded animal bones show rings similar to growth rings in trees. The bones of some dinosaurs don't show this ring structure. Instead, they are similar to bones found in modern mammals, as you can see in **Figure 21.**

Figure 21
Some dinosaur bones show structural features that are like mammal bones, leading some paleontologists to think that dinosaurs were warm blooded like mammals. **A** This shows the structure of a dinosaur bone. **B** This shows the structure of a mammal bone.

> ✔ **Reading Check** *Why do some paleontologists think that dinosaurs were warm blooded?*

These observations indicate that some dinosaurs might have been warm-blooded, fast-moving animals somewhat like present-day mammals and birds. They might have been quite different from present-day reptiles.

Good Mother Dinosaurs The fossil record also indicates that some dinosaurs nurtured their young and traveled in herds in which the adults surrounded their young.

One such dinosaur is *Maiasaura.* This dinosaur built nests in which it laid its eggs and raised its offspring. Nests have been found in relatively close clusters, indicating that more than one family of dinosaurs built in the same area. Some fossils of hatchlings have been found near adult animals, leading paleontologists to think that some dinosaurs nurtured their young. In fact, *Maiasaura* hatchlings might have stayed in the nest while they grew in length from about 35 cm to more than 1 m.

Research Visit the Glencoe Science Web site at **science.glencoe.com** for more information about dinosaurs. Communicate to your class what you learn.

Figure 22
Birds might have evolved from dinosaurs.

A *Bambiraptor feinberger* is a 75-million-year-old member of a family of meat-eating dinosaurs thought by some paleontologists to be closely related to birds.

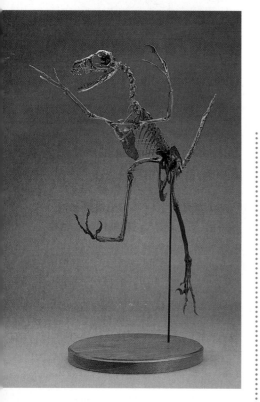

B Considered one of the world's most priceless fossils, *Archaeopteryx* was first found in a limestone quarry in Germany in 1861.

Birds Birds appeared during the Jurassic Period. Some paleontologists think that birds evolved from small, meat-eating dinosaurs much like *Bambiraptor feinberger* in **Figure 22A.** The earliest bird, *Archaeopteryx,* shown in **Figure 22B,** had wings and feathers. However, because *Archaeopteryx* had features not shared with modern birds, scientists know it was not a direct ancestor of today's birds.

Mammals Mammals first appeared in the Triassic Period. The earliest mammals were small, mouselike creatures, as shown in **Figure 23.** Mammals are warm-blooded vertebrates that have hair or fur covering their bodies. The females produce milk to feed their young. These two characteristics have enabled mammals to survive in many changing environments.

Gymnosperms During most of the Mesozoic Era, gymnosperms (JIHM nuh spurmz), which first appeared in the Paleozoic Era, dominated the land. Gymnosperms are plants that produce seeds but not flowers. Many gymnosperms are still around today. These include pines and ginkgo trees.

Angiosperms Angiosperms (AN jee uh spurmz), or flowering plants, first evolved during the Cretaceous Period. Angiosperms produce seeds with hard outer coverings.

Because their seeds are enclosed and protected, angiosperms can live in many environments. Angiosperms are the most diverse and abundant land plants today. Present-day angiosperms that evolved during the Mesozoic Era include magnolia and oak trees.

Figure 23
The earliest mammals were small creatures that resembled today's mice and shrews.

End of an Era The Mesozoic Era ended about 65 million years ago with a major extinction of land and marine species. Many groups of animals, including the dinosaurs, disappeared suddenly at this time. Many paleontologists hypothesize that a comet or asteroid collided with Earth, causing a huge cloud of dust and smoke to rise into the atmosphere, blocking out the Sun. Without sunlight the plants died, and all the animals that depended on these plants also died. Not everything died, however. All the organisms that you see around you today are descendants of the survivors of the great extinction at the end of the Mesozoic Era.

Math Skills Activity

Calculating Extinction Using Percentages

Example Problem

At the end of the Cretaceous Period, large numbers of animals became extinct. Scientists still are trying to understand why some types of animals survived while others died. Looking at data about amphibians, reptiles, and mammals that lived during the Cretaceous Period, can you determine what percentage of amphibians survived the extinction?

Solution

1 *This is what you know:*

Animal Extinctions

Animal Type	Groups Living Before Extinction Event (n)	Groups Left After Extinction Event (t)
Amphibians	12	4
Reptiles	63	30
Mammals	24	8

2 *This is what you need to find out:* p = the percentage of amphibian groups that survived the Cretaceous extinction

3 *This is the equation you need to use:* $p = t / n \times 100$
Both t and n are shown on the above chart.

4 *Substitute the known values:* $p = 4 / 12 \times 100 = 33.3\%$

Practice Problem

Using the same equation, calculate the percentage of reptiles and then the percentage of mammals that survived. Which type of animal was least affected by the extinction?

For more help, refer to the Math Skill Handbook.

TRY AT HOME

Mini LAB

Calculating the Age of the Atlantic Ocean

Procedure

1. On a **world map** or **globe**, measure the distance in kilometers between a point on the east coast of South America and a point on the west coast of Africa.
2. Measure in SI several times and take the average of your results.
3. Assuming that Africa has been moving away from South America at a rate of 3.5 cm per year, calculate how many years it took to create the Atlantic Ocean.

Analysis

1. Did the values used to obtain your average value vary much?
2. How close did your age come to the accepted estimate for the beginning of the breakup of Pangaea in the Triassic Period?

The Cenozoic Era

The **Cenozoic** (sen uh ZOH ihk) **Era,** or era of recent life, began about 65 million years ago and continues today. Many mountain ranges in North and South America and Europe began to form in the Cenozoic Era. In the late Cenozoic, the climate became much cooler and ice ages occurred. The Cenozoic Era is subdivided into two periods. The first of these is the Tertiary Period. The present-day period is the Quaternary Period. It began about 1.8 million years ago.

✓ **Reading Check** *What happened to the climate during the late Cenozoic Era?*

Times of Mountain Building Many mountain ranges formed during the Cenozoic Era. These include the Alps in Europe and the Andes in South America. The Himalaya, shown in **Figure 24,** formed as India moved northward and collided with Asia. The collision crumpled and thickened Earth's crust, raising the highest mountains presently on Earth. Many people think the growth of these mountains has helped create cooler climates worldwide.

Figure 24
Ⓐ The Himalaya extend along the India-Tibet border and contain some of the world's tallest mountains. Ⓑ India drifted north and finally collided with Asia, forming the Himalaya.

Further Evolution of Mammals

Throughout much of the Cenozoic Era, expanding grasslands favored grazing plant eaters like horses, camels, deer, and some elephants. Many kinds of mammals became larger. Horses evolved from small, multi-toed animals into the large, hoofed animals of today. However, not all mammals remained on land. Ancestors of the present-day whales and dolphins evolved to make their lives in the sea.

As Australia and South America separated from Antarctica during the continuing breakup of the continents, many species became isolated. They evolved separately from life-forms in other parts of the world. Evidence of this can be seen today in Australia's marsupials. Marsupials are mammals such as kangaroos, koalas, and wombats (shown in **Figure 25**) that carry their young in a pouch.

Your species, *Homo sapiens*, probably appeared about 140,000 years ago. Some people suggest that the appearance of humans could have led to the extinction of many other mammals. As their numbers grew, humans competed for food that other animals relied upon. Also, fossil bones and other evidence indicate that early humans were hunters.

Figure 25
The wombat is one of many Australian marsupials. As a result of human activities, the number and range of wombats have diminished.

Section 3 Assessment

1. In which era, period, and epoch did *Homo sapiens* first appear?

2. Did mammals become more or less abundant after the extinction of the dinosaurs? Explain why.

3. How did the development of seeds with a hard outer covering enable angiosperms to survive in a wide variety of climates?

4. Give two reasons why some paleontologists hypothesize that dinosaurs were warm-blooded animals.

5. **Think Critically** How could two species that evolved on separate continents have many similarities?

Skill Builder Activities

6. **Researching Information** Arrange these organisms in sequence according to when they first appeared on Earth: *mammals, reptiles, dinosaurs, fish, angiosperms, birds, insects, amphibians, land plants,* and *bacteria.* **For more help, refer to the** Science Skill Handbook.

7. **Converting Units** A fossil mosasaur, a giant marine reptile, measured 9 m in lenth and had a skull that measured 45 cm in length. What fraction of the mosasaur's total length did the skull account for? Compare your length with the mosasaur's length. **For more help, refer to the** Math Skill Handbook.

Activity

Use the Internet

Discovering the Past

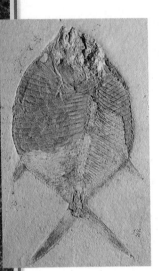

Imagine what your state was like millions of years ago. What animals might have been roaming around the spot where you now sit? Can you picture a *Tyrannosaurus rex* roaming the area that is now your school? The animals and plants that once inhabited your region might have left some clues to their identity—fossils. Scientists use fossils to piece together what Earth looked like in the geologic past. Fossils can help determine whether an area used to be dry land or was underwater. Fossils can help uncover clues about how plants and animals have evolved over the course of time. Using the resources of the Internet and by sharing data with your peers, you can start to discover how North America has changed through time.

Recognize the Problem

How has your area changed over geologic time?

Form a Hypothesis

How might the area where you are now living have looked thousands or millions of years ago? Do you think that the types of animals and plants have changed much over time? Form a hypothesis concerning the change in organisms and geography from long ago to the present day in your area.

Goals
- **Gather** information about fossils found in your area.
- **Communicate** details about fossils found in your area.
- **Synthesize** information from sources about the fossil record and the changes in your area over time.

SCIENCE *Online* Go to the Glencoe Science Web site at **science.glencoe.com** to get more information about fossils and changes over geologic time and for data collected by other students.

Fossils in Your Area					
Fossil Name	Plant or Animal Fossil	Age of Fossils	Details About Plant or Animal Fossil	Location of Fossil	Additional Information

420 CHAPTER 14 Geologic Time

Test Your Hypothesis

Plan

1. **Determine** the age of the rocks that make up your area. Were they formed during Precambrian time, the Paleozoic Era, the Mesozoic Era, or the Cenozoic Era?

2. Gather information about the fossil plants and animals found in your area during one of the above geologic time intervals. Find specific information on when, where and how the fossil organisms lived. If no fossils are known from your area, find out information about the fossils found nearest your area.

Do

1. Make sure your teacher approves your plan before you start.

2. Go to the Glencoe Science Web site at **science.glencoe.com** to post your data in the table. Add any additional information you think is important to understanding the fossils found in your area.

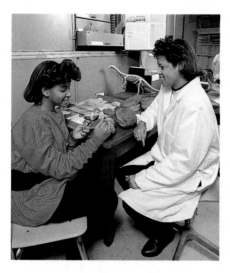

Analyze Your Data

1. What present-day relatives of prehistoric animals or plants exist in your area?

2. How have the organisms in your area changed over time? Is your hypothesis supported? Why or why not?

3. What other information did you discover about your area's climate or environment from the geologic time period you investigated?

Draw Conclusions

1. Find this *Use the Internet* activity on the Glencoe Science Web site at **science.glencoe.com.** Compare your data to those of other students. Study other students' data to compare information about the geologic time periods and fossils that you investigated. Review data that other students have entered about fossils they have researched.

2. **Describe** the plant and animal fossils that have been discovered in your area. What clues did you discover about the environment of your fossil organisms? How do these compare to the environment of your area today?

3. **Infer** from the fossil organisms found in your area what the geography and climate were like during the geologic time period you chose.

Science Stats

Extinct!

Did you know...

...In 1865, when Lewis Carroll wrote about the dodo in his famous book *Alice in Wonderland*, the bird had been gone for almost two hundred years from the island of Mauritius in the Indian Ocean. First seen by European settlers in 1507, the dodo was hunted for food. The birds were extinct by 1681.

...The earliest known gliding reptile had "wings" with a span of 30 cm that were wide flaps of flesh not attached to its limbs. From fossils found in Madagascar, scientists believe that *Coelurosauravus* (see lor oh SOR uh vuhs) lived between 260 and 246 million years ago.

Coelurosauravus

Saber-toothed cat

...The saber-toothed cat lived in the Americas from about 1.6 million to 8,000 years ago. *Smilodon*, the best-known saber-toothed cat, was among the most ferocious carnivores. It had large canine teeth, about 15 cm long, which it used to pierce the flesh of its prey.

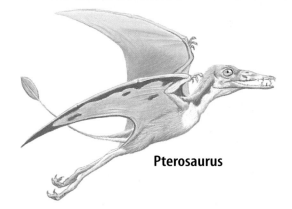

Pterosaurus

...The first vertebrates to fly were reptiles called pterosaurs (TER uh sawrz). The front limbs of these reptiles developed into wings during the Triassic Period. Their wingspans ranged from 1.8 m to more than 12 m. Some pterosaurs were fish eaters, flying low over water to catch fish with their narrow jaws.

Great Mass Extinctions of Species

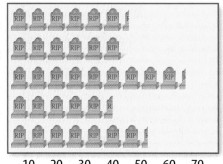

...The woolly mammoth lived in the cold tundra regions during the Ice Age. It looked rather like an elephant with long hair, had a mass between 5,300 kg and 7,300 kg, and was between 3 m and 4 m tall.

Woolly mammoth

Do the Math

1. If 75 species of organisms become extinct each day, how many would become extinct during a leap year?
2. How many years did it take from the first sighting of the dodo bird by Europeans until it became extinct?
3. What was the range between the smallest and largest wingspans among pterosaurs?

Go Further

Use the Glencoe Science Web site at **science.glencoe.com** to research extinct animals, fish, birds, plants, and other life-forms. Trace the origins of each of the species and learn how long its kind existed on Earth.

Chapter 14 Study Guide

Reviewing Main Ideas

Section 1 Life and Geologic Time

1. Geologic time is divided into eons, eras, periods, and epochs.

2. Divisions within the geologic time scale are based largely on major evolutionary changes in organisms.

3. Plate movements cause changes that affect organic evolution. *How can the building of mountains like those shown here affect the evolution of species?*

Section 2 Early Earth History

1. Cyanobacteria were an early form of life that evolved during Precambrian time. Trilobites, fish, and corals were abundant during the Paleozoic Era.

2. Plants and animals began to move onto land during the middle of the Paleozoic Era. Land plants and animals then evolved rapidly and colonized the land.

3. The Paleozoic Era was a time of mountain building. The Appalachian Mountains formed when several islands and finally Africa collided with North America.

4. At the end of the Paleozoic Era, many marine invertebrates became extinct. *What kind of marine organism is shown here?*

Section 3 Middle and Recent Earth History

1. Reptiles and gymnosperms were dominant land life-forms in the Mesozoic Era. Mammals and angiosperms began to dominate the land in the Cenozoic Era.

2. Pangaea broke apart during the Mesozoic Era. Many mountain ranges formed during the Cenozoic.

3. *Homo sapiens* appeared during the Pleistocene Epoch. *How might* Homo sapiens *have contributed to the extinction of animals like the one shown here?*

FOLDABLES
Reading & Study Skills

After You Read

To help you review geological time, use the Organizational Study Fold you made at the beginning of the chapter.

Visualizing Main Ideas

Complete the concept map on geologic time using the following choices: Cenozoic, Trilobites in oceans, mammals common, Paleozoic, Dinosaurs roam Earth, and Abundant gymnosperms.

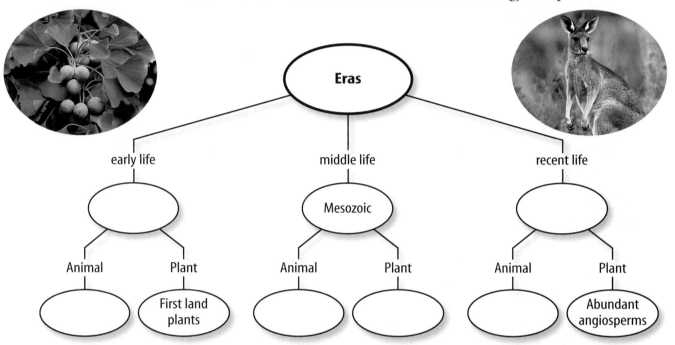

Vocabulary Review

Vocabulary Words

a. Cenozoic Era
b. cyanobacteria
c. eon
d. epoch
e. era
f. geologic time scale
g. Mesozoic Era
h. natural selection
i. organic evolution
j. Paleozoic Era
k. Pangaea
l. period
m. Precambrian time
n. species
o. trilobite

Study Tip

Outline the chapters to make sure that you're understanding the key ideas. Writing down the main points of the chapter will help you remember important details and understand larger themes.

Using Vocabulary

The sentences below include vocabulary words that have been used incorrectly. Change the incorrect word so that the sentence reads correctly.

1. A change in the hereditary features of a species over a long period is extinction.

2. A record of events in Earth history is natural selection.

3. The largest subdivision of geologic time is the period.

4. The process by which the best-suited individuals survive in their environment is organic evolution.

5. A group of individuals that normally breed only among themselves is an epoch.

Checking Concepts

Choose the word or phrase that best completes the sentence.

1. How many millions of years ago did the era in which you live begin?
   A) 650   C) 1.6
   B) 245   D) 65

2. What is the process by which better-suited organisms survive and reproduce?
   A) endangerment   C) gymnosperm
   B) extinction   D) natural selection

3. What is the next smaller division of geologic time after the era?
   A) period   C) epoch
   B) stage   D) eon

4. When did the most recent ice age occur?
   A) Pennsylvanian   C) Tertiary
   B) Triassic   D) Quaternary

5. What was one of the earliest forms of life on Earth?
   A) gymnosperm   C) angiosperm
   B) cyanobacterium   D) dinosaur

6. Which group of organisms evolved from the same ancestors as amphibians?
   A) trilobites   C) angiosperms
   B) lungfish   D) gymnosperms

7. During which era did the dinosaurs live?
   A) Mesozoic   C) Miocene
   B) Paleozoic   D) Cenozoic

8. Which type of plant has seeds without protective coverings?
   A) angiosperms   C) gymnosperms
   B) flowering plants D) magnolias

9. Which group of plants evolved during the Mesozoic Era and is the dominant plant group today?
   A) gymnosperms   C) ginkgoes
   B) angiosperms   D) algae

10. When did the Ediacaran fauna live?
    A) Precambrian time   C) Mesozoic Era
    B) Paleozoic Era   D) Cenozoic Era

Thinking Critically

11. Why couldn't plants move onto land until an ozone layer formed?

12. Why are trilobites classified as index fossils?

13. What is the most significant difference between Precambrian life-forms and Paleozoic life-forms?

14. How is natural selection related to organic evolution?

15. In the early 1800s, a naturalist proposed that the giraffe species has a long neck as a result of years of stretching their necks to reach leaves in tall trees. Explain why this isn't true.

Developing Skills

16. **Interpreting Scientific Illustrations** The circle graph below represents geologic time. Determine which interval of geologic time is represented by each portion of the graph. Which interval was longest? Which do we know the least about?

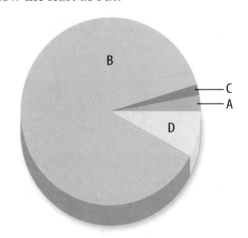

17. Interpreting Data The Cenozoic Era has lasted 65 million years. What percentage of Earth's 4.5-billion-year history is that?

18. Comparing and Contrasting Use the outlines of the present-day continents to make a sketch of Pangaea.

19. Forming Hypotheses Suggest some reasons why trilobites might have become extinct at the end of the Paleozoic Era.

20. Interpreting Data A student found what he thought was a piece of dinosaur bone in Pleistocene sediment. How likely is it that he is right? Explain.

Performance Assessment

21. Make a Model In the Activity, you learned how a particular characteristic might evolve within a species. Modify the experimental model by using color instead of height as a characteristic. Design your activity with the understanding that varimals live in a dark-colored forest environment.

TECHNOLOGY

Go to the Glencoe Science Web site at **science.glencoe.com** or use the **Glencoe Science CD-ROM** for additional chapter assessment.

Test Practice

Many animals have inhabited Earth. Some of these animals have been divided into groups below.

Group A	Group B	Group C
Saber-toothed cat	*Apatosaurus*	Frog
Giant ground sloth	*Maiasaura*	Turtle
Woolly mammoth	*Tyrannosaurus*	Salamander

Study the table and answer the following questions.

1. The animals in Group A are different from the animals in Group B because only the animals in Group A are _____ .
  A) mammals **C)** amphibians
  B) reptiles **D)** birds

2. The animals in Group B are different from the animals in Group C because only the animals in Group B are _____ .
  F) dinosaurs **H)** amphibians
  G) reptiles **J)** herbivores

3. The animals in all three groups are similar because _____ .
  A) They make their own food.
  B) They have hair.
  C) They must eat other organisms.
  D) They are meat eaters.

Reading Comprehension

Read the passage. Then read each question that follows the passage. Decide which is the best answer to each question.

Life on Earth

The oldest rocks found on Earth have been estimated to be about 4 billion years old. But these rocks do not contain any fossil remains. The oldest rocks that have been found with any evidence of life are about 3.5 billion years old. These rocks have fossils of single-celled organisms that once lived in the oceans.

No life could have survived on land at that time because there was so much ultraviolet radiation coming down to Earth's surface from the Sun. Early bacteria did something very important that made it possible for other species to inhabit Earth.

Scientists believe there was little oxygen in Earth's early atmosphere. Nearly 3 billion years ago, some early bacteria produced oxygen during photosynthesis. This oxygen bubbled into the oceans, and after millions of years, entered the atmosphere.

This oxygen from photosynthesis was very important. As it entered the atmosphere, some of it absorbed energy from the Sun's radiation. The radiation caused chemical reactions, which produced ozone gas. This ozone began acting as a protective shield, blocking Earth's surface from most of the incoming ultraviolet radiation. As a result, it became possible for life to survive on land without being damaged by ultraviolet radiation.

The fossil record provides clues about past life on Earth. The earliest multicellular organisms lived in the oceans during the late part of the Precambrian portion of Earth's history. Fossil evidence indicates that animals did not inhabit land until much later.

> **Test-Taking Tip** Number the paragraphs in the passage to make sure that you are referring to the correct paragraph when answering a question.

1. According to the passage, what type of organism added oxygen to Earth's early atmosphere?
   A) vertebrates
   B) mammals
   C) bacteria
   D) fungi

2. Which of these is the best summary of this passage?
   F) The earliest organisms lived in the ocean.
   G) The oldest rocks that have been found with any evidence of life are about 3.5 billion years old.
   H) The production of oxygen by bacteria, which began nearly 3 billion years ago, helped make life on land possible.
   J) Oxygen can act like a protective shield, blocking Earth's surface from most of the incoming ultraviolet radiation.

3. What is the main idea of the fourth paragraph of this passage?
   A) Oxygen entered the atmosphere.
   B) Oxygen from photosynthesis was very important.
   C) Oxygen absorbed radiation from the Sun to make ozone.
   D) The oldest rocks on Earth do not have any fossils in them.

Reasoning and Skills

Read each question and choose the best answer.

1. Which of these facts best explains why the oldest fossils have been discovered in sediments that were deposited in ancient oceans?
- **A)** Only aquatic animals leave fossils.
- **B)** Fossils from land organisms have been destroyed.
- **C)** The first life on Earth began in the oceans.
- **D)** All living organisms need water.

Test-Taking Tip Some answer choices might be true statements but not the *best* explanation to a particular question.

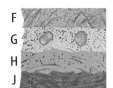
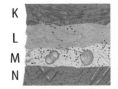

Rock layers at Site X   Rock layers at Site Z

2. The fossils found in layer G are considered index fossils by paleontologists. This means that _____.
- **F)** they are not as old as the fossils found in layer M
- **G)** layer G is most likely the same age as layer M
- **H)** they are the same age as the rock in layer L
- **J)** the fossils are the oldest types of fossils found on Earth

Test-Taking Tip Index fossils appeared during a specific block of time in Earth's history. They can be used to date layers of rock.

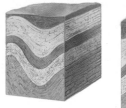

3. Which of the sequences below could fill the gap in the picture above?

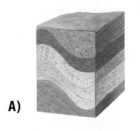
A)   **B)**

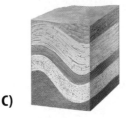
C)   **D)**

Test-Taking Tip Sedimentary rock layers can be distinguished from each other in pictures by using different patterns to represent each layer.

Consider this question carefully before writing your answer on a separate sheet of paper.

4. The Himalaya are found at the boundary between two large continental plates. These mountains were formed as the result of what geologic process?

Test-Taking Tip Think about what happens as the continental plates move slowly over Earth's plastic-like asthenosphere.

How Are
Air & Advertising
Connected?

HAND LAUNDRY

In the late 1800s, two scientists were studying the composition of air when they discovered an element that hadn't been known before. They named it "neon," and soon this new element, represented by the chemical symbol Ne, had been assigned a spot in the periodic table (right). It took a few years for people to figure out something useful to do with neon! In 1910, a French engineer experimented with passing an electrical current through neon gas in a vacuum tube. The result was a spectacular orange-red light. Neon's advertising possibilities were quickly realized, and soon the first neon sign blazed on a boulevard in Paris. Today, neon signs in a wide range of colors advertise shops and services all over the world. The other colors are made by mixing neon with other gases and by using tinted tubes.

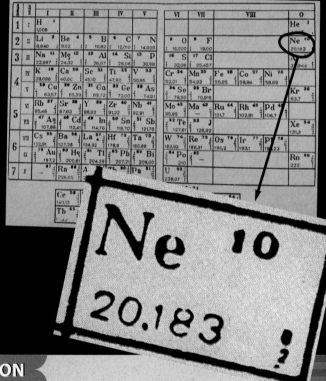

SCIENCE CONNECTION

NOBLE GASES Neon is one of six elements known as the noble gases. These gases make up a tiny fraction of the atmosphere. Conduct research to find out more about this group. What other gases belong to the group? What do they all have in common? Why are they called "noble"? Create a chart that shows when each of the noble gases was discovered and what distinctive characteristics and uses it has.

15 Atmosphere

Why is it difficult to breathe at high elevations? Why are some mountain peaks permanently covered with snow? These mountain climbers aren't supplementing oxygen just because the activity is physically demanding. At elevations like this, the amount of oxygen available in the air is so small that the climbers' bodily functions might not be supported. In this chapter, you'll learn about the composition and structure of the atmosphere. You also will learn how energy is transferred in the atmosphere. In addition, you'll examine the water cycle and major wind systems.

What do you think?

Science Journal Look at the picture below with a classmate. Discuss what this might be. Here's a hint: *It "pops" in thin air.* Write your answer or best guess in your Science Journal.

The air around you is made of billions of molecules. These molecules are constantly moving in all directions and bouncing into every object in the room, including you. Air pressure is the result of the billions of collisions of molecules into these objects. Because you usually do not feel molecules in air hitting you, do the activity below to see the effect of air pressure.

Observe air pressure

1. Cut out a square of cardboard about 10 cm on a side from a cereal box.
2. Fill a glass to the brim with water.
3. Hold the cardboard firmly over the top of the glass, covering the water, and invert the glass.
4. Slowly remove your hand holding the cardboard in place and observe.

Observe

Write a paragraph in your Science Journal describing what happened to the cardboard when you inverted the glass and removed your hand. How does air pressure explain what happened?

Before You Read

FOLDABLES
Reading & Study Skills

Making a Sequence Study Fold Make the following Foldable to help you visualize the layers of Earth's atmosphere.

1. Stack three sheets of paper in front of you so the short sides are at the top.
2. Slide the top sheet up so that about four centimeters of the middle sheet show. Slide the middle sheet up so that about four centimeters of the bottom sheet show.
3. Fold the sheets top to bottom to form six tabs and staple along the topfold. Turn the Foldable so the staples are at the bottom.
4. Label the flaps *Earth's Atmosphere, Troposphere, Stratosphere, Mesosphere, Thermosphere,* and *Exosphere,* as shown.
5. As you read the chapter, write information about each layer of Earth's atmosphere under the tabs.

Exosphere
Thermosphere
Mesosphere
Stratosphere
Troposphere
Earth's Atmosphere

Earth's Atmosphere

Figure 1
Earth's atmosphere, as viewed from space, is a thin layer of gases. The atmosphere keeps Earth's temperature in a range that can support life.

Importance of the Atmosphere

Earth's **atmosphere,** shown in **Figure 1,** is a thin layer of air that forms a protective covering around the planet. If Earth had no atmosphere, days would be extremely hot and nights would be extremely cold. Earth's atmosphere maintains a balance between the amount of heat absorbed from the Sun and the amount of heat that escapes back into space. It also protects life-forms from some of the Sun's harmful rays.

Makeup of the Atmosphere

Earth's atmosphere is a mixture of gases, solids, and liquids that surround the planet. It extends from Earth's surface to outer space. The atmosphere is much different today from what it was when Earth was young.

Earth's early atmosphere, produced by erupting volcanoes, contained nitrogen and carbon dioxide, but little oxygen. Then, more than 2 billon years ago, Earth's early organisms released oxygen into the atmosphere as they made food with the aid of sunlight. These early organisms, however, were limited to layers of ocean water deep enough to be shielded from the Sun's harmful rays, yet close enough to the surface to receive sunlight. Eventually, a layer rich in ozone (O_3) that protects Earth from the Sun's harmful rays formed in the upper atmosphere. This protective layer eventually allowed green plants to flourish all over Earth, releasing even more oxygen. Today, a variety of life forms, including you, depends on a certain amount of oxygen in Earth's atmosphere.

Gases in the Atmosphere Today's atmosphere is a mixture of the gases shown in **Figure 2**. Nitrogen is the most abundant gas, making up 78 percent of the atmosphere. Oxygen actually makes up only 21 percent of Earth's atmosphere. As much as four percent of the atmosphere is water vapor. Other gases that make up Earth's atmosphere include argon and carbon dioxide.

The composition of the atmosphere is changing in small but important ways. For example, car exhaust emits gases into the air. These pollutants mix with oxygen and other chemicals in the presence of sunlight and form a brown haze called smog. Humans burn fuel for energy. As fuel is burned, carbon dioxide is released as a by-product into Earth's atmosphere. Increasing energy use may increase the amount of carbon dioxide in the atmosphere.

Solids and Liquids in Earth's Atmosphere In addition to gases, Earth's atmosphere contains small, solid particles such as dust, salt, and pollen. Dust particles get into the atmosphere when wind picks them up off the ground and carries them along. Salt is picked up from ocean spray. Plants give off pollen that becomes mixed throughout part of the atmosphere.

The atmosphere also contains small liquid droplets other than water droplets in clouds. The atmosphere constantly moves these liquid droplets and solids from one region to another. For example, the atmosphere above you may contain liquid droplets and solids from an erupting volcano thousands of kilometers from your home, as illustrated in **Figure 3.**

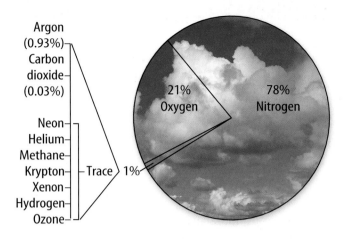

Argon (0.93%)
Carbon dioxide (0.03%)

Neon
Helium
Methane
Krypton — Trace 1%
Xenon
Hydrogen
Ozone

21% Oxygen   78% Nitrogen

Figure 2
This graph shows the percentages of the gases, excluding water vapor, that make up Earth's atmosphere.

Figure 3
Solids and liquids can travel large distances in Earth's atmosphere, affecting regions far from their source.

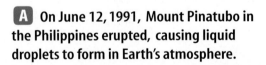
A On June 12, 1991, Mount Pinatubo in the Philippines erupted, causing liquid droplets to form in Earth's atmosphere.

B Droplets of sulfuric acid from volcanoes can produce spectacular sunrises.

Layers of the Atmosphere

What would happen if you left a glass of chocolate milk on the kitchen counter for a while? Eventually, you would see a lower layer with more chocolate separating from upper layers with less chocolate. Like a glass of chocolate milk, Earth's atmosphere has layers. There are five layers in Earth's atmosphere, each with its own properties, as shown in **Figure 4.** The lower layers include the troposphere and stratosphere. The upper atmospheric layers are the mesosphere, thermosphere, and exosphere. The troposphere and stratosphere contain most of the air.

Lower Layers of the Atmosphere You study, eat, sleep, and play in the **troposphere,** which is the lowest of Earth's atmospheric layers. It contains 99 percent of the water vapor and 75 percent of the atmospheric gases. Rain, snow, and clouds occur in the troposphere, which extends up to about 10 km.

The stratosphere, the layer directly above the troposphere, extends from 10 km above Earth's surface to about 50 km. As **Figure 4** shows, a portion of the stratosphere contains higher levels of a gas called ozone. Each molecule of ozone is made up of three oxygen atoms bonded together. Later in this section you will learn how ozone protects Earth from the Sun's harmful rays.

Figure 4
Earth's atmosphere is divided into five layers. *Which layer of the atmosphere do you live in?*

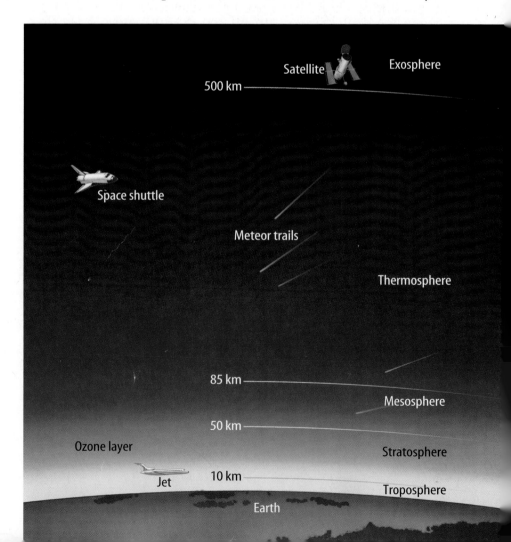

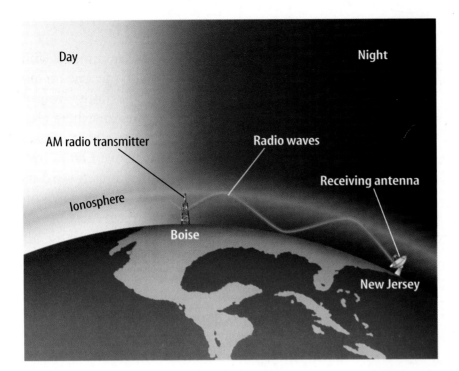

Day

Night

Ionosphere

AM radio transmitter

Radio waves

Receiving antenna

Boise

New Jersey

Figure 5
During the day, the ionosphere absorbs radio transmissions. This prevents you from hearing distant radio stations. At night, the ionosphere reflects radio waves. The reflected waves can travel to distant cities.

Upper Layers of the Atmosphere Beyond the stratosphere are the mesosphere, thermosphere, and exosphere. The mesosphere extends from the top of the stratosphere to about 85 km above Earth. If you've ever seen a shooting star, you might have witnessed a meteor in the mesosphere.

The thermosphere is named for its high temperatures. This is the thickest atmospheric layer and is found between 85 km and 500 km above Earth's surface.

Within the mesosphere and thermosphere is a layer of electrically charged particles called the **ionosphere** (i AHN uh sfir). If you live in New Jersey and listen to the radio at night, you might pick up a station from Boise, Idaho. The ionosphere allows radio waves to travel across the country to another city, as shown in **Figure 5.** During the day, energy from the Sun interacts with the particles in the ionosphere, causing them to absorb AM radio frequencies. At night, without solar energy, AM radio transmissions reflect off the ionosphere, allowing radio transmissions to be received at greater distances.

The space shuttle in **Figure 6** orbits Earth in the exosphere. In contrast to the troposphere, the layer you live in, the exosphere has so few molecules that the wings of the shuttle are useless. In the exosphere, the spacecraft relies on bursts from small rocket thrusters to move around. Beyond the exosphere is outer space.

✓ Reading Check *How does the space shuttle maneuver in the exosphere?*

Figure 6
Wings help move aircraft in lower layers of the atmosphere. The space shuttle can't use its wings to maneuver in the exosphere because so few molecules are present.

Atmospheric Pressure

Imagine you're a football player running with the ball. Six players tackle you and pile one on top of the other. Who feels the weight more—you or the player on top? Like molecules anywhere else, atmospheric gases have mass. Atmospheric gases extend hundreds of kilometers above Earth's surface. As Earth's gravity pulls the gases toward its surface, the weight of these gases presses down on the air below. As a result, the molecules nearer Earth's surface are closer together. This dense air exerts more force than the less dense air near the top of the atmosphere. Force exerted on an area is known as pressure.

Like the pile of football players, air pressure is greater near Earth's surface and decreases higher in the atmosphere, as shown in **Figure 7.** People find it difficult to breathe in high mountains because fewer molecules of air exist there. Jets that fly in the stratosphere must maintain pressurized cabins so that people can breathe.

Figure 7
Air pressure decreases as you go higher in Earth's atmosphere.

✓ Reading Check *Where is air pressure greater—in the exosphere or in the troposphere?*

Problem-Solving Activity

How does altitude affect air pressure?

Atmospheric gases extend hundreds of kilometers above Earth's surface, but the molecules that make up these gases are fewer and fewer in number as you go higher. This means that air pressure decreases with altitude.

Identifying the Problem

The graph on the right shows these changes in air pressure. Note that altitude on the graph goes up only to 50 km. The troposphere and the stratosphere are represented on the graph, but other layers of the atmosphere are not. By examining the graph, can you understand the relationship between altitude and pressure?

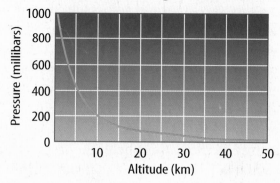

Solving the Problem

1. Estimate the air pressure at an altitude of 5 km.
2. Does air pressure change more quickly at higher altitudes or at lower altitudes?

Temperature in Atmospheric Layers

The Sun is the source of most of the energy on Earth. Before it reaches Earth's surface, energy from the Sun must pass through the atmosphere. Because some layers contain gases that easily absorb the Sun's energy while other layers do not, the various layers have different temperatures, illustrated by the red line in **Figure 8.**

Molecules that make up air in the troposphere are warmed mostly by heat from Earth's surface. The Sun warms Earth's surface, which then warms the air above it. When you climb a mountain, the air at the top is usually cooler than the air at the bottom. Every kilometer you climb, the air temperature decreases about 6.5°C.

Molecules of ozone in the stratosphere absorb some of the Sun's energy. Energy absorbed by ozone molecules raises the temperature. Because more ozone molecules are in the upper portion of the stratosphere, the temperature in this layer rises with increasing altitude.

Like the troposphere, the temperature in the mesosphere decreases with altitude. The thermosphere and exosphere are the first layers to receive the Sun's rays. Few molecules are in these layers, but each molecule has a great deal of energy. Temperatures here are high.

Mini LAB

Determining if air has mass

Procedure

1. On a **pan balance,** find the mass of an **inflatable ball** that is completely deflated.
2. Hypothesize about the change in the mass of the ball when it is inflated.
3. Inflate the ball to its maximum recommended inflation pressure.
4. Determine the mass of the fully inflated ball.

Analysis

1. What change occurs in the mass of the ball when it is inflated?
2. Infer from your data whether air has mass.

Temperature of the Atmosphere at Various Altitudes

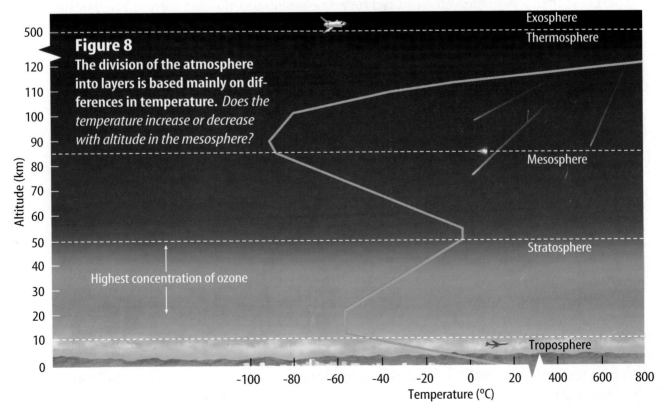

Figure 8
The division of the atmosphere into layers is based mainly on differences in temperature. *Does the temperature increase or decrease with altitude in the mesosphere?*

The Ozone Layer

Within the stratosphere, about 19 km to 48 km above your head, lies an atmospheric layer called the **ozone layer.** Ozone is made of oxygen. Although you cannot see the ozone layer, your life depends on it.

The oxygen you breathe has two atoms per molecule, but an ozone molecule is made up of three oxygen atoms bound together. The ozone layer contains a high concentration of ozone and shields you from the Sun's harmful energy. Ozone absorbs most of the ultraviolet radiation that enters the atmosphere. **Ultraviolet radiation** is one of the many types of energy that come to Earth from the Sun. Too much exposure to ultraviolet radiation can damage your skin and cause cancer.

CFCs Evidence exists that some air pollutants are destroying the ozone layer. Blame has fallen on **chlorofluorocarbons** (CFCs), chemical compounds used in some refrigerators, air conditioners, and aerosol sprays, and in the production of some foam packaging. CFCs can enter the atmosphere if these appliances leak or if they and other products containing CFCs are improperly discarded.

Recall that an ozone molecule is made of three oxygen atoms bonded together. Chlorofluorocarbon molecules, shown in **Figure 9,** destroy ozone. When a chlorine atom from a chlorofluorocarbon molecule comes near a molecule of ozone, the ozone molecule breaks apart. One of the oxygen atoms combines with the chlorine atom, and the rest form a regular, two-atom molecule. These compounds don't absorb ultraviolet radiation the way ozone can. In addition, the original chlorine atom can continue to break apart thousands of ozone molecules. The result is that more ultraviolet radiation reaches Earth's surface.

Figure 9
Chlorofluorocarbon (CFC) molecules were used in refrigerators and air conditioners. Each CFC molecule has three chlorine atoms. One atom of chlorine can destroy approximately 100,000 ozone molecules.

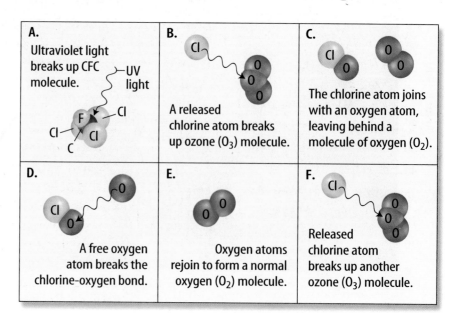

A. Ultraviolet light breaks up CFC molecule. — UV light

B. A released chlorine atom breaks up ozone (O_3) molecule.

C. The chlorine atom joins with an oxygen atom, leaving behind a molecule of oxygen (O_2).

D. A free oxygen atom breaks the chlorine-oxygen bond.

E. Oxygen atoms rejoin to form a normal oxygen (O_2) molecule.

F. Released chlorine atom breaks up another ozone (O_3) molecule.

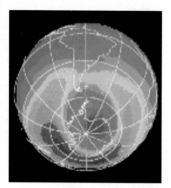
October 1980

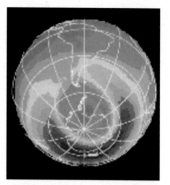
October 1988

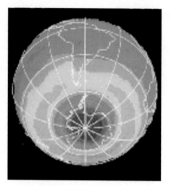
October 1990

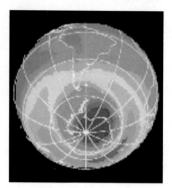
September 1999

Health
INTEGRATION

Ozone Holes Each year, more than 1.3 million Americans develop skin cancer, and more than 9,500 die from it. Exposure to ultraviolet radiation can cause skin cancer. If the ozone layer disappeared, skin cancer rates might increase. In 1986, scientists found areas in the stratosphere with extremely low amounts of ozone. One large hole was found over Antarctica. A smaller hole was discovered over the north pole. **Figure 10** shows how the ozone layer has thinned and developed holes.

In the mid 1990s, many governments banned the production and use of CFCs. Perhaps over time, the areas where the ozone layer is thinning will recover.

Figure 10
These images of Antarctica were produced using data from a NASA satellite. The purple color shows how the ozone hole has grown bigger over time.

Section 1 Assessment

1. Earth's early atmosphere had little oxygen. How did oxygen come to make up 21 percent of Earth's present atmosphere?

2. List the layers of the atmosphere in order beginning at Earth's surface.

3. While hiking in the mountains, you notice that it is harder to breathe as you climb higher. Explain why this is so.

4. What are some effects from a thinning ozone layer?

5. **Think Critically** During the day, the radio only receives AM stations from a city near you. At night, you are able to listen to an AM radio station from a distant city. Explain why this is possible.

Skill Builder Activities

6. **Interpreting Scientific Illustrations** Using **Figure 2,** determine the total percentage of nitrogen and oxygen in the atmosphere. What is the total percentage of argon and carbon dioxide? **For more help, refer to the** Science Skill Handbook.

7. **Communicating** The names of the atmospheric layers end with the suffix -*sphere* a word that means "ball." Use a dictionary to find out what *tropo-, meso-, thermo-,* and *exo-* mean. In your Science Journal, write the meaning of these prefixes and explain if the layers are appropriately named. **For more help, refer to the** Science Skill Handbook.

Activity

Evaluating Sunscreens

Without protection, sun exposure can damage your health. Sunscreens protect your skin from ultraviolet radiation. In this activity, you will draw inferences using the labels of different sunscreens.

What You'll Investigate
How effective are various brands of sunscreens?

Materials
variety of sunscreens of different brand names

Goals
- **Draw inferences** based on labels on sunscreen brands.
- **Compare** the effectiveness of different sunscreen brands for protection against the Sun.
- **Compare** the cost of several sunscreen brands.

Safety Precautions

Sunscreen Assessment			
Brand Name			
SPF			
Cost per Milliliter			
Misleading Terms			

Procedure

1. Make a data table in your Science Journal using the following terms: *brand name, SPF, misleading terms,* and *cost per milliliter.*

2. The Sun Protection Factor (SPF) tells you how long the sunscreen will protect you. For example, an SPF of 4 allows you to stay in the Sun four times longer than if you did not use sunscreen. Record the SPF of each sunscreen on your data table.

3. **Calculate** the cost per milliliter of each sunscreen brand.

4. Government guidelines say that terms like *sunblock* and *waterproof* are misleading because sunscreens cannot block the Sun, and they wash off in water. List the misleading terms in your data table for each brand.

Conclude and Apply

1. **Explain** why you need to use sunscreen.

2. A minimum of SPF 15 is considered adequate protection for a sunscreen. Sunscreens with an SPF greater than 30 are considered by government guidelines to be misleading because sunscreens will wash or wear off. Evaluate the SPF of each brand of sunscreen.

3. Considering the cost and effectiveness of all the sunscreen brands, discuss which brand you consider to be the best buy.

Communicating Your Data

Create a poster on the proper use of sunscreens, and provide guidelines for selecting the safest product. **For more help, refer to the** Science Skill Handbook.

Energy Transfer in the Atmosphere

Energy from the Sun

The Sun provides most of the energy on Earth. This energy drives winds and ocean currents and allows plants to grow and produce food, providing nutrition for many animals. When Earth receives energy from the Sun, three different things can happen to that energy, as shown in **Figure 11.** Some energy is reflected back into space by clouds, atmospheric particles, and Earth's surface. Some is absorbed by the atmosphere. The rest is absorbed by land and water on Earth's surface.

Heat

Heat is energy that flows from an object with a higher temperature to an object with a lower temperature. Energy from the Sun reaches Earth's surface and heats objects such as roads, rocks, and water. Heat then is transferred through the atmosphere in three ways—radiation, conduction, and convection, as shown in **Figure 12.**

As You Read

What You'll Learn

■ **Describe** what happens to the energy Earth receives from the Sun.
■ **Compare and contrast** radiation, conduction, and convection.
■ **Explain** the water cycle.

Vocabulary

radiation   hydrosphere
conduction   condensation
convection

Why It's Important

The Sun provides energy to Earth's atmosphere, allowing life to exist.

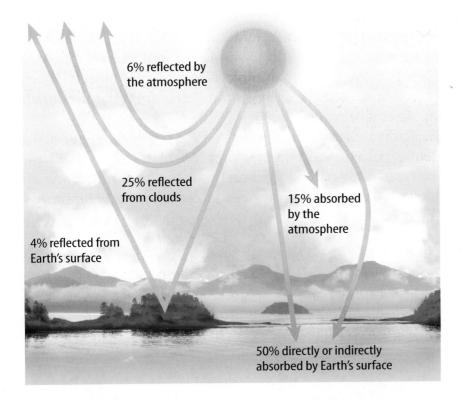

6% reflected by the atmosphere

25% reflected from clouds

15% absorbed by the atmosphere

4% reflected from Earth's surface

50% directly or indirectly absorbed by Earth's surface

Figure 11
The Sun is the source of energy for Earth's atmosphere. Thirty-five percent of incoming solar radiation is reflected back into space. *How much is absorbed by Earth's surface and atmosphere?*

443

Radiation warms the surface.

The air near Earth's surface is heated by conduction.

Cooler air pushes warm air upward, creating a convection current.

Figure 12
Heat is transferred within Earth's atmosphere by radiation, conduction, and convection.

Physics
INTEGRATION

Specific heat is the amount of heat required to change the temperature of a substance one degree. Substances with high specific heat absorb a lot of heat for a small increase in temperature. Land warms faster than water does. Infer whether soil or water has a higher specific heat value.

Radiation Sitting on the beach, you feel the Sun's warmth on your face. How can you feel the Sun's heat even though you aren't in direct contact with it? Energy from the Sun reaches Earth in the form of radiant energy, or radiation. **Radiation** is energy that is transferred in the form of rays or waves. Earth radiates some of the energy it absorbs from the Sun back toward space. Radiant energy from the Sun warms your face.

✔ Reading Check *How does the Sun warm your skin?*

Conduction If you walk barefoot on a hot beach, your feet heat up because of conduction. **Conduction** is the transfer of energy that occurs when molecules bump into one another. Molecules are always in motion, but molecules in warmer objects move faster than molecules in cooler objects. When objects are in contact, energy is transferred from warmer objects to cooler objects.

Radiation from the Sun heated the beach sand, but direct contact with the sand warmed your feet. In a similar way, Earth's surface conducts energy directly to the atmosphere. As air moves over warm land or water, molecules in air are heated by direct contact.

Convection After the atmosphere is warmed by radiation or conduction, the heat is transferred by a third process called convection. **Convection** is the transfer of heat by the flow of material. Convection circulates heat throughout the atmosphere. How does this happen?

When air is warmed, the molecules in it move apart and the air becomes less dense. Air pressure decreases because fewer molecules are in the same space. In cold air, molecules move closer together. The air becomes more dense and air pressure increases. Cooler, denser air sinks while warmer, less dense air rises, forming a convection current. As **Figure 12** shows, radiation, conduction, and convection together distribute the Sun's heat throughout Earth's atmosphere.

The Water Cycle

Hydrosphere is a term that describes all the water on Earth's surface. Water moves constantly between the atmosphere and the hydrosphere in the water cycle, shown in **Figure 13.**

If you watch a puddle in the Sun, you'll notice that over time the puddle gets smaller and smaller. Energy from the Sun causes the water in the puddle to change from a liquid to a gas by a process called evaporation. Water that evaporates from lakes, streams, and oceans enters Earth's atmosphere.

If water vapor in the atmosphere cools enough, it changes back into a liquid. This process of water vapor changing to a liquid is called **condensation.**

Clouds form when condensation occurs high in the atmosphere. Clouds are made up of tiny water droplets that can collide to form larger drops. As the drops grow, they fall to Earth as precipitation, which completes the cycle by returning water to the hydrosphere.

TRY AT HOME
Mini LAB

Modeling Heat Transfer

Procedure
1. Cover the outside of an empty **soup can** with **black construction paper.**
2. Fill the can with **cold water** and feel it with your fingers.
3. Place the can in sunlight for 1 h then pour the water over your fingers.

Analysis
1. Does the water in the can feel warmer or cooler after placing the can in sunlight?
2. What types of heat transfer did you model?

Figure 13
In the water cycle, water moves from Earth to the atmosphere and back to Earth again.

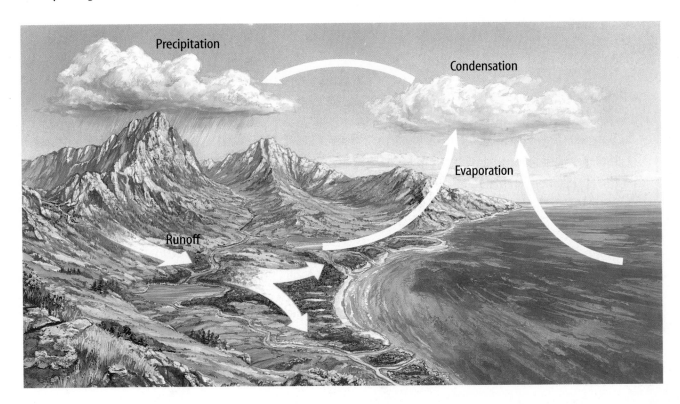

Precipitation

Condensation

Evaporation

Runoff

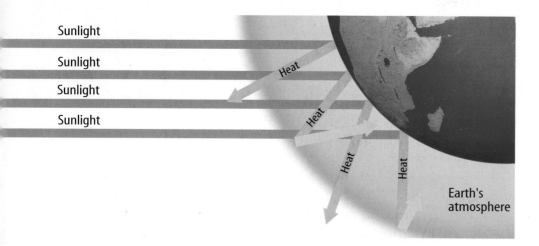

Sunlight

Sunlight

Sunlight

Sunlight

Heat

Heat

Heat

Heat

Earth's
atmosphere

Earth's Atmosphere is Unique

On Earth, radiation from the Sun can be reflected into space, absorbed by the atmosphere, or absorbed by land and water. Once it is absorbed, heat can be transferred by radiation, conduction, or convection. Earth's atmosphere, shown in **Figure 14,** helps control how much of the Sun's radiation is absorbed or lost.

Figure 14
Earth's atmosphere creates a delicate balance between energy received and energy lost.

 Reading Check *What helps control how much of the Sun's radiation is absorbed on Earth?*

Why doesn't life exist on Mars or Venus? Mars is a cold, lifeless world because its atmosphere is too thin to support life or to hold much of the Sun's heat. Temperatures on the surface of Mars range from 35°C to −170°C. On the other hand, gases in Venus's dense atmosphere trap heat coming from the Sun. The temperature on the surface of Venus is 470°C. Living things would burn instantly if they were placed on Venus's surface. Life on Earth exists because the atmosphere holds just the right amount of the Sun's energy.

Section 2 Assessment

1. How does the Sun transfer energy to Earth?
2. How is Earth's atmosphere different from the atmosphere on Mars?
3. How is heat transferred from the stove to the water when you boil a pot of water?
4. Briefly describe the steps included in the water cycle.
5. **Think Critically** What would happen to temperatures on Earth if the Sun's heat were not distributed throughout the atmosphere?

Skill Builder Activities

6. **Concept Mapping** Make a concept map that explains what happens to radiant energy that reaches Earth. **For more help, refer to the** Science Skill Handbook.

7. **Solving One-Step Equations** Earth is about 150 million km from the Sun. The radiation coming from the Sun travels at 300,000 km/s. How long does it take for radiation from the Sun to reach Earth? **For more help, refer to the** Math Skill Handbook.

Air Movement

Forming Wind

Uneven heating of Earth's surface by the Sun causes some areas to be warmer than others. Recall from Section 2 that warmer air expands, becoming less dense than colder air. This causes air pressure to be generally lower where air is heated. Wind is the movement of air from an area of higher pressure to an area of lower pressure.

Heated Air Areas of Earth receive different amounts of radiation from the Sun because Earth is curved. **Figure 15** illustrates why the equator receives more radiation than areas to the north or south. The heated air at the equator is less dense, so it is displaced by denser, colder air, creating convection currents.

This cold, denser air comes from the poles, which receive less radiation from the Sun, making air at the poles much cooler. The resulting dense, high-pressure air sinks and moves along Earth's surface. However, dense air sinking as less-dense air rises does not explain everything about wind.

As You Read

What **You'll Learn**

- **Explain** why different latitudes on Earth receive different amounts of solar energy.
- **Describe** the Coriolis effect.
- **Locate** doldrums, trade winds, prevailing westerlies, polar easterlies, and jet streams.

Vocabulary

Coriolis effect   sea breeze
jet stream   land breeze

Why **It's Important**

Wind systems determine major weather patterns on Earth.

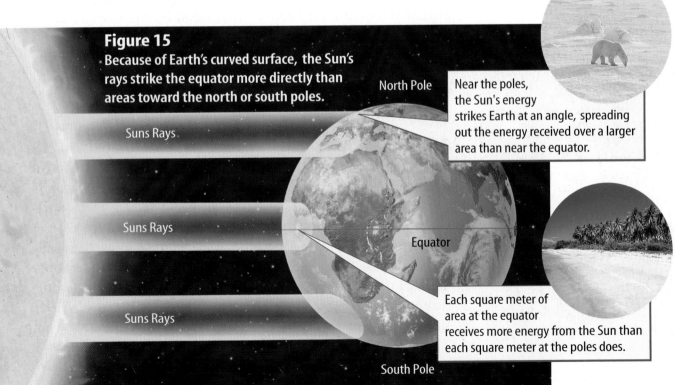

Figure 15
Because of Earth's curved surface, the Sun's rays strike the equator more directly than areas toward the north or south poles.

Suns Rays

Suns Rays

Suns Rays

North Pole

Equator

South Pole

Near the poles, the Sun's energy strikes Earth at an angle, spreading out the energy received over a larger area than near the equator.

Each square meter of area at the equator receives more energy from the Sun than each square meter at the poles does.

Figure 16
The Coriolis effect causes moving air to turn to the right in the northern hemisphere and to the left in the southern hemisphere.

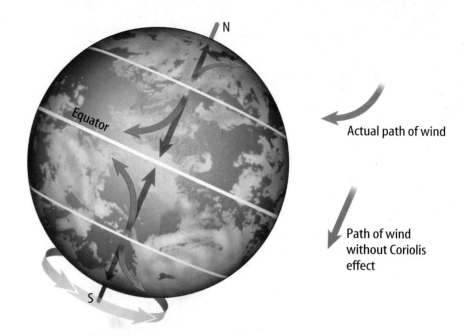

Actual path of wind

Path of wind without Coriolis effect

The Coriolis Effect What would happen if you threw a ball to someone sitting directly across from you on a moving merry-go-round? Would the ball go to your friend? By the time the ball got to the opposite side, your friend would have moved and the ball would appear to have curved.

Like the merry-go-round, the rotation of Earth causes moving air and water to appear to turn to the right north of the equator and to the left south of the equator. This is called the **Coriolis** (kohr ee OH lus) **effect.** It is illustrated in **Figure 16.** The flow of air caused by differences in the amount of solar radiation received on Earth's surface and by the Coriolis effect creates distinct wind patterns on Earth's surface. These wind systems not only influence the weather, they also determine when and where ships and planes travel most efficiently.

Global Winds

How did Christopher Columbus get from Spain to the Americas? The *Nina,* the *Pinta,* and the *Santa Maria* had no source of power other than the wind in their sails. Early sailors discovered that the wind patterns on Earth helped them navigate the oceans. These wind systems are shown in **Figure 17.**

Sometimes sailors found little or no wind to move their sailing ships near the equator. It also rained nearly every afternoon. This windless, rainy zone near the equator is called the doldrums. Look again at **Figure 17.** Near the equator, the Sun heats the air and causes it to rise, creating low pressure and little wind. The rising air then cools, causing rain.

Research Visit the Glencoe Science Web site at **science.glencoe.com** to learn more about global winds. Communicate to your class what you've learned.

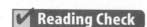 **Reading Check** *What are the doldrums?*

VISUALIZING GLOBAL WINDS

Figure 17

The Sun's uneven heating of Earth's surface forms giant loops, or cells, of moving air. The Coriolis effect deflects the surface winds to the west or east, setting up belts of prevailing winds that distribute heat and moisture around the globe.

A WESTERLIES Near 30° north and south latitude, Earth's rotation deflects air from west to east as air moves toward the polar regions. In the United States, the westerlies move weather systems, such as this one along the Oklahoma-Texas border, from west to east.

B DOLDRUMS Along the equator, heating causes air to expand, creating a zone of low pressure. Cloudy, rainy weather, as shown here, develops almost every afternoon.

C TRADE WINDS Air warmed near the equator travels toward the poles but gradually cools and sinks. As the air flows back toward the low pressure of the doldrums, the Coriolis effect

deflects the surface wind to the west. Early sailors, in ships like the one above, relied on these winds to navigate global trade routes.

D POLAR EASTERLIES In the polar regions, cold, dense air sinks and moves away from the poles. Earth's rotation deflects this wind from east to west.

60° N — Polar easterlies

Westerlies

30° N —

Trade winds

0° — Equatorial doldrums

Trade winds

30° S —

Westerlies

60°S —

Polar easterlies

Surface Winds Air descending to Earth's surface near 30° north and south latitude creates steady winds that blow in tropical regions. These are called trade winds because early sailors used their dependability to establish trade routes.

Between 30° and 60° latitude, winds called the prevailing westerlies blow in the opposite direction from the trade winds. Prevailing westerlies are responsible for much of the movement of weather across North America.

Polar easterlies are found near the poles. Near the north pole, easterlies blow from northeast to southwest. Near the south pole, polar easterlies blow from the southeast to the northwest.

Winds in the Upper Troposphere Narrow belts of strong winds, called **jet streams,** blow near the top of the troposphere. The polar jet stream forms at the boundary of cold, dry polar air to the north and warmer, more moist tropical air to the south, as shown in **Figure 18.** The jet stream moves faster in the winter because the difference between cold air and warm air is greater. The jet stream helps move storms across the country.

Jet pilots take advantage of the jet streams. When flying eastward, planes save time and fuel. Going west, planes fly at different altitudes to avoid the jet streams.

Local Wind Systems

Global wind systems determine the major weather patterns for the entire planet. Smaller wind systems affect local weather. If you live near a large body of water, you're familiar with two such wind systems—sea breezes and land breezes.

Figure 18
A strong current of air, called the jet stream, forms between cold, polar air and warm, tropical air.

A Flying from Boston to Seattle may take 30 min longer than flying from Seattle to Boston.

B The polar jet stream in North America usually is found between 10 km and 15 km above Earth's surface.

Cold air

Polar jet stream

Warm air

A

Warm air

Cool air

Sea breeze

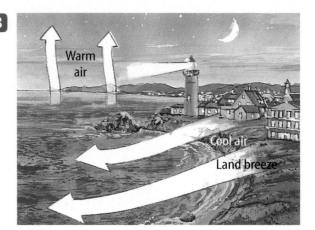

B

Warm air

Cool air

Land breeze

Sea Breezes Convection currents over areas where the land meets the sea can cause wind. A **sea breeze,** shown in **Figure 19,** is created during the day because solar radiation warms the land more than the water. Air over the land is heated by conduction. This heated air is less dense and has lower pressure. Cooler, denser air over the water has higher pressure and flows toward the warmer, less dense air. A convection current results, and wind blows from the sea toward the land.

 Reading Check *How does a sea breeze form?*

Land Breezes At night, land cools much more rapidly than ocean water. Air over the land becomes cooler than air over the ocean. Cooler, denser air above the land moves over the water, as the warm air over the water rises. Movement of air toward the water from the land is called a **land breeze.**

Figure 19
These daily winds occur because land heats up and cools off faster than water does. **A** During the day, cool air from the water moves over the land, creating a sea breeze. **B** At night, cool air over the land moves toward the warmer air over the water, creating a land breeze.

Section 3 Assessment

1. Why do some parts of Earth's surface, such as the equator, receive more of the Sun's heat than other regions?
2. How does the Coriolis effect influence wind circulation on Earth?
3. Why does little wind and lots of afternoon rain occur in the doldrums?
4. Which wind system helped early sailors navigate Earth's oceans?
5. **Think Critically** How does the jet stream help move storms across North America?

Skill Builder Activities

6. **Comparing and Contrasting** Compare and contrast sea breezes and land breezes. **For more help,** refer to the Science Skill Handbook.
7. **Using Graphics Software** Use graphics software and **Figure 17** to draw the wind systems on Earth. Make a separate graphic of major wind circulation cells shown by black arrows. On another graphic, show major surface winds. Print your graphics and share them with your class. **For more help,** refer to the Technology Skill Handbook.

The Heat Is On

Sometimes, a plunge in a pool or lake on a hot summer day feels cool and refreshing. Why does the beach sand get so hot when the water remains cool? A few hours later, the water feels warmer than the land does. In this activity, you'll explore how water and land absorb heat.

Recognize the Problem

How do soil and water compare in their abilities to absorb and emit heat?

Form a Hypothesis

Form a hypothesis about how soil and water compare in their abilities to absorb and release heat. Write another hypothesis about how air temperatures above soil and above water differ during the day and night.

Safety Precautions

WARNING: *Be careful when handling the hot overhead light. Do not let the light or its cord make contact with water.*

Possible Materials

ring stand	clear plastic boxes (2)
soil	overhead light
metric ruler	with reflector
water	thermometers (4)
masking tape	colored pencils (4)

Goals

■ **Design** an experiment to compare heat absorption and release for soil and water.

■ **Observe** how heat release affects the air above soil and above water.

Test Your Hypothesis

Plan

1. As a group, agree upon and write your hypothesis.

2. **List** the steps that you need to take to test your hypothesis. Include in your plan a description of how you will use your equipment to compare heat absorption and release for water and soil.

3. **Design** a data table in your Science Journal for both parts of your experiment—when the light is on and energy can be absorbed and when the light is off and energy is released to the environment.

Do

1. Make sure your teacher approves your plan and your data table before you start.

2. Carry out the experiment as planned.

3. During the experiment, record your observations and complete the data table in your Science Journal.

4. Include the temperatures of the soil and the water in your measurements. Also compare heat release for water and soil. Include the temperatures of the air immediately above both of the substances. Allow 15 min for each test.

Analyze Your Data

1. Use your colored pencils and the information in your data tables to make line graphs. Show the rate of temperature increase for soil and water. Graph the rate of temperature decrease for soil and water after you turn the light off.

2. **Analyze** your graphs. When the light was on, which heated up faster—the soil or the water?

3. **Compare** how fast the air temperature over the water changed with how fast the temperature over the land changed after the light was turned off.

Draw Conclusions

1. Were your hypotheses supported or not? Explain.

2. **Infer** from your graphs which cooled faster—the water or the soil.

3. **Compare** the temperatures of the air above the water and above the soil 15 minutes after the light was turned off. How do water and soil compare in their abilities to absorb and release heat?

𝒞ommunicating
Your Data

Make a poster showing the steps you followed for your experiment. Include graphs of your data. **Display** your poster in the classroom. **For more help, refer to the Science Skill Handbook.**

Song of the Sky Loom[1]
Brian Swann, ed.

This Native American prayer probably comes from the Tewa-speaking Pueblo village of San Juan, New Mexico. The poem is actually a chanted prayer used in ceremonial rituals.

Mother Earth   Father Sky

    we are your children

With tired backs we bring you gifts you love

Then weave for us a garment of brightness
    its warp[2] the white light of morning,
      weft[3] the red light of evening,
        fringes the falling rain,
      its border the standing rainbow.

Thus weave for us a garment of brightness
So we may walk fittingly where birds sing,
So we may walk fittingly where grass is green.

Mother Earth   Father Sky

Respond to the Reading

1. Why do the words *Mother Earth* and *Father Sky* appear on either side and above and below the rest of the words?

2. Why does the song use the image of a garment to describe Earth's atmosphere?

[1] a machine or device from which cloth is produced

[2] threads that run lengthwise in a piece of cloth

[3] horizontal threads interlaced through the warp in a piece of cloth

Understanding Literature

Metaphor A metaphor is a figure of speech that compares seemingly unlike things. Unlike a simile, a metaphor does not use the connecting words *like* or *as*. For instance, in the song you just read, Father Sky is a loom. A loom is a machine or device that weaves cloth. The song describes the relationship between Earth and sky as being a woven garment. Lines such as "weave for us a garment of brightness" serve as metaphors for how Mother Earth and Father Sky together create an atmosphere in which their "children," or humans, can thrive.

Science Connection In this chapter, you learned about the composition of Earth's atmosphere. The atmosphere maintains the proper balance between the amount of heat absorbed from the Sun and the amount of heat that escapes back into space. You also learned about the water cycle and how water evaporates from Earth's surface back into the atmosphere. Using metaphor instead of scientific facts, the Tewa song conveys to the reader how the relationship between Earth and its atmosphere is important to all living things.

Linking Science and Writing

Creating a Metaphor Write a four-line poem that uses a metaphor to describe rain. You can choose to write about a gentle spring rain or a thunderous rainstorm. Remember that a metaphor does not use the words *like* or *as*. Therefore, your poem should begin with something like "Rain is …" or "Heavy rain is …"

Career Connection

Meteorologist

Kim Perez is an on-air meteorologist for The Weather Channel, a national cable television network. She became interested in the weather when she was living in Cincinnati, Ohio. There, in 1974, she witnessed the largest tornado on record. Ms. Perez now broadcasts weather reports to millions of television viewers. Meteorologists study computer models of Earth's atmosphere. These models help them predict short-term and long-term weather conditions for the United States and the world.

SCIENCE *Online* To learn more about careers in meteorology, visit the Glencoe Science Web site at **science.glencoe.com.**

Reviewing Main Ideas

Section 1 Earth's Atmosphere

1. Earth's atmosphere is made up mostly of gases, with some suspended solids and liquids. The unique atmosphere allows life on Earth to exist.

2. The atmosphere is divided into five layers with different characteristics.

3. The ozone layer protects Earth from too much ultraviolet radiation, which can be harmful. *How do chlorofluorocarbon molecules destroy ozone?*

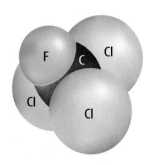

Section 2 Energy Transfer in the Atmosphere

1. Earth receives its energy from the Sun. Some of this energy is reflected back into space, and some is absorbed.

2. Heat is distributed in Earth's atmosphere by radiation, conduction, and convection.

3. Energy from the Sun powers the water cycle between the atmosphere and Earth's surface. *Clouds form during which part of the water cycle?*

4. Unlike the atmosphere on Mars or Venus, Earth's unique atmosphere maintains a balance between energy received and energy lost that keeps temperatures mild. This delicate balance allows life on Earth to exist.

Section 3 Air Movement

1. Because Earth's surface is curved, not all areas receive the same amount of solar radiation. This uneven heating causes temperature differences at Earth's surface.

2. Convection currents modified by the Coriolis effect produce Earth's global winds.

3. The polar jet stream is a strong current of wind found in the upper troposphere. It forms at the boundary between cold, polar air and warm, tropical air.

4. Land breezes and sea breezes occur near the ocean. *Why do winds change direction from day to night?*

After You Read

FOLDABLES Reading & Study Skills

Draw pictures on the front of your Foldable of things that you might find in each layer of Earth's atmosphere.

Visualizing Main Ideas

Complete the following cycle map on the water cycle.

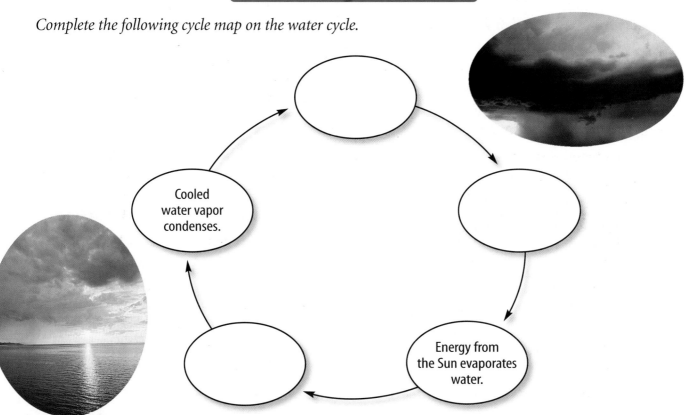

Cooled water vapor condenses.

Energy from the Sun evaporates water.

Vocabulary Review

Vocabulary Review

a. atmosphere
b. chlorofluorocarbon
c. condensation
d. conduction
e. convection
f. Coriolis effect
g. hydrosphere
h. ionosphere
i. jet stream
j. land breeze
k. ozone layer
l. radiation
m. sea breeze
n. troposphere
o. ultraviolet radiation

THE PRINCETON REVIEW Study Tip

Describe ways that you might design an experiment to prove scientific principles.

Using Vocabulary

The sentences below include terms that have been used incorrectly. Change the incorrect terms so that the sentence reads correctly.

1. Chlorofluorocarbons are dangerous because they destroy the hydrosphere.

2. Narrow belts of strong winds called sea breezes blow near the top of the troposphere.

3. The thin layer of air that surrounds Earth is called the troposphere.

4. Heat energy transferred in the form of waves is called condensation.

5. The ozone layer helps protect us from the Coriolis effect.

Chapter **15** Assessment

Checking Concepts

Choose the word or phrase that best answers the question.

1. What is the most abundant gas in the atmosphere?
   A) oxygen   **C)** argon
   B) water vapor **D)** nitrogen

2. What causes a brown haze near cities?
   A) conduction   **C)** car exhaust
   B) mud   **D)** wind

3. Which is the uppermost layer of the atmosphere?
   A) troposphere **C)** exosphere
   B) stratosphere **D)** thermosphere

4. What layer of the atmosphere has the most water?
   A) troposphere **C)** mesosphere
   B) stratosphere **D)** exosphere

5. What protects living things from too much ultraviolet radiation?
   A) the ozone layer **C)** nitrogen
   B) oxygen   **D)** argon

6. Where is air pressure least?
   A) troposphere **C)** exosphere
   B) stratosphere **D)** thermosphere

7. How is energy transferred when objects are in contact?
   A) trade winds **C)** radiation
   B) convection   **D)** conduction

8. Which surface winds are responsible for most of the weather movement across the United States?
   A) polar easterlies **C)** prevailing westerlies
   B) sea breeze   **D)** trade winds

9. What type of wind is a movement of air toward water?
   A) sea breeze   **C)** land breeze
   B) polar easterlies **D)** trade winds

10. What are narrow belts of strong winds near the top of the troposphere called?
    A) doldrums   **C)** polar easterlies
    B) jet streams **D)** trade winds

Thinking Critically

11. Why are there few or no clouds in the stratosphere?

12. It is thought that life could not have existed on land until the ozone layer formed about 2 billion years ago. Why does life on land require an ozone layer?

13. Why do sea breezes occur during the day but not at night?

14. Describe what happens when water vapor rises and cools.

15. Why does air pressure decrease with an increase in altitude?

Developing Skills

16. **Concept Mapping** Complete the cycle concept map below using the following phrases to explain how air moves to form a convection current: *Cool air moves toward warm air, warm air is lifted and cools,* and *cool air sinks.*

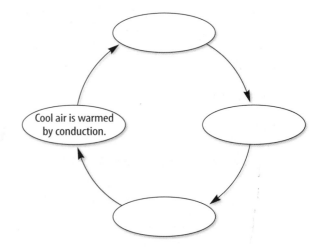

Cool air is warmed by conduction.

17. Drawing Conclusions In an experiment, a student measured the air temperature 1 m above the ground on a sunny afternoon and again in the same spot 1h after sunset. The second reading was lower than the first. What can you infer from this?

18. Forming Hypotheses Carbon dioxide in the atmosphere prevents some radiation from Earth's surface from escaping to space. Hypothesize how the temperature on Earth might change if more carbon dioxide were released from burning fossil fuels.

19. Identifying and Manipulating Variables and Controls Design an experiment to find out how plants are affected by differing amounts of ultraviolet radiation. In the design, use filtering film made for car windows. What is the variable you are testing? What are your constants? Your controls?

20. Recognizing Cause and Effect Why is the inside of a car hotter than the outdoor temperature on a sunny summer day?

Performance Assessment

21. Poster Illustrate or find magazine photos of convection currents that occur in everyday life.

22. Experiment Design and conduct an experiment to find out how different surfaces such as asphalt, soil, sand, and grass absorb and reflect solar energy. Share the results with your class.

TECHNOLOGY

Go to the Glencoe Science Web site at **science.glencoe.com** or use the **Glencoe Science CD-ROM** for additional chapter assessment.

 **Test Practice**

Each layer of Earth's atmosphere has a unique composition and temperature. The four layers closest to Earth's surface are shown in the diagram below.

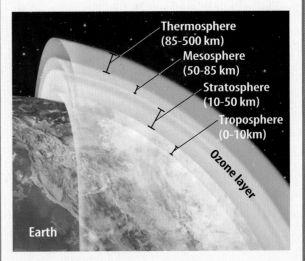

Thermosphere (85-500 km)
Mesosphere (50-85 km)
Stratosphere (10-50 km)
Troposphere (0-10km)
Ozone layer
Earth

Study the diagram and answer the following questions.

1. In which part of the atmosphere is ozone located.
A) Thermosphere   **C)** Stratosphere
B) Troposphere   **D)** Mesosphere

2. According to the diagram, how far does the mesosphere extend above Earth's surface?
F) 10 km   **H)** 50 km
G) 85 km   **I)** 60 km

3. What is the correct order of atmospheric layers that the space shuttle goes through when landing on Earth?
A) Mesosphere   **C)** Stratosphere
    Stratosphere   Troposphere
    Troposphere   Mesosphere
B) Troposphere   **D)** Mesosphere
    Stratosphere   Troposphere
    Mesosphere   Stratosphere

Weather

It's summer and you've gone to your aunt's house in the country. You're playing baseball with your cousins, getting ready to bat, when suddenly you feel a strange sensation. The hot, humid air has suddenly turned cooler, and a strong breeze has kicked up. To the west, tall, black clouds are rapidly advancing. You see a flash of lightning and hear a loud clap of thunder. In this chapter, you'll learn how to measure weather conditions, interpret weather information, and make predictions.

What do you think?

Science Journal Look at the picture below with a classmate. Discuss what this might be or what is happening. Here's a hint: *It's calm in the center and rough around the edges.* Write your answer or best guess in your Science Journal.

EXPLORE ACTIVITY

How can it rain one day and be sunny the next? Powered by heat from the Sun, the air that surrounds you stirs and swirls. This constant mixing produces storms, calm weather, and everything in between. What causes rain and where does the water come from? Do the activity below to find out.

Demonstrate how rain forms

WARNING: *Boiling water and steam can cause burns.*

1. Bring a pan of water to a boil on a hot plate.

2. Carefully hold another pan containing ice cubes about 20 cm above the boiling water. Be sure to keep your hands and face away from the steam.

3. Keep the pan with the ice cubes in place until you see drops of water dripping from the bottom.

Observe

In your Science Journal, describe how the droplets formed. Infer where the water on the bottom of the pan came from.

Before You Read

FOLDABLES
Reading & Study Skills

Making an Organizational Study Fold When information is grouped into clear categories, it is easier to make sense of what you are learning. Make the following Foldable to help you organize your thoughts about weather.

1. Stack two sheets of paper in front of you so the short side of both sheets is at the top.

2. Slide the top sheet up so that about 4 cm of the bottom sheet show.

3. Fold both sheets top to bottom to form four tabs and staple along the top fold, as shown.

4. Label the flaps *Weather, What is Weather?, Weather Patterns,* and *Forecasting Weather,* as shown.

5. As you read the chapter, list what you learn under the appropriate flaps.

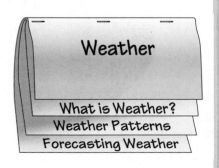

1 What is weather?

What You'll Learn

- **Explain** how solar heating and water vapor in the atmosphere affect weather.
- **Discuss** how clouds form and how they are classified.
- **Describe** how rain, hail, sleet, and snow develop.

Vocabulary

weather
humidity
relative humidity
dew point
fog
precipitation

Why It's Important

Weather changes affect your daily activities.

Weather Factors

It might seem like small talk to you, but for farmers, truck drivers, pilots, and construction workers, the weather can have a huge impact on their livelihoods. Even professional athletes, especially golfers, follow weather patterns closely. You can describe what happens in different kinds of weather, but can you explain how it happens?

Weather refers to the state of the atmosphere at a specific time and place. Weather describes conditions such as air pressure, wind, temperature, and the amount of moisture in the air.

The Sun provides almost all of Earth's energy. Energy from the Sun evaporates water into the atmosphere where it forms clouds. Eventually, the water falls back to Earth as rain or snow. However, the Sun does more than evaporate water. It is also a source of heat energy. Heat from the Sun is absorbed by Earth's surface, which then heats the air above it. Weather, as shown in **Figure 1,** is the result of heat and Earth's air and water.

Figure 1
The Sun provides the energy that drives Earth's weather.
Can you find any storms in this image?

Molecules in air

Wind

Molecules in air

Temperature   Pressure

Temperature   Pressure

A When air is heated, it expands and becomes less dense. This creates lower pressure.

B Molecules making up air are closer together in cooler temperatures, creating high pressure. Wind blows from higher pressure toward lower pressure.

Air Temperature During the summer when the Sun is hot and the air is still, a swim can be refreshing. But would a swim seem refreshing on a cold, winter day? The temperature of air influences your daily activities.

Air is made up of molecules that are always moving randomly, even when there's no wind. Temperature is a measure of the average amount of motion of molecules. When the temperature is high, molecules in air move rapidly and it feels warm. When the temperature is low, molecules in air move less rapidly, and it feels cold.

Wind Why can you fly a kite on some days but not others? Kites fly because air is moving. Air moving in a specific direction is called wind. As the Sun warms the air, the air expands and becomes less dense. Warm, expanding air has low atmospheric pressure. Cooler air is denser and tends to sink, bringing about high atmospheric pressure. Wind results because air moves from regions of high pressure to regions of low pressure. You may have experienced this on a small scale if you've ever spent time along a beach, as in **Figure 2.**

Many instruments are used to measure wind direction and speed. Wind direction can be measured using a wind vane. A wind vane has an arrow that points in the direction from which the wind is blowing. A wind sock has one open end that catches the wind, causing the sock to point in the direction toward which the wind is blowing. Wind speed can be measured using an anemometer (a nuh MAH muh tur). Anemometers have rotating cups that spin faster when the wind is strong.

Figure 2
The temperature of air can affect air pressure. Wind is air moving from high pressure to low pressure.

Life Science
INTEGRATION

Birds and mammals maintain a fairly constant internal temperature, even when the temperature outside their bodies changes. On the other hand, the internal temperature of fish and reptiles changes when the temperature around them changes. Infer from this which group is more likely to survive a quick change in the weather.

Figure 3

Warmer air can have more water vapor than cooler air can because water vapor doesn't easily condense in warm air.

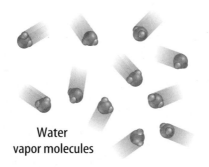

Water vapor molecules

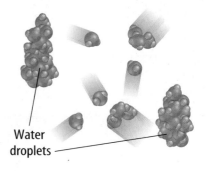

Water droplets

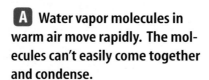

A Water vapor molecules in warm air move rapidly. The molecules can't easily come together and condense.

B As air cools, water molecules in air move closer together. Some of them collide, allowing condensation to take place.

Determining Dew Point

Procedure

1. Partially fill a **metal can** with room-temperature **water.** Dry the outer surface of the can.
2. Place a **stirring rod** in the water.
3. Slowly stir the water and add small amounts of **ice.**
4. On a data table in your **Science Journal,** with a thermometer, note the exact water temperature at which a thin film of moisture first begins to form on the outside of the metal can.
5. Repeat steps 1 through 4 two more times.
6. The average of the three temperatures at which the moisture begins to appear is the dew point temperature of the air surrounding the metal container.

Analysis

1. What determines the dew point temperature?
2. Will the dew point change with increasing temperature if the amount of moisture in the air doesn't change? Explain.

Humidity Heat evaporates water into the atmosphere. Where does the water go? Water vapor molecules fit into spaces among the molecules that make up air. The amount of water vapor present in the air is called **humidity.**

Air doesn't always contain the same amount of water vapor. As you can see in **Figure 3,** more water vapor can be present when the air is warm than when it is cool. At warmer temperatures, the molecules of water vapor in air move quickly and don't easily come together. At cooler temperatures, molecules in air move more slowly. The slower movement allows water vapor molecules to stick together and form droplets of liquid water. The formation of liquid water from water vapor is called condensation. When enough water vapor is present in air for condensation to take place, the air is saturated.

✓ Reading Check *Why can more water vapor be present in warm air than in cold air?*

Relative Humidity On a hot, sticky afternoon, the weather forecaster reports that the humidity is 50 percent. How can the humidity be low when it feels so humid? Weather forecasters report the amount of moisture in the air as relative humidity. **Relative humidity** is a measure of the amount of water vapor present in the air compared to the amount needed for saturation at a specific temperature.

If you hear a weather forecaster say that the relative humidity is 50 percent, it means that the air contains 50 percent of the water needed for the air to be saturated.

As shown in **Figure 4,** air at 25°C is saturated when it contains 22 g of water vapor per cubic meter of air. The relative humidity is 100 percent. If air at 25°C contains 11 g of water vapor per cubic meter, the relative humidity is 50 percent.

Dew Point

When the temperature drops, less water vapor can be present in air. The water vapor in air will condense to a liquid or form ice crystals. The temperature at which air is saturated and condensation forms is the **dew point.** The dew point changes with the amount of water vapor in the air.

You've probably seen water droplets form on the outside of a glass of cold milk. The cold glass cooled the air next to it to its dew point. The water vapor in the surrounding air condensed and formed water droplets on the glass. In a similar way, when air near the ground cools to its dew point, water vapor condenses and forms dew. Frost may form when temperatures are near 0°C.

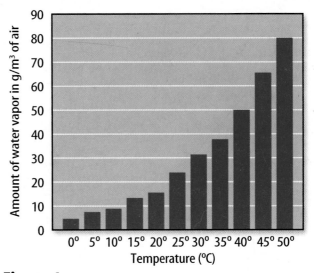

Figure 4
This graph shows that as the temperature of air increases, more water vapor can be present in the air.

Math Skills Activity

Calculating Whether Dew Will Form

Example Problem

One summer day, the relative humidity is 80 percent and the temperature is 35°C. Will the dew point be reached if the temperature falls to 25°C?

Solution

1 *This is what you know:*

From Figure 4

Air Temperature (°C)	Amount of Water Vapor Needed for Saturation (g/m³)
35	37
25	22

2 *This is what you need to find:* x = amount of water vapor in 35°C air at 80 percent relative humidity. Is $x > 22$ g/m³ or is $x < 22$ g/m³?

3 *This is how you solve the problem:* $x = .80 \ (37 \text{ g/m}^3)$
$x = 29.6$ g/m³ of water vapor
29.6 g/m³ $>$ 22 g/m³, so the dew point is reached and dew will form.

Practice Problem

If the relative humidity is 50 percent and the air temperature is 30°C, will the dew point be reached if the temperature falls to 20°C?

What clouds are in the sky today? To find out more about clouds, see the **Clouds Field Guide** at the back of this book.

Forming Clouds

Why are there clouds in the sky? Clouds form as warm air is forced upward, expands, and cools. **Figure 5** shows several ways that warm, moist air forms clouds. As the air cools, the amount of water vapor needed for saturation decreases and the relative humidity increases. When the relative humidity reaches 100 percent, the air is saturated. Water vapor soon begins to condense in tiny droplets around small particles such as dust and salt. These droplets of water are so small that they remain suspended in the air. Billions of these droplets form a cloud.

Classifying Clouds

Clouds are classified mainly by shape and height. Some clouds extend high into the sky, and others are low and flat. Some dense clouds bring rain or snow, while thin, wispy clouds appear on mostly sunny days. The shape and height of clouds vary with temperature, pressure, and the amount of water vapor in the atmosphere.

Figure 5
Clouds form when moist air is lifted and cools. This occurs where air is heated, at mountain ranges, and where cold air meets warm air.

A Rays from the Sun heat the ground and the air next to it. The warm air rises and cools. If the air is moist, some water vapor condenses and forms clouds.

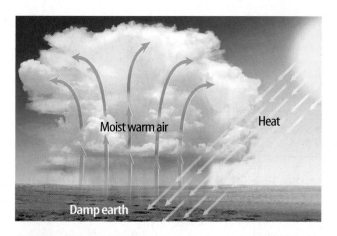

B As moist air moves over mountains, it is lifted and cools. Clouds formed in this way can cover mountains for long periods of time.

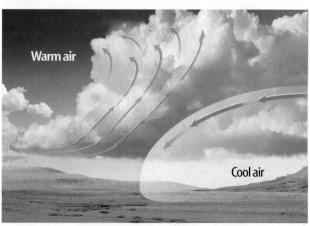

C When cool air meets warm, moist air, the warm air is lifted and cools. *What happens to the water vapor when the dew point is reached?*

Shape The three main cloud types are stratus, cumulus, and cirrus. Stratus clouds form layers, or smooth, even sheets in the sky. Stratus clouds usually form at low altitudes and may be associated with fair weather or rain or snow. When air is cooled to its dew point near the ground, it forms a stratus cloud called **fog**, as shown in **Figure 6.**

Cumulus (KYEW myuh lus) clouds are masses of puffy, white clouds, often with flat bases. They sometimes tower to great heights and can be associated with fair weather or thunderstorms.

Cirrus (SIHR us) clouds appear fibrous or curly. They are high, thin, white, feathery clouds made of ice crystals. Cirrus clouds are associated with fair weather, but they can indicate approaching storms.

Height Some prefixes of cloud names describe the height of the cloud base. The prefix *cirro-* describes high clouds, *alto-* describes middle-elevation clouds, and *strato-* refers to clouds at low elevations. Some clouds' names combine the altitude prefix with the term *stratus* or *cumulus*.

Cirrostratus clouds are high clouds, like those in **Figure 7.** Usually, cirrostratus clouds indicate fair weather, but they also can signal an approaching storm. Altostratus clouds form at middle levels. If the clouds are not too thick, sunlight can filter through them.

Figure 6
Fog surrounds the Golden Gate Bridge, San Francisco. Fog is a stratus cloud near the ground.

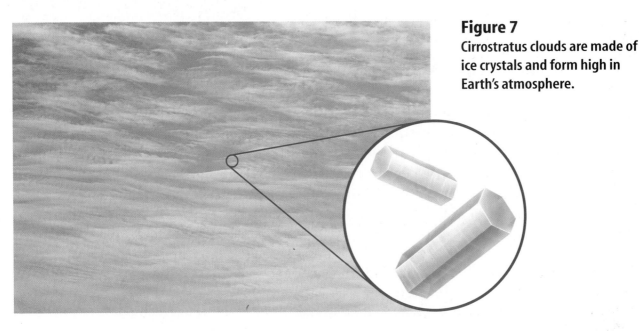

Figure 7
Cirrostratus clouds are made of ice crystals and form high in Earth's atmosphere.

Figure 8
Water vapor in air collects on particles to form water droplets or ice crystals. The type of precipitation that is received on the ground depends on the temperature of the air.

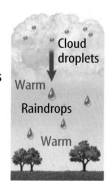

A When the air is warm, water vapor forms raindrops that fall as rain.

B When the air is cold, water vapor forms snowflakes.

Rain- or Snow-Producing Clouds Clouds associated with rain or snow often have the word nimbus attached to them. The term *nimbus* is Latin for "dark rain cloud" and this is a good description, because the water content of these clouds is so high that little sunlight can pass through them. When a cumulus cloud grows into a thunderstorm, it is called a cumulonimbus (kyew myuh loh NIHM bus) cloud. These clouds can tower to nearly 18,000 km. Nimbostratus clouds are layered clouds that can bring long, steady rain or snowfall.

Precipitation

Water falling from clouds is called **precipitation.** Precipitation occurs when cloud droplets combine and grow large enough to fall to Earth. The cloud droplets form around small particles, such as salt and dust. These particles are so small that a puff of smoke can contain millions of them.

You might have noticed that raindrops are not all the same size. The size of raindrops depends on several factors. One factor is the strength of updrafts in a cloud. Strong updrafts can keep drops suspended in the air where they can combine with other drops and grow larger. The rate of evaporation as a drop falls to Earth also can affect its size. If the air is dry, the size of raindrops can be reduced or they can completely evaporate before reaching the ground. Air temperature determines whether water forms rain, snow, sleet, or hail—the four main types of precipitation. **Figure 8** shows these different types of precipitation. Drops of water falling in temperatures above freezing fall as rain. Snow forms when the air temperature is so cold that water vapor changes directly to a solid. Sleet forms when raindrops pass through a layer of freezing air near Earth's surface, forming ice pellets.

✔ **Reading Check** *What are the four main types of precipitation?*

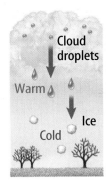

C When the air near the ground is cold, sleet, which is made up of many small ice pellets, falls.

D Hailstones are pellets of ice that form inside a cloud.

Hail Hail is precipitation in the form of lumps of ice. Hail forms in cumulonimbus clouds of a thunderstorm when water freezes in layers around a small nucleus of ice. Hailstones grow larger as they're tossed up and down by rising and falling air. Most hailstones are smaller than 2.5 cm but can grow larger than a softball. Of all forms of precipitation, hail produces the most damage immediately, especially if winds blow during a hailstorm. Falling hailstones can break windows and destroy crops.

If you understand the role of water vapor in the atmosphere, you can begin to understand weather. The relative humidity of the air helps determine whether a location will have a dry day or experience some form of precipitation. The temperature of the atmosphere determines the form of precipitation. Studying clouds can add to your ability to forecast weather.

Section Assessment

1. When does water vapor in air condense?
2. What is the difference between humidity and relative humidity?
3. How do clouds form?
4. How does precipitation occur and what determines the type of precipitation that falls to Earth?
5. **Think Critically** Cumulonimbus clouds form when warm, moist air is suddenly lifted. How can the same cumulonimbus cloud produce rain and hail?

Skill Builder Activities

6. **Concept Mapping** Make a network-tree concept map that compares clouds and their descriptions. Use these terms: *cirrus, cumulus, stratus, feathery, fair weather, puffy, layered, precipitation, clouds, dark,* and *steady precipitation.* **For more help, refer to the** Science Skill Handbook.

7. **Making and Using Graphs** Use **Figure 4** to determine how much water vapor can be present in air when the temperature is 40°C. **For more help, refer to the** Science Skill Handbook.

Weather Patterns

As You Read

What You'll Learn

- **Describe** how weather is associated with fronts and high- and low-pressure areas.
- **Explain** how tornadoes develop from thunderstorms.
- **Discuss** the dangers of severe weather.

Vocabulary

air mass   hurricane
front   blizzard
tornado

Why It's Important

Air masses, pressure systems, and fronts cause weather to change.

Weather Changes

When you leave for school in the morning, the weather might be different from what it is when you head home in the afternoon. Because of the movement of air and moisture in the atmosphere, weather constantly changes.

Air Masses An **air mass** is a large body of air that has properties similar to the part of Earth's surface over which it develops. For example, an air mass that develops over land is dry compared with one that develops over water. An air mass that develops in the tropics is warmer than one that develops over northern regions. An air mass can cover thousands of square kilometers. When you observe a change in the weather from one day to the next, it is due to the movement of air masses. **Figure 9** shows air masses that affect the United States.

Figure 9
Six major air masses affect weather in the United States. Each air mass has the same characteristics of temperature and moisture content as the area over which it formed.

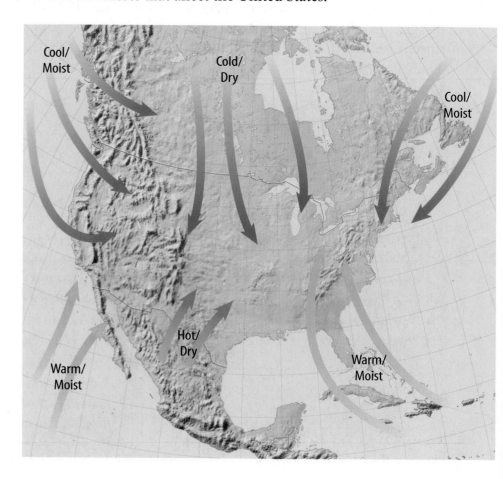

Cool/
Moist

Cold/
Dry

Cool/
Moist

Cool/
Moist

Hot/
Dry

Warm/
Moist

Warm/
Moist

Physics
INTEGRATION

Highs and Lows

Atmospheric pressure varies over Earth's surface. Anyone who has watched a weather report on television has heard about high- and low-pressure systems. Recall that winds blow from areas of high pressure to areas of low pressure. As winds blow into a low-pressure area in the northern hemisphere, Earth's rotation causes these winds to swirl in a counterclockwise direction. Large, swirling areas of low pressure are called cyclones and are associated with stormy weather.

✔ **Reading Check** *How do winds move in a cyclone?*

Winds blow away from a center of high pressure. Earth's rotation causes these winds to spiral clockwise in the northern hemisphere. High-pressure areas are associated with fair weather and are called anticyclones. Air pressure is measured using a barometer, like the one shown in **Figure 10.**

Variation in atmospheric pressure affects the weather. Low pressure systems at Earth's surface are regions of rising air. In Section 1, you learned that clouds form when air is lifted and cools. Areas of low pressure usually have cloudy weather. Sinking motion in high-pressure air masses makes it difficult for air to rise and clouds to form. That's why high pressure usually means good weather.

Figure 10
A barometer measures atmospheric pressure. The red pointer points to the current pressure. Watch how atmospheric pressure changes over time when you line up the white pointer to the one indicating the current pressure each day.

Fronts

A boundary between two air masses of different density, moisture, or temperature is called a **front.** If you've seen a weather map in the newspaper or on the evening news, you've seen fronts represented by various types of curving lines.

Cloudiness, precipitation, and storms sometimes occur at frontal boundaries. Four types of fronts include cold, warm, occluded, and stationary.

Cold and Warm Fronts A cold front, shown on a map as a blue line with triangles, occurs when colder air advances toward warm air. The cold air wedges under the warm air like a plow. As the warm air is lifted, it cools and water vapor condenses, forming clouds. When the temperature difference between the cold and warm air is large, thunderstorms and even tornadoes may form.

Warm fronts form when lighter, warmer air advances over heavier, colder air. A warm front is drawn on weather maps as a red line with red semicircles.

Data Update Visit the Glencoe Science Web site at **science.glencoe.com** to find the current atmospheric pressure, temperature, and wind direction in your town or nearest city. Look up these weather conditions for a city west of your town. Forecast the weather for your town based on your research.

Occluded and Stationary Fronts An occluded front involves three air masses of different temperatures—colder air, cool air, and warm air. An occluded front may form when a cold air mass moves toward cool air with warm air between the two. The colder air forces the warm air upward, closing off the warm air from the surface. Occluded fronts are shown on maps as purple lines with triangles and semicircles.

A stationary front occurs when a boundary between air masses stops advancing. Stationary fronts may remain in the same place for several days, producing light wind and precipitation. A stationary front is drawn on a weather map as an alternating red and blue line. Red semicircles point toward the cold air and blue triangles point toward the warm air. **Figure 11** summarizes the four types of fronts.

Figure 11
Cold, warm, occluded, and stationary fronts occur at the boundaries of air masses. Cloudiness and precipitation occur at front boundaries.

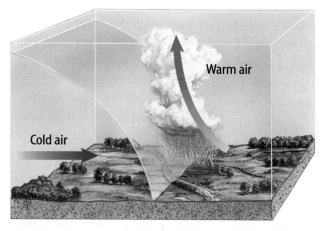

A A cold front can advance rapidly. Thunderstorms often form as warm air is suddenly lifted up over the cold air.

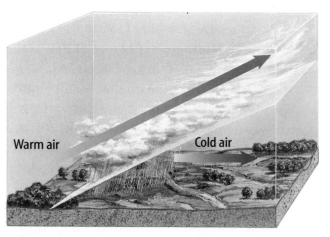

B Warm air slides over colder air along a warm front, forming a boundary with a gentle slope. This can lead to hours, if not days, of wet weather.

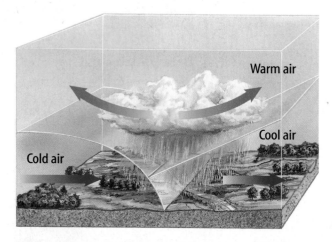

C The term *occlusion* means "closure." Colder air forces warm air upward, forming an occluded front that closes off the warm air from the surface.

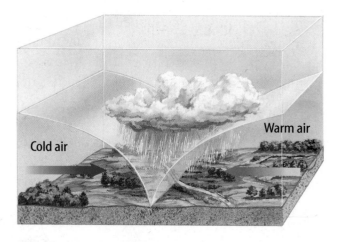

D A stationary front results when neither cold air nor warm air advances.

Severe Weather

Despite the weather, you usually can do your daily activities. If it's raining, you still go to school. You can still get there even if it snows a little. However, some weather conditions, such as those caused by thunderstorms, tornadoes, and blizzards, prevent you from going about your normal routine. Severe weather poses danger to people, structures, and animals.

Thunderstorms In a thunderstorm, heavy rain falls, lightning flashes, thunder roars, and hail might fall. What forces cause such extreme weather conditions? Thunderstorms occur in warm, moist air masses and along fronts. Warm, moist air can be forced upward where it cools and condensation occurs, forming cumulonimbus clouds that can reach heights of 18 km, like the one in **Figure 12.** When rising air cools, water vapor condenses into water droplets or ice crystals. Smaller droplets collide to form larger ones, and the droplets fall through the cloud toward Earth's surface. The falling droplets collide with still more droplets and grow larger. Raindrops cool the air around them. This cool, dense air then sinks and spreads over Earth's surface. Sinking, rain-cooled air and strong updrafts of warmer air cause the strong winds associated with thunderstorms. Hail also may form as ice crystals alternately fall to warmer layers and are lifted into colder layers by the strong updrafts inside cumulonimbus clouds.

Thunderstorm damage Sometimes thunderstorms can stall over a region, causing rain to fall heavily for a period of time. When streams cannot contain all the water running into them, flash flooding can occur. Flash floods can be dangerous because they occur with little warning.

Strong winds generated by thunderstorms also can cause damage. If a thunderstorm is accompanied by winds traveling faster than 89 km/h, it is classified as a severe thunderstorm. Hail from a thunderstorm can dent cars and the aluminum siding on houses. Although rain from thunderstorms helps crops grow, hail has been known to flatten and destroy entire crops in a matter of minutes.

Figure 12
Tall cumulonimbus clouds may form quickly as warm, moist air rapidly rises.

Figure 13
This time-elapsed photo shows a thunderstorm over Arizona.

Lightning and Thunder

What are lightning and thunder? Inside a storm cloud, warm air is lifted rapidly as cooler air sinks. This movement of air can cause different parts of a cloud to become oppositely charged. When current flows between regions of opposite electrical charge, lightning flashes. Lightning, as shown in **Figure 13,** can occur within a cloud, between clouds, or between a cloud and the ground.

Thunder results from the rapid heating of air around a bolt of lightning. Lightning can reach temperatures of about 30,000°C, which is more than five times the temperature of the surface of the Sun. This extreme heat causes air around the lightning to expand rapidly. Then it cools quickly and contracts. The rapid movement of the molecules forms sound waves heard as thunder.

Tornadoes   Some of the most severe thunderstorms produce tornadoes. A **tornado** is a violent, whirling wind that moves in a narrow path over land. In severe thunderstorms, wind at different heights blows in different directions and at different speeds. This difference in wind speed and direction, called wind shear, creates a rotating column parallel to the ground. A thunderstorm's updraft can tilt the rotating column upward into the thunderstorm creating a funnel cloud. If the funnel comes into contact with Earth's surface, it is called a tornado.

✔ **Reading Check**   *What causes a tornado to form?*

A tornado's destructive winds can rip apart buildings and uproot trees. High winds can blow through broken windows. When winds blow inside a house, they can lift off the roof and blow out the walls, making it look as though the building exploded. The updraft in the center of a powerful tornado can lift animals, cars, and even houses into the air. Although tornadoes rarely exceed 200 m in diameter and usually last only a few minutes, they often are extremely destructive. In May 1999, multiple thunderstorms produced more than 70 tornadoes in Kansas, Oklahoma, and Texas. This severe tornado outbreak caused 40 deaths, 100 injuries, and more than $1.2 billion in property damage.

SCIENCE
Online

Research Visit the Glencoe Science Web site at **science.glencoe.com** to research the number of lightning strikes in your state during the last year. Compare your findings with previous years. Communicate to your class what you learn.

NATIONAL GEOGRAPHIC VISUALIZING TORNADOES

Figure 14

Tornadoes are extremely rapid, rotating winds that form at the base of cumulonimbus clouds. Smaller tornadoes may even form inside larger ones. Luckily, most tornadoes remain on the ground for just a few minutes. During that time, however, they can cause considerable—and sometimes strange—damage, such as driving a fork into a tree.

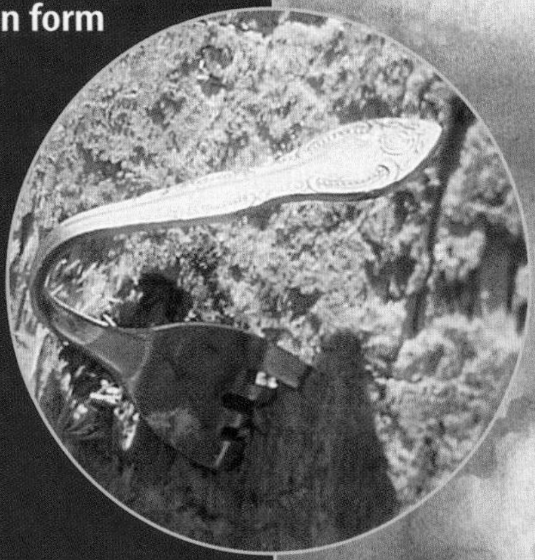

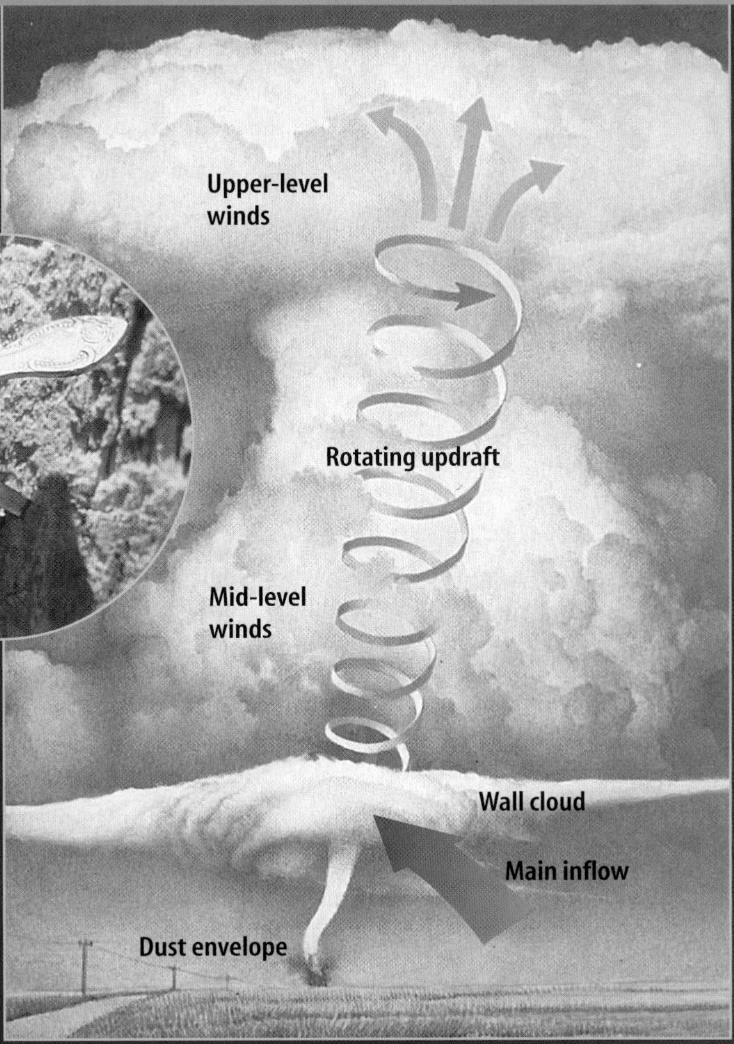

Upper-level winds
Rotating updraft
Mid-level winds
Wall cloud
Main inflow
Dust envelope

Tornadoes often form from a type of cumulonimbus cloud called a wall cloud. Strong, spiraling updrafts of warm, moist air may form in these clouds. As air spins upward, a low-pressure area forms, and the cloud descends to the ground in a funnel. The tornado sucks up debris as it moves along the ground, forming a dust envelope.

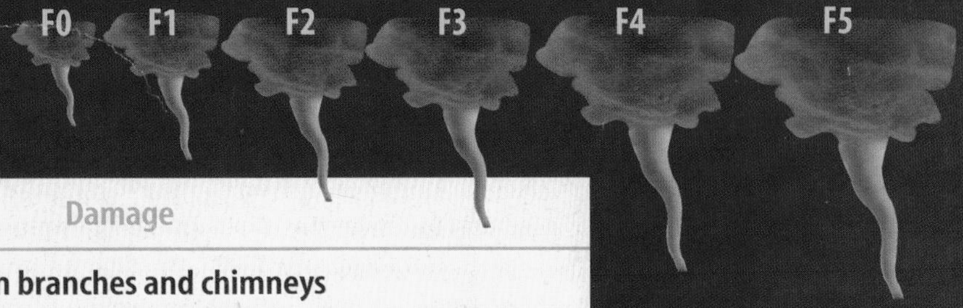

F0
F1
F2
F3
F4
F5

The Fujita Scale

	Wind speed (km/h)	Damage
F0	<116	Light: broken branches and chimneys
F1	116–180	Moderate: roofs damaged, mobile homes upturned
F2	181–253	Considerable: roofs torn off homes, large trees uprooted
F3	254–332	Severe: trains overturned, roofs and walls torn off
F4	333–419	Devastating: houses completely destroyed, cars picked up and carried elsewhere
F5	420–512	Incredible: total demolition

The Fujita scale, named after tornado expert Theodore Fujita, ranks tornadoes according to how much damage they cause. Fortunately, only one percent of tornadoes are classified as violent (F4 and F5).

Figure 15
In this hurricane cross section, the small, red arrows indicate rising, warm, moist air. This air forms cumulus and cumulonimbus clouds in bands around the eye. The green arrows indicate cool, dry air sinking in the eye and between the cloud bands.

Hurricanes The most powerful storm is the hurricane. A **hurricane,** illustrated in **Figure 15,** is a large, swirling, low-pressure system that forms over the warm Atlantic Ocean. It is like a machine that turns heat energy from the ocean into wind. A storm must have winds of at least 119 km/h to be called a hurricane. Similar storms are called typhoons in the Pacific Ocean and cyclones in the Indian Ocean.

Hurricanes are similar to low-pressure systems on land, but they are much stronger. In the Atlantic and Pacific Oceans, low pressure sometimes develops near the equator. In the northern hemisphere, winds around this low pressure begin rotating counterclockwise. The strongest hurricanes affecting North America usually begin as a low-pressure system west of Africa. Steered by surface winds, these storms can travel west, gaining strength from the heat and moisture of warm ocean water.

When a hurricane strikes land, high winds, tornadoes, heavy rains, and high waves can cause a lot of damage. Floods from the heavy rains can cause additional damage. Hurricane weather can destroy crops, demolish buildings, and kill people and other animals. As long as a hurricane is over water, the warm, moist air rises and provides energy for the storm. When a hurricane reaches land, however, its supply of energy disappears and the storm loses power.

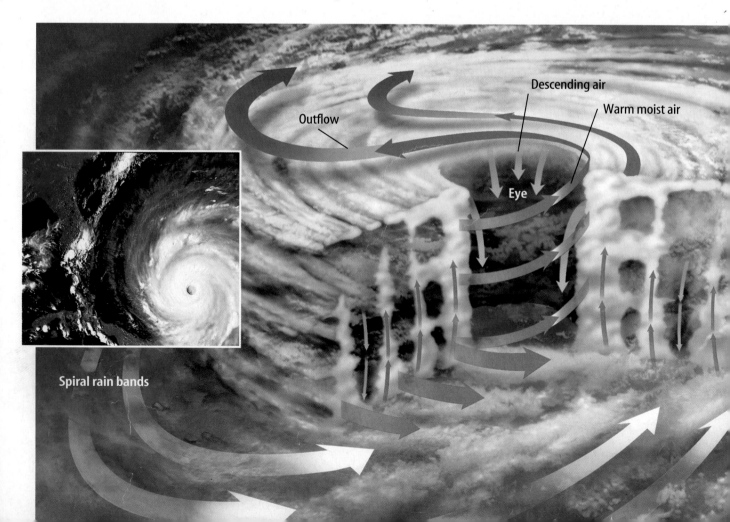

Descending air

Warm moist air

Outflow

Eye

Spiral rain bands

Blizzards Severe storms also can occur in winter. If you live in the northern United States, you may have awakened from a winter night's sleep to a cold, howling wind and blowing snow, like the storm in **Figure 16.** The National Weather Service classifies a winter storm as a **blizzard** if the winds are 56 km/h, the temperature is low, the visibility is less than 400 m in falling or blowing snow, and if these conditions persist for three hours or more.

Severe Weather Safety When severe weather threatens, the National Weather Service issues a watch or warning. Watches are issued when conditions are favorable for severe thunderstorms, tornadoes, floods, blizzards, and hurricanes. During a watch, stay tuned to a radio or television station reporting the weather. When a warning is issued, severe weather conditions already exist. You should take immediate action. During a severe thunderstorm or tornado warning, take shelter in the basement or a room in the middle of the house away from windows. When a hurricane or flood watch is issued, be prepared to leave your home and move farther inland.

Blizzards can be blinding and have dangerously low temperatures with high winds. During a blizzard, stay indoors. Spending too much time outside can result in severe frostbite.

Figure 16
Blizzards can be extremely dangerous because of their high winds, low temperatures, and poor visibility.

Section 2 Assessment

1. Why is fair weather common during periods of high pressure?
2. How does a cold front form? What effect does a cold front have on weather?
3. What causes lightning and thunder in a thunderstorm?
4. What is the difference between a watch and a warning? How can you keep safe during a tornado warning?
5. **Think Critically** Explain why some fronts produce stronger storms than others.

Skill Builder Activities

6. **Recognizing Cause and Effect** Describe how an occluded front may form over your city and what effects it can have on the weather. **For more help, refer to the** Science Skill Handbook.
7. **Using an Electronic Spreadsheet** Make a spreadsheet comparing warm fronts, cold fronts, occluded fronts, and stationary fronts. Indicate what kind of clouds and weather systems form with each. **For more help, refer to the** Technology Skill Handbook.

3 Weather Forecasts

Weather Observations

You can determine current weather conditions by checking the thermometer and looking to see whether clouds are in the sky. You know when it's raining. You have a general idea of the weather because you are familiar with the typical weather where you live. If you live in Florida, you don't expect snow in the forecast. If you live in Maine, you assume it will snow every winter. What weather concerns do you have in your region?

A **meteorologist** (meet ee uh RAH luh jist) is a person who studies the weather. Meteorologists take measurements of temperature, air pressure, winds, humidity, and precipitation. Computers, weather satellites, Doppler radar shown in **Figure 17,** and instruments attached to balloons are used to gather data. Such instruments improve meteorologists' ability to predict the weather. Meteorologists use the information provided by weather instruments to make weather maps. These maps are used to make weather forecasts.

Forecasting Weather Meteorologists gather information about current weather and use computers to make predictions about future weather patterns. Because storms can be dangerous, you do not want to be unprepared for threatening weather. However, meteorologists cannot always predict the weather exactly because conditions can change rapidly.

The National Weather Service depends on two sources for its information—data collected from the upper atmosphere and data collected on Earth's surface. Meteorologists of the National Weather Service collect information recorded by satellites, instruments attached to weather balloons, and from radar. This information is used to describe weather conditions in the atmosphere above Earth's surface.

Figure 17
A meteorologist uses Doppler radar to track a tornado. Since the nineteenth century, technology has greatly improved weather forecasting.

Station Models When meteorologists gather data from Earth's surface, it is recorded on a map using a combination of symbols, forming a **station model.** A station model, like the one in **Figure 18,** shows the weather conditions at a specific location on Earth's surface. Information provided by station models and instruments in the upper atmosphere is entered into computers and used to forecast weather.

Temperature and Pressure In addition to station models, weather maps have lines that connect locations of equal temperature or pressure. A line that connects points of equal temperature is called an **isotherm** (I suh thurm). *Iso* means "same" and *therm* means "temperature." You probably have seen isotherms on weather maps on TV or in the newspaper.

An **isobar** is a line drawn to connect points of equal atmospheric pressure. You can tell how fast wind is blowing in an area by noting how closely isobars are spaced. Isobars that are close together indicate a large pressure difference over a small area. A large pressure difference causes strong winds. Isobars that are spread apart indicate a smaller difference in pressure. Winds in this area are gentler. Isobars also indicate the locations of high- and low-pressure areas.

✔ **Reading Check** *How do isobars indicate wind speed?*

Figure 18
A station model shows the weather conditions at one specific location.

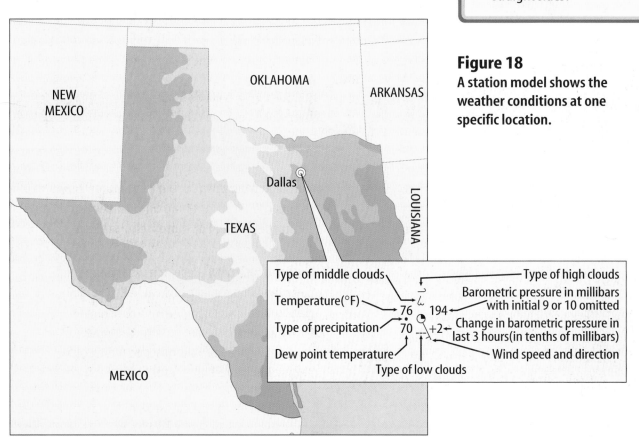

Type of middle clouds
Temperature(°F)
Type of precipitation
Dew point temperature
76
70
194
+2
Type of high clouds
Barometric pressure in millibars with initial 9 or 10 omitted
Change in barometric pressure in last 3 hours(in tenths of millibars)
Wind speed and direction
Type of low clouds

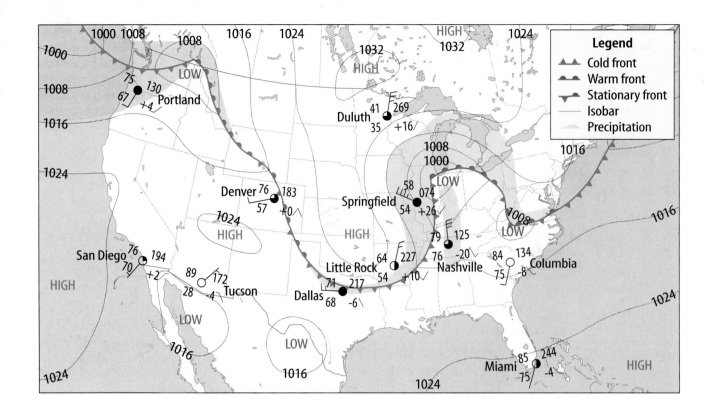

Figure 19
Highs, lows, isobars, and fronts on this weather map help meteorologists forecast the weather.

Weather Maps On a weather map like the one in **Figure 19,** pressure areas are drawn as circles with the word High or Low in the middle of the circle. Fronts are drawn as lines and symbols. When you watch weather forecasts on television, notice how weather fronts move from west to east. This is a pattern that meteorologists depend on to forecast weather.

Section 3 Assessment

1. What instruments do meteorologists use to collect weather data?

2. What is a station model?

3. How does the National Weather Service make weather maps?

4. What do closely spaced isobars on a weather map indicate?

5. **Think Critically** In the morning you hear a meteorologist forecast today's weather as sunny and warm. After school, it is raining. Why is the weather so hard to predict?

Skill Builder Activities

6. **Concept Mapping** Using a computer, make an events chain concept map for how a weather forecast is made. **For more help, refer to the** Science Skill Handbook.

7. **Communicating** Research what happened to American colonial troops at Valley Forge during the winter of 1777–1778. Imagine that you were a soldier during that winter. In your Science Journal, describe your experiences. **For more help, refer to the** Science Skill Handbook.

Activity

Reading a Weather Map

Meteorologists use a series of symbols to provide a picture of local and national weather conditions. With what you know, can you interpret weather information from weather map symbols?

What You'll Investigate
How do you read a weather map?

Materials
hand lens
Weather Map Symbols Appendix
Figure 19

Goals
- ■ **Learn** how to read a weather map.
- ■ **Use** information from a station model and a weather map to forecast weather.

Procedure

Use the information provided in the questions below and the Weather Map Symbols Appendix to learn how to read a weather map.

1. Find the station models on the map for Portland, Oregon, and Miami, Florida. Find the dew point, wind direction, barometric pressure, and temperature at each location.

2. Looking at the placement of the isobars, determine whether the wind would be stronger at Springfield, Illinois, or at San Diego, California. Record your answer. What is another way to determine the wind speed at these locations?

3. **Determine** the type of front near Dallas, Texas. Record your answer.

4. The triangles or half-circles are on the side of the line toward the direction the front is moving. In which direction is the cold front located over Washington state moving?

Conclude and Apply

1. Locate the pressure system over southeast Kansas. Predict what will happen to the weather of Nashville, Tennessee, if this pressure system moves there.

2. Prevailing westerlies are winds responsible for the movement of much of the weather across the United States. Based on this, would you expect Columbia, South Carolina, to continue to have clear skies? Explain.

3. The direction line on the station model indicates the direction from which the wind blows. The wind is named for that direction. Infer from this the name of the wind blowing at Little Rock, Arkansas.

*C*ommunicating
Your Data

Pretend you are a meteorologist for a local TV news station. Make a poster of your weather data and present a weather forecast to your class. **For more help, refer to the** Science Skill Handbook.

Measuring Wind Speed

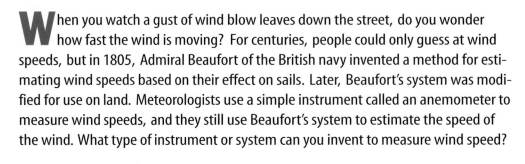

When you watch a gust of wind blow leaves down the street, do you wonder how fast the wind is moving? For centuries, people could only guess at wind speeds, but in 1805, Admiral Beaufort of the British navy invented a method for estimating wind speeds based on their effect on sails. Later, Beaufort's system was modified for use on land. Meteorologists use a simple instrument called an anemometer to measure wind speeds, and they still use Beaufort's system to estimate the speed of the wind. What type of instrument or system can you invent to measure wind speed?

Recognize the Problem

How could you use simple materials to invent an instrument or system for measuring wind speeds?

Thinking Critically

What observations do you use to estimate the speed of the wind?

Goals
- **Invent** an instrument or devise a system for measuring wind speeds using common materials.
- **Devise** a method for using your invention or system to compare different wind speeds.

Possible Materials 🖐
paper
scissors
confetti
grass clippings
meterstick
*measuring tape
Alternate materials

Data Source
Refer to Section 1 for more information about anemometers and other wind speed instruments. Consult the data table for information about Beaufort's wind speed scale.

Planning the Model

1. Scan the list of possible materials and choose the materials you will need to devise your system.

2. **Devise** a system to measure different wind speeds. Be certain the materials you use are light enough to be moved by slight breezes.

Check the Model Plans

1. **Describe** your plan to your teacher. Provide a sketch of your instrument or system and ask your teacher how you might improve its design.

2. Present your idea for measuring wind speed to the class in the form of a diagram or poster. Ask your classmates to suggest improvements in your design that will make your system more accurate or easier to use.

Making the Model

1. Confetti or grass clippings that are all the same size can be used to measure wind speed by dropping them from a specific height. Measuring the distances they travel in different strength winds will provide data for devising a wind speed scale.

2. Different sizes and shapes of paper also could be dropped into the wind, and the strength of the wind would be determined by measuring the distances traveled by these different types of paper.

Beufort's Wind Speed Scale	
Description	Wind Speed (km/h)
calm—smoke drifts up	less than 1
light air— smoke drifts with wind	1–5
light breeze— leaves rustle	6–11
gentle breeze— leaves move constantly	12–19
moderate breeze— branches move	20–29
fresh breeze— small trees sway	30–39
strong breeze— large branches move	40–50
moderate gale— whole trees move	51–61
fresh gale—twigs break	62–74
strong gale— slight damage to houses	75–87
whole gale— much damage to houses	88–101
storm— extensive damage	102–120
hurricane— extreme damage	more than 120

Analyzing and Applying Results

1. **Explain** why it is important for meteorologists to measure wind speeds.

2. **Compare** your results with Beaufort's wind speed scale.

3. **Develop** a scale for your method.

4. **Evaluate** how well your system worked in gentle breezes and strong winds.

5. **Analyze** what problems may exist in the design of your system and suggest steps you could take to improve your design.

*C*ommunicating Your Data

Demonstrate your system for the class. **Compare** your results and measurements with the results of other classmates.

Rain

Y̶ou listen to a meteorologist give the long-term weather forecast. Another week with no rain in sight. As a farmer, you are concerned that your crops are withering in the fields. Home owners' lawns are turning brown. Wildfires are possible. Cattle are starving. And, if farmers' crops die, there could be a shortage of food and prices will go up for consumers.

Cloud seeding is an inexact science

makers

Flares contain chemicals which will seed clouds.

Meanwhile, several states away, another farmer is listening to the weather report calling for another week of rain. Her crops are getting so water soaked that they are beginning to rot.

Weather. Can't scientists find a way to better control it? The answer is...not exactly. Scientists have been experimenting with methods to control our weather since the 1940s. And nothing really works.

Cloud seeding is one such attempt. It uses technology to enhance the natural rainfall process. The idea has been used to create rain where it is needed or to reduce hail damage. Government officials also use cloud seeding or weather modification to try to reduce the force of a severe storm.

Some people seed a cloud by flying a plane above it and releasing highway-type flares with chemicals, such as silver iodide. Another method is to fly beneath the cloud and spray a chemical that can be carried into the cloud by air currents.

Flares are lodged under a plane. The pilot will drop them into potential rain clouds.

Cloud seeding doesn't work with clouds that have little water vapor or are not near the dew point. Seeding chemicals must be released into potential rain clouds. The chemicals provide nuclei for water molecules to cluster around. Water then falls to Earth as precipitation.

Cloud seeding does have its critics. If you seed clouds and cause rain for your area, aren't you preventing rain from falling in another area? Would that be considered "rain theft" by people who live in places where the cloudburst would naturally occur? What about those cloud-seeding agents? Could the cloud-seeding chemicals, such as silver iodide and acetone, affect the environment in a harmful way? Are humans meddling with nature and creating problems in ways that haven't been determined?

Currently, Montana, Pennsylvania, and New Mexico are states that don't allow cloud seeding within their state boundaries. But officials in Texas and California, the two states with the largest number of cloud-seeding programs, feel strongly that cloud seeding is an important technology when it comes to dealing with weather.

CONNECTIONS Debate Learn more about cloud seeding and other methods of changing weather. Then debate whether or not cloud seeding can be considered "rain theft."

SCIENCE *Online*

For more information, visit science.glencoe.com

Reviewing Main Ideas

Section 1   What is weather?

1. Important factors that determine weather include air pressure, wind, temperature, and the amount of moisture in the air.

2. More water vapor can be present in warm air than in cold air. Water vapor condenses when the dew point is reached. Clouds are formed when warm, moist air rises and cools to its dew point.

3. Rain, hail, sleet, and snow are types of precipitation. *What causes hail to form during severe thunderstorms?*

Section 2   Weather Patterns

1. Fronts form when air masses with different characteristics, such as temperature, moisture, or density, meet. Types of fronts include cold fronts, warm fronts, occluded fronts, and stationary fronts.

2. High atmospheric pressure at Earth's surface usually means good weather. Cloudy and stormy weather occurs under low pressure.

3. Tornadoes are intense, whirling windstorms that can result from wind shears inside a thunderstorm.

4. Hurricanes and blizzards are large, severe storms with strong winds. *Why does a hurricane, shown below, lose strength as it moves over land?*

Section 3   Weather Forecasts

1. Meteorologists use information from radar, satellites, computers, and other weather instruments to make weather maps and forecasts.

2. Symbols on a station model indicate the weather at a particular location. *What is the dew point temperature on the station model shown here?*

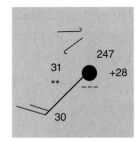

3. Weather maps include information about temperature and air pressure.

FOLDABLES
Reading & Study Skills

After You Read

To help you review facts about weather, use the Foldable you made at the beginning of the chapter.

Visualizing Main Ideas

Complete the following concept map about air temperature, water vapor, and pressure.

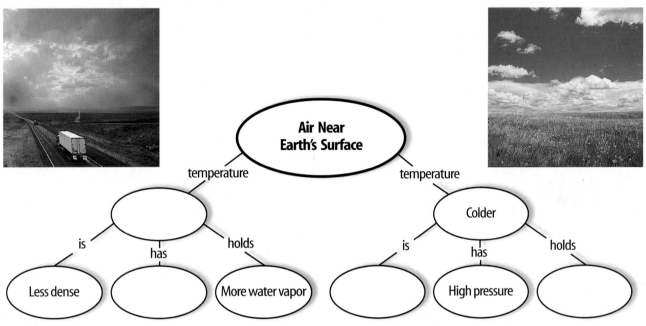

Vocabulary Review

Vocabulary Words

a. air mass
b. blizzard
c. dew point
d. fog
e. front
f. humidity
g. hurricane
h. isobar
i. isotherm
j. meteorologist
k. precipitation
l. relative humidity
m. station model
n. tornado
o. weather

Study Tip

After each day's lesson, make a practice quiz for yourself. Later, when you're studying for the test, take the practice quizzes that you created.

Using Vocabulary

Explain the differences between the vocabulary words in each of the following sets.

1. air mass, front
2. humidity, relative humidity
3. relative humidity, dew point
4. dew point, precipitation
5. hurricane, tornado
6. blizzard, fog
7. meteorologist, station model
8. precipitation, fog
9. isobar, isotherm
10. isobar, front

Checking Concepts

Choose the word or phrase that best answers the question.

1. Which type of air has a relative humidity of 100 percent?
   A) humid   C) dry
   B) temperate   D) saturated

2. What is a large body of air that has the same properties as the area over which it formed called?
   A) air mass   C) front
   B) station model D) isotherm

3. At what temperature does water vapor in air condense?
   A) dew point   C) front
   B) station model D) isobar

4. Which type of precipitation forms when water vapor changes directly into a solid?
   A) rain   C) sleet
   B) hail   D) snow

5. Which type of the following clouds are high feathery clouds made of ice crystals?
   A) cirrus   C) cumulus
   B) nimbus   D) stratus

6. Which type of front may form when cool air, cold air and warm air meet?
   A) warm   C) stationary
   B) cold   D) occluded

7. Which is issued when severe weather conditions exist and immediate action should be taken?
   A) front   C) station model
   B) watch   D) warning

8. Which term means the amount of water vapor in the air?
   A) dew point   C) humidity
   B) precipitation D) relative humidity

9. What does an anemometer measure?
   A) air pressure   C) wind speed
   B) relative humidity D) precipitation

10. What is a large, swirling storm that forms over warm, tropical water called?
    A) hurricane   C) blizzard
    B) tornado   D) hailstorm

Thinking Critically

11. Explain the relationship between temperature and relative humidity.

12. Describe how air, water, and the Sun interact to cause weather.

13. Explain why northwest Washington often has rainy weather and southwest Texas is dry.

14. What does it mean if the relative humidity is 79 percent?

15. Why don't hurricanes form in Earth's polar regions?

Developing Skills

16. **Comparing and Contrasting** Compare and contrast the weather at a cold front to that at a warm front.

17. **Observing and Inferring** You take a hot shower. The mirror in the bathroom fogs up, like the one below. Infer from this information what has happened.

18. **Interpreting Scientific Illustrations** Use the cloud descriptions in the **Cloud Field Guide** at the back of the book to describe the weather at your location today. Then try to predict tomorrow's weather.

19. **Concept Mapping** Complete the sequence map below showing how precipitation forms.

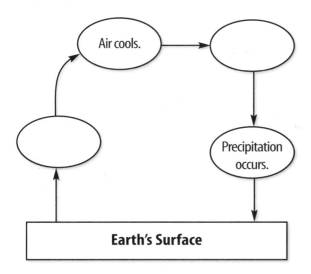

20. **Comparing and Contrasting** Compare and contrast tornadoes and thunderstorms.

Performance Assessment

21. **Board Game** Make a board game using weather terms. You could make cards to advance or retreat a token.

22. **Design** your own weather station. Record temperature, precipitation, and wind speed for one week.

TECHNOLOGY

Go to the Glencoe Science Web site at **science.glencoe.com** or use the **Glencoe Science CD-ROM** for additional chapter assessment.

Test Practice

Hurricanes are rated on a scale based on their wind speed and barometric pressure. The table below lists the hurricane category by the wind speed and pressure of the storm.

Hurricane Rating Scale

Category	Wind Speed (km/h)	Barometric Pressure (millibars)
1	119–154	>980
2	155–178	965–980
3	179–210	945–964
4	211–250	920–944
5	>250	<920

Study the table and answer the following questions.

1. In 1992, Hurricane Andrew, with winds of 233 km/hr and a pressure of 922 mb, struck southeast Florida. What category was Hurricane Andrew?
   A) 1
   B) 2
   C) 3
   D) 4

2. Which of the following best describes the pressure and wind when categorizing a hurricane?
   F) Storm category increases as wind increases and pressure decreases.
   G) Storm category increases as wind decreases and pressure increases.
   H) Storm category increases as wind and pressure increase.
   J) Storm category decreases as wind and pressure decrease.

17 Climate

As summer fades, trees take on the beautiful colors of autumn. As the temperature continues to drop throughout the season, those beautiful leaves will fall, and the bare branches will signify winter. What causes the change in seasons? Why do some places have four distinct seasons, while others have only a wet and dry season? In this chapter, you will learn what climate is and how climates are classified. You also will learn what causes climate changes and how humans and animals adapt to different climates.

What do you think?

Science Journal Look at the picture below with a classmate. Discuss what you think this might be or what is happening. Here's a hint: *These are frozen, but they aren't trees.* Write your best guess in your Science Journal.

EXPLORE ACTIVITY

You wouldn't go to Alaska to swim or to Jamaica to snow ski. You know the climates in these places aren't suited for these sports. In this activity, you'll explore the climates in different parts of the world.

Track the climates of the world

1. Obtain a world atlas, globe, or large classroom map. Select several cities from as many different parts of the world as possible.

2. Record the longitude and latitude of your cities. Note if they are near mountains or an ocean.

3. Research the average temperature of your cities. In what months are they hottest? Coldest? What is the average yearly rainfall? What kinds of plants and animals live in the region? Record your findings.

4. Compare your findings with those of the rest of your class. Can you see any relationship between latitude and climate? Do cities near an ocean or a mountain range have different climatic characteristics?

Observe

As you read the chapter, keep track of the daily weather conditions in your cities. Are these representative of the kind of climates your cities are supposed to have? Suggest reasons why day-to-day weather conditions may vary.

Before You Read

FOLDABLES Reading & Study Skills

Making a Classify Study Fold Make the following Foldable to help you organize objects or events into groups based on their common features.

1. Place a sheet of paper in front of you so the short side is at the top. Fold the paper in half from top to bottom. Then fold it in half again top to bottom two more times. Unfold all the folds.

2. Using the fold lines as a guide, refold the paper into a fan. Unfold all the folds again.

3. Title your Foldable *Climate Classification* and label the sections *Tropical, Mild, Dry, Continental, Polar,* and *High Elevation.*

4. As you read the chapter, define each and write notes on weather patterns on the back of each fold.

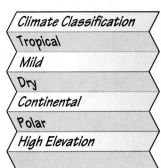

What is climate?

As You Read

What You'll Learn

- **Describe** what determines climate.
- **Explain** how latitude and other factors affect the climate of a region.

Vocabulary

climate
tropics
polar zone
temperate zone

Why It's Important

Climate affects the way you live.

Climate

If you wandered through a tropical rain forest, you would see beautiful plants flowering in shades of pink and purple beneath a canopy of towering trees. A variety of exotic birds and other animals would dart among the tree branches and across the forest floor. The sounds of singing birds and croaking frogs would surround you. All of these organisms thrive in hot temperatures and abundant rainfall. Rain forests have a hot, wet climate. **Climate** is the pattern of weather that occurs in an area over many years. It determines the types of plants or animals that can survive, and it influences how people live.

Climate is determined by averaging the weather of a region over a long period of time, such as 30 years. Scientists average temperature, precipitation, air pressure, humidity, and number of days of sunshine to determine an area's climate. Some factors that affect the climate of a region include latitude, landforms, location of lakes and oceans, and ocean currents.

Latitude and Climate

As you can see in **Figure 1,** regions close to the equator receive the most solar radiation. Latitude, a measure of distance north or south of the equator, affects climate. **Figure 2** compares cities at different latitudes. The **tropics**—the region between latitudes 23.5°N and 23.5°S—receive the most solar radiation because the Sun shines almost directly over these areas. The tropics have temperatures that are always hot, except at high elevations. The **polar zones** extend from 66.5°N and 66.5°S latitude to the poles. Solar radiation hits these zones at a low angle, spreading energy over a large area. During winter, polar regions receive little or no solar radiation. Polar regions are never warm.

✔ Reading Check *How does latitude affect climate?*

Between the tropics and the polar zones are the **temperate zones.** Temperatures here are moderate. Most of the United States is in a temperate zone.

Figure 1
The tropics are warmer than the temperate zones and the polar zones because the tropics receive the most direct solar energy.

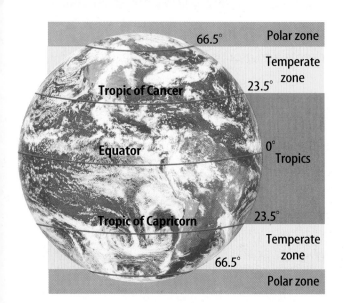

Other Factors

In addition to the general climate divisions of polar, temperate, and tropical, natural features such as large bodies of water, ocean currents, and mountains affect climate within each zone. Large cities also change weather patterns and influence the local climate.

Large Bodies of Water If you live or have vacationed near an ocean, you may have noticed that water heats up and cools down more slowly than land does. This is because it takes a lot more heat to increase the temperature of water than it takes to increase the temperature of land. In addition, water must give up more heat than land does for it to cool. Large bodies of water can affect the climate of coastal areas by absorbing or giving off heat. This causes many coastal regions to be warmer in the winter and cooler in the summer than inland areas at similar latitude. Look at **Figure 2** again. You can see the effect of an ocean on climate by comparing the average temperatures in a coastal city and an inland city, both located at 37°N latitude.

Figure 2
This map shows average daily low temperatures in four cities during January and July. It also shows average yearly precipitation.

Mini LAB

Observing Solar Radiation

Procedure
1. Darken the room.
2. Hold a **flashlight** about 30 cm from a **globe.** Shine the light directly on the equator. With your finger, trace around the light.
3. Now, tilt the flashlight to shine on 30°N latitude. The size of the lighted area should increase. Repeat at 60°N latitude.

Analysis
1. How did the size and shape of the light beam change as you directed the light toward higher latitudes?
2. How does Earth's tilt affect the solar radiation received by different latitudes?

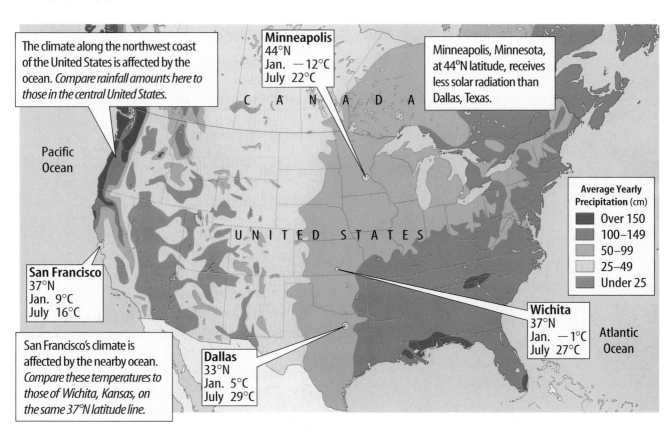

The climate along the northwest coast of the United States is affected by the ocean. *Compare rainfall amounts here to those in the central United States.*

Minneapolis
44°N
Jan. −12°C
July 22°C

Minneapolis, Minnesota, at 44°N latitude, receives less solar radiation than Dallas, Texas.

Pacific Ocean

C A N A D A

U N I T E D   S T A T E S

Average Yearly Precipitation (cm)
- Over 150
- 100–149
- 50–99
- 25–49
- Under 25

San Francisco
37°N
Jan. 9°C
July 16°C

San Francisco's climate is affected by the nearby ocean. *Compare these temperatures to those of Wichita, Kansas, on the same 37°N latitude line.*

Dallas
33°N
Jan. 5°C
July 29°C

Wichita
37°N
Jan. −1°C
July 27°C

Atlantic Ocean

When air rises over a mountain, the air expands and its temperature decreases, causing water vapor to condense and form rain. Temperature changes caused by air expanding or contracting also occur in some machines. Why does the air coming out of a bicycle pump feel cold?

Ocean Currents Ocean currents affect coastal climates. Warm currents begin near the equator and flow toward higher latitudes, warming the land regions they pass. When the currents cool off and flow back toward the equator, they cool the air and climates of nearby land.

✔ **Reading Check** *How do ocean currents affect climate?*

Winds blowing from the sea are often moister than those blowing from land. Therefore, some coastal areas have wetter climates than places farther inland. Look at the northwest coast of the United States shown in **Figure 2.** The large amounts of precipitation in Washington, Oregon, and northern California can be explained by this moist ocean air.

Mountains At the same latitude, the climate is colder in the mountains than at sea level. When radiation from the Sun is absorbed by Earth's surface, it heats the land. Heat from Earth then warms the atmosphere. In the mountains, the air has fewer molecules to absorb this heat. This is because Earth's atmosphere gets thinner at higher altitudes.

Problem-Solving Activity

How do large cities influence temperature?

The temperature of rural areas surrounding a large city can differ from the temperature in the city's downtown by several degrees. This difference in temperature is called the heat-island effect. It is partly a result of the land in cities being covered with buildings, roads, and pavement, while rural areas have far more open space and vegetation. Do you think the heat-island effect can change as a city grows? Use your ability to interpret data to find out.

Identifying the Problem

The table lists the average summer high temperatures in and around a large metropolitan city in 1978 and 1993. By examining the data, can you tell if the heat-island effect has changed over time?

Average Summer Temperatures		
Distance from Downtown (km)	Temperatures (°C)	
	1978	1993
0	32.5	35
8	28.0	32.0
16	27.5	30.5
24	27.5	29.5

Solving the Problem

1. What distances from downtown experienced the greatest and smallest temperature change between 1978 and 1993? Can you account for the differences?
2. Suppose that in 1999 the average temperature in the downtown area was 36°C, but the temperature 24 km away was 33°C. Would you expect this? Explain.

A The windward side of a large mountain range—the side facing the wind—often receives heavy precipitation.

B Deserts are common on the leeward side—the side away from the wind—of mountain ranges.

Rain Shadows Mountains also affect regional climates, as shown in **Figure 3.** On the windward side of a mountain range, air rises, cools, and drops its moisture. On the leeward side of a mountain range air descends, heats up, and dries the land. Deserts are common behind mountains.

Cities Large cities affect local climates. Streets, parking lots, and buildings heat up, in turn heating the air. Air pollution traps this heat, creating what is known as the heat-island effect. Temperatures in a city can be 5°C higher than in surrounding rural areas.

Figure 3
Large mountain ranges can affect climate by forcing air to rise over the windward side and to descend on the leeward side. *How can this cause the two sides of the Sierra Nevada Mountain Range to be so different?*

Section 1 Assessment

1. What factors determine the climate of a region?
2. Explain how two cities located at the same latitude can have different climates.
3. How do mountains affect climate?
4. What is the heat island effect?
5. **Think Critically** Explain why types of plants and animals found at different elevations on a mountain range might differ. Would latitude affect the elevation at which some organisms are found?

Skill Builder Activities

6. **Comparing and Contrasting** Compare and contrast tropical and polar climates. **For more help, refer to the** Science Skill Handbook.
7. **Solving One-Step Equations** The coolest average summer temperature in the United States is 2°C at Barrow, Alaska, and the warmest is 37°C at Death Valley, California. Calculate the range of average summer temperatures in the United States. **For more help, refer to the** Math Skill Handbook.

Climate Types

Figure 4
The type of vegetation in a region depends on the climate.
What do these plants tell you about the climate shown here?

Classifying Climates

What is the climate like where you live? Would you call it generally warm? Usually wet and cold? Or different depending on the time of year? How would you classify the climate in your region? Life is full of familiar classification systems—from musical categories to food groups. Classifications help to organize your thoughts and to make your life more efficient. That's why Earth's climates also are classified and are organized into the various types that exist. Climatologists—people who study climates—usually use a system developed in 1918 by Wladimir Köppen to classify climates. Köppen observed that the types of plants found in a region depended on the climate of the area. **Figure 4** shows one region Köppen might have observed. He classified climates by using the temperature and precipitation of regions that had different plant types.

The climate classification system shown in **Figure 5** separates climates into six groups—tropical, mild, dry, continental, polar, and high elevation. These groups are further separated into types. For example, the dry climate classification is separated into semiarid and arid.

Adaptations

Climates vary around the world, and as Köppen observed, the type of climate that exists in an area determines the vegetation found there. Fir trees aren't found in deserts, nor are cacti found in rain forests. In fact, all organisms are best suited for certain climates. Organisms are adapted to their environment. An **adaptation** is any structure or behavior that helps an organism survive in its environment. Adaptations are inherited. They develop in a population over a long period of time. Once adapted to a particular climate, organisms may not be able to survive in other climates.

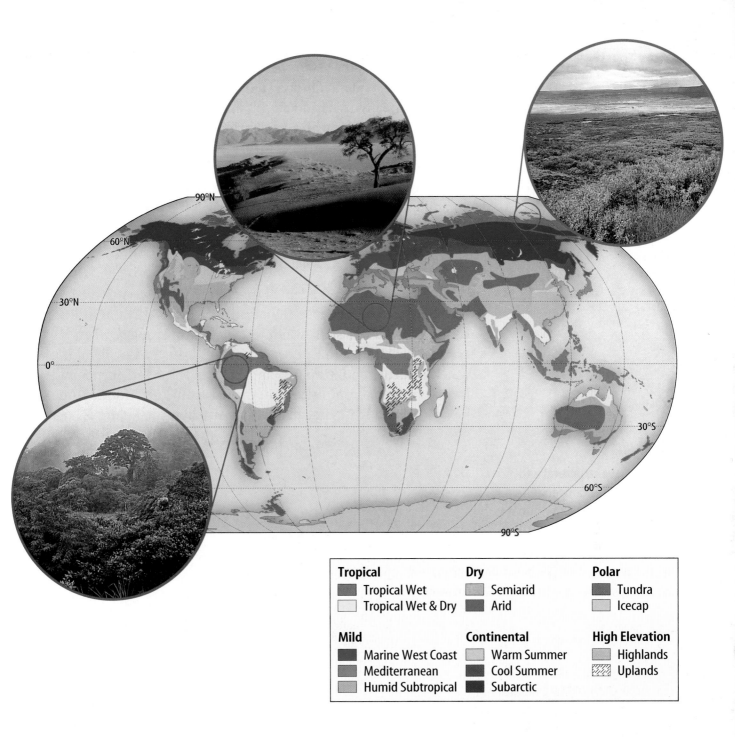

Tropical
- Tropical Wet
- Tropical Wet & Dry

Dry
- Semiarid
- Arid

Polar
- Tundra
- Icecap

Mild
- Marine West Coast
- Mediterranean
- Humid Subtropical

Continental
- Warm Summer
- Cool Summer
- Subarctic

High Elevation
- Highlands
- Uplands

Structural Adaptations Some organisms have body structures that help them survive in certain climates. The fur of mammals is really hair that insulates them from cold temperatures. A cactus has a thick, fleshy stem. This structural adaptation helps a cactus hold water. The waxy stem covering prevents water inside the cactus from evaporating. Instead of broad leaves, these plants have spiny leaves that further reduce water loss.

Reading Check *How do cacti conserve water?*

Figure 5
This map shows a climate classification system similar to the one developed by Köppen. *What patterns can you see in the locations of certain climate types?*

Behavioral Adaptations Some organisms display behavioral adaptations that help them survive in a particular climate. For example, rodents and certain other mammals undergo a period of greatly reduced activity in winter called **hibernation.** During hibernation, body temperature drops and body processes are reduced to a minimum. Some of the factors thought to trigger hibernation include cooler temperatures, shorter days, and lack of adequate food. The length of time that an animal hibernates varies depending on the particular species of animal and the environmental conditions.

> ✔ **Reading Check** *What is hibernation?*

Other animals have adapted differently. During cold weather, bees cluster together in a tight ball to conserve heat. On hot, sunny days, desert snakes hide under rocks. At night when it's cooler, they slither out in search of food. Instead of drinking water as turtles and lizards do in wet climates, desert turtles and lizards obtain the moisture they need from their food. Some behavioral and structural adaptations are shown in **Figure 6.**

Figure 6
Organisms have structural and behavioral adaptations that help them survive in particular climates.

A These hibernating bats have adapted their behavior to survive winter.

B The needles and the waxy skin of a cactus are structural adaptations to a desert climate. *How do these adaptations help cacti conserve water?*

C Polar bears have structural adaptations to keep them warm. The hairs of their fur trap air and heat.

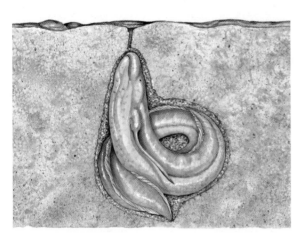

Figure 7
Lungfish survive periods of intense heat and drought by going into an inactive state called estivation. **A** During the dry season when water evaporates, lungfish dig into the mud and curl up in a small chamber they make at the lake's bottom. **B** During the wet season, lungfish reemerge to live in small lakes and pools.

Estivation Lungfish, shown in **Figure 7,** survive periods of intense heat by entering an inactive state called estivation (es tuh VAY shun). As the weather gets hot and water evaporates, the fish burrows into mud and covers itself in a leathery mixture of mud and mucus. It lives this way until the warm, dry months pass.

Like other organisms, you have adaptations that help you adjust to climate. In hot weather, your sweat glands release water onto your skin. The water evaporates, taking some heat with it. As a result, you become cooler. In cold weather, you may shiver to help your body stay warm. When you shiver, the rapid muscle movements produce some heat. What other adaptations to climate do people have?

Section Assessment

1. How can climates be classified? What type of climate do you live in?

2. Use **Figure 5** and a world map to identify the climate type for each of the following locations: Cuba, North Korea, Egypt, and Uruguay.

3. What are some behavioral adaptations that allow animals to stay warm? What structural adaptations help keep animals warm?

4. How is hibernation different from estivation? How is it similar?

5. **Think Critically** What adaptations help dogs keep cool during hot weather?

Skill Builder Activities

6. **Forming Hypotheses** Some scientists have suggested that Earth's climate is getting warmer. What effects might this have on vegetation and animal life in various parts of the United States? **For more help, refer to the** Science Skill Handbook.

7. **Communicating** Research the types of vegetation found in the six climate regions shown in **Figure 5.** Write a paragraph in your Science Journal describing why vegetation can be used to help define climate boundaries. **For more help, refer to the** Science Skill Handbook.

Climatic Changes

What You'll Learn

- **Explain** what causes seasons.
- **Describe** how El Niño affects climate.
- **Explore** possible causes of climatic change.

Vocabulary

season
El Niño
greenhouse effect
global warming
deforestation

Why It's Important

Changing climates could affect sea level and life on Earth.

Earth's Seasons

In temperate zones, you can play softball under the summer Sun and in the winter go sledding with friends. Weather changes with the season. **Seasons** are short periods of climatic change caused by changes in the amount of solar radiation an area receives. **Figure 8** shows Earth revolving around the Sun. Because Earth is tilted, different areas of Earth receive changing amounts of solar radiation throughout the year.

Seasonal Changes Because of fairly constant solar radiation near the equator, the tropics do not have much seasonal temperature change. However, they do experience dry and rainy seasons. The middle latitudes, or temperate zones, have warm summers and cool winters. Spring and fall are usually mild.

✔ **Reading Check** *What are seasons like in the tropics?*

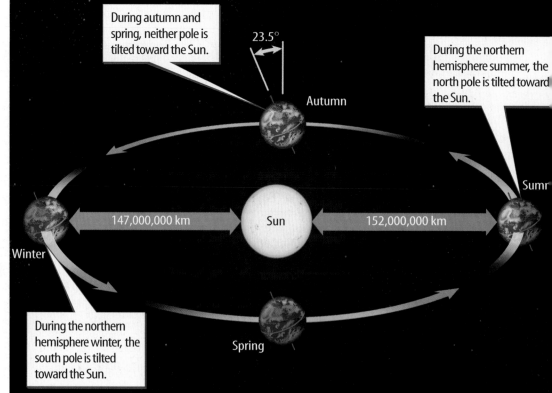

Figure 8
As Earth revolves around the Sun, different areas of Earth are tilted toward the Sun, which causes different seasons. *During which northern hemisphere season is Earth closer to the Sun?*

During autumn and spring, neither pole is tilted toward the Sun.

23.5°

During the northern hemisphere summer, the north pole is tilted toward the Sun.

Autumn

Sumr

147,000,000 km   Sun   152,000,000 km

Winter

During the northern hemisphere winter, the south pole is tilted toward the Sun.

Spring

Figure 9
A strong El Niño, like the one that occurred in 1998, can affect weather patterns around the world.

A A severe drought struck Indonesia, contributing to forest fires.

B California was plagued by large storms that produced pounding surf and shoreline erosion.

High Latitudes During the year, the high latitudes near the poles have great differences in temperature and number of daylight hours. As shown in **Figure 8,** during summer in the northern hemisphere, the north pole is tilted toward the Sun. During summer at the north pole, the Sun doesn't set for nearly six months. During that same time, the Sun never rises at the south pole. At the equator days are about the same length all year long.

El Niño and La Niña

El Niño (el NEEN yoh) is a climatic event that involves the tropical Pacific Ocean and the atmosphere. During normal years, strong trade winds that blow east to west along the equator push warm surface water toward the western Pacific Ocean. Cold, deep water then is forced up from below along the coast of South America. During El Niño years, these winds weaken and sometimes reverse. The change in the winds allows warm, tropical water in the upper layers of the Pacific to flow back eastward to South America. Cold, deep water is no longer forced up from below. Ocean temperatures increase by 1°C to 7°C off the coast of Peru.

El Niño can affect weather patterns. It can alter the position and strength of one of the jet streams. This changes the atmospheric pressure off California and wind and precipitation patterns around the world. This can cause drought in Australia and Africa. This also affects monsoon rains in Indonesia and causes storms in California, as shown in **Figure 9.**

The opposite of El Niño is La Niña, shown in **Figure 10.** During La Niña, the winds blowing across the Pacific are stronger than normal, causing warm water to accumulate in the western Pacific. The water in the eastern Pacific near Peru is cooler than normal. La Niña may cause droughts in the southern United States and excess rainfall in the northwestern United States.

TRY AT HOME

Mini LAB

Modeling El Niño

Procedure
1. During El Niño, trade winds blowing across the Pacific Ocean from east to west slacken or even reverse. Surface waters move back toward the coast of Peru.
2. Add **warm water** to a **9 in by 13 in baking pan** until it is two-thirds full. Place the pan on a smooth countertop.
3. Blow as hard as you can across the surface of the water along the length of the pan. Next, blow with less force. Then, blow in the opposite direction.

Analysis
1. What happened to the water as you blew across its surface? What was different when you blew with less force and when you blew from the opposite direction?
2. Explain how this is similar to what happens during an El Niño event.

Figure 10

Weather in the United States can be affected by changes that occur thousands of kilometers away. Out in the middle of the Pacific Ocean, periodic warming and cooling of a huge mass of seawater—phenomena known as El Niño and La Niña, respectively—can impact weather across North America. During normal years (right), when neither El Niño nor La Niña is in effect, strong winds tend to keep warm surface waters contained in the western Pacific while cooler water wells up to the surface in the eastern Pacific.

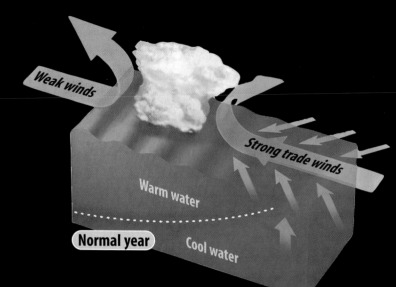

Weak winds

Strong trade winds

Warm water

Normal year

Cool water

EL NIÑO During El Niño years, winds blowing west weaken and may even reverse. When this happens, warm waters in the western Pacific move eastward, preventing cold water from upwelling. These changes can alter global weather patterns and trigger heavier-than-normal precipitation across much of the United States.

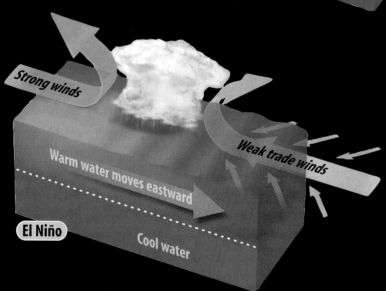

Strong winds

Weak trade winds

Warm water moves eastward

El Niño

Cool water

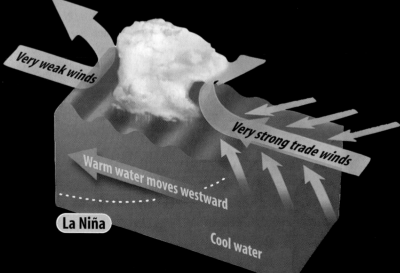

Very weak winds

Very strong trade winds

Warm water moves westward

La Niña

Cool water

LA NIÑA During La Niña years, stronger-than-normal winds push warm Pacific waters farther west, toward Asia. Cold, deep-sea waters then well up strongly in the eastern Pacific, bringing cooler and often drier weather to many parts of the United States.

El Niño

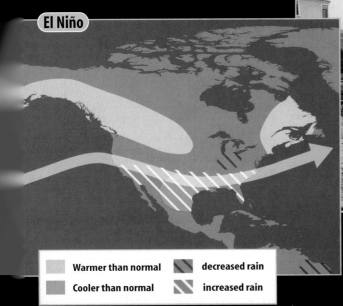

Warmer than normal	decreased rain
Cooler than normal	increased rain

Sun-warmed surface water spans the Pacific Ocean during El Niño years. Clouds form above the warm ocean, carrying moisture aloft. The jet stream, shown by the white arrow above, helps bring some of this warm, moist air to the United States.

▲ LANDSLIDE Heavy rains in California resulting from El Niño can lead to landslides. This upended house in Laguna Niguel, California, took a ride downhill during the El Niño storms of 1998.

La Niña

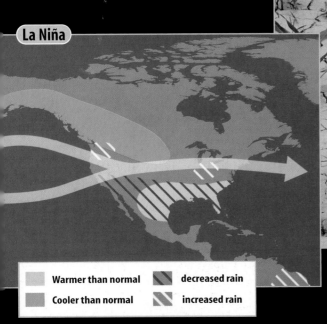

Warmer than normal	decreased rain
Cooler than normal	increased rain

During a typical La Niña year, warm ocean waters, clouds, and moisture are pushed away from North America. A weaker jet stream often brings cooler weather to the northern parts of the continent and hot, dry weather to southern areas.

▲ PARCHED LAND The Southeast may experience drought conditions, like those that struck the cornfields of Montgomery County, Maryland, during the La Niña summer of 1988.

Climatic Change

If you were exploring in Antarctica near Earth's south pole and found a 3-million-year-old fossil of a warm-weather plant or animal, what would it tell you? You might conclude that the climate of that region changed because Antarctica is much too cold for similar plants and animals to survive today. Some warm-weather fossils found in polar regions indicate that at times in Earth's past, worldwide climate was much warmer than at present. At other times Earth's climate has been much colder than it is today.

Sediments in many parts of the world show that at several different times in the past 2 million years, glaciers covered large parts of Earth's surface. These times are called ice ages. During the past 2 million years, ice ages have alternated with warm periods called interglacial intervals. Ice ages seem to last 60,000 to 100,000 years. Most interglacial periods are shorter, lasting 10,000 to 15,000 years. We are now in an interglacial interval that began about 11,500 years ago. Additional evidence suggests that climate can change even more quickly. Ice cores record climate in a way similar to tree rings. Cores drilled in Greenland show that during the last ice age, colder times lasting 1,000 to 2,000 years changed quickly to warmer spells that lasted about as long. **Figure 11** shows a scientist working with ice cores.

Figure 11
Each layer of ice in a core records detailed climate information for a single year. Some cores cover 300,000 years.

What causes climatic change?

Climatic change has many varied causes. These causes of climatic change can operate over short periods of time or very long periods of time. Catastrophic events, including meteorite collisions and large volcanic eruptions, can affect climate over short periods of time, such as a year or several years. These events add solid particles and liquid droplets to the upper atmosphere, which can change climate. Another factor that can alter Earth's climate is short- or long-term changes in solar output, which is the amount of energy given off by the Sun. Changes in Earth's movements in space affect climate over many thousands of years, and movement of Earth's crustal plates can change climate over millions of years. All of these things can work separately or together to alter Earth's climate.

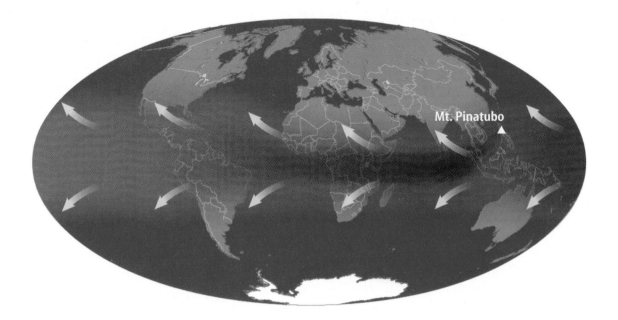

Atmospheric Solids and Liquids

Atmospheric Solids and Liquids Small solid and liquid particles always are present in Earth's atmosphere. These particles can enter the atmosphere naturally or be added to the atmosphere by humans as pollution. Some ways that particles enter the atmosphere naturally include volcanic eruptions, soot from fires, and wind erosion of soil particles. Humans add particles to the atmosphere through automobile exhaust and smokestack emissions. These small particles can affect climate.

Catastrophic events such as meteorite collisions and volcanic eruptions put enormous volumes of dust, ash, and other particles into the atmosphere. These particles block so much solar radiation that they can cool the planet. **Figure 12** shows how a major volcanic eruption affected Earth's atmosphere.

In cities, particles put into the atmosphere as pollution can change the local climate. These particles can increase the amount of cloud cover downwind from the city. Some studies have even suggested that rainfall amounts can be reduced in these areas. This may happen because many small cloud droplets form rather than larger droplets that could produce rain.

Energy from the Sun Solar radiation provides Earth's energy. If the output of radiation from the Sun varies, Earth's climate could change. Some changes in the amount of energy given off by the Sun seem to be related to the presence of sunspots. Sunspots are dark spots on the surface of the Sun. **WARNING:** *Never look directly at the Sun.* Evidence supporting the link between sunspots and climate includes an extremely cold period in Europe between 1645 and 1715. During this time, very few sunspots appeared on the Sun.

Figure 12
Mount Pinatubo in the Philippines erupted in 1991. During the eruption, particles were spread high into the atmosphere and circled the globe. Over time, particles spread around the world, blocking some of the Sun's energy from reaching Earth. The gray areas show how particles from the eruption moved around the world.

Health
INTEGRATION

Atmospheric particles produced as pollution can affect human health as well as climate. These small particles, often called particulates, can enter the lungs and cause tissue damage. People who have existing lung or heart problems are affected most seriously. Do research to find out what types of laws have been written to reduce particulate pollution.

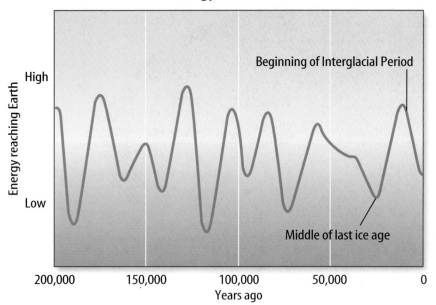

Energy From the Sun

Energy reaching Earth

High

Low

Beginning of Interglacial Period

Middle of last ice age

200,000   150,000   100,000   50,000   0
Years ago

Figure 13
The curving line shows how the amount of the Sun's energy that strikes the northern hemisphere changed over the last 200,000 years. *Describe the amount of energy that reached the northern hemisphere during the last ice age.*

Earth Movements Another explanation for some climatic changes involves Earth's movements in space. Earth's axis currently is tilted 23.5° from perpendicular to the plane of its orbit around the Sun. In the past, this tilt has increased to 24.5° and has decreased to 21.5°. When this tilt is at its maximum, the change between summer and winter is probably greater. Earth's tilt changes about every 41,000 years. Some scientists hypothesize that the change in tilt affects climate.

Two additional Earth movements also cause climatic change. Earth's axis wobbles in space just like the axis of a top wobbles when it begins to spin more slowly. This can affect the amount of solar energy received by different parts of Earth. Also, the shape of Earth's orbit changes. Sometimes it is more circular than at present and sometimes it is more flattened. The shape of Earth's orbit changes over a 100,000-year cycle.

Amount of Solar Energy These movements of Earth cause the amount of solar energy reaching different parts of Earth to vary over time, as shown in **Figure 13.** These changes might have caused glaciers to grow and shrink over the last few million years. However, they do not explain why glaciers have occurred so rarely over long spans of geologic time.

Crustal Plate Movement Another explanation for major climatic change over tens or hundreds of millions of years concerns the movement of Earth's crustal plates. The movement of continents and oceans affects the transfer of heat on Earth, which in turn affects wind and precipitation patterns. Through time, these altered patterns can change climate. One example of this is when movement of Earth's plates created the Himalaya about 40 million years ago. The growth of these mountains changed climate over much of Earth.

As you've learned, many theories attempt to answer questions about why Earth's climate has changed through the ages. Probably all of these things play some role in changing climates. More study needs to be done before all the factors that affect climate will be understood.

Climatic Changes Today

Beginning in 1992, representatives from many countries have met to discuss the greenhouse effect and global warming. These subjects also have appeared frequently in the headlines of newspapers and magazines. Some people are concerned that the greenhouse effect could be responsible for some present-day warming of Earth's atmosphere and oceans.

The **greenhouse effect** is a natural heating that occurs when certain gases in Earth's atmosphere trap heat. Radiation from the Sun strikes Earth's surface and causes it to warm. Some of this heat then is radiated back toward space. Greenhouse gases in the atmosphere absorb a portion of this heat. This keeps Earth warmer than it would be otherwise. The greenhouse effect is illustrated in **Figure 14.**

There are many natural greenhouse gases in Earth's atmosphere. Water vapor, carbon dioxide, and methane are some of the most important ones. Without these greenhouse gases, life would not be possible on Earth. Like Mars, Earth would be too cold. However, if the greenhouse effect is too strong, Earth could get too warm. High levels of carbon dioxide in its atmosphere indicate that this has happened on the planet Venus.

SCIENCE *Online*

Research Visit the Glencoe Science Web site at **science.glencoe.com** for more information about the greenhouse effect.

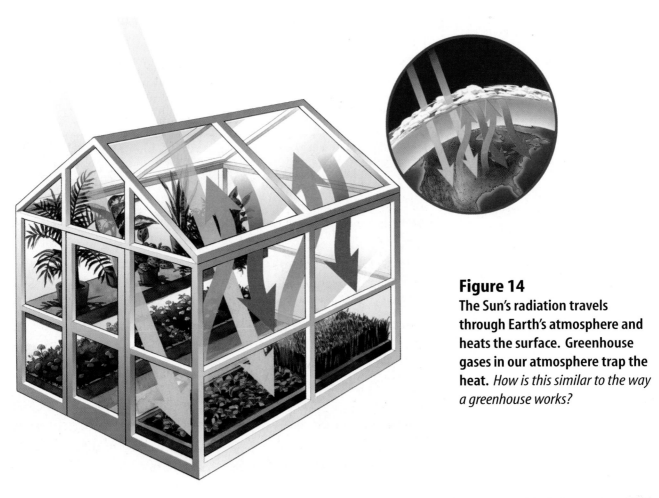

Figure 14
The Sun's radiation travels through Earth's atmosphere and heats the surface. Greenhouse gases in our atmosphere trap the heat. *How is this similar to the way a greenhouse works?*

Global Warming

Global warming means that the average global temperature of Earth is rising. One likely reason for global warming is the increase of greenhouse gases in our atmosphere. An increase in greenhouse gases increases the greenhouse effect. In the last 100 years, the surface temperature on Earth has increased 0.5°C. This might be a result of more greenhouse gases.

✔ **Reading Check** *How can greenhouse gases affect Earth's climate?*

If Earth's average temperature continues to rise, many glaciers could melt. When glaciers melt, the extra water causes sea levels to rise. Low-lying coastal areas would experience increased flooding. Already some ice caps and small glaciers are beginning to melt and recede, as shown in **Figure 15.** Sea level is rising in some places. Some scientific studies show that these events are related to Earth's increased temperature.

You learned in the previous section that organisms are adapted to their environments. When environments change, can organisms cope? In some tropical waters around the world, corals are dying. Many people think these deaths are caused by warmer water to which the corals are not adapted.

Some climate models show that in the future, Earth's temperatures will increase faster than they have in the last 100 years. Next, you will learn how human activity may add to global warming, and you will find out what you can do to help decrease this problem.

Figure 15
This glacier in Greenland might have receded from its previous position because of global warming. The pile of sediment in front shows how far the glacier once reached.

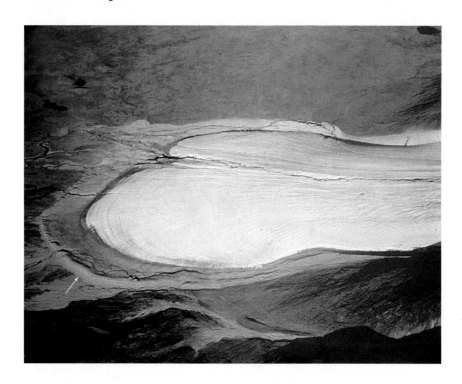

Human Activities

Human activities affect the air in Earth's atmosphere. Burning fossil fuels and removing vegetation increase the amount of carbon dioxide (CO$_2$) in the atmosphere. Because carbon dioxide is a greenhouse gas, it might contribute to global warming. Each year, the amount of carbon dioxide in the atmosphere continues to increase.

Burning Fossil Fuels When natural gas, oil, and coal are burned for energy, the carbon in these fossil fuels combines with atmospheric oxygen to form carbon dioxide. This increases the amount of carbon dioxide in Earth's atmosphere. Studies indicate that humans have increased carbon dioxide levels in the atmosphere by about 25 percent over the last 150 years.

Deforestation Destroying and cutting down forests, called **deforestation,** also affects the amount of carbon dioxide in the atmosphere. Forests, such as the one shown in **Figure 16,** are cleared for mining, roads, buildings, and grazing cattle. Large tracts of forest have been cleared in every country on Earth. Tropical forests have been decreasing at a rate of about one percent each year for the past two decades.

As trees grow, they take in carbon dioxide from the atmosphere. Trees use this carbon dioxide to produce wood and leaves. When trees are cut down, the carbon dioxide they could have removed from the atmosphere remains. Cut-down trees often are burned for fuel or to clear the land. Burning trees produces even more carbon dioxide.

SCIENCE *Online*

Research Visit the Glencoe Science Web site at **science.glencoe.com** for more information about deforestation. Communicate to your class what you learn.

Figure 17
Solar energy is beginning to fill some human energy needs.
A These solar panels in California generate electricity without adding CO_2 to the atmosphere.
B Solar-powered cars may be common in the future.

How to Reduce CO_2

What can you do to help reduce the amount of CO_2 in the atmosphere? Conserving electricity is one answer. When you conserve electricity, you reduce the amount of fossil fuel that is burned. One way to save fuel is to change daily activities that rely on energy from burning fossil fuels. Turn off the TV, for instance, when no one is watching it. Walk or ride a bike to the store, if possible, instead of using a car. People also can use different energy sources to meet their energy needs, as shown in **Figure 17.**

Another way to reduce CO_2 is to plant vegetation. As you've learned, plants remove carbon dioxide from the atmosphere. Correctly planted vegetation also can shelter homes from cold winds or the blazing Sun and reduce the use of electricity.

Section Assessment

1. How is Earth's tilted axis responsible for seasons?

2. In what way does El Niño change the climate? What climate changes are caused by La Niña?

3. What factors can cause Earth's climate to change?

4. How are people adding carbon dioxide to the atmosphere?

5. **Think Critically** If Earth's climate continues to warm, how might your community be affected?

Skill Builder Activities

6. **Recognizing Cause and Effect** Using a globe, model the three movements of Earth in space that can cause climatic change. **For more help, refer to the** Science Skill Handbook.

7. **Using a Word Processor** Use word processing software to design a pamphlet to inform people why Earth's climate changes. What are your predictions for the future? On what evidence do you base these predictions? Include this evidence in your pamphlet. **For more help, refer to the** Technology Skill Handbook.

Activity

The Greenhouse Effect

Do you remember climbing into the car on a warm, sunny day? Why was it so hot inside the car when it wasn't that hot outside? It was hotter in the car because the car functioned like a greenhouse. You experienced the greenhouse effect.

What You'll Investigate
How can you demonstrate the greenhouse effect?

Materials
identical large, empty glass jars (2)
lid for one jar
nonmercury thermometers (3)

Goals
- **Model** the greenhouse effect.
- **Measure and graph** temperature changes.

Safety Precautions
Be careful when you handle glass thermometers. If a thermometer breaks, do not touch it. Have your teacher dispose of the glass safely.

Procedure

1. Lay a thermometer inside each jar.

2. Place the jars next to each other by a sunny window. Lay the third thermometer between the jars.

3. **Record** the temperatures of the three thermometers. They should be the same.

4. Place the lid on one jar.

5. **Record** the temperatures of all three thermometers at the end of 5, 10, and 15 min.

6. Make a line graph that shows the temperatures of the three thermometers for the 15 min of the experiment.

Conclude and Apply

1. **Explain** why you placed a thermometer between the two jars.

2. What were the constants in this experiment? What was the variable?

3. Which thermometer experienced the greatest temperature change during your experiment?

4. **Analyze** what occurred in this experiment. How was the lid in this experiment like the greenhouse gases in the atmosphere?

5. **Infer** from this experiment why you should never leave a pet inside a closed car in warm weather.

Communicating Your Data

Give a brief speech describing your conclusions to your class. **For more help, refer to the** Science Skill Handbook.

Activity

Microclimates

A microclimate is a localized climate that differs from the main climate of a region. Buildings in a city, for instance, can affect the climate of the surrounding area. Large buildings can create microclimates by blocking the Sun or changing wind patterns.

What You'll Investigate

Does your school create microclimates?

Materials

thermometers
psychrometer
paper strip or wind sock
large cans (4 or 5)
beakers or rain gauges (4 or 5)
unlined paper

Alternate materials

Goals

- **Observe** temperature, wind speed, relative humidity, and precipitation in areas outside your school.
- **Identify** local microclimates.

Safety Precautions

WARNING: *If a thermometer breaks, do not touch it. Have your teacher dispose of the glass safely.*

Relative Humidity										
Dry Bulb Temperature (°C)	**Dry Bulb Temperature Minus Wet Bulb Temperature (°C)**									
	1	2	3	4	5	6	7	8	9	10
14	90	79	70	60	51	42	34	26	18	10
15	90	80	71	61	53	44	36	27	20	13
16	90	81	71	63	54	46	38	30	23	15
17	90	81	72	64	55	47	40	32	25	18
18	91	82	73	65	57	49	41	34	27	20
19	91	82	74	65	58	50	43	36	29	22
20	91	83	74	66	59	51	44	37	31	24
21	91	83	75	67	60	53	46	39	32	26
22	92	83	76	68	61	54	47	40	34	28
23	92	84	76	69	62	55	48	42	36	30
24	92	84	77	69	62	56	49	43	37	31
25	92	84	77	70	63	57	50	44	39	33

Procedure

1. Select four or five sites around your school building. Also, select a control site well away from the school.

2. Attach a thermometer to an object near each of the locations you selected. Set up a rain gauge, beaker, or can to collect precipitation.

3. Visit each site at two predetermined times, one in the morning and one in the afternoon, each day for a week. Record the temperature and measure any precipitation that might have fallen. Use a wind sock or paper strip to determine wind direction.

4. To find relative humidity, you'll need to use a psychrometer. A psychrometer is an instrument with two thermometers—one wet and one dry. As moisture from the wet thermometer evaporates, it takes

heat energy from its environment, and the environment immediately around the wet thermometer cools. The thermometer records a lower temperature. Relative humidity can be found by finding the difference between the wet thermometer and the dry thermometer and by using the chart on the previous page. Record all of your weather data.

5. **Analyze** your data to find patterns. Make separate line graphs for temperature, relative humidity, and precipitation for your morning and afternoon data. Make a table showing wind direction data.

Conclude and Apply

1. Why did you take weather data at a control site away from the school building? How did the control help you analyze and interpret your data?

2. **Compare and contrast** weather data for each of your sites. What microclimates did you identify around your school building? How did these climates differ from the control site? How did they differ from each other?

3. **Infer** what conditions could have caused the microclimates that you identified. Are your microclimates similar to those that might exist in a large city? Explain.

Communicating
Your Data

Use your graphs to make a large poster explaining your conclusions. Display your poster in the school building. **For more help, refer to the Science Skill Handbook.**

The Year There Was

Mount St. Helens in Washington state erupted in May 1980. Its ash cloud temporarily darkened the sky and was carried thousands of kilometers away by upper-air currents.

Y ou've seen pictures of erupting volcanoes. One kind of volcano sends smoke, rock, and ash high into the air above the crater. Another kind of volcano erupts with fiery, red-hot rivers of lava snaking down its sides. Erupting volcanoes are nature's forces at their mightiest, causing destruction and death. But not everyone realizes how far-reaching the destruction can be.

Large volcanic eruptions can affect people thousands of kilometers away. In fact, major volcanic eruptions can have effects that reach around the globe.

An erupting volcano can temporarily change Earth's climate. The ash a volcano ejects into the atmosphere can create day after day without sunshine. Other particles move high into the atmosphere and are carried all the way around Earth, sometimes causing global temperatures to drop for several months.

The Summer That Never Came

An example of a volcanic eruption with wide-ranging effects occurred in 1783 in Iceland, an island nation in the North Atlantic Ocean. Winds carried a black cloud of ash from an erupting volcano in Iceland westward across northern Canada, Alaska, and across the Pacific Ocean to Japan. The summer turned bitterly cold in these places. Water froze, and heavy snowstorms pelted the land. Sulfurous gases from the erupting volcano combined with water to form particles of acid that reflected solar energy back into space. This "blanket" in the atmosphere kept the Sun's rays from heating up part of Earth.

The most tragic result of this eruption was the death of many Kauwerak people, who lived in western Alaska. Only a handful of Kauwerak survived the summer that never came. They had no opportunity to catch needed foods to keep them alive through the following winter.

No Summer

Erupting volcanoes can cause unusual changes in climate

A truck is dwarfed by clouds of volcanic ash during the eruption of Mt. Pinatubo in the Philippines. The volcano had been dormant for 611 years.

Another Year Without a Summer

In 1815, the eruption of Tambora, on the island of Sumbawa in Indonesia, is blamed for North America and Europe having "the year without a summer" in 1816. Once again, the blanket of acid particles, pushed by upper-level winds, caused areas far removed from the volcanic activity to face a drastic change in weather patterns. In New England in 1816, frosts occurred and snow fell well into June. Massive crop failures occurred in North America and Europe.

Only volcanoes that throw debris above the troposphere cause this kind of climatic change. Melissa Free, a researcher for the National Oceanographic and Atmospheric Administration, says, "If [debris] is in the troposphere, it is taken out and dissipated more rapidly by rainfall." But, Free says, to cause climatic change, "...the eruption has to be strong enough to push material above the troposphere and into the stratosphere." And that's what happened when Tambora erupted, leading to a very unusual summer.

CONNECTIONS Locate Using an atlas, locate Indonesia and Iceland. Using reference materials, find five facts about each place. Make a map of each nation and illustrate the map with your five facts.

SCIENCE *Online*
For more information, visit
science.glencoe.com

Reviewing Main Ideas

Section 1 What is climate?

1. An area's climate is the average weather over a long period of time, such as 30 years.

2. The three main climate zones are tropical, polar, and temperate.

3. Features such as oceans, mountains, and even large cities affect climate. *What differences might you find in the climates on the opposite sides of this mountain range?*

Section 2 Climate Types

1. Climates can be classified by various characteristics, such as temperature, precipitation, and vegetation. World climates commonly are separated into six major groups.

2. Organisms have structural and behavioral adaptations that help them survive in particular climates. Many organisms can survive only in the climate they are adapted to.

3. Adaptations develop in a population over a long period of time.

Section 3 Climatic Changes

1. Seasons are caused by the tilt of Earth's axis as Earth revolves around the Sun.

2. El Niño disrupts normal temperature and precipitation patterns around the world. *How could El Niño help cause California mudflows like this one?*

3. Geological records show that over the past few million years, Earth's climate has alternated between ice ages and warmer periods called interglacials.

4. The greenhouse effect occurs when certain gases trap heat in Earth's atmosphere.

5. Humans may be contributing to global warming by producing greenhouse gases. Carbon dioxide enters the atmosphere when fossil fuels such as oil and coal are burned. *How can planting vegetation like the tree shown here help decrease greenhouse gases in the atmosphere?*

FOLDABLES
Reading & Study
Skills

After You Read

To help you review the various aspects of the climate classification system, use the Classify Study Fold you made at the beginning of the chapter.

Visualizing Main Ideas

Complete the following concept map on climate.

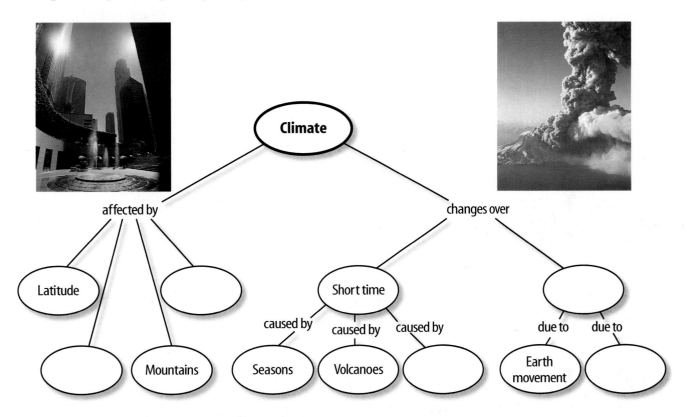

Vocabulary Review

Vocabulary Words

a. adaptation
b. climate
c. deforestation
d. El Niño
e. global warming
f. greenhouse effect
g. hibernation
h. polar zone
i. season
j. temperate zone
k. tropics

Using Vocabulary

The sentences below include vocabulary words that have been used incorrectly. Change the incorrect words so that the sentence reads correctly. Underline your change.

1. Earth's north pole is in the temperate zone.

2. Deforestation causes the Pacific Ocean to become warmer off the coast of Peru.

3. During adaptation, an animal's body temperature drops.

4. Season is the pattern of weather that occurs over many years.

5. Greenhouse effect means global temperatures are rising.

Study Tip

Ask what kinds of questions to expect on the test. Ask for practice tests so that you can become familiar with the test-taking materials.

Chapter 17 Assessment

Checking Concepts

Choose the word or phrase that best answers the question.

1. What is commonly found in places where warm air crosses a mountain and descends?
   A) lakes
   C) deserts
   B) rain forests
   D) glaciers

2. During which of the following is the eastern Pacific warmer than normal?
   A) El Niño
   C) summer
   B) La Niña
   D) spring

3. Which of the following is a greenhouse gas in Earth's atmosphere?
   A) helium
   C) hydrogen
   B) carbon dioxide
   D) oxygen

4. Which latitude receives the most direct rays of the Sun year-round?
   A) 60°N
   C) 30°S
   B) 90°N
   D) 0°

5. What happens as you climb a mountain?
   A) temperature decreases
   B) temperature increases
   C) air pressure increases
   D) air pressure remains constant

6. Which of the following is true of El Niño?
   A) It cools the Pacific Ocean near Peru.
   B) It causes flooding in Australia.
   C) It cools the waters off Alaska.
   D) It may occur when the trade winds slacken or reverse.

7. What do changes in Earth's orbit affect?
   A) Earth's shape
   C) Earth's rotation
   B) Earth's climate
   D) Earth's tilt

8. The Köppen climate classification system includes categories based on precipitation and what other factor?
   A) temperature
   C) winds
   B) air pressure
   D) latitude

9. Which of the following is an example of structural adaptation?
   A) hibernation
   C) fur
   B) migration
   D) estivation

10. Which of these can people do in order to help reduce global warming?
    A) conserve energy
    C) produce methane
    B) burn coal
    D) remove trees

Thinking Critically

11. Why might global warming lead to the extinction of some organisms?

12. What might you infer if you find fossils of tropical plants in a desert?

13. On a summer day, why would a Florida beach be cooler than an orange grove that is 2 km inland?

14. What would happen to global climates if the Sun emitted more energy?

15. Why will it be cooler if you climb to a higher elevation in a desert?

Developing Skills

16. **Making and Using Graphs** The following table gives average precipitation amounts for Phoenix, Arizona. Make a bar graph of these data. Which climate type do you think Phoenix represents?

Precipitation in Phoenix, Arizona	
Season	Precipitation (cm)
Winter	5.7
Spring	1.2
Summer	6.7
Autumn	5.9
Total	19.5

17. Communicating Explain how atmospheric pressure over the Pacific Ocean might affect how the trade winds blow.

18. Predicting Make a chain-of-events chart to explain the effect of a major volcanic eruption on climate.

19. Forming Hypotheses A mountain glacier in South America has been getting smaller over several decades. What hypotheses should a scientist consider to explain why this is occurring?

20. Concept Mapping Complete the concept map using the following: *tropics, 0°–23.5° latitude, polar, temperate, 23.5°–66.5° latitude, 66.5° latitude to poles.*

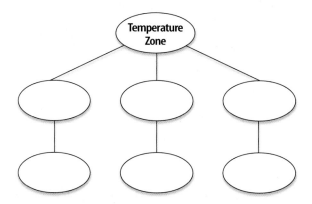

Performance Assessment

21. Science Display Make a display illustrating different factors that can affect climate. Be sure to include detailed diagrams and descriptions for each factor in your display. Present your display to the class.

 **Test Practice**

The graph below shows long-term variations in atmospheric carbon dioxide levels and global temperature.

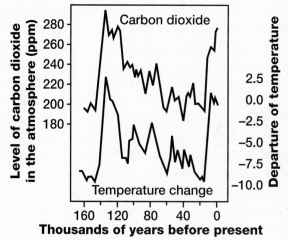

Study the graph and answer the following questions.

1. Which of these statements is TRUE according to the graph?
  A) Earth's mean temperature has never been hotter than it is today.
  B) The level of CO_2 has never been higher than today.
  C) The mean global temperature 60,000 years ago was less than today.
  D) The level of CO_2 in the atmosphere 80,000 years ago was 280 parts per million.

2. Which of the following statements BEST describes this graph?
  F) As CO_2 levels have increased, so has global temperature.
  G) As CO_2 levels have increased, global temperature has decreased.
  H) As global temperature has increased, CO_2 levels have decreased.
  J) No relationship exists between CO_2 and global temperatures.

CHAPTER 18

Ocean Motion

Surfers in Hawaii experience firsthand the enormous power of moving water. It surprises people to learn that wind causes most waves, from small ripples to the giant waves of hurricanes, some more than 30 m high. Wind also creates surface currents. Other types of currents move through the ocean, too. In this chapter, you'll learn about the composition of ocean water, the interaction between the atmosphere and the oceans, and how waves, currents, and tides are created.

What do you think?

Science Journal Look at the picture below with a classmate. Discuss what you think this might be or what is happening. Here's a hint: *Daily fluctuations make this happen.* Write your answer or best guess in your Science Journal.

Surface currents are caused by wind, but wind cannot cause currents deep in the ocean. Instead, deep-water currents are created by differences in the density of ocean water. Several factors affect water density. One is temperature. Do the activity below to see how temperature differences create these kinds of currents.

Explore how currents work

1. In a bowl, mix ice and cold water to make ice water.
2. Fill a beaker with warm tap water.
3. Add a few drops of food coloring to the ice water and stir the mixture.
4. Use a dropper to place some of this ice water on top of the warm water.

Observe

In your Science Journal, describe what happened. Did adding cold water on top produce a current? Look up the word *convection* in a dictionary. Infer why the current you created is called a convection current.

Before You Read

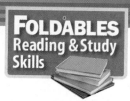

FOLDABLES
Reading & Study Skills

Making a Cause and Effect Study Fold Make the following Foldable to help you understand the cause and effect relationship of ocean motion.

1. Place a sheet of paper in front of you so the long side is at the top. Fold the paper in half from the left side to the right side. Fold top to bottom and crease. Then unfold.
2. Through the top thickness of paper, cut along the middle fold line to form two tabs. Label the tabs *Causes of Ocean Motion* and *Effects of Ocean Motion*.
3. As you read the chapter, write what you learn about why the ocean moves and the types of ocean motion under the top tab.

① Ocean Water

Importance of Oceans

Imagine yourself lying on a beach and listening to the waves gently roll onto shore. A warm breeze blows off the water, making it seem as if you're in a tropical paradise. It's easy to appreciate the oceans under these circumstances, but the oceans affect your life in other ways, too.

Varied Resources   Oceans are important for food, minerals, transportation, and weather. Today, as shown in **Figure 1,** a huge variety of different resources comes from the oceans of the world. Oceans also allow efficient transportation. Can you think of a product you own that was transported by a ship? Energy and mineral resources also are found in oceans. Oil wells often are drilled in shallow ocean water. Oceans affect weather and climate. Hurricanes develop in some tropical waters, and moist air masses move onto land from oceans. Ocean currents keep some places warm while creating cool, foggy days elsewhere.

✓ **Reading Check**   *What resources come from oceans?*

Figure 1
People depend on the oceans for many resources.

A Krill are tiny, shrimplike animals that live in the Antarctic Ocean. Some cultures use krill in noodles and rice cakes.

B Kelp is a fast-growing seaweed that is a source of algin, used in making ice cream, salad dressing, medicines, and cosmetics.

Origin of Oceans

During Earth's first billion years, its surface, shown in **Figure 2A,** was much more volcanically active than it is today. When volcanoes erupt, they spew lava and ash, and they give off water vapor, carbon dioxide, and other gases. Scientists hypothesize that about 4 billion years ago, this water vapor began to be stored in Earth's early atmosphere. Over millions of years, it cooled enough to condense into storm clouds. Torrential rains began to fall. Shown in **Figure 2B,** oceans were formed as this water filled low areas on Earth called **basins.** Today, 70 percent of Earth's surface is covered by ocean water.

Composition of Oceans

Ocean water contains dissolved gases such as oxygen, carbon dioxide, and nitrogen. Oxygen is the gas that almost all organisms need for respiration. It enters the oceans in two ways—directly from the atmosphere and from organisms that photosynthesize. Carbon dioxide enters the ocean from the atmosphere and from organisms when they respire. The atmosphere is the only important source of nitrogen gas. Bacteria combine nitrogen and oxygen to create nitrates, which are important nutrients for plants.

If you've ever tasted ocean water, you know that it is salty. Ocean water contains many dissolved salts. Chloride, sodium, sulfate, magnesium, calcium, and potassium are some of the ions in seawater. An ion is a charged atom or group of atoms. Some of these ions come from rocks that are dissolved slowly by rivers and groundwater. These include calcium, magnesium, and sodium. Rivers carry these chemicals to the oceans. Erupting volcanoes add other ions, such as sulfate and chloride.

Figure 2
Earth's oceans formed from water vapor.

A Water vapor was released into the atmosphere by volcanoes that also gave off other gases, such as carbon dioxide and nitrogen.

B Condensed water vapor formed storm clouds. Oceans formed when basins filled with water from torrential rains.

 **Reading Check** *How do sodium and chloride ions get into seawater?*

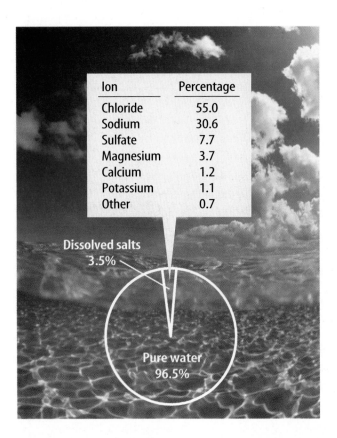

Ion	Percentage
Chloride	55.0
Sodium	30.6
Sulfate	7.7
Magnesium	3.7
Calcium	1.2
Potassium	1.1
Other	0.7

Dissolved salts
3.5%

Pure water
96.5%

Figure 3
Ocean water contains about 3.5 percent dissolved salts. *If you evaporated 1,000 g of seawater, how many grams of salt would be left?*

Salts The most abundant elements in seawater are the hydrogen and oxygen that make up water. Ions of many other elements are found dissolved in seawater. When seawater is evaporated, these ions combine to form materials called salts. Sodium and chloride make up most of the ions in seawater. If seawater evaporates, the sodium and chloride ions combine to form a salt called halite. Halite is the common table salt you use to season food. It is this dissolved salt and similar ones that give ocean water its salty taste.

Salinity (say LIH nuh tee) is a measure of the amount of salts dissolved in seawater. It usually is measured in grams of dissolved salt per kilogram of seawater. One kilogram of ocean water contains about 35 g of dissolved salts, or 3.5 percent. The chart in **Figure 3** shows the most abundant ions in ocean water. The proportion and amount of dissolved salts in seawater remain nearly constant and have stayed about the same for hundreds of millions of years. This tells you that the composition of the oceans is in balance. Evidence that scientists have gathered indicates that Earth's oceans are not growing saltier.

Life Science INTEGRATION

Removal of Elements Although rivers, volcanoes, and the atmosphere constantly add material to the oceans, the oceans are considered to be in a steady state. This means that elements are added to the oceans at about the same rate that they are removed. They are removed when ocean water evaporates, leaving salt behind, and when organisms use the dissolved salts to make shells. Some marine animals remove calcium ions from the water to form bones. Other animals, such as oysters and clams, use the dissolved calcium to form shells. Some algae, called diatoms, have silica shells. Because many organisms use calcium and silicon, these elements are removed more quickly from seawater than elements such as chlorine or sodium.

Desalination Salt can be removed from ocean water by a process called desalination (dee sa luh NAY shun). If you have ever swum in the ocean, you know what happens when your skin dries. The white, flaky substance on your skin is salt. As seawater evaporates, salt is left behind. As demand for freshwater increases throughout the world, scientists are working on technology to remove salt to make seawater drinkable.

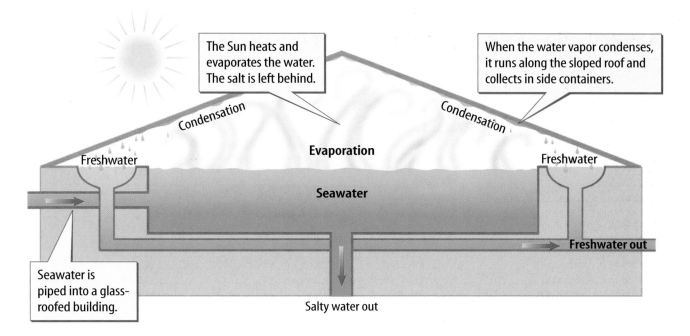

The Sun heats and evaporates the water. The salt is left behind.

When the water vapor condenses, it runs along the sloped roof and collects in side containers.

Condensation

Condensation

Freshwater

Evaporation

Freshwater

Seawater

Freshwater out

Seawater is piped into a glass-roofed building.

Salty water out

Desalination Plants Some methods of desalination include evaporating seawater and collecting the freshwater as it condenses on a glass roof. **Figure 4** shows how a desalination plant that uses solar energy works. Other plants desalinate water by passing it through a membrane that removes the dissolved salts. Freshwater also can be obtained by melting frozen seawater. As seawater freezes, the ice crystals that form contain much less salt than the remaining water. The salty, unfrozen water then can be separated from the ice. The ice can be washed and melted to produce freshwater.

Figure 4
This desalination plant uses solar energy to produce freshwater.

Section Assessment

1. Describe at least five ways that Earth's oceans affect your life.

2. According to scientific hypothesis, how were Earth's oceans formed? When do scientists hypothesize they formed?

3. Where do the dissolved salts in ocean water come from?

4. How does oxygen get into oceans?

5. **Think Critically** Organisms in the oceans are important sources of food and medicine. What steps can humans take to ensure that these resources are available for future generations?

Skill Builder Activities

6. **Concept Mapping** Make a concept map that shows how sodium and chloride become dissolved in ocean water and what happens when the seawater evaporates. Use the terms *rivers, volcanoes, halite, source of, sodium, chloride,* and *combine to form*. **For more help, refer to the** Science Skill Handbook.

7. **Using Proportions** If the average salinity of seawater is 35 parts per thousand, how many grams of dissolved salts will 500 g of seawater contain? **For more help, refer to the** Math Skill Handbook.

Ocean Currents

As You Read

What You'll Learn

- **Explain** how winds and the Coriolis effect influence surface currents.
- **Discuss** the temperatures of coastal waters.
- **Describe** density currents.

Vocabulary

surface current   upwelling
Coriolis effect   density current

Why It's Important

Ocean currents and the atmosphere transfer heat that creates the climate you live in.

Surface Currents

When you stir chocolate into a glass of milk, do you notice the milk swirling around in the glass in a circle? If so, you've observed something similar to an ocean current. Ocean currents are a mass movement, or flow, of ocean water. An ocean current is like a river within the ocean.

Surface currents move water horizontally—parallel to Earth's surface. These currents are powered by wind. The wind forces the ocean to move in huge, circular patterns. **Figure 5** shows these major surface currents. Notice that some currents are shown with red arrows and some are shown with blue arrows. Red arrows indicate warm currents. Blue arrows indicate cold currents. The currents on the ocean's surface are related to the general circulation of winds on Earth.

Surface currents move only the upper few hundred meters of seawater. Some seeds and plants are carried between continents by surface currents. Sailors take advantage of these currents along with winds to sail more efficiently from place to place.

Figure 5
These are the major surface currents of Earth's oceans.

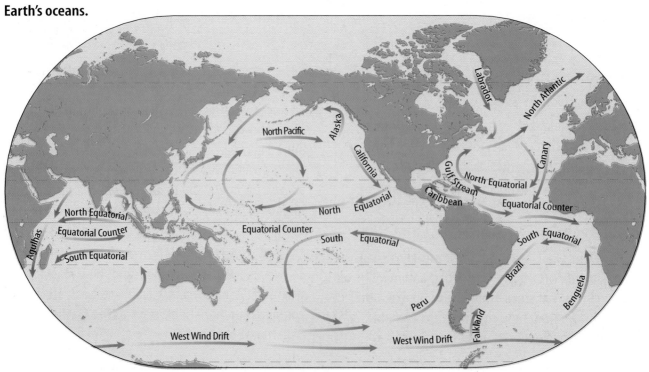

How Surface Currents
Form Surface ocean currents and surface winds are affected by the Coriolis (kor ee OH lus) effect. The **Coriolis effect** is the shifting of winds and surface currents from their expected paths that is caused by Earth's rotation. Imagine that you try to draw a line straight out from the center of a disk to the edge of the disk. You probably could do that with no problem. But what would happen if the disk were slowly spinning like the one in

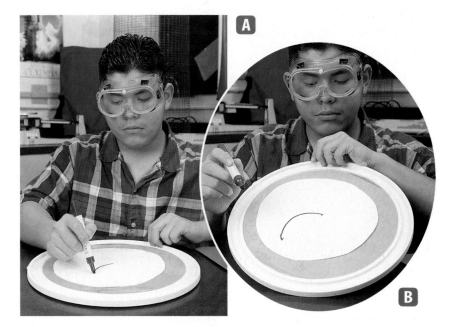

Figure 6A? As the student tried to draw a straight line, the disk rotated and, as shown in **Figure 6B,** the line curved.

A similar thing happens to wind and surface currents. Because Earth rotates toward the east, winds appear to curve to the right in the northern hemisphere and to the left in the southern hemisphere. These surface winds can cause water to pile up in certain parts of the ocean. When gravity pulls water off the pile, the Coriolis effect turns the water. This causes surface water in the oceans to spiral around the piles of water. The Coriolis effect causes currents north of the equator to turn to the right. Currents south of the equator are turned to the left. Look again at the map of surface currents in **Figure 5** to see the results of the Coriolis effect.

The Gulf Stream Although satellites provide new information about ocean movements, much of what is known about surface currents comes from records that were kept by sailors of the nineteenth century. Sailors always have used surface currents to help them travel quickly. Sailing ships depend on some surface currents to carry them to the west and others to carry them east. During the American colonial era, ships floated on the 100-km-wide Gulf Stream current to go quickly from North America to England. Find the Gulf Stream current in the Atlantic Ocean on the map in **Figure 5.**

In the late 1700s, Deputy Postmaster General Benjamin Franklin received complaints about why it took longer to receive a letter from England than it did to send one there. Upon investigation, Franklin found that a Nantucket whaling captain's map furnished the answer. Going against the Gulf Stream delayed ships sailing west from England by up to 110 km per day.

Figure 6
A The student draws a line straight out from the center of the disk. **B** Because the disk was spinning, the line is curved.

Research Visit the Glencoe Science Web site at **science.glencoe.com** for more information about ocean currents. Communicate to your class what you learn.

Tracking Surface Currents Items that wash up on beaches, such as the bottle shown in **Figure 7,** provide information about ocean currents. Drift bottles containing messages and numbered cards are released from a variety of coastal locations. The bottles are carried by surface currents and might end up on a beach. The person who finds a bottle writes down the date and the location where the bottle was found. Then the card is sent back to the institution that launched the bottle. By doing this, valuable information is provided about the current that carried the bottle.

Figure 7
Bottles and other floating objects that enter the ocean are used to gain information about surface currents.

Warm and Cold Surface Currents Notice in **Figure 5** that currents on the west coasts of continents begin near the poles where the water is colder. The California Current that flows along the west coast of the United States is a cold surface current. East-coast currents originate near the equator where the water is warmer. Warm surface currents, such as the Gulf Stream, distribute heat from equatorial regions to other areas of Earth. **Figure 8** shows the warm water of the Gulf Stream in red and orange. Cooler water appears in blue and green.

As warm water flows away from the equator, heat is released to the atmosphere. The atmosphere is warmed. This transfer of heat influences climate.

Figure 8
Data about ocean temperature collected by a satellite were used to make this surface-temperature image of the Atlantic Ocean.

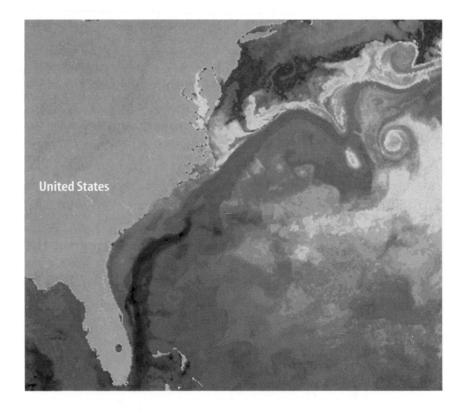

United States

Figure 9
Winds push surface water away from the coast of Peru, causing upwelling. This process brings colder water to the surface.

Water movement

Southerly wind

Upwelling

Upwelling

Upwelling is a circulation in the ocean that brings deep, cold water to the ocean surface. Along some coasts of continents, wind blowing parallel to the coast carries water away from the land because of the Coriolis effect, as shown in **Figure 9.** Cold, deep ocean water rises to the surface and replaces water that has moved away from shore. This water contains high concentrations of nutrients from organisms that died, sank to the bottom, and decayed. Nutrients promote plankton growth, which attracts fish. Areas of upwelling occur along the coasts of Oregon, Washington, and Peru and create important fishing grounds.

Density Currents

Deep in the ocean, waters circulate not because of wind but because of density differences. A **density current** forms when a mass of seawater becomes more dense than the surrounding water. Gravity causes more dense seawater to sink beneath less dense seawater. This deep, dense water then slowly spreads to the rest of the ocean.

The density of seawater can be increased if salinity increases, as you can see if you perform the MiniLAB on this page. It also can be increased by a decrease in temperature. In the Explore Activity, the cold water was more dense than the warm water in the beaker. The cold water sank to the bottom. This created a density current that moved the food coloring.

Changes in temperature and salinity work together to create density currents. Density currents circulate ocean water slowly—moving as little as a few meters per month.

TRY AT HOME

Mini LAB

Modeling a Density Current

Procedure
1. Fill a **clear plastic storage box** (shoe-box size) with room-temperature **water.**
2. Mix several spoonfuls of table **salt** into a **glass** of water at room temperature.
3. Add a few drops of **food coloring** to the saltwater solution. Pour the solution slowly into the freshwater in the large container.

Analysis
1. Describe what happened when you added salt water to freshwater.
2. How does this lab relate to density currents?

Chemistry INTEGRATION

When salt dissolves in water, the freezing point of the mixture is lowered. The greater the number of particles dissolved in the water is, the more the freezing point is lowered. How does this help ocean water near the poles get colder than freshwater could?

Deep Waters An important density current begins in Antarctica where the most dense ocean water forms during the winter. As ice forms, seawater freezes, but the salt is left behind in the unfrozen water. This extra salt increases the salinity and, therefore, the density of the ocean water until it is very dense. This dense water sinks and slowly spreads along the ocean bottom toward the equator, forming a density current. In the Pacific Ocean, this water could take 1,000 years to reach the equator.

In the North Atlantic Ocean, cold, dense water forms around Norway, Greenland, and Labrador. These waters sink, forming North Atlantic Deep Water. In about the northern one third to one half of the Atlantic Ocean, North Atlantic Deep Water forms the bottom layer of ocean water. In the southern part of the Atlantic Ocean, it flows at depths of about 3,000 m, just above the denser water formed near Antarctica. The dense waters circulate more quickly in the Atlantic Ocean than in the Pacific Ocean. In the Atlantic, a density current could circulate in 275 years.

Math Skills Activity

Calculating Density

Example Problem

You have an aquarium full of freshwater in which you have dissolved salt. If the mass of the salt water is 123,000 g and its volume is 120,000 cm^3, what is the density of the salt water?

Solution

1. *This is what you know:* volume: $v = 120,000$ cm^3
   mass of salt water: $m = 123,000$ g

2. *This is what you need to find:* density of water: d

3. *This is the equation you need to use:* $d = m/v$

4. *Substitute the known values:* $d = 123,000\text{g} /120,000\text{cm}^3 = 1.025$ g/cm^3

   Check your answer by multiplying your answer by the volume.
   Do you calculate the same mass of salt water that was given?

┌─ **Practice Problems** ─────────────────────────────────

1. Calculate the density of 78,000 cm^3 of salt water with a mass of 79,000 g.
2. If a sample of ocean water has a density of 1.03 g/cm^3 and a mass of 50,000 g, what is the volume of the water?

For more help, refer to the Math Skill Handbook.

Intermediate Waters A density current also occurs in the Mediterranean Sea, a nearly enclosed body of water. The warm temperatures and dry air in the region cause large amounts of water to evaporate from the surface of the sea. This evaporation increases the salinity and density of the water. This dense water from the Mediterranean flows through the narrow Straits of Gibraltar into the Atlantic Ocean at a depth of about 320 m. When it reaches the Atlantic, it flows to depths of 1,000 m to 2,000 m because it is more dense than the water in the upper parts of the North Atlantic Ocean. However, the water from the Mediterranean is less dense than the very cold, salty water flowing from the North Atlantic Ocean around Greenland, Norway, and Labrador. Therefore, as shown in **Figure 10,** the Mediterranean water forms a middle layer of water—the Mediterranean Intermediate Water.

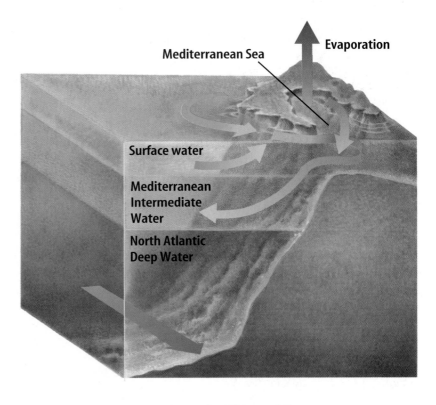

Mediterranean Sea

Evaporation

Surface water

Mediterranean Intermediate Water

North Atlantic Deep Water

 Reading Check *What causes the Mediterranean Intermediate Water to form?*

Figure 10
Dense layers of North Atlantic Deep Water form in the Greenland, Labrador, and Norwegian Seas. This water flows southward along the North Atlantic seafloor. Less dense water from the Mediterranean Sea forms Mediterranean Intermediate Water.

Section 2 Assessment

1. What factors create surface currents?
2. What is the Coriolis effect?
3. How do density currents circulate water?
4. What is upwelling?
5. **Think Critically** The latitudes of San Diego, California, and Charleston, South Carolina, are exactly the same. However, the average yearly water temperature in the ocean off Charleston is much higher than the water temperature off San Diego. Explain why.

Skill Builder Activities

6. **Predicting** A river flows into the ocean. Predict what will happen to this layer of freshwater. Explain your prediction. **For more help, refer to the** Science Skill Handbook.

7. **Using an Electronic Spreadsheet** Make a spreadsheet that compares surface and density currents. Focus on characteristics such as wind, horizontal and vertical movement, temperature, and density. **For more help, refer to the** Technology Skill Handbook.

Ocean Waves and Tides

Figure 11
Ocean waves carry energy through seawater.

A Identify the crests and troughs in this picture.

Waves

If you've been to the seashore or seen a beach on TV, you've watched waves roll in. There is something hypnotic about ocean waves. They keep coming and coming, one after another. But what is an ocean wave? A **wave** is a rhythmic movement that carries energy through matter or space. In the ocean, waves like those in **Figure 11A** move through seawater.

Describing Waves Several terms are used to describe waves, as shown in **Figure 11B.** Notice that waves look like hills and valleys. The **crest** is the highest point of the wave. The **trough** (TRAWF) is the lowest point of the wave. Wavelength is the horizontal distance between the crests or between the troughs of two adjacent waves. Wave height is the vertical distance between crest and trough.

Half the distance of the wave height is called the amplitude (AM pluh tewd) of the wave. The amplitude squared is proportional to the amount of energy the wave carries. For example, a wave with twice the amplitude of the wave in **Figure 11** carries four times ($2 \times 2 = 4$) the energy. On a calm day, the amplitude of ocean waves is small. But during a storm, wave amplitude increases and the waves carry a lot more energy. Large waves can damage ships and coastal property.

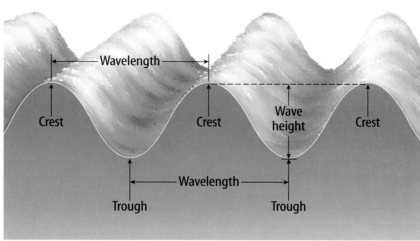

B The crest, trough, wavelength, and wave height describe a wave.

Wave Movement You might have noticed that if you throw a pebble into a pond, a circular wave moves outward from where the pebble entered the water, as shown in **Figure 12.** A bobber on a fishing line floating in the water will bob up and down as the wave passes, but it will not move outward with the wave. Notice that the bobber's position doesn't change.

When you watch an ocean wave, it looks as though the water is moving forward. But unless the wave is breaking onto shore, the water does not move forward. Each molecule of water stays in about the same place as the wave passes. **Figure 13** shows this. Water in a wave moves around in circles. Only the energy moves forward while the water remains in about the same place. Below a depth equal to about half the wavelength, water movement stops. Below that depth, water is not affected by waves. Submarines that travel below this level usually are not affected by surface storms.

Breakers A wave changes shape in the shallow area near shore. Near the shoreline, friction with the ocean bottom slows water at the bottom of the wave. As the wave slows, its crest and trough come closer together. The wave height increases. The top of a wave, not slowed by friction, moves faster than the bottom. Eventually, the top of the wave outruns the bottom and it collapses. The wave crest falls as water tumbles over on itself. The wave breaks onto the shore. **Figure 13** also shows this process. This collapsing wave is a **breaker.** It is the collapse of this wave that propels a surfer and surfboard onto shore. After a wave breaks onto shore, gravity pulls the water back into the sea.

 *What causes an ocean wave to slow down?*

Mini LAB

Modeling Water Particle Movement

Procedure 🥽 👕
1. Put a piece of **tape** on the outside bottom of a clear, rectangular **plastic storage box.** Fill the box with **water.**
2. Float a **cork** in the container above the piece of tape.
3. Use a **spoon** to make gentle waves in the container.
4. Observe the movement of the waves and the cork.

Analysis
1. Describe the movement of the waves and the motion of the cork.
2. Compare the movement of the cork in the water with the movement of water particles in a wave.

Figure 13

As ocean waves move toward the shore, they seem to be traveling in from a great distance, hurrying toward land. Actually, the water in waves moves relatively little, as shown here. It's the energy in the waves that moves across the ocean surface. Eventually that energy is transferred—in a crash of foam and spray—to the land.

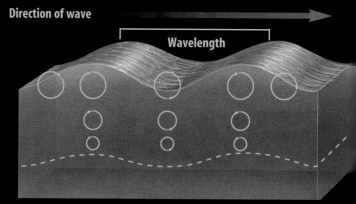

Direction of wave

Wavelength

A Particles of water move around in circles rather than forward. Near the water's surface, the circles are relatively large. Below the surface, the circles become progressively smaller. Little water movement occurs below a depth equal to about one-half of a wave's length.

B The energy in waves, however, does move forward. One way to visualize this energy movement is to imagine a line of dominoes. Knock over the first domino, and the others fall in sequence. As they fall, individual dominoes—like water particles in waves—remain close to where they started. But each transfers its energy to the next one down the line.

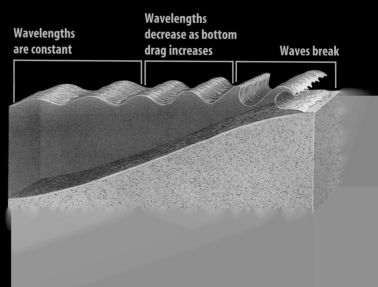

Wavelengths are constant

Wavelengths decrease as bottom drag increases

Waves break

C As waves approach shore, wavelength decreases and wave height increases. This causes breakers to form. Where ocean floor rises steeply to beach, incoming waves break quickly at a great height, forming huge arching waves.

How Water Waves Form On a windy day, waves form on a lake or ocean. When wind blows across a body of water, friction causes the water to move along with the wind. If the wind speed is great enough, the water begins to pile up, forming a wave. As the wind continues to blow, the wave increases in height. Some waves reach tremendous heights, as shown in **Figure 14.** Storm winds have been known to produce waves more than 30 m high—taller than a six-story building.

The height of waves depends on the speed of the wind, the distance over which the wind blows, and the length of time the wind blows. When the wind stops blowing, waves stop forming. But once set in motion, waves continue moving for long distances, even if the wind stops. The waves you see lapping at a beach could have formed halfway around the world.

☑ **Reading Check** *What factors affect the height of waves?*

Tides

When you go to a beach, you probably notice the level of the sea rise and fall during the day. This rise and fall in sea level is called a **tide.** A tide is caused by a giant wave produced by the gravitational pull of the Sun and the Moon. This wave has a wave height of only 1 m or 2 m, but it has a wavelength that is thousands of kilometers long. As the crest of this wave approaches the shore, sea level appears to rise. This rise in sea level is called high tide. Later, as the trough of the wave approaches, sea level appears to drop. This drop in sea level is referred to as low tide.

Research Visit the Glencoe Science Web site at **science.glencoe.com** for more information about tides. Communicate to your class what you learn.

Figure 15
A large difference between high tide and low tide can be seen at Mont-Saint-Michel off the northwestern coast of France.

A Mont-Saint-Michel lies about 1.6 km offshore and is connected to the mainland at low tide.

B Incoming tides move very quickly, making Mont-Saint-Michel an island at high tide.

Tidal Range As Earth rotates, different locations on Earth's surface pass through the high and low positions. Many coastal locations, such as the Atlantic and Pacific coasts of the United States, experience two high tides and two low tides each day. One low-tide/high-tide cycle takes 12 h, 25 min. A daily cycle of two high tides and two low tides takes 24 h, 50 min—slightly more than a day. But because ocean basins vary in size and shape, some coastal locations, such as many along the Gulf of Mexico, have only one high and one low tide each day. The **tidal range** is the difference between the level of the ocean at high tide and low tide. Notice the tidal range in the photos in **Figure 15.**

Extreme Tidal Ranges The shape of the seacoast and the shape of the ocean floor affect the ranges of tides. Along a smooth, wide beach, the incoming water can spread over a large area. There the water level might rise only a few centimeters at high tide. In a narrow gulf or bay, however, the water might rise many meters at high tide.

Most shorelines have tidal ranges between 1 m and 2 m. Some places, such as those on the Mediterranean Sea, have tidal ranges of only about 30 cm. Other places have large tidal ranges. Mont-Saint-Michel, shown in **Figure 15,** lies in the Gulf of Saint-Malo off the northwestern coast of France. There the tidal range reaches about 13.5 m.

The dock shown in **Figure 16** is in Digby, Nova Scotia in the Bay of Fundy. This bay is extremely narrow, which contributes to large tidal ranges. The difference between water levels at high tide and low tide can be as much as 15 m.

Figure 16
The Bay of Fundy has the greatest tidal range in the world. *Was this picture taken at high tide or low tide?*

Tidal Bores In some areas when a rising tide enters a shallow, narrow river from a wide area of the sea, a wave called a tidal bore forms. A tidal bore can have a breaking crest or it can be a smooth wave. Tidal bores tend to be found in places with large tidal ranges. The Amazon River in Brazil, the Tsientang River in China, and rivers that empty into the Bay of Fundy in Nova Scotia have tidal bores.

When a tidal bore enters a river, it causes surface water to reverse its flow. In the Amazon River, the tidal bore rushes 650 km upstream at speeds of 65 km/h, causing a wave more than 5 m in height. Four rivers that empty into the Bay of Fundy have tidal bores. In those rivers, bore rafting is a popular sport.

The Gravitational Effect of the Moon For the most part, tides are caused by the interaction of gravity in the Earth-Moon system. The Moon's gravity exerts a strong pull on Earth. Earth and the water in Earth's oceans respond to this pull. The water bulges outward as Earth and the Moon revolve around a common center of mass. These events are explained in **Figure 17.**

Two bulges of water form, one on the side of Earth closest to the Moon and one on the opposite side of Earth. The reason two bulges form is because the Moon's gravity pulls harder on parts of Earth closer to the Moon than on parts farther away. You can imagine this if you think about pulling a large ball of dough in the same direction but harder on one side than the other side. The ball of dough will stretch and form two bulges. Earth does the same thing. The ocean bulges are the high tides, and the areas of Earth's oceans that are not toward or away from the Moon are the low tides. As Earth rotates, different locations on its surface pass through high and low tide.

Life Science
INTEGRATION

Limpets are sea snails that live on rocky shores. When the tide comes in, they glide over the rocks to graze on seaweed. When the tide goes out, they use strong muscles to pull their shells tight against the rocks. Find out how other organisms survive in the zone between high and low tides.

Figure 17
The Moon and Earth revolve around a common center of mass. Because the Moon's gravity pulls harder on parts of Earth closer to the Moon, a bulge of water forms on the side of Earth facing the Moon and the side of Earth opposite the Moon.

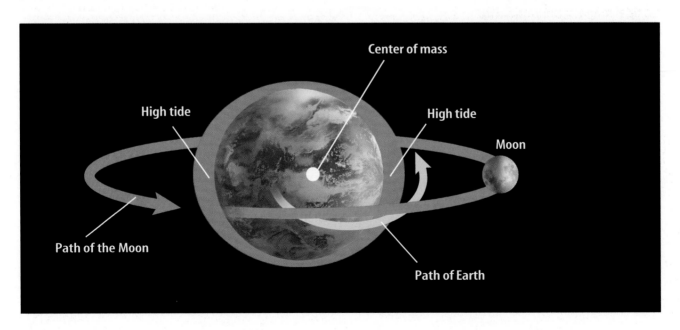

Figure 18
The gravitational attraction of the Sun causes spring tides and neap tides.

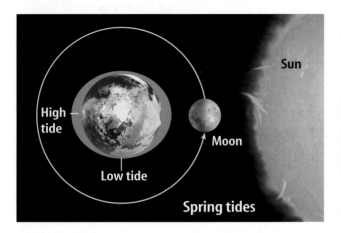

A When the Sun, the Moon, and Earth are aligned, spring tides occur.

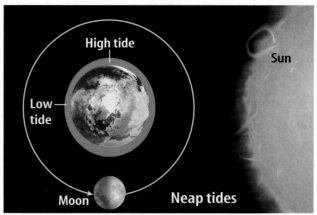

B When the Sun, Earth, and the Moon form a right angle, neap tides occur.

The Gravitational Effect of the Sun The Sun also affects tides. The Sun can strengthen or weaken the Moon's effects. When the Moon, Earth, and the Sun are lined up together, the combined pull of the Sun and the Moon causes spring tides, shown in **Figure 18A.** During spring tides, high tides are higher and low tides are lower than normal. The name *spring tide* has nothing to do with the season of spring. It comes from the German word *springen,* which means "to jump." When the Sun, Earth, and the Moon form a right angle, as shown in **Figure 18B,** high tides are lower and low tides are higher than normal. These are called neap tides.

Section 3 Assessment

1. Describe the parts of an ocean wave.
2. How does wind create water waves?
3. What causes high tides? Spring tides?
4. Compare water and wave movement.
5. **Think Critically** At the ocean, you spot a wave about 200 m from shore. A few seconds later, the wave breaks on the beach. Explain why the water in the breaker is not the same water that was in the wave 200 m away.

Skill Builder Activities

6. **Comparing and Contrasting** Compare and contrast the effects of the Sun and the Moon on Earth's tides. **For more help, refer to the** Science Skill Handbook.
7. **Communicating** Many planets have more than one moon. In your Science Journal, write a description of what tides might be like if Earth had two moons. **For more help, refer to the** Science Skill Handbook.

Activity

Making Waves

W ind generates some waves. The energy of motion is transferred from the wind to the surface water of the ocean. What factors influence the generation of waves?

What You'll Investigate
How do the speed of the wind and the length of time the wind blows affect the height of a wave?

Materials
11" × 14" white paper	water
3-speed electric fan	metric ruler
gooseneck lamp	
clock or watch	
rectangular, clear-plastic storage box	

Goals
■ **Observe** how wind speed and duration affect wave height.

Safety Precautions
Do not allow any part of the light or cord to come in contact with the water.

Procedure

1. Position the box on white paper beside the lamp.

2. Fill the plastic box with water to within 3 cm of the top. Direct light from the lamp onto the box.

3. Place the fan at one end of the box to create waves. Start the fan on its slowest speed. Keep the fan on during measuring.

4. After 3 min, measure the height of the waves caused by the fan. Record your observations in a table similar to the one shown. Through the plastic box, observe the shadows of the waves on the white paper.

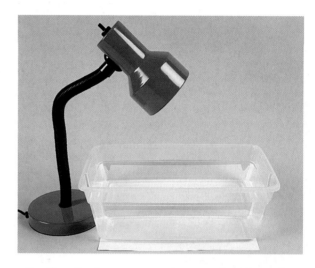

5. After 5 min, measure the wave height and record your observations.

6. Repeat steps 3 to 5 with the fan on medium, then on high.

7. Turn off the fan. Unplug it. Observe what happens.

Wave Data

Fan Speed	Time (min)	Wave Height (mm)	Observations
Low	3		
Low	5		
Medium	3		
Medium	5		
High	3		
High	5		

Conclude and Apply

1. **Analyze** your data to determine whether the wave height is affected by the length of time that the wind blows. Explain.

2. **Analyze** your data to determine whether the height of the waves is affected by the speed of the wind. Explain.

Sink or Float

As you know, ocean water contains many dissolved salts. How does this affect objects within the oceans? Why do certain objects float on top of the ocean's waves, while others sink directly to the bottom? Density is a measurement of mass per volume. You can use density to determine whether an object will float within a certain volume of water of a specific salinity. In this activity you will investigate the effect of salinity on whether an object floats or sinks.

Recognizing the Problem

How does salinity affect whether a potato will float or sink?

Form a Hypothesis

Based on what you know so far about salinity, why things float or sink, and the density of a potato, plus what it looks and feels like, formulate a hypothesis. Do you think the salinity of water has any effect on objects that are floating in water? What kind of effect? Will they float or sink? How would a dense object like a potato be different from a less dense object like a cork?

Goals
■ **Design** an experiment to identify how increasing salinity affects the ability of a potato to float in water.

Possible Materials
small, uncooked potato
teaspoon
salt
large glass bowl
water
balance
large graduated cylinder
metric ruler

Safety

Test Your Hypothesis

Plan

1. As a group, agree upon and write your hypothesis statement.

2. Devise a method to test how salinity affects whether a potato floats in water.

3. **List** the steps you need to take to test your hypothesis. Be specific, describing exactly what you will do at each step.

4. Read over your plan for testing your hypothesis.

5. How will you determine the densities of the potato and the different water samples? How you will measure the salinity of the water? How will you change the salinity of the water? Will you add teaspoons of salt one at a time?

6. How you will measure the ability of an object to float? Could you somehow measure the displacement of the water? Perhaps you could draw a line somewhere on your bowl and see how the position of the potato changes.

7. **Design** a data table where you can record your results. Include columns/rows for the salinity and float/sink measurements. What else should you include?

Do

1. Make sure your teacher approves your plan before you start.

2. Carry out the experiment.

3. While conducting the experiment, record your data and any observations that you or other group members make in your Science Journal.

Analyze Your Data

1. **Compare** how the potato floated in water with different salinities.

2. How does the ability of an object to float change with changing salinity?

Draw Conclusions

1. Did your experiment support the hypothesis you made?

2. A heavily loaded ship barely floats in the Gulf of Mexico. Based on what you learned, infer what might happen to the ship if it travels into the freshwater of the Mississippi River.

Communicating

Your Data

Prepare a chart showing the results of your experiment. Share the chart with members of your class. **For more help, refer to the** Science Skill Handbook.

"The Jungle of Ceylon"
from Passions and Impressions
by Pablo Neruda

Respond to the Reading

1. What were his impressions of the island on arrival?
2. What words does the author choose to describe waves?
3. How would you describe the climate of Ceylon?

The following passage is part of a travel chronicle describing the Chilean poet Pablo Neruda's visit to the island of Ceylon, now called Sri Lanka, which is located southeast of India. The author considered himself so connected to Earth that he wrote in green ink.

Felicitous[1] shore! A coral reef stretches parallel to the beach; there the ocean interposes in its blues the perpetual white of a rippling ruff[2] of feathers and foam; the triangular red sails of sampans[3]; the unmarred line of the coast on which the straight trunks of the coconut palms rise like explosions, their brilliant green Spanish combs nearly touching the sky.

… In the deep jungle, there is a silence like that of libraries: abstract and humid.

1 Happy
2 round collar made of layers of lace
3 East Asian boats

Understanding Literature

Imagery Imagery is a series of words that evoke pictures to the reader. Poets use imagery to connect images to abstract concepts. The poet, here, wants to capture a particular feature of the reef and does so by describing it as a "ruff of feathers and foam," invoking the image of a gentle place, without the author saying so. Imagery also gives the reader more information about the story or chronicle. The poet further describes the shore as "happy", which helps us learn that the poet is arriving on the island on a clear, calm day.

Where else in the poem does the poet use imagery to convey a mood or feeling?

Science Connection In the poem there are several indicators that the wind, which causes waves and currents, is light and the waves are small, using imagery as discussed above.

Sri Lanka, however, often is plagued by monsoons, which affect ocean conditions and local climate. Monsoons are seasonal reversals of the regional winds. During the wet season, moist winds blow in from the sea, causing storms and producing waves. During the dry season, winds blow from the land and sunny days are common.

Linking Science and Writing

Weather Report Write a weather report for fishers and others who work at sea. Pick a geographic location to focus on. If possible, do research on wave conditions in this area. Include the times for low and high tides in your report. Add any additional weather information that you think might be important to people who work at or around the sea.

Career Connection

Oceanographer

Oceanographer Dr. Robert D. Ballard is an American oceanographer who revolutionized deep-sea archaeology. He developed several high-tech vessels that can explore ocean bottoms previously out of reach. Dr. Ballard discovered the location of the wreckage of the *Titanic, Lusitania,* and *Bismark.* He also discovered the wreckage of eight ancient ships in the Mediterranean Sea. Dr. Ballard has degrees in chemistry and geology as well as doctorate degrees in marine geology and geophysics.

SCIENCE *Online* To learn more about careers in oceanography, visit the Glencoe Science Web site at **science.glencoe.com.**

Reviewing Main Ideas

Section 1 Ocean Water

1. Earth's ocean water might have originated from water vapor released from volcanoes. Over millions of years, the water condensed and rain fell, filling basins.

2. The oceans are a mixture of water, dissolved salts, and dissolved gases that are in constant motion.

3. Groundwater and rivers weather rock and dissolve some minerals to form ions. The ions are carried to the oceans where they give seawater its salty taste. *What kind of salt, shown here, makes up most of the salt left when seawater is evaporated?*

Section 2 Ocean Currents

1. Wind causes surface currents. Surface currents are affected by the Coriolis effect. The Coriolis effect turns currents north of the equator clockwise and turns currents south of the equator counterclockwise.

2. Surface currents can greatly affect climate and economic activity such as fishing. Upwelling brings deep, cold water to the ocean's surface.

3. Cool currents off western coasts originate far from the equator. Warmer currents along eastern coasts begin near the equator.

4. Differences in temperature and salinity between water masses in the oceans set up circulation patterns called density currents. *How do density currents originate in the Mediterranean Sea, shown here?*

Section 3 Ocean Waves and Tides

1. A wave is a rhythmic movement that carries energy. The crest is the highest point of a wave. The trough is the lowest point.

2. In a wave, energy moves forward while water particles move around in small circles.

3. Wind causes water to pile up and form most water waves. Tides are not caused by wind. *What causes the high and low tides shown here?*

FOLDABLES
Reading & Study Skills

After You Read

Under the bottom tab of your Foldable, write about the effects of ocean motion on climate, world economics, and ocean water.

Visualizing Main Ideas

Complete the following concept map on ocean motions.

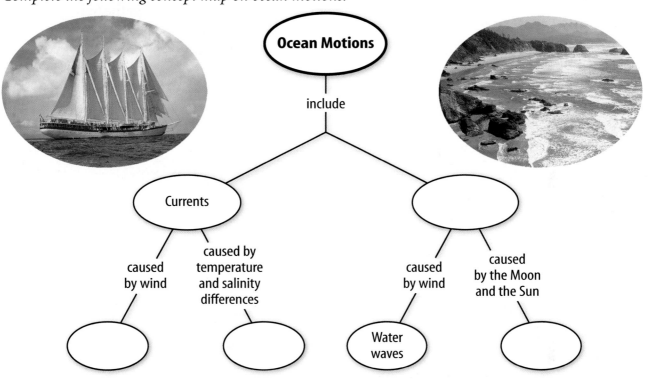

Vocabulary Review

Vocabulary Words

a. basin
b. breaker
c. Coriolis effect
d. crest
e. density current
f. salinity

g. surface current
h. tidal range
i. tide
j. trough
k. upwelling
l. wave

Study Tip

After you've read a chapter, go back to the beginning and speed-read through what you've just read. This will help you better remember what you have read.

Using Vocabulary

Replace the underlined words with the vocabulary words that have the same meaning.

1. The <u>amount of dissolved salts</u> in seawater has stayed about the same for hundreds of millions of years.

2. An <u>area where nutrient-rich water comes to the surface</u> is a good place to catch fish.

3. Wind creates a <u>horizontal current at the top of the ocean</u>.

4. Along most ocean beaches, a <u>rise and fall of the ocean related to gravitational pull</u> is easy to see.

5. Wind pushes on water to make a <u>movement of energy through the water</u>.

Checking Concepts

Choose the word or phrase that best answers the question.

1. Where might ocean water have originated?
   A) salt marshes   C) basins
   B) volcanoes   D) surface currents

2. How does chlorine enter the oceans?
   A) volcanoes   C) density currents
   B) rivers   D) groundwater

3. What is the most common ion found in ocean water?
   A) chloride   C) boron
   B) calcium   D) sulfate

4. What causes most surface currents?
   A) density differences   C) salinity
   B) the Gulf Stream   D) wind

5. What is the highest point on a wave called?
   A) wave height   C) crest
   B) trough   D) wavelength

6. In the ocean, what is the rhythmic movement that carries energy through seawater?
   A) current   C) crest
   B) wave   D) upwelling

7. Which of the following causes the density of seawater to increase?
   A) a decrease in temperature
   B) a decrease in salinity
   C) an increase in temperature
   D) a decrease in pressure

8. In which direction does the Coriolis effect cause currents in the northern hemisphere to turn?
   A) east
   B) south
   C) counterclockwise
   D) clockwise

9. Tides are affected by the positions of which celestial bodies?
   A) Earth and the Moon
   B) Earth, the Moon, and the Sun
   C) Venus, Earth, and Mars
   D) the Sun, Earth, and Mars

10. What affects surface currents?
    A) crests   C) the Coriolis effect
    B) upwellings   D) tides

Thinking Critically

11. If a sealed bottle is dropped into the ocean off the coast of Florida, where do you think it might wash up? Explain.

12. Why do silicon and calcium remain in seawater for a shorter time than sodium?

13. Describe the Antarctic density current.

14. How would the density of seawater at the mouth of the Mississippi River and in the Mediterranean Sea compare? Explain.

15. Refer to the graph below. On which day is the high tide highest? Lowest? On which day(s) is the low tide lowest? Highest? On which day(s) would Earth, the Moon, and the Sun be lined up? On which day(s) would the Moon, Earth, and the Sun form a right angle?

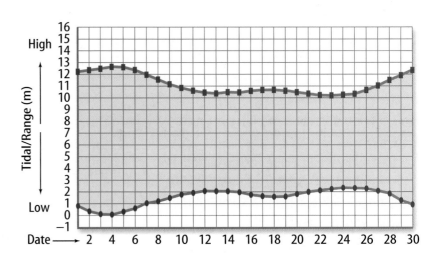

Developing Skills

16. Recognizing Cause and Effect What causes upwelling? What effect does it have? What can happen when upwelling stops?

17. Comparing and Contrasting Compare and contrast ocean waves and ocean currents.

18. Predicting Predict how drift bottles that are dropped into the ocean at points A and B will move. Explain.

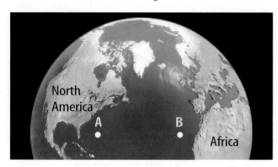

19. Recognizing Cause and Effect In the Mediterranean Sea, a density current forms because of the high rate of evaporation of water from the surface. How can evaporation cause a density current?

Performance Assessment

20. Invention Design a method for desalinating water that does not use solar energy. Draw it, and display it for your class.

21. Design and Perform an Experiment Create an experiment to test the density of water at different temperatures.

TECHNOLOGY

Go to the Glencoe Science Web site at **science.glencoe.com** or use the **Glencoe Science CD-ROM** for additional chapter assessment.

Test Practice

A marine scientist used the following graphic to support a lecture about ocean currents.

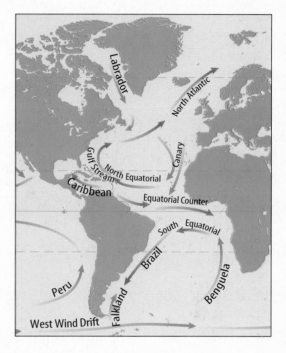

Study the graphic and answer the following questions.

1. The direction of ocean currents in the northern hemisphere is _____.
   A) counterclockwise
   B) north to south only
   C) clockwise
   D) east to west only

2. A reasonable conclusion based on the information in the graphic is that _____.
   F) the ocean's currents only flow in one direction
   G) the ocean's waters are constantly in motion
   H) the Gulf Stream flows east to west
   J) the Atlantic Ocean is deep

Oceanography

W hat would it be like to live in the ocean? Imagine all the different organisms you might see. What would the ocean floor look like? In this chapter, you will learn about ocean basin features and the structures located along ocean plate boundaries. You'll learn about the mineral resources found in the ocean. You also will read about many different kinds of ocean inhabitants. Finally, you'll learn how the oceans are being polluted, the effects of this pollution, and what you can do to help clean up the oceans.

What do you think?

Science Journal Examine the photo below with a classmate. Discuss what this might be or what is happening. Here's a hint: *It lives in the dark depths.* Write your answer or best guess in your Science Journal.

EXPLORE ACTIVITY

Scientists use sonar to map the sea bottom. Using the speed of sound waves and the time it takes them to travel, scientists can calculate the depth of the ocean bottom. You will model sonar in this activity.

Model sonar

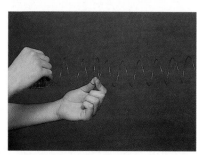

1. With one person holding each end, stretch a spring until it is taut. Measure the distance between the ends.

2. Pinch two coils together. When the spring is steady, release the coils to create a wave.

3. Record the time it takes the wave to travel back and forth five times. Divide this number by five to calculate the time of one round trip.

4. Calculate the speed of the wave by multiplying the distance by two and dividing this number by the time.

5. Move closer to your partner. Take in coils to keep the spring at the same tension. Repeat steps 2 and 3.

6. Calculate the new distance by multiplying the new time by the speed from step 4, and then dividing this number by two.

Observe

Write in your Science Journal about how this activity models sonar.

Before You Read

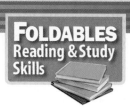

FOLDABLES
Reading & Study Skills

Making a Main Ideas Study Fold
Make the following Foldable to help you identify the major topics about oceanography.

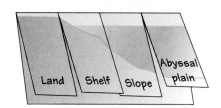

1. Place a sheet of paper in front of you so the long side is at the top. Fold the paper in half from top to bottom.

2. Now fold the paper in half from the left side to the right side two times. Unfold the last two folds.

3. Through the top thickness of paper, cut along each of the fold lines to form four tabs. Label the tabs *Land, Shelf, Slope*, and *Abyssal Plain*.

4. Before you read the chapter, draw the regions of the ocean on the front of the tabs. As you read the chapter, write information about each region under the tabs.

The Seafloor

The Ocean Basins

Imagine yourself driving a deep-sea submersible along the ocean floor. Surrounded by cold, black water, the lights of your vessel reflect off of what looks like a mountain range ahead. As you continue, you find a huge opening in the seafloor—so deep you can't even see the bottom. What other ocean floor features can you find in **Figure 1**?

Ocean basins, which are low areas of Earth that are filled with water, have many different features. Beginning at the ocean shoreline is the continental shelf. The **continental shelf** is the gradually sloping end of a continent that extends under the ocean. On some coasts, the continental shelf extends a long distance. For instance, on North America's Atlantic and Gulf coasts, it extends 100 km to 350 km into the sea. On the Pacific Coast, where the coastal range mountains are close to the shore, the shelf is only 10 km to 30 km wide. The ocean covering the continental shelf can be as deep as 350 m.

Figure 1
This map shows features of the ocean basins. Locate a trench and a mid-ocean ridge.

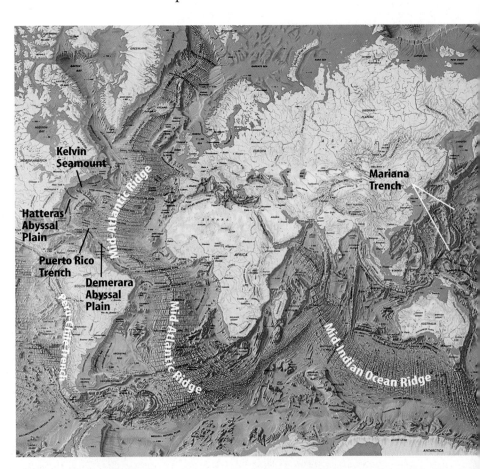

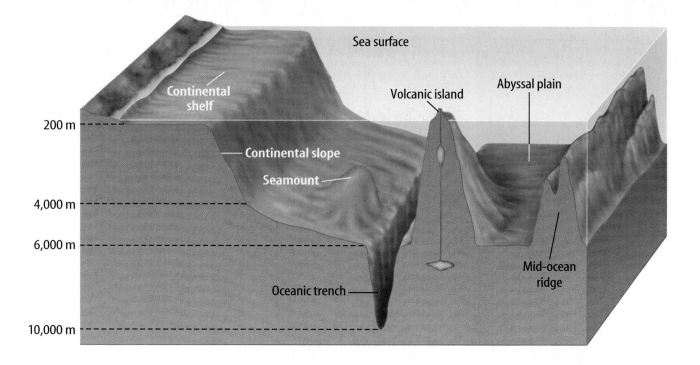

Sea surface

Continental shelf

Continental slope

Seamount

200 m

4,000 m

6,000 m

Oceanic trench

10,000 m

Volcanic island

Abyssal plain

Mid-ocean ridge

Figure 2 shows that beyond the shelf, the ocean floor drops more steeply, forming the continental slope. The **continental slope** extends from the outer edge of the continental shelf down to the ocean floor. Beyond the continental slope lie the trenches, valleys, plains, mountains, and ridges of the ocean basin.

In the deep ocean, sediment, dervied mostly from land, settles constantly on the ocean floor. These deposits fill in valleys and create flat seafloor areas called **abyssal** (uh BIH sul) **plains.**

Abyssal plains are from 4,000 m to 6,000 m below the ocean surface. Can you locate the abyssal plain shown in **Figure 2?**

In the Atlantic Ocean, areas of extremely flat abyssal plains can be large. One example is the Canary Abyssal Plain, which has an area of approximately 900,000 km^2. Other abyssal plains found in the Atlantic Ocean include the Hatteras and Demerara Abyssal Plains, both shown in **Figure 1.** Some areas of abyssal plains have small hills and seamounts. Seamounts are underwater, inactive volcanic peaks. They most commonly are found in the Pacific Ocean. Can you locate a seamount in **Figure 1?**

Reading Check *What are seamounts?*

Figure 2
Ocean basin features are continuous from shore to shore. (Features in this diagram are not to scale.) *Where does the continental shelf end and the continental slope begin?*

Aleutian Abyssal Plain

East Pacific Rise

Peru-Chile Trench

Ridges and Trenches

Locate the Mid-Atlantic Ridge in **Figure 1.** Mid-ocean ridges can be found at the bottom of all ocean basins. They form a continuous underwater ridge approximately 70,000 km long. A **mid-ocean ridge** is the area in an ocean basin where new ocean floor is formed. Crustal plates, which are large sections of Earth's uppermost mantle and crust, are moving constantly. As they move, the ocean floor changes. When ocean plates separate, hot magma from Earth's interior forms new ocean crust. This is the process of seafloor spreading. New ocean floor is being formed at a rate of approximately 2.5 cm per year along the Mid-Atlantic Ridge.

New ocean floor forms along mid-ocean ridges as lava erupts through cracks in Earth's crust. **Figure 3** shows newly erupted lava on the seafloor. When the lava hits the water, it cools quickly into solid rock, forming new seafloor. While seafloor is being formed in some parts of the oceans, it is being destroyed in others. Areas where old ocean floor slides beneath another plate and descends into Earth's mantle are called subduction zones.

Figure 3
New seafloor forms at mid-ocean ridges. A type of lava called pillow lava lies newly formed at this ridge on the ocean floor.

Reading Check *How does new ocean floor form?*

Math Skills Activity

Calculating a Feature's Slope

Example Problem

If the width of a continental shelf is 320 km and it increases in depth a total of 300 m in that distance, what is its slope?

Solution

1. *This is what you know:*   width = 320 km
   increase in depth = 300 m

2. *This is what you need to find:*   slope: s
3. *This is the equation you need to use:* s = increase in depth ÷ width
4. *Solve the equation by substituting in known values:* s = 300 m ÷ 320 km = 0.94 m/km

> **Practice Problem**
>
> The width of a continental slope is 40 km. It increases in depth by 2,000 m. What is the slope of the continental slope?

For more help, refer to the Math Skill Handbook.

Figure 4
Located at subduction zones, trenches are important ocean basin features.

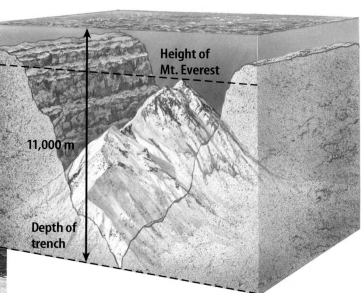

Height of Mt. Everest

11,000 m

Depth of trench

A In 1960, the world's deepest dive was made in the Mariana Trench. The *Trieste* carried Jacque Piccard and Donald Walsh to a depth of almost 11 km.

B If Earth's tallest mountain, Mount Everest, were set in the bottom of the Mariana Trench of the Pacific Basin, it would be covered with more than 2,000 m of water.

Subduction Zones On the ocean floor, subduction zones are marked by deep ocean trenches. A **trench** is a long, narrow, steep-sided depression where one crustal plate sinks beneath another. Most trenches are found in the Pacific Basin. Ocean trenches are usually longer and deeper than any valley on any continent. One trench, famous for its depth, is the Mariana Trench. It is located to the south and east of Japan in the Pacific Basin. This trench reaches 11 km below the surface of the water, and it is the deepest place in the Pacific. **Figure 4A** shows the deep-sea vessel, the *Trieste*, that descended into the trench in 1960. **Figure 4B** shows that the Mariana Trench is so deep that Mount Everest could easily fit into it.

Mineral Resources from the Seafloor

Resources can be found in many places in the ocean. Some deposits on the continental shelf are relatively easy to extract. Others can be found only in the deep abyssal regions on the ocean floor. People still are trying to figure out how to get these valuable resources to the surface. As you read, suggest some methods that could be used to retrieve hard-to-reach resources.

TRY AT HOME
Mini LAB

Modeling the Mid-Atlantic Ridge

Procedure
1. Set two **tray tables** 2 cm apart.
2. Gather ten **paper towels** that are still connected. Lay one end of the paper towels on each table so the towels hang down into the space between the tables.
3. Slowly pull each end of the paper towels away from each other.

Analysis
1. Explain how this models the Mid-Atlantic Ridge.
2. How long does it take for 2.5 cm of new ocean crust to form at the Mid-Atlantic Ridge? How long does it take 25 cm to form?

Figure 5
The ocean is rich with mineral resources. *In what areas of the world can phosphorite be found? Diamonds?*

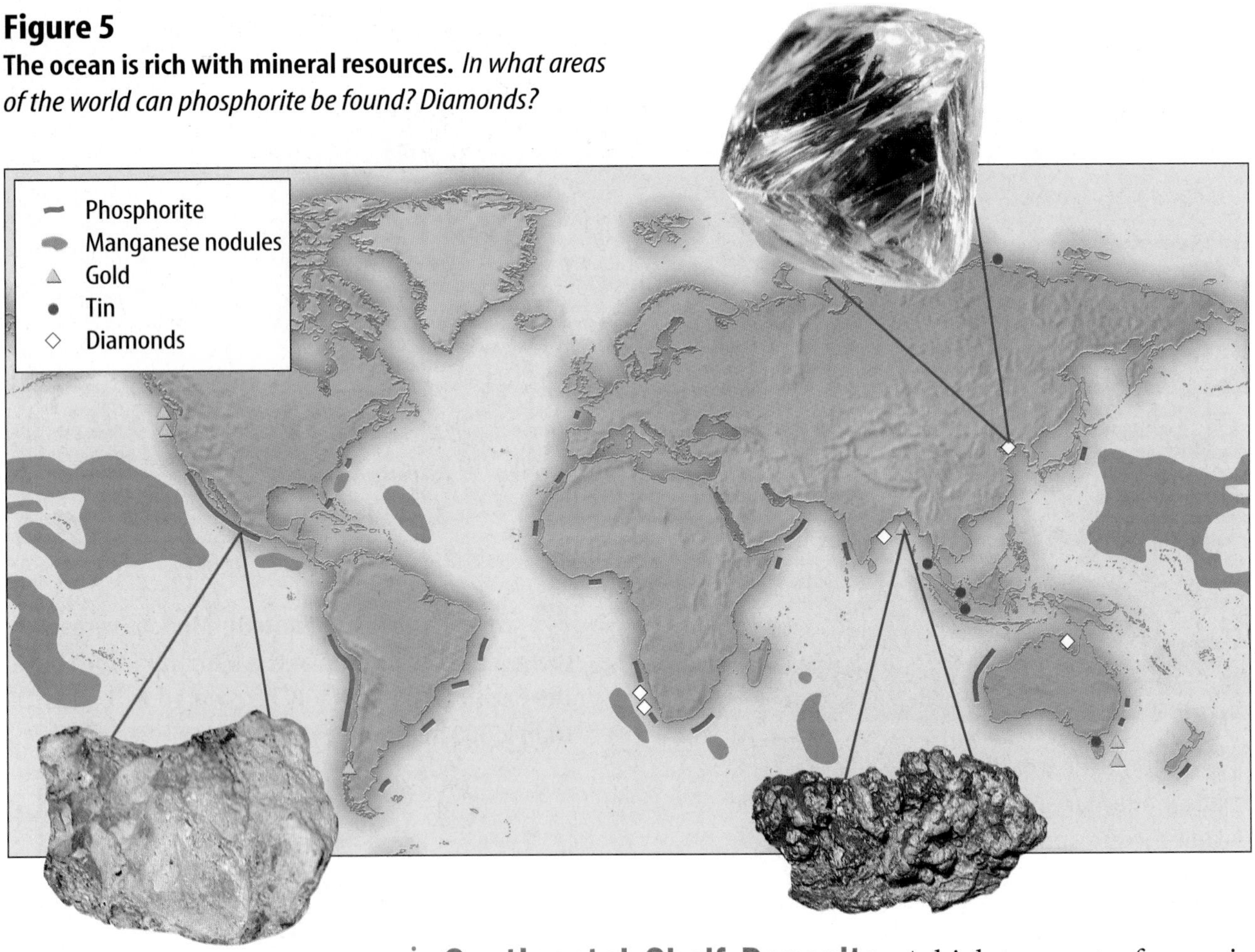

Continental Shelf Deposits A high amount of organic activity occurs in the waters above the continental shelf, and sediment accumulates to great thickness on the ocean floor. This is why many different kinds of resources can be found there, such as petroleum and natural gas deposits. Approximately 20 percent of the world's oil comes from under the seabed. To extract these substances, wells are drilled into the seafloor from floating vessels and fixed platforms.

Other deposits on the continental shelf include phosphorite, which is used to make fertilizer, and limestone, which is used to make cement. Sand and gravel, both economically important, also can be dredged from the continental shelf.

Rivers that flow into oceans transport important minerals to the continental shelf from land. Sometimes the energy of ocean waves and currents can cause denser mineral grains that have been brought in by rivers to concentrate in one place. These deposits, called placer (PLAHS ur) deposits, can occur in coastal regions where rivers entering the ocean suddenly lose energy, slow down, and drop their sediment. Metals such as gold and titanium and gemstones such as diamonds are mined from placer deposits in some coastal regions. **Figure 5** shows where some resources in the ocean can be found.

Deep-Water Deposits Through the holes and cracks along mid-ocean ridges, plumes of hot water billow out into surrounding seawater. As the superheated water cools, mineral deposits sometimes form. As a result, elements such as sulfur and metals like iron, copper, zinc, and silver can be concentrated in these areas. Today, no one is mining these valuable materials from the depths because it would be too expensive to recover them. However, in the future, these deposits could become important.

Other mineral deposits can precipitate from seawater. In this process, minerals that are dissolved in ocean water come out of solution and form solids on the ocean floor. Manganese nodules are small, darkly colored lumps strewn across large areas of the ocean basins. **Figure 6** shows these nodules. Manganese nodules form by a chemical process that is not fully understood. They form around nuclei such as discarded sharks' teeth, growing slowly, perhaps as little as 1 mm to 10 mm per million years. These nodules are rich in manganese, copper, iron, nickel, and cobalt, which are used in the manufacture of steel, paint, and batteries. Most of the nodules lie thousands of meters deep in the ocean and are not currently being mined, although suction devices similar to huge vacuum cleaners have been tested to collect them.

Figure 6
These manganese nodules were found on the floor of the Pacific Ocean. *Can you think of an efficient way to gather them from a depth of 4 km?*

Section 1 Assessment

1. Compare and contrast continental shelves and continental slopes.

2. Contrast mid-ocean ridges and trenches.

3. Describe what an abyssal plain looks like and how it forms.

4. How do placer deposits form? Name two examples of placer deposits that are mined. How do manganese nodules form?

5. **Think Critically** Depth soundings, taken as a ship moves across an ocean, are consistently between 4,000 m and 4,500 m. Infer over which area of seafloor the ship is passing.

Skill Builder Activities

6. **Comparing and Contrasting** Compare and contrast the Atlantic Ocean Basin with the Pacific Ocean Basin. Which basin contains many deep-sea trenches? Which basin is getting larger with time? **For more help, refer to the** Science Skill Handbook.

7. **Using Statistics** Each year for three years, the distance between two locations across an ocean basin increases by 1.8 cm, 4.1 cm, and 3.2 cm, respectively. What is the average rate of separation of these locations during this time? **For more help, refer to the** Math Skill Handbook.

Activity

Mapping the Ocean Floor

How do people know what the ocean floor looks like? Scientists use sonar and radar to identify structures on the ocean floor. Data from these instruments are used to calculate the distance to the bottom. You will use data collected in the Atlantic Ocean to make a bottom profile.

What You'll Investigate
What does the ocean floor look like?

Materials
graph paper

Goals
- **Make** a profile of the ocean floor.
- **Identify** seafloor structures.

Procedure

1. Set up a graph as shown.

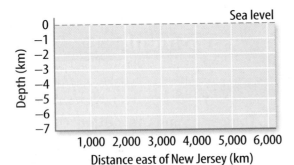

2. **Plot** each data point and connect the points with a smooth line.

3. **Color** water blue and the seafloor brown.

Conclude and Apply

1. What ocean floor structures occur between 160 km and 1,050 km east of New Jersey? Between 2,000 km and 4,500 km? Between 5,300 km and 5,650 km?

2. When a profile of a feature is drawn to scale, the horizontal and vertical scales must be the same. Does your profile give an accurate picture of the ocean floor? Explain.

Ocean Floor Data		
Station Number	Distance from New Jersey (km)	Depth to Ocean Floor (m)
1	0	0
2	160	165
3	200	1,800
4	500	3,500
5	1,050	5,450
6	1,450	5,100
7	1,800	5,300
8	2,000	5,600
9	2,300	4,750
10	2,400	3,500
11	2,600	3,100
12	3,000	4,300
13	3,200	3,900
14	3,450	3,400
15	3,550	2,100
16	3,700	1,275
17	3,950	1,000
18	4,000	0
19	4,100	1,800
20	4,350	3,650
21	4,500	5,100
22	5,000	5,000
23	5,300	4,200
24	5,450	1,800
25	5,500	920
26	5,650	0

Life in the Ocean

Life Processes

Life processes such as breathing oxygen, eating and digesting food, making new cells, and growing take place in your body every day. It takes energy to do this plus walk from one classroom to another or play soccer. Organisms that live in the ocean also carry out life processes every day. The octopus shown in **Figure 7** will get the oxygen it needs from the water. It will have to eat, and it will use energy to capture prey and to escape predators. It will make new cells and eventually reproduce. Like other marine organisms, it is adapted to accomplish these processes in the salty water of the ocean.

One of the most important processes in the ocean, as it is on land, is that organisms obtain food to use for energy. Obtaining the food necessary to survive can be done in several ways.

Life Science
INTEGRATION

Photosynthesis Nearly all of the energy used by organisms in the ocean ultimately comes from the Sun. Radiant energy from the Sun penetrates seawater to an average depth of 100 m. Marine organisms such as plants and algae use energy from the Sun to build their tissues and produce their own food. This process of making food is called **photosynthesis**. During photosynthesis, carbon dioxide and water are changed to sugar and oxygen in the presence of sunlight. Organisms that undergo photosynthesis are called producers. Marine producers include sea grasses, seaweeds, and microscopic algae. Although they might seem unimportant because they are small, microscopic algae are responsible for approximately 90 percent of all marine production. Seaweeds account for only about two percent to five percent of total marine production. Organisms that feed on producers are called consumers. Consumers in the marine environment include shrimp, fish, dolphins, whales, and sharks.

As You Read

***What* You'll Learn**
- **Describe** photosynthesis and chemosynthesis in the oceans.
- **List** the key characteristics of plankton, nekton, and benthos.
- **Compare and contrast** ocean margin habitats.

Vocabulary

photosynthesis	benthos
chemosynthesis	estuary
plankton	reef
nekton	

***Why* It's Important**
The ocean environment is fragile, and many organisms, including humans, depend on it for their survival.

Figure 7
Hunting at night, this Pacific octopus feeds on snails and crabs. It uses camouflage, ink, and speed to avoid predators.

Energy Relationships Energy from the Sun is transferred through food chains. Although the organisms of the ocean capture only a small part of the Sun's energy, this energy is passed from producer to consumer, then to other consumers. In **Figure 8,** notice that in one food chain, a large whale shark consumes small, shrimplike organisms as its basic food. In the other chain, microscopic algae found in water are eaten by microscopic animals called copepods (KOH pah pahdz). The copepods are, in turn, eaten by herring. Cod eat the herring, seals eat the cod, and eventually great white sharks eat the seals. At each stage in the food chain, energy obtained by one organism is used by other organisms to move, grow, repair cells, reproduce, and eliminate waste.

☑ **Reading Check**   *What is passed on at each stage in a food chain?*

In an ecosystem—a community of organisms and their environment—many complex feeding relationships exist. Most organisms depend on more than one species for food. For example, herring eat more than copepods, cod eat more than herring, seals eat more than cod, and white sharks eat more than seals. In an ecosystem, food chains overlap and are connected much like the threads of a spider's web. These highly complex systems are called food webs.

Figure 8
Numerous food chains exist in the ocean. Some food chains are simple and some are complex.

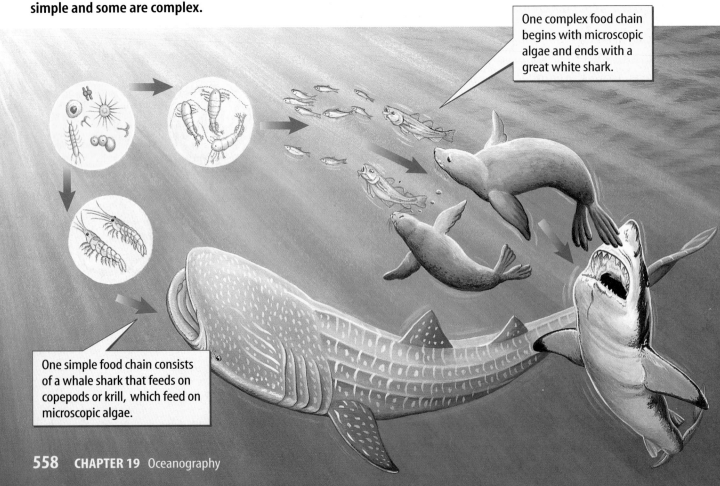

One complex food chain begins with microscopic algae and ends with a great white shark.

One simple food chain consists of a whale shark that feeds on copepods or krill, which feed on microscopic algae.

Chemosynthesis Other types of food webs do not depend on the Sun and photosynthesis. These food webs depend on bacteria that perform chemosynthesis. **Chemosynthesis** (kee moh SIHN thuh sus) involves using sulfur or nitrogen compounds as an energy source, instead of light from the Sun, to produce food. Bacteria that perform chemosynthesis using sulfur compounds live along mid-ocean ridges near hydrothermal vents where no light is available. Recall that superheated water from the crust contains high amounts of sulfur. The bacteria found here form the base of a food chain and support a host of highly specialized organisms such as giant tube worms, clams, crabs, and shrimp.

Other Life Processes Reproduction also is a vital life process. Some organisms, such as corals and sponges, depend on ocean currents for successful reproduction. Shown in **Figure 9,** these organisms release reproductive cells into the water where they unite to form more organisms of the same type. Other organisms, such as salmon and the Atlantic eel, travel long distances across the ocean in order to reproduce in a specific location. One important aspect of successful reproduction is finding a safe place for eggs and newly hatched larvae to develop. You will learn later in this section that some places in the ocean are used by marine organisms for this purpose.

Ocean Life

Many varieties of plants and animals live in the ocean. Although some organisms live in the open ocean or on the deep ocean floor, most marine organisms live in the waters above or on the floor of the continental shelf. In this relatively shallow water, the Sun penetrates to the bottom, allowing for photosynthesis. Because light is available for photosynthesis, large numbers of producers live in the waters above the continental shelf. These waters also contain many nutrients that producers use to carry out life processes. As a result, the greatest source of food is located in the waters of the continental shelf.

Figure 9
Because sponges live attached to the ocean bottom, they depend on currents to carry their reproductive cells to nearby sponges. *What would happen if a sponge settled in an area without strong currents?*

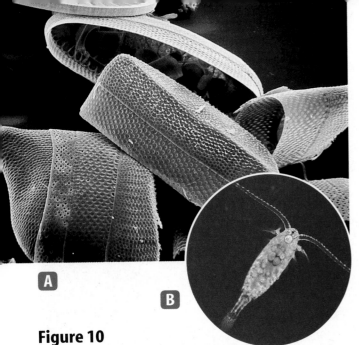

A

B

Figure 10
A Diatoms are phytoplankton that live in freshwater and ocean water. **B** The zooplankton shown here is a copepod. Although it has reached its adult size, it is still microscopic.

Plankton Marine organisms that drift with the currents are called **plankton.** Plankton range from microscopic algae and animals to organisms as large as jellyfish. Most phytoplankton—plankton that are producers—are one-celled organisms that float in the upper layers of the ocean where light needed for photosynthesis is available. One abundant form of phytoplankton is a one-celled organism called a diatom. Diatoms are shown in **Figure 10A.** Diatoms and other phytoplankton are the source of food for zooplankton, animals that drift with ocean currents.

Examples of zooplankton include newly hatched fish and crabs, jellyfish, and tiny adults of some organisms like the one shown in **Figure 10B.** These organisms feed on phytoplankton and are usually the second step in ocean food chains. Most animal plankton depend on surface currents to move them, but some can swim short distances.

Nekton Animals that actively swim, rather than drift with the currents in the ocean, are called **nekton.** Nekton include all swimming forms of fish and other animals, from tiny herring to huge whales. Nekton can be found from polar regions to the tropics and from shallow water to the deepest parts of the ocean. In **Figure 11,** the Greenland shark, the manatee, and the deep-ocean fish are all nekton. As nekton move throughout the oceans, it is important that they are able to control their buoyancy, or how easily they float or sink. What happens when you hold your breath underwater, then let all of the air out of your lungs at once? The air held in your lungs provides buoyancy and helps you float. As the air is released, you sink. Many fish have a special organ filled with gas that helps them control their buoyancy. By changing their buoyancy, organisms can change their depth in the ocean. The ability to move between different depths allows animals to search more areas for food.

Reading Check *What are nekton?*

Some deep-dwelling nekton are adapted with special light-generating organs. The light has several uses for these organisms. The deep-sea fish, shown in **Figure 11C,** dangles a luminous lure from beneath its jaw. When prey attracted by the lure are close enough, they are swallowed quickly. Some deep-sea organisms use this light to momentarily blind predators so they can escape. Others use it to attract mates.

Figure 11
Nekton are found living in all areas of the ocean, warm or cold, shallow or deep.

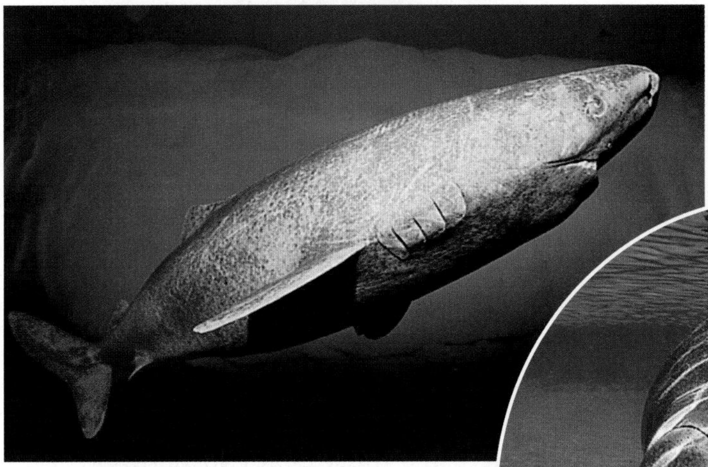

A This Greenland shark lives in the cold waters of the North Atlantic.

B Manatees are found in tropical regions around the world.

C This deep-sea fish deals with living under pressure at a depth of 4 km.

Bottom Dwellers The plants and animals living on or in the seafloor are the **benthos** (BEN thohs). Benthic animals include crabs, snails, sea urchins, and bottom-dwelling fish such as flounder. These organisms move or swim across the bottom to find food. Other benthic animals that live permanently attached to the bottom, such as sea anemones and sponges, filter out food particles from seawater. Certain types of worms live burrowed in the sediment of the ocean floor. Bottom-dwelling animals can be found living from the shallow water of the continental shelf to the deepest areas of the ocean. Benthic plants and algae, however, are limited to the shallow areas of the ocean where enough sunlight penetrates the water to allow for photosynthesis. One example of a benthic algae is kelp, which is anchored to the bottom and grows toward the surface from depths of up to 30 m.

Ocean Margin Habitats

The area of the environment where a plant or animal normally lives is called a habitat. Along the near-shore areas of the continental shelf, called ocean margins, a variety of habitats exist. Beaches, rocky shores, estuaries, and coral reefs are some examples of the different habitats found along ocean margins.

Mini LAB

Observing Plankton

Procedure
1. Place one or two drops of **pond, lake, or ocean water** onto a **microscope slide.**
2. With light coming through the sample from beneath, use a **microscope** to observe your sample. Look for microscopic life such as plankton.
3. Find at least three different types of plankton.

Analysis
1. Draw detailed pictures of three types of plankton.
2. Classify the plankton as phytoplankton or zooplankton.

Beaches At the edge of a sandy beach where the waves splash, you can find some microscopic organisms and worms that spend their entire lives between moist grains of sand. Burrowing animals such as small clams and mole crabs make holes in the sand. When water covers the holes, these animals rise to the surface to filter food from the water. Where sand is covered constantly by water, larger animals like horseshoe crabs, snails, fish, turtles, and sand dollars reside. **Figure 12** shows some of the organisms that are found living on sandy beaches.

Although the beach is great fun for people, it is a very stressful environment for the plants and animals that live there. They constantly deal with waves, changing tides, and storms, all of which redistribute large amounts of sand. Large waves produced by storms, such as hurricanes, can cause damage to beaches as they crash onto shore. These organisms must adapt to natural changes as well as changes created by humans. Damming rivers, building harbors, and constructing homes and hotels near the shoreline disrupts natural processes on the beach.

Rocky Shore Areas In some regions the shoreline is rocky, as shown in **Figure 13.** Algae, sea anemones, mussels, and barnacles encrust submerged rocks. Sea stars, sea urchins, octopuses, and hermit crabs crawl along the rock surfaces, looking for food.

Tide pools are formed when water remains onshore, trapped by the rocks during low tide. Tide pools are an important habitat for many marine organisms. They serve as protected areas where many animals such as octopuses and fish can develop safely from juveniles to adults. Tide pools contain an abundance of food and offer protection from larger predators.

SCIENCE Online

Research Visit the Glencoe Science Web site at **science.glencoe.com** for more information about beach erosion. Communicate to your class what you learn.

Figure 12
Organisms inhabit many different shore areas. *If you were a marine organism, where would you want to live? Why?*

Figure 13

Life is tough in the intertidal zone—the coastal area between the highest high tide and the lowest low tide. Organisms here are pounded by waves and alternately covered and uncovered by water as tides rise and fall. These organisms tend to cluster into three general zones along the shore. Where they live depends on how well they tolerate being washed by waves, submerged at high tide, or exposed to air and sunlight when the tide is low.

Upper intertidal zone

Mid-intertidal zone

Lower intertidal zone

UPPER INTERTIDAL ZONE This part of the intertidal zone is splashed by high waves and is usually covered by water only during the highest tides each month. It is home to crabs that scuttle among periwinkle snails, limpets, and a few kinds of algae that can withstand long periods of dryness.

Wavy turban snail

Stone crab

Periwinkle

Algae

MID-INTERTIDAL ZONE Submerged at most high tides and exposed at most low tides, this zone is populated by brown algae, sponges, barnacles, mussels, chitons, snails, and sea stars. These creatures are resistant to drying out and good at clinging to slippery surfaces.

Gooseneck barnacles

Lined chiton

Blue mussels

LOWER INTERTIDAL ZONE This section of the intertidal zone is exposed only during the lowest tides each month. It contains the most diverse collection of living things. Here you find sea urchins, large sea stars, brittle stars, nudibranchs, sea cucumbers, anemones, and many kinds of fish.

Sea lemon nudibranch

African coaster

Sea urchins

Estuaries An **estuary** is an area where the mouth of a river opens into an ocean. Because estuaries receive freshwater from rivers, they are not as salty as the ocean. Rivers also bring nutrients to estuaries. Areas with many nutrients usually have many phytoplankton, which form the base of the food chain. Shown in **Figure 14,** estuaries are full of life from salt-tolerant grasses to oysters, clams, shrimps, fish, and even manatees.

Estuaries are an important habitat to many marine organisms. Newly hatched fish, shrimps, crabs, and other animals enter estuaries as microscopic organisms and remain there until adulthood. For these vulnerable animals, fewer predators and more food are found in estuaries.

Figure 14
Estuaries are called the nurseries of the oceans because many creatures spend their early lives there.

Coral Reefs Corals thrive in clear, warm water that receives a lot of sunlight. This means that they generally live in warm latitudes, between 30°N and 30°S, and in water that is no deeper than 40 m. Each coral animal builds a hard capsule around its body from the calcium it removes from seawater. Each capsule is cemented to others to form a large colony called a reef. A **reef** is a rigid, wave-resistant structure built by corals from skeletal material. As a coral reef forms, other benthos such as sea stars and sponges and nekton such as fish and turtles begin living on it.

In all ocean margin habitats, nutrients, food, and energy are cycled among organisms in complex food webs. Plankton, nekton, and benthos depend on each other for survival.

Section Assessment

1. Describe the processes of photosynthesis and chemosynthesis.
2. Give an example of a marine producer and a marine consumer.
3. List the key characteristics of plankton, nekton, and benthos.
4. Compare and contrast the characteristics of coral reef and estuary habitats.
5. **Think Critically** The amount of nutrients in the water decreases as the distance from the continental shelf increases. What effect does this have on open-ocean food chains?

Skill Builder Activities

6. **Identifying and Manipulating Variables and Controls** Describe how you could set up an experiment to test the effects of different amounts of light on marine producers. **For more help, refer to the** Science Skill Handbook.

7. **Using Graphics Software** Design a creative poster that shows energy relationships in a food chain. Begin with photosynthesis. Use clip art, scanned photographs, or computer graphics. **For more help, refer to the** Technology Skill Handbook.

Ocean Pollution

Sources of Pollution

How would you feel if someone came into your bedroom; spilled oil on your carpet; littered your room with plastic bags, cans, bottles, and newspapers; then sprayed insect killer and scattered sand all over? Organisms in the ocean experience these things when people pollute seawater.

Pollution is the introduction of harmful waste products, chemicals, and other substances not native to an environment. A pollutant is a substance that causes damage to organisms by interfering with life processes.

Pollutants from land eventually will reach the ocean in one of four main ways. They can be dumped deliberately and directly into the ocean. Material can be lost overboard accidentally during storms or shipwrecks. Some pollutants begin in the air and enter the ocean through rain. Other pollutants will reach the ocean by being carried in rivers that empty into the ocean. **Figure 15** illustrates how pollutants from land enter the oceans.

✔ **Reading Check** *How do pollutants reach the ocean?*

As You Read

What You'll Learn
- **List** five types of ocean pollution.
- **Explain** how ocean pollution affects the entire world.
- **Describe** how ocean pollution can be controlled.

Vocabulary
pollution

Why It's Important
Earth's health depends on the oceans being unpolluted.

Figure 15
Ocean pollution comes from many sources.

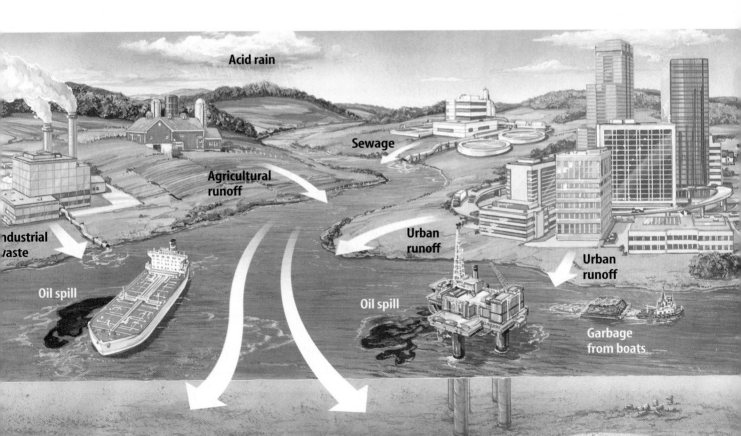

Acid rain

Sewage

Agricultural runoff

Industrial waste

Oil spill

Urban runoff

Urban runoff

Oil spill

Garbage from boats

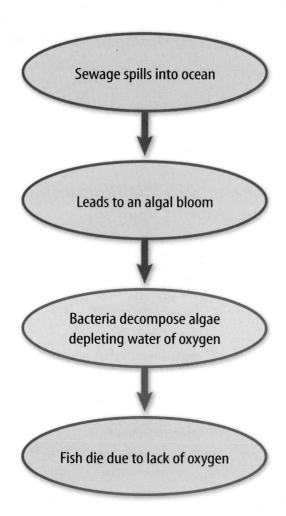

Figure 16
Fish kills occur when the oxygen supply is low. *How does a fish kill affect the food web?*

Sewage In some regions, human sewage leaks from septic tanks or is pumped directly into oceans or into rivers leading to an ocean. The introduction of sewage to an area of the ocean can cause immediate changes in the ecosystem, as shown by the following example. Sewage is a pollutant that acts like fertilizer. It is rich in nutrients that cause some types of algae to reproduce rapidly, creating what is called a bloom. The problem occurs when the algae die. As huge numbers of bacteria reproduce and decompose the algae, much of the oxygen in the water is used up. Other organisms, such as fish, cannot get enough oxygen. As a result, fish die in a phenomenon called a fish kill, as illustrated in **Figure 16.**

Reading Check *What is an algal bloom?*

When sewage is dumped routinely into the same area year after year, changes take place. Entire ecosystems have been altered drastically as a result of long-term, repeated exposure to sewage and fertilizer runoff. In some areas of the world, sewage is dumped directly onto coral reefs. When this happens algae can outgrow the coral because the sewage acts like a fertilizer. Eventually, the coral organisms die. If this occurs, other organisms that depend on the reef for food and shelter also can be affected.

Chemical Pollution Industrial wastes from land can harm marine organisms. When it rains, the herbicides (weed killers) and insecticides (insect killers) used in farming and on lawns are carried to streams. Eventually, they can reach the ocean and kill other organisms far from where they were applied originally. Sometimes industrial wastes are released directly into streams that eventually empty into oceans. Other chemicals are released into the air, where they later settle into the ocean. Industrial chemicals include metals like mercury and lead and chemicals like polychlorinated biphenyls (PCBs). In a process called biological amplification (am plah fah KAY shun), harmful chemicals can build up in the tissues of organisms that are at the top of the food chain. Higher consumers like dolphins and seabirds accumulate greater amounts of a toxin as they continue to feed on smaller organisms. At high concentrations, some chemicals can damage an organism's immune and reproductive systems. Explosives and nuclear wastes also have been dumped, by accident and on purpose, into some regions of the oceans.

Oil Pollution Although oil spills from tankers that have collided or are leaking are usually highly publicized, they are not the biggest source of oil pollution in the ocean. As much as 44 percent of oil that reaches the ocean comes from land. Oil that washes from cars and streets, or that is poured down drains or into soil, flows into streams. Eventually, this oil reaches the ocean. Other sources of oil pollution are leaks at offshore oil wells and oil mixed with wastewater that is pumped out of ships. **Figure 17** shows the percentage of different sources of oil entering the oceans each year.

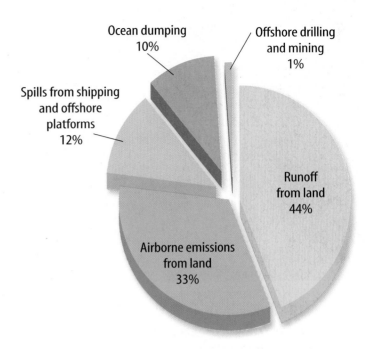

Ocean dumping 10%

Offshore drilling and mining 1%

Spills from shipping and offshore platforms 12%

Runoff from land 44%

Airborne emissions from land 33%

Figure 17
Although oil spills are highly publicized and tragic, the same harmful oil enters the ocean every day from many other sources. *What can be done to reduce the amount of oil entering the oceans?*

Solid-Waste Pollution Even in the most remote areas of the world, such as uninhabited islands that are thousands of miles from any major city, large amounts of trash wash up on the beach. **Figure 18** shows the amount of debris collected by a scientist on an island in the Pacific Ocean, 8,000 km east of Australia in just one day. The presence of trash ruins a beautiful beach, and solid wastes, such as plastic bags and fishing line, can entangle animals. Animals such as sea turtles mistakenly eat plastic bags, because they look so much like their normal prey, floating jellyfish. Illegally dumped medical waste such as needles, plastic tubing, and bags also are a threat to humans and other animals.

6 Lightbulbs   **7 Aerosol cans**   **25 Shoes**

71 Plastic bottles   **171 Glass bottles**   **268 Plastic pieces**

Figure 18
These items are like the ones found washed ashore on one of the Pitcairn Islands in the South Pacific. The number of each item found is shown below the figure. Also among the rubble were broken toys, two pairs of gloves, and an asthma inhaler.

Sediment Silt also can pollute the ocean. Human activities such as agriculture, deforestation, and construction tear up the soil. Rain washes soil into streams and eventually into an ocean nearby, as shown in **Figure 19.** This causes huge amounts of silt to accumulate in many coastal areas. Coral reefs and saltwater marshes are safe, protected places where young marine organisms grow to adults. When large amounts of silt cover coral reefs and fill marshes, these habitats are destroyed. Without a safe place to grow larger, many organisms will not survive.

Figure 19
When large amounts of silt enter seawater, the filter-feeding systems of animals such as oysters and clams can be clogged.

Effects of Pollution

You already have learned some examples of how pollution affects the ocean and the organisms that live there. Today, there is not a single area of the ocean that is not polluted in some way. As pollution from land continues to reach the ocean, scientists are recording dramatic changes in this environment.

Estuaries and the rivers that feed into them from Delaware to North Carolina have suffered from toxic blooms of *Pfiesteria* since the late 1980s. These blooms have killed billions of fish. *Pfiesteria*, a type of plankton, also has caused rashes, nausea, memory loss, and the weakening of the immune system in humans. The cause of these blooms is thought to be runoff contaminated by fertilizers and other waste materials. In Florida, toxic red tides kill fish and manatees. Some people also blame these red tides on sewage releases and fertilizer runoff. **Figure 20** shows an increase in the number of harmful algal blooms since the early 1970s.

Figure 20
Some scientists hypothesize that a relationship exists between increased pollution in the ocean and the number of harmful algal blooms in the last 30 years.

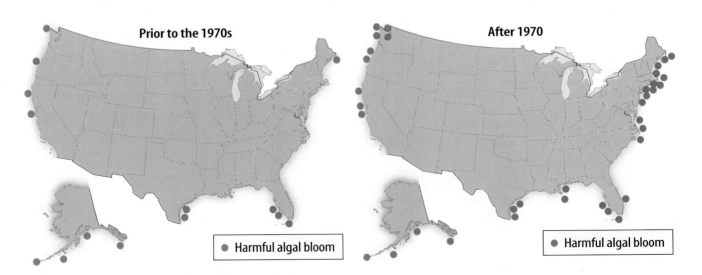

Controlling Pollution

Some people believe that oceans take care of themselves because they are large. However, other people view ocean pollution as a serious problem. Many international organizations have met to develop ways of reducing ocean pollution. Treaties prohibit the dumping of some kinds of hazardous wastes from vessels, aircraft, and platforms. One treaty requires that some ships and operators of offshore platforms have oil pollution emergency plans. This includes having the proper equipment to combat oil spills and practicing what to do if a spill takes place. Recall that a large amount of pollution enters the ocean from land. Although the idea of reducing land pollution to better protect the ocean has been discussed, no international agreement exists to prevent and control land-based activities that affect the oceans.

 **Reading Check** *What has been done to help control ocean pollution?*

What You Can Do Current international and U.S. laws aren't effective enough. Further cooperation is needed to reduce ocean pollution. You can help by disposing of wastes properly and volunteering for beach or community cleanups, like the one shown in **Figure 21.** You can recycle materials such as newspapers, glass, and plastics and never dump chemicals like oil or paint onto soil or into water. One of the best things you can do is continue to learn about marine pollution and how people affect the oceans. What other things will help reduce ocean pollution?

Figure 21
Under careful supervision, picking up trash is an easy way to help reduce ocean pollution.

Section Assessment

1. List five human activities that pollute the oceans. Suggest a solution to each.

2. How does pollution of the oceans affect the entire world?

3. In what ways have international treaties helped reduce pollution?

4. What can you do to help prevent ocean pollution?

5. **Think Critically** To widen beaches, some cities pump offshore sediment onto them. How might this affect organisms that live in coastal waters?

Skill Builder Activities

6. **Concept Mapping** Make an events-chain concept map that describes how runoff from land can reach the ocean. Include examples of pollution that could be in the runoff. **For more help, refer to the** Science Skill Handbook.

7. **Communicating** Submit a letter to the editor of a newspaper. In your letter, explain why ocean pollution is a problem that people can help prevent. List examples of things people can do to help. **For more help, refer to the** Science Skill Handbook.

Resources from the Oceans

Oceans cover most of Earth's surface. Humans get many things from oceans such as seafood, medicines, oil, and diamonds. Humans also use oceans for recreation and to transport materials from place to place. What else comes from oceans? Scientists continue to discover and research new uses for ocean resources. You might not realize that you probably use many products every day that are made from organisms that live in oceans.

Recognize the Problem

What products do you use that come from the oceans?

Form a Hypothesis

Think about the plants and animals that live in the oceans. How could these organisms be used to make everyday products? Form a hypothesis about the types of products that could be manufactured from these organisms.

Goals

- **Research and identify** organisms that are used to make products.
- **Explain** why it is important to keep oceans clean.

Data Source

SCIENCE *Online* Go to the Glencoe Science Web site at **science.glencoe.com** for more information about resources from the oceans, hints on which products come from the oceans, and data from other students.

Test Your Hypothesis

Plan

1. **Identify** web sites listed on the Glencoe Science Web site and identify other resources that will help you complete the data table shown on the right.

2. Notice that to complete the table you must identify products made from marine organisms, where the organisms are collected or harvested, and alternative products.

3. Plan how and when you will locate the information.

Do

1. Make sure your teacher approves your plan and your resource list before you begin.

Ocean Resources Data			
Organism	Location Where Collected or Harvested	Product (name and use)	Alternatives

2. Find at least three ocean organisms that are used to make products you use every day.

3. **Identify** the name and any uses of the product.

4. **Research** where each organism lives and the method by which it is collected or harvested.

5. **Identify** alternative products.

Analyze Your Data

1. **Describe** the different ways in which ocean organisms are useful to humans.

2. Are there any substitutes or alternatives available for the ocean organisms in the products?

Draw Conclusions

1. How might the activities of humans affect any of the ocean organisms you researched?

2. Are the substitute or alternative products more or less expensive?

3. Can you tell whether the ocean-made product is better than the substitute product? Explain.

4. Why is it important to conserve ocean resources and keep oceans clean?

Communicating **Your Data**

SCIENCE *Online* Find this *Use the Internet* activity on the Glencoe Science Web site at **science.glencoe.com.** Post your data in the table provided. Compare your data to those of other students.

Filter-feeding tube worm

Strange Creatures from the

In 1977, the *Alvin*, a small submersible craft specially designed to explore the ocean depths, took three geologists down about 2,200 m below the sea surface. They wanted to be the first to observe and study the formations of the Galápagos Rift deep in the Pacific Ocean. What they saw was totally unexpected. Instead of barren rock, the geologists found life—a lot of life. And they had never even considered having a life scientist as part of the research team!

The crew of the *Alvin* discovered hydrothermal vents—underwater openings where hot water (400°C) spurts from cracks in the rocks on the ocean floor. Some organisms thrive there because of the hydrogen sulfide that exists at the vents. Many of these organisms are like nothing humans had ever seen before. They are organisms that live in extremely hot temperatures and use hydrogen sulfide as their food supply.

Some of these organisms begin as nothing more than a mouth and grow into long, tube-like shapes. These organisms don't need sunlight to live but thrive in total darkness and can withstand pressure hundreds of times greater than at sea level.

The worm, *Alvinella pompejana*, can tolerate very high temperatures.

The discovery and study of hydrothermal vents almost has been overshadowed by the amazing variety of life that was found there. But scientists think these openings on the ocean floor (many located along the Mid-Atlantic Ridge) control the temperature and movement of nearby ocean waters, as well as have a significant effect on the ocean's chemical content. These vents also act as outlets for Earth's inner heat.

The *Alvin* enters the water.

Scientists also have discovered that the vent communities are temporary. Each vent eventually shuts down and the organisms somehow disperse to other vents. Exactly how this happens is an area of ongoing research. One idea is that "whale falls" provide stepping stones from one vent to another. A whale fall is the remains of a whale that has died and sunk to the seafloor. The offspring of vent creatures might take refuge and grow to adulthood at a whale fall, releasing their own young to seek out a new vent community.

Ocean Floor

PACIFIC OCEAN

COSTA RICA
PANAMA
COLOMBIA
ECUADOR
PERU

Galápagos Rift

GALÁPAGOS IS.

North America

South America

Australia

Blood-red tube worms live deep beneath the sea.

CONNECTIONS Creative Writing Imagine you were a passenger on the *Alvin*. Write about your adventure as you came upon the hydrothermal communities. Describe in detail some of the things you saw. Draw pictures of some of the unique creatures that live there.

SCIENCE *Online*
For more information, visit science.glencoe.com

Reviewing Main Ideas

Section 1 The Seafloor

1. The continental shelf is a gently sloping part of the continent that extends into the oceans. The continental slope extends beyond the continental shelf to the ocean floor. The abyssal plain is a flat area of the ocean floor. *Can you identify the continental shelf and slope in the figure below?*

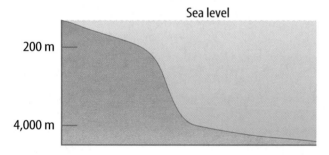

Sea level

200 m

4,000 m

2. Cracks form where the seafloor is spreading apart. Along mid-ocean ridges, new seafloor forms. Seafloor slips beneath another crustal plate at a trench.

3. Petroleum, natural gas, and placer deposits are mined from continental shelves. In the future, perhaps manganese nodules and other deep-water deposits will be mined.

Section 2 Life in the Ocean

1. Marine organisms are specially adapted to live in salt water. These organisms consume food, produce energy, and reproduce in the oceans.

2. Photosynthesis is the basis of most of the food chains in the ocean. Chemosynthesis is a special process of making food using chemical energy. Energy is transferred through food webs. Organisms that can make their own food are called producers. Organisms that feed on producers are called consumers.

3. Organisms that drift in ocean currents are called plankton. Nekton are marine organisms that swim. Benthos are plants and animals that live on or near the ocean floor. *Is this sponge classified as plankton, nekton, or benthos?*

4. Ocean margin habitats, found along the continental shelf, include sandy beaches, rocky shores, estuaries, and coral reefs.

Section 3 Ocean Pollution

1. Sources of pollution include sewage, chemical pollution, oil spills, solid waste pollution, and sediment. *Can you name the source of pollution shown here?*

2. Ocean pollution can disrupt food webs and threaten marine organisms.

3. International treaties and U.S. laws have been made to help reduce ocean pollution. Everyone can help reduce ocean pollution.

FOLDABLES
Reading & Study
Skills

After You Read

Using the information in your Foldable, explain the effects of ocean pollution. Predict whether pollution will increase or decrease.

Visualizing Main Ideas

Complete the following chart on ocean-margin organisms.

Ocean Margin Organisms

Organism	Ocean-Margin Environments			
	Sandy Beach	Rocky Shore	Estuary	Coral Reef
Plankton	phytoplankton zooplankton		phytoplankton, zooplankton	
Nekton		octopuses, fish		fish, turtles
Benthos			grasses, snails, clams	

Vocabulary Review

Vocabulary Words

a. abyssal plain
b. benthos
c. chemosynthesis
d. continental shelf
e. continental slope
f. estuary
g. mid-ocean ridge
h. nekton
i. photosynthesis
j. plankton
k. pollution
l. reef
m. trench

THE PRINCETON REVIEW **Study Tip**

Copy your notes from class. Explaining each concept in detail will help you remember it and understand it completely.

Using Vocabulary

Replace the underlined word(s) with the correct vocabulary word(s).

1. Animals such as whales, sea turtles, and fish are examples of <u>plankton</u>.

2. <u>Photosynthesis</u> occurs in areas where organisms use sulfur as energy to produce their own food.

3. The <u>continental shelf</u> drops from the edge of a continent out to the deep abyssal plains of the ocean floor.

4. New ocean floor is formed at <u>trenches</u>.

5. An area where the mouth of a river opens into an ocean is a <u>reef</u>.

Checking Concepts

Choose the word or phrase that best answers the question.

1. What is the flattest feature of the ocean floor?
   A) rift valley
   B) continental slope
   C) seamount
   D) abyssal plain

2. What might be found in areas where rivers enter oceans?
   A) rift valleys
   B) manganese nodules
   C) abyssal plains
   D) placer deposits

3. What are formed along subduction zones?
   A) mid-ocean ridges
   B) continental shelves
   C) trenches
   D) density currents

4. What are ocean organisms that drift in ocean currents called?
   A) nekton
   B) pollutants
   C) benthos
   D) plankton

5. How do some deep-water bacteria in the ocean make food?
   A) photosynthesis
   B) chemosynthesis
   C) respiration
   D) rifting

6. Which organisms reproduce rapidly, resulting eventually in a lack of oxygen?
   A) fish
   B) animal plankton
   C) algae
   D) corals

7. In which area of the ocean is the greatest source of food found?
   A) on abyssal plains
   B) in trenches
   C) along continental shelves
   D) along the mid-ocean ridge

8. Where does most oil pollution originate?
   A) tanker collisions
   B) runoff from land
   C) leaks at offshore wells
   D) in wastewater pumped from ships

9. Where does new seafloor form?
   A) trenches
   B) mid-ocean ridges
   C) abyssal plains
   D) continental shelves

10. What do organisms do to produce their own food?
    A) photosynthesize
    B) excrete wastes
    C) consume
    D) respire

Thinking Critically

11. Why might some industries be interested in mining manganese nodules?

12. Explain why ocean pollution is considered to be a serious international problem.

13. Discuss how agricultural chemicals can kill marine organisms.

14. Would you expect to find coral reefs growing around the bases of underwater volcanoes off the coast of Alaska? Explain.

15. Scientists currently are researching the use of chemicals produced by marine organisms to help fight diseases that affect humans including viruses and certain types of cancer. How would ocean pollution affect the ability to discover and research new drugs?

Developing Skills

16. **Using Scientific Illustrations** Use **Figure 8** to determine which organisms would starve if phytoplankton became extinct.

17. **Classifying** Classify each of these sea creatures as plankton, nekton, or benthos: shrimps, dolphins, sea stars, krill, coral, manatees, and algae.

18. Drawing Conclusions At point A an echo, a sound wave bounced off the ocean floor, took 2 s to reach the ship. It took 2.4 s at point B. Which point is deeper?

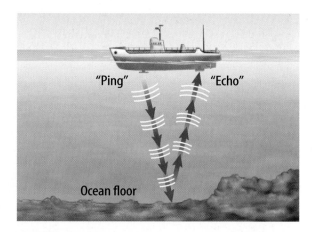

"Ping" "Echo"

Ocean floor

19. Measuring in SI If kelp grows at a steady rate of 30 cm per day, how long would it take to reach a length of 25 m?

20. Comparing and Contrasting Compare and contrast beach and rocky shore habitats.

Performance Assessment

21. Graph a Profile Previously, you made a profile of the ocean bottom along the 38° N latitude line, but it was not drawn to scale. Make a scale profile of the area between 3,600 km and 4,100 km from New Jersey. Use the scale 1 mm = 1,000 m.

22. Poster Choose a sea animal and research its life processes. Classify it as plankton, nekton, or benthos. Design a poster that includes all of this information.

TECHNOLOGY

Go to the Glencoe Science Web site at **science.glencoe.com** or use the **Glencoe Science CD-ROM** for additional chapter assessment.

THE PRINCETON REVIEW **Test Practice**

The table shows the year and location of oil being spilled into the ocean as a result of fire, ships running aground, or leakage.

Oil Spills Around the World		
Year	Location	Spill Size (millions of liters)
1967	Land's End, England	144.7
1972	Gulf of Oman, Oman	143.5
1978	Brittany, France	260.2
1979	Bay of Campeche, Mexico	530.3
1983	South Africa	297.3
1988	Newfoundland, Canada	163.2
1991	Persian Gulf	909.0
2001	Galapagos Islands	0.6

Use the information given in the table to answer the following questions.

1. At which location was the largest spill?
A) Brittany, France
B) South Africa
C) Persian Gulf
D) Lands End, England

2. Oil entering the ocean is a major concern because:
F) the presence of oil can reduce water quality
G) the presence of oil in the water can be harmful to marine life
H) large oils spills can be difficult to clean up
J) all of the above

Reading Comprehension

Read the passage. Then read each question that follows the passage. Decide which is the best answer to each question.

Hurricanes: An Exchange Between Ocean and Atmosphere

Hurricanes are among the most feared of all weather storms in the Atlantic region. In the western Pacific they are known as typhoons, and in Australia and the Indian Ocean they are called cyclones. Hurricane season in the United States and Caribbean occurs each year between June and November.

Hurricanes over the Atlantic Ocean begin as low-pressure systems, usually in the tropical seas west of Africa. The trade winds blow these storms westward. Heat from the warm ocean water gives the system energy. Once the water temperature rises to 27°C, a hurricane can form.

To be classified as a hurricane, the wind speed of the storm must exceed 119 km/h. Hurricanes can last for several days and can reach heights up to 16 km above the water.

The Atlantic has about ten tropical storms each year. Of these, six of them might develop into full-blown hurricanes.

Hurricanes can cause severe damage to anything near them on the water. After a hurricane reaches land, it loses its source of energy—warm ocean water—and begins to weaken. Even though the strength of these storms fades as they reach shore, hurricanes frequently are responsible for billions of dollars of damage and loss of lives.

Test-Taking Tip Take your time and read the passage carefully.

1. Which of the following was described first in the passage?
   A) Hurricanes can reach heights up to 16 km above the surface of the water.
   B) Hurricane season in the Atlantic region occurs between June and November.
   C) Hurricanes start to lose energy upon reaching land.
   D) Hurricanes can be suppressed by the high, strong winds of El Niño.

2. To classify as a hurricane, the wind speed of the storm must be more than _____.
   F) 27 km/h
   G) 16 km/h
   H) 119 km/h
   J) 140 km/h

3. What is the energy source of hurricanes?
   A) high winds
   B) warm ocean water
   C) trade winds
   D) cold air from land

Reasoning and Skills

Read each question and choose the best answer.

1. During El Niño years, upwelling of ocean water off the coast of Peru is greatly reduced, the water warms, and the number of fish decreases. Which of the following best explains why this occurs?
   A) The Coriolis effect shifts currents toward land.
   B) High concentrations of nutrients in the surface water increase its density.
   C) Winds blowing water from the coast slacken.
   D) The Eastern Pacific warms, thus changing the direction of currents off Peru.

Test-Taking Tip Think about what causes upwelling, then choose the answer that offers the most reasonable explanation for why it might stop.

Nature of Earth's Atmosphere	
Layer	**Characteristic**
Troposphere	Contains most of the water vapor
Stratosphere	Includes ozone layer
Mesosphere	Falling temperatures
Thermosphere	Very high temperatures

2. According to the information in the table, rain would most likely originate in the _____.
   F) troposphere   H) mesosphere
   G) stratosphere   J) thermosphere

Test-Taking Tip Examine the table carefully. Find the words you would most closely associate with rain and follow the row back to the correct answer.

Tornado Damage	
Type of Tornado	**Damage**
F0	Light: Broken branches and chimneys
F1	Moderate: Roof damage
F2	Considerable: Roofs torn off, trees uprooted
F3	Severe: Heavy roofs and walls torn off
F4	Devastating: Houses leveled
F5	Incredible: Houses picked up
F6	Total demolition

3. What type of tornado occurred near this home?
   A) light   C) considerable
   B) moderate   D) severe

Test-Taking Tip Obtain a copy of the photograph above and circle all of the things damaged in the picture. Compare what you found to the information given in the table.

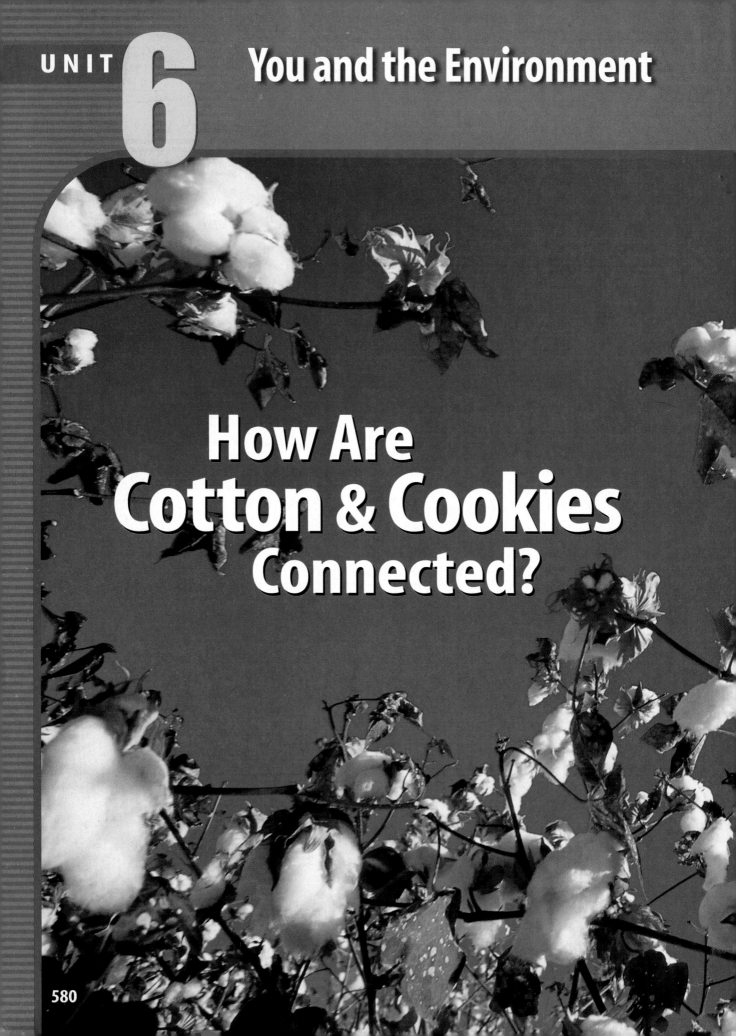

How Are
Cotton & Cookies
Connected?

In the 1800s, the economy of the South depended heavily on cotton and tobacco—two crops that rob the soil of nutrients, especially nitrogen. By the late 1800s, the soil was in poor shape. A scientist named George Washington Carver set out to change that. He promoted the technique of crop rotation—alternating soil-depleting crops such as cotton with soil-enriching crops such as peanuts. Many farmers listened to Carver and began planting peanuts. However, there was little market for the crop. So Carver poured his energy into developing uses for peanuts. Ultimately, he came up with more than 300 products made from peanuts—everything from soap to axle grease. He also created the first recipe for peanut butter cookies, which have become an American favorite.

SCIENCE CONNECTION

NITROGEN FIXATION Plants need nitrogen to grow well. Peanuts are legumes, plants that—with the help of certain bacteria—take nitrogen from the air and convert it into a form that plants can use. Conduct research to identify some other legumes and to discover how they "fix" nitrogen. In a one-page report, describe nitrogen fixation. If you were a farmer, what crops would you choose and how would you plant them to make your farm more productive?

Our Impact on Land

People are everywhere. You see people while at school, while waiting in traffic, or while shopping at the mall. There are a lot of people on Earth, and more are added every second. How do you think the human population affects natural resources on Earth? In this chapter, you'll learn about the impact that human population has on land. You also will read about ways humans use the land and how recycling helps conserve energy and materials.

What do you think?

Science Journal Look at the picture below with a classmate. Discuss what you think these might be. Here's a hint: *They will be here for a very long time.* Write your answer or best guess in your Science Journal.

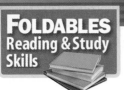

You're the first one on the school bus in the morning. After a few more stops, you notice that the bus is rather noisy. By the time you get to school, every seat is taken. It's loud, and it's hot. Like the school bus, space on Earth is limited. Do the activity below to see what happens when the number of people on Earth increases over time.

Model human population growth

1. On a piece of paper, draw a square that is 10 cm on each side. This square represents 1 km² of land.
2. In 1965, an average of 22 people lived on 1 km² of land. Draw 22 small circles inside your square to represent this.
3. In 1990, the average was 35. Add 14 circles to illustrate this increase.
4. In 2025, the estimated number of people will be 53. Add enough circles to represent this increase.
5. Prepare a bar graph that shows population density for these years.

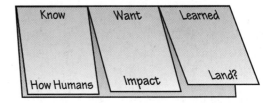

Observe

In your Science Journal, use the bar graph you made to explain how Earth's human population has changed over time.

Before You Read

Making a Know-Want-Learn Study Fold Make the following Foldable to help you identify what you already know and what you want to know about how human activities impact land.

1. Place a sheet of paper in front of you so the long side is at the top. Fold the paper in half from top to bottom.
2. Fold both sides in. Unfold the paper so three sections show.
3. Through the top thickness of paper, cut along each of the fold lines to the topfold, forming three tabs. Label each tab *Know*, *Want*, and *Learned*, as shown and then, *How humans impact land* across the front of the paper.
4. Before you read the chapter, write what you know and what you want to know about how humans impact land under the first two tabs.
5. As you read the chapter, add to or correct what you have written under the tabs.

SECTION

Population Impact on the Environment

As You Read

What You'll Learn
- **Describe** how fast the human population is increasing.
- **Identify** reasons for Earth's rapid increase in human population.
- **List** several ways each person can affect the environment.

Vocabulary
population
carrying capacity
pollutant

Why It's Important
As the human population grows, resources are depleted and more waste is produced.

Population and Carrying Capacity

Look around and identify the kinds of living things you see. You might see students, fish in an aquarium, or squirrels in the trees. Perhaps plants are on the windowsill. These are examples of populations. A **population** is all of the individuals of one species occupying a particular area. As you can see in **Figure 1,** the area can be small or large. For example, a human population can be of one community, such as Los Angeles, or the entire planet.

Earth's Increasing Population Do you ever wonder how many people live on Earth? The global population in 2000 was 6.05 billion. Each day, the number of humans increases by approximately 260,000. Earth is now experiencing a population explosion. The word *explosion* is used because the rate at which the population is growing has increased rapidly in recent history.

✔ **Reading Check** *Why is the increasing number of humans on Earth called a population explosion?*

Figure 1
A population is the number of individuals occupying an area. **A** The population of this classroom is 24. **B** The population of Cranford includes the population of the classroom.

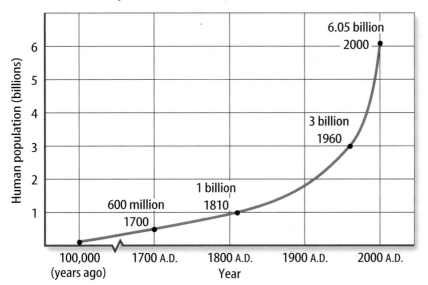

Population Growth of Modern Humans

Human population (billions)

6.05 billion
2000

3 billion
1960

1 billion
1810

600 million
1700

100,000
(years ago)   1700 A.D.   1800 A.D.   1900 A.D.   2000 A.D.

Year

Figure 2
Human population remained relatively steady until the beginning of the nineteeth century.
Why has the human population experienced such a sharp increase in growth rate since about 1800?

Population Growth Many years ago, few people lived on Earth. You can see in **Figure 2** that it took thousands of years for the population to reach one billion people. After that, the population increased much faster. The human population has increased because modern medicine, clean water, and better nutrition have decreased the death rate. This means that more people are living longer. In addition, the number of births has increased because more people survive to the age at which they can have children.

By 2100, the population is predicted to be more than twice what it is now. Imagine the effect such a large human population will have on the environment. Will the things that have helped the population grow, such as improved health care and clean water, be maintainable? Will Earth have enough natural resources to support such a large population?

Population Limits Each person uses space and resources. Population size depends on the amount of available resources and how members of the population use them. If resources become scarce or if the environment is damaged, members of the population can suffer and population size could decrease.

People once thought that Earth had an endless supply of resources such as fossil fuels, metals, and rich soils. It's now known that this isn't true. Earth's resources are limited. The planet has a carrying capacity. **Carrying capacity** is the largest number of individuals of a particular species that the environment can support. Unless Earth's resources are treated with care, they could disappear and the human population might reach its carrying capacity.

SCIENCE *Online*

Data Update For an online population update, visit the Glencoe Science Web site at **science.glencoe.com.** Communicate to your class what you've learned.

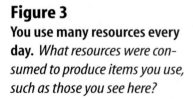

Life Science

INTEGRATION

Bald eagles are fish-eating birds whose population in the United States declined rapidly during the 1950s and early 1960s. One of the reasons this occurred was the use of pesticides, which affected reproduction. Research how pesticides applied to land can end up in an eagle's body.

People and the Environment

How will you affect the environment over your lifetime? By the time you're 75 years old, you will have produced enough garbage to equal the mass of thirteen African elephants (55,000 kg). You will have consumed enough water to fill 68,000 bathtubs (18 million L). If you live in the United States, you will have used five times as much energy as an average person living elsewhere in the world.

Daily Activities Every day you affect the environment. The electricity you use might be generated by burning fossil fuels. The environment changes when fuels are mined, and again later when they are burned. The water that you use must be treated to make it as clean as possible before being returned to the environment. You eat food, which needs soil to grow. Much of the food you eat is grown using chemical substances, such as pesticides and herbicides, to kill insects and weeds. These chemicals can get into water supplies and threaten the health of living things if the chemicals become too concentrated. How else do you and other people affect the environment?

As you can see in **Figure 3,** many of the products you use are made of plastic and paper. Plastic begins as oil. The process of refining oil can produce **pollutants**—substances that contaminate the environment. In the process of changing trees to paper, several things happen that impact the environment. Trees are cut down. Oil is used to transport the trees to the paper mill, and water and air pollutants are given off in the papermaking process.

✔ **Reading Check** *How do the products you use affect the environment?*

Figure 3
You use many resources every day. *What resources were consumed to produce items you use, such as those you see here?*

Packaging Produces Waste

The land is changed when resources are removed from it. The environment is further impacted when those resources are shaped into usable products. After the products are produced and consumed, they must be discarded. Look at **Figure 4.** Unnecessary packaging is only one of the problems associated with waste disposal.

Figure 4
Packaging foods for single servings uses more paper and plastic than buying food in bulk does.

The Future

As the population continues to grow, more resources are used and more waste is created. Traffic-choked highways, overflowing garbage dumps, shrinking forests, and vanishing wildlife are possible. What can be done to prevent these problems? As you learn more about how you affect the environment, you'll discover what you can do to help make the future world one that everyone can live in and enjoy. An important step that you can take is to think carefully about your use of natural resources. If you conserve resources, you can lessen the impact on the environment.

Section Assessment

1. Using **Figure 2,** estimate the human population increase from 1800 to 1960. How many people were added to the population from 1960 to 2000?

2. List three reasons why the human population is increasing rapidly.

3. What might happen if the human population reaches its carrying capacity?

4. How do your daily activities affect Earth's available resources?

5. **Think Critically** In some areas of the world, individuals have less negative impact on the environment than citizens in the United States do. In your Science Journal, explain why this is so.

Skill Builder Activities

6. **Researching Information** Some areas of the world are experiencing a decrease in population. Find out where this is happening and some reasons for the decrease. **For more help, refer to the** Science Skill Handbook.

7. **Making and Using Graphs** Make a line graph of the data shown below. Plot years on the x-axis and population on the y-axis. Use your graph to infer the population of humans in the year 2040. **For more help, refer to the** Math Skill Handbook.

Human Population (billions)
1998 — 5.86
2010 — 6.97
2025 — 8.66

Using Land

As You Read

What You'll Learn

- **Identify** ways that land is used.
- **Explain** how land use creates environmental problems.
- **Identify** things you can do to help protect the environment.

Vocabulary

stream discharge
sanitary landfill
hazardous waste
enzyme

Why It's Important

Land is a resource to use responsibly.

Land Usage

You may not think of land as a natural resource. Yet it is as important to people as oil, clean air, and clean water. Through agriculture, logging, garbage disposal, and urban development, we use land—and sometimes abuse it.

Agriculture About 16 million km^2 of Earth's total land surface is used as farmland. To feed the growing world population, some farmers use higher-yielding seeds and chemical fertilizers. These methods help increase the amount of food grown on each km^2 of land. Herbicides and pesticides also are used to reduce weeds, insects, and other pests that can damage crops.

Organic farming techniques, as shown in **Figure 5,** use natural fertilizers, crop rotation, and biological pest controls. These methods help crops thrive without using chemicals. You can buy organically grown fruits and vegetables in many places today.

Whenever vegetation is removed from an area, such as a construction site or tilled farmland, soil is exposed. Without plant roots to hold soil in place, nothing prevents the soil from being carried away by running water and wind. Several centimeters of topsoil may be lost in one year. In some places, it can take more than 1,000 years for new topsoil to develop.

Figure 5
Organic farms such as this one reduce the environmental impact of chemicals on land. *How does organic farming differ from other techniques?*

Reducing Erosion Some farmers practice no-till farming. They don't plow the soil from harvest until planting. Instead, farmers plant seed between the stubble left from the previous year.

Another way to reduce soil loss is by contour plowing. The rows are tilled across hills and valleys. When it rains, water and soil are captured by the plowed rows, reducing erosion.

Feeding Livestock Land also is used for feeding livestock. Animals such as cattle eat vegetation and then are used as food for humans. In the United States, the majority of land used for grazing is unsuitable for crops. However, about 20 percent of the total cropland in our country is used to grow feed for livestock.

Look at **Figure 6.** A square kilometer of vegetable crops can feed many more people than a square kilometer of land used to raise livestock. Some people argue that a more efficient use of the land would be to grow crops directly for human consumption, rather than for consumption by livestock. However, many consider meat and dairy products an important part of their diet. They argue that these foods are important sources of protein.

Figure 6
Land is used for feeding livestock. **A** About half the corn raised in the United States, such as the corn grown on this Iowa farm, is used to feed livestock. **B** Land is used more efficiently when vegetable crops are grown directly for human consumption.

A

B

Amount of Food Grown on 1 km² of Land

5.6 million kg of Tomatoes

4.5 million kg of Potatoes

28,000 kg of Beef

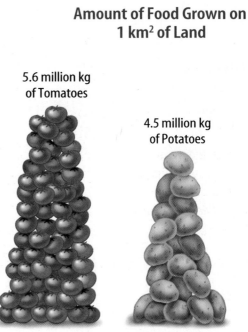
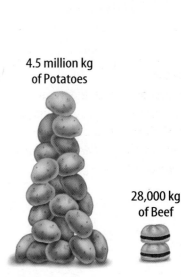

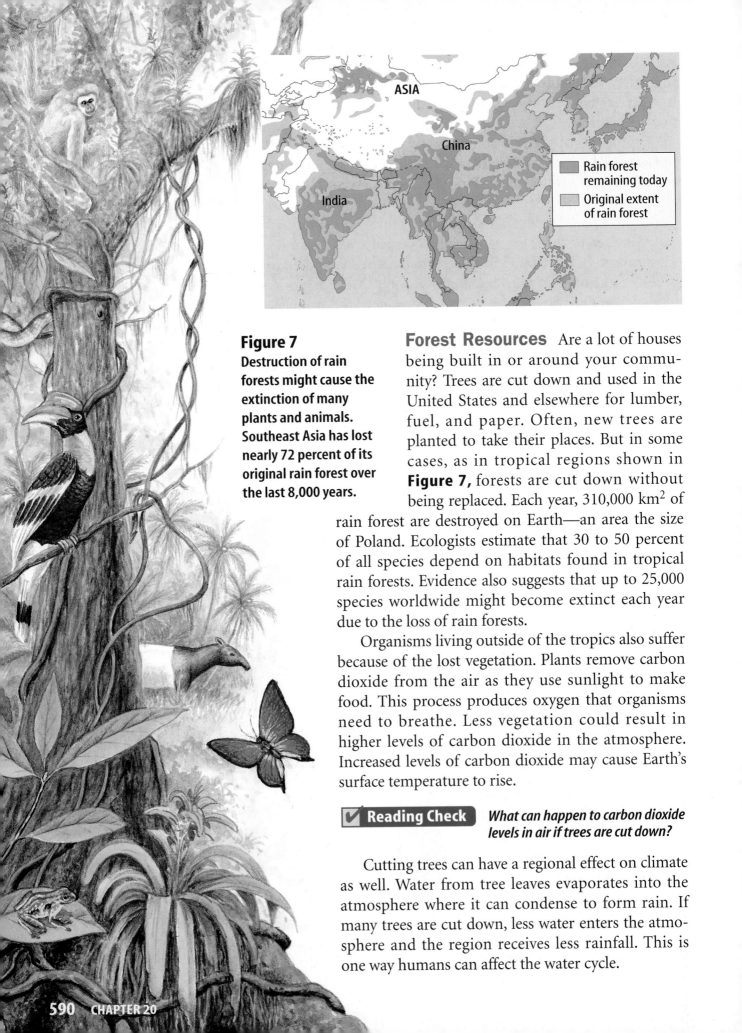

ASIA

China

India

Rain forest remaining today

Original extent of rain forest

Figure 7
Destruction of rain forests might cause the extinction of many plants and animals. Southeast Asia has lost nearly 72 percent of its original rain forest over the last 8,000 years.

Forest Resources Are a lot of houses being built in or around your community? Trees are cut down and used in the United States and elsewhere for lumber, fuel, and paper. Often, new trees are planted to take their places. But in some cases, as in tropical regions shown in **Figure 7,** forests are cut down without being replaced. Each year, 310,000 km² of rain forest are destroyed on Earth—an area the size of Poland. Ecologists estimate that 30 to 50 percent of all species depend on habitats found in tropical rain forests. Evidence also suggests that up to 25,000 species worldwide might become extinct each year due to the loss of rain forests.

Organisms living outside of the tropics also suffer because of the lost vegetation. Plants remove carbon dioxide from the air as they use sunlight to make food. This process produces oxygen that organisms need to breathe. Less vegetation could result in higher levels of carbon dioxide in the atmosphere. Increased levels of carbon dioxide may cause Earth's surface temperature to rise.

Reading Check *What can happen to carbon dioxide levels in air if trees are cut down?*

Cutting trees can have a regional effect on climate as well. Water from tree leaves evaporates into the atmosphere where it can condense to form rain. If many trees are cut down, less water enters the atmosphere and the region receives less rainfall. This is one way humans can affect the water cycle.

Development Since 1960, nearly 640,000 km of highway pavement have been added in the United States. Highway building often leads to more paving as office buildings, stores, and parking lots are constructed.

Paving land prevents water from soaking into the soil. Instead, it runs off into sewers or streams. A stream's discharge increases when more water enters its channel. **Stream discharge** is the volume of water flowing past a point per unit of time. During heavy rainstorms in paved areas, rainwater flows directly into streams, increasing stream discharge and the risk of flooding.

Many communities use underground water supplies for drinking. Covering land with roads, sidewalks, and parking lots reduces the amount of rainwater that soaks into the ground to refill underground water supplies.

Some communities preserve areas that cannot be paved. Land is set aside for environmental protection, as shown in **Figure 8.** Preserving space beautifies the environment, increases the area into which water can soak, and provides space for recreation and other outdoor activities.

Figure 8
Some communities set aside land that cannot be developed, such as this area near Portland, Oregon.
How does preserving green space near cities protect the environment?

Problem-Solving Activity

How does land use affect stream discharge?

It's not unusual for streams and rivers to flood after heavy rain. The amount of water flowing quickly into waterways may be more than streams and rivers can carry. Land use can affect how much runoff enters a waterway. Would changing the landscape increase flooding? Use your ability to interpret a data table to find out.

Land Use	Runoff to Streams (%)
Commercial (offices and stores)	75
Residential (houses)	40
Natural areas (forest and grassland)	29

Identifying the Problem

The table at the right lists the percentage of rainfall that runs off land. Compare the amount of runoff for each of the land uses listed. Assume that all of the regions are the same size and have the same slope. Looking at the table, do you see a relationship between what is on the land and how much water runs off of it?

Solving the Problem

1. Two years after construction of a commercial development near a stream, houses downstream flooded after a heavy rain. What contributed to the flooding?
2. What are some ways that developers can help reduce the risk of flooding?

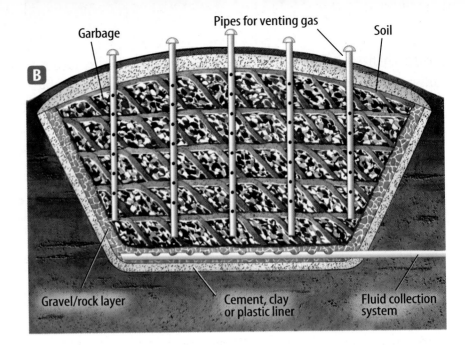

Garbage Pipes for venting gas Soil

B

Gravel/rock layer Cement, clay or plastic liner Fluid collection system

Figure 9

A The majority of garbage is deposited in sanitary landfills. *What are some problems associated with landfill disposal?*

B Sanitary landfills are designed to contain garbage and prevent contamination of the surrounding land and water.

Physics

INTEGRATION

Wastes from nuclear power plants must be stored safely because radioactivity is dangerous. The U.S. government is currently studying a site in Nevada for nuclear waste disposal because the area is remote, little rain falls, and the underground water supply is far below the proposed storage facility. What is radioactivity and how can it harm the environment?

Sanitary Landfills Land also is used when consumed products are thrown away. About 60 percent of our garbage goes into sanitary landfills. A **sanitary landfill,** like the one illustrated in **Figure 9,** is an area where each day's garbage is deposited and covered with soil. The soil prevents the deposit from blowing away, helps decompose some materials, and reduces the odor produced by the decaying waste.

Sanitary landfills also are designed to prevent liquid wastes from draining into the soil and groundwater below. New sanitary landfills are lined with plastic, concrete, or clay-rich soils that trap the liquid waste. Because of these linings, sanitary landfills greatly reduce the chance that pollutants will leak into the surrounding soil and groundwater.

Since many materials do not decompose in landfills, or they decompose slowly, landfills fill with garbage, and new ones must be built. Locating an acceptable area to build a landfill can be difficult. Type of soil, the depth to groundwater, and neighborhood concerns must be considered.

Hazardous Wastes

Some of the wastes that are thrown away are dangerous to organisms. Wastes that are poisonous, that cause cancer, or that can catch fire are called **hazardous wastes.** Previously, everyone—industries and individuals alike—put hazardous wastes into landfills, along with household garbage. In the 1980s, many states passed environmental laws that prohibit industries from disposing of hazardous wastes in sanitary landfills. New technologies which help recycle hazardous wastes have decreased the need to dispose of them.

Household Hazardous Waste Unlike most industries, individuals discard hazardous wastes such as insect sprays, batteries, drain cleaners, bleaches, medicines, and paints in the trash. It may seem that when you throw something in the garbage, it's gone and you don't need to be concerned with it anymore. Unfortunately, some garbage can remain unchanged in a landfill for hundreds of years. You can help by disposing of hazardous wastes at special hazardous waste-collection sites. Contact your local government to find out about collections in your area.

Phytoremediation Hazardous substances can contaminate soil. These contaminants may come from nearby industries or leaking landfills. Water contaminated from such a source can filter into the ground and leave the toxic substances in the soil. Some plants can help fix this problem in a method called phytoremediation (FITE uh rem ee dee AY shun). *Phyto* means "plant" and *remediation* means "to fix or remedy a problem."

During phytoremediation, roots of certain plants such as alfalfa, grasses, and pine trees can absorb metals, including copper, lead, and zinc from contaminated soil just as they absorb other nutrients. **Figure 10** shows how metals are absorbed from the soil and taken into plant tissue.

What happens to these plants after they absorb metals? If livestock were to eat contaminated alfalfa, the harmful metals could end up in your milk or meat. Plants that become concentrated with metals from soil eventually must be harvested and either composted to recycle the metals or burned. If these plants are destroyed by burning, the ash residue contains the hazardous waste that was in the plant tissue and must be disposed of at a hazardous waste site.

Breaking Down Organic Pollutants Living things also can clean up pollutants other than metals. Substances that contain carbon and other elements like hydrogen, oxygen, and nitrogen, are called organic compounds. Examples of organic pollutants are gasoline, oil, and solvents.

Organic pollutants can be broken down into simpler, harmless substances, some of which plants use for growth. Some plant roots release enzymes (EN zimez) into the soil. **Enzymes** are substances that make chemical reactions go faster. Enzymes from plant roots increase the rate at which organic pollutants are broken down into simpler substances. Plants use these substances for growth.

✓ Reading Check *What is the role of enzymes in the breakdown of organic pollutants?*

Figure 10
Metals such as copper can be removed from soil and be absorbed by plant tissues.
Why can't this vegetation be fed to livestock?

Metal absorbed

Composting   Burning

Metal recovery

Ash disposal

Figure 11
Many countries set aside land in the form of national parks as natural preserves. *How do these natural preserves benefit humans and other living things?*

Natural Preserves

Not all land on Earth is being utilized to produce usable materials or for storing waste. As shown in **Figure 11,** some land remains mostly uninhabited by people. National forestlands, grasslands, and national parks in the United States are protected from many problems that you've read about in this section. In many other countries throughout the world, land also is set aside for natural preserves. As the world population continues to rise, the strain on the environment is likely to worsen. Preserving some land in its natural state will benefit future generations.

Section Assessment

1. In your Science Journal, list six ways that people use land.
2. Discuss environmental problems that can be created by agriculture and trash disposal.
3. What can you do that would benefit the environment?
4. How can development increase flooding?
5. **Think Critically** Preserving land beautifies the environment, provides recreational space, and benefits future generations. Are there any disadvantages to setting aside large areas of land as natural preserves?

Skill Builder Activities

6. **Forming a Hypothesis** Develop a hypothesis about how migrating birds might be affected by cutting down forests. **For more help, refer to the** Science Skill Handbook.

7. **Using a Word Processor** Suppose that a new landfill is needed in your community. Where do you think it should be located? Try to convince people that you've selected the best place for the landfill. Write a letter to the editor of the local newspaper listing reasons for your choice. Map the location of your choice. **For more help, refer to the** Technology Skill Handbook.

Activity

What to wear?

List some items in your house that will end up in a landfill. You might include milk jugs, cereal boxes, and food scraps. But what about old worn-out clothing? In this activity, you'll observe what happens to different types of clothes that are buried in a landfill.

What You'll Investigate
Do different materials decompose at the same rate?

Materials 🔧 🔌 🚫 📄
identical baking trays (2)
garden soil
clothing made of natural fibers (linen tablecloth, cotton shirt, wool socks, silk scarf)
clothing made of artificial materials (fleece jacket, polyester shirt, acrylic sweater, rayon dress, nylon stockings)
toothpicks
transparent tape
scissors
spray bottle filled with water

Goals
■ **Compare** the decomposing rates of natural and artificial clothing materials.
■ **Infer** the effect of these materials on landfills.

Procedure
1. Collect several articles of clothing and separate those made with natural fibers from those made from artificial materials.
2. Cut 3-cm squares of each type of clothing.
3. Cut 1-cm by 3-cm labels from a sheet of notebook paper, and write one label for each of your clothing squares. Tape each label to the tip of a toothpick.
4. Fill each tray halfway with garden soil. Lay your artificial cloth squares in one tray and your natural cloth squares in the other tray. Be certain the squares don't overlap. Thoroughly moisten all squares using the spray bottle.
5. Identify each clothing square by attaching a toothpick label.
6. Cover your squares with soil. Moisten the soil and place the trays in a dark place. Keep the soil equally moist for three weeks.
7. After three weeks, dig up your samples and observe each square. Record your observations in your Science Journal.

Conclude and Apply:
1. **Compare** the amount of decomposition of the two types of materials.
2. **Infer** the effects of clothing made with natural materials on landfills.
3. **Infer** the effects of clothing made with artificial materials on landfills.
4. **Research** materials used to manufacture clothing. Determine if the material is made from recycled products such as plastic bottles.

Communicating Your Data
Compare the types of clothing worn by your classmates with the types you used in your experiment. Contrast the results of their experiments with your observations. **For more help, refer to the** Science Skill Handbook.

Conserving Resources

Field
GUIDE

What happens to the material you recycle? To find out how your trash is reused, see the **Waste Management Field Guide** at the back of the book.

A Disposable Lifestyle

In the United States and other industrialized countries, people have a throwaway lifestyle. When they are done with something, they throw it away. This means more products must be made to replace what's been thrown away, more land is used, and landfills become filled. You can help by conserving resources. **Conservation** is the careful use of Earth materials to reduce damage to the environment.

Reduce, Reuse, Recycle

The United States makes up only five percent of the world's human population, yet it consumes 30 percent of the world's natural resources, as shown in **Figure 12.** Ways to conserve resources include reducing the use of materials, and reusing and recycling materials. You can reduce the consumption of materials in simple ways, such as using both sides of notebook paper or carrying lunch to school in a nondisposable container. Reusing an item means finding another use for it instead of throwing it away. You can reuse old clothes by giving them to someone else or by cutting them into rags. The rags can be used in place of paper towels for cleaning jobs around your home.

Figure 12
A person in the United States consumes more resources than the average consumption per person elsewhere.

Yearly Consumption Per Person

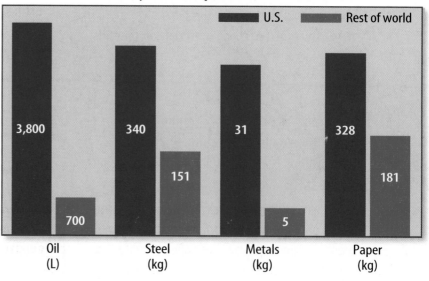

	U.S.	Rest of world
Oil (L)	3,800	700
Steel (kg)	340	151
Metals (kg)	31	5
Paper (kg)	328	181

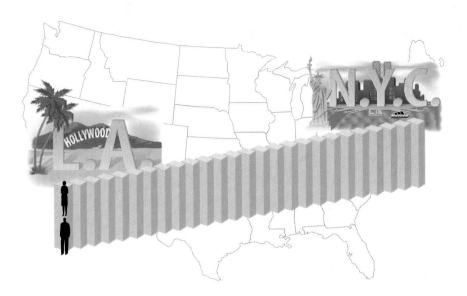

Figure 13
People in the United States throw away enough office and writing paper each year to build a wall 3.6 m high stretching from New York City to Los Angeles.

Reusing Yard Waste

Outdoors, you can do helpful things, too. If you cut grass or rake leaves, you can compost these items instead of putting them into the trash. **Composting** means piling yard wastes where they can decompose gradually. Decomposed material provides needed nutrients for your garden or flower bed. Some cities no longer pick up yard waste to take to landfills. In these places, composting is common. If everyone in the United States composted, it would reduce the trash put into landfills by 20 percent.

Recycling Materials

Recycling means using materials again. When you recycle wastes such as glass, paper, plastic, steel, or tires, you help conserve Earth's resources, energy, and landfill space.

Paper makes up about 40 percent of the mass of trash. As shown in **Figure 13,** Americans throw away a large amount of paper each year. Recycling this paper would use 58 percent less water and generate 74 percent less air pollution than producing new paper from trees. The paper shown in the figure doesn't even include newspapers. More than 500,000 trees are cut every week just to print newspapers.

How much energy do you think is saved when you recycle aluminum cans? Twenty aluminum cans can be recycled with the energy that is needed to produce a single new can from aluminum ore. If you recycle, you will reduce the trash you generate in your lifetime by 60 percent. If you don't recycle, you'll generate trash equal to at least 600 times your mass.

Figure 14 shows that the amount of material deposited in landfills has decreased since 1980. In addition to saving landfill space, reducing, reusing and recycling can reduce energy use and minimize the need to extract raw materials from Earth.

TRY AT HOME

Classifying Your Trash for One Day

Procedure
1. Label a table with the following columns: Paper, Plastic, Glass, Metal, and Food Waste.
2. Record items you throw out in one day. At the end of the day, count the number of **trash items** in each column.
3. Rank each column by number from the fewest trash items to the most trash items.

Analysis
1. Compare your rankings with those of others in your household.
2. What activities can you change to decrease the amount of trash you produce?

Figure 14

Although trash production in the United States is increasing, the amount of trash deposited in landfills is decreasing. In 1980, 82 percent of discarded trash ended up in a landfill. Today, only 55 percent is taken to the dump—thanks to the use of waste-reducing methods such as those shown below.

Landfill Use in the United States

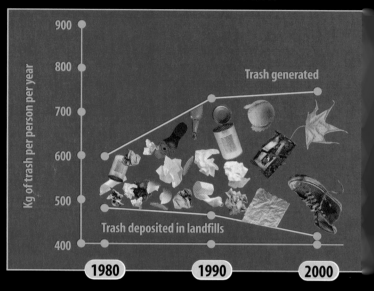

Kg of trash per person per year

900
800
700
600
500
400

Trash generated

Trash deposited in landfills

1980   1990   2000

CANS AND GLASS

MIXED PAPER

◀ COMPOSTING Yard trimmings placed in a pile will decompose and form a substance called compost. Compost then can be used on flowers and vegetables to help them grow.

▲ RECYCLING In 1980, about nine percent of trash was recycled. Now nearly 30 percent of America's trash is reused.

▶ WASTE TO ENERGY Some waste material can be burned to produce electricity. This plant in Rochester, Massachusetts, burns trash to generate electricity for a local paper company.

Recycling Methods What types of recycling programs does your state have? Many states or cities have some form of recycling laws. For example, in some places people who recycle pay lower trash-collection fees. In other places a refundable deposit is made on all beverage containers. This means paying extra money at the store for a drink, but you get your money back if you return the container to the store for recycling.

✓ Reading Check *How have states and cities encouraged people to recycle?*

There are several disadvantages to recycling. More people and trucks are needed to haul materials separately from your trash. The materials then must be separated at special facilities like the one shown in **Figure 15.** In addition, demand for things made from recycled materials must exist, and items made from recycled materials often cost more.

The Population Outlook The human population explosion already has had an effect on the environment and the organisms that inhabit Earth. It's unlikely that the population will begin to decline in the near future. To make up for this, resources must be used wisely. Conserving resources by reducing, reusing, and recycling is an important way that you can make a difference.

Figure 15
In recycling facilities like this one, materials must be separated before they can be reused.

Section Assessment

1. List four advantages and two disadvantages of recycling.

2. What is the difference between reducing and reusing materials?

3. How does compost benefit your garden or flower bed?

4. List two simple ways that you and your classmates can reduce your consumption of Earth materials.

5. **Think Critically** Why is it more important to conserve resources as the human population increases?

Skill Builder Activities

6. **Collecting and Organizing Data** Contact a local sanitary landfill. Find out how long it will take for your community's landfill to be full. How will waste be disposed of after the landfill is full? **For more help, refer to the** Science Skill Handbook.

7. **Using a Word Processor** Find out the email address of your local chamber of commerce. Compose a letter suggesting ways to encourage businesses to recycle. **For more help, refer to the** Technology Skill Handbook.

Activity

A World Full of People

Every second, five people are born on Earth and two people die. As a result, there is a net increase of three people in the world every second of every day. That amounts to about 95 million new people every year. This is nearly equal to the population of Central Africa. What effects will this rapid increase in human population have on Earth? How crowded will Earth become in your lifetime?

What You'll Investigate

How crowded will different regions of Earth become in the next ten years?

Materials
small objects such as popcorn kernels or dried beans (1,000)
large map of the world (the map must show the countries of the world)
clock or watch
calculator

Goals
■ **Demonstrate** the world's human population increase in the next decade.
■ **Predict** the world's population in 50 years.
■ **Record, graph,** and **interpret** population data.

Safety Precautions 🥽
Never eat or taste anything from a lab, even if you are confident that you know what it is.

Procedure

1. Copy the data table below in your Science Journal.

2. Lay the map out on a table. The map represents Earth and the 6 billion people already living here.

3. Each minute of time will represent one year. During your first minute, place 95 popcorn kernels on the continents of your map. Each kernel represents 1 million new people.

4. Place nine of your kernels inside the borders of developed countries such as the United States, Canada, Japan, Australia, and countries in Europe. Place 86 kernels inside the borders of developing nations located in South America, Africa, and Asia.

5. Continue adding 95 kernels to your map in the same fashion each minute for 10 min. Record the total population increase for each year (each minute of the lab) in your data table.

Population Data	
Time (years)	**Total Population Increase**

Conclude and Apply

1. Make a graph of your data showing the time in years on the horizontal axis and the world population on the vertical axis.

2. How many new people will be added to Earth in the next 10 years? Determine the world's population in 10 years.

3. At the current rate of population growth, calculate the world's population in 50 years.

4. **Determine** world population in ten years if only 4.5 million people are added each year.

5. **Compare** the population growth in developed countries to the growth of developing countries.

6. **Discuss** ways the increase in the human population will affect Earth's resources in the future.

Communicating
Your Data

Draw your graph on a computer and present your findings to the class. **For more help, refer to the** Science Skill Handbook.

Hazardous

Danger: Hazardous Waste Area. Unauthorized Persons Keep Out.

During much of the 1980s, this sign greeted visitors to Love Canal, a housing project in Niagara Falls, New York. The housing project was closed because it had been built on a hazardous waste dump and people were getting sick. Hazardous wastes are substances that are poisonous, that cause cancer, or that can catch fire. Some human health effects of exposure to hazardous waste include nerve damage, birth defects, and lowered resistance to disease.

Many types of manufacturing companies produce hazardous waste. In addition, some common household items, such as paints and oven cleaners, contain hazardous wastes.

The Environmental Protection Agency (EPA) estimates that U.S. industries produce about 265 million metric tons of hazardous wastes each year. Much of this waste is handled by the industries themselves—they recycle the waste or convert it to harmless substances. About 60 million tons of hazardous waste, however, must be disposed of in a safe manner. Incineration, or burning, is one way to dispose of hazardous wastes. However, the safety of this method is hotly debated.

For Incineration

People in favor of incineration note that the EPA has strict rules that govern the effectiveness of incinerators. Done correctly, it destroys 99.99 percent of toxic materials.

A danger sign in a garbage dump alerts visitors to the presence of hazardous waste.

The Love Canal housing development was closed because of toxic waste.

Waste

It causes health risks, but how to safely get rid of it?

Although the remaining ash must still be disposed of, it is often less hazardous than the original waste material. Supporters also note that incineration is safer than simply storing the hazardous wastes or dumping them in landfills.

Against Incineration

Other people say that incinerators fail to destroy all hazardous wastes and that some toxins are released in the process. They also note that new substances are generated during incineration, and that scientists don't yet know how these new substances will impact the environment or human health.

Lastly, they say that incineration may reduce efforts to reuse or recycle hazardous wastes.

While the debate goes on, scientists continue to develop better methods for dealing with hazardous wastes. As Roberta Crowell Barbalace, an environmental scientist, wrote in an article, "In an ideal environment there would be no hazardous waste facilities. The problem is that we don't live in an ideal environment ... Until some new technology is found for dealing with or eliminating hazardous waste, disposal facilities will be necessary to protect both humans and the environment."

CONNECTIONS Research Find out more about incineration. Then use this feature and your research to conduct a class debate about the advantages and disadvantages of incineration.

SCIENCE *Online*

For more information, visit
science.glencoe.com

Reviewing Main Ideas

Section 1 Population Impact on the Environment

1. Modern medicine, clean water, and better nutrition have contributed to the human population explosion on Earth.

2. Earth's resources are limited.

3. Our daily activities use resources and produce waste. *What resources were consumed to produce the items shown here?*

Section 2 Using Land

1. Land is used for farming, grazing livestock, lumber, development, and disposal.

2. Farming and development are ways that using land can impact the environment. Chemicals are used in farming. *What are some of the impacts caused by paving land?*

3. Using forest resources can impact organisms and Earth's climate.

4. Plants are sometimes used to break down and absorb pollutants from contaminated land.

5. New technologies have reduced greatly the need for hazardous waste disposal.

6. One way to preserve our land is to set aside natural areas.

Section 3 Conserving Resources

1. Recycling, reducing, and reusing materials are important ways to conserve natural resources. *How can you recycle yard waste, such as these grass clippings?*

2. Recycling saves energy and much-needed space in land-fills.

3. Different methods can be used to encourage recycling.

4. Reducing, reusing, and recycling consumed items has decreased the annual deposits of trash in landfills since 1980.

FOLDABLES
Reading & Study Skills

After You Read

Write what you learned about how human activities impact land under the right tab of your Foldable. Explain the importance of human activities on land.

Visualizing Main Ideas

Complete the concept map about using land.

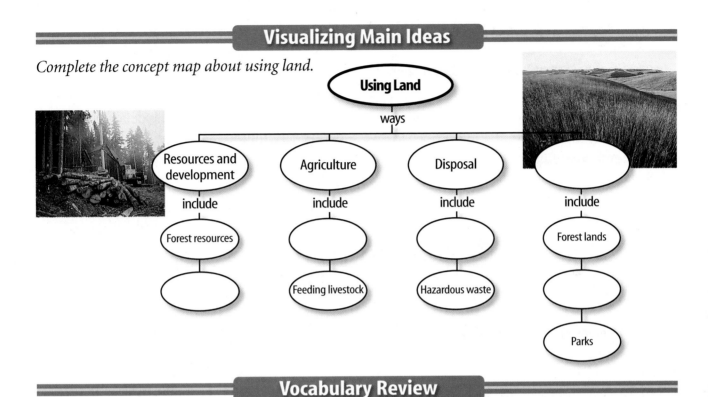

Using Land

ways

Resources and development — include — Forest resources — ◯

Agriculture — include — ◯ — Feeding livestock

Disposal — include — ◯ — Hazardous waste

◯ — include — Forest lands — ◯ — Parks

Vocabulary Review

Vocabulary Words

a. carrying capacity
b. composting
c. conservation
d. enzyme
e. hazardous waste
f. pollutant
g. population
h. recycling
i. sanitary landfill
j. stream discharge

Using Vocabulary

Using the vocabulary list above, replace the underlined word or phrase with the correct vocabulary word.

1. The total number of individuals of a particular species in an area is called the <u>carrying capacity</u>.

Study Tip

Get together with a friend. Quiz each other from your textbook and class material.

2. <u>Composting</u> means using resources carefully to reduce damage to the environment.

3. Substances that can make chemical reactions go faster are called <u>pollutants</u>.

4. <u>Hazardous wastes</u> means using materials again.

5. A <u>population</u> is the maximum number of individuals of a particular type that the planet will support.

6. An area where waste is deposited is called <u>stream discharge</u>.

7. Poisonous, cancer-causing, or ignitable wastes are called <u>enzymes</u>.

8. <u>Conservation</u> means piling yard wastes where they can decompose gradually.

9. A <u>sanitary landfill</u> is a substance that contaminates the environment.

10. The volume of river water flowing past a point per unit of time is called <u>recycling</u>.

Checking Concepts

Choose the word or phrase that best answers the question.

1. Where is most of the trash in the United States disposed of?
   A) recycling centers
   B) landfills
   C) hazardous waste sites
   D) compost piles

2. Between 1960 and 2000, world population increased by how many billions of people?
   A) 5.9   C) 1.0
   B) 4.2   D) 3.1

3. What percentage of Earth's resources does the United States use?
   A) 5   C) 30
   B) 10   D) 50

4. About what percentage of U.S. cropland is used to grow feed for livestock?
   A) 100   C) 50
   B) 1   D) 20

5. What do we call an object that can be processed in some way so that it can be used again?
   A) trash   C) disposable
   B) recyclable   D) hazardous

6. What is about 40 percent of the mass of our trash made up of?
   A) glass   C) yard waste
   B) aluminum   D) paper

7. What term is used to describe using plants to clean up contaminated soil?
   A) recycling   C) sanitary landfill
   B) phytoremediation   D) composting

8. Which of the following increases a stream's discharge?
   A) planting trees   C) paving land
   B) preserving land   D) organic farming

9. How many thousand square kilometers of rain forest disappear each year?
   A) 3   C) 31
   B) 310   D) 150

10. What is used to cover daily deposits of trash in a landfill?
    A) soil   C) gravel
    B) plastic   D) hazardous waste

Thinking Critically

11. How would reducing materials used for packaging products affect our disposal of solid wastes?

12. Developing land can change stream discharge. What are some ways that land can be developed without this impact?

13. Although land is farmable in several developing countries, hunger is a major problem in many of these places. Give some reasons why this might be so.

14. Forests in Germany are dying due to acid rain. What effects might this loss of trees have on the environment?

15. Describe how you could encourage your neighbors to recycle their aluminum cans.

Developing Skills

16. **Classifying** Group the following materials as hazardous or nonhazardous: gasoline, newspaper, leaves, lead, can of paint, glass.

17. **Interpreting Scientific Illustrations** One hectare, shown here, is a square of land measuring 100 meters by 100 meters. How many hectares are in one km^2 of land?

100 m

100 m

1 hectare
or
10,000 m²

18. Concept Mapping Complete this concept map about phytoremediation.

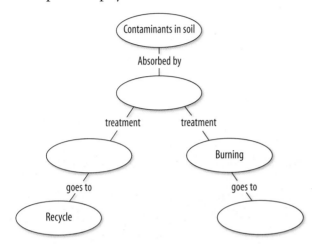

19. Comparing and Contrasting Compare and contrast farming and developing land. How do these activities affect stream discharge?

20. Researching Find out whether your community excludes yard waste from landfills.

Performance Assessment

21. Using Math in Science Collect your family's junk mail for one week and weigh it. Divide this weight by the number of people in your home. Multiply this number by 300 million (the U.S. population). If 17 trees are cut to make each metric ton of paper, calculate how many trees are cut each year to make junk mail for the entire U.S. population.

22. Evaluating a Hypothesis Design an experiment to determine factors that decrease the time it takes for newspapers or yard wastes to decompose.

TECHNOLOGY

Go to the Glencoe Science Web site at **science.glencoe.com** or use the **Glencoe Science CD-ROM** for additional chapter assessment.

 **Test Practice**

In 1782, when the Bald Eagle was named the national symbol of the U.S., there were about 75,000 pairs in the lower 48 states. By 1963, there were only 450 pairs. Alarmed, Americans supported laws banning the use of pesticides, such as DDT, which had been determined to be harmful to the eagle.

Study the graph and answer the following questions.

1. According to the graph, the number of pairs of eagles in 1973 was _____.
   A) fewer than 500
   B) fewer than 1,000
   C) greater than 1,000
   D) greater than 2,000

2. How many more pairs of bald eagles were in the lower 48 states in 1999 than in 1973?
   F) just over 4,000
   G) just over 3,000
   H) about 5,000
   J) about 2,000

Our Impact on Water and Air

The canoe glides peacefully through the water. A gentle breeze invites you to breathe deeply, almost tasting the fresh air. You look down as your paddle pulls through the clear water and see fish swimming. Once upon a time all the water on Earth was like this. In this chapter, you'll investigate water and air pollution sources, ways of reducing water and air pollution, and the importance of clean water and air.

What do you think?

Science Journal Look at the picture below with a classmate and discuss what this might be. Here's a hint: *They're small and difficult to detect in water.* Write your answer or best guess in your Science Journal.

EXPLORE ACTIVITY

Some water pollution is easy to see. Where pollutants have directly entered a stream, the water can be discolored, have an odor, or contain dead fish. Suppose the water appears to be clean. Does that mean it's free of pollution? Do the activity below to find out.

Model water pollution

1. Pour 125 mL of water into a large jar.
2. Add one drop (0.05 ml) of food coloring to the water and stir.
3. Add an additional 125 mL of water to the jar and stir.
4. Repeat step 3 until you cannot see the food coloring.

Observe

In your Science Journal, calculate the concentration of food coloring in your jar with each 125-mL addition of water. Will the concentration of food coloring ever become zero by diluting the solution?

Before You Read

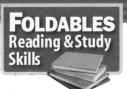

FOLDABLES
Reading & Study Skills

Making a Venn Diagram Study Fold **Make the following Foldable to compare and contrast the characteristics of water pollution and air pollution.**

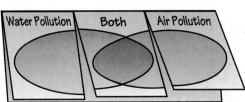

1. Place a sheet of paper in front of you so the short side is at the top. Fold the paper in half from top to bottom.
2. Fold both sides in. Unfold the paper so three sections show.
3. Through the top thickness of paper, cut along each of the fold lines to the top fold, forming three tabs. Label the tabs *Water Pollution, Both,* and *Air Pollution* and draw ovals on the front of the foldable as shown.
4. As you read the chapter, write information about each type of pollution under the tabs.

Water Pollution

As You Read

What You'll Learn

- **Identify** types of water pollutants and their effects.
- **Discuss** ways to reduce water pollution.
- **List** ways that you can help reduce water pollution.

Vocabulary
point source pollution
nonpoint source pollution
pesticide
fertilizer
sewage

Why It's Important
All organisms on Earth depend on water for life.

Importance of Clean Water

All organisms need water. Plants need water to make food from sunlight. Some animals such as fish, frogs, and whales live in water. What about you? You cannot live without drinking water. What happens if water isn't clean? Polluted water contains chemicals and organisms that can cause disease or bring death to many living things. Water also can be polluted with sediments, such as silt and clay.

Sources of Water Pollution

If you were hiking along a stream or lake and became thirsty, would it be safe to drink the stream water? Many streams and lakes in the United States are polluted in some way. Even streams that look clear and sparkling might not be safe for drinking.

Point source pollution is pollution that enters water from a specific location, such as drainpipes or ditches, as shown in **Figure 1.** Pollution from point sources can be controlled or treated before the water is released to a body of water.

However, many times bodies of water become polluted and no one knows exactly where the pollution comes from. Pollution that enters a body of water from a large area, such as lawns, construction sites, and roads, is called **nonpoint source pollution.** Nonpoint sources also include pollutants in rain or snow. More than 75 percent of water pollution in the United States comes from nonpoint sources.

Figure 1
A Point sources include industrial wastes. **B** Nonpoint sources cannot be traced to a single location.

Sediment The largest source of water pollution in the United States is sediment. Sediment is loose material, such as rock fragments and mineral grains, that is moved by erosion. Rivers always have carried sediment to oceans, but human activities can increase the amount of sediment in rivers, lakes, and oceans. Each year, about 25 billion metric tons of sediment are carried from farm fields to bodies of water on Earth. At least 50 billion additional tons run off of construction sites, cleared forests, and land used to graze livestock. Sediment makes water cloudy and blocks sunlight that underwater plants need to make food. Sediment also covers the eggs of organisms that live in water, preventing organisms from receiving the oxygen they need to develop.

Agriculture and Lawn Care Farmers and home owners apply **pesticides,** which are substances that destroy pests, to keep insects and weeds from destroying their crops and lawns. When farmers and home owners apply pesticides to their crops and lawns, some of the chemicals run off into water. These chemicals might be harmful to people and other organisms, such as the frog in **Figure 2.**

Fertilizers are chemicals that help plants grow. However, rain washes away as much as 25 percent of the fertilizers applied to farms and yards into ponds, streams, and rivers. Fertilizers contain nitrogen and phosphorus that algae, living in water, use to grow and multiply. Lakes or ponds with high nitrogen and phosphorous levels, such as the one shown in **Figure 3,** can be choked with algae. When algae die and decompose, oxygen in the lake is used up more rapidly. This can cause fish and other organisms to die. Earth's nitrogen cycle is modified when fertilizers enter the water system.

Reading Check *How do fertilizers cause water pollution?*

Figure 2
Research suggests that some pesticides in the environment could lead to deformities in frogs, such as missing legs.

Figure 3
Nitrogen and phosphorus in fertilizer cause algae to grow and multiply. Fish can die when algae decompose, using up oxygen.

B Algae grow and multiply.

D Without enough oxygen, fish may die.

C Algae die and decay, using up oxygen.

A Fertilizer applied to lawns or farms runs off.

Human Waste Another source of water pollution is human waste. When you flush a toilet or take a shower, the water that goes into drains, called **sewage,** contains human waste, household detergents, and other chemicals. Human waste contains harmful organisms that can make people sick.

In most cities and towns in the United States, underground pipes take the water you use from your home to a sewage treatment plant. Sewage treatment plants, such as the one in **Figure 4,** remove pollution using several steps. These steps purify the water by removing solid materials from the sewage, killing harmful bacteria, and reducing the amount of nitrogen and phosphorus in the water before it is returned to the environment.

Math Skills Activity

Calculating with Percentages

The following table lists surface water pollutants in rivers and streams. It shows the number of sampling stations that have an increased or a decreased level of pollution in a 10-year period.

Trends in River and Stream Water Quality 1980–1989				
Measured Pollutant	Total No. of Stations Examined	No. of Stations With Decrease in Pollutant Level	No. of Stations With Increase in Pollutant Level	No Change
Sediments	324	36	6	282
Bacteria From Sewage	313	41	9	263
Total Phosphorus	410	90	21	299
Nitrogen	344	27	21	296

Example Problem

What percent of stations has shown an increase in nitrogen over a 10-year period?

Solution

1. *This is what you know:*   Nitrogen: number of stations with an increase = 21
                                                     total number of stations examined = 344

2. *This is what you need to find:*   percentage: _____%

3. *This is the equation you need to use:*   % = (stations with increase)/(total stations) $\times$ 100

4. *Substitute the known values:* $(21)/(344) \times 100 = 6.1\%$

   Check your answer by multiplying the total stations by the percent in decimal form to obtain the number of stations with an increase.

Practice Problems

1. What percent of stations has shown a decrease in bacteria?

2. What percent of stations has shown an increase in sediment?

For more help, refer to the Math Skill Handbook.

Figure 4

Sewage from most towns and cities is treated at municipal sewage treatment plants. Wastewater entering a sewage plant contains organic matter, paper, grease, bacteria, nitrogen, and phosphorus. As shown below, the wastewater from homes and businesses is purified in three stages—primary, secondary, and tertiary—before it is pumped back into a stream or river.

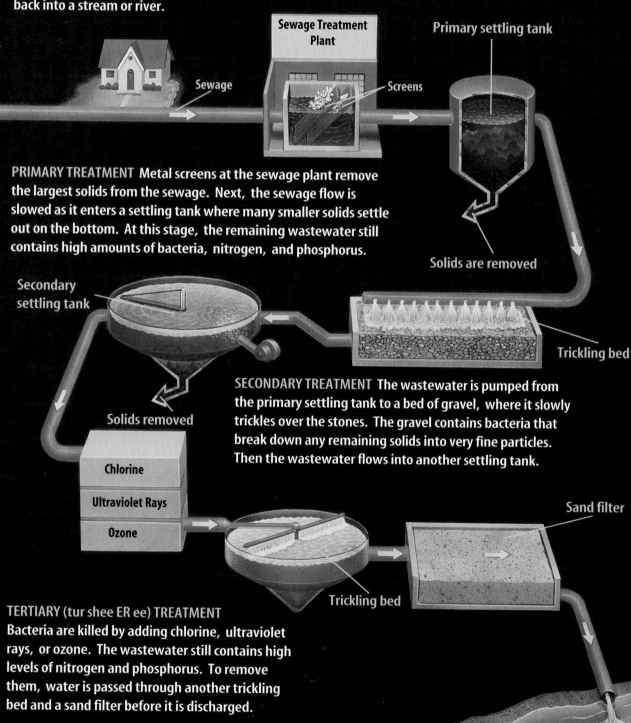

Sewage Treatment Plant

Primary settling tank

Sewage

Screens

PRIMARY TREATMENT Metal screens at the sewage plant remove the largest solids from the sewage. Next, the sewage flow is slowed as it enters a settling tank where many smaller solids settle out on the bottom. At this stage, the remaining wastewater still contains high amounts of bacteria, nitrogen, and phosphorus.

Solids are removed

Secondary settling tank

Trickling bed

Solids removed

Chlorine

Ultraviolet Rays

Ozone

SECONDARY TREATMENT The wastewater is pumped from the primary settling tank to a bed of gravel, where it slowly trickles over the stones. The gravel contains bacteria that break down any remaining solids into very fine particles. Then the wastewater flows into another settling tank.

Sand filter

Trickling bed

TERTIARY (tur shee ER ee) TREATMENT
Bacteria are killed by adding chlorine, ultraviolet rays, or ozone. The wastewater still contains high levels of nitrogen and phosphorus. To remove them, water is passed through another trickling bed and a sand filter before it is discharged.

Pollutants, such as mercury, that are removed from wastewater before the water is returned to the environment don't just disappear. They are either recycled or disposed of in a safe manner. What law of physics states that matter can't be created or destroyed? How can removing pollutants from the water cause pollution on land?

Metals Many metals such as mercury, lead, nickel, and cadmium can be poisonous, even in small amounts. For example, lead and mercury in drinking water can damage the nervous system. However, metals such as these are valuable in making items you use such as paper, paints, and stereos. Before environmental laws were written, a large amount of metal was released with wastewater from factories. Today, laws control how much metal can be released. Because metals remain in the environment for a long time, metals released many years ago still are polluting bodies of water today.

Mining also releases metals into water. For example, in the state of Tennessee, more than 43 percent of all streams and lakes contain metals from mining activities. In the mid 1980s, gold was found near the Amazon River in South America. Miners use mercury to trap the gold and separate it from sediments. Each year, more than 130 tons of mercury end up in the Amazon River.

Oil and Gasoline Oil and gasoline run off roads and parking lots and into streams and rivers when it rains. These compounds contain pollutants that might cause cancer. Gasoline is stored at gas stations in tanks below the ground. In the past, the tanks were made of steel. Some of these tanks rusted and leaked gasoline into the surrounding soil and groundwater. As little as one gallon of gasoline can make an entire city's water supply unsafe for drinking.

Federal laws passed in 1988 require all new gasoline tanks to have a double layer of steel or fiberglass. In addition, by 1998, all new and old underground tanks must have had equipment installed that detects spills and must be made of materials that will not develop holes. These laws help protect soil and groundwater from gasoline and oil stored in underground tanks.

Figure 5
During the manufacture of many products, such as electricity from this power plant, water is needed for cooling the machinery. Heated water remains in large towers and ponds until it has cooled to a temperature that is safe for fish and other organisms.

Heat When a factory makes a product, heat often is released. Sometimes, cool water from a nearby ocean, river, lake, or underground supply is used to cool factory machines. The heated water then is released. This water can pollute because it contains less oxygen than cool water does. In addition, organisms that live in water are sensitive to changes in temperature. A sudden release of heated water can kill a large number of fish in a short time. Water can be cooled before it is released into a river by using a cooling tower or pond, as shown in **Figure 5.**

Reducing Water Pollution

One way to reduce water pollution is by treating water before it enters a stream, lake, or river. In 1972, the United States Congress amended the Water Pollution Control Act. This law provided funds to build sewage-treatment facilities. It required industries to remove or treat pollution in water discharged to a lake or stream. The Clean Water Act of 1987 made additional money available for sewage treatment and set goals for reducing point source and nonpoint source pollution.

Another law, the Safe Drinking Water Act of 1996, strengthens health standards for drinking water. This legislation also protects rivers, lakes, and streams that are sources of drinking water.

International Cooperation Several countries have worked together to reduce water pollution. Lake Erie is on the border between the United States and Canada. Prior to the 1970s, phosphorus and nitrogen from sewage, soaps, and fertilizers entered Lake Erie from homes, yards, and farms, causing algae to grow and reproduce. The lake became a green, soupy mess. In the summer, the algae died and sank to the lake bottom. As the dead algae decayed, large areas of the lake bottom no longer had oxygen and, therefore, no life.

Pollutants also were discharged from many steel, automobile, and other factories along Lake Erie. **Figure 6** shows how on June 22, 1969, greasy debris on a large river flowing through Cleveland, Ohio, caught fire. This event was a wake-up call for everyone concerned about the quality of water in the United States and around the world.

In the 1970s, the United States and Canada made two water-quality agreements. These agreements set goals for reducing pollution in the Great Lakes. As a result of these agreements, limits were placed on the amount of phosphorus and other pollutants allowed into Lake Erie.

✓ Reading Check *Which countries worked together to control water pollution in Lake Erie?*

Today, the green slime is gone and the fish are back. However, more than 300 human-made chemicals still can be found in Lake Erie, and some of them are hazardous. The United States and Canada are studying ways to remove them from the lake.

Figure 6
Because of laws passed since 1972, Lake Erie's water has improved.

A Firefighters battled debris burning on the Cuyahoga River in Cleveland, Ohio. This alerted people in the United States to water pollution problems.

B Today, millions of people enjoy this natural resource.

How can you help?

Through laws and regulations, the quality of many streams, rivers, and lakes in the United States has improved. However, as **Figure 7** shows, much remains to be done. Individuals and industries alike need to continue to work to reduce water pollution. You easily can help by keeping contaminants out of Earth's water supply and by conserving water.

Dispose of Wastes Safely When you dispose of household chemicals such as paint and motor oil, don't pour them onto the ground or down the drain. Hazardous wastes that are poured directly onto the ground move through the soil and eventually reach the groundwater below. Pouring them down the drain is no better because they flow through the sewer, through the wastewater-treatment plant, and into a stream or river where they can harm the organisms living there.

What should you do with these wastes? First, read the label on the container for instructions on disposal. Don't throw the container into the trash if the label tells you not to. Store chemical wastes so that they can't leak. Call your local government officials and ask how to dispose of these wastes in your area safely. Many communities have specific times each year when they collect hazardous wastes. These wastes then are disposed of at special hazardous waste sites.

SCIENCE *Online*

Research Visit the Glencoe Science Web site at **science.glencoe.com** for more information about water conservation. Communicate to your class what you learn.

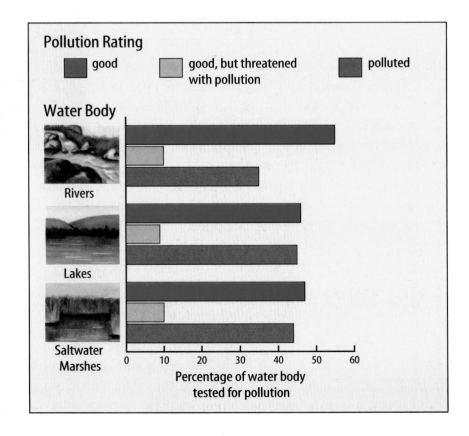

Figure 7
This graph shows that water pollution is still a problem in the United States. *What is the percentage of rivers listed as polluted?*

Figure 8
Water pollution can be reduced if less water is used.

B One drip every 5s from a leaky faucet will waste nearly 2,400 L of water per year.

A Toilets made before 1994 use nearly 76 L of water per person per day. Replacing your old toilet with a new one can save your family up to 82,000 L of water per year.

C Turning off the water while brushing your teeth will save more than 19 L per day.

Conserve Water How much water do you use every day? You use water every time you flush a toilet, take a bath, clean your clothes, wash dishes, wash a car, or use a hose or lawn sprinkler. A typical U.S. citizen uses an average of 692 L of water per day. Unless it comes from a home well, this water must be purified before it reaches your home. After you use it, it must be treated again. **Figure 8** shows how using simple conservation methods can save water. Conserving water reduces the need for water treatment and reduces water pollution.

Section 1 Assessment

1. Name six sources of water pollution.

2. What is the difference between point source and nonpoint source pollution?

3. How have U.S. laws helped reduce water pollution?

4. What are two things you can do to help reduce water pollution?

5. **Think Critically** Southern Florida has many dairy farms and sugarcane fields. It also contains Everglades National Park—a shallow river system with polluted waters. What kinds of pollutants are in the Everglades? How did they get there?

Skill Builder Activities

6. **Testing a Hypothesis** A stream near your home is polluted. You hypothesize that a large factory is responsible. Design an experiment that will help you test whether the factory is the pollution source. **For more help, refer to the** Science Skill Handbook.

7. **Using Graphics Software** Use graphics software to design a pamphlet that informs people how to reduce the amount of water they use. Be creative and include graphics in your pamphlet. **For more help, refer to the** Technology Skill Handbook.

Activity

Elements in Water

When you look at water, it is often clear and looks as if there is not much in it. However, there are many compounds, microscopic organisms, and other substances that can be in the water, but aren't easily visible. How can you find out what else might be in the water? Can you also find out how much of it is in the water?

What You'll Investigate
What is the nitrate and phosphate content of water?

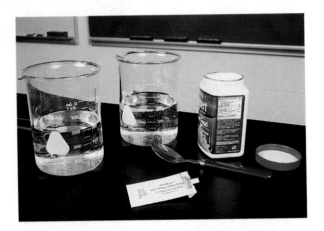

Materials
beakers (2)	nitrate test kit
tap water	phosphate test kit
plant fertilizer	stirrer
teaspoon	

Goals
- **Determine** the nitrate and phosphate content of two samples of water.
- **Compare** the levels and explain any differences you find.

Safety Precautions
Never eat or drink anything in the lab. Use gloves and goggles when handling fertilizer.

Procedure
1. Half-fill two large beakers with tap water.
2. Add a teaspoon of plant fertilizer to one of the beakers and stir well.
3. **Predict** which beaker might have a greater level of nitrate.
4. Using an appropriate kit, measure the nitrate content of each beaker of water.
5. Clean the test kit between measurements. Record your measurements.
6. **Predict** which beaker might have a greater level of phosphates.
7. Using an appropriate kit, measure the phosphate content of each beaker of water. Be sure to clean the kit between measurements. Record your measurements.

Conclude and Apply
1. **Describe** your results. Were the levels of each compound you measured the same in both samples?
2. Were your predictions correct?
3. **Explain** any differences that you found.
4. **Explain** how the use of fertilizers can cause problems in lakes and streams.

Communicating Your Data

Compare your results with those of others in your class. Discuss any differences found in your measurements.

Air Pollution

Causes of Air Pollution

Cities can be exciting because they are centers of business, culture, and entertainment. Unfortunately, cities also have many cars, buses, and trucks that burn fuel for energy. The brown haze you sometimes see forms from the exhaust of these vehicles. Air pollution comes from burning fuels in factories, generating electricity, and burning trash. Dust from plowed fields, construction sites, and mines also contributes to air pollution.

Natural sources add pollutants to the air, too. For example, radon is a naturally occurring gas given off by certain kinds of rock. This gas can seep into basements of homes built on these rocks. Exposure to radon can increase the risk of lung cancer. Natural sources of pollution also include particles and gases emitted into air from erupting volcanoes and fires.

What is smog?

One type of air pollution found in urban areas is called smog, a term originally used to describe the combination of smoke and fog. Major sources of smog, shown in **Figure 9,** include cars, factories, and power plants.

Sources of Smog

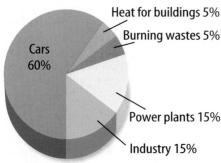

Heat for buildings 5%
Burning wastes 5%
Cars 60%
Power plants 15%
Industry 15%

Figure 9
Cars are one of the main sources of air pollution in the United States. *What percentage of smog comes from power plants and industry combined?*

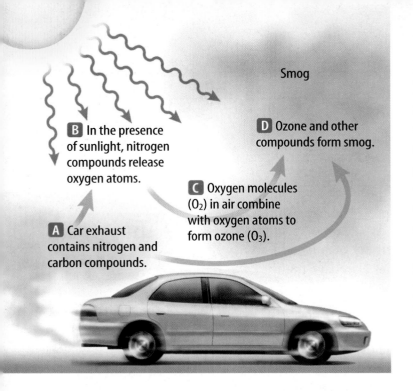

B In the presence of sunlight, nitrogen compounds release oxygen atoms.

Smog

D Ozone and other compounds form smog.

C Oxygen molecules (O_2) in air combine with oxygen atoms to form ozone (O_3).

A Car exhaust contains nitrogen and carbon compounds.

Figure 10
In the presence of sunlight, car exhaust can form smog.

Figure 11
Conditions in the atmosphere can worsen air pollution.

How Smog Forms The hazy, yellowish brown blanket of smog that is sometimes found over cities is called **photochemical smog** because it forms with the help of sunlight. Pollutants get into the air when gasoline is burned, releasing nitrogen and carbon compounds. These compounds, as shown in **Figure 10,** react in the presence of sunlight to produce other substances. One of the substances produced is ozone. Ozone high in the atmosphere protects you from the Sun's ultraviolet radiation. However, ozone near Earth's surface is a major component of smog. Smog can damage sensitive tissues, like plants or your lungs.

Nature and Smog Certain natural conditions contribute to smoggy air. For example, in many places, smog is not a problem because winds disperse the pollutants that cause smog to form. In other areas, landforms add to smog development. The mountains surrounding Los Angeles, for example, can prevent smog from being carried away by winds.

Figure 11 shows how the atmosphere also can influence the formation of smog. Normally, warmer air is found near Earth's surface. However, sometimes warm air traps cool air near the ground. This is called a temperature inversion, and it reduces the amount of mixing in the atmosphere, causing pollutants to accumulate near Earth's surface.

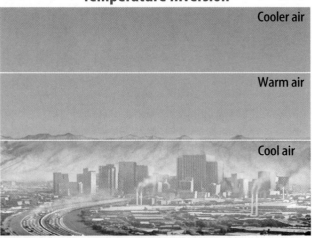

Normal Conditions

Cooler air

Cool air

Warm air

A Usually, air temperature decreases with distance above Earth's surface. Air pollutants can be carried far away from their source.

Temperature Inversion

Cooler air

Warm air

Cool air

B During a temperature inversion, warm air overlies cool air. Air pollutants can't be dispersed and can accumulate to unhealthy levels.

Acid Rain

When sulfur oxides from coal-burning power plants and nitrogen oxides from cars combine with moisture in the air, they form acids. When acidic moisture falls to Earth as rain or snow, it is called **acid rain.** Acid rain can corrode structures, damage forests, and harm organisms. The amount of acid is measured using the **pH scale,** shown in **Figure 12.** A lower number means greater acidity. Substances with a pH lower than 7 are **acids.** Substances with a pH above 7 are **bases.**

Natural lakes and streams have a pH between 6 and 8. When rain is acidic, the pH of streams and lakes may decrease. As **Figure 12** shows, certain organisms, like snails, can't live in acidic water.

CFCs

In the atmosphere, a layer containing a higher concentration of ozone molecules protects Earth from some of the Sun's harmful rays. Chlorofluorocarbons (CFCs) from air conditioners and refrigerators might be destroying this ozone layer. Even though the use of CFCs is declining worldwide, these compounds can remain in the upper atmosphere for decades.

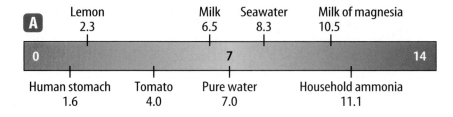

| A | Lemon 2.3 | | Milk 6.5 | Seawater 8.3 | Milk of magnesia 10.5 |

| 0 | 7 | 14 |

Human stomach 1.6   Tomato 4.0   Pure water 7.0   Household ammonia 11.1

B   **pH Tolerance of Various Organisms**

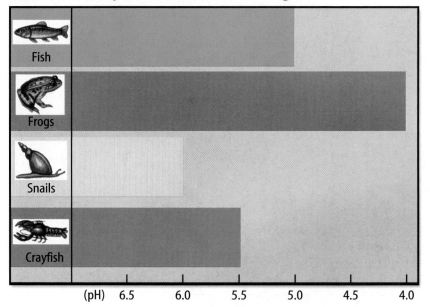

Fish

Frogs

Snails

Crayfish

(pH)   6.5   6.0   5.5   5.0   4.5   4.0

Mini LAB

Identifying Acid Rain

Procedure

1. Use a clean **glass or plastic container** to collect a **sample of precipitation.**
2. Use **pH paper** or a **pH computer probe** to determine the acidity level of your sample. If you have collected snow, allow it to melt before measuring its pH.
3. Record the indicated pH of your sample and compare it with the results of other classmates who have followed the same procedure.

Analysis

1. What is the average pH of the samples obtained from this precipitation?
2. Compare and contrast the pH of your samples with those of the substances shown on the pH scale in **Figure 12.**

Figure 12

A The natural pH of rainwater is about 5.6. Acid rain is precipitation with a pH below 5.6.

B This chart shows the different levels of acidity in water that certain organisms can live in. *Which types of organisms might you find in a pond with a pH of 5.5 or greater? Which organisms would be in the pond if the pH dropped to 4.5?*

Air Pollution and Your Health

Suppose you're an athlete in a large city training for a competition. You might have to get up at 4:30 A.M. to exercise. Later in the day, the smog levels might be so high that it wouldn't be safe for you to exercise outdoors. In some large cities, athletes adjust their training schedules to avoid exposure to ozone and other pollutants. Schools schedule football games for Saturday afternoons when smog levels are lower. Parents are warned to keep their children indoors when smog exceeds certain levels.

Health Disorders How hazardous is dirty air? Approximately 250,000 people in the United States suffer from pollution-related breathing disorders. About 60,000 deaths each year in the United States are blamed on air pollution. **Figure 13** illustrates some of the health problems caused by air pollution. Ozone damages lung tissue, making people more susceptible to diseases such as pneumonia and asthma. Less severe symptoms of ozone include burning eyes, dry throat, and headache.

How do you know if ozone levels in your community are safe? You may have seen the Air Quality Index reported in your newspaper. **Table 1** shows the index along with ways to protect your health when ozone levels are high.

Carbon monoxide, a colorless, odorless gas found in car exhaust, also contributes to air pollution. This gas can make people ill, even in small concentrations because it replaces oxygen in your blood.

SCIENCE Online

Collect Data Air quality is reported daily for major cities. Visit the Glencoe Science Web site at **science.glencoe.com** and look up the air quality index of a city near you each day for seven days. Graph the results.

Figure 13
Air pollution can be a health hazard. Compounds in the air can affect your body.

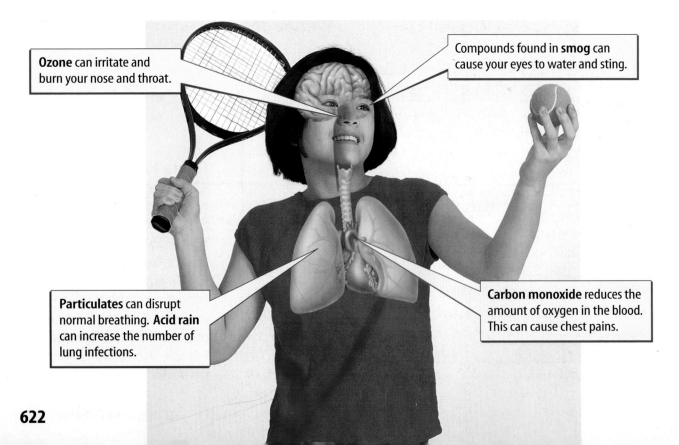

Ozone can irritate and burn your nose and throat.

Compounds found in **smog** can cause your eyes to water and sting.

Particulates can disrupt normal breathing. **Acid rain** can increase the number of lung infections.

Carbon monoxide reduces the amount of oxygen in the blood. This can cause chest pains.

Acid Rain What do you suppose happens when you inhale the humid air from acid rain? Acid is breathed deep inside your lungs. This may cause irritation and reduce your ability to fight respiratory infections. When you breathe, oxygen travels from the air to your lungs. Lungs damaged by acid rain cannot move oxygen to the blood easily. This puts stress on your heart.

Particulates Thick, black smoke from a forest fire, exhaust from school buses and large trucks, smoke billowing from a factory, and dust picked up by the wind all contain **particulate** (par TIH kyuh luht) **matter**. Particulate matter consists of fine particles such as dust, pollen, mold, ash, and soot that are in the air.

Particulate matter ranges in size from large, visible solids like dust and soil particles to microscopic particles that form when substances are burned. Smaller particles are more dangerous, because they can travel deeper into the lungs. When particulate matter is breathed in, it can irritate and damage the lungs, causing breathing problems.

✓ Reading Check *Why are small particles dangerous to your health?*

Reducing Air Pollution

Pollutants moving through the atmosphere don't stop when they reach the borders between states and countries. They go wherever the wind carries them. This makes them difficult to control. Even if one state or country reduces its air pollution levels, pollutants from another state or country can blow across the border. For example, burning coal in midwestern states might cause acid rain in the northeast and Canada.

When states and nations cooperate, pollution problems can be reduced. People from around the world have met on several occasions to try to eliminate some kinds of air pollution. At one meeting in Montreal, Canada, an agreement called the Montreal Protocol was written to phase out the manufacture and use of CFCs by 2000. In 1989, 29 countries that consumed 82 percent of CFCs signed the agreement. By 1999, 184 countries signed it.

Table 1 Air-Quality Guide for Ozone

Air Quality	Air Quality Index	Protect Your Health
Good	0–50	No health impacts occur.
Moderate	51–100	People with breathing problems should limit outdoor exercise.
Unhealthy for Certain People	101–150	Everyone, especially children and elderly, should not exercise outside for long periods of time.
Unhealthy	151–200	People with breathing problems should avoid outdoor activities.

TRY AT HOME
Mini LAB

Examining the Content of Air

Use caution when reaching high places. Students with dust allergies should not perform this activity.

Procedure
1. Find a **high shelf or the top of a tall cabinet** in your home—someplace that hasn't been cleaned for a while.
2. Using a **white cloth,** thoroughly dust the surface.
3. Observe the cloth under a **magnifying lens.**

Analysis
1. What did you see on your cloth? Where did these particles come from?
2. Explain what you think happens when you breathe in these particles.

Table 2 Clean Air Regulations

Urban Air Pollution	All cars manufactured since 1996 must reduce nitrogen oxide emissions by 60 percent and hydrocarbons by 35 percent from their 1990 levels.
Acid Rain	Sulfur dioxide emissions had to be reduced by 14 million tons from 1990 levels by the year 2000.
Airborne Toxins	Industries must limit the emission of 200 compounds that cause cancer and birth defects.
Ozone-Depleting Chemicals	Industries were required to immediately cease production of many ozone-depleting substances in 1996.

Air Pollution in the United States The United States Congress passed several laws to protect the air. The Clean Air Act of 1990, summarized in **Table 2,** addressed some air pollution problems by regulating emissions from cars, energy production, and other industries. In 1997, new levels for ozone and particulate matter were proposed.

Since the passage of the Clean Air Act, the amount of some pollutants released into the air has decreased, as shown in **Figure 14.** However, one out of four people in the United States still breathes unhealthy air.

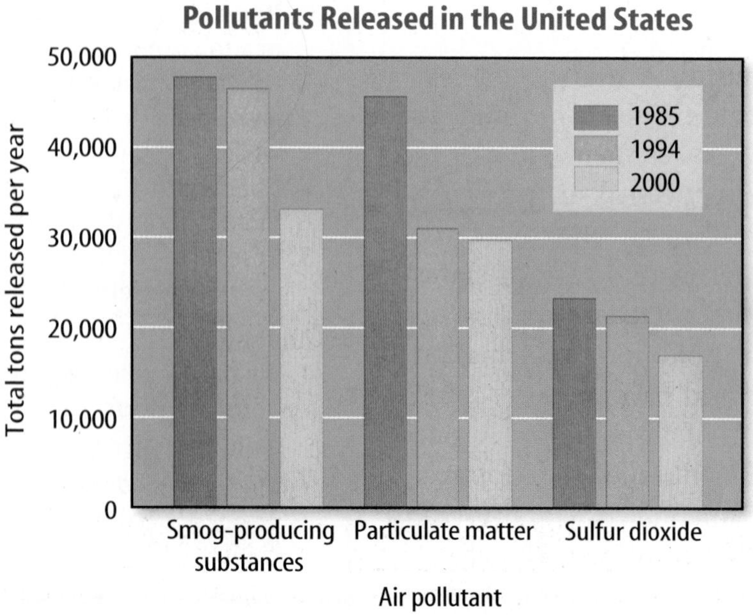

Reducing Emissions More than 80 percent of sulfur dioxide emissions comes from coal-burning power plants. Coal from some parts of the United States contains a lot of sulfur. When this coal is burned, sulfur oxides combine with moisture in the air to form sulfuric acid, causing acid rain. Sulfur dioxide can be removed by passing the smoke through a scrubber. A **scrubber** lets the gases react with a limestone and water mixture. Another way to decrease the amount of sulfur dioxide is by burning low-sulfur coal.

Electric power plants that burn fossil fuels emit particulates into the atmosphere. **Figure 15** shows how particulates can be removed using an electrostatic separator.

Figure 14
This graph shows that some air pollutants have decreased since the passage of the Clean Air Act.
How many tons of particulates were released to the air in 1994?

Pollutants Released in the United States

Getting Around Cars produce more than 80 percent of the carbon monoxide and 40 percent of the nitrogen oxides that enter the atmosphere in the United States. Better emission-control devices on cars will help, but will that be enough to solve smog problems? Today, Americans drive more per year than they ever did before. More time spent driving and more cars on the roads lead to more traffic congestion. Cars produce even more pollutants when they are stopped in traffic or at traffic lights.

The Clean Air Act can work only if we all cooperate. Cleaning the air takes money, time, and effort. How might you take part in this cleanup? You can change your lifestyle. For example, you can walk, ride a bike, or use public transportation to get to a friend's house instead of asking for a ride. You also can set the thermostat in your house lower in the winter and higher in the summer.

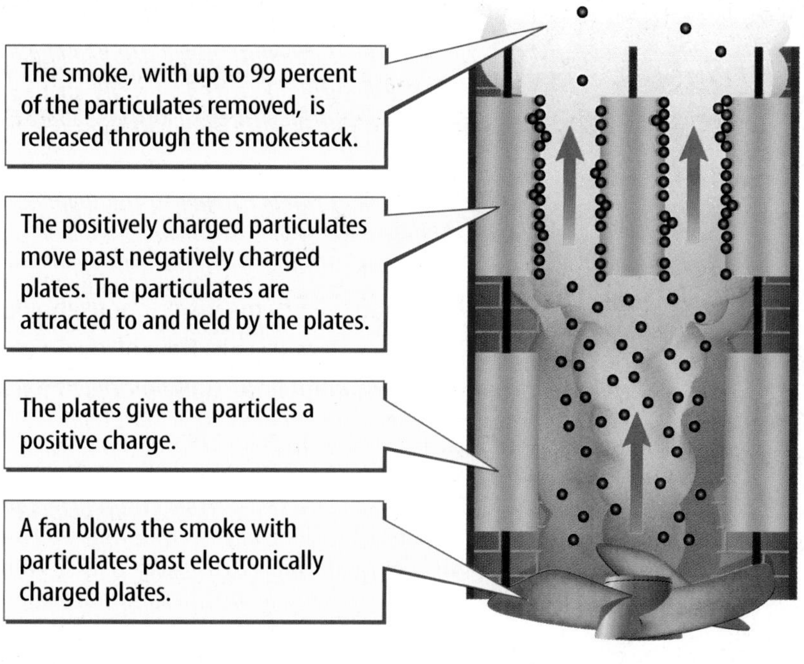

The smoke, with up to 99 percent of the particulates removed, is released through the smokestack.

The positively charged particulates move past negatively charged plates. The particulates are attracted to and held by the plates.

The plates give the particles a positive charge.

A fan blows the smoke with particulates past electronically charged plates.

Figure 15
Electrostatic separators can remove almost all of the particulates in industrial smoke.

Reading Check *What can you do to prevent air pollution?*

Section 2 Assessment

1. Name four sources of air pollution.
2. What are three pollutants released into the air when fuels are burned?
3. Name several ways air pollution can affect your health.
4. How can people reduce air pollution?
5. **Think Critically** Laws were passed in 1970 requiring coal-burning power plants to use tall smokestacks to disperse pollutants. Power plants in the midwestern states complied with that law, and people in eastern Canada began complaining about acid rain. Explain the connection.

Skill Builder Activities

6. **Classifying** Use the information in **Table 1** to classify the following air quality indices: 43, 152, 7, 52, 147, and 98. What category of air quality would you like to have in your neighborhood? Why is it important to have limits on pollutants from cars and factories as the U.S. population grows and people drive more? **For more help, refer to the** Science Skill Handbook.

7. **Communicating** In your Science Journal, create a crossword puzzle using at least 12 important terms found in this section. **For more help, refer to the** Science Skill Handbook.

Activity

Design Your Own Experiment

What's in the air?

When you dust items in your household, you are cleaning up particles that settled out of the air. How often do you have to dust to keep your furniture clean? Just imagine how many pieces of particulate matter the air must hold.

Recognize the Problem

Do some areas of your environment have more particulates than other areas?

Form a Hypothesis

Based on your knowledge of your neighborhood, hypothesize what kinds of particulate matter you will find in your environment. Will all areas in your community contain the same types and amounts of particulate matter?

Goals
- **Design** an experiment to collect and analyze particulate matter in the air in your community.
- **Observe and describe** the particulate matter you collect.

Safety Precautions

Wear a thermal mitt, safety goggles, and an apron while working with a hot plate and while pouring the gelatin from the pan or pot into the lids. Never eat anything in the lab.

Possible Materials
small box of plain gelatin
hot plate
pan or pot
water
marker
refrigerator
plastic lids (4)
microscope
Hand lens
Alternate materials

Test Your Hypothesis

Plan

1. As a group, agree upon your hypothesis and decide how you will test it. Identify which results will confirm or refute the hypothesis.

2. **List** the steps you need to take to test your hypothesis. Be specific. Describe exactly what you will do at each step. List your materials.

3. Prepare a data table in your Science Journal to record your observations.

4. Label your lids with the location where you decide to place them.

5. Mix the gelatin according to the directions on the box. Carefully pour a thin layer of gelatin into each lid. Use this to collect air particulate matter.

6. Read over your entire experiment to make sure that all steps are in a logical order.

7. Identify any constants, variables, and controls of the experiment.

Do

1. Make sure your teacher approves your plan.

2. Carry out the experiment as planned.

3. While the experiment is going on, record any observations that you make and complete the data table in your Science Journal.

Analyze Your Data

1. **Describe** the types of materials you collected in each lid.

2. **Calculate** the number of particles on each lid.

3. What did you use as a control in this experiment?

4. What were your variables?

5. **Graph** your results using a bar graph. Place the number of particulates on the *y*-axis and the test site location on the *x*-axis.

Draw Conclusions

1. Did the results support your hypothesis? Explain.

2. **Explain** why different sizes of particulate matter may be found at different locations.

3. **Infer** why some test-site locations showed more particulates than other sites did.

*C*ommunicating Your Data

Give an oral presentation to another class on air pollution in your community. Demonstrate your experiment and graphically display the results.

MEET

In 1958, retired biologist Rachel Carson (1907–1964) received a letter from a worried friend. Several songbirds had died immediately after the pesticide DDT was sprayed over an area of woods. In the 1940s and 1950s, DDT was sprayed over large areas of land to kill insects that caused crop damage and to eliminate diseases such as malaria. DDT was considered to be a scientific miracle. The letter Carson received, however, indicated something she had long suspected—there was a downside to the miracle.

Her friend's letter troubled Carson so much that she launched an investigation into the possible harmful effects of pesticides—a study that took four years of research, interviews, and analysis. The result was Carson's famous book, *Silent Spring*. In it, she stated her findings that pesticide use was destroying the food supply of many animals, killing birds and fish, and poisoning human food supplies. She wrote that unless action was taken, an eerie stillness would settle over the world, a world without songbirds—a silent spring.

A biologist and writer who made people aware of the fragility of nature

RACHEL CARSON

The publication of *Silent Spring* led to a heated debate in the United States over the use of pesticides. But it also led to a change in how people thought about the natural world. Before Carson's book, few people thought about nature and how human activities might affect Earth's organisms. Thanks to *Silent Spring*, many people began to realize that Earth and the organisms living on Earth are closely connected.

"The more clearly we can focus our attention on the wonders and realities of the universe about us, the less taste we have for its destruction," Carson wrote.

Carson's findings were verified and DDT was banned. Many species of birds owe their continuing existence to her efforts. The most famous example is the national symbol of the United States—the bald eagle. DDT caused the bald eagles' eggs to weaken and break, bringing the species close to extinction. Since the ban on DDT, bald eagles have been making a slow comeback. Their recovery was affirmed in 1999, when the U.S. Fish & Wildlife Service proposed to remove the bald eagle from the endangered species list.

CONNECTIONS Make Posters Research an environmental issue you are concerned about. Make posters to educate others in your school or community about the issue. Look for quotes you'd like to use to help illustrate your poster. You might find some in Carson's book.

SCIENCE *Online*
For more information, visit
science.glencoe.com

Reviewing Main Ideas

Section 1 Water Pollution

1. Water pollution comes from industrial discharge; runoff of pesticides and fertilizers from lawns and farms; and water from your home. *How can oil dripping onto pavement eventually pollute a stream?*

2. National and international cooperation is necessary if water pollution is to be reduced. In the United States, the 1990 Clean Water Act set up standards for sewage and wastewater-treatment facilities and for runoff from roadways and farms.

3. Conserving water in your daily activities can help reduce water pollution. *How can doing laundry conserve water and electricity if only full loads are washed?*

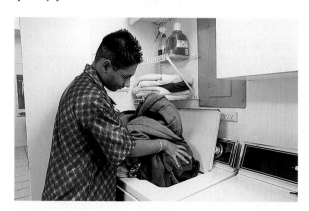

4. The water quality of many lakes and rivers in the United States is improving.

Section 2 Air Pollution

1. Exhaust from vehicles pollutes the air. Other sources of air pollution include power plants, fires and volcanoes.

2. Natural conditions, such as landforms and temperature inversions, can affect air quality.

3. Polluted air can affect human health. Breathing particles, ozone, and acid rain can damage your lungs.

4. Air pollutants don't have boundaries. They drift between states and countries. National and international cooperation is necessary to reduce the problem. *Which type of air pollution can damage structures like the statue below?*

FOLDABLES
Reading & Study Skills

After You Read

Compare and contrast characteristics of water and air pollution. List things in common under the middle tab of your Foldable.

Visualizing Main Ideas

Complete the following concept map on types of pollution.

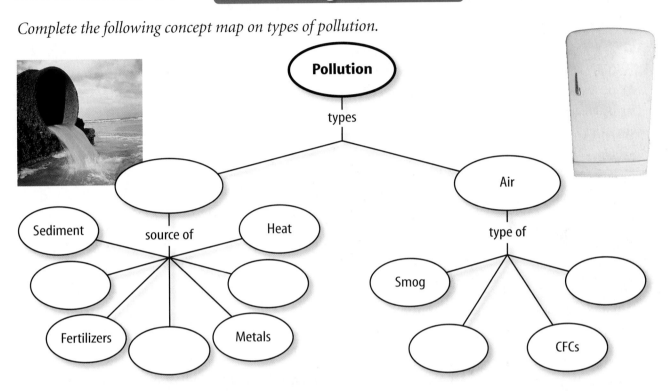

Vocabulary Review

Vocabulary words

a. acid
b. acid rain
c. base
d. carbon monoxide
e. fertilizer
f. nonpoint source pollution
g. particulate matter
h. pesticide
i. pH scale
j. photochemical smog
k. point source pollution
l. scrubber
m. sewage

Using Vocabulary

Replace the underlined phrase with the correct vocabulary words.

1. Smog that forms with the aid of sunlight contains ozone near Earth's surface.

2. Pollution that enters water from a specific location can be controlled or treated before it enters a body of water.

3. Chemicals that help plants and other organisms grow can reduce the amount of oxygen in lakes.

4. A device that lowers sulfur emissions from coal-burning power plants can help reduce acid rain.

5. When fine solids such as dust, ash, and soot that are in the air are inhaled, they can irritate and damage the lungs.

THE PRINCETON REVIEW **Study Tip**

Write the questions as well as the answers to end-of-chapter quizzes. This will help you form complete responses to important questions.

Checking Concepts

Choose the word or phrase that best answers the question.

1. Which is most responsible for city smog?
   A) power plants   C) industries
   B) burning wastes   D) cars

2. What forms when pollutants react in the presence of sunlight?
   A) pH   C) particulates
   B) photochemical smog D) acid rain

3. Which of the following describes substances with a low pH?
   A) neutral   C) dense
   B) acidic   D) basic

4. What combines with moisture in the air to form acid rain?
   A) ozone   C) lead
   B) sulfur oxides D) oxygen

5. Which of the following is a nonpoint source?
   A) runoff from a golf course
   B) sewage
   C) wastewater from industry
   D) discharge from a ditch into a river

6. What is the largest source of water pollution in the United States?
   A) sediment C) heat
   B) metals   D) gasoline

7. What is the pH of acid rain?
   A) less than 5.6   C) greater than 7.0
   B) between 5.6 and 7.0 D) greater than 9.5

8. What kind of pollution are airborne solids that range in size from large grains to microscopic?
   A) pH   C) particulate
   B) ozone D) acid rain

9. Which of the following causes algae to grow?
   A) pesticides C) metals
   B) sediment   D) fertilizers

10. Which act gives money to local governments to treat wastewater?
    A) Water Pollution Control Act
    B) Clean Air Act
    C) Montreal Protocol
    D) Safe Drinking Water Act

Thinking Critically

11. How might cities with smog problems lessen the dangers to people who live and work in the cities?

12. What are some ways to control nonpoint pollution sources?

13. Why is it important to place stricter limits on pollutants from cars and factories as the U.S. population grows and people drive more?

14. Pollution occurs when heated water is released into a nearby body of water. What effects does this type of pollution have on organisms living in the water?

15. How might a community in a desert cope with water-supply problems?

Developing Skills

16. **Making and Using Graphs** This table lists the total phosphorus entering Lake Erie for the listed year. Make a line graph of these data.

17. **Recognizing Cause and Effect** What effect will an increase in the human population have on the need for freshwater?

Phosphorus Entering Lake Erie	
Year	Metric Tons
1976	15,000
1982	12,000
1988	8,000
1995	7,000

18. Concept Mapping Complete this concept map about sewage treatment.

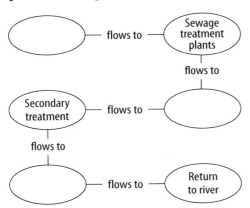

19. Relating Cause and Effect Explain how air and water are polluted when automobiles are used for transportation.

20. Communicating Explain what you personally can do to reduce air pollution.

Performance Assessment

21. Design and Perform an Experiment Design and perform an experiment to test the effects of acid rain on different kinds of vegetation. You might choose to use different types of vegetation as your variable and use acidity level as your constant. You also might want to vary the pH of the solution.

22. Letter to the Editor Survey your town for evidence of air and water pollution. Write a letter to the editor of your local newspaper communicating what you have observed. Include suggestions for reducing pollution.

TECHNOLOGY

Go to the Glencoe Science Web site at **science.glencoe.com** or use the **Glencoe Science CD-ROM** for additional chapter assessment.

Test Practice

Rafael went to the library to do some research on acid rain. One of the pages he photocopied was the pH Scale shown in the diagram below.

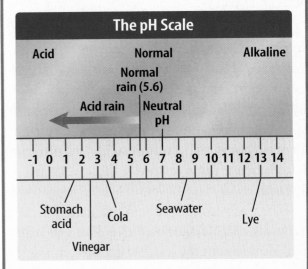

Study the diagram and answer the following questions.

1. According to the diagram, which substance has a pH greater than 10?
A) cola
C) seawater
B) lye
D) vinegar

2. Which of the following statements is correct?
F) Stomach acid has a higher pH than seawater.
G) Normal rain has a neutral pH.
H) Vinegar has a lower pH than cola.
J) Seawater has a higher pH than lye.

3. Which of the following best describes normal rain?
A) strongly alkaline
B) slightly alkaline
C) strongly acid
D) slightly acid

Reading Comprehension

Read the passage. Then read each question that follows the passage. Decide which is the best answer to each question.

Acid Rain

The United States has become increasingly dependent on fossil fuels as sources of energy. When fossil fuels such as coal, oil, and natural gas are burned, they emit oxides of carbon, sulfur, and nitrogen into the air. These pollutants combine with the moisture in the air to form carbonic acid, sulfuric acid, and nitric acid. When it rains or snows, these acids fall to Earth.

The problem of acid rain originated during the Industrial Revolution more than 150 years ago. However, in the last few decades, it has become more <u>severe</u>. Recent increases are due mainly to industrial coal and oil combustion and automobile exhaust emissions.

Acid rain upsets the delicate balance of the ecosystems in streams, rivers, and lakes. It also increases soil acidity, reducing crop production and forest growth. Acid rain also contributes to the destruction of buildings and monuments made of stone.

The problem of acid rain is not limited to industrial and urban areas. Winds can carry polluted clouds hundreds of miles before they release their acidic moisture. Consequently, it has become a global problem requiring a global solution.

Efforts to reduce the production of acid rain have concentrated on removing dangerous emissions at their source. Power plants are permitted to burn only low-sulfur coal. Anti-pollution devices called "scrubbers" are used to trap gases emitted from smokestacks before they can enter the atmosphere. Cars must pass strict emissions tests before they can be sold.

The average citizen can help by conserving energy at home and by buying more fuel-efficient automobiles. Nevertheless, the only long-range solution to the problems of air pollution in general and acid rain in particular is the development of alternative sources of energy other than fossil fuels.

> **Test-Taking Tip** Try drawing a picture or making a concept map to help you organize the main ideas in this passage.

1. In this passage, the word <u>severe</u> means _____.
   A) apparent
   B) complicated
   C) obvious
   D) serious

2. Which conclusion is best supported by information given in the passage?
   F) The problem of acid rain has no quick solution.
   G) The problem of acid rain appeared only recently.
   H) Scientists do not know what chemicals make up acid rain.
   J) There is little the average citizen can do to help reduce the incidence of acid rain.

3. Which of the following would NOT help reduce air pollution?
   A) conserving more energy at home
   B) burning high-sulfur coal
   C) requiring cars to pass stricter emissions tests
   D) removing more of the dangerous emissions at their source

Reasoning and Skills

Read each question and choose the best answer.

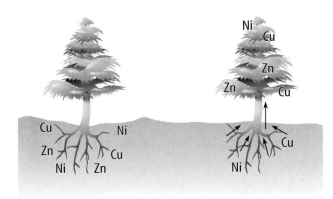

1. Phytoremediation occurs when certain types of plants remove harmful metals from soil. According to the picture, all of the following metals were removed from the soil EXCEPT _____.
   A) copper
   B) zirconium
   C) nickel
   D) zinc

Test-Taking Tip   Review the chemical element symbols for different metals. Refer to the periodic table at the back of the book.

2. A scientist is going to study the effects of phytoremediation. She wants to measure the rates at which certain types of trees remove different metals. When chosing an area to study, which would be the best choice?
   F) along the banks of a stream
   G) a closed auto factory
   H) a new housing development
   J) along a sea coastline

3. Information about enzymes would most likely be found under which heading in a table of contents?
   A) Carrying Capacity
   B) Yard Waste and You
   C) Breaking Down Pollutants
   D) Lining Our Landfills

Test-Taking Tip   Review the characteristics and uses of enzymes.

**PESTICIDES
SEDIMENT
FERTILIZERS**

4. Which if these belongs with the group above?
   F) smog
   G) carbon monoxide
   H) human waste
   J) particulates

Test-Taking Tip   Review the different types of water and air pollution.

Consider this question carefully before writing your answer on a separate sheet of paper.

5. There are different types of smog. Explain the causes and characteristics of each different type of smog.

Test-Taking Tip   Review the causes and characteristics of photochemical smog and sulfurous smog.

How Are Thunderstorms & Neutron Stars Connected?

In 1931, an engineer built an antenna to study thunderstorm static that was interfering with radio communication. The antenna did detect static from storms, but it also picked up something else: radio signals coming from beyond our solar system. That discovery marked the birth of radio astronomy. By the 1960s, radio astronomy was thriving. In 1967, astronomer Jocelyn Bell Burnell detected a peculiar series of radio pulses coming from far out in space. At first, she and her colleagues theorized that the signals might be a message from a distant civilization. Soon, however, scientists determined that the signals must be coming from something called a neutron star (below)—a rapidly spinning star that gives off a radio beam from its magnetic pole.

Axis

Direction of spin

Radio beam

SCIENCE CONNECTION

NEUTRON STARS Find out more about what neutron stars are and how they form. You might begin your research by visiting the Glencoe Science Web site at **science.glencoe.com** or by consulting an encyclopedia or astronomy textbook. Then work with a partner to design a demonstration that uses a flashlight to show how a spinning neutron star emits a radio signal that sweeps past Earth like the rotating beam of a lighthouse.

22 Exploring Space

Stars and planets have always fascinated humans. We admire their beauty, and our nearest star—the Sun—provides energy that enables life to exist on Earth. For centuries, people have studied space from the ground. But, in the last few decades, space travel has allowed us to get a closer look. In this chapter, you'll learn how space is explored with telescopes, rockets, probes, satellites, and space shuttles. You'll see how astronauts like Shannon Lucid, shown here, now can spend months living and working aboard space stations.

What do you think?

Science Journal Look at the picture below with a classmate. Discuss what you think this might be or what might be happening. Here's a hint: *It's part of a dusty trail that's far, far away.* Write your answer or best guess in your Science Journal.

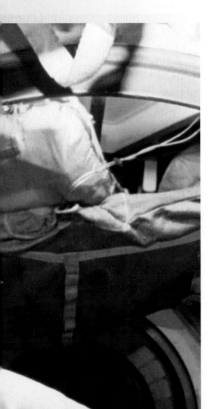

EXPLORE ACTIVITY

You might think exploring space with a telescope is easy because the visible light coming from stars is so bright and space is dark. But space contains massive clouds of gases, dust, and other debris called nebulae that block part of the starlight traveling to Earth making it more difficult for astronomers to observe deep space. What does visible light look like when viewed through clouds of dust or gas?

Model visible light seen through nebulae

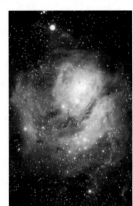

1. Turn on a lightbulb and darken the room.
2. View the lightbulb through a sheet of dark plastic.
3. View the lightbulb through different-colored plastic sheets.
4. View the light through a variety of different-colored balloons such as yellow, blue, red, and purple. Observe how the light changes when you slowly let the air out of each balloon.

Observe

Write a paragraph in your Science Journal describing how this activity modeled the difficulty astronomers have when viewing stars through thick nebulae?

Before You Read

FOLDABLES
Reading & Study Skills

Making a Sequence Study Fold Identifying a sequence helps you understand what you are experiencing and predict what might occur next. Before you read this chapter, make the following Foldable to prepare you to learn about the sequence of space exploration.

1. Place a sheet of paper in front of you so the short side is at the top. Fold the paper in half from the left side to the right side.
2. Fold the top and bottom in to divide the paper into thirds. Unfold the paper so three sections show.
3. Through the top thickness of paper, cut along each of the fold lines to the left fold, forming three tabs. Label the tabs "Past", "Present", and "Future", as shown.
4. As you read the chapter, write what you learn under the tabs.

Radiation from Space

1

As You Read

What You'll Learn

- **Explain** the electromagnetic spectrum.
- **Identify** the differences between refracting and reflecting telescopes.
- **Recognize** the differences between optical and radio telescopes.

Vocabulary

electromagnetic spectrum
refracting telescope
reflecting telescope
observatory
radio telescope

Why It's Important

You can learn much about space without traveling there.

Electromagnetic Waves

As you just read, living in space now is possible. The same can't be said, though, for space travel to distant galaxies. If you've dreamed about racing toward distant parts of the universe—think again. Even at the speed of light, it would take years and years to reach even the nearest stars.

Light from the Past When you look at a star, the light that you see left the star many years ago. Although light travels fast, distances between objects in space are so great that it sometimes takes millions of years for the light to reach Earth.

The light and other energy leaving a star are forms of radiation. Radiation is energy that is transmitted from one place to another by electromagnetic waves. Because of the electric and magnetic properties of this radiation, it's called electromagnetic radiation. Electromagnetic waves carry energy through empty space and through matter.

Electromagnetic radiation is everywhere around you. When you turn on the radio, peer down a microscope, or have an X ray taken—you're using various forms of electromagnetic radiation.

Figure 1
The electromagnetic spectrum ranges from gamma rays with wavelengths of less than 0.000 000 000 01 m to radio waves more than 100,000 m long. *How does frequency change as wavelength shortens?*

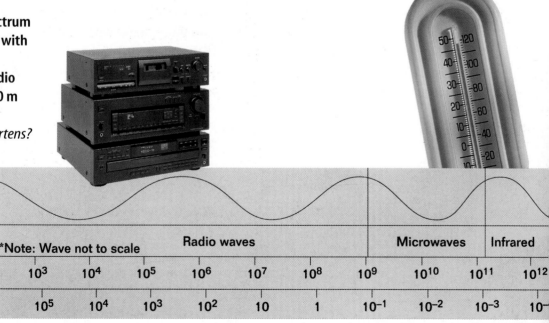

*Note: Wave not to scale	Radio waves					Microwaves		Infrared
10^3	10^4	10^5	10^6	10^7	10^8	10^9	10^{10}	10^{11}	10^{12}
10^5	10^4	10^3	10^2	10	1	10^{-1}	10^{-2}	10^{-3}	10^-

Electromagnetic Radiation Sound waves, which are a type of mechanical wave, can't travel through empty space. How, then, do we hear the voices of the astronauts while they're in space? When astronauts speak into a microphone, the sound waves are converted into electromagnetic waves called radio waves. The radio waves travel through space and through Earth's atmosphere. They're then converted back into sound waves by electronic equipment and audio speakers.

Radio waves and visible light from the Sun are just two types of electromagnetic radiation. Other types include gamma rays, X rays, ultraviolet waves, infrared waves, and microwaves. **Figure 1** shows these forms of electromagnetic radiation arranged according to their wavelengths. This arrangement of electromagnetic radiation is called the **electromagnetic spectrum.** Forms of electromagnetic radiation also differ in their frequencies. Frequency is the number of times a wave vibrates per unit of time. The shorter the wavelength is, the more vibrations will occur, as shown in **Figure 1.**

Speed of Light Although the various electromagnetic waves differ in their wavelengths, they all travel at 300,000 km/s in a vacuum. This is called the speed of light. Visible light and other forms of electromagnetic radiation travel at this incredible speed, but the universe is so large that it takes millions of years for the light from some stars to reach Earth.

When electromagnetic radiation from stars and other objects reaches Earth, scientists use it to learn about its source. One tool for studying electromagnetic radiation from distant sources is a telescope.

Health
INTEGRATION

Many newspapers include an ultraviolet (UV) index to urge people to minimize their exposure to the Sun. Compare the wavelengths and frequencies of red and violet light, shown below in **Figure 1.** Infer what properties of UV light cause damage to tissues of organisms.

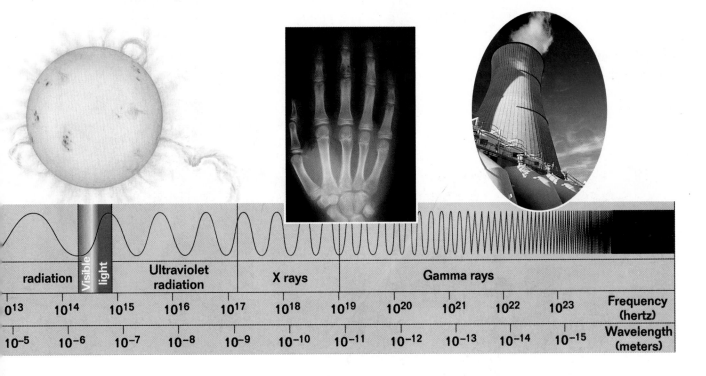

radiation	Visible light	Ultraviolet radiation		X rays		Gamma rays				

0^{13}	10^{14}	10^{15}	10^{16}	10^{17}	10^{18}	10^{19}	10^{20}	10^{21}	10^{22}	10^{23}	**Frequency (hertz)**
10^{-5}	10^{-6}	10^{-7}	10^{-8}	10^{-9}	10^{-10}	10^{-11}	10^{-12}	10^{-13}	10^{-14}	10^{-15}	**Wavelength (meters)**

Optical Telescopes

Optical telescopes use light, which is a form of electromagnetic radiation, to produce magnified images of objects. Light is collected by an objective lens or mirror, which then forms an image at the focal point of the telescope. The focal point is where light that is bent by the lens or reflected by the mirror comes together to form a point. The eyepiece lens then magnifies the image. The two types of optical telescopes are shown in **Figure 2.**

A **refracting telescope** uses convex lenses, which are curved outward like the surface of a ball. Light from an object passes through a convex objective lens and is bent to form an image at the focal point. The eyepiece magnifies the image.

A **reflecting telescope** uses a curved mirror to direct light. Light from the object being viewed passes through the open end of a reflecting telescope. This light strikes a concave mirror, which is curved inward like a bowl and located at the base of the telescope. The light is reflected off the interior surface of the bowl to the focal point where it forms an image. Sometimes, a smaller mirror is used to reflect light into the eyepiece lens, where it is magnified for viewing.

Using Optical Telescopes Most optical telescopes used by professional astronomers are housed in buildings called **observatories.** Observatories often have dome-shaped roofs that can be opened up for viewing. However, not all telescopes are located in observatories. The *Hubble Space Telescope* is an example.

Figure 2
These diagrams show how each type of optical telescope collects light and forms an image.

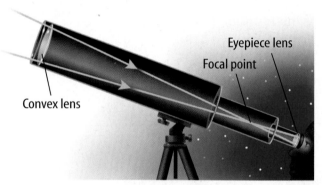

Eyepiece lens
Focal point
Convex lens

A In a refracting telescope, a double convex lens focuses light to form an image at the focal point.

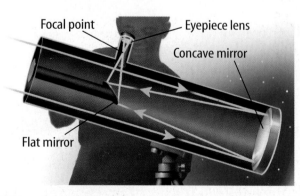

Focal point
Eyepiece lens
Concave mirror
Flat mirror

B In a reflecting telescope, a concave mirror focuses light to form an image at the focal point.

C Optical telescopes are widely available for use by individuals.

Hubble Space Telescope The *Hubble Space Telescope* was launched in 1990 by the space shuttle *Discovery*. Because *Hubble* is located outside Earth's atmosphere, which absorbs and distorts some of the energy received from space, it should have produced clear images. However, when the largest mirror of this reflecting telescope was shaped, a mistake was made. As a result, images obtained by the telescope were not as clear as expected. In December 1993, a team of astronauts repaired the *Hubble Space Telescope* by installing a set of small mirrors designed to correct images obtained by the faulty mirror. Two more missions to service *Hubble* were carried out in 1997 and 1999, shown in **Figure 3.** Among the objects viewed by *Hubble* after it was repaired in 1999 was a large cluster of galaxies known as Abell 2218.

✔ Reading Check *Why is* Hubble *located outside Earth's atmosphere?*

Figure 3
The *Hubble Space Telescope* was serviced at the end of 1999. Astronauts replaced devices on *Hubble* that are used to stabilize the telescope.

Large Reflecting Telescopes Since the early 1600s, when the Italian scientist Galileo Galilei first turned a telescope toward the stars, people have been searching for better ways to study what lies beyond Earth's atmosphere. For example, the twin Keck reflecting telescopes, shown in **Figure 4,** have segmented mirrors 10 m wide. Until 2000, these mirrors were the largest reflectors ever used. To cope with the difficulty of building such huge mirrors, the Keck telescope mirrors are built out of many small mirrors that are pieced together. In 2000, the European Southern Observatory's telescope, in Chile, consisted of four 8.2-m reflectors, making it the largest optical telescope in use.

✓ Reading Check *About how long have people been using telescopes?*

Active and Adaptive Optics The most recent innovations in optical telescopes involve active and adaptive optics. With active optics, a computer corrects for changes in temperature, mirror distortions, and bad viewing conditions. Adaptive optics is even more ambitious. Adaptive optics uses a laser to probe the atmosphere and relay information to a computer about air turbulence. The computer then adjusts the telescope's mirror thousands of times per second, which lessens the effects of atmospheric turbulence. Telescope images are clearer when corrections for air turbulence, temperature changes, and mirror-shape changes are made.

Figure 4

The twin Keck telescopes on Mauna Kea in Hawaii can be used together, more than doubling their ability to distinguish objects. A Keck reflector is shown in the inset photo. Currently, plans include using these telescopes, along with four others to obtain images that will help answer questions about the origin of planetary systems.

Radio Telescopes

As shown in the spectrum illustrated in **Figure 1,** stars and other objects radiate electromagnetic energy of various types. Radio waves are an example of long-wavelength energy in the electromagnetic spectrum. A **radio telescope,** such as the one shown in **Figure 5,** is used to study radio waves traveling through space. Unlike visible light, radio waves pass freely through Earth's atmosphere. Because of this, radio telescopes are useful 24 hours per day under most weather conditions.

Radio waves reaching Earth's surface strike the large, concave dish of a radio telescope. This dish reflects the waves to a focal point where a receiver is located. The information allows scientists to detect objects in space, to map the universe, and to search for signs of intelligent life on other planets.

Later in this chapter, you'll learn about the instruments that travel into space and send back information that telescopes on Earth's surface cannot obtain.

Figure 5
This radio telescope is used to study radio waves traveling through space.

Section Assessment

1. What is the difference between radio telescopes and optical telescopes?

2. If red light has a longer wavelength than blue light, which has a greater frequency?

3. Compare and contrast refracting and reflecting telescopes.

4. How does adaptive optics in a telescope help solve problems caused by atmospheric turbulence?

5. **Think Critically** It takes light from the closest star to Earth (other than the Sun) about four years to reach Earth. If intelligent life were on a planet circling that star, how long would it take for scientists on Earth to send them a radio transmission and for the scientists to receive their reply?

Skill Builder Activities

6. **Sequencing** Sequence these electromagnetic waves from longest wavelength to shortest wavelength: *gamma rays, visible light, X rays, radio waves, infrared waves, ultraviolet waves,* and *microwaves.* **For more help, refer to the** Science Skill Handbook.

7. **Solving One-Step Equations** The magnifying power (*Mp*) of a telescope is determined by dividing the focal length of the objective lens (FL_{obj}) by the focal length of the eyepiece lens (FL_{eye}) using the following equation:

$$Mp = FL_{obj}/FL_{eye}$$

If FL_{obj} = 1,200 mm and FL_{eye} = 6 mm, what is the telescope's magnifying power? **For more help, refer to the** Math Skill Handbook.

Building a Reflecting Telescope

Nearly four hundred years ago, scientist Galileo Galilei saw what no human had ever seen before. Using the telescope he built, Galileo discovered moons revolving around Jupiter, observed craters on the Moon in detail, and saw sunspots on the surface of the Sun. What was it like to make these discoveries? You will find out as you make your own reflecting telescope.

What You'll Investigate
How do you construct a reflecting telescope?

Materials
flat mirror
shaving or cosmetic mirror (a curved, concave mirror)
magnifying lenses of different magnifications (3–4)

Goals
- **Construct** a reflecting telescope.
- **Observe** magnified images using the telescope and different magnifying lenses.

Safety Precautions
WARNING: *Never observe the Sun directly or with mirrors.*

Procedure
1. Position the cosmetic mirror so that you can see the reflection of the object you want to look at. Choose an object such as the Moon, a planet, or an artificial light source.
2. Place the flat mirror so that it is facing the cosmetic mirror.
3. Adjust the position of the flat mirror until you can see the reflection of the object in it.

4. View the image of the object in the flat mirror with one of your magnifying glasses. Observe how the lens magnifies the image.
5. Use your other magnifying lenses to view the image of the object in the flat mirror. Observe how the different lenses change the image of the object.

Conclude and Apply
1. **Describe** how the image of the object changed when you used different magnifying lenses.
2. **Identify** the part or parts of your telescope that reflected the light of the image.
3. **Identify** the part or parts of your telescope that magnified the image of the object.
4. **Explain** how the three parts of your telescope worked to reflect and magnify the light of the object.
5. **Infer** how the materials you used would have differed if you had constructed a refracting telescope instead of a reflecting telescope.

*C*ommunicating
Your Data
Write an instructional pamphlet for amateur astronomers about how to construct a reflecting telescope. **For more help, refer to the** Science Skill Handbook.

Early Space Missions

The First Missions into Space

You're offered a choice—front-row-center seats for this weekend's rock concert, or a copy of the video when it's released. Wouldn't you rather be right next to the action? Astronomers feel the same way about space. Even though telescopes have taught them a great deal about the Moon and planets, they want to learn more by going to those places or by sending spacecraft where humans can't go.

Rockets The space program would not have gotten far off the ground using ordinary airplane engines. To break free of gravity and enter Earth's orbit, spacecraft must travel at speeds greater than 11 km/s. The space shuttle and several other spacecrafts are equipped with special engines that carry their own fuel. **Rockets,** like the one in **Figure 6,** are engines that have everything they need for the burning of fuel. They don't even require air to carry out the process. Therefore, they can work in space, which has no air. The simplest rocket engine is made of a burning chamber and a nozzle. More complex rockets have more than one burning chamber.

Rocket Types Two types of rockets are distinguished by the type of fuel they use. One type is the liquid-propellant rocket and the other is the solid-propellant rocket. Solid-propellant rockets are generally simpler but they can't be shut down after they are ignited. Liquid-propellant rockets can be shut down after they are ignited and can be restarted. The space shuttle uses liquid-propellant rockets and solid-propellant rockets.

As You Read

What You'll Learn

- **Compare and contrast** natural and artificial satellites.
- **Identify** the differences between artificial satellites and space probes.
- **Explain** the history of the race to the Moon.

Vocabulary

rocket	Project Mercury
satellite	Project Gemini
orbit	Project Apollo
space probe	

Why It's Important

Early missions that sent objects and people into space began a new era of human exploration.

Figure 6
Rockets differ according to the types of fuel used to launch them. This rocket uses liquid oxygen for fuel.

Figure 7
In this view of the shuttle, a red-colored external liquid fuel tank is behind a white, solid rocket booster.

Rocket Launching Solid-propellant rockets use a powdery or rubberlike fuel and a liquid such as liquid oxygen. The burning chamber of a rocket is a tube that has a nozzle at one end. As the solid propellant burns, hot gases exert pressure on all inner surfaces of the tube. The tube pushes back on the gas except at the nozzle where hot gases escape. Thrust builds up and pushes the rocket forward.

Liquid-propellant rockets use a liquid fuel and, commonly, liquid oxygen, stored in separate tanks. To ignite the rocket, the liquid oxygen is mixed with the liquid fuel in the burning chamber. As the mixture burns, forces are exerted and the rocket is propelled forward. **Figure 7** shows the space shuttle, with both types of rockets, being launched.

Math Skills Activity

Using a Grid to Draw

Points are defined by two coordinates, called an ordered pair. To plot an ordered pair, find the first number on the horizontal *x*-axis and the second on the vertical *y*-axis. The point is placed where these two coordinates intersect. Line segments are drawn to connect points.

Example Problem
Using an *x-y* grid and point coordinates, draw a symmetrical house.

Solution

1. On a piece of graph paper, label and number the x-axis 0 to 6 and the y-axis 0 to 6, as shown here.

2. Plot the following points and connect them with straight line segments, as shown here. (1,1), (5,1), (5,4), (3,6), (1,4)

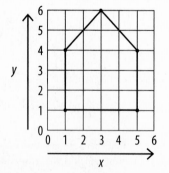

Section	Points
1	(1,−8) (3,−13) (6,−21) (9,−21) (9,−17) (8,−15) (8,−12), (6,−8) (5,−4) (4,−3) (4,−1) (5,1) (6,3) (8,3) (9,4) (9,7) (7,11) (4,14) (4,22) (−9,22) (−9,10) (−10,5) (−11,−1) (−11,−7) (−9,−8) (−8,−7) (−8,−1) (−6,3) (−6,−3) (−6,−9) (−7,−20) (−8,−21) (−4,−21) (−4,−18) (−3,−14) (−1,−8)
2	(0,11) (2,13) (2,17) (0,19) (−4,19) (−6,17) (−6,13) (−4,11)
3	(−4,9) (1,9) (1,5) (−1,5) (−2,6) (−4,6)

Practice Problem

Label and number the *x*-axis −12 to 10 and the *y*-axis −22 to 23. Draw an astronaut by plotting and connecting the points in each section. Do not draw segments to connect points in different sections.

For more help, refer to the Math Skill Handbook.

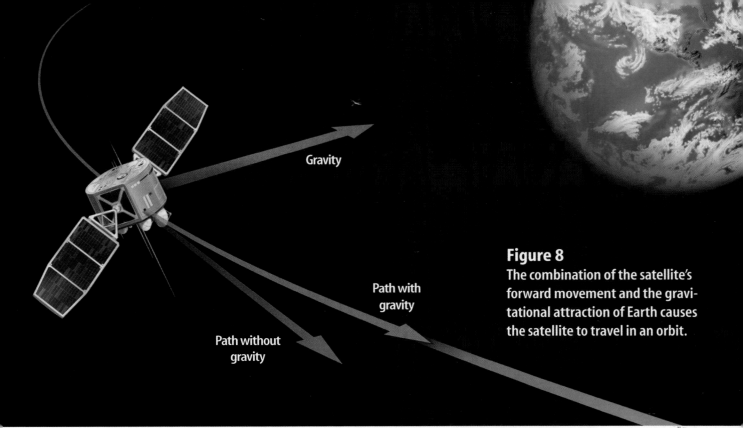

Gravity

Path with
gravity

Path without
gravity

Figure 8
The combination of the satellite's forward movement and the gravitational attraction of Earth causes the satellite to travel in an orbit.

Satellites The space age began in 1957 when the former Soviet Union used a rocket to send *Sputnik I* into space. *Sputnik I* was the first artificial satellite. A **satellite** is any object that revolves around another object. When an object enters space, it travels in a straight line unless a force, such as gravity, makes it turn. Earth's gravity pulls a satellite toward Earth. The result of the satellite traveling forward while at the same time being pulled toward Earth is a curved path, called an **orbit,** around Earth. This is shown in **Figure 8.** *Sputnik I* orbited Earth for 57 days before gravity pulled it back into the atmosphere, where it burned up.

Figure 9
Data obtained from the satellite *Terra,* launched in 1999, illustrates the use of space technology to study Earth. This false-color image includes data on spring growth, sea-surface temperature, carbon monoxide concentrations, and reflected sunlight, among others.

Satellite Uses *Sputnik I* was an experiment to show that artificial satellites could be made and placed into orbit around Earth.

Today, thousands of artificial satellites orbit Earth. Communication satellites transmit radio and television programs to locations around the world. Other satellites gather scientific data, like those shown in **Figure 9,** which can't be obtained from Earth, and weather satellites constantly monitor Earth's global weather patterns.

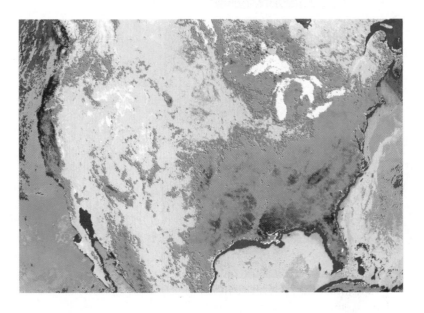

Space Probes

Not all objects carried into space by rockets become satellites. Rockets also can be used to send instruments into space to collect data. A **space probe** is an instrument that gathers information and sends it back to Earth. Unlike satellites that orbit Earth, space probes travel far into the solar system as illustrated in **Figure 10.** Some even have traveled out of the solar system. Space probes, like many satellites, carry cameras and other data-gathering equipment, as well as radio transmitters and receivers that allow them to communicate with scientists on Earth. **Table 1** shows some of the early space probes launched by the National Aeronautics and Space Administration (NASA).

Table 1 Some Early Space Missions

Mission Name		Date Launched	Destination	Data Obtained
Mariner 2		August 1962	Venus	verified high temperatures in Venus's atmosphere
Pioneer 10		March 1972	Jupiter	sent back photos of Jupiter—first probe to encounter an outer planet
Viking 1		August 1975	Mars	orbiter mapped the surface of Mars; lander searched for life on Mars
Magellan		May 1989	Venus	mapped Venus's surface and returned data on the composition of Venus's atmosphere

Figure 10

Probes have taught us much about the solar system. As they travel through space, these car-size craft gather data with their onboard instruments and send results back to Earth via radio waves. Some data collected during these missions are made into pictures, a selection of which is shown here.

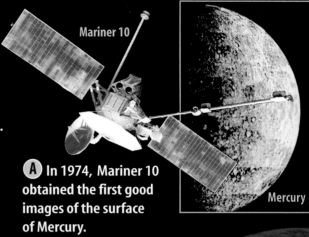

Mariner 10

Mercury

A In 1974, Mariner 10 obtained the first good images of the surface of Mercury.

Venera 8

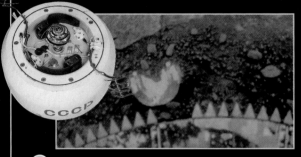

B A Soviet Venera probe took this picture of the surface of Venus on March 1, 1982. Parts of the spacecraft's landing gear are visible at the bottom of the photograph.

Magellan

D In 1990, Magellan imaged craters, lava domes, and great rifts, or cracks, on the surface of Venus.

Venus

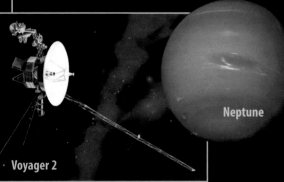

Neptune

Voyager 2

C The Voyager 2 mission included flybys of the outer planets Jupiter, Saturn, Uranus, and Neptune. Voyager took this photograph of Neptune in 1989 as the craft sped toward the edge of the solar system.

Jupiter

Galileo

E NASA's veteran space traveler Galileo nears Jupiter in this artist's drawing. The craft arrived at Jupiter in 1995 and sent back data, including images of Europa, one of Jupiter's 16 moons, seen below in a color-enhanced view.

Europa

Voyager and Pioneer Probes Space probes *Voyager 1* and *Voyager 2* were launched in 1977 and now are heading toward deep space. *Voyager 1* flew past Jupiter and Saturn. *Voyager 2* flew past Jupiter, Saturn, Uranus, and Neptune. These probes now are exploring beyond the solar system as part of the Voyager Interstellar Mission. Scientists expect these probes to continue to transmit data to Earth for at least 20 more years.

Pioneer 10, launched in 1972, was the first probe to survive a trip through the asteroid belt and encounter an outer planet, Jupiter. As of 2000, *Pioneer 10* is more than 11 billion km from Earth, and will continue beyond the solar system. The probe carries a gold medallion with an engraving of a man, a woman, and Earth's position in the galaxy.

Galileo Launched in 1989, *Galileo* reached Jupiter in 1995. In July 1995, *Galileo* released a smaller probe that began a five-month approach to Jupiter. The small probe took a parachute ride through Jupiter's violent atmosphere in December 1995.

Before being crushed by the atmospheric pressure, it transmitted information about Jupiter's composition, temperature, and pressure to the satellite orbiting above. *Galileo* studied Jupiter's moons, rings, and magnetic fields and then relayed this information to scientists who were waiting eagerly for it on Earth.

Life Science INTEGRATION

Studies of Jupiter's moon Europa by *Galileo* indicate that an ocean of water may exist under the surface of Europa. A cracked outer layer of ice makes up Europa's surface, shown in **Figure 11.** The cracks in the surface may be caused by geologic activity that heats the ocean underneath the surface. Sunlight penetrates these cracks, further heating the ocean and setting the stage for the possible existence of life on Europa. *Galileo* ended its study of Europa in 2000. More advanced probes will be needed to determine whether life exists on this icy moon.

✔ **Reading Check** *What features on Europa suggest the possibility of life existing on this moon?*

In October and November of 1999, *Galileo* approached Io, another one of Jupiter's moons. It came within 300 km and took photographs of a volcanic vent named Loki, which emits more energy than all of Earth's volcanoes combined. *Galileo* also discovered eruption plumes that shoot gas made of sulfur and oxygen.

SCIENCE *Online*

Collect Data Visit the Glencoe Science Web site at **science.glencoe.com** to get the latest information on Galileo's discoveries. Record your information in your Science Journal.

Figure 11
Future missions will be needed to determine whether life exists on Europa.

Moon Quest

Throughout the world, people were shocked when they turned on their radios and television sets in 1957 and heard the radio transmissions from *Sputnik I* as it orbited Earth. All that *Sputnik I* transmitted was a sort of beeping sound, but people quickly realized that launching a human into space wasn't far off.

In 1961, Soviet cosmonaut Yuri A. Gagarin became the first human in space. He orbited Earth and returned safely. Soon, President John F. Kennedy called for the United States to send humans to the Moon and return them safely to Earth. His goal was to achieve this by the end of the 1960s. The race for space was underway.

The U.S. program to reach the Moon began with **Project Mercury.** The goals of Project Mercury were to orbit a piloted spacecraft around Earth and to bring it back safely. The program provided data and experience in the basics of space flight. On May 5, 1961, Alan B. Shepard became the first U.S. citizen in space. In 1962, *Mercury* astronaut John Glenn became the first U.S. citizen to orbit Earth. **Figure 12** shows Glenn preparing for liftoff.

Reading Check *What were the goals of Project Mercury?*

Project Gemini The next step in reaching the Moon was called **Project Gemini.** Teams of two astronauts in the same *Gemini* spacecraft orbited Earth. One *Gemini* team met and connected with another spacecraft in orbit—a skill that would be needed on a voyage to the Moon.

The *Gemini* spacecraft was much like the *Mercury* spacecraft, except it was larger and easier for the astronauts to maintain. It was launched by a rocket known as a *Titan II,* which was a liquid fuel rocket.

In addition to connecting spacecraft in orbit, another goal of Project *Gemini* was to investigate the effects of space travel on the human body.

Along with the Mercury and Gemini programs, a series of robotic probes was sent to the Moon. *Ranger* proved that a spacecraft could be sent to the Moon. In 1966, *Surveyor* landed gently on the Moon's surface, indicating that the Moon's surface could support spacecraft and humans. The mission of *Lunar Orbiter* was to take pictures of the Moon's surface that would help determine the best future lunar landing sites.

Figure 12
An important step in the attempt to reach the Moon was John Glenn's first orbit around Earth.

Modeling a Satellite

WARNING: *Stand a safe distance away from classmates. Use heavy string.*

Procedure
1. Tie one end of a 2-m-long **string** to a **rubber stopper.**
2. Thread the string through a 15-cm piece of **hose.**
3. Tie the other end of the string securely to several large **steel nuts.**
4. Swing the rubber stopper in a circle above your head. Swing the stopper at different speeds.

Analysis
Based upon your observations, explain how a satellite stays in orbit above Earth.

Figure 13
The Lunar Rover vehicle was first used during the *Apollo 15* mission. Riding in the moon buggy, *Apollo 15, 16,* and *17* astronauts explored large areas of the lunar surface.

Project Apollo The final stage of the U.S. program to reach the Moon was **Project Apollo.** On July 20, 1969, *Apollo 11* landed on the Moon's surface. Neil Armstrong was the first human to set foot on the Moon. His first words as he stepped onto its surface were, "That's one small step for man, one giant leap for mankind." Edwin Aldrin, the second of the three *Apollo 11* astronauts, joined Armstrong on the Moon, and they explored its surface for two hours. While they were exploring, Michael Collins remained in the Command Module; Armstrong and Aldrin then returned to the Command Module before beginning the journey home. A total of six lunar landings brought back more than 2,000 samples of moon rock and soil for study before the program ended in 1972. **Figure 13** shows an astronaut exploring the Moon's surface from the Lunar Rover vehicle.

Sharing Knowledge During the past three decades, most missions in space have been carried out by individual countries, often competing to be the first or the best. Today, countries of the world cooperate more and work together, sharing what each has learned. Projects are being planned for cooperative missions to Mars and elsewhere. As you read the next section, you'll see how the U.S. program has progressed since the days of Project Apollo and what may be planned for the future.

Section Assessment

1. Explain why Neptune has eight satellites even though it is not orbited by human-made objects.

2. *Galileo* was considered a space probe as it traveled to Jupiter. Once there, however, it became an artificial satellite. Explain.

3. List several discoveries made by the *Voyager 1* and *Voyager 2* space probes.

4. Draw a time line beginning with *Sputnik* and ending with Project Apollo. Include descriptions of important missions.

5. **Think Critically** Is Earth a satellite of any other body in space? Explain.

Skill Builder Activities

6. **Using an Electronic Spreadsheet** Use a spreadsheet program to generate a table of recent successful satellites and space probes launched by the United States. Include a description of the craft, the date it was launched, and its mission. **For more help, refer to the** Technology Skill Handbook.

7. **Solving One-Step Equations** Suppose a spacecraft were launched at a speed of 40,200 km/h. Express this speed in kilometers per second. **For more help, refer to the** Math Skill Handbook.

Current and Future Space Missions

The Space Shuttle

Imagine spending millions of dollars to build a machine, sending it off into space, and watching its 3,000 metric tons of metal and other materials burn up after only a few minutes of work. That's exactly what NASA did with the rocket portions of spacecraft for many years. The early rockets were used only to launch a small capsule holding astronauts into orbit. Then sections of the rocket separated from the rest and burned when reentering the atmosphere.

A Reusable Spacecraft NASA administrators, like many others, realized that it would be less expensive and less wasteful to reuse resources. The reusable spacecraft that transports astronauts, satellites, and other materials to and from space is called the **space shuttle,** shown in **Figure 14,** as it is landing.

At launch, the space shuttle stands on end and is connected to an external liquid-fuel tank and two solid-fuel booster rockets. When the shuttle reaches an altitude of about 45 km, the emptied, solid-fuel booster rockets drop off and parachute back to Earth. These are recovered and used again. The external liquid-fuel tank separates and falls back to Earth, but it isn't recovered.

Work on the Shuttle After the space shuttle reaches space, it begins to orbit Earth. There, astronauts perform many different tasks. In the cargo bay, astronauts can conduct scientific experiments and determine the effects of spaceflight on the human body. When the cargo bay isn't used as a laboratory, the shuttle can launch, repair, and retrieve satellites. Then the satellites can be returned to Earth or repaired onboard and returned to space. After a mission, the shuttle glides back to Earth and lands like an airplane. A large landing field is needed as the gliding speed of the shuttle is 335 km/h.

As You Read

What **You'll Learn**
- **Explain** the benefits of the space shuttle.
- **Identify** the usefulness of orbital space stations.
- **Explore** future space missions.

Vocabulary
space shuttle
space station

Why **It's Important**
Many future space missions have planned experiments that may benefit you.

Figure 14
The space shuttle is designed to make many trips into space.

Figure 15
Astronauts performed a variety of tasks while living and working in space onboard *Skylab*.

Figure 16
Russian and American scientists have worked together to further space exploration.

Space Stations

Astronauts can spend only a short time living in the space shuttle. Its living area is small, and the crew needs more room to live, exercise, and work. A **space station** has living quarters, work and exercise areas, and all the equipment and support systems needed for humans to live and work in space.

In 1973, the United States launched the space station *Skylab*, shown in **Figure 15.** Crews of astronauts spent up to 84 days there, performing experiments and collecting data on the effects on humans of living in space. In 1979, the abandoned *Skylab* fell out of orbit and burned up as it entered Earth's atmosphere.

Crews from the former Soviet Union have spent more time in space, onboard the space station *Mir,* than crews from any other country. Cosmonaut Dr. Valery Polyakov returned to Earth after 438 days in space studying the long-term effects of weightlessness.

Cooperation in Space

In 1995, the United States and Russia began an era of cooperation and trust in exploring space. Early in the year, American Dr. Norman Thagard was launched into orbit aboard the Russian *Soyuz* spacecraft, along with two Russian cosmonaut crewmates. Dr. Thagard was the first U.S. astronaut launched into space by a Russian booster and the first American resident of the Russian space station *Mir.*

In June 1995, Russian cosmonauts rode into orbit onboard the space shuttle *Atlantis,* America's 100th crewed launch. The mission of *Atlantis* involved, among other studies, a rendezvous and docking with the space station *Mir.* The cooperation that existed on this mission, as shown in **Figure 16,** continued through eight more space shuttle-*Mir* docking missions. Each of the eight missions was an important step toward building and operating the *International Space Station.* In 2001, the abandoned *Mir* space station fell out of orbit and burned up upon reentering the atmosphere. Cooperation continued as the *International Space Station* began to take form.

The International Space Station

The *International Space Station (ISS)* will be a permanent laboratory designed for long-term research projects. Diverse topics will be studied, including research on the growth of protein crystals. This particular project will help scientists determine protein structure and function, which is expected to enhance work on drug design and the treatment of many diseases.

The *ISS* will draw on the resources of 16 nations. These nations will build units for the space station, which then will be transported into space onboard the space shuttle and Russian launch rockets. The station will be constructed in space. **Figure 17** shows what the completed station will look like.

Reading Check *What is the purpose of the* International Space Station?

Phases of ISS NASA is planning the *ISS* program in phases. Phase One, now concluded, involved the space shuttle-*Mir* docking missions. Phase Two began in 1998 with the launch of the Russian-built *Zarya Module,* also known as the Functional Cargo Block, and will end with the delivery of a U.S. laboratory onboard the space shuttle. The first assembly of *ISS* occurred in December of 1998 when a space shuttle mission attached the Unity module to *Zarya.* During Phase Two, crews of three people were delivered to the space station.

Living in Space The project will continue with Phase Three when the Japanese Experiment Module, the European Columbus Orbiting Facility, and another Russian lab will be delivered.

It is hoped that the *International Space Station* will be completed in 2005. A three- or four-person crew then should be able to work comfortably onboard the station. A total of 47 separate launches will be required to take all the components of the *ISS* into space and prepare it for permanent habitation. NASA plans for crews of astronauts to stay onboard the station for several months at a time. NASA already has conducted numerous tests to prepare crews of astronauts for extended space missions. One day, the station could be a construction site for ships that will travel to the Moon and Mars.

Figure 17
This is a picture of what the proposed *International Space Station* will look like when it is completed in 2006.

SCIENCE Online

Research Visit the Glencoe Science Web site at **science.glencoe.com** for more information on the *International Space Station.* Share your information with the class.

Figure 18
Gulleys, channels, and aprons of sediment imaged by the *Mars Global Surveyor* are similar to features on Earth known to be caused by flowing water. This water is thought to seep out from beneath the surface of Mars.

Exploring Mars

Two of the most successful missions in recent years were the 1996 launchings of the *Mars Global Surveyor* and the *Mars Pathfinder*. *Surveyor* orbited Mars, taking high-quality photos of the planet's surface as shown in **Figure 18.** *Pathfinder* descended to the Martian surface, using rockets and a parachute system to slow its descent. Large balloons absorbed the shock of landing. *Pathfinder* carried technology to study the surface of the planet, including a remote-controlled robot rover called Sojourner. Using information gathered by studying photographs taken by *Surveyor,* scientists determined that water recently had seeped to the surface of Mars in some areas.

✓ Reading Check *What type of data were obtained by the* **Mars Global Surveyor?**

Although the *Mars Global Surveyor* and the *Mars Pathfinder* missions were successful, not all the missions to Mars have met with the same success. The *Mars Climate Orbiter,* launched in 1998, was lost in September of 1999. An incorrect calculation of the force that the thrusters were to exert caused the spacecraft to be lost. Engineers had used English units instead of metric units. Then, in December of 1999, the *Mars Polar Lander* was lost just as it was making its descent to the planet. This time, it is believed that the spacecraft thought it had landed and shut off its thrusters too soon. NASA tried to make contact with the lander but never had any success.

New Millennium Program

To continue space missions into the future, NASA has created the New Millennium Program (NMP). The goal of the NMP is to develop advanced technology that will let NASA send smart spacecraft into the solar system. This will reduce the amount of ground control needed. They also hope to reduce the size of future spacecraft to keep the cost of launching them under control. NASA's challenge is to prove that certain cutting-edge technologies, as well as mission concepts, work in space.

Exploring the Moon

Does water exist in the craters of the Moon's poles? This is one question NASA intends to explore with data gathered from the *Lunar Prospector* spacecraft shown in **Figure 19.** Launched in 1998, the *Lunar Prospector's* one-year mission was to orbit the Moon, mapping its structure and composition. Early data obtained from the spacecraft indicate that hydrogen might be present in the rocks of the Moon's poles. Hydrogen is one of the elements found in water. Scientists now hypothesize that ice on the floors of the Moon's polar craters may be the source of this hydrogen. Ice might survive indefinitely at the bottom of these craters because it would always be shaded from the Sun.

Other things could account for the presence of hydrogen. It could be from solar wind or certain minerals. The *Lunar Prospector* was directed to crash into a crater at the Moon's south pole when its mission ended in July 1999. Scientists hoped that any water vapor thrown up by the collision could be detected using special telescopes. However, it didn't work. Further studies are needed to determine if water exists on the Moon.

SCIENCE *Online*

Data Update For an online update on the New Millenium Program, visit the Glencoe Science Web site at **science.glencoe.com**

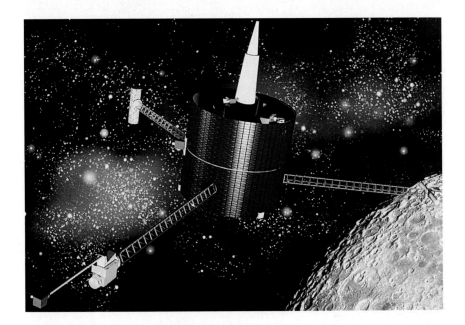

Figure 19
The *Lunar Prospector* analyzed the Moon's composition during its one-year mission.

Cassini

In October 1997, NASA launched the space probe *Cassini*. This probe's destination is Saturn. *Cassini*, shown in **Figure 20,** will not reach its goal until 2004. At that time, the space probe will explore Saturn and surrounding areas for four years. One part of its mission is to deliver the European Space Agency's *Huygens* probe to Saturn's largest moon, Titan. Some scientists theorize that Titan's atmosphere may be similar to the atmosphere of early Earth.

Figure 20
Cassini is currently on its way to Saturn. After it arrives, it will spend four years studying Saturn and its surrounding area.

The Next Generation Space Telescope Not all space missions involve sending astronauts or probes into space. Plans are being made to launch a new space telescope that is capable of observing the first stars and galaxies in the universe. The *Next Generation Space Telescope*, shown in **Figure 21,** will be the successor to the *Hubble Space Telescope*. As part of the Origins project, it will provide scientists with the opportunity to study the evolution of galaxies, the production of elements by stars, and the process of star and planet formation. To accomplish these tasks, the telescope will have to be able to see objects 400 times fainter than those currently studied with ground-based telescopes such as the twin Keck telescopes. NASA hopes to launch the *Next Generation Space Telescope* as early as 2009.

Figure 21
The *Next Generation Space Telescope* will attempt to observe stars and galaxies that formed early in the history of the universe.

Everyday Space Technology Items developed for space exploration don't always stay in space. In fact, many of today's cutting-edge technologies are modifications of research or technology used in the space program. For example, NASA space suit technology, shown in **Figure 22,** was used to give a child with a skin disorder the opportunity to play outside. Without the suit, the child could have been seriously hurt by the Sun's rays.

Space technology also has been used to understand, diagnose, and treat heart disease. Programmable pacemakers, developed through space technology, have given doctors more programming possibilities and more detailed information about their patients' health.

Other advances include ribbed swimsuits that reduce water resistance. Also, badges have been designed that warn workers of toxic chemicals in the air by turning color when the wearer is exposed to a particular chemical. Jet engines capable of much higher speeds than current jet engines are being developed as well.

A new technology that may prevent many accidents also has been developed. Equipment on emergency vehicles causes traffic lights to turn yellow and then red for other vehicles approaching the same intersections. The equipment activates the traffic lights when fast-moving emergency vehicles come close to such an intersection, preventing crashes.

Figure 22
Space technology has helped children go places and do things that they otherwise wouldn't be able to do.

Section 3 Assessment

1. What is the main advantage of the space shuttle?

2. Why were the space shuttle-*Mir* docking missions so important?

3. What is the *International Space Station* used for? Describe how the *ISS* could help future space missions.

4. Describe Phase Three of the *International Space Station* program.

5. **Think Critically** What makes the space shuttle more versatile than earlier spacecraft?

Skill Builder Activities

6. **Making and Using Tables** Make a table of the discoveries from missions to the Moon and Mars. **For more help, refer to the** Science Skill Handbook.

7. **Communicating** Suppose you're in charge of assembling a crew for a new space station. Select 50 people to do a variety of jobs, such as farming, maintenance, scientific experimentation, and so on. In your Science Journal, list and explain your choices. **For more help, refer to the** Science Skill Handbook.

Star Sightings

Polaris ——————

For thousands of years, humans have used the stars to learn about Earth. From star sightings, you can map the change of seasons, navigate the oceans, and even determine the size of Earth.

Polaris, or the North Star, has occupied an important place in human history. The location of Polaris is not affected by Earth's rotation. At any given observation point, it always appears at the same angle above the horizon. At Earth's north pole, Polaris appears directly overhead. At the equator, it is just above the northern horizon. Polaris provides a standard from which other locations can be measured. Such star sightings can be made using the astrolabe, an instrument used to measure the height of a star above the horizon.

Recognize the Problem

How can you determine the size of Earth?

Form a Hypothesis

Think about what you have learned about sightings of Polaris. How does this tell you that Earth is round? Knowing that Earth is round, form a hypothesis about how you can estimate the circumference of Earth based on star sightings.

Goals
- **Record** your sightings of Polaris.
- **Share** the data with other students to calculate the circumference of Earth.

Safety Precautions
WARNING: *Do not use the astrolabe during the daytime to observe the Sun.*

Data Sources
SCIENCE*Online* Go to the Glencoe Science Web site at **science.glencoe.com** to obtain instructions on how to make an astrolabe. Also visit the Web site for more information about the location of Polaris, and for data from other students.

Test Your Hypothesis

Plan

1. Obtain an astrolabe or construct one using the instructions posted on the Glencoe Science Web site.

2. **Design** a data table in your Science Journal similar to the one below.

Polaris Observations		
Your Location:		
Date	**Time**	**Astrolabe Reading**

3. Decide as a group how you will make your observations. Does it take more than one person to make each observation? When will it be easiest to see Polaris?

Do

1. Make sure your teacher approves your plan before you start.

2. Carry out your observations.

3. **Record** your observations in your data table.

4. Average your readings and post them in the table provided on the Glencoe Science Web site.

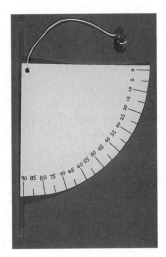

Analyze Your Data

1. **Research** the names of cities that are at approximately the same longitude as your hometown. Gather astrolabe readings at the Glencoe Science Web site from students in one of those cities.

2. **Compare** your astrolabe readings. Subtract the smaller reading from the larger one.

3. Determine the distance between your star sighting location and the other city.

4. **Calculate** the circumference of Earth using the following relationship.
Circumference = (360°) × (distance between locations)/ difference between readings

Draw Conclusions

1. How does the circumference of Earth that you calculated compare with the accepted value of 40,079 km?

2. What are some possible sources of error in this method of determining the size of Earth? What improvements would you suggest?

Communicating Your Data

SCIENCE *Online* Find this *Use the Internet* activity on the Glencoe Science Web site at **science.glencoe.com** Create a poster that includes a table of your data and data from students in other cities. Perform a sample circumference calculation for your class.

TIME

SCIENCE AND
Society

SCIENCE
ISSUES
THAT AFFECT
YOU!

Cities in Space

Should the U.S. spend money to colonize space?

Humans have landed on the Moon, and spacecrafts have landed on Mars. But these space missions are just small steps that may lead to a giant new space program. As technology improves, humans may be able to visit and even live on other planets. The twenty-first century may turn science fiction into science fact. But is it worth the time and money involved?

Those in favor of living in space point to the International Space Station that already is orbiting Earth. It's an early step toward establishing floating cities where astronauts can live and work. The 94 billion dollar station may pave the way for "ordinary people" to live in space, too. As Earth's population continues to increase and there is less room on this planet, why not create ideal cities on another planet or a floating city in space? That reason, combined with the fact that there is little pollution in space, makes the idea appealing to many.

Critics of colonizing space think we should spend the hundreds of billions of dollars that it would cost to colonize space on projects to help improve people's lives here on Earth. Building better housing, developing ways to feed the hungry, finding cures for diseases, and increasing funds for education should come first, these people say. And, critics continue, if people want to explore, why not explore right here on Earth? "The ocean floor is Earth's last frontier," says one person. "Why not explore that?"

If humans were to move permanently to space, the two most likely destinations would be Mars and the Moon, both bleak places.

But those in favor of moving to these places say humans could find a way to make them livable. They argue that humans have made homes in harsh climates and in rugged areas, and people can meet the challenges of living in space.

Choosing Mars

Mars may be the best place to live. Photos suggest that the planet once had liquid water on its surface. If that water is now frozen underground, humans may someday be able to tap into it.

NASA is studying whether it makes sense to send astronauts and scientists to explore Mars. An international team would live there for about 500 days, collecting and studying soil and rock samples for clues as to whether Mars is a planet that could be settled. NASA says this journey could begin as early as 2009.

But a longer-range dream to transform Mars into an Earthlike place with breathable air and usable water is just that—a dream. Some small steps are being taken to make that dream more realistic. Experimental plants are being developed that could absorb Mars' excess carbon dioxide and release oxygen. Solar mirrors, already available, could warm Mars' surface.

Those for and against colonizing space agree on one thing—setting up colonies on Mars or the Moon will take large amounts of money, research, and planning. It also will take the same spirit of adventure that has led history's pioneers into so many bold frontiers—deserts, the Poles, and the sky.

Is the International Space Station a small step toward colonizing space?

An early Mars colony might look something like this. Settlers would live in air-filled domes and even grow crops.

Reviewing Main Ideas

Section 1 Radiation from Space

1. The arrangement of electromagnetic waves according to their wavelengths is the electromagnetic spectrum.

2. Optical telescopes produce magnified images of objects. *What does this reflecting telescope use to focus light that produces an image?*

3. Radio telescopes collect and record radio waves given off by some space objects.

Section 2 Early Space Missions

1. A satellite is an object that revolves around another object. The moons of planets are natural satellites. Artificial satellites are those made by people.

2. A space probe travels into the solar system, gathers data, and sends them back to Earth. *How far can space probes, like the one pictured here, travel?*

3. Early American piloted space programs included the Gemini and Apollo Projects.

Section 3 Recent and Future Space Missions

1. Space stations provide the opportunity to conduct research not possible on Earth. The *International Space Station* is being constructed in space with the cooperation of more than a dozen nations.

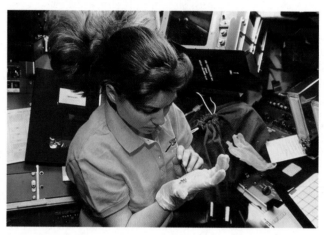

2. The space shuttle is a reusable spacecraft that carries astronauts, satellites, and other cargo to and from space. *What special obstacles must astronauts overcome when they conduct research in space?*

3. Space technology is used to solve problems on Earth not related to space travel. Advances in engineering related to space travel have led to problem solving in medicine and environmental sciences, among other fields.

FOLDABLES
Reading & Study Skills

After You Read

Use what you've learned to predict the future of space exploration. Record your predictions under the Future tab of your Foldable.

Visualizing Main Ideas

Complete the following concept map about the race to the Moon. Use the following phrases: first satellite, Project Gemini, Project Mercury, team of two astronauts orbits Earth, Project Apollo.

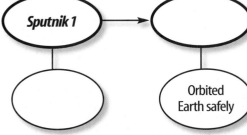
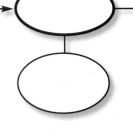

Sputnik 1 →　○　→　○　→　○

○　　Orbited Earth safely　　○　　First human on the Moon

Vocabulary Review

Vocabulary Words

a. electromagnetic spectrum
b. observatory
c. orbit
d. Project Apollo
e. Project Gemini
f. Project Mercury
g. radio telescope
h. reflecting telescope
i. refracting telescope
j. rocket
k. satellite
l. space probe
m. space shuttle
n. space station

Using Vocabulary

Each of the following sentences is false. Make each sentence true by replacing the underlined word(s) with the correct vocabulary word(s).

1. A <u>radio</u> telescope uses lenses to bend light.

2. A <u>space probe</u> is an object that revolves around another object in space.

3. <u>Project Apollo</u> was the first piloted U.S. space program.

4. A <u>satellite</u> carries people and tools to and from space.

5. In the <u>space station</u>, electromagnetic waves are arranged according to their wavelengths.

Checking Concepts

Choose the word or phrase that best answers the question.

1. Which spacecraft has sent images of Venus to scientists on Earth?
   A) *Voyager*
   B) *Viking*
   C) *Apollo 11*
   D) *Magellan*

2. Which kind of telescope uses mirrors to collect light?
   A) radio
   B) electromagnetic
   C) refracting
   D) reflecting

3. What was *Sputnik I?*
   A) the first telescope
   B) the first artificial satellite
   C) the first observatory
   D) the first U.S. space probe

4. Which kind of telescope can be used during the day or night and during bad weather?
   A) radio
   B) electromagnetic
   C) refracting
   D) reflecting

5. When fully operational, what is the maximum number of people who will crew the *International Space Station?*
   A) 3
   B) 7
   C) 15
   D) 50

6. Which space mission's goal was to put a spacecraft into orbit and bring it back safely?
   A) Project Mercury
   B) Project Apollo
   C) Project Gemini
   D) *Viking I*

7. Which of the following is a natural satellite of Earth?
   A) *Skylab*
   B) the space shuttle
   C) the Sun
   D) the Moon

8. What does the space shuttle use to place a satellite into space?
   A) liquid-fuel tank
   B) booster rocket
   C) mechanical arm
   D) cargo bay

9. What was *Skylab?*
   A) a space probe
   B) a space station
   C) a space shuttle
   D) an optical telescope

10. What part of the space shuttle is reused?
    A) liquid-fuel tanks
    B) *Gemini* rockets
    C) booster engines
    D) Saturn rockets

Thinking Critically

11. Describe any advantages that a Moon-based telescope would have over an Earth-based telescope?

12. Would a space probe to the Sun's surface be useful? Explain.

13. Which do you think is a wiser method of exploration—space missions with people onboard or robotic space probes? Why?

14. Suppose two astronauts are outside the space shuttle orbiting Earth. The audio speaker in the helmet of one astronaut quits working. The other astronaut is 1 m away and shouts a message. Can the first astronaut hear the message? Explain.

15. Space probes have crossed Pluto's orbit, but never have visited the planet. Explain.

Developing Skills

16. **Making and Using Tables** Copy and complete the table below. Use information from several resources.

United States Space Probes		
Probe	Launch Date(s)	Planets or Objects Visited
Vikings 1 and *2*		
Galileo		
Lunar Prospector		
Pathfinder		

17. Concept Mapping Use the following phrases to complete the concept map about rocket launching: *thrust pushes rocket forward, rocket breaks free of Earth's gravity, propellant is ignited,* and *hot gases exert pressure on walls of burning chamber.*

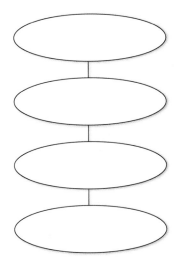

18. Classifying Classify the following as a satellite or a space probe: *Cassini, Sputnik I, Hubble Space Telescope,* space shuttle, and *Voyager 2.*

19. Comparing and Contrasting Compare and contrast space probes and satellites.

Performance Assessment

20. Poem Research a space probe launched within the last five years. Write a poem that includes its destination, the goals for its mission and something about the individuals who crewed the flight.

TECHNOLOGY

Go to the Glencoe Science Web site at **science.glencoe.com** or use the **Glencoe Science CD-ROM** for additional chapter assessment.

Test Practice

Scientists use several different kinds of telescopes to make observations about space. Information about some types of telescopes is listed in the table below.

1. According to the table, a refracting telescope focuses light using a _____.
   A) convex lens
   B) mirror
   C) dish
   D) receiver

Types of Telescopes		
Telescope	**Use**	**How it Works**
Optical Refracting Telescope	Produces magnified images of distant objects	Uses a convex lens to bend and focus light
Optical Reflecting Telescope	Produces magnified images of distant objects	Uses mirrors to reflect and focus light
Radio Telescope	Collects radio waves from space	Large dish reflects and focuses waves to receiver.

2. While in space, the *Hubble Space Telescope* needed its largest mirror repaired in 1993. While costly, the repair mission was a huge success. According to the chart, the *Hubble Space Telescope* is _____.
   F) an optical refracting telescope
   G) an optical reflecting telescope
   H) a radio telescope
   J) an optical radio receiver

The Sun-Earth-Moon System

Can you say with certainty that Earth is round? Why do you feel coldest when Earth is closest to the Sun? In this chapter, you will find the answers to these questions. You'll learn why the lengths of day and night change and why seasons occur. You'll also learn why the Moon's appearance changes throughout the month and why scientists think its surface might be a home for humans one day.

What do you think?

Science Journal Look at the picture below with a classmate. Discuss what this might be. Here's a hint: *It's behind the man in the Moon.* Write your best guess in your Science Journal.

EXPLORE ACTIVITY

The Sun rises in the morning. This occurs because Earth is moving through space. The movements of Earth cause day and night, as well as the seasons. In this activity, you will explore Earth's movements.

Model rotation and revolution

1. Hold a basketball with one finger at the top and one at the bottom. Have a classmate gently spin the ball.

2. Explain how this models rotation.

3. Continue to hold the basketball and walk one complete circle around another student in your class.

4. How does this model revolution?

Observe

In your Science Journal, compare and contrast rotation and revolution.

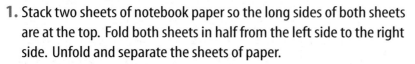

Before You Read

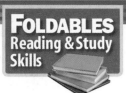

FOLDABLES Reading & Study Skills

Making a Question Study Fold Make the following Foldable to organize and answer questions as you read this chapter.

1. Stack two sheets of notebook paper so the long sides of both sheets are at the top. Fold both sheets in half from the left side to the right side. Unfold and separate the sheets of paper.

2. Take one sheet of paper and cut along the fold line, starting in the center of the fold and stopping at both margin lines as shown.

3. Place the second sheet in front so the long side is at the top. Cut along the fold line from the bottom and from the top, stopping at the margin lines.

4. Insert the second sheet into the cut of the first sheet. Unfold the inserted sheet; align the cuts along the fold of the other.

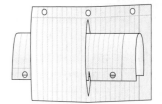

5. Fold both sheets in half to make a book as shown. On the first two pages, record questions you have about the Sun, on the middle two pages, questions about Earth, and on the last two pages, questions about the Moon. As you read the chapter, answer your questions.

Earth

As You Read

What You'll Learn

- **Examine** Earth's physical characteristics.
- **Differentiate** between rotation and revolution.
- **Discuss** what causes seasons to change.

Vocabulary

sphere   ellipse
axis   solstice
rotation   equinox
revolution

Why It's Important

Earth's movements cause night and day and the changing of seasons.

Figure 1
For many years, sailors have observed that the tops of ships coming across the horizon appear first. This suggests that Earth is spherical, not flat, as was once widely believed.

Properties of Earth

You awaken at daybreak to catch the Sun "rising" from the dark horizon. Then it begins its daily "journey" from east to west across the sky. Finally the Sun "sinks" out of view as night falls. Is the Sun moving—or are you?

It wasn't long ago that people thought Earth was the center of the universe. It was widely believed that the Sun revolved around Earth, which stood still. It is now common knowledge that the Sun only appears to be moving around Earth. Because Earth spins as it revolves around the Sun, it creates the illusion that the Sun is moving across the sky.

Another mistaken idea about Earth concerned its shape. Even as recently as the days of Christopher Columbus, many people believed Earth to be flat. Because of this, they were afraid that if they sailed far enough out to sea, they would fall off the edge of the world. How do you know this isn't true? How have scientists determined the true shape of Earth?

Spherical Shape   A round, three-dimensional object is called a **sphere** (SFIHR). Its surface is the same distance from its center at all points. Some common examples of spheres are basketballs and tennis balls.

In the late twentieth century, artificial satellites and space probes sent back pictures showing that Earth is spherical. Much earlier, Aristotle, a Greek astronomer and philosopher who lived around 350 B.C., suspected that Earth was spherical. He observed that Earth cast a curved shadow on the Moon during an eclipse.

In addition to Aristotle, other individuals made observations that indicated Earth's spherical shape. Early sailors, for example, noticed that approaching ships came into view a little at a time, as shown in **Figure 1.**

Additional Evidence Sailors also noticed changes in how the night sky looked. As they sailed north or south, the North Star moved higher or lower in the sky. The best explanation was a spherical Earth.

Today, most people know that Earth is spherical. They also know all objects are attracted by gravity to the center of a spherical Earth. Astronauts have clearly seen the spherical shape of Earth. However, it bulges slightly at the equator and is somewhat flattened at the poles, so it is not a perfect sphere.

Rotation Earth's **axis** is the imaginary vertical line around which Earth spins. This line cuts directly through the center of Earth, as shown in the illustration accompanying **Table 1.** The poles are located at the north and south ends of Earth's axis. The spinning of Earth on its axis, called **rotation,** causes day and night to occur. Here is how it works. As Earth rotates, you can see the Sun come into view at daybreak. Earth continues to spin, making it seem as if the Sun moves across the sky until it sets at night. During night, your area of Earth has rotated so that it is on the opposite side as the Sun. Because of this, the Sun is no longer visible to you. Earth continues to rotate steadily, and eventually the Sun comes into view again the next morning. One complete rotation takes about 24 h, or one day. How many rotations does Earth complete during one year? As you can infer from **Table 1,** it completes about 365 rotations during its one-year journey around the Sun.

Reading Check *Why does the Sun seem to rise and set?*

Life Science INTEGRATION

Suppose that Earth's rotation took twice as long as it does now. In your Science Journal, predict how conditions such as global temperatures, work schedules, plant growth, and other factors might change under these circumstances.

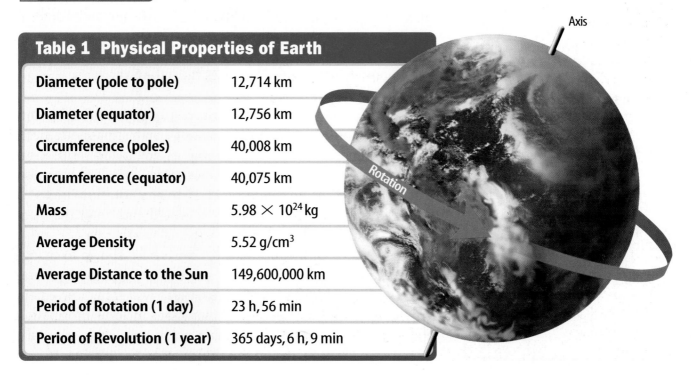

Table 1 Physical Properties of Earth	
Diameter (pole to pole)	12,714 km
Diameter (equator)	12,756 km
Circumference (poles)	40,008 km
Circumference (equator)	40,075 km
Mass	5.98×10^{24} kg
Average Density	5.52 g/cm^3
Average Distance to the Sun	149,600,000 km
Period of Rotation (1 day)	23 h, 56 min
Period of Revolution (1 year)	365 days, 6 h, 9 min

Axis

Rotation

Figure 2

Earth's magnetic field is similar to that of a bar magnet, almost as if Earth contained a giant magnet. Earth's magnetic axis is angled 11.5 degrees from its rotational axis.

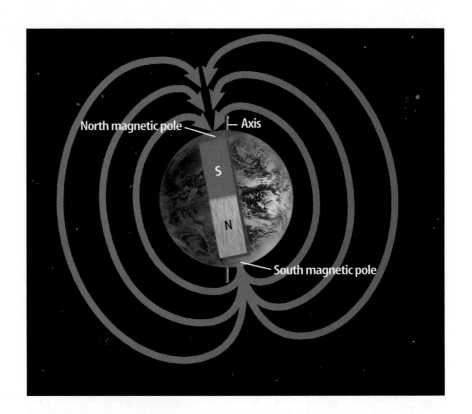

North magnetic pole — Axis
S
N
South magnetic pole

TRY AT HOME
Mini LAB

Making Your Own Compass

Procedure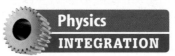
WARNING: *Use care when handling sharp objects.*

1. Cut off the bottom of a **plastic foam cup** to make a polystyrene disk.
2. Magnetize a **sewing needle** by continuously stroking the needle in the same direction with a **magnet** for 1 min.
3. **Tape** the needle to the center of the foam disk.
4. Fill a **plate** with **water** and float the disk, needle side up, in the water.

Analysis

1. What happened to the needle and disk when you placed them in the water? Why did this happen?
2. Infer how ancient sailors might have used magnets to help them navigate on the open seas.

Physics INTEGRATION

Magnetic Field Scientists hypothesize that the movement of material inside Earth's core, along with Earth's rotation, generates a magnetic field. This magnetic field is much like that of a bar magnet. Earth has a north and a south magnetic pole, just as a bar magnet has opposite magnetic poles at each of its ends. When you sprinkle iron shavings over a bar magnet, the shavings align with the magnetic field of the magnet. As you can see in **Figure 2,** Earth's magnetic field is similar—almost as if Earth contained a giant bar magnet. Earth's magnetic field protects you from harmful solar radiation by trapping many charged particles from the Sun.

Magnetic Axis When you observe a compass needle pointing north, you are seeing evidence of Earth's magnetic field. Earth's magnetic axis, the line joining its north and south magnetic poles, does not align with its rotational axis. The magnetic axis is inclined at an angle of 11.5° to the rotational axis. If you followed a compass needle, you would end up at the magnetic north pole rather than the rotational north pole.

The location of the magnetic poles has been shown to change slowly over time. The magnetic poles move around the rotational (geographic) poles in an irregular way. This movement can be significant over decades. Many maps include information about the position of the magnetic north pole at the time the map was made. Why would this information be important?

What causes changing seasons?

Flowers bloom as the days get warmer. The Sun appears higher in the sky, and daylight lasts longer. Spring seems like a fresh, new beginning. What causes these wonderful changes?

Orbiting the Sun You learned earlier that Earth's rotation causes day and night. Another important motion is **revolution,** which is Earth's yearly orbit around the Sun. Just as the Moon is Earth's satellite, Earth is a satellite of the Sun. If Earth's orbit were a circle with the Sun at the center, Earth would maintain a constant distance from the Sun. However, this is not the case. Earth's orbit is an **ellipse** (ee LIHPS)—an elongated, closed curve. The Sun is not at the center of the ellipse but is a little toward one end. Because of this, the distance between Earth and the Sun changes during Earth's yearlong orbit. Earth gets closest to the Sun—about 147 million km away—around January 3. The farthest Earth gets from the Sun is about 152 million km away. This happens around July 4 each year.

Reading Check *What is an ellipse?*

Does this elliptical orbit cause seasonal temperatures on Earth? If it did, you would expect the warmest days to be in January. You know this isn't the case in the northern hemisphere, something else must cause the change.

Even though Earth is closest to the Sun in January, the change in distance is small. Earth is exposed to almost the same amount of Sun all year. But the amount of solar energy any one place on Earth receives varies greatly during the year. Next, you will learn why.

A Tilted Axis Earth's axis is tilted 23.5° from a line drawn perpendicular to the plane of its orbit. It is this tilt that causes seasons. Daylight hours are longer for the hemisphere, or half of Earth, that is tilted toward the Sun. Think of how early it gets dark in the winter compared to the summer. As shown in **Figure 3,** the hemisphere that is tilted toward the Sun receives more hours of sunlight each day than the hemisphere that is tilted away from the Sun. The longer period of sunlight is one reason summer is warmer than winter, but it is not the only reason.

Research Visit the Glencoe Science Web site at **science.glencoe.com** for more information about ellipses. What special properties do they have? What effect does the shape of Earth's orbit have on incoming solar energy throughout the year? Present your findings to the class.

Figure 3
In summer, the northern hemisphere is tilted toward the Sun. Notice that the north pole is always lit during the summer. *Why are daylight hours longer in the summer than in the winter?*

North Pole

Radiation from the Sun Earth's tilt also causes the Sun's radiation to strike the hemispheres at different angles. The hemisphere tilted toward the Sun receives more direct rays, thus more total solar radiation than the hemisphere tilted away from the Sun. In the hemisphere tilted away from the Sun, the Sun appears low in the sky, and its rays are slanted.

Summer occurs in the hemisphere tilted toward the Sun, where the Sun appears high in the sky. Its radiation strikes Earth at a higher angle and for longer periods of time. The hemisphere receiving less radiation experiences winter.

Solstices

The **solstice** is the day when the Sun reaches its greatest distance north or south of the equator. In the northern hemisphere, the summer solstice occurs on June 21 or 22, and the winter solstice occurs on December 21 or 22. Both solstices are illustrated in **Figure 4.** In the southern hemisphere, the winter solstice is in June and the summer solstice is in December. Summer solstice is nearly the longest day of the year. After the summer solstice, days begin to get shorter. The winter solstice is nearly the shortest day of the year, but after the winter solstice, the period of sunlight grows longer each day.

Figure 4
During the summer solstice in the northern hemisphere, the Sun is directly over the tropic of Cancer, the latitude line at 23.5° N latitude. During the winter solstice, the Sun is directly over the tropic of Capricorn, the latitude line at 23.5° S latitude. At fall and spring equinoxes, the Sun is directly over the equator.

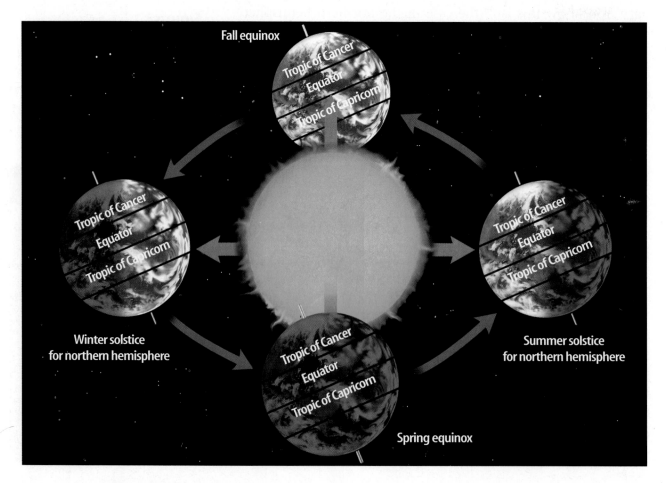

Equinoxes

An **equinox** (EE kwuh nahks) occurs when the Sun is directly above Earth's equator. Because of the tilt of Earth's axis, the Sun's position relative to the equator changes constantly. Most of the time, the Sun is either north or south of the equator, but two times each year it is directly over it, resulting in the spring and fall equinoxes. As you can see in **Figure 4,** on the equinox the Sun's most direct rays shine on the equator.

During an equinox, the number of daylight hours and night-time hours is nearly equal all over the world. Also at this time, neither the northern hemisphere nor the southern hemisphere is tilted toward the Sun.

In the northern hemisphere, the Sun reaches the spring equinox on March 20 or 21, and the fall equinox occurs on September 22 or 23. In the southern hemisphere, the equinoxes are reversed. Spring occurs in September and fall occurs in March.

Earth Data Review As you have learned, Earth is a sphere that rotates on a tilted axis. This rotation causes day and night. Earth's tilted axis and its revolution around the Sun cause the seasons. One Earth revolution takes one year. In the next section, you will read how the Moon rotates on its axis and revolves around Earth.

SCIENCE *Online*

Collect Data Visit the Glencoe Science Web site at **science.glencoe.com** for data about seasons. How are seasons different in other parts of the world? Make a poster summarizing what you learn.

Section 1 Assessment

1. Why did Aristotle think Earth was spherical?

2. Compare and contrast rotation and revolution.

3. Describe how Earth's distance from the Sun changes throughout the year. When is Earth closest to the Sun?

4. Why is it summer in Earth's northern hemisphere at the same time it is winter in the southern hemisphere?

5. **Think Critically** **Table 1** lists Earth's distance from the Sun as an average. Why isn't an exact measurement available for this distance?

Skill Builder Activities

6. **Recognizing Cause and Effect** Answer these questions about the Sun-Earth-Moon relationship. **For more help, refer to the** Science Skill Handbook.
   a. What causes seasons on Earth?
   b. Why does the Sun appear to rise in the east and set in the west each day?

7. **Using an Electronic Spreadsheet** Create a table of Earth's physical properties. Show the following: *diameter, mass, period of rotation,* and *other data.* Then, write a description of Earth based on your table. **For more help, refer to the** Technology Skill Handbook.

The Moon— Earth's Satellite

As You Read

What You'll Learn

■ **Identify** phases of the Moon and their cause.
■ **Explain** why solar and lunar eclipses occur.
■ **Infer** what the Moon's surface features may reveal about its history.

Vocabulary

moon phase	waning
new moon	solar eclipse
waxing	lunar eclipse
full moon	maria

Why It's Important

Learning about the Moon can teach you about Earth.

Motions of the Moon

Doesn't it seem as if the Moon's shape changes night after night? Sometimes, just after sunset, you can see a full, round Moon low in the sky. Other times, only half of the Moon is visible, and it's high in the sky at sunset. At times, the Moon is even visible during the day. What causes the Moon to change in appearance and position in the sky?

Rotation and Revolution Just as Earth rotates on its axis and revolves around the Sun, the Moon rotates on its axis and revolves around Earth. The Moon's revolution around Earth is responsible for the changes in its appearance. If the Moon rotates on its axis, why can't you see it spin around in space? The reason is that the Moon's rotation takes 27.3 days—the same amount of time it takes to revolve once around Earth. Because these two motions take the same amount of time, the same side of the Moon always faces Earth, as shown in **Figure 5.**

You can demonstrate this by having a friend hold a ball in front of you. Direct your friend to move the ball in a circle around you while keeping the same side of it facing you. Everyone else in the room will see all sides of the ball. You will see only one side.

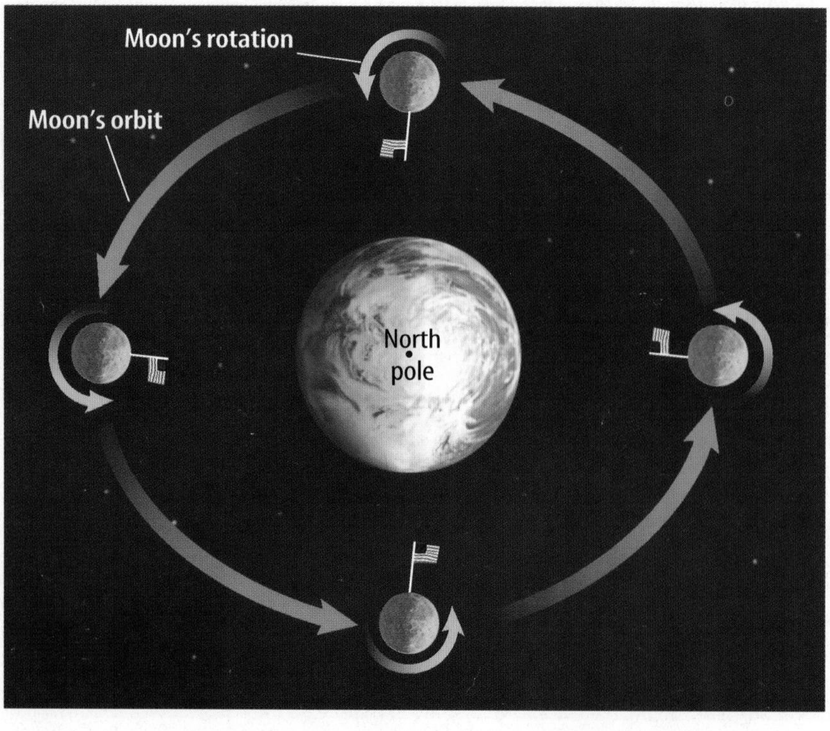

Figure 5
In about one month, the Moon orbits Earth. It also completes one rotation on its axis during the same period. *Does this affect which side of the Moon faces Earth? Explain.*

Reflection of the Sun The Moon seems to shine because its surface reflects sunlight. Just as half of Earth experiences day as the other half experiences night, half of the Moon is lighted while the other half is dark. As the Moon revolves around Earth, you see different portions of its lighted side, causing the Moon's appearance to change.

Phases of the Moon

Moon phases are the different forms that the Moon takes in its appearance from Earth. The phase depends on the relative positions of the Moon, Earth, and the Sun, as seen in **Figure 6** on the next page. A **new moon** occurs when the Moon is between Earth and the Sun. During a new moon, the lighted half of the Moon is facing the Sun and the dark side faces Earth. The Moon is in the sky, but it cannot be seen. The new moon rises and sets with the Sun.

✔ **Reading Check** *Why can't you see a new moon?*

Waxing Phases After a new moon, the phases begin waxing. **Waxing** means that more of the illuminated half of the Moon can be seen each night. About 24 h after a new moon, you can see a thin slice of the Moon. This phase is called the waxing crescent. About a week after a new moon, you can see half of the lighted side of the Moon, or one quarter of the Moon's surface. This is the first quarter phase.

The phases continue to wax. When more than one quarter is visible, it is called waxing *gibbous* after the Latin word for "humpbacked." A **full moon** occurs when all of the Moon's surface facing Earth reflects light.

Waning Phases After a full moon, the phases are said to be waning. When the Moon's phases are **waning,** you see less of its illuminated half each night. Waning gibbous begins just after a full moon. When you can see only half of the lighted side, it is the third-quarter phase. The Moon continues to appear to shrink. Waning crescent occurs just before another new moon. Once again, you can see only a small slice of the Moon.

It takes about 29.5 days for the Moon to complete its cycle of phases. Recall that it takes about 27.3 days for the Moon to revolve around Earth. The discrepancy between these two numbers is due to Earth's revolution. The roughly two extra days are what it takes for the Moon to keep up constantly with Earth as it orbits around the Sun.

Mini LAB

Comparing the Sun and the Moon

Procedure
1. Find an area where you can make a chalk mark on **pavement or another surface.**
2. Tie a piece of **chalk** to one end of a 400-cm-long **string.**
3. Hold the other end of the string to the pavement.
4. Have a friend pull the string tight and walk around you, leaving a mark (the Sun) on the pavement.
5. Draw a 1-cm-diameter circle in the middle of the larger circle (the Moon).

Analysis
1. How big is the Sun compared to the Moon?
2. The diameter of the Sun is 1.39 million km. The diameter of Earth is 12,756 km. Draw two new circles modeling the sizes of the Sun and Earth. What scale did you use?

Figure 6
The phases of the Moon change during a cycle that lasts about 29.5 days.

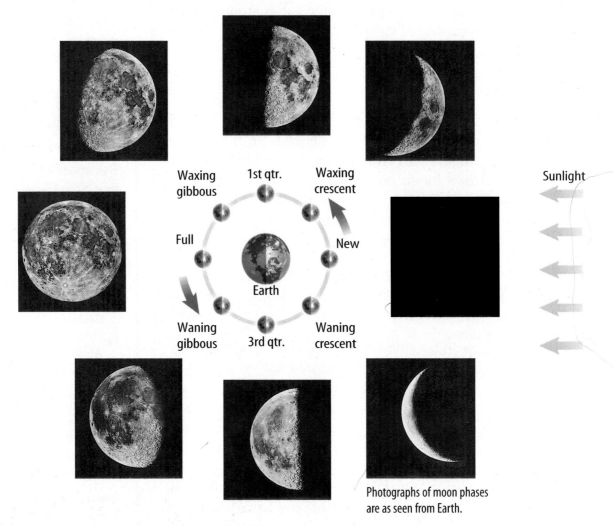

Waxing gibbous · 1st qtr. · Waxing crescent · Full · Earth · New · Waning gibbous · 3rd qtr. · Waning crescent · Sunlight

Photographs of moon phases are as seen from Earth.

Figure 7
Only the outer portion of the Sun's atmosphere is visible during a total solar eclipse. It looks like a halo around the Moon.

Eclipses

Imagine living 10,000 years ago. You are foraging for nuts and fruit when unexpectedly the Sun disappears from the sky. The darkness lasts only a short time, and the Sun soon returns to full brightness. You know something strange has happened, but you don't know why. It will be almost 8,000 years before anyone can explain what you just experienced.

The event just described was a total solar eclipse (ih KLIPS), shown in **Figure 7.** Today, most people know what causes such eclipses, but without this knowledge, they would have been terrifying events. During a solar eclipse, many animals act as if it is nighttime. Cows return to their barns and chickens go to sleep. What causes the day to become night and then change back into day?

✓ **Reading Check**   *What happens during a total solar eclipse?*

What causes an eclipse? The revolution of the Moon causes eclipses. Eclipses occur when Earth or the Moon temporarily blocks the sunlight from reaching the other. Sometimes, during a new moon, the Moon's shadow falls on Earth and causes a solar eclipse. During a full moon, Earth's shadow can be cast on the Moon, resulting in a lunar eclipse.

An eclipse can occur only when the Sun, the Moon, and Earth are lined up perfectly. Because the Moon's orbit is not in the same plane as Earth's orbit around the Sun, eclipses occur only a few times each year.

Eclipses of the Sun A **solar eclipse** occurs when the Moon moves directly between the Sun and Earth and casts its shadow over part of Earth, as seen in **Figure 8.** Depending on where you are on Earth, you may experience a total eclipse or a partial eclipse. The darkest portion of the Moon's shadow is called the umbra (UM bruh). A person standing within the umbra experiences a total solar eclipse. During a total solar eclipse, the only visible portion of the Sun is a pearly white glow around the edge of the eclipsing Moon.

Surrounding the umbra is a lighter shadow on Earth's surface called the penumbra (puh NUM bruh). Persons standing in the penumbra experience a partial solar eclipse. **WARNING:** *Regardless of where you stand, never look directly at the Sun during an eclipse. The light can permanently damage your eyes.*

SCIENCE *Online*

Collect Data Visit the Glencoe Science Web site at **science.glencoe.com** for more information about solar and lunar eclipses due to occur over the next several years. Make a chart showing when and where they will occur.

Figure 8
Only a small area of Earth experiences a total solar eclipse during the eclipse event.

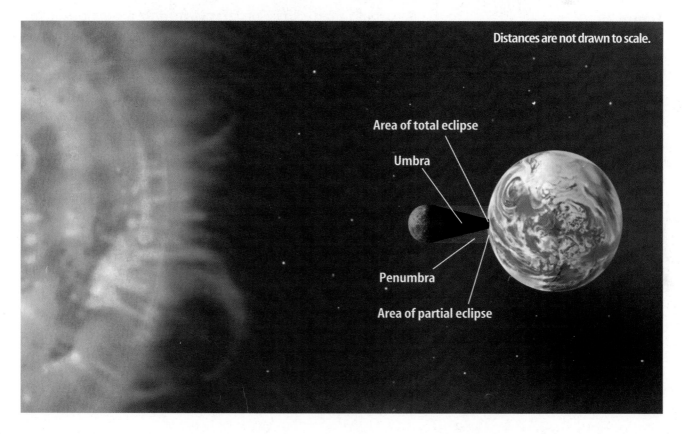

Distances are not drawn to scale.

Area of total eclipse

Umbra

Penumbra

Area of partial eclipse

Figure 9
These photographs show the Moon moving from right to left into Earth's umbra, then out again.

Figure 10
During a total lunar eclipse, Earth's shadow blocks light coming from the Sun.

Eclipses of the Moon When Earth's shadow falls on the Moon, a **lunar eclipse** occurs. A lunar eclipse begins when the Moon moves into Earth's penumbra. As the Moon continues to move, it enters Earth's umbra and you see a curved shadow on the Moon's surface, as in **Figure 9.** Upon moving completely into Earth's umbra, as shown in **Figure 10,** the Moon goes dark, signaling that a total lunar eclipse has occurred. Sometimes sunlight bent through Earth's atmosphere causes the eclipsed Moon to appear red.

A partial lunar eclipse occurs when only a portion of the Moon moves into Earth's umbra. The remainder of the Moon is in Earth's penumbra and, therefore, receives some direct sunlight. A penumbral lunar eclipse occurs when the Moon is totally within Earth's penumbra. However, it is difficult to tell when a penumbral lunar eclipse occurs because some direct sunlight is falling on the side of the Moon facing Earth.

A total lunar eclipse can be seen by anyone on the nighttime side of Earth where the Moon is not hidden by clouds. In contrast, only a lucky few people get to witness a total solar eclipse. Only those people in the small region where the Moon's umbra strikes Earth can witness one.

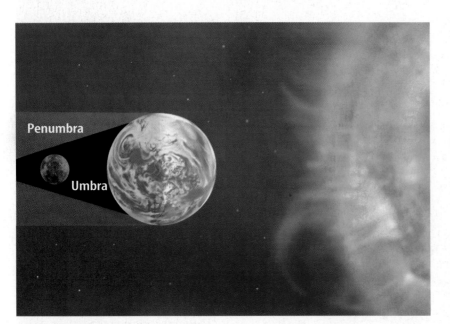

Penumbra

Umbra

The Moon's Surface

When you look at the Moon, as shown in **Figure 12** on the next page, you can see many depressions called craters. Meteorites, asteroids, and comets striking the Moon's surface created most of these craters, which formed early in the Moon's history. Upon impact, cracks may have formed in the Moon's crust, allowing lava to reach the surface and fill up the large craters. The resulting dark, flat regions are called **maria** (MAHR ee uh). The igneous rocks of the maria are 3 billion to 4 billion years old. So far, they are the youngest rocks to be found on the Moon. This indicates that craters formed after the Moon's surface originally cooled. The maria formed early enough in the Moon's history that molten material still remained in the Moon's interior. The Moon once must have been as geologically active as Earth is today. Before the Moon cooled to the current condition, the interior separated into distinct layers.

Inside the Moon

Earthquakes allow scientists to learn about Earth's interior. In a similar way, scientists use instruments such as the one in **Figure 11A** to study moonquakes. The data they have received have led to the construction of several models of the Moon's interior. One such model, shown in **Figure 11B,** suggests that the Moon's crust is about 60 km thick on the side facing Earth. On the far side, it is thought to be about 150 km thick. Under the crust, a solid mantle may extend to a depth of 1,000 km. A partly molten zone of the mantle may extend even farther down. Below this mantle may lie a solid, iron-rich core.

Physics
INTEGRATION

Waves generated by earthquakes and moonquakes change speeds when encountering materials with different physical properties. Research these waves to find out what factors affect their speed and direction. Infer how scientists might use this information to learn about the interiors of the Moon and Earth.

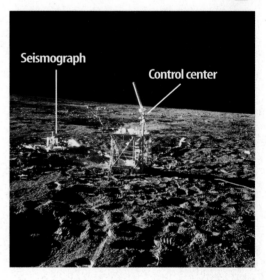

Seismograph

Control center

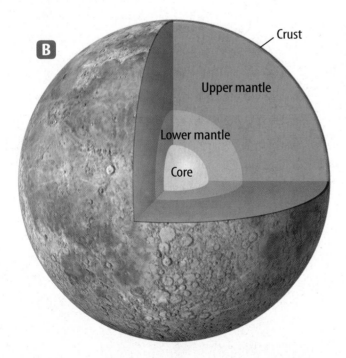

Crust

Upper mantle

Lower mantle

Core

Figure 11
A Equipment, such as the seismograph left on the Moon by the *Apollo 12* mission, helps scientists study moonquakes.
B Models of the Moon's interior were created from data obtained by scientists studying moonquakes.

Figure 12

By looking through binoculars, you can see many of the features on the surface of the Moon. These include craters that are hundreds of kilometers wide, light-colored mountains, and darker patches that early astronomers called maria (Latin for "seas"). However, as the NASA Apollo missions discovered, these so-called seas do not contain water. In fact, maria (singular, mare) are flat, dry areas formed by ancient lava flows. Some of the Moon's geographic features are shown below, along with the landing sites of Apollo missions sent to investigate Earth's closest neighbor in space.

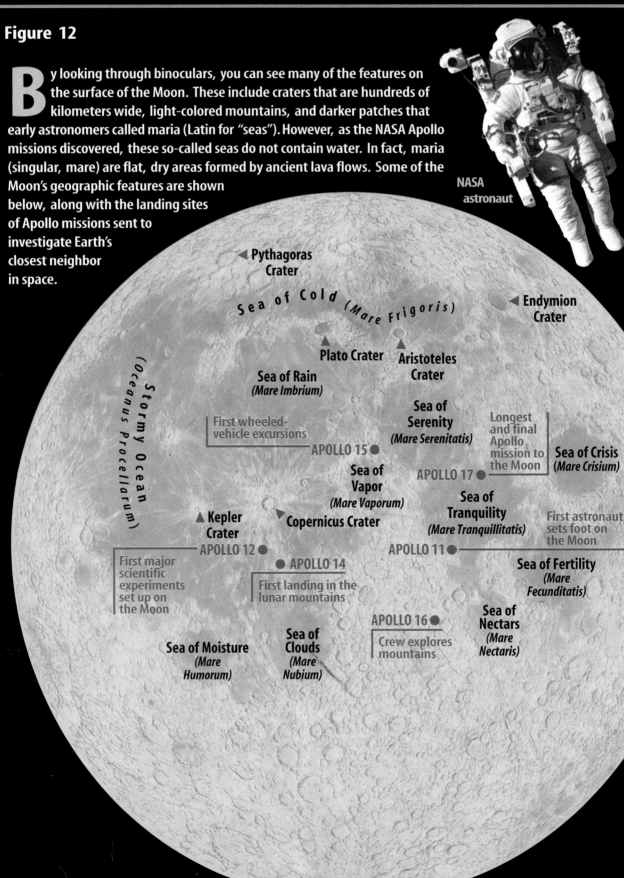

NASA astronaut

A A Mars-sized
object collided
with Earth.

B The blast ejected
material from both
objects into space.

C A ring of gas
and debris formed
around Earth.

D Particles in the
ring joined together
to form the Moon.

The Moon's Origin

Before the *Apollo* space missions in the 1960s and 1970s, there were three leading theories about the Moon's origin. According to one theory, the Moon was captured by Earth's gravity. Another held that material surrounding Earth condensed to produce the Moon. An alternative theory proposed that Earth ejected molten material that became the Moon.

The Impact Theory The data gathered by the *Apollo* missions have led many scientists to support a new theory, known as the impact theory. It states that the Moon formed billions of years ago from condensing gas and debris thrown off when Earth collided with a Mars-sized object as shown in **Figure 13.**

Figure 13
According to the impact theory, a Mars-sized object collided with Earth around 4.6 billion years ago. Vaporized materials ejected by the collision began orbiting Earth and quickly consolidated into the Moon.

Problem-Solving Activity

What will you use to survive on the Moon?

You have crash-landed on the Moon. It will take one day to reach a moon colony on foot. The side of the Moon that you are on will be facing away from the Sun during your entire trip. You manage to salvage the following items from your wrecked ship: food, rope, solar-powered heating unit, battery-operated heating unit, oxygen tanks, map of the constellations, compass, matches, water, solar-powered radio transmitter, three flashlights, signal mirror, and binoculars.

Identifying the Problem

The Moon lacks a magnetic field and has no atmosphere. How do the Moon's physical properties and the lack of sunlight affect your decisions?

Solving the Problem

1. Which items will be of no use to you? Which items will you take with you?
2. Describe why each of the salvaged items is useful or not useful.

Figure 14
Moon rocks collected by astronauts provide scientists with information about the Moon and Earth.

The Moon in History Regardless of how the Moon formed, it has played an important role. Studying the Moon's phases and eclipses led to the conclusion that both Earth and the Moon were in motion around the Sun. The curved shadow Earth casts on the Moon indicated to early scientists that Earth was spherical. When Galileo first turned his telescope toward the Moon, he found a surface scarred by craters and maria. Before that time, many people believed that all planetary bodies were perfectly smooth and lacking surface features.

 Reading Check *How has observing the Moon been important to science?*

More recently, actual Moon rocks became available for scientists to study, as seen in **Figure 14.** By doing so, they hope to learn more about Earth. The Moon was important in the past and promises to be important in the future, as well.

Section ② Assessment

1. How are the Sun, the Moon, and Earth positioned relative to each other during a new moon?

2. What do maria look like? How did they form?

3. What are the umbra and penumbra? How do they relate to eclipses?

4. What is the difference between a solar and a lunar eclipse? Explain what causes each and why more people see a lunar eclipse.

5. **Think Critically** What do the surface features of the Moon tell you about its history?

Skill Builder Activities

6. **Predicting** Look at a calendar or almanac to find out when the next new moon will occur. Using this information, predict when the first-quarter phase will begin. **For more help, refer to the** Science Skill Handbook.

7. **Communicating** Research the various theories about the Moon's origin in astronomy books and magazines. In your Science Journal, report on and make a diagram of each, including the impact theory. Evaluate the strengths and weaknesses of each theory. **For more help, refer to the** Science Skill Handbook.

Activity

Moon Phases and Eclipses

You have learned that Moon phases and eclipses result from the relative positions of the Sun, the Moon, and Earth. In this activity, you will demonstrate the positions of these bodies during certain phases and eclipses. You also will see why only a small portion of the people on Earth witness a total solar eclipse during a particular eclipse event.

What You'll Investigate

Can a model be devised to show the positions of the Sun, the Moon, and Earth during various phases and eclipses?

Materials

light source (unshaded)   globe
polystyrene ball   pencil

Goals

■ **Model** moon phases.
■ **Model** solar and lunar eclipses.

Procedure

1. Review the illustrations of Moon phases and eclipses shown in Section 2.

2. Use the light source as a Sun model and a polystyrene ball on a pencil as a Moon model. Move the Moon around the globe to duplicate the exact position that would have to occur for a lunar eclipse to take place.

3. Move the Moon to the position that would cause a solar eclipse.

4. Place the Moon at each of the following phases: first quarter, full moon, third quarter, and new moon. Identify which, if any, type of eclipse could occur during each phase. Record your data.

Moon Phase Observations	
Moon Phase	**Observations**
first quarter	
full moon	
third quarter	
new moon	

5. Place the Moon at the location where a lunar eclipse could occur. Move it slightly toward Earth, then away from Earth. Note the amount of change in the size of the shadow.

6. Repeat step 5 with the Moon in a position where a solar eclipse could occur.

Conclude and Apply

1. During which phase(s) of the Moon is it possible for an eclipse to occur?

2. **Describe** the effect of a small change in distance between Earth and the Moon on the size of the umbra and penumbra.

3. **Infer** why a lunar and a solar eclipse do not occur every month.

4. Why have only a few people experienced a total solar eclipse?

5. **Diagram** the positions of the Sun, Earth, and the Moon during a first quarter moon.

6. Why might it be better to call a full moon a half moon?

Communicating Your Data

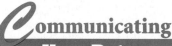

Communicate your answers to other students. **For more help, refer to the** Science Skill Handbook.

3 Exploring Earth's Moon

As You Read

What You'll Learn

- **Describe** recent discoveries about the Moon.
- **Examine** facts about the Moon that might influence future space travel.

Vocabulary

impact basin

Why It's Important

Continuing Moon missions may result in discoveries about Earth's origin.

Figure 15
This time line illustrates some of the most important events in the history of Moon exploration.

Missions to the Moon

The Moon has always fascinated humanity. People have made up stories about how it formed. Children's stories even suggested it was made of cheese. Of course, for centuries astronomers also have studied the Moon for clues to its makeup and origin. In 1959, the former Soviet Union launched the first *Luna* spacecraft, enabling up-close study of the Moon. Two years later, the United States began a similar program with the first *Ranger* spacecraft. Following the uncrewed *Ranger* missions, the United States launched a series of *Lunar Orbiters.* The spacecraft in these early missions took detailed photographs of the Moon. There also were seven *Surveyor* spacecraft designed to land on the Moon. Five of these spacecraft successfully touched down on the lunar surface. The *Surveyor* probes took detailed photographs and performed the first analysis of lunar soil. The goal of the program was to prepare for landing astronauts on the Moon. This goal was achieved in 1969 by the astronauts of *Apollo 11.* By 1972, when the *Apollo* missions ended, 12 U.S. astronauts had walked on the Moon. A time line of these important moon missions can be seen in **Figure 15.**

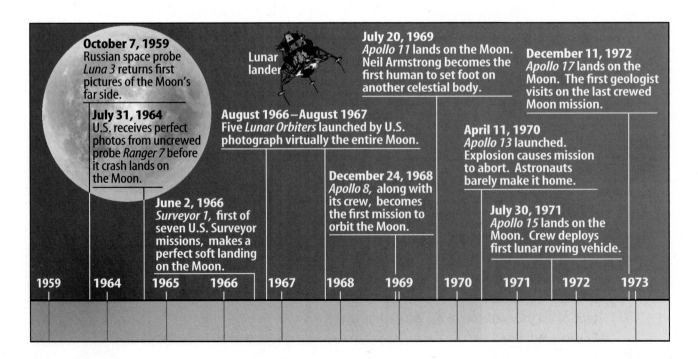

October 7, 1959
Russian space probe *Luna 3* returns first pictures of the Moon's far side.

July 31, 1964
U.S. receives perfect photos from uncrewed probe *Ranger 7* before it crash lands on the Moon.

June 2, 1966
Surveyor 1, first of seven U.S. Surveyor missions, makes a perfect soft landing on the Moon.

Lunar lander

August 1966—August 1967
Five *Lunar Orbiters* launched by U.S. photograph virtually the entire Moon.

December 24, 1968
Apollo 8, along with its crew, becomes the first mission to orbit the Moon.

July 20, 1969
Apollo 11 lands on the Moon. Neil Armstrong becomes the first human to set foot on another celestial body.

April 11, 1970
Apollo 13 launched. Explosion causes mission to abort. Astronauts barely make it home.

July 30, 1971
Apollo 15 lands on the Moon. Crew deploys first lunar roving vehicle.

December 11, 1972
Apollo 17 lands on the Moon. The first geologist visits on the last crewed Moon mission.

1959 | 1964 | 1965 | 1966 | 1967 | 1968 | 1969 | 1970 | 1971 | 1972 | 1973

Surveying the Moon More than 20 years passed before the United States resumed its studies of the Moon from space. In 1994, the *Clementine* was placed into lunar orbit. Its goal was to conduct a two-month survey of the Moon's surface. An important aspect of this study was collecting data on the mineral content of Moon rocks. In fact, this part of its mission was instrumental in naming the spacecraft. Clementine was the daughter of a miner in the ballad *My Darlin' Clementine*. While in orbit, *Clementine* also mapped features on the Moon's surface, including huge impact basins.

✔ **Reading Check** *Why was* **Clementine** *placed in lunar orbit?*

Impact Basins When meteorites and other objects strike the Moon, they leave behind depressions in the Moon's surface. The depression left behind by an object striking the Moon is known as an **impact basin,** or impact crater. The South Pole-Aitken Basin is the oldest identifiable impact feature on the Moon's surface. At 12 km in depth and 2,500 km in diameter, it is also the largest and deepest impact basin in the solar system. Data returned by *Clementine* gave scientists the first set of high-resolution photographs of this area of the Moon. Because much of this basin stays in shadow throughout the Moon's rotation, a cold area has formed where ice deposits from impacting comets might have collected, as shown in **Figure 16.** A large plateau that is always in sunlight also was discovered in this area. If ice truly is near this plateau, as indicated by radio signals that *Clementine* reflected off the Moon to Earth, it would be the ideal location to build a moon colony powered by solar energy.

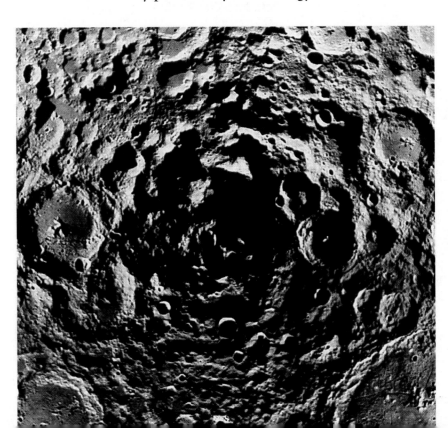

Figure 16
The South Pole-Aitken Basin is the largest of its kind found anywhere in the solar system. The deepest craters in the basin stay in shadow throughout the Moon's rotation. Ice deposits from impacting comets are thought to have collected at the bottom of these craters. This image shows the possible location of such ice.

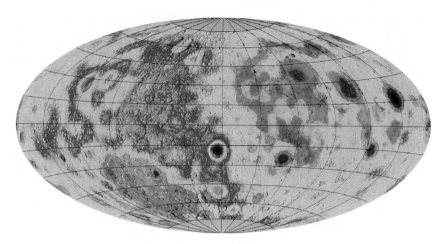

Figure 17
This computer-enhanced map based on *Clementine* data indicates the thickness of the Moon's crust. The crust of the side of the Moon facing Earth, shown mostly in red, is thinner than the crust on the far side of the Moon.

Mapping the Moon

A large part of *Clementine's* mission included taking high-resolution photographs so a detailed map of the Moon's surface could be compiled. Because the five cameras mounted on *Clementine* were able to resolve features as small as 200 m across, human knowledge of the Moon's surface increased immensely. One image resulting from other *Clementine* data is shown in **Figure 17.** It shows that the crust on the side of the Moon that faces Earth is much thinner than the crust on the far side. Additional information shows that the Moon's crust is thinnest under impact basins. Based on analysis of the light data received from *Clementine,* a global map of the Moon also was created that shows its composition, as seen in **Figure 18.**

✔ **Reading Check** *What information about the Moon did scientists learn from* Clementine?

The Lunar Prospector The success of *Clementine* at a relatively low cost opened the door for further Moon missions. In 1998, NASA launched the desk-sized *Lunar Prospector* into lunar orbit to collect more information about the lunar surface. The spacecraft spent a year orbiting the Moon from pole to pole, once every two hours. The resulting maps confirmed the *Clementine* data. The *Lunar Prospector* was scheduled to conduct a detailed study of the Moon from 100 km above the surface and to look for clues about its origin and makeup.

Figure 18
One of *Clementine's* main missions was to map the Moon. The different colors in this map represent the different types of lunar surface material, because not all parts of the Moon are made up of the same materials. This is the side of the Moon that faces Earth.

Icy Poles Early data from the *Lunar Prospector,* shown in **Figure 19,** indicate hydrogen is present in crater rocks at the Moon's poles. Hydrogen is one of the two elements in water. This information, combined with data from *Clementine,* has led scientists to hypothesize that ice may exist in the floors of craters at both Moon poles. These deep craters are cold because sunlight never reaches their floors. Temperatures are as low as −233°C. They are definitely cold enough to have preserved any ice that may have collected from colliding comets. Scientists estimate that 6 billion tons of ice might lie buried under 40 cm of crushed rock at the Moon's poles.

When the *Lunar Prospector* mission ended in July of 1999, it still was unknown whether the hydrogen was from water on the Moon or some other source, such as solar wind. NASA decided to crash the spacecraft into a crater at the Moon's south pole that might contain ice. The scientists hoped that the crash would release large quantities of water vapor that might be detected with special telescopes on Earth. The chance of this experiment working was considered slight, and the results were inconclusive. Water might exist on the Moon, but additional research will be needed before definite conclusions can be drawn. However, data from the *Lunar Prospector* have enabled scientists to confirm that the Moon has a small, iron-rich core about 600 km in diameter. The fact that the Moon has such a small core supports the impact theory of its origin because only small amounts of iron would have been blasted off the primitive Earth. Most of Earth's iron would have remained deep in the planet's interior.

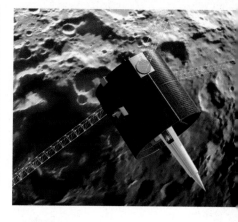

Figure 19
The *Lunar Prospector* provided data indicating that ice might exist at the Moon's poles. Further investigation will reveal whether or not this is true.

Section Assessment

1. List two discoveries about the Moon made by *Clementine.*

2. What was the main mission of the *Lunar Prospector?*

3. How did studies of the Moon change after the 1950s?

4. What was the goal of the *Surveyor* and *Apollo* missions? How did these missions further scientists' studies of the Moon?

5. **Think Critically** Why would the discovery of ice at the Moon's poles be important to future space flights?

Skill Builder Activities

6. **Concept Mapping** Sequence the following Moon missions in the order they occurred: *Lunar Prospector, Apollo, Lunar Orbiter, Ranger,* and *Clementine.* **For more help, refer to the** Science Skill Handbook.

7. **Using Fractions** The Moon's orbit is tilted at an angle of 5° to Earth's orbit around the Sun. Using a protractor draw the Moon's orbit around Earth. What fraction of a full circle (360°) is 5°? **For more help, refer to the** Math Skill Handbook.

Activity

Tilt and Temperature

If you walk barefeet on blacktop pavement at noon, you can feel the effect of solar energy. The Sun's rays hit most directly at midday. Now consider the fact that Earth is tilted on its axis. How does this tilt affect how directly light rays strike an area on Earth? How is the angle of the light rays related to the amount of heat energy and the changing seasons?

What You'll Investigate

How does the angle at which light strikes Earth affect the amount of heat energy received by any area on Earth?

Materials

tape
black construction paper (one sheet)
gooseneck lamp with 75-watt bulb
celsius thermometer
watch
protractor

Goals

- **Measure** the temperature change in a surface after light strikes it at different angles.
- **Describe** how the angle of light relates to seasons on Earth.

Safety Precautions

Do not touch the lamp without safety gloves. The lightbulb and shade can be hot even when the lamp has been turned off. Handle the thermometer carefully. If it breaks, do not touch anything. Inform your teacher immediately.

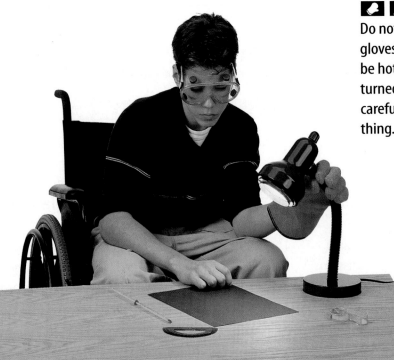

Procedure

1. Choose three angles that you will use to aim the light at the paper.

2. **Determine** how long you will shine the light at each angle before you measure the temperature. Measure the temperature at two times for each angle. Use the same time periods for each angle.

3. Copy the following data table into your Science Journal and fill in the temperature the paper reached at each angle and time.

4. Tape a sheet of black construction paper to a desk or the floor.

5. Using the protractor, set the gooseneck lamp so that it will shine on the paper at one of the angles you chose.

6. Turn on the lamp. Use the thermometer to measure the temperature of the paper at the end of the first time period. Continue shining the lamp on the paper until the second time period has passed. Measure the temperature again. Record your data in your data table.

7. Turn off the lamp until the paper cools to room temperature. Repeat steps 5 and 6 using your other two angles.

Temperature Data			
Angle of Lamp	Initial Temperature	Temperature at ____ Minutes/Seconds	Temperature at ____ Minutes/Seconds
First angle			
Second angle			
Third angle			

Conclude and Apply

1. **Describe** your experiment. Identify the variables in your experiment. Which were your independent and dependent variables?

2. **Graph** your data using a line graph. Describe what your graph tells you about the data.

3. What happened to the temperature of the paper as you changed the angle of light?

4. **Predict** how your results might have been different if you used white paper. Explain why.

5. **Describe** how the results of this experiment apply to seasons on Earth.

Communicating Your Data

Compare your results with those of other students in your class. **Discuss** how the different angles and time periods affected the temperatures.

UNITED STATES

GULF OF MEXICO

North America

MEXICO

CUBA

Chichén Itzá

South America

Australia

An ancient people used many calendars to track time and to help them in everyday life

THE Mayan Calendar

Most people don't give the setup of calendars a second thought. They take for granted that a week is seven days, and that a year is 12 months. But this wasn't always the case. Roughly 1,750 years ago, in what is now south Mexico and Central America, the Mayan people invented a sophisticated calendar system. Their system used observations of Sun and Moon cycles to help them figure out time. They also developed a sophisticated mathematical system where single units are written with dots, and bars are used to represent five single units.

The Mayans had several calendars that they used at the same time. Two were most important. One was based on 260 days, the other on 365 days. The calendars were so accurate and useful that later civilizations, including the Aztecs, adopted them.

These glyphs represent four different days of the *Tzolkin* calendar.

The 260-Day Calendar

One calendar, called the *Tzolkin* (tz uhl KIN), was based on the planting, harvesting, drying, and storing of corn—the main crop of the Mayans. Each day of the *Tzolkin* had one of 20 names, as well as a number from 1 to 13. Each day also had a Mayan god associated with it.

Mayan priests used this calendar to determine the dates of religious festivals. The priests also used it to help decide when to go to war and what to name children. Just as important, the Mayans used the *Tzolkin* for timing their planting and harvesting.

The 365-Day Calendar

Another Mayan calendar was called the *Haab* (HAHB) and was based on the orbit of Earth around the Sun. It was divided into 18 months with 20 days each, plus five extra days at the end of each year. The 260-day *Tzolkin* calendar fit into the 365-day *Haab* year, so the Mayans used both calendars at once.

Used together, these calendars made the Mayans the most accurate reckoners of time before the modern period. In fact, they were only one day off every 6,000 years. Their idea of measuring time by Earth moving around the Sun is the basis for our calendar today.

This calendar was carved more than 1,000 years ago by an unknown Mayan.

Kukulkan (*left*), built around the year 1050, was used by the Mayans as a calendar. It had four stairways, each with 91 steps. Including the platform on top, the total number of steps is 365, the number of days in a year. You can see this pyramid if you visit Chichén Itzá in Mexico.

CONNECTIONS Drawing Symbols **The Mayans created picture symbols for each day of their week. Historians call these symbols glyphs. Collaborate with another student to invent seven glyphs—one for each day of the week. Compare them with other glyphs on the Glencoe Science Web site.**

SCIENCE *Online*
For more information, visit science.glencoe.com

Reviewing Main Ideas

Section 1 Earth

1. Earth is a sphere that bulges slightly at its equator.

2. Earth rotates once per day and completes its orbit around the Sun in a little more than 365 days.

3. Earth has a magnetic field like that of a bar magnet.

4. Seasons on Earth are caused by the tilt of Earth's axis and its revolution around the Sun. Because Earth's axis is tilted, the amount of solar energy each hemisphere receives varies throughout the year, causing the seasons.

Section 2 The Moon—Earth's Satellite

1. Earth's Moon goes through phases that depend on the relative positions of the Sun, the Moon, and Earth. *What are the relative positions of the Sun, Earth, and the Moon during a new moon?*

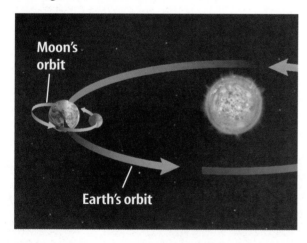

Moon's orbit

Earth's orbit

2. Eclipses occur when Earth or the Moon temporarily blocks sunlight from reaching the other. A solar eclipse occurs when the Moon moves directly between the Sun and Earth. A lunar eclipse occurs when Earth's shadow falls on the Moon.

3. The Moon's maria are the result of ancient lava flows. Craters on the Moon's surface formed from impacts with meteorites, asteroids, and comets. *What is the Moon's surface like?*

Section 3 Exploring Earth's Moon

1. The *Clementine* spacecraft took detailed, high-resolution photographs of the Moon's surface.

2. Data from *Clementine* indicate that the Moon's South Pole-Aitken Basin may contain ice deposits that could supply water for a Moon colony. *What could people living on the Moon do with water?*

3. NASA continued exploring the Moon with the *Lunar Prospector*. Data from *Lunar Prospector* lend support to the ice theory.

FOLDABLES
Reading & Study Skills

After You Read

Exchange Foldables with a classmate and quiz each other. Ask questions that can be answered with the information in your classmate's Foldable.

Visualizing Main Ideas

Complete the following concept map on Moon formation.

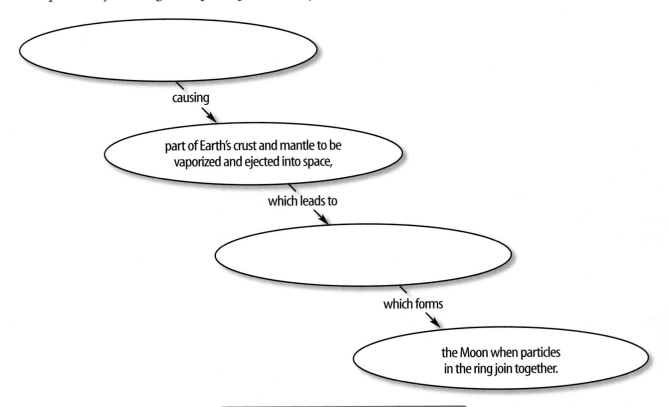

causing

part of Earth's crust and mantle to be vaporized and ejected into space,

which leads to

which forms

the Moon when particles in the ring join together.

Vocabulary Review

Vocabulary Words

a. axis
b. ellipse
c. equinox
d. full moon
e. impact basin
f. lunar eclipse
g. maria
h. moon phase
i. new moon
j. revolution
k. rotation
l. solar eclipse
m. solstice
n. sphere
o. waning
p. waxing

THE PRINCETON REVIEW — Study Tip

Practice reading graphs and charts. Make a table that contains the same information that a graph does.

Using Vocabulary

Replace each underlined word with the correct vocabulary word.

1. The spinning of Earth around its axis is called <u>revolution</u>.

2. The <u>equinox</u> is the point at which the Sun reaches its greatest distance north or south of the equator.

3. The Moon is said to be <u>waxing</u> when the portion of the lighted side that can be seen becomes smaller.

4. The depression left behind by an object striking the Moon is called an <u>ellipse</u>.

5. Earth's orbit is an <u>eclipse</u>.

Checking Concepts

Choose the word or phrase that best answers the question.

1. How long does it take for the Moon to rotate once?
   A) 24 hours   C) 27.3 hours
   B) 365 days   D) 27.3 days

2. Where is Earth's circumference greatest?
   A) equator   C) poles
   B) mantle   D) axis

3. During an equinox, the Sun is directly over what part of Earth?
   A) southern hemisphere
   B) northern hemisphere
   C) equator
   D) pole

4. What causes the Sun to appear to rise and set?
   A) Earth's revolution
   B) the Sun's revolution
   C) Earth's rotation
   D) Earth's elliptical orbit

5. How long does it take for the Moon to revolve once around Earth?
   A) 24 hours   C) 27.3 hours
   B) 365 days   D) 27.3 days

6. What is it called when the lighted portion of the Moon appears to get larger?
   A) waning   C) rotating
   B) waxing   D) crescent shaped

7. What kind of eclipse occurs when the Moon blocks sunlight from reaching Earth?
   A) solar   C) full
   B) new   D) lunar

8. What is the darkest part of the shadow during an eclipse?
   A) waxing gibbous   C) waning gibbous
   B) umbra   D) penumbra

9. What is the name for a depression on the Moon caused by an object striking its surface?
   A) eclipse   C) phase
   B) moonquake   D) impact basin

10. What fact does data gathered from the *Clementine* spacecraft support?
    A) The Moon rotates once in 29.5 days.
    B) The Moon has a thinner crust on the side facing Earth.
    C) The Moon revolves once in 29.5 days.
    D) The Moon has no crust.

Thinking Critically

11. How would the Moon appear to an observer in space during its revolution? Would phases be observable? Explain.

12. What would be the effect on Earth's seasons if the axis were tilted at 28.5° instead of 23.5°?

13. Seasons in the two hemispheres are opposite. Explain how this supports the statement that seasons are NOT caused by Earth's changing distance from the Sun.

14. How would solar eclipses be different if the Moon were twice as far from Earth? Explain.

15. Which observed motions of the Moon are real? Which are apparent? Explain.

Developing Skills

16. **Predicting** Predict how the information gathered by Moon missions could be helpful in the future for people wanting to establish a colony on the Moon.

17. **Using Variables, Constants, and Controls** Describe a simple activity to show how the Moon's rotation and revolution work to keep the same side facing Earth at all times.

18. Comparing and Contrasting Compare and contrast a waning moon and a waxing moon.

19. Concept Mapping Copy and complete the cycle concept map shown on this page. Show the sequence of the Moon's phases.

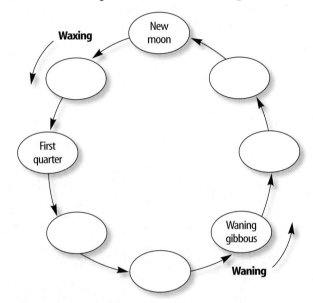

20. Hypothesizing Gravity is weaker on the Moon than it is on Earth. Use this fact to help you form a hypothesis as to why more craters are present on the far side of the Moon than on the side facing Earth.

Performance Assessment

21. Poem Write a poem in which you describe the various surface features of the Moon. Be sure to include information on how these features formed. Share your poem with others.

TECHNOLOGY

Go to the Glencoe Science Web site at **science.glencoe.com** or use the **Glencoe Science CD-ROM** for additional chapter assessment.

THE PRINCETON REVIEW **Test Practice**

The diagram below shows one complete lunar cycle.

1   **2**   **3**   **4**   **5**

Study the picture and answer the following questions.

1. About how long does it take the Moon to complete the cycle shown above?
  A) one hour
  B) one day
  C) one month
  D) one year

2. Which phase of the Moon shown above does not reflect any sunlight toward Earth?
  A) 1
  B) 2
  C) 3
  D) 4

3. At which phase could a solar eclipse occur?
  A) 1
  B) 2
  C) 3
  D) 4

4. Which of the above phases is first quarter?
  A) 1
  B) 2
  C) 3
  D) 4

The Solar System

Did you know that some of the brightest objects in the night sky are not stars at all but other planets in the solar system? A planetarium, such as New York's Hayden Planetarium pictured here, is a place where people can go to learn about objects in space. In this chapter, you will learn about the solar system's planets and how they are being explored. You also will learn about other objects such as comets, meteoroids, and asteroids.

What do you think?

Science Journal Look at the picture below with a classmate. Discuss what you think this might be or what is happening. Here's a hint: *Death sometimes can foster new birth.* Write down your answer or best guess in your Science Journal.

EXPLORE ACTIVITY

The planets of the solar system are like neighbors in space, but to humans on Earth, they look like tiny points of light among the thousands of others visible on a clear night. With the help of telescopes and space probes, the points of light become giant spheres, some with rings and moons and others pitted with countless craters. In this activity, you'll explore how craters are formed on the surfaces of planets and moons.

Model crater formation

1. Place white flour into a metal cake pan to a depth of 3 cm, completely covering the bottom of the pan.

2. Cover the flour with 1 cm of colored powdered drink mix or different colors of gelatin powder.

3. From different heights ranging from 10 cm to 25 cm, drop various-sized objects into the pan. Use marbles, bolts, and nuts.

Observe

In your Science Journal, make a drawing that shows what happened to the surface of the powder in the pan when each object was dropped from different heights.

Before You Read

FOLDABLES
Reading & Study Skills

Making a Compare and Contrast Study Fold As you study this chapter, use this Foldable to compare and contrast inner planets and outer planets. When you compare two things, you say how they are similar. When you contrast two things, you say how they are different.

INNER PLANETS

OUTER PLANETS

1. Place a sheet of paper in front of you so the long side is at the top. Fold the paper in half from the left side to the right side and then unfold.

2. Fold each side in to the centerfold line to divide the paper into fourths.

3. Write "Inner Planets" on one flap and "Outer Planets" on the other.

4. On the back of each flap, contrast inner planets and outer planets.

5. Under the flaps in the center section, compare inner planets and outer planets.

The Solar System

As You Read

What **You'll Learn**

■ **Compare** the Sun-centered and Earth-centered models of the solar system.
■ **Describe** current models of the formation of the solar system.

Vocabulary
solar system

Why **It's Important**
The solar system is your neighborhood in space.

Ideas About the Solar System

On a clear night, gazing at the sky can be an awe-inspiring experience. Early observers who noted the changing positions of the planets presented differing ideas about the solar system based on their observations and beliefs. Today, people know that the Sun and the stars only appear to move through the sky because Earth is moving. This wasn't always an accepted fact.

Earth-Centered Model Many early Greek scientists thought the planets, the Sun, and the Moon were fixed in separate spheres that rotated around Earth. The stars were thought to be fixed in another sphere that also rotated around Earth.

This is called the Earth-centered model of the solar system. It included Earth, the Moon, the Sun, five planets—Mercury, Venus, Mars, Jupiter, and Saturn—and the sphere of stars.

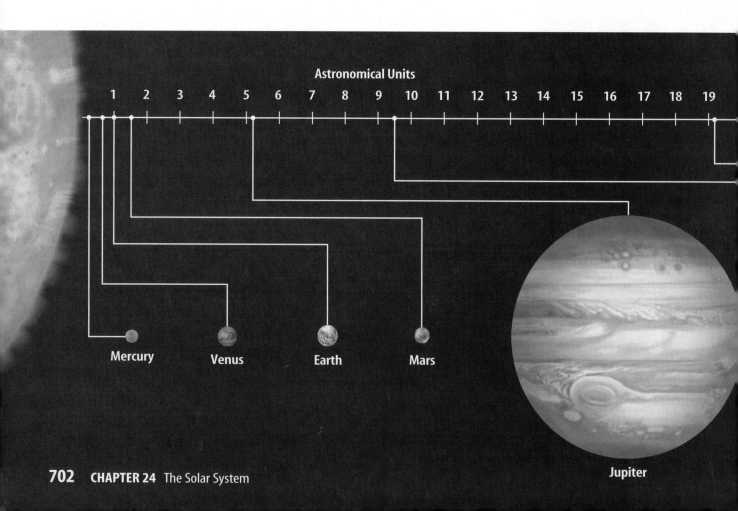

Astronomical Units

1 2 3 4 5 6 7 8 9 10 11 12 13 14 15 16 17 18 19

Mercury   Venus   Earth   Mars

Jupiter

Sun-Centered Model

People believed the idea of an Earth-centered solar system for centuries. Then in 1543, Polish astronomer Nicholas Copernicus published a different view.

Copernicus stated that the Moon revolved around Earth and that Earth and the other planets revolved around the Sun. He also stated that the daily movement of the planets and the stars was due to Earth's rotation. This is the Sun-centered model of the solar system.

Using his telescope, Italian astronomer Galileo Galilei observed that Venus went through a full cycle of phases like the Moon's, which could be explained only if Venus were orbiting the Sun. From this, he concluded that Venus revolves around the Sun and that the Sun is the center of the solar system.

Modern View of the Solar System

We now know that the **solar system** is made up of nine planets, including Earth, and many smaller objects that orbit the Sun. The nine planets and the Sun are shown in **Figure 1.** Notice how small Earth is compared with some of the other planets and the Sun.

The solar system includes a huge volume of space that stretches in all directions from the Sun. The Sun contains 99.86 percent of the mass of the solar system. Because of its gravitational pull, the Sun is the central object in the solar system. All other objects in the solar system revolve around the Sun.

SCIENCE *Online*

Research Visit the Glencoe Science Web site at **science.glencoe.com** for more information about the Solar System. Communicate to your class what you learn.

Figure 1
Each of the nine planets in the solar system is unique. The distances between the planets and the Sun are shown on the scale. One astronomical unit (AU) is the average distance between Earth and the Sun.

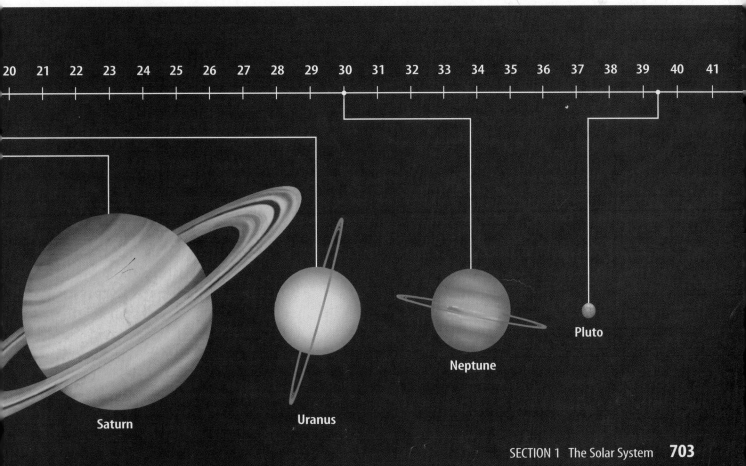

Saturn   Uranus   Neptune   Pluto

In 1915, Albert Einstein developed a theory that explains why gravity exists. According to his theory, a mass makes a dent in space, just as a heavy object dents a rubber sheet. A ball placed on the sheet tends to fall into the dent, just as an object tends to fall into a dent in space. This tendency to fall toward the dent is what is called gravity. Do research to find out more about Einstein and his work.

How the Solar System Formed

Scientists hypothesize that the solar system formed from part of a nebula of gas, ice, and dust, like the one shown in **Figure 2,** about 4.6 billion years ago. Follow the steps shown in **Figures 3A** through **3D,** which illustrate how this might have happened. A cloud of material in this nebula was rotating slowly in space. A nearby star might have exploded, and the shock waves from this event could have caused the cloud to start contracting. As it contracted, the matter in the cloud was squeezed into less space. The cloud's density became greater, and the attraction of gravity pulled more gas and dust toward the cloud center. This caused the cloud to rotate faster, which in turn caused it to flatten into a disk with a dense center.

As the cloud contracted, its temperature began to increase. Eventually, the temperature in the core of the cloud reached about 10 million degrees Celsius and nuclear fusion began. A star was born—the beginning of the Sun.

Nuclear fusion occurs when atoms with low mass, such as hydrogen, combine to form heavier elements, such as helium. The new, heavier element contains slightly less mass than the sum of the lighter atoms that formed it. The "lost" mass is converted into energy.

Reading Check *At what temperature does nuclear fusion begin?*

Planet Formation Not all of the nearby gas, ice, and dust was drawn into the core of the cloud. The matter that did not get pulled into the cloud's center collided and stuck together to form the planets and asteroids. Close to the Sun, the temperature was hot, and the easily vaporized elements could not condense into solids. This is why lighter elements are scarcer in the planets near the Sun than in planets farther out in the solar system.

The inner planets of the solar system—Mercury, Venus, Earth, and Mars—are small, rocky planets with iron cores. The outer planets are Jupiter, Saturn, Uranus, Neptune, and Pluto. Pluto, a small planet, is the only outer planet made mostly of rock and ice. The other outer planets are much larger and are made mostly of lighter substances such as hydrogen, helium, methane, and ammonia.

Figure 2
Systems of planets such as the solar system form in areas of space like this, called a nebula.

Figure 3

Through careful observations, astronomers have found clues that help explain how the solar system may have formed. **A** About 4.6 billion years ago, the solar system was a vast, swirling cloud of gas, ice, and dust. **B** Gradually, part of the nebula contracted into a large, tightly packed, spinning disk. The disk's center was so hot and dense that nuclear fusion reactions began to occur, and the Sun was born. **C** Eventually, the rest of the material in the disk cooled enough to clump into scattered solids. **D** Finally, these clumps collided and combined to become the nine planets that make up the solar system today.

Table 1 Average Orbital Speed	
Planet	Average Orbital Speed (km/s)
Mercury	48
Venus	35
Earth	30
Mars	24
Jupiter	13
Saturn	9.7
Uranus	6.8
Neptune	5.4
Pluto	4.7

Motions of the Planets

When Nicholas Copernicus developed his Sun-centered model of the solar system, he thought that the planets orbited the Sun in circles. In the early 1600s, German mathematician Johannes Kepler began studying the orbits of the planets. He discovered that the shapes of the orbits are not circular. They are oval shaped, or elliptical.

His calculations further showed that the Sun is not at the center of the orbits but is slightly offset.

Kepler also discovered that the planets travel at different speeds in their orbits around the Sun, as shown in **Table 1.** By studying these speeds, you can see that the planets closer to the Sun travel faster than planets farther away from the Sun. Because of their slower speeds and the longer distance they must travel, the outer planets take much longer to orbit the Sun than the inner planets do.

Johannes Kepler

Copernicus's ideas, considered radical at the time, led to the birth of modern astronomy. Early scientists didn't have technology such as computers and space probes to perform rapid calculations and learn about the planets. Nevertheless, they developed theories about the solar system that still are used today.

Section ① Assessment

1. Describe the Sun-centered model of the solar system.

2. How do most scientists hypothesize that the solar system formed?

3. The outer planets are rich in water, methane, and ammonia—the materials needed for life. Yet life is unlikely on these planets. Explain.

4. Why do the outer planets take longer to orbit the Sun than the inner planets do?

5. **Think Critically** Would a year on the planet Neptune be longer or shorter than an Earth year? Explain.

Skill Builder Activities

6. **Concept Mapping** Make a concept map that compares and contrasts the Earth-centered model with the Sun-centered model of the solar system. **For more help, refer to the** Science Skill Handbook.

7. **Solving One-Step Equations** Use the average-orbital-speed data in **Table 1** to determine how much faster Mercury travels in its orbit than Earth travels in its orbit. How much faster does Mars travel than Neptune? How much faster does Earth travel than Pluto? **For more help, refer to the** Math Skill Handbook.

Activity

Planetary Orbits

Planets travel around the Sun along fixed paths called orbits. As you construct a model of a planetary orbit, you will observe that the shape of planetary orbits is an ellipse.

What You'll Investigate
How can you model planetary orbits?

Materials
thumbtacks or pins (2)   metric ruler
cardboard (23 cm × 30 cm)   string (25 cm)
paper (21.5 cm × 28 cm)   pencil

Goals
■ **Model** planetary orbits.
■ **Calculate** changes in ellipses.

Safety Precautions

Procedure

1. Place a blank sheet of paper on top of the cardboard and insert two thumbtacks or pins about 3 cm apart.

2. Tie the string into a circle with a circumference of 15 cm to 20 cm. Loop the string around the thumbtacks. With someone holding the tacks or pins, place your pencil inside the loop and pull it tight.

3. Moving the pencil around the tacks and keeping the string tight, mark a line until you have completed a smooth, closed curve, called an ellipse.

4. Repeat steps 1 through 3 several times. First, vary the distance between the tacks, then vary the length of the string. However, change only one of these each time. Make a data table to record the changes in the sizes and shapes of the ellipses.

5. Orbits usually are described in terms of eccentricity, e. The eccentricity of any ellipse is determined by dividing the distance, d, between the foci (fixed points—here, the tacks) by the length, l, of the major axis. See the diagram below.

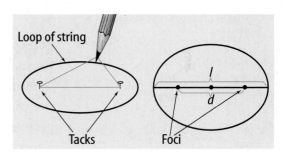

Loop of string

Tacks   Foci

6. **Calculate** and record the eccentricity of the ellipses that you constructed.

7. **Research** the eccentricities of planetary orbits. Construct an ellipse with the same eccentricity as Earth's orbit.

Conclude and Apply

1. **Analyze** the effect that a change in the length of the string or the distance between the tacks has on the shape of the ellipse.

2. **Hypothesize** what must be done to the string or placement of tacks to decrease the eccentricity of a constructed ellipse.

3. **Describe** the shape of Earth's orbit. Where is the Sun located within the orbit?

*C*ommunicating
Your Data

Compare your results with those of other students. **For more help, refer to the** Science Skill Handbook.

The Inner Planets

As You Read

What You'll Learn

- **List** the inner planets in their relative order from the Sun.
- **Describe** important characteristics of each inner planet.
- **Compare and contrast** Venus and Earth.

Vocabulary

Mercury   Earth
Venus   Mars

Why It's Important

The planet you live on is uniquely capable of sustaining life.

Inner Planets

Today, people know more about the solar system than ever before. Better telescopes allow astronomers to observe the planets from Earth and space. In addition, space probes have explored much of the solar system. Prepare to take a tour of the solar system through the eyes of some space probes.

Mercury The closest planet to the Sun is **Mercury.** It is also the second-smallest planet. The first American spacecraft mission to Mercury was in 1974–1975 by *Mariner 10.* The spacecraft flew by the planet and sent pictures back to Earth. *Mariner 10* photographed only 45 percent of Mercury's surface, so scientists don't know what the other 55 percent looks like. What they do know is that the surface of Mercury has many craters and looks much like Earth's Moon. It also has cliffs as high as 3 km on its surface. These cliffs may have formed at a time when Mercury apparently shrank in diameter, as seen in **Figure 4.**

Why would Mercury have shrunk? *Mariner 10* detected a weak magnetic field around Mercury. This indicates that the planet has an iron core. Some scientists hypothesize that the crust of Mercury solidified while the iron core was still hot and molten.

As the core cooled and solidified, it contracted. The cliffs might have resulted from breaks in the crust caused by this contraction.

Figure 4
Large cliffs on Mercury might have formed when the crust of the planet broke as the planet contracted.

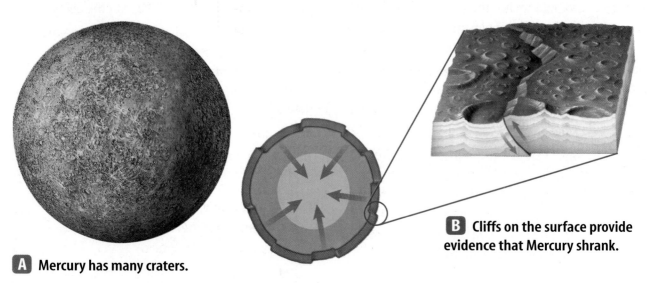

A Mercury has many craters.

B Cliffs on the surface provide evidence that Mercury shrank.

Does Mercury have an atmosphere? Because of Mercury's small size and low gravitational pull, most gases that could form an atmosphere escape into space. *Mariner 10* found traces of gases that were first thought to be an atmosphere. However, these gases are now known to be temporarily trapped hydrogen and helium from the solar wind. Mercury traps these gases and holds them for just a few weeks.

Earth-based observations have also found traces of sodium and potassium around Mercury. Scientists think that these atoms come from rocks in the planet's crust. Therefore, Mercury has no true atmosphere. This lack of atmosphere and the nearness of Mercury to the Sun cause this planet to have great extremes in temperature. Mercury's surface temperature can reach 425°C during the day and it can drop to −170°C at night.

Venus The second planet from the Sun is **Venus,** shown in **Figure 5A.** Venus is sometimes called Earth's twin because its size and mass are similar to Earth's. In 1962, *Mariner 2* flew within 34,400 km of Venus and sent back information about Venus's atmosphere and rotation. The former Soviet Union landed the first probe on the surface of Venus in 1970. *Venera 7,* however, stopped working in less than an hour because of the high temperature and pressure. Additional *Venera* probes photographed and mapped the surface of Venus using cameras and radar. Between 1990 and 1994, the U.S. *Magellan* probe used its radar to make the most detailed maps yet of Venus's surface. It collected radar images of 98 percent of Venus's surface. Notice the huge volcano with visible lava flows shown in **Figure 5B.**

Clouds on Venus are so dense that only a small percentage of the sunlight that strikes the top of the clouds reaches the planet's surface. Much of the solar energy that does reach the surface is trapped by carbon dioxide gas in Venus's atmosphere. This causes a greenhouse effect similar to, but more intense than, Earth's greenhouse effect. Due to this intense greenhouse effect, the temperature on the surface of Venus is between 450°C and 475°C.

Figure 5
A This radar image of Venus's surface was made from data acquired by *Magellan*. **B** Maat Mons is the highest volcano on Venus. Lava flows extend for hundreds of kilometers across the plains.

Earth **Figure 6** shows **Earth,** the third planet from the Sun. The average distance from Earth to the Sun is 150 million km, or one astronomical unit (AU). Unlike other planets, surface temperatures on Earth allow water to exist as a solid, liquid, and gas. Earth's atmosphere causes most meteors to burn up before they reach the surface, and it protects life-forms from the effects of the Sun's intense radiation.

Mars Look at **Figure 7A.** Can you guess why **Mars,** the fourth planet from the Sun, has been called the red planet? Iron oxide in the weathered rocks on its surface gives it a reddish-yellow color. Other features of Mars visible from Earth are its polar ice caps and changes in the coloring of the planet's surface. The ice caps are made mostly of frozen carbon dioxide and frozen water.

Most of the information scientists have about Mars came from *Mariner 9,* the *Viking* probes, *Mars Global Surveyor,* and *Mars Pathfinder. Mariner 9* orbited Mars in 1971 and 1972. It revealed long channels on the planet that might have been carved by flowing water. *Mariner 9* also discovered the largest volcano in the solar system, Olympus Mons, shown in **Figure 7B.** Olympus Mons is probably extinct. Large rift valleys that formed in the Martian crust also were discovered. One such valley, Valles Marineris, is shown in **Figure 7C.**

Figure 6
More than 70 percent of Earth's surface is covered by liquid water. *What is unique about surface temperatures on Earth?*

Figure 7
Many features on Mars are similar to those on Earth.

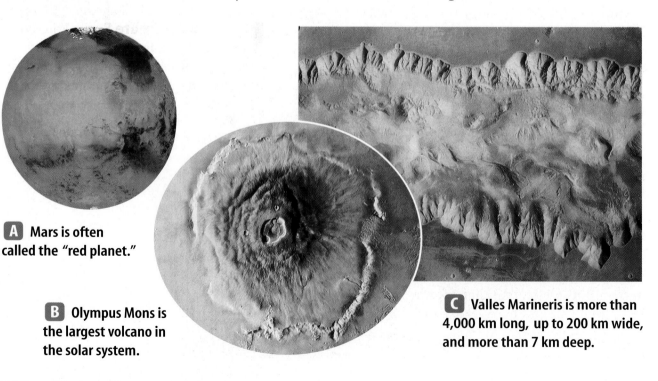

A Mars is often called the "red planet."

B Olympus Mons is the largest volcano in the solar system.

C Valles Marineris is more than 4,000 km long, up to 200 km wide, and more than 7 km deep.

The Viking Probes The *Viking 1* and *2* probes arrived at Mars in 1976. Each spacecraft consisted of an orbiter and a lander. The *Viking 1* and *2* orbiters photographed the entire surface of Mars from their orbits, while the *Viking 1* and *2* landers touched down on the planet's surface. The landers carried equipment to detect possible life on Mars. Some of this equipment was designed to analyze gases given off by Martian soil. These experiments found no conclusive evidence of gaseous life by-products coming from the soil. The *Viking* landers also sent back pictures of a reddish-colored, windswept surface.

Global Surveyor and Pathfinder The *Mars Pathfinder* carried a robot rover named Sojourner with equipment that allowed it to analyze samples of Martian rock and soil. Data from these tests indicated that iron in Mars's crust might have been leached out by groundwater. *Mars Pathfinder* also gathered data on Mars's weather. In addition, cameras onboard *Global Surveyor* showed features that looked like gullies formed by flowing water and deposits of soil and rock carried by water flows. The features, shown in **Figure 8,** are young enough that scientists are considering the idea that liquid groundwater may exist on Mars and that it sometimes reaches the surface. It is also possible that recent volcanic activity could have melted frost beneath the Martian surface. The features compare to those left by flash floods on Earth, such as on Mount St. Helens. Scientists also believe that water is frozen into Mars's crust at the poles.

Reading Check *What evidence indicates that Mars has water?*

Mini LAB

Inferring Effects of Gravity

Procedure
1. Suppose you are a crane operator who is sent to Mars to help build a Mars colony.
2. You know that your crane can lift 44,500 N on Earth and Mars, but the gravity on Mars is only 40 percent of Earth's gravity.
3. Determine how much mass your crane could lift on Mars.

Analysis
1. How can what you have discovered be an advantage over construction on Earth?
2. How might construction advantages change the overall design of the Mars colony?

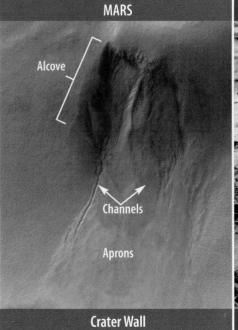

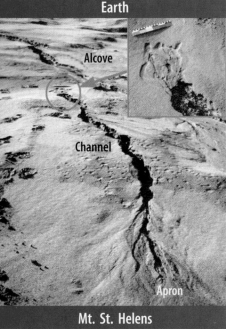

Figure 8
Compare the features found on Mars with those found on an area of Mount St. Helens in Washington state that experienced a flash flood.

Mars's Atmosphere The *Viking* and *Global Surveyor* probes analyzed gases in the Martian atmosphere and determined atmospheric pressure and temperature. They found that Mars's atmosphere is much thinner than Earth's. It is composed mostly of carbon dioxide, with some nitrogen and argon. Surface temperatures range from −125°C to 35°C. The temperature difference between day and night results in strong winds on the planet, which can cause global dust storms during certain seasons. This information will help in planning possible human exploration of Mars in the future.

Martian Seasons Mars is tilted on its axis by 25°, which is close to Earth's tilt of 23.5°. Because of this, Mars goes through seasons as it orbits the Sun, just like Earth does. The polar ice caps get larger during the Martian winter as ice collects on their surface. The ice caps shrink during the summer. As one ice cap shrinks, the other expands, and both their surfaces change color during different seasons. Wind causes this seasonal change in the coloration of the Martian surface. When the seasons change, winds blow the dust around on the planet's surface. As dust blows off one area, it might look darker.

Math Skills Activity

Calculating with Percentages

Example Problem

The diameter of Earth is 12,756 km. The diameter of Mars is 53.3 percent of the diameter of Earth. Calculate the diameter of Mars.

Solution

1 *This is what you know:* diameter of Earth: 12,756 km
percent of Earth's diameter: 53.3%
decimal equivalent: 0.533 (53.3% ÷ 100)

2 *This is what you need to find:* diameter of Mars

3 *This is the equation you need to use:* (diameter of Earth) × (decimal equivalent)
= diameter of Mars

4 *Solve the equation for the diameter of Mars:* (12,756 km) × (0.533) = 6,799 km

Practice Problem

Use the same procedure to calculate the diameter of Venus. Its diameter is 94.9 percent of the diameter of Earth.

For more help, refer to the Math Skill Handbook.

Martian Moons Mars has two small, irregularly shaped moons that are heavily cratered. Phobos, shown in **Figure 9,** is about 25 km in length, and Deimos is about 13 km in length. Deimos orbits Mars once every 31 h, while Phobos speeds around Mars once every 7 h.

Phobos has grooves on its surface that seem to radiate out in all directions from the giant Stickney Crater. Some of the grooves are 700 m across and 90 m deep. Phobos's orbit is spiraling slowly inward toward Mars. It is expected to crash into the Martian surface in about 50 million years.

Deimos is the outer of Mars's two moons. It is among the smallest known moons in the solar system. Its surface is smoother in appearance than that of Phobos because some of its craters have partially filled with soil and rock.

As you toured the inner planets through the eyes of the space probes, you saw how each planet is unique. Refer to **Table 3** following Section 3 for a summary of the planets. Mercury, Venus, Earth, and Mars are different from the outer planets, which you'll explore in the next section.

Figure 9
Phobos orbits Mars once every 7 h. *Why does Phobos have so many craters?*

Section 2 Assessment

1. How are Mercury and Earth's Moon similar?
2. List one important characteristic of each inner planet.
3. Although Venus often is called Earth's twin, why would life as you know it be unlikely on Venus?
4. Name the inner planets in order from the Sun.
5. **Think Critically** Do the closest planets to the Sun always have the hottest surface temperatures? Explain.

Skill Builder Activities

6. **Interpreting Data** Using the information in this section, explain how Mars is like Earth. How are they different? **For more help, refer to the** Science Skill Handbook.
7. **Communicating** Use textbooks and NASA materials to investigate NASA's missions to Mars. In your Science Journal, report on the possibility of life on Mars and the tests that have been conducted to see whether life-forms exist. **For more help, refer to the** Science Skill Handbook.

3 The Outer Planets

As You Read

What You'll Learn

- **Describe** the major characteristics of Jupiter, Saturn, Uranus, and Neptune.
- **Explain** how Pluto differs from the other outer planets.

Vocabulary

Jupiter   Uranus
Great Red Spot   Neptune
Saturn   Pluto

Why It's Important

Studying the outer planets might help scientists better understand Earth.

Outer Planets

You may have heard about the *Voyager* and *Galileo* space-crafts. They were not the first probes to the outer planets, but they gathered a lot of new information about them. Follow the spacecrafts as you read about their journeys to the outer planets.

Jupiter In 1979, *Voyager 1* and *Voyager 2* flew past **Jupiter,** the largest planet and the fifth planet from the Sun. *Galileo* reached Jupiter in 1995. The spacecrafts gathered new information about Jupiter's atmosphere and discovered three new moons. *Voyager* probes also revealed that Jupiter has faint dust rings around it and that one of its moons has volcanoes on it.

Jupiter's Atmosphere Jupiter is composed mostly of hydrogen and helium, with some ammonia, methane, and water vapor. Scientists hypothesize that the atmosphere of hydrogen and helium gradually changes to a planetwide ocean of liquid hydrogen and helium toward the middle of the planet. Below this liquid layer might be a solid rocky core. The extreme pressure and temperature, however, would make the core different from any rock on Earth.

You've probably seen pictures from the probes of Jupiter's colorful clouds. In **Figure 10,** you can see bands of white, red, tan, and brown clouds in its atmosphere. Continuous storms of swirling, high-pressure gas have been observed on Jupiter. The **Great Red Spot** is the most spectacular of these storms. Lightning also has been observed within Jupiter's clouds.

Figure 10
A Jupiter is the largest planet in the solar system, containing more mass than all of the other planets combined. **B** The Great Red Spot is a giant storm about 25,000 km in size from east to west.

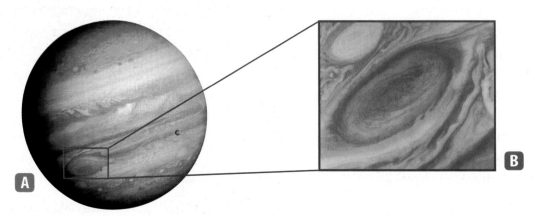

Table 2 Large Moons of Jupiter

	Io The most volcanically active object in the solar system; sulfurous compounds give it its distinctive reddish and orange colors; has a thin oxygen, sulfur, and sulfur dioxide atmosphere.
	Europa Rocky interior is covered by a 100-km-thick crust of ice, which has a network of cracks, indicating tectonic activity; an ocean might exist under the ice crust; has a thin oxygen atmosphere.
	Ganymede Has a crust of ice about 100 km thick, covered with grooves; crust might surround an ocean of water or slushy ice; has a rocky core and a thin oxygen atmosphere.
	Callisto Has a heavily cratered crust of ice and rock several hundred kilometers thick; crust might surround a salty ocean around a rock core; has a thin atmosphere of carbon dioxide.

Moons of Jupiter At least 28 moons orbit Jupiter. In 1610, the astronomer Galileo Galilei was the first person to see Jupiter's four largest moons, shown in **Table 2.** Io (I oh) is the closest large moon to Jupiter. Jupiter's tremendous gravitational force and the gravity of Europa, Jupiter's next large moon, pull on Io. This force heats up Io, causing it to be the most volcanically active object in the solar system. You can see a volcano erupting on Io in **Figure 11.** Europa is composed mostly of rock with a thick, smooth crust of ice. Under the ice might be an ocean as deep as 50 km. If this ocean of water exists, it will be the only place in the solar system, other than Earth and possibly Ganymede and Callisto, where liquid water exists in large quantities. Next is Ganymede, the largest moon in the solar system—larger even than the planet Mercury. Callisto, the last of Jupiter's large moons, is composed mostly of ice and rock. Studying these moons adds to knowledge about the origin of Earth and the rest of the solar system.

Figure 11
Voyager 2 photographed the eruption of this volcano on Io in July 1979.

Figure 12
Saturn's rings are composed of pieces of rock and ice.

Saturn The *Voyager* probes next surveyed Saturn in 1980 and 1981. **Saturn** is the sixth planet from the Sun. It is the second-largest planet in the solar system, but it has the lowest density. Its density is so low that the planet would float in water.

Saturn's Atmosphere Similar to Jupiter, Saturn is a large, gaseous planet. It has a thick outer atmosphere composed mostly of hydrogen and helium. Saturn's atmosphere also contains ammonia, methane, and water vapor. As you go deeper into Saturn's atmosphere, the gases gradually change to liquid hydrogen and helium. Below its atmosphere and liquid layer, Saturn might have a small, rocky core.

Rings and Moons The *Voyager* probes gathered new information about Saturn's ring system and its moons. The probes showed that Saturn has several broad rings. Each large ring is composed of thousands of thin ringlets. **Figure 12** shows that they are composed of countless ice and rock particles. These particles range in size from a speck of dust to tens of meters across. This makes Saturn's ring system the most complex of all the outer gaseous planets.

At least 30 moons orbit Saturn. The latest moons to be discovered orbiting Saturn were found using telescopes in Chile and Hawaii. Some scientists suggest that Saturn's gravity captured several of these moons as they passed nearby. The largest of Saturn's moons, Titan, is larger than the planet Mercury. It has an atmosphere of nitrogen, argon, and methane. Thick clouds prevent scientists from seeing the surface of Titan.

Uranus Beyond Saturn, Voyager 2 flew by Uranus in 1986. Uranus (YOOR uh nus) is the seventh planet from the Sun and was discovered in 1781. It is a large, gaseous planet with 21 satellites and a system of thin, dark rings. Three of its satellites were identified in 1999 from Earth-based observations. Most of the rings of Uranus are less than 10 km across. Two of Uranus's moons, Cordelia and Ophelia, keep the outer ring from dispersing into space.

Uranus's Characteristics The atmosphere of Uranus is composed of hydrogen, helium, and some methane. Methane gives the planet the bluish-green color that you see in **Figure 13.** Methane absorbs the red and yellow light, and the clouds reflect the green and blue. Few cloud bands and storm systems can be seen on Uranus. Evidence suggests that under its atmosphere, Uranus has a mantle of liquid and solid water, methane, and ammonia surrounding a rocky core.

Figure 14 shows one of the most unusual features of Uranus. Its axis of rotation is tilted on its side compared with the other planets. The axes of rotation of the other planets, except Pluto, are nearly perpendicular to the planes of their orbits. However, Uranus's axis of rotation is nearly parallel to the plane of its orbit. Some scientists believe a collision with another object tipped Uranus on its side.

Figure 13
The atmosphere of Uranus gives the planet its distinct bluish-green color.

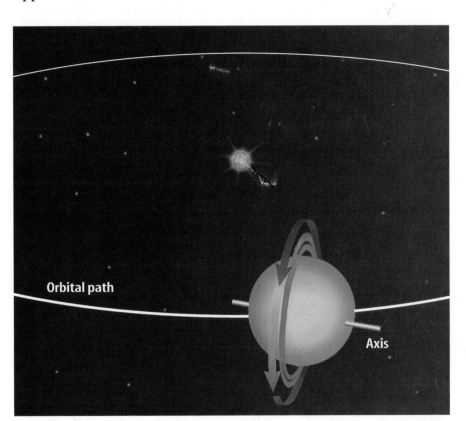

Orbital path

Axis

Figure 14
Uranus rotates on an axis nearly parallel to the plane of its orbit. During its revolution around the Sun, one of the poles, at times, points almost directly at the Sun.

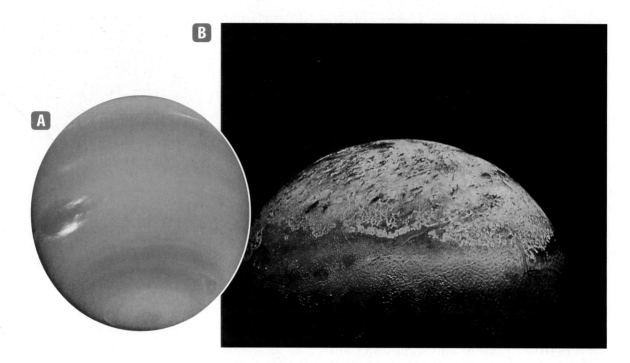

Figure 15

A Neptune has a distinctive bluish-green color. **B** The pinkish hue of Neptune's largest moon, Triton, is thought to come from an evaporating layer of nitrogen ice.

Physics
INTEGRATION

All change requires energy. On Earth energy from the Sun powers hurricanes. However, hurricanes on Neptune probably get most of their energy from Neptune's interior. Do research and write a short paragraph comparing and contrasting hurricanes on Earth and storms on Neptune.

Neptune Passing Uranus, *Voyager 2* traveled to Neptune, another large, gaseous planet. Discovered in 1846, **Neptune** is usually the eighth planet from the Sun. However, Pluto's orbit crosses inside Neptune's during part of its voyage around the Sun. Between 1979 and 1999, Pluto was closer to the Sun than was Neptune. Early in 1999, Pluto once again became the farthest planet from the Sun.

Neptune's Characteristics Neptune's atmosphere is similar to Uranus's atmosphere. The methane content gives Neptune, shown in **Figure 15A,** its distinctive bluish-green color, just as it does for Uranus.

✔ **Reading Check** *What gives Neptune its bluish-green color?*

Neptune has dark-colored storms in its atmosphere that are similar to the Great Red Spot on Jupiter. One discovered by *Voyager 2* in 1989 was called the Great Dark Spot. It was about the size of Earth. However, observations by the *Hubble Space Telescope* in 1994 showed that the Great Dark Spot had disappeared. Bright clouds also form, then disappear in several hours. This shows that Neptune's atmosphere is active and changes rapidly.

Under its atmosphere, Neptune is thought to have a layer of liquid water, methane, and ammonia that might change to solid ice. Neptune probably has a rocky core.

Voyager 2 detected six new moons, so the total number of Neptune's known moons is now eight. Triton, shown in **Figure 15B,** is the largest. It has a thin atmosphere composed mostly of nitrogen.

Pluto The smallest planet in the solar system, and the one scientists know the least about, is Pluto. For 20 years of its 249-year orbit, Pluto is closer to the Sun than Neptune. However, because **Pluto** is farther from the Sun than Neptune during most of its orbit, it is considered to be the ninth planet from the Sun. From Pluto's surface, the Sun would appear as only a bright star. Pluto is vastly different from the other outer planets. It's surrounded by only a thin atmosphere, and it's the only outer planet with a solid, icy-rock surface.

Pluto's Moon Pluto's single moon, Charon, has a diameter about half the size of Pluto's. It was discovered in 1978 when a bulge was noticed on a photograph of the planet. Later photographs, taken with improved telescopes, showed that the bulge was a moon. Pluto and Charon are shown in **Figure 16.** Because of their close size and orbit, some scientists consider them to be a double planet.

Data from the *Hubble Space Telescope* indicate the presence of a vast disk of icy comets called the Kuiper Belt near Neptune's orbit. Some of the comets are hundreds of kilometers in diameter. Are Pluto and Charon members of this belt? Or did they simply form at the distance where they are? Scientists might not find out until a probe is sent to Pluto.

Figure 16
The *Hubble Space Telescope* gave astronomers their first clear view of Pluto and Charon as distinct objects.

Section 3 Assessment

1. Describe the main differences between the outer planets and the inner planets.

2. Are any moons in the solar system larger than any planets? If so, which ones?

3. How does Pluto differ from the other outer planets? How is it similar to the inner planets?

4. Why are Pluto and Charon sometimes considered a double planet?

5. **Think Critically** Some scientists think life could exist on one of Jupiter's moons. On which moons would you look for life? Why?

Skill Builder Activities

6. **Recognizing Cause and Effect** Answer the following questions about Jupiter. **For more help, refer to the Science Skill Handbook.**
   a. How is the Great Red Spot affected by Jupiter's rotation?
   b. How does Jupiter's mass affect its gravitational force?

7. **Using Graphics Software** Use graphing software to plot each planet's average speed against its distance from the Sun. Describe the type of curve that is produced. **For more help, refer to the Technology Skill Handbook.**

Table 3 Planets

Mercury

- closest to the Sun
- second-smallest planet
- surface has many craters and high cliffs
- no atmosphere
- temperatures range from 425°C during the day to −170°C at night
- has no moons

Venus

- similar to Earth in size and mass
- thick atmosphere made mostly of carbon dioxide
- droplets of sulfuric acid in atmosphere give clouds a yellowish color
- surface has craters, faultlike cracks, and volcanoes
- greenhouse effect causes surface temperatures of 450°C to 475°C
- has no moons

Earth

- atmosphere protects life
- surface temperatures allow water to exist as solid, liquid, and gas
- only planet where life is known to exist
- has one large moon

Mars

- surface appears reddish-yellow because of iron oxide in rocks
- ice caps are made of frozen carbon dioxide and water
- channels indicate that water has flowed on the surface; has large volcanoes and valleys
- has a thin atmosphere composed mostly of carbon dioxide
- surface temperatures range from −125°C to 35°C
- huge dust storms often blanket the planet
- has two small moons

Jupiter

- largest planet
- has faint rings
- atmosphere is mostly hydrogen and helium; continuous storms swirl on the planet—the largest is the Great Red Spot
- has four large moons and at least 24 smaller moons; one of its moons, Io, has active volcanoes

Saturn

- second-largest planet
- thick atmosphere is mostly hydrogen and helium
- has a complex ring system
- has at least 30 moons—the largest, Titan, is larger than Mercury

Uranus

- large, gaseous planet with thin, dark rings
- atmosphere is hydrogen, helium, and methane
- axis of rotation is parallel to plane of orbit
- has at least 21 moons

Neptune

- large, gaseous planet with rings that vary in thickness
- is sometimes farther from the Sun than Pluto is
- methane in atmosphere causes its bluish-green color
- has dark-colored storms in atmosphere
- has eight moons

Pluto

- small, icy-rock planet with thin atmosphere
- single moon, Charon, is half the diameter of the planet

Other Objects in the Solar System

As You Read

What You'll Learn

- **Describe** where comets come from and how a comet develops as it approaches the Sun.
- **Distinguish** among comets, meteoroids, and asteroids.

Vocabulary

comet   meteorite
meteor   asteroid

Why It's Important

Comets, meteoroids, and asteroids might be composed of material that formed early in the history of the solar system.

Figure 17
Comet Hale-Bopp was most visible in March and April 1997.

Comets

The planets and their moons are the most noticeable members of the Sun's family, but many other objects also orbit the Sun. Comets, meteoroids, and asteroids are important other objects in the solar system.

You might have heard of Halley's comet. A **comet** is composed of dust and rock particles mixed with frozen water, methane, and ammonia. Halley's comet was last seen from Earth in 1986. English astronomer Edmund Halley realized that comet sightings that had taken place about every 76 years were really sightings of the same comet. This comet, which takes about 76 years to orbit the Sun, was named after him. Halley's comet is just one example of the many other objects in the solar system besides the planets.

Oort Cloud Dutch astronomer Jan Oort proposed the idea that a large collection of comets lies in a cloud that completely surrounds the solar system. This cloud, called the Oort Cloud, is located beyond the orbit of Pluto. Oort suggested that the gravities of the Sun and nearby stars interact with comets in the Oort Cloud. Comets either escape from the solar system or get captured into smaller orbits.

Comet Hale-Bopp On July 23, 1995, two amateur astronomers made an exciting discovery. A new comet, Comet Hale-Bopp, was headed toward the Sun. Larger than most that approach the Sun, it was the brightest comet visible from Earth in 20 years. Shown in **Figure 17**, Comet Hale-Bopp was at its brightest in March and April 1997.

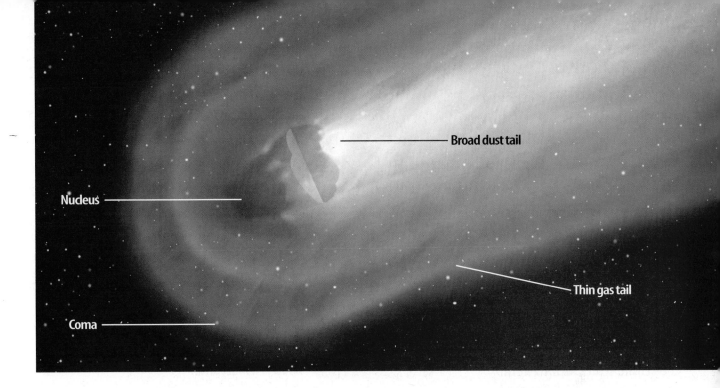

Nucleus

Coma

Broad dust tail

Thin gas tail

Structure of Comets The *Hubble Space Telescope,* satellites orbiting Earth, and spacecrafts such as the *International Cometary Explorer* have all gathered information about comets. Notice the structure of a comet shown in **Figure 18.** It is like a large, dirty snowball or a mass of frozen ice and rock.

As the comet approaches the Sun, it starts to change. Ices of water, methane, and ammonia begin to vaporize because of the heat from the Sun, releasing dust and bits of rock. The vaporized gases and released dust form a bright cloud called a coma around the nucleus, or solid part, of the comet. The solar wind pushes on the gases and released dust in the coma, causing the particles to form tails that always point away from the Sun.

After many trips around the Sun, most of the frozen ice in a comet's nucleus has vaporized. All that's left are the small particles, which are spread throughout the orbit of the original comet.

Meteoroids, Meteors, and Meteorites

You learned that comets tend to vaporize and break up after they have passed close to the Sun many times. The small pieces of the comet's nucleus spread out into a loose group within the original orbit of the comet. These smaller pieces, along with particles derived from other sources, are called meteoroids.

Sometimes the path of a meteoroid crosses the position of Earth, and it enters Earth's atmosphere at speeds of 15 km/s to 70 km/s. Most meteoroids are so small that they completely burn up in Earth's atmosphere. A meteoroid that burns up in Earth's atmosphere is called a **meteor.** People often see meteors like the one in **Figure 19** and call them shooting stars.

Figure 18
A comet consists of a nucleus, a coma, a dust tail, and a gas tail.

Figure 19
A meteoroid that burns up in Earth's atmosphere is called a meteor.

Figure 20
Meteorites occasionally strike Earth's surface. A large meteorite struck Arizona, forming a crater about 1.2 km in diameter and about 200 m deep.

Meteor Showers Each time Earth passes through the loose group of particles within the old orbit of a comet, many small particles of rock and dust enter the atmosphere. Because more meteors than usual are seen, the event is called a meteor shower.

When a meteoroid is large enough, it might not burn up completely in the atmosphere. If it strikes Earth, it is called a **meteorite.** Barringer Crater in Arizona, shown in **Figure 20,** was formed when a large meteorite struck Earth about 50,000 years ago. Most meteorites are probably debris from asteroid collisions or broken-up comets, but some originate from the Moon and Mars.

✔ **Reading Check** *What is a meteorite?*

Asteroids

An **asteroid** is a piece of rock similar to the material that formed into the planets. Most asteroids are located in an area between the orbits of Mars and Jupiter called the asteroid belt. Find the asteroid belt in **Figure 21.** Why are they located there? The gravity of Jupiter might have kept a planet from forming in the area where the asteroid belt is located now.

Other asteroids are scattered throughout the solar system. They might have been thrown out of the belt by Jupiter's gravity. Some may have since been captured as moons around other planets. Some scientists think that Mars's moons are captured asteroids.

Figure 21
The asteroid belt lies between the orbits of Mars and Jupiter.

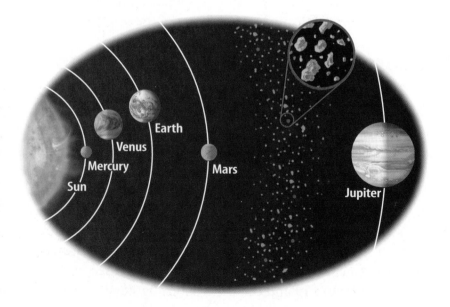

Exploring Asteroids The sizes of the asteroids in the asteroid belt range from tiny particles to objects 940 km in diameter. Ceres is the largest and the first one discovered. The next three in order of size are Vesta (530 km), Pallas (522 km), and 10 Hygiea (430 km). Two asteroids, Ida and Gaspra, shown in **Figure 22,** were photographed by *Galileo* on its way to Jupiter.

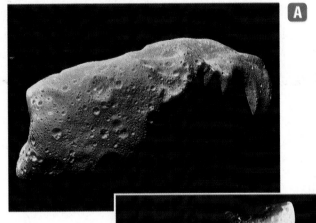

NEAR On February 14, 2000, the *Near Earth Asteroid Rendezvous (NEAR)* spacecraft went into orbit around the asteroid 433 Eros and successfully began its one-year mission of data gathering. Data from the spacecraft show that Eros's surface has a large number of craters. Other data indicate that Eros might be similar to the most common type of meteorite that strikes Earth. On February 12, 2001, *NEAR* ended its mission by becoming the first spacecraft to land softly on an asteroid.

Comets, meteoroids, and asteroids probably are composed of material that formed early in the history of the solar system. Scientists study the structure and composition of these space objects in order to learn what the solar system might have been like long ago. Understanding this could help scientists better understand how Earth formed.

Figure 22
A The asteroid Ida is about 56 km long. **B** Gaspra is about 20 km long.

Section 4 Assessment

1. Why does a comet's tail form as it approaches the Sun?
2. What type of feature might be formed on Earth if a large meteorite reached its surface?
3. Describe differences among comets, meteoroids, and asteroids.
4. What was the mission of the *NEAR* spacecraft?
5. **Think Critically** What is the chemical composition of comets? Are comets more similar to the inner or the outer planets?

Skill Builder Activities

6. **Forming a Hypothesis** A meteorite found in Antarctica in 1979 is thought to have originated on Mars about 16 million years ago. Write a hypothesis to explain how a piece of Mars might have ended up on Earth. **For more help, refer to the** Science Skill Handbook.

7. **Communicating** The asteroid belt contains many objects—from tiny particles to objects 940 km in diameter. In your Science Journal, describe how mining the asteroids for valuable minerals might be accomplished. **For more help, refer to the** Science Skill Handbook.

Solar System Distance Model

Distances between the planets of the solar system are large. Can you design a model that will demonstrate these large distances?

Recognize the Problem

Can a model be designed to show relative distances in the solar system?

Thinking Critically

How can you design a model of a reasonable size that will demonstrate the distances between and among the Sun and planets of the solar system?

Possible Materials

meterstick	string (several meters)
scissors	notebook paper
pencil	(several sheets)

Safety Precautions

Use care when handling scissors.

Data Source

SCIENCE *Online* Go to the Glencoe Science Web site at **science.glencoe.com** to find information about distances in the solar system.

Goals

■ **Design** a table of scale distances and model the distances between and among the Sun and the planets.

Planetary Distances				
Planet	Distance to Sun (km)	Distance to Sun (AU)	Scale Distance (1 AU = 10 cm)	Scale Distance (1 AU = 2 m)
Mercury				
Venus				
Earth				
Mars				
Jupiter				
Saturn				
Uranus				
Neptune				
Pluto				

Test Your Hypothesis

Planning the Model

1. **List** the steps that you need to take in making your model. Be specific, describing exactly what you will do at each step.

2. **List** the materials that you will need to complete your model.

3. **Make a table** of scale distances you will use in your model.

4. **Write** a description of how you will build your model, explaining how it will demonstrate relative distances between and among the Sun and planets of the solar system.

Check the Model Plans

1. **Compare** your scale distances with those of other students. Discuss why each of you chose the scale you did.

2. Make sure your teacher approves your plan before you start.

Making the Model

1. **Construct** the model using your scale distances.

2. While constructing the model, write any observations that you or other members of your group make, and complete the data table in your Science Journal. Calculate the scale distances that would be used in your model if 1 AU = 2 m.

Analyzing and Applying Results

1. **Explain** how a scale distance is determined.

2. Was it possible to work with your scale? Explain why or why not.

3. How much string would be required to construct a model with a scale distance of 1 AU = 2 m?

4. Proxima Centauri, the closest star to the Sun, is about 270,000 AU from the Sun. Based on your scale, how much string would you need to place this star on your model?

Compare your scale model with those of other students. Discuss any differences. **For more help,** refer to the Science Skill Handbook.

IT CAME FROM OUTER

An unexpected visitor crashes into

"Old Kentucky Home"

On September 4, 1990, Frances Pegg had just returned from grocery shopping and was unloading bags of groceries in her kitchen in Burnwell, Kentucky. Suddenly and unexpectedly, she heard a loud crashing sound. Her husband, Arthur, had heard the same sound. Before it, however, he had heard another sound—one that was similar to a noise a helicopter makes. The sound frightened the couple's goat and horse. The noise had come from an object that had crashed through the Peggs' roof, their ceiling, and the floor of their porch. They couldn't see what the object was, but the noise sounded like a gunshot, and pieces of wood from their home flew everywhere.

The Burnwell meteorite crashed into the Peggs' home and landed in their basement on the right.

They thought that perhaps a part of an airplane had fallen off as it flew overhead. The next day the couple looked under their front porch and found the culprit—a chunk of rock from outer space. It was a meteorite!

Frances and Arthur Pegg were not hurt, but they certainly were puzzled.

SPACE!

Actual size

When the Burnwell meteorite entered Earth's atmosphere, it became covered in a shiny black crust. Its mass is about 2 kilograms.

For seven years, the Peggs kept their "space rock" at home, making them local celebrities. The rock appeared on TV, and the couple were interviewed by newspaper reporters. In 1997, the Peggs sold the meteorite to the National Museum of Natural History in Washington, D.C., which has a collection of over 9,000 meteorites. Scientists there study meteorites to learn more about space and the beginnings of the solar system. One astronomer explained, "Meteorites were formed at about the same time as the solar system, about 4.6 billion years ago, though some are younger."

Scientists especially are interested in the Burnwell meteorite because its chemical make up is different from other meteorites previously studied. Meteorites are made of various amounts of minerals and metals, including iron, cobalt, and nickel. The Burnwell meteorite is richer in metallic iron and nickel than other known meteorites and is less rich in some metals such as cobalt. Scientists are comparing the rare Burnwell rock with data from NASA space probes to find out if there are more meteorites like the one that fell on the Peggs' roof. But so far, it seems the Peggs' visitor from outer space is one-of-a-kind.

CONNECTIONS **Research** the latest information on meteorites. How do they give clues to how our solar system was formed? Visit the Glencoe Science Web site. Report to the class.

SCIENCE *Online*

For more information, visit science.glencoe.com

Chapter **24** Study Guide

Section 1 The Solar System

1. Early astronomers thought that the planets, the Moon, the Sun, and the stars were fixed in separate spheres that rotated around Earth.

2. The Sun-centered model of the solar system states that the Sun is the center of the solar system. Earth and the other planets revolve around the Sun.

3. Scientists think the solar system formed from a cloud of gas, ice, and dust about 4.6 billion years ago.

4. Johannes Kepler discovered that the planets orbit the Sun in elliptical, not circular, orbits.

Section 2 The Inner Planets

1. The moonlike planet Mercury has craters and cliffs on its surface.

2. Venus and Earth are similar in size and mass. Venus has a dense atmosphere.

3. On Earth, surface temperatures allow water to exist as a solid, liquid, and gas. Earth's atmosphere protects life-forms from the Sun's radiation.

4. Mars appears to be reddish-yellow due to the iron oxide content of its weathered rocks. Recent studies by *Global Surveyor* indicate that Mars's surface once had large amounts of water flowing over it. *Why do the polar ice caps of Mars, shown here, change size?*

Section 3 The Outer Planets

1. Faint rings and at least 28 moons orbit the gaseous planet Jupiter.

2. Saturn has pronounced rings. *What are Saturn's rings, shown here, made of?*

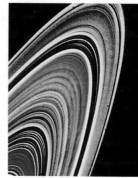

3. Uranus is a large, gaseous planet with many moons and several rings. Neptune is similar to Uranus in size, composition, and storm-like features.

4. Pluto has a surface of icy rock.

Section 4 Other Objects in the Solar System

1. As a comet approaches the Sun, a bright coma of dust and gas forms. *Why do comet tails always point away from the Sun?*

2. Meteoroids and asteroids are relatively small objects within the solar system. Meteoroids form when asteroids collide, when comets break up, or when meteorites collide with the Moon or planets.

FOLDABLES
Reading & Study Skills

After You Read

To help you review the similarities and differences among inner planets and outer planets, use the Foldable you made at the beginning of the chapter.

Visualizing Main Ideas

Complete the following concept map on the solar system.

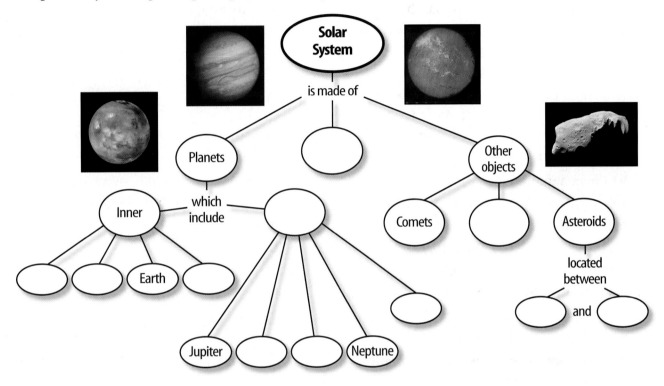

Vocabulary Review

Vocabulary Words

a. asteroid
b. comet
c. Earth
d. Great Red Spot
e. Jupiter
f. Mars
g. Mercury
h. meteor

i. meteorite
j. Neptune
k. Pluto
l. Saturn
m. solar system
n. Uranus
o. Venus

THE PRINCETON REVIEW **Study Tip**

Practice reading tables. See if you can devise a graph that shows the same information that a table does.

Using Vocabulary

Replace the underlined words with the correct vocabulary words.

1. A meteoroid that burns up in Earth's atmosphere is called a <u>meteorite</u>.

2. The axis of rotation of <u>Neptune</u> is tilted on its side compared with the other planets.

3. <u>Uranus</u> is the second-largest planet in the solar system, but it has the lowest density.

4. The *Viking 1* and *2* orbiters photographed the entire surface of <u>Venus</u> from orbit.

5. One of the moons of <u>Mars</u> has volcanoes on it.

Checking Concepts

Choose the word or phrase that best answers the question.

1. Who proposed a Sun-centered solar system?
   A) Ptolemy   **C)** Galileo
   B) Copernicus   **D)** Oort

2. How does the Sun produce energy?
   A) magnetism
   B) nuclear fission
   C) nuclear fusion
   D) the greenhouse effect

3. What is the shape of planetary orbits?
   A) circles   **C)** squares
   B) ellipses   **D)** rectangles

4. Which planet has extreme temperatures because it has no atmosphere?
   A) Earth   **C)** Saturn
   B) Jupiter   **D)** Mercury

5. Water is a solid, liquid, and gas on which planet?
   A) Pluto   **C)** Saturn
   B) Uranus   **D)** Earth

6. Where is the largest known volcano in the solar system?
   A) Earth   **C)** Mars
   B) Jupiter   **D)** Uranus

7. What do scientists call a rock that strikes Earth's surface?
   A) asteroid   **C)** meteorite
   B) comet   **D)** meteoroid

8. Which planet has a complex ring system made of thousands of ringlets?
   A) Pluto   **C)** Uranus
   B) Saturn   **D)** Mars

9. Which planet has a Great Red Spot?
   A) Uranus   **C)** Jupiter
   B) Earth   **D)** Pluto

10. In what direction do comet tails always point?
    A) toward the Sun
    B) away from the Sun
    C) toward Earth
    D) away from the Oort Cloud

Thinking Critically

11. Why is the surface temperature on Venus so much higher than that on Earth?

12. Describe a relationship between a planet's mass and the number of satellites it has.

13. Why are probe landings on Jupiter or Saturn unlikely events?

14. What evidence suggests that water is or once was present on Mars?

15. An observer on Earth can watch Venus go through phases much like Earth's Moon does. Explain.

Developing Skills

16. **Making and Using Graphs** Use the graph below to explain how the time of a planet's revolution is related to its distance from the Sun.

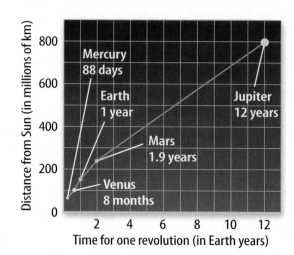

17. **Forming a Hypothesis** Mercury is the closest planet to the Sun, yet it does not reflect much of the Sun's light. What can you say about Mercury's color?

18. **Concept Mapping** Complete the concept map on this page to show how a comet changes as it travels through space.

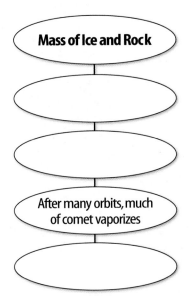

Mass of Ice and Rock

After many orbits, much of comet vaporizes

Performance Assessment

19. **Display** Mercury, Venus, Mars, Jupiter, and Saturn can be observed with the unaided eye. Research where in the sky these planets can be observed in the next year. Construct a display with your findings. Include time of day, day of the year, and locations with respect to known landmarks in your area on your display.

THE PRINCETON REVIEW

Test Practice

The table below presents data about the four inner planets of the solar system.

The Inner Planets			
Planet	Diameter (km)	Distance from the Sun (millions of km)	Period of Revolution (days)
Mercury	4,880	58	88
Venus	12,104	108	225
Earth	12,756	150	365
Mars	6,799	228	687

Study the table and answer the following questions.

1. How many millions of kilometers farther from the Sun is Venus than Mercury?
   - **A)** 150
   - **B)** 108
   - **C)** 58
   - **D)** 50

2. The best conclusion to be made from these data is that _____.
   - **F)** Mercury is a larger planet than Mars
   - **G)** the period of a planet's revolution increases with increasing distance from the Sun
   - **H)** over time, the planets are gradually moving farther and farther away from the Sun
   - **J)** the size of a planet increases with increasing period of revolution

Stars and Galaxies

When you look at stars, do you ever wonder why some stars appear brighter than others do? How do you know how far stars are from Earth and what they are made of? How does the Sun compare to the other stars you see? In this chapter, you will find the answers to these questions. You'll also learn how stars are classified and about the different kinds of galaxies that stars are grouped into. In addition, you'll learn about the Big Bang theory, which most astronomers believe is the most likely way that the universe began.

What do you think?

Science Journal Look at the picture below with a classmate. Discuss what this might be or what is happening. Here's a hint: *When it comes to space, two is sometimes better than one.* Write down your answer or best guess in your Science Journal.

EXPLORE **A**CTIVITY

This photo may look like a scene from the latest science fiction movie, but it shows a real event—two galaxies colliding. However, don't be worried. Most clusters of galaxies are not moving toward each other. They are moving apart, and astronomers have concluded that the universe is expanding in all directions. In the following activity, model how the universe might be expanding.

Model the universe

1. Partially inflate a balloon. Clip the neck shut with a clothespin.

2. Draw six evenly spaced dots on the balloon with a felt-tip marker. Label the dots A through F.

3. Use a string and ruler to measure the distance, in millimeters, from dot A to each of the other dots.

4. Inflate the balloon some more.

5. Measure the distances from dot A again.

6. Inflate the balloon again and take new measurements.

Observe

If each dot represents a cluster of galaxies and the balloon represents the universe, describe in your Science Journal the motion of the clusters relative to one another.

Before You Read

FOLDABLES
Reading & Study Skills

Making a Concept Map Study Fold
A concept map is a diagram that shows how concepts or ideas are related and organizes information. Make this Foldable to show what you already know about stars, galaxies, and the Universe.

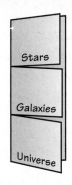

1. Place a sheet of paper in front of you so the short side is at the top. Fold the paper in half from the left side to the right side.

2. Fold the top and bottom in to divide the paper into thirds. Unfold the paper so three columns show.

3. Through the top thickness of paper, cut along each of the fold lines to the left fold, forming three tabs. Label the tabs "Stars," "Galaxies," and "Universe," as shown.

4. Before you read the chapter, write what you already know about stars, galaxies, and the Universe under the tabs.

5. As you read the chapter, add to or correct what you have written under the tabs.

Stars

Constellations

It's fun to look at cloud formations and find ones that remind you of animals, people, or objects that you recognize. It takes much more imagination to play this game with celestial bodies. Ancient Greeks, Romans, and other early cultures observed patterns of stars in the sky called **constellations** and imagined that they represented mythological characters, animals, or familiar objects.

From Earth, a constellation looks like spots of light arranged in a particular shape against the dark night sky. **Figure 1** shows how the constellation of the mythological Greek hunter Orion appears from Earth. It also shows how stars in the constellation have no relationship to each other in space.

Stars in the sky can be found at specific locations within a constellation. For example, you can find the star Betelgeuse (BEE tul jooz) in the shoulder of the mighty hunter Orion. Orion's faithful companion is his dog, Canis Major. Sirius, the brightest star visible from the northern hemisphere, is in Canis Major.

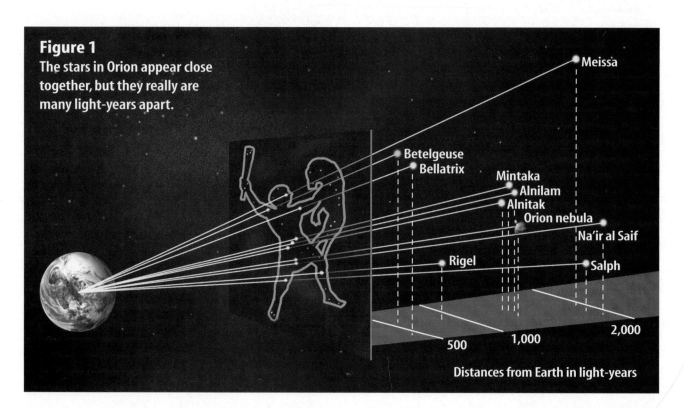

Figure 1
The stars in Orion appear close together, but they really are many light-years apart.

Meissa

Betelgeuse
Bellatrix

Mintaka
Alnilam
Alnitak
Orion nebula
Na'ir al Saif

Rigel
Salph

500   1,000   2,000

Distances from Earth in light-years

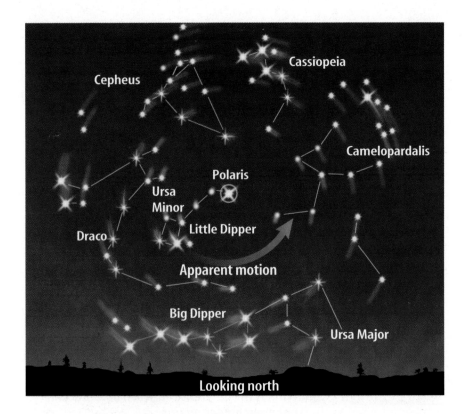

Cepheus **Cassiopeia**

Camelopardalis

Polaris

Ursa Minor

Little Dipper

Draco

Apparent motion

Big Dipper

Ursa Major

Looking north

Figure 2
The Big Dipper, in red, is part of the constellation Ursa Major. It is visible year-round in the northern hemisphere. Constellations close to Polaris rotate around Polaris, which is almost directly over the north pole.

Modern Constellations Modern astronomy divides the sky into 88 constellations, many of which were named by early astronomers. You probably know some of them. Can you recognize the Big Dipper? It's part of the constellation Ursa Major, shown in **Figure 2.** Notice how the front two stars of the Big Dipper point almost directly at the North Star, Polaris, which is located at the end of the Little Dipper in the constellation Ursa Minor. Polaris is positioned almost directly over Earth's north pole.

Circumpolar Constellations As Earth rotates, Ursa Major, Ursa Minor, and other constellations in the northern sky circle around Polaris. Because of this, they are called circumpolar constellations. The constellations appear to move, as shown in **Figure 2,** because Earth is in motion. The stars appear to complete one full circle in the sky in less than 24 h as Earth rotates on its axis. One circumpolar constellation that's easy to find is Cassiopeia (kas ee uh PEE uh). You can look for five bright stars that form a big W or a big M in the northern sky, depending on the season. In spring and summer, Cassiopeia forms an M, and in fall and winter, it forms a W.

As Earth orbits the Sun, different constellations come into view while others disappear. Because of their unique position, circumpolar constellations are visible all year long. Other constellations are not. Orion, which is visible in the winter in the northern hemisphere, can't be seen there in the summer because the daytime side of Earth is facing it.

TRY AT HOME

Mini LAB

Observing Star Patterns

Procedure
1. On a clear night, go outside after dark and study the stars. Take an adult with you and help each other find some common constellations.
2. Let your imagination flow to find patterns of stars that look like something familiar.
3. Draw the stars you see, note their positions, and include a drawing of what you think each star pattern resembles.

Analysis
1. Which of your constellations match those observed by your classmates?
2. How can recognizing star patterns be useful?

Absolute and Apparent Magnitudes

When you look at constellations, you'll notice that some stars are brighter than others. For example, Sirius looks much brighter than Rigel. Is Sirius a brighter star, or is it just closer to Earth, making it appear to be brighter? As it turns out, Sirius is 100 times closer to Earth than Rigel is. If Sirius and Rigel were the same distance from Earth, Rigel would appear much brighter in the night sky than Sirius would.

When you refer to the brightness of a star, you can refer to its absolute magnitude or its apparent magnitude. The **absolute magnitude** of a star is a measure of the amount of light it gives off. A measure of the amount of light received on Earth is called the **apparent magnitude.** A star that's rather dim can appear bright in the sky if it's close to Earth, and a star that's bright can appear dim if it's far away. If two stars are the same distance away, what might cause one of them to be brighter than the other?

Which constellations are visible during different seasons? To find out, see the **Backyard Astronomy Field Guide** at the back of the book.

✔ **Reading Check** *What is the difference between absolute and apparent magnitude?*

Problem-Solving Activity

Are distance and brightness related?

The apparent magnitude of a star is affected by its distance from Earth. This activity will help you determine the relationship between distance and brightness.

Identifying the Problem

Luisa conducted an experiment to determine the relationship between distance and the brightness of stars. She used a meterstick, a light meter, and a lightbulb. She placed the bulb at the zero end of the meterstick, then placed the light meter at the 20-cm mark and recorded the distance and the light-meter reading in her data table. Readings are in luxes, which are units for measuring light intensity. Luisa then increased the distance from the bulb to the light meter and took more readings. By examining the data in the table, can you see a relationship between the two variables?

Effect of Distance on Light	
Distance (cm)	Meter Reading (luxes)
20	4150.0
40	1037.5
60	461.1
80	259.4

Solving the Problem

1. What happened to the amount of light recorded when the distance was increased from 20 cm to 40 cm? When the distance was increased from 20 cm to 60 cm?

2. What does this indicate about the relationship between light intensity and distance? What would the light intensity be at 100 cm? Would making a graph help you visualize the relationship?

Measurement in Space

How do scientists determine distance to stars from the solar system that Earth is part of? One way is to measure its parallax—the apparent shift in the position of an object when viewed from two different positions. Extend your arm and look at your thumb first with your left eye closed and then with your right eye closed, as the girl in **Figure 3A** is doing. Your thumb appears to change position with respect to the background. Now do the same experiment with your thumb closer to your face, as shown in **Figure 3B.** What do you observe? The nearer an object is to the observer, the greater its parallax is.

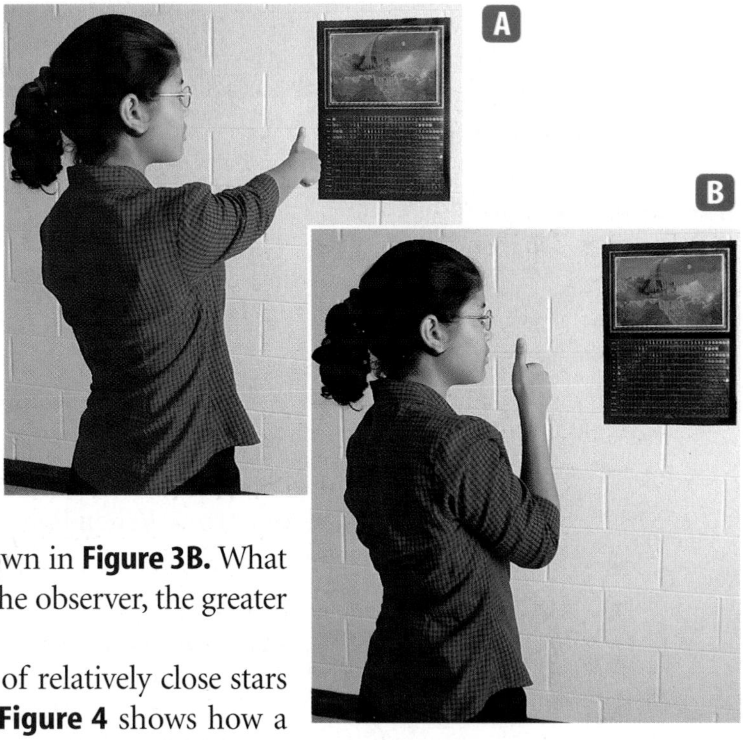

Astronomers can measure the parallax of relatively close stars to determine their distances from Earth. **Figure 4** shows how a close star's position appears to change. Knowing the angle that the star's position changes and the size of Earth's orbit, astronomers can calculate the distance of the star from Earth.

Because space is so vast, a special unit of measure is needed to record distances. Distances between stars and galaxies are measured in light-years. A **light-year** is the distance that light travels in one year. Light travels at 300,000 km/s, or about 9.5 trillion km in one year. The nearest star to Earth, other than the Sun, is Proxima Centauri. Proxima Centauri is a mere 4.3 light-years away, or about 40 trillion km.

Figure 3
A Your thumb appears to move less against the background when it is farther away.
B It appears to move more when it is closer.

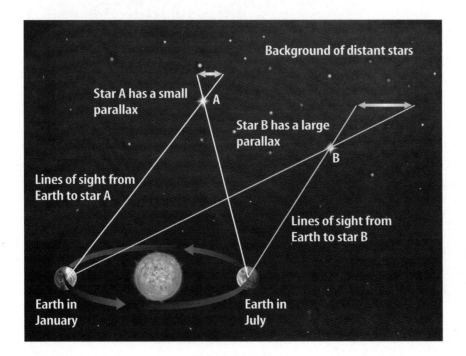

Figure 4
Parallax can be seen if you observe the same star when Earth is at two different points during its orbit around the Sun. The star's position relative to more distant background stars will appear to change. *Is star* A *or* B *farther from Earth?*

Figure 5
These star spectra were made by placing a diffraction grating over a telescope's objective lens. A diffraction grating produces a spectrum by causing interference of light waves. *What causes the lines in spectra?*

Properties of Stars

The color of a star indicates its temperature. For example, hot stars are a blue-white color. A relatively cool star looks orange or red. Stars that have the same temperature as the Sun have a yellow color.

Astronomers study the composition of stars by observing their spectra. When fitted into a telescope, a spectroscope acts like a prism. It spreads light out in the rainbow band called a spectrum. When light from a star passes through a spectroscope, it breaks into its component colors. Look at the spectrum of a star in **Figure 5.** Notice the dark lines caused by elements in the star's atmosphere. Light radiated from a star passes through the star's atmosphere. As it does, elements in the atmosphere absorb some of this light. The wavelengths of visible light that are absorbed appear as dark lines in the spectrum. Each element absorbs certain wavelengths, producing a certain pattern of dark lines. Every chemical element produces a unique pattern of dark lines. Like a fingerprint, the patterns of lines can be used to identify which elements are in a star's atmosphere.

Section 1 Assessment

1. What is a constellation?
2. How does Earth's revolution affect the viewing of constellations throughout the year?
3. If two stars give off equal amounts of light, why might one look brighter?
4. If the spectrum of a star shows the same absorption lines as the Sun, what can be said about the star's composition?
5. **Think Critically** Several thousand stars have large enough parallaxes that their distances can be studied using parallax. Most of these stars are invisible to the naked eye. What does this indicate about their absolute magnitudes?

Skill Builder Activities

6. **Recognizing Cause and Effect** Suppose you viewed Proxima Centauri through a telescope today. How old were you when the light that you see left Proxima Centauri? Why might Proxima Centauri look dimmer than Betelgeuse, a large star that is 520 light-years away? **For more help, refer to the** Science Skill Handbook.

7. **Using Graphics Software** Use graphics software on a computer to make a star chart of major constellations visible from where you live during the current season. Include several reference points to help others find the charted constellations. **For more help, refer to the** Technology Skill Handbook.

The Sun

The Sun's Layers

Within the universe, the Sun is an ordinary star—not too spectacular. However, to you it's important. The Sun is the center of the solar system, and it makes life possible on Earth. More than 99 percent of all the matter in the solar system is in the Sun.

Notice the different layers of the Sun, shown in **Figure 6,** as you read about them. Like other stars, the Sun is an enormous ball of gas that produces energy by fusing hydrogen into helium in its core. This energy travels outward through the radiation zone and the convection zone. In the convection zone, gases circulate in giant swirls. Finally, energy passes into the Sun's atmosphere.

The Sun's Atmosphere

The lowest layer of the Sun's atmosphere and the layer from which light is given off is the **photosphere.** The photosphere often is called the surface of the Sun, although the surface is not a smooth feature. Temperatures there are about 6,000 K. Above the photosphere is the **chromosphere.** This layer extends upward about 2,000 km above the photosphere. A transition zone occurs between 2,000 km and 10,000 km above the photosphere. Above the transition zone is the **corona.** This is the largest layer of the Sun's atmosphere and extends millions of kilometers into space. Temperatures in the corona are as high as 2 million K. Charged particles continually escape from the corona and move through space as solar wind.

Figure 6
Energy produced by fusion in the Sun's core travels outward by radiation and convection. The Sun's atmosphere— the photosphere, chromosphere, and corona—shines by the energy produced in the core.

As You Read

What You'll Learn
- **Describe** the strucure of the Sun.
- **Explain** how sunspots, prominences, and solar flares are related.
- **Explain** why the Sun is considered an average star and how it differs from stars in binary systems.

Vocabulary

photosphere	corona
chromosphere	sunspot

Why It's Important
The Sun is the source of most energy on Earth.

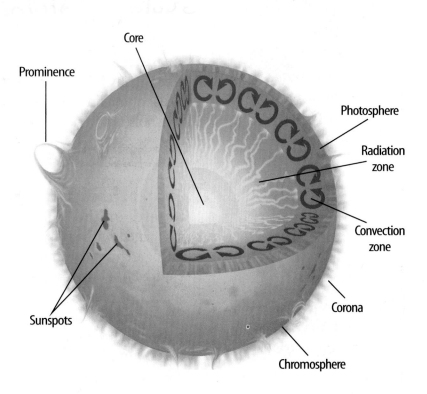

Core

Prominence

Photosphere

Radiation zone

Convection zone

Corona

Sunspots

Chromosphere

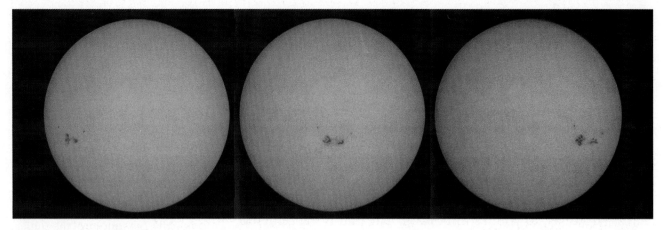

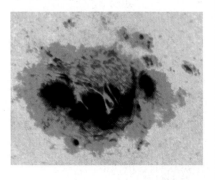

A Notice how these sunspots move as the Sun rotates.

Figure 7
Sunspots are bright, but when viewed against the rest of the photosphere, they appear dark.

B This is a close-up photo of a large sunspot.

Surface Features

From the viewpoint that you observe the Sun, its surface appears to be a smooth layer. But the Sun's surface has many features, including sunspots, prominences, flares, and CMEs.

Sunspots Areas of the Sun's surface that appear dark because they are cooler than surrounding areas are called **sunspots.** Ever since Galileo Galilei viewed sunspots with a telescope, scientists have been studying them. Because scientists could observe the movement of individual sunspots, shown in **Figure 7,** they concluded that the Sun rotates. However, the Sun doesn't rotate as a solid body, as Earth does. It rotates faster at its equator than at its poles. Sunspots at the equator take about 25 days to complete one rotation. Near the poles, they take about 33 days.

Sunspots aren't permanent features on the Sun. They appear and disappear over a period of several days, weeks, or months. The number of sunspots increases and decreases in a fairly regular pattern called the sunspot, or solar activity, cycle. Times when many large sunspots occur are called sunspot maximums. Sunspot maximums occur about every 10 to 11 years. Periods of sunspot minimum occur in between.

 **Reading Check** *What is a sunspot cycle?*

Prominences and Flares Sunspots are related to several features on the Sun's surface. The intense magnetic fields associated with sunspots might cause prominences, which are huge, arching columns of gas. Notice the huge prominence in **Figure 8A.** Some prominences blast material from the Sun into space at speeds ranging from 600 km/s to more than 1,000 km/s.

Gases near a sunspot sometimes brighten suddenly, shooting outward at high speed. These violent eruptions are called solar flares. You can see a solar flare in **Figure 8B.**

CMEs During a sunspot maximum, like the one that occurred in 2000, brilliant coronal mass ejections (CMEs) are emitted from the Sun. When a CME is released in the direction of Earth, it appears as a halo around the Sun.

CMEs present little danger to life on Earth, but the highly charged solar wind material, along with ultraviolet light and X rays from solar flares, can reach Earth and cause disruption of radio signals. High-energy particles contained in CMEs are carried past Earth's magnetic field. This sets up electrical currents that flow toward the poles. The currents of electricity ionize gas in Earth's atmosphere. The ionized gases produce the light of an aurora, shown in **Figure 8C.** Power distribution equipment also can be affected.

SCIENCE *Online*

Research Visit the Glencoe Science Web site at **science.glencoe.com** for more information about sunspots, solar flares, prominences, and CMEs. Communicate to your class what you learned.

Figure 8

Features such as solar prominences and solar flares can reach hundreds of thousands of kilometers into space. CMEs are generated as magnetic fields above sunspot groups rearrange. CMEs can trigger events that produce auroras.

A Solar prominence

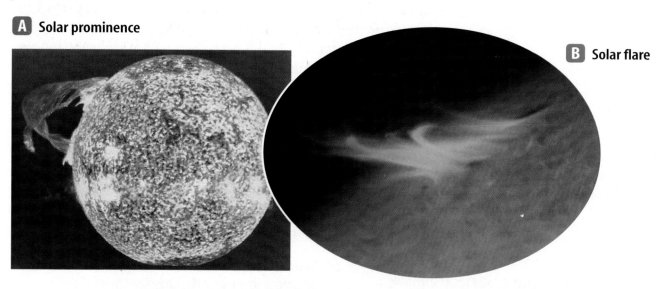

B Solar flare

C Aurora

The Sun—An Average Star

The Sun is a middle-aged star. Its absolute magnitude is typical, and it shines with a yellow light. Although the Sun is an average star, it is somewhat unusual in one way. Most stars are part of a system in which two or more stars orbit each other. When two stars orbit each other, they make up a binary system.

In some cases, astronomers can detect binary systems because one star occasionally eclipses the other. Algol, in the constellation Perseus, is an example of this. The total amount of light from the star system becomes dim and then bright again on a regular cycle.

Figure 9
Most stars were formed originally in large clusters containing hundreds, or even thousands, of stars.

In other cases, three stars orbit around each other, forming a triple star system. The closest star system to the Sun—the Alpha Centauri system, including Proxima Centauri—is a triple star.

Stars also can move through space together as a cluster. In a star cluster, many stars are relatively close to one another, so their gravitational attraction to each other is strong. Most star clusters are far from the solar system, and each appears as a fuzzy patch in the night sky. The double cluster in the northern part of the constellation Perseus is shown in **Figure 9.** On a dark night in autumn, you can see the double cluster with binoculars, but you can't see its individual stars. The Pleiades star cluster can be seen in the constellation of Taurus in the winter sky. On a clear, dark night, you might be able to see seven of the stars in this cluster.

Section 2 Assessment

1. What are the different layers that make up the Sun?
2. Describe the characteristics of sunspots.
3. How are sunspots, prominences, solar flares, and CMEs related? How does each affect Earth?
4. What characteristics does the Sun have in common with other stars? What characteristic makes it different from most other stars?
5. **Think Critically** Because most stars are found in multiple-star systems, what might explain why the Sun is a single star?

Skill Builder Activities

6. **Interpreting Scientific Illustrations** Use **Figure 6** to answer the questions below. **For more help, refer to the** Science Skill Handbook.
   a. Which layers make up the Sun's atmosphere?
   b. What process circulates gas in the Sun's convection zone?

7. **Communicating** Explain how the Sun generates energy. In your Science Journal, write a short paragraph hypothesizing what might happen to the Sun when it depletes its supply of hydrogen. **For more help, refer to the** Science Skill Handbook.

Activity

Sunspots

Sunspots can be observed moving across the face of the Sun as it rotates. Measure the movement of sunspots, and use your data to determine the Sun's period of rotation.

What You'll Investigate
Can sunspot motion be used to determine the Sun's period of rotation?

Materials
several books   clipboard
piece of cardboard   small tripod
drawing paper   scissors
refracting telescope

Goals
- **Observe** sunspots and estimate their size.
- **Estimate** the rate of apparent motion of sunspots.

Safety Precautions
Handle scissors with care.

Procedure

1. Find a location where the Sun can be viewed at the same time of day for a minimum of five days. **WARNING:** *Do not look directly at the Sun. Do not look through the telescope at the Sun. You could damage your eyes.*

2. If the telescope has a small finder scope attached, remove it or keep it covered.

3. Set up the telescope with the eyepiece facing away from the Sun, as shown. Align the telescope so that the shadow it casts on the ground is the smallest size possible. Cut and attach the cardboard as shown in the photo.

4. Use books to prop the clipboard upright. Point the eyepiece at the drawing paper.

5. Move the clipboard back and forth until you have the largest image of the Sun on the paper. Adjust the telescope to form a clear image. Trace the outline of the Sun on the paper.

6. Trace any sunspots that appear as dark areas on the Sun's image. Repeat this step at the same time each day for a week.

7. Using the Sun's diameter (approximately 1,390,000 km), estimate the size of the largest sunspots that you observed.

8. **Calculate** how many kilometers the sunspots appear to move each day.

9. **Predict** how many days it will take for the same group of sunspots to return to the same position in which they appeared on day 1.

Conclude and Apply

1. What was the estimated size and rate of apparent motion of the largest sunspots?

2. **Infer** how sunspots can be used to determine that the Sun's surface is not solid like Earth's.

*C*ommunicating Your Data

Compare your conclusions with those of other students in your class. **For more help, refer to the** Science Skill Handbook.

Evolution of Stars

As You Read

What You'll Learn

- **Describe** how stars are classified.
- **Explain** how the temperature of a star relates to its color.
- **Describe** how a star evolves.

Vocabulary

nebula	supergiant
giant	neutron star
white dwarf	black hole

Why It's Important

Like humans, stars are born, mature, grow old, and die.

Classifying Stars

When you look at the night sky, all stars might appear to be similar, but they are quite different. Like people, they vary in age and size, but stars also vary in temperature.

In the early 1900s, Ejnar Hertzsprung and Henry Russell made some important observations. They noticed that in general, stars with higher temperatures also have brighter absolute magnitudes.

Hertzsprung and Russell developed a graph, shown in **Figure 10,** to show this relationship. They placed temperatures across the bottom and absolute magnitudes up one side. A graph that shows the relationship of a star's temperature to its absolute magnitude is called a Hertzsprung-Russell (H-R) diagram.

The Main Sequence As you can see, stars seem to fit into specific areas of the graph. Most stars fit into a diagonal band that runs from the upper left to the lower right of the chart. This band, called the main sequence, contains hot, blue, bright stars in the upper left and cool, red, dim stars in the lower right. Yellow, main sequence stars, like the Sun, fall in between.

Figure 10
The relationships among a star's color, temperature, and brightness are shown in this Hertzsprung-Russell diagram. Main sequence stars run from the upper-left corner to the lower-right corner. Stars in the upper left are hot, bright stars, and stars in the lower right are cool, faint stars. *What type of star shown in the diagram is the coolest, brightest star?*

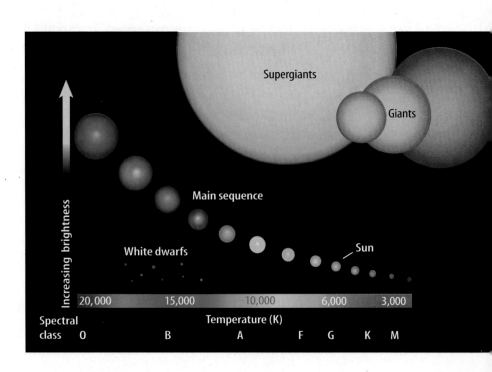

Dwarfs and Giants About 90 percent of all stars are main sequence stars. Most of these are small, red stars found in the lower right of the H-R diagram. Among main sequence stars, the hottest stars generate the most light and the coolest ones generate the least. What about the ten percent of stars that are not part of the main sequence? Some of these stars are hot but not bright. These small stars are located on the lower left of the H-R diagram and are called white dwarfs. Other stars are extremely bright but not hot. These large stars on the upper right of the H-R diagram are called giants, or red giants because they are usually red in color. The largest giants are called super-giants. **Figure 11** shows the supergiant, Antares—a star 300 times the Sun's diameter—in the constellation Scorpius. It is 5,600 times as bright as the Sun.

 **Reading Check** *What kinds of stars are on the main sequence?*

How do stars shine?

When the H-R diagram was developed, scientists didn't know what caused stars to shine. Hertzsprung and Russell developed their diagram without knowing what produced the light and heat of stars.

For centuries, people were puzzled by the questions of what stars were made of and how they produced light. Many people had estimated that Earth was only a few thousand years old. The Sun could have been made of coal and shined for that long. However, when people realized that Earth was much older, they wondered what material possibly could burn for so many years. Early in the twentieth century, scientists began to understand the process that keeps stars shining for billions of years.

Figure 11
Antares is a bright, supergiant located 400 light-years from Earth. Although its temperature is only about 3,500 K, it is the 16th brightest star in the sky.

Generating Energy In the 1930s, scientists discovered reactions between the nuclei of atoms. They hypothesized that temperatures in the center of the Sun must be high enough to cause hydrogen to fuse to make helium. That reaction would release tremendous amounts of energy. In this reaction, four hydrogen nuclei combine to create one helium nucleus. The mass of one helium nucleus is less than the mass of four hydrogen nuclei, so some mass is lost in the reaction.

Years earlier, in 1905, Albert Einstein had proposed a theory stating that mass can be converted into energy. This was stated as the famous equation $E = mc^2$. In this equation, E is the energy produced, m is the mass, and c is the speed of light. The small amount of mass "lost" when hydrogen atoms fuse to form a helium atom is converted to a large amount of energy.

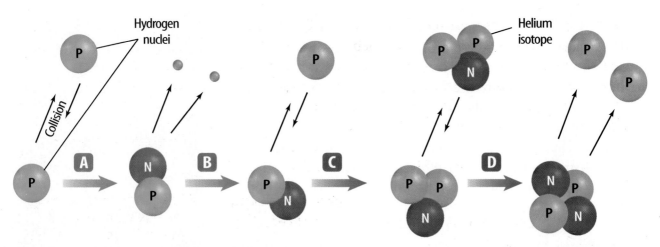

A One proton decays to a neutron, releasing subatomic particles and some energy.

B Another proton fuses with a proton and neutron to form an isotope of helium. Energy is given off again.

C Two helium isotopes fuse.

D A helium nucleus (two protons and two neutrons) forms as two protons break away. In the process, still more energy is released.

Figure 12
Fusion begins in a star's core as protons (hydrogen nuclei) collide.
What happens to the "lost" mass during this process?

Research Visit the Glencoe Science Web site at **science.glencoe.com** to find out more about the evolution of stars. Record this information in your Science Journal.

Fusion Shown in **Figure 12,** fusion occurs in the cores of stars. Only in the core are temperatures high enough to cause atoms to fuse. Normally, they would repel each other, but in the core of a star where temperatures can exceed 15,000,000 K, atoms can move so fast that some of them fuse upon colliding.

Evolution of Stars

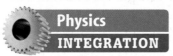

The H-R diagram explained a lot about stars. However, it also led to more questions. Many wondered why some stars didn't fit in the main sequence group and what happened when a star depleted its supply of hydrogen fuel. Today, scientists have theories of how stars evolve, what makes them different from one another, and what happens when they die. **Figure 13** illustrates the lives of different types of stars.

When hydrogen fuel is depleted, a star loses its main sequence status. This can take less than 1 million years for the brightest stars to many billions of years for the faintest stars. The Sun has a main sequence life span of about 10 billion years. Half of its life span is still in the future.

Nebula Stars begin as a large cloud of gas and dust called a **nebula.** As the particles of gas and dust exert a gravitational force on each other, the nebula begins to contract. Gravitational forces cause instability within the nebula. The nebula can break apart into smaller pieces. Each piece eventually will collapse to form a star.

A Star Is Born As the particles in the smaller clouds move closer together, the temperatures in each nebula increase. When the temperature inside the core of a nebula reaches 10 million K, fusion begins. The energy released radiates outward through the condensing ball of gas. As the energy radiates into space, stars are born.

✔ Reading Check *How are stars born?*

Main Sequence to Giant Stars In the newly formed star, the heat from fusion causes pressure that balances the attraction due to gravity. The star becomes a main sequence star. It continues to use up its hydrogen fuel.

When hydrogen in the core of the star is depleted, a balance no longer exists between pressure and gravity. The core contracts, and temperatures inside the star increase. This causes the outer layers of the star to expand and cool. In this late stage of its life cycle, a star is called a **giant.**

After the core temperature reaches 100 million K, helium nuclei fuse to form carbon in the giant's core. By this time, the star has expanded to an enormous size, and its outer layers are much cooler than they were when it was a main sequence star. In about 5 billion years, the Sun will become a giant.

White Dwarfs After the star's core uses up its helium, it contracts even more and its outer layers escape into space. This leaves behind the hot, dense core. The core contracts under the force of gravity. At this stage in a star's evolution, it becomes a **white dwarf.** A white dwarf is about the size of Earth.

Chemistry
INTEGRATION

The spectrum of a star shows dark absorption lines of helium and hydrogen and is bright in the blue end. Describe as much as you can about the star's composition and surface temperature.

Figure 13
The life of a star depends greatly on its mass. Massive stars eventually become neutron stars or possibly black holes. *What happens to stars that are the size of the Sun?*

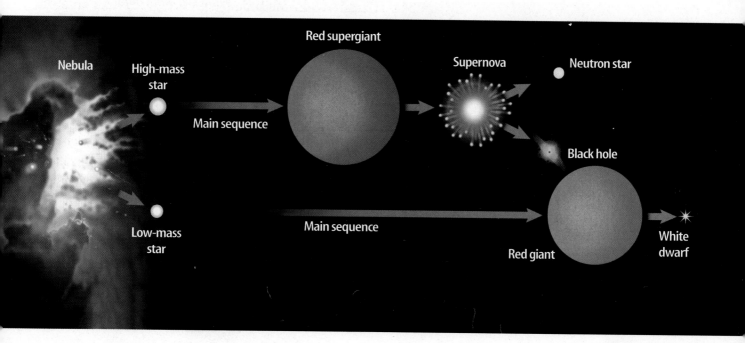

Nebula   High-mass star   Main sequence   Red supergiant   Supernova   Neutron star   Black hole

Low-mass star   Main sequence   Red giant   White dwarf

About forty years ago, radio waves were detected coming from some small regions of space. These radio sources were called quasi-stellar radio sources, which was shortened to quasars. Today, quasars are known to be distant galaxies with supermassive, rotating black holes in their centers. Radio waves and other types of radiation are produced as matter falls into the black holes. Research to learn how scientists solved the mystery of quasars.

Figure 14
The black hole at the center of galaxy M87 pulls matter into it at extremely high velocities. Some matter is ejected to produce a jet of gas that streams away from the center of the galaxy at nearly light speed.

Supergiants and Supernovas In stars that are more than ten times more massive than the Sun, the stages of evolution occur more quickly and more violently. Look back at **Figure 13.** In massive stars, the core heats up to much higher temperatures. Heavier and heavier elements form by fusion, and the star expands into a **supergiant.** Eventually, iron forms in the core. Much of the star's energy is no longer used to produce light and heat, and the core collapses violently, sending a shock wave outward through the star. The outer portion of the star explodes, producing a supernova. A supernova can be millions of times brighter than the original star was.

Neutron Stars If the collapsed core of a supernova is about twice as massive as the Sun, it may shrink to approximately 20 km in diameter. Only neutrons can exist in the dense core, and it becomes a **neutron star.** Neutron stars are so dense that a teaspoonful would weigh about 100 million metric tons in Earth's gravity. As dense as neutron stars are, they can contract only so far because the neutrons resist the inward pull of gravity.

Black Holes If the remaining dense core from a supernova is more than three times more massive than the Sun, probably nothing can stop the core's collapse. Under these conditions, all of the core's mass collapses to a point that has no volume. The gravity from this mass is so strong that nothing can escape from it, not even light. Because light cannot escape, the region is called a **black hole.** If you could shine a flashlight on a black hole, the light simply would disappear into it.

 **Reading Check**   *What is a black hole?*

Black holes, however, are not like giant vacuum cleaners, sucking in distant objects. A black hole has an event horizon, which is a region inside of which nothing can escape. If something—including light—crosses the event horizon, it will be pulled into the black hole. Beyond the event horizon, the black hole's gravity pulls on objects just as it would if the mass had not collapsed. Stars and planets can orbit around a black hole.

The photograph in **Figure 14** was taken by the *Hubble Space Telescope*. It shows a jet of gas streaming out of the center of galaxy M87. This jet of gas formed as matter flowed toward a black hole, and some of the gas was ejected along the polar axis without falling in.

Recycling Matter A star begins its life as a nebula, such as the one shown in **Figure 15.** Where does the matter in a nebula come from? Nebulas form partly from the matter that was once in other stars. A star ejects enormous amounts of matter during its lifetime. Some of this matter is incorporated into nebulas, which can evolve to form new stars. The matter in stars is recycled many times.

What about the matter created in the cores of stars? Are elements such as carbon and iron also recycled? These elements can become parts of new stars. In fact, spectrographs have shown that the Sun contains some carbon, iron, and other such elements. Because the Sun is an average, main sequence star, it is too young and its mass is too small to have formed these elements itself. The Sun condensed from material that was created in stars that died many billions of years ago.

Some elements condense to form planets and other bodies rather than stars. In fact, your body contains many atoms that were fused in the cores of ancient stars. Evidence suggests that the first stars formed from hydrogen and helium and that all the other elements have formed in the cores of stars or as stars explode.

Figure 15
Stars are forming in the Orion Nebula and other similar nebulae.

Section 3 Assessment

1. Explain why giants are not in the main sequence on the H-R diagram. How do their temperatures and absolute magnitudes compare with those of main sequence stars?

2. What can be said about the absolute magnitudes of two equal-sized stars whose colors are blue and yellow?

3. How do stars produce energy?

4. Outline the history and probable future of the Sun.

5. **Think Critically** Why doesn't the helium currently in the Sun's core undergo fusion?

Skill Builder Activities

6. **Sequencing** Sequence the following in order of most evolved to least evolved: *main sequence star, supergiant, neutron star,* and *nebula.* **For more help, refer to the** Science Skill Handbook.

7. **Solving One-Step Equations** Assume that a star's core has shrunk to a diameter of 12 km. What would be the circumference of the shrunken stellar core? Use the equation $C = \pi d$. How does this compare with the circumference of Earth with a diameter of 12,756 km? **For more help, refer to the** Math Skill Handbook.

Galaxies and the Universe

As You Read

What You'll Learn

- **Identify** the three main types of galaxies.
- **List** several characteristics of the Milky Way Galaxy.
- **Describe** evidence that supports the Big Bang theory.

Vocabulary

galaxy
Big Bang theory

Why It's Important

Studying the universe could help scientists determine whether life is possible elsewhere.

Galaxies

Long ago, people believed that Earth was the center of the universe. Today you know that the Sun is the center of the solar system. But where is the solar system in relation to the galaxy? Where is the galaxy located in the universe?

You are on Earth, and Earth orbits the Sun. Does the Sun orbit anything? How does it interact with other objects in the universe? The Sun is one star among many in a **galaxy**—a large group of stars, gas, and dust held together by gravity. The galaxy in which Earth is found is called the Milky Way. It might contain as many as one trillion stars, including the Sun. Galaxies are separated by huge distances—often millions of light-years.

In the same way that stars are grouped together within galaxies, galaxies are grouped into clusters. The cluster that the Milky Way belongs to is called the Local Group. It contains more than 30 galaxies of various sizes and types. The three major types of galaxies are spiral, elliptical, and irregular.

Spiral Galaxies The Milky Way is a spiral galaxy, as shown in **Figure 16.** Notice that spiral galaxies have spiral arms that wind outward from inner regions. These arms are made up of bright stars and dust. The fuzzy patch seen in the constellation of Andromeda is a spiral galaxy. It's so far away that you can't see its individual stars. Instead, its combined light appears as a hazy spot in the sky. The Andromeda Galaxy is about 2 million light-years away and is a member of the Local Group.

Arms in a normal spiral start close to the center of the galaxy. Barred spirals have spiral arms extending from a large bar of stars and gas that passes through the center of the galaxy.

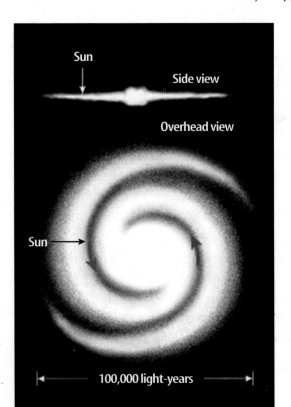

Sun
Side view

Overhead view

Sun

|← 100,000 light-years →|

Figure 16
This illustration shows a side view and an overhead view of the Milky Way. *What group of galaxies is the Milky Way part of?*

Elliptical Galaxies A common type of galaxy is the elliptical galaxy. **Figure 17** shows an elliptical galaxy in the constellation Andromeda. These galaxies are shaped like large, three-dimensional ellipses. Many are football shaped, but others are round. Some elliptical galaxies are small, while others are so large that several galaxies the size of the Milky Way would fit inside one of them.

Irregular Galaxies The third type—an irregular galaxy—includes most of those galaxies that don't fit into the other classifications. Irregular galaxies have many different shapes. They are smaller than the other types of galaxies. Two irregular galaxies called the Clouds of Magellan orbit the Milky Way. The Large Magellanic Cloud is shown in **Figure 18.**

✔ **Reading Check** *How do the three different types of galaxies differ?*

The Milky Way Galaxy

The Milky Way might contain one trillion stars. The visible disk of stars shown in **Figure 16** is about 100,000 light-years across. Find the location of the Sun. Notice that it is located about 30,000 light-years from the galaxy's center in one of the spiral arms. In the galaxy, all stars orbit around a central region, or core. Based on a distance of 30,000 light-years and a speed of 235 km/s, the Sun orbits the center of the Milky Way once every 240 million years.

The Milky Way usually is classified as a normal spiral galaxy. However, some evidence suggests that it might be a barred spiral. It is difficult to know for sure because astronomers have limited data about how the galaxy looks from the outside.

You can't see the normal spiral or barred shape of the Milky Way because you are located within one of its spiral arms. You can, however, see the Milky Way stretching across the sky as a misty band of faint light. You can see the brightest part of the Milky Way if you look low in the southern sky on a moonless summer night. All the stars you can see in the night sky belong to the Milky Way.

Figure 17
This photo shows an example of an elliptical galaxy. *What are the two other types of galaxies?*

Figure 18
The Large Magellanic Cloud is an irregular galaxy. It's a member of the Local Group, and it orbits the Milky Way.

Origin of the Universe

People have long wondered how the universe formed. Several models of its origin have been proposed. One model was the steady state theory. It proposed that the universe always has been the same as it is now. The universe always has existed and always will. As matter in the universe expands outward, new matter is created to keep the overall density of the universe the same or in a steady state. However, evidence indicates that the universe was much different in the past from what it is today.

A second idea is called the oscillating model. In this model, the universe began with expansion occurring in all areas of the universe. Over time, the expansion slowed and the matter in the universe contracted. Then the process began again—repeating over and over, oscillating back and forth.

Neither of these theories can be proven or disproven, but evidence suggests that a third idea is more likely to be correct. The universe started with a big bang and has been expanding ever since. This theory will be described later.

Expansion of the Universe

What does it sound like when a train is blowing its whistle while it travels past you? The whistle has a higher pitch as the train approaches you. Then the whistle seems to drop in pitch as the train moves away. This effect is called the Doppler shift. The Doppler shift occurs with light as well as with sound. **Figure 19** shows how the Doppler shift causes changes in the light coming from distant stars and galaxies. If a star is moving toward Earth, its wavelengths of light are compressed. If a star is moving away from Earth, its wavelengths of light are stretched.

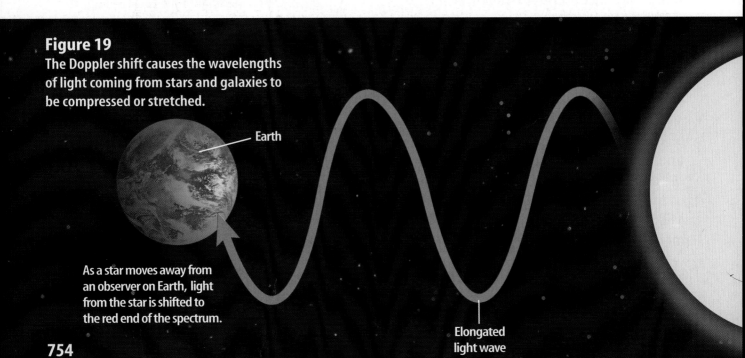

Figure 19
The Doppler shift causes the wavelengths of light coming from stars and galaxies to be compressed or stretched.

Earth

As a star moves away from an observer on Earth, light from the star is shifted to the red end of the spectrum.

Elongated light wave

The Doppler Shift Look at the spectrum of a star in **Figure 20A.** Note the position of the dark lines. How do they compare with the lines in **Figures 20B** and **20C?** They have shifted in position. What caused this shift? As you just read, when a star is moving toward Earth, its wavelengths of light are compressed, just as the sound waves from the train's whistle are. This causes the dark lines in the spectrum to shift toward the blue-violet end of the spectrum. A red shift in the spectrum occurs when a star is moving away from Earth. In a red shift, the dark lines shift toward the red end of the spectrum.

Red Shift In 1929, Edwin Hubble published an interesting fact about the light coming from most galaxies. When a spectrograph is used to study light from galaxies beyond the Local Group, a red shift occurs in the light. What does this red shift tell you about the universe?

Because all galaxies beyond the Local Group show a red shift in their spectra, they must be moving away from Earth. If all galaxies outside the Local Group are moving away from Earth, then the entire universe must be expanding. Remember the Explore Activity at the beginning of the chapter? The dots on the balloon moved apart as the model universe expanded. Regardless of which dot you picked, all the other dots moved away from it. In a similar way, galaxies beyond the Local Group are moving away from Earth.

Figure 20
A This spectrum shows dark absorption lines. **B** The dark lines shift toward the blue-violet end for a star moving toward Earth. **C** The lines shift toward the red end for a star moving away from Earth.

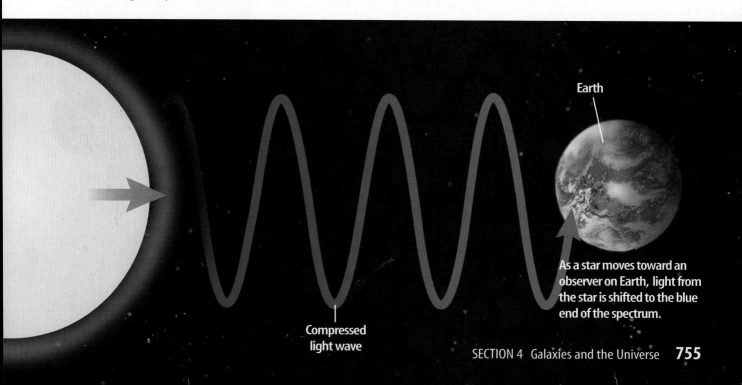

Earth

As a star moves toward an observer on Earth, light from the star is shifted to the blue end of the spectrum.

Compressed light wave

NATIONAL GEOGRAPHIC

VISUALIZING THE BIG BANG THEORY

Figure 21

The Big Bang theory states that the universe probably began 12 billion to 15 billion years ago with an enormous explosion. Even today, galaxies are rushing apart from this explosion.

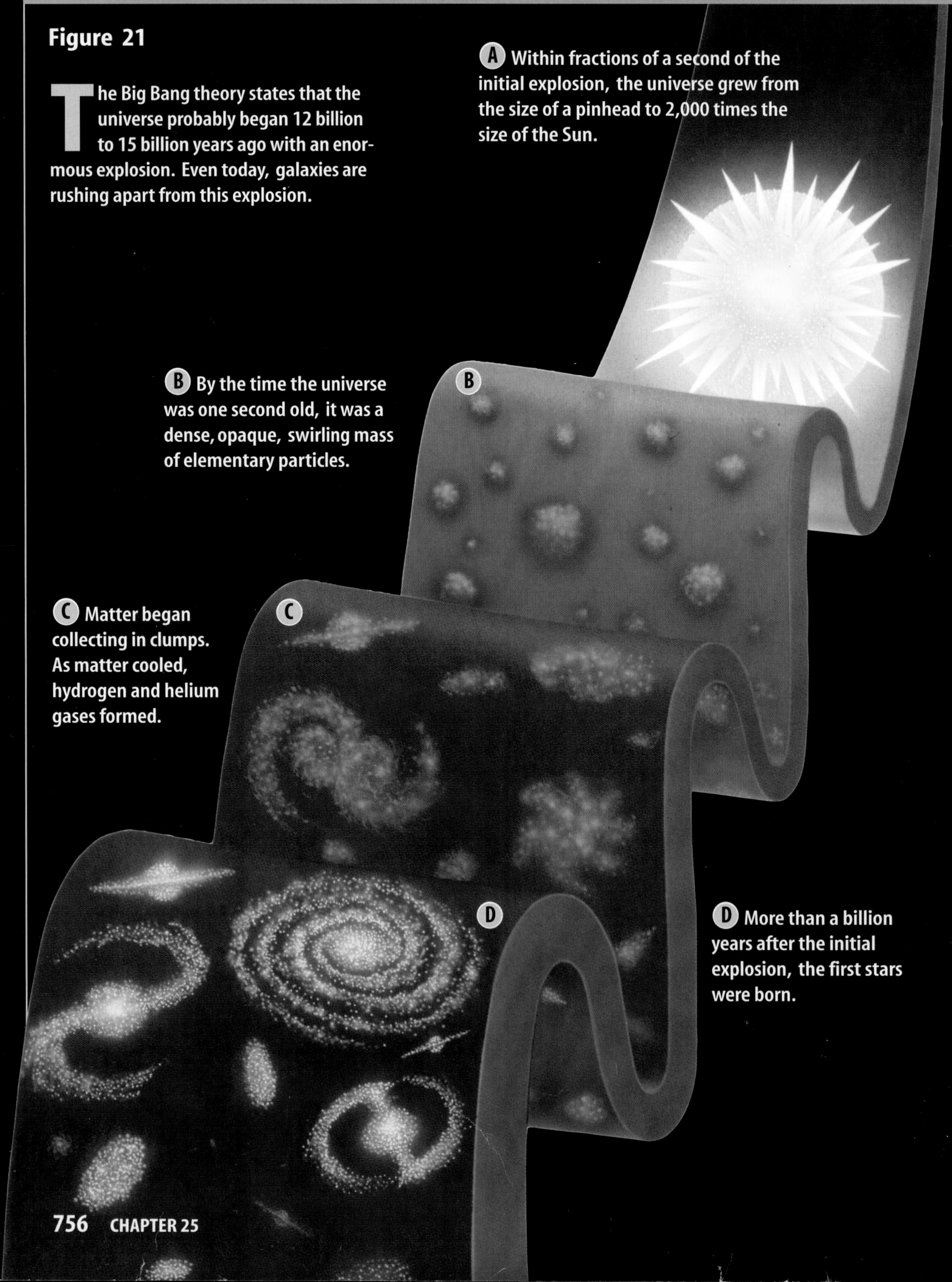

A Within fractions of a second of the initial explosion, the universe grew from the size of a pinhead to 2,000 times the size of the Sun.

B By the time the universe was one second old, it was a dense, opaque, swirling mass of elementary particles.

C Matter began collecting in clumps. As matter cooled, hydrogen and helium gases formed.

D More than a billion years after the initial explosion, the first stars were born.

The Big Bang Theory

When scientists determined that the universe was expanding, they realized that galaxy clusters must have been closer together in the past. The leading theory about the formation of the universe, called the **Big Bang theory,** is based on this explanation. **Figure 21** illustrates the Big Bang theory. According to this theory, approximately 12 billion to 15 billion years ago, the universe began with an enormous explosion. The entire universe began to expand everywhere at the same time.

Looking Back in Time The time-exposure photograph shown in **Figure 22** was taken by the *Hubble Space Telescope.* It shows more than 1,500 galaxies at distances of more than 10 billion light-years. These galaxies could date back to when the universe was no more than 1 billion years old and are in various stages of development. One astronomer says humans might be looking back to a time when the Milky Way was forming. Studies like this eventually will allow astronomers to determine the approximate age of the universe.

Whether the universe will expand forever or stop expanding is still unknown. If enough matter exists, gravity might halt the expansion, and the universe will contract until everything comes to a single point. However, recent studies of distant supernovas indicate that some energy might be causing the universe to expand faster. If this is correct, the universe might expand forever.

Figure 22
The light from the galaxies in this photo mosaic took billions of years to reach Earth.

Section 4 Assessment

1. List the three major types of galaxies. What do they have in common?

2. What is the name of the galaxy that you live in? What motion do its stars exhibit?

3. What is the Doppler shift?

4. How far away are the most distant galaxies?

5. **Think Critically** All galaxies outside the Local Group show a red shift. Within the Local Group, some galaxies show a red shift and some show a blue shift. What does this tell you about the galaxies in the Local Group?

Skill Builder Activities

6. **Comparing and Contrasting** Compare and contrast the three models of the origin of the universe. **For more help, refer to the** Science Skill Handbook.

7. **Communicating** Research and write a report in your Science Journal about the most recent evidence supporting or disputing the Big Bang theory. Describe how the Big Bang theory explains observations of galaxies made with spectrometers. **For more help, refer to the** Science Skill Handbook.

Measuring Parallax

Parallax is the apparent shift in the position of an object when viewed from two locations. The nearer an object is to the observer, the greater its parallax is. Do this activity to design a model and use it in an experiment that will show how distance affects the amount of observed parallax.

Recognizing the Problem

How can you build a model to show the relationship between distance and parallax?

Form a Hypothesis

State a hypothesis about how a model must be built in order for it to be used to show how distance affects the amount of observed parallax.

Possible Materials
meterstick	masking tape
metric ruler	pencil

Goals
- **Design** a model to show how the distance from an observer to an object affects the object's parallax shift.
- **Design** an experiment that shows how distance affects the amount of observed parallax.

Safety Precautions

WARNING: *Be sure to wear goggles to protect your eyes.*

Test Your Hypothesis

Plan

1. As a group, agree upon and write your hypothesis statement.

2. **List** the steps you need to take to build your model. Be specific, describing exactly what you will do at each step.

3. Devise a method to test how distance from an observer to an object, such as a pencil, affects the relative position of the object.

4. **List** the steps you will take to test your hypothesis. Be specific, describing exactly what you will do at each step.

5. Read over your plan for the model to be used in this experiment.

6. How will you determine changes in observed parallax? Remember, these changes should occur when the distance from the observer to the object is changed.

7. You should measure shifts in parallax from several different positions. How will these positions differ?

8. How will you measure distances accurately and compare relative position shift?

Do

1. Make sure your teacher approves your plan before you start.

2. **Construct** the model your team has planned.

3. Carry out the experiment as planned.

4. While conducting the experiment, record any observations that you or other members of your group make in your Science Journal.

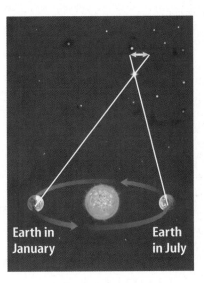

Earth in January

Earth in July

Analyze Your Data

1. **Compare** what happened to the object when it was viewed with one eye closed, then the other.

2. At what distance from the observer did the object appear to shift the most?

Draw Conclusions

1. **Infer** what happened to the apparent shift of the object's location as the distance from the observer was increased or decreased.

2. **Describe** how astronomers might use parallax to study stars.

ommunicating

Your Data

Prepare a chart showing the results of your experiment. **Share** the chart with members of your class. **For more help, refer to the** Science Skill Handbook.

Stars and Galaxies

Did you know...

... The 11,000-kg Hubble Space Telescope orbits Earth every 95 minutes—that's a speed of about 27,250 km/h. The heavy telescope moves more than 750 times faster than the fastest human does on Earth.

... A star in Earth's galaxy explodes as a supernova about once a century. The most famous supernova of this galaxy occurred in 1054 and was recorded by the ancient Chinese and Koreans. The explosion was so powerful that it could be seen during the day, and its brightness lasted for weeks. Other major supernovas in the Milky Way that were observed from Earth occurred in 185, 393, 1006, 1181, 1572, and 1604.

... The large loops of material called solar prominences can extend more than 320,000 km above the Sun's surface. This is so high that two Jupiters and three Earths could fit under the arch.

... **Some of the most famous stars** on Earth can be found on the Hollywood Walk of Fame. The walk contains more than 2,000 stars in honor of various Hollywood film actors and actresses and is located in Hollywood, California.

... **The red giant star** Betelgeuse has a diameter larger than that of Earth's Sun. This gigantic star measures 450,520,000 km in diameter, while the Sun's diameter is a mere 1,390,176 km.

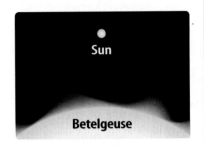

Sun

Betelgeuse

Distance (light-years) from Earth to the Brightest Stars Visible from the Northern Hemisphere in January

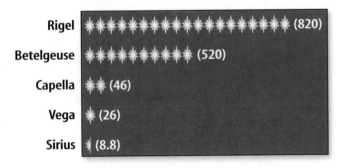

Rigel ✶✶✶✶✶✶✶✶✶✶✶✶✶✶✶✶✶✶✶✶✶ (820)

Betelgeuse ✶✶✶✶✶✶✶✶✶ (520)

Capella ✶✶ (46)

Vega ✶ (26)

Sirius ✶ (8.8)

... **Just as Earth goes around the Sun,** the whole solar system is going around the center of the Milky Way. It takes Earth one year to go around the Sun, but it takes the solar system about 240 million years to go around the center of the Milky Way.

Do the Math

1. Compare the diameter of Betelgeuse to that of the Sun.
2. The Sun orbits the Milky Way once every 240 million years. About how many revolutions of the galaxy has it made since its formation?
3. How many kilometers does the *Hubble Space Telescope* travel in one minute? About how long would it take the fastest human to travel that distance?

Go Further

Is it possible for Earth astronauts to travel to the nearest stars? For more information go to the Glencoe Science Web site at **science.glencoe.com.** How long would such a trip take? What problems would have to be overcome? Write a brief report on what you find.

Chapter 25 Study Guide

Reviewing Main Ideas

Section 1 Stars

1. Constellations are groups of stars that change positions throughout the year because Earth moves. The constellations seem to move because Earth rotates on its axis and revolves around the Sun.

2. The magnitude of a star is a measure of the star's brightness. Absolute magnitude is a measure of the light actually given off by a star. Apparent magnitude is a measure of the amount of light received on Earth.

3. Parallax is the apparent shift in the position of an object when viewed from two different positions. The closer to Earth a star is, the greater its shift in parallax is.

4. A star's composition can be determined from the star's spectrum.

Section 2 The Sun

1. The Sun produces energy by fusing hydrogen into helium in its core. Light is given off from the photosphere, which is the lowest layer of the Sun's atmosphere.

2. Sunspots are areas of the Sun that are cooler and less bright than surrounding areas.

3. Sunspots, prominences, flares, and CMEs are caused by the intense magnetic field of the Sun, which is a main sequence star. *Why is the Sun, shown here, considered an average star?*

Section 3 Evolution of Stars

1. Stars are classified by their positions on the H-R diagram. Most stars are main sequence stars.

2. When hydrogen is depleted in a main sequence star, the star's core collapses and its temperature increases. The star becomes a giant or a supergiant. As the star evolves, it might become a white dwarf.

3. Stars with large masses can explode to form a supernova. The outer layers are blown away and the core evolves into a neutron star or black hole. *At what temperature does fusion begin inside a nebula like the one shown here?*

Section 4 Galaxies and the Universe

1. A galaxy is a large group of stars, gas, and dust held together by gravity. Galaxies can be spiral, elliptical, or irregular in shape.

2. The Milky Way is a spiral galaxy containing about 200 billion stars.

3. The most accepted theory about the origin of the universe is the Big Bang theory.

FOLDABLES Reading & Study Skills

After You Read

To help you review what you have read in this chapter, use your Foldable to explain the relationship among stars, galaxies, and the universe.

Visualizing Main Ideas

Complete the following concept map that shows the evolution of a main sequence star with a mass similar to that of the Sun.

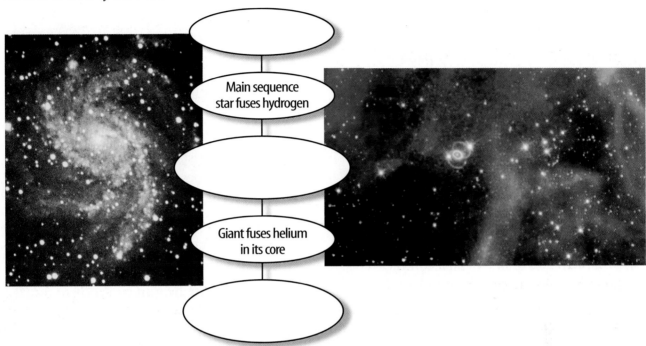

Main sequence star fuses hydrogen

Giant fuses helium in its core

Vocabulary Review

Vocabulary Review

a. absolute magnitude
b. apparent magnitude
c. Big Bang theory
d. black hole
e. chromosphere
f. constellation
g. corona
h. galaxy

i. giant
j. light-year
k. nebula
l. neutron star
m. photosphere
n. sunspot
o. supergiant
p. white dwarf

THE PRINCETON REVIEW **Study Tip**

As you read, look up the definition of any prefixes you do not recognize. Once you know the meaning of a prefix, you'll be able to figure out the definitions of many new words.

Using Vocabulary

Explain the difference between the terms in each of the following sets.

1. absolute magnitude, apparent magnitude

2. black hole, neutron star

3. chromosphere, photosphere

4. galaxy, constellation

5. light-year, galaxy

6. giant, white dwarf

7. corona, sunspot

8. giant, supergiant

9. apparent magnitude, light-year

10. galaxy, Big Bang theory

Checking Concepts

Choose the word or phrase that best answers the question.

1. What are constellations?
   A) clusters   C) black holes
   B) giants   D) patterns

2. What is a measure of the amount of a star's light received on Earth?
   A) absolute magnitude   C) fusion
   B) apparent magnitude   D) parallax

3. What increases as an object comes closer to an observer?
   A) absolute magnitude   C) parallax
   B) red shift   D) blue shift

4. What happens after a nebula contracts and temperatures increase to 10 million K?
   A) a black hole forms
   B) a supernova forms
   C) fusion begins
   D) white dwarfs form

5. What is about 20 km in size?
   A) giant   C) black hole
   B) white dwarf   D) neutron star

6. What does the Sun fuse hydrogen into?
   A) carbon   C) iron
   B) oxygen   D) helium

7. What are loops of matter flowing from the Sun called?
   A) sunspots   C) coronas
   B) auroras   D) prominences

8. What are groups of galaxies called?
   A) clusters   C) giants
   B) supergiants   D) binary systems

9. Which galaxies are sometimes shaped like footballs?
   A) spiral   C) barred
   B) elliptical   D) irregular

10. What do scientists study to determine shifts in wavelengths of light?
    A) spectrum   C) corona
    B) surface   D) chromosphere

Thinking Critically

11. What is significant about the 1995 discovery by the *Hubble Space Telescope* of more than 1,500 galaxies, some at distances of more than 10 billion light-years?

12. How do scientists know that black holes exist if these objects don't emit any visible light?

13. Why can parallax be used only to measure distances to stars that are relatively close to the solar system?

14. Why doesn't the helium currently in the Sun's core undergo fusion?

15. Betelgeuse is the brightest star in the constellation Orion. However, Betelgeuse is a cool, red star that is 520 light-years from Earth. How can Betelgeuse look so bright if it is far away and has a cool surface?

Developing Skills

16. **Concept Mapping** Make a concept map that shows the relationship of temperature, color, and brightness of stars and their positions on the H-R diagram.

17. **Comparing and Contrasting** Compare and contrast the Sun with other stars on the H-R diagram.

18. **Measuring in SI** The Milky Way is 100,000 light-years in diameter. What scale would you use if you were to construct a model of the Milky Way with a diameter of 20 cm? What scale would you use to construct a model with a diameter of 5 m?

19. Interpreting Data Use the chart below to answer the following questions.

Magnitude and Distance of Stars			
Star	Apparent Magnitude	Absolute Magnitude	Distance in Light-Years
A	-26	4.8	0.000,02
B	-1.5	1.4	8.7
C	0.1	4.4	4.3
D	0.1	-7.0	815
E	0.4	-5.9	520
F	1.0	-0.6	45

a. Which star appears brightest from Earth? Which star appears dimmest from Earth?

b. Which star would appear brightest from a distance of 10 light-years? Which star would appear dimmest from this distance?

20. Making a Model Design and construct scale models of a spiral and a barred spiral Milky Way. Show the approximate position of the Sun in each.

Performance Assessment

21. Story Write a short science-fiction story about an astronaut traveling through the universe. In your story, tell what the astronaut observes. Use as many of the vocabulary terms from this chapter as you can.

TECHNOLOGY

Go to the Glencoe Science Web site at **science.glencoe.com** or use the **Glencoe Science CD-ROM** for additional chapter assessment.

Test Practice

The diagram below shows the distance from Earth to Proxima Centauri.

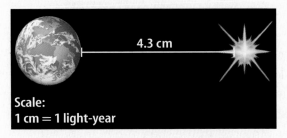

Scale:
1 cm = 1 light-year

Study the picture and answer the following questions.

1. Which of these instruments should you use to view Proxima Centauri?
   A) binoculars
   B) magnifying glass
   C) telescope
   D) microscope

2. What is the approximate distance of Proxima Centauri from Earth?
   F) 3.2 light-years
   G) 4.3 light-years
   H) 5.2 light-years
   J) 6.2 light-years

3. If astronomers shined a beam of laser light from Earth, how long would it take to reach Proxima Centauri?
   A) 3.2 years
   B) 4.3 years
   C) 5.2 years
   D) 6.2 years

4. At this scale, how far from Earth would you draw a star that is 100 light-years away?
   F) 50 cm   **H)** 100 cm
   G) 75 cm   **J)** 200 cm

Standardized Test Practice

Reading Comprehension

Read the passage. Then read each question following the passage. Determine the best answer.

Miss Mitchell's Comet

Maria Mitchell, born in 1818, grew up on the island of Nantucket, off the coast of Massachusetts. Maria Mitchell's father believed girls should aspire to high academic goals. An astronomer himself, he encouraged his daughter's interest in the stars and planets. Mr. Mitchell built a small observatory including a telescope on the roof of the family home.

One night in 1847, Maria spotted a star that she had never noticed before. It was five degrees above the North Star. The next night, Maria looked for the star, and it had moved. Maria realized that what she saw was not a star but a comet. Maria told her father the news. Mr. Mitchell wrote a letter to Professor William Bond of the Harvard University Observatory. Professor Bond sent Maria's name to the King of Denmark, who had offered a prize to the first

person who discovered a comet using a telescope instead of binoculars or the unaided eye. Maria became acknowledged as a scholar, and in 1848 she became the first woman admitted to the American Academy of Arts and Sciences. After studying, Maria traveled the world. She lectured about astronomy and encouraged girls to get involved in science. She was Professor of Astronomy at Vassar College in Poughkeepsie, New York from 1865 until a year before her death in 1889.

Today, the Maria Mitchell Association continues her work. Every year, it awards a prize to a group or person who encourages the advancement of females in science. The association is at Maria's house on Nantucket where there is a large observatory.

> **Test-Taking Tip** When asked the order of events in a story, find each answer choice in the passage and number them 1 to 4, from first to last.

1. Which of these happened first in Maria Mitchell's life?
   A) Maria joined the American Academy of Arts and Sciences.
   B) Maria began teaching at Vassar College.
   C) Maria discovered a comet while looking through a telescope.
   D) Maria traveled the world, lecturing about astronomy.

2. The main idea of this passage is that Maria _____.
   F) was not sure if she saw a star or a comet
   G) used her success to encourage others
   H) grew up on Nantucket Island
   J) received credit for her comet because of her father's help

Standardized Test Practice

THE PRINCETON REVIEW

Reasoning and Skills

Read each question and choose the best answer.

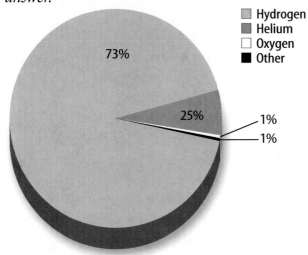

- Hydrogen
- Helium
- Oxygen
- Other

73%

25%

1%

1%

1. The circle graph shows the percentages of elements found in the Sun. What element is most abundant in the Sun?
  - **A)** Helium
  - **B)** Other
  - **C)** Hydrogen
  - **D)** Oxygen

Test-Taking Tip The key next to the graph is shaded to represent the elements found in the Sun.

2. The Sun produces energy by a process called fusion. During fusion, hydrogen nuclei combine to form helium. How would the circle graph shown above change as the Sun ages?
  - **F)** The helium slice would get larger.
  - **G)** The hydrogen slice would get larger.
  - **H)** The helium slice would get smaller.
  - **J)** The circle graph would not change.

Test-Taking Tip Think about what happens during fusion. Then examine the circle graph and select the best answer.

Consider this question carefully before writing your answer on a separate sheet of paper.

3. Astronomers indicate the brightness of a star in two ways—absolute magnitude and apparent magnitude. What do these terms mean, and why is it helpful to talk about the stars in two different ways?

Test-Taking Tip After defining each term on a separate sheet of paper, decide what the major difference between the two is. Then decide why this difference is important and carefully write out your answer.

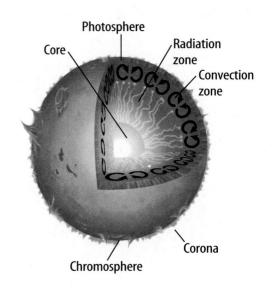

Photosphere, Core, Radiation zone, Convection zone, Corona, Chromosphere

4. The figure above represents a cross-section of ____.
  - **A)** Earth
  - **B)** the Moon
  - **C)** the Sun
  - **D)** Saturn

Test-Taking Tip Use key words from the graphic, such as Photosphere, to identify the correct answer.

Student Resources

Student Resources

Building Stones

Since ancient times, people have used naturally available materials such as stone and wood to construct their homes, places of worship, palaces, and other buildings and monuments. As early as 2500 B.C., the Egyptians had learned how to cut and transport large blocks of limestone from mountainsides for use in the construction of the pyramids. The Maya also used stone to construct their magnificent cities. Since those times, stone has remained a popular building material not only for its beauty but also for its durability and strength.

Types of Building Stone and Some Famous Stone Monuments

The choice of stone used in construction depends on the purpose, availability, cost, and properties of the particular stone. On the following pages are some of the most common building stones, one or more of which was used in the construction of many of the famous buildings in the world. By using this Field Guide, you can learn which stones were used to build these structures. You also can learn to identify the stones used to construct other monuments, buildings, and structures in your neighborhood or town.

Granite

Granite's mineral composition determines its color. For example, some granites are pink because they contain the mineral potassium feldspar. The minerals in granite resist wear and tear caused by wind and precipitation.

The pink color of granite in Enchanted Rock is caused by the presence of microcline, a variety of potassium feldspar.

Marischal College

Built in the late 1890s in Aberdeen, Scotland, Marischal College is one of the largest granite buildings in the world. Constructed in a gothic style, the building has an ornate façade and represents unusually intricate sculpture work for granite.

Field Activity

Take a walk around your town and find a public building or structure made of stone. Then use this field guide to identify the stones used to construct that building. Record your findings in your Science Journal. Visit the Glencoe Science Web site at **science.glencoe.com** to find out which stones were used to build other famous buildings.

Indiana limestone

Limestone

Limestone commonly ranges in color from white to gray, but other colors are possible. Compared to other stones, limestone is not resistant when exposed to air and water in humid climates. However, it still is used as a building stone, especially where it is available locally.

Marble

In its purest form marble is white, but impurities produce other colors. The impurities often weave throughout marble, producing a beautiful mosaic appearance. Many marbles are composed mainly of calcite or dolomite. The softness of these minerals allows marble to be carved easily.

Italian marble

Lincoln Memorial

The Lincoln Memorial in Washington, D.C., was built to honor Abraham Lincoln. The exterior of the building, which is constructed in the Greek style, is made of marble quarried from Colorado and Tennessee. The walls and columns inside are made of Indiana limestone, and the ceiling, floor, and platform are made of marble from Tennessee and Alabama. Lincoln's statue was carved using white marble from Georgia.

Sandstone

The sand grains in sandstone give it a coarse, rustic appearance. Although sandstone often is white or tan, mineral cements can color the rock vivid shades of orange or red. When sand grains are well-cemented and composed mainly of resistant minerals such as quartz, sandstones can be excellent building materials.

Uluru (OO lew rew), the world's largest outcropping of sandstone, was previously called Ayers Rock.

Piece of sandstone

The Red Fort

Built by the Shah Jahan in the seventeenth century, the Red Fort in Delhi, India, is made of red sandstone. The octagonal, or eight-sided, fort contains palaces, gardens, army barracks, and other buildings, including a stable for the king's horses and elephants. The entire fort is surrounded by a large wall.

Slate

The color of slate varies, but it is usually gray, black, green, or red. The color is determined by the stones' organic and mineral content. Slate contains fine-grained, well-compacted clay minerals and mica which make it water-tight and easy to separate into layers. These properties make slate ideal to use for roofing and paving.

The Gloddfa Ganol Slate Mine in Wales is the world's largest slate mine.

Mud Island Park

The flat and compact nature of slate make it a natural choice for this River Walk in Mud Island Park, Memphis, Tennessee. A street map of Memphis was etched into the surface of the slate.

Mud Island Park River Walk

Field GUIDE

I f you've ever built a sandcastle on a beach, you know that shoreline features are often temporary. Shorelines are constantly changing. As the tide comes in, waves wash higher and higher onto the beach, getting closer to your sandcastle. Finally, your sandcastle disappears as the waves rush over it. However, the ocean water that carried away the sand in your castle simply deposited it somewhere else.

Earth's Shorelines

Every landmass on Earth has a shoreline—an area where the ocean meets the land. The United States has approximately 153,000 km of shoreline. Natural forces, such as wind, waves, and currents constantly shape and sculpt these shoreline regions. Along rocky shores, rock-laden waves batter the land away. The chemical action of seawater in waves slowly dissolves some of the minerals coastal rock contains. Yet, along gently sloping shores, fragments that were eroded elsewhere are deposited to form beaches and barrier islands. Along shorelines, the ocean both erodes and builds the land.

Shorelines vary in shape and they have different features. This field guide offers a look at some features you might see along various shorelines.

Shorelines

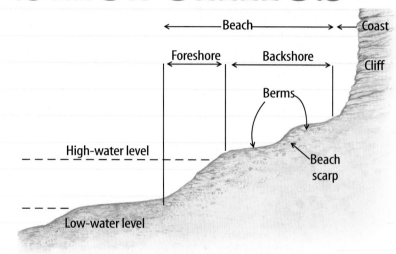

General Shoreline Features

Along a shoreline is an area of loose materials that extends from the low-water line landward to the place where permanent vegetation occurs. This area is called the beach or shore. The beach includes the **foreshore**—the part which lies between high-water and low-water marks at ordinary tide—and the **backshore**—the part of the beach that is usually dry. **Berms** are nearly horizontal parts of the backshore formed by materials deposited by wave action. Some beaches have no berms, but others might have several. On beaches with more than one berm, the berms are separated by **scarps**, which are steep slopes. Some beaches also have cliffs—areas where the land rises sharply a large distance above the water or surrounding land.

Field Activity

Research a location along the shoreline of the United States. Identify the shoreline features in that location. In your Science Journal, describe the features and the patterns of erosion and deposition that created them.

Rocky Shores: Cliffs

Where cliffs or rocks line the shore, the land is probably retreating. Cliffs are sometimes made of layers of different kinds of rock. You often can see these layers in the face of a cliff. The steepness of a cliff varies with the rock hardness and the angle of rock layers. Usually, the harder the rock is, the steeper the cliff is. Sometimes, when lower layers of rock are eroded, an overhanging cliff can form. If the overhang becomes too great, it can collapse and send rocks tumbling down, leaving a pile of rubble at the base of the cliff.

Rocky Shores: Sea Caves, Arches, and Stacks

As waves pound against a sea cliff, the rock near the base of a cliff gradually wears down to form sea caves. Sometimes waves enlarge a cave so much that a crack appears in the roof. During rough weather, water can be seen spraying out of the opening, forming what is called a blowhole.

Sometimes headlands extend into the water like the prow of a boat. If waves erode both sides of a headland, arches will form. Eventually, arches collapse, leaving isolated rock columns called stacks.

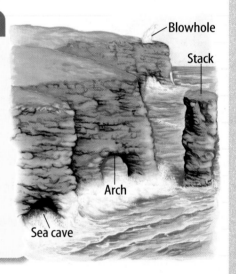

Blowhole

Stack

Arch

Sea cave

Rocky Shores: Boulder and Bay-head Beaches

The ocean transports loose rock that it wears away from shorelines. Waves grind the rock into smaller particles and deposit them onto the shore, forming beaches. Boulder beaches are narrow areas of rocks, stones, and gravel, called shingle, at the base of sea cliffs. Bay-head beaches are small, sandy, crescent-shaped beaches that form in coves between two rocky headlands.

Sandy Beaches

Lowland beaches are broad, gently sloping sandy beaches, usually with a strip of stones and gravel along the upper part. They often have dunes of sand blown inland by onshore winds.

The **beach proper** is the exposed, sandy area of the shoreline. Waves breaking at an angle to the land produce longshore drift. As a wave approaches, it slides up the beach at an angle, then drains straight back into the ocean. This process shifts sand along the beach, moving it from one area to another, often forming cusps that give the beach a scalloped look.

Sandy Beaches: Swash Zone

The **swash zone,** which separates the exposed beach from the ocean, is where waves lap onto the shore. The upper limit of the beach is the swash line, which is the highest level that the sea reaches during large storms. On a gently sloping shore, the ocean erodes beach sediment on the lower part of the swash zone and deposits it on the upper part. On a steeply sloping shore, the opposite occurs.

Sandy Beaches: Dunes

Dunes are large mounds of sand behind ocean beaches. Dunes form as wind carries dry sand farther inland away from the beach proper. A dune begins when an obstruction, such as a rock, a shell, or a piece of driftwood, breaks the wind stream and causes sand to drop on the side of the object that is facing away from the wind direction.

Barrier Islands

Barrier islands are elongated islands of sediments—usually sand—that typically form parallel to the shore. They are separated from the mainland by bodies of water that vary in size. Barrier islands are found along the Atlantic Coast and the Gulf of Mexico. For barrier islands to form, a large supply of sand and moderate wave energy are needed. Barrier islands are constantly changing and moving. Winds, waves, currents, and tides build and erode island shorelines in a never-ending process.

Barrier Islands: Structure

Barrier islands have several distinct zones. Beaches on barrier islands are usually sandy, but some are composed of gravel. Behind the beaches are dunes that help protect the islands from winds, waves, and storms. Sea oats and beach grass growing on the dunes help stabilize the island. On some islands, maritime forests grow behind the dunes. Trees and bushes that are salt tolerant grow in this region. Some islands also have inland ponds that serve as nurseries for young fish, crabs, and shrimp.

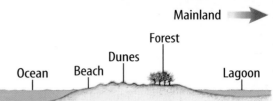

Mainland →
Forest
Dunes
Ocean   Beach   Lagoon

Deltas

Some rivers build up huge deposits of soil or silt where they empty into the ocean, forming coastal lands called deltas. The Mississippi River flows south and empties into the Gulf of Mexico. As the river meets the Gulf, it loses speed and dumps its sediments in an expanding fan-shaped zone. The sediments dropped at the river's mouth tend to build rapidly, causing the river's mouth to grow outward toward the open waters of the Gulf.

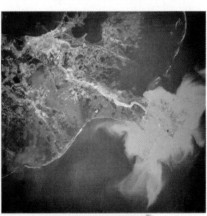

Estuaries

Estuaries are partially enclosed bodies of water where freshwater from rivers and streams runs into and mixes with salty ocean water. Water flowing from the land carries

sediments, nutrients, and pollutants. As water flows through freshwater and saltwater marshes in estuaries, much of the sediment and pollution are filtered out. Wetland plants and soils also protect the areas further inland by absorbing floodwaters and lessening storm surges. Saltwater marsh grasses and other plants help prevent erosion and stabilize the shoreline. Over long periods of time, unless sea levels rise, estuaries tend to fill with sediment and become much smaller.

Field GUIDE

Earth's crust is squeezed and pulled as tectonic plates move. Energy builds up in rocks and when it is released, it can dramatically and visibly change the structure of the land surface. Faults are fractures in Earth produced when sections of Earth's crust move past each other and release energy, sometimes during earthquakes. Small faults can be nearly invisible on the surface. Large faults, caused by intense, long-time crustal movement, can be huge cracks in the ground that extend for hundreds of miles. Types of faults include normal; reverse, or thrust; and strike-slip.

When tectonic plates collide, great mountain chains can be uplifted out of Earth. When rocks are subjected to stress, they do not always break and form faults—under some conditions, rocks will bend and fold. These folded rocks can reach Earth's surface. Upward-arching folds are known as anticlines. Downward-sagging folds are known as synclines. Folds can be large enough to form mountains or small enough to hold in your hand.

Earth's tectonic plates have been moving for hundreds of millions of years. Although they move very slowly— about 10 cm per year—they have moved across great distances over time.

Faults and Folds

Normal Fault: The Rio Grande Rift

Extending across a distance of more than 1,200 km, the Rio Grande Rift stretches from Colorado to Northern Mexico and separates the Colorado Plateau from the Great Plains. The Rio Grande Rift is an example of how normal faults can affect Earth's surface. It is the result of fractures in Earth's crust that are formed as the crust is pulled apart. The Rio Grande Rift is active and the faults in the rift release their built-up energy in the form of frequent, small tremors rather than as large earthquakes. The region also contains many volcanoes.

The Rio Grande Rift

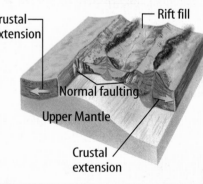

Crustal extension — Rift fill

Normal faulting

Upper Mantle

Crustal extension

Formation of ri

Field Activity

Go to **science.glencoe.com** to take an online geology field trip. Click on the links provided to view different rock formations throughout the United States. In your Science Journal, write or draw the overall structure and attitude, or positioning, of the rock formations.

Thrust Fault: The Appalachian Mountains

The Appalachian Mountains were uplifted more than 300 million years ago during a plate collision. As the African and North American Plates collided, huge slices of Earth's crust were moved great distances along thrust faults. These slices of crust make up Earth's surface in some parts of the Appalachian Mountains today.

Fault Surface

Thrust fault

Hayward Fault

Tectonic Plate collision

Location of Hayward Fault

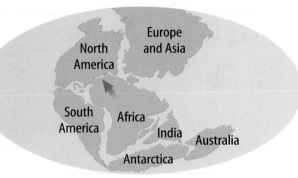

Strike-Slip Fault: The Hayward Fault

This strike-slip fault—part of the 900-km San Andreas Fault system—extends about 120 km and is considered to be the state's second-strongest fault. It is a fracture in Earth's crust: the Pacific Plate to the west is moving northwest in relation to the North American Plate. Energy from this movement continues to build until it is released in the form of an earthquake. In 1868, an earthquake occurred along the Hayward Fault, damaging San Francisco.

Fold: Anticline at Hartland Quay

Layers of sedimentary rock are folded in this spectacular anticline at Hartland Quay in Devon, England. The anticline has been eroded in the foreground exposing the ends of the tilted rock layers.

Anticline at Hartland Qua

Anticlines and synclines in sequence

Anticlines

Synclines

Fold: Syncline on Mount Kidd

This syncline is part of Mount Kidd in Alberta, Canada, in the Kananaskis Valley of the Canadian Rockies. Unlike the American Rockies, which were originally large chunks of granite, the Canadian Rockies began as piles of sediment that were compressed when plates collided.

The syncline on Mount Kidd

Location of the Alps

Fold: The Alps

The Alps, which are a beautiful example of folded mountains, were formed as the African Plate began to move closer to the Eurasian Plate. The pressure exerted by this movement caused the land to fold, creating these peaks more than 15 million years ago.

The folded Alps

Erosion of the Blue Ridge Mountains

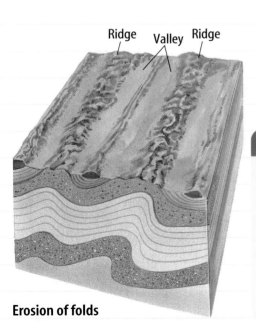

Erosion of folds

Fold: The Blue Ridge Mountains

Folded mountains derive their name from the abundance of folds in the rocks. The Blue Ridge Mountains in the Valley and Ridge region of the Appalachians are an excellent example of mountains formed by rocks that folded, then began to erode. The harder rocks formed ridges, and the softer rocks eroded to form valleys.

Field GUIDE

Suppose you live hundreds of miles from the ocean. One day, you find a rock with a clear imprint of a seashell on its surface. What might this imprint, or fossil, tell you about the area in which you live?

Fossils are the traces or remains of organisms that lived long ago. Fossils provide a record of the vast array of life-forms that have lived on Earth, and they show how these life-forms have changed over time. Fossils also can give clues about how the rock formed over millions of years.

Common Fossils

The most easily found fossils are shells, or casts of shells, from the sea animals that once lived in them. Some fossils might have been well preserved, but most are only fragments of the ancient organism. Fossils of soft-bodied animals and plants are not so easily found because they decay more quickly. However, some rare fossils of soft-bodied organisms have been found and have provided a wealth of information about the organisms.

This field guide shows examples of common fossils belonging to some of the major fossil groups. When you find a fossil, you can use this guide to help identify the group to which the fossil belongs.

Fossils

Fossil Groups

Coral fossil

Coral

Corals are small sea creatures that live attached to the ocean floor in warm, shallow waters. The animals build structures that have a honeycomb-like appearance. These structures commonly are found as fossils. The first corals appeared during the Paleozoic Era of geologic time. Use the key on the fourth page of this field guide to identify these symbols.

Field Activity

Go on an Internet fossil hunt at **science.glencoe.com**. Find five different fossils and make a sketch of each in your Science Journal. Then, below each illustration, write a description of the fossil and explain what kind of organism the fossil came from.

Trilobite

Trilobites (TRI loh bites) are now extinct, but they once were common in ancient seas. Their bodies were divided into three parts—one slightly raised center lobe flanked by two flatter lobes. Trilobites measured from less than 1 cm long to more than 20 cm long.

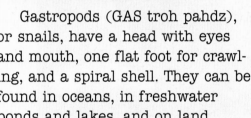

Trilobite fossil

Gastropod or Snail

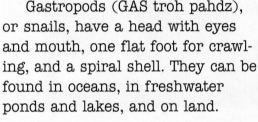

Gastropods (GAS troh pahdz), or snails, have a head with eyes and mouth, one flat foot for crawling, and a spiral shell. They can be found in oceans, in freshwater ponds and lakes, and on land.

Snail fossil

Crinoid or Sea Lily

In spite of its name, the sea lily is an animal. Most crinoids (KRI noydz) attach themselves to the soft, muddy floor of the ocean and then use their long arms to catch prey.

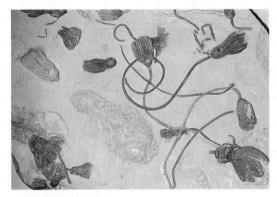

Multiple sea lily fossils

Ammonoid

Ammonoids (A muh noydz) were free-swimming shellfish related to squid. Ammonites appeared during the Devonian Era, long before dinosaurs roamed Earth. They became extinct when dinosaurs did.

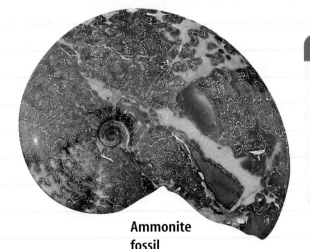

Ammonite fossil

Echinoid

Echinoids (ih KI noydz)—sea urchins and sand dollars—can be ball shaped, cone shaped, or flat. The sea urchin that left this fossil was cone shaped. It used its mouth, located on the flat bottom side, to swallow sand from which it took its nourishment.

Echinoid fossil

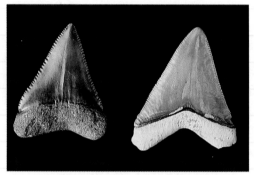

Shark's tooth fossils

Shark

The soft cartilage of a shark's body decays rapidly, so shark fossils are usually teeth or scales. Sharks are abundant in today's oceans.

Plant (Leaves)

Some plant fossils show leaves, fronds, or stems. Tree ferns thrived during the Carboniferous Period (360 million to 286 million years ago). Seed ferns can be identified by their fronds, or branches, which are divided into smaller leaflets.

Seed fern fossils

Petrified wood fossil

Plant (Wood)

Some plant fossils are pieces of wood. This petrified wood fossil is from the Petrified Forest in Arizona. Growth rings in wood indicate changing growth rates during each year. Growth rings often are clear enough to be counted.

FOSSIL MAP OF THE UNITED STATES

The key below provides a symbol for each fossil group shown in this guide. Use the key to find out which fossils can be found in your region.

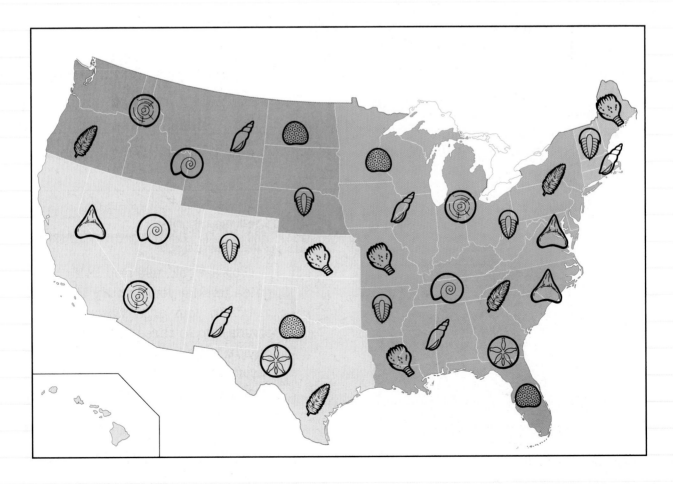

Field GUIDE

Clouds

Clouds are like people—they come in many different sizes and shapes. Some tower thousands of meters in the sky. Others are like fragile wisps of cotton candy floating in the air. All clouds are formed by atmospheric conditions that in turn form Earth's weather. Using this field guide, you can learn to identify different types of clouds and try your hand at weather forecasting.

How Clouds Are Classified

Clouds are classified based on their shape and height. The height of a cloud is represented by the prefix used in its name. For example, a cirrocumulus (*cirro + cumulus*) cloud is a high cloud with a puffy shape.

Key to Cloud Classification

The following symbols are used in this field guide to represent the height and shape of common clouds.

Key to Cloud Classification

Height		Shape	
symbol	prefix	symbol	prefix
	Cirro Describes high clouds with bases starting above 6000 m.		**Cirrus** Latin meaning: curl Describes wispy, stringy clouds
	Alto Describes middle clouds with bases between 2000 m to 6000 m.		**Cumulus** Latin meaning: pile or heap. Describes puffy, lumpy-looking clouds
			Stratus Latin meaning: layer Describes featureless sheets of clouds
	Strato Refers to low clouds below 2000 m.		**Nimbus** Latin meaning: cloud Describes low, gray rain clouds

Field Activity

For a week, use this field guide to help you identify the clouds in your area. Observe the clouds two to three times each day. In your Science Journal, record the date, time, types of clouds observed, and the general weather conditions. What relationships can you infer between the weather and types of clouds that are present?

Cirrus

Feathery cirrus clouds are the highest clouds. They are formed of ice crystals. They usually signal fair weather, but they also can be a sign of changing weather.

Cirrostratus

These thin, sheetlike clouds often form ahead of advancing storms, particularly if they're followed by middle clouds.

Cirrocumulus

Cirrocumulus clouds are small, rounded, white puffs. They appear individually or in long rows. Their rippled pattern resembles the scales of fish. Hence, a sky full of cirrocumulus clouds is called a mackerel sky.

Altostratus

Gray or blue-gray altostratus clouds—they're never white—often cover the entire sky. They are a sign of widespread, steady rain ahead.

Altocumulus

These puffy, white or gray clouds look like rows of soft cotton balls. Randomly scattered altocumulus clouds can mean several days of fair weather. When the clouds resemble little castles, expect a thunderstorm by day's end.

Nimbostratus

Dark-gray nimbostratus clouds are associated with steady rain or snow. This precipitation is light to moderate—rarely heavy. Nimbostratus clouds often have streaks that extend to the ground.

Stratus

Low-lying stratus clouds cover the sky in a blanket of gray. Light rain or drizzle usually accompanies these clouds.

Stratocumulus

Low, lumpy stratocumulus clouds are often a sign of fair weather. To distinguish them from altocumulus clouds, extend your arm toward the cloud. An altocumulus cloud will be roughly the size of your thumbnail. A stratocumulus cloud will be about the size of your fist.

Cumulus

Small, scattered cumulus clouds with slight vertical growth signal fair weather. They have dome-shaped or tower-shaped tops, like cauliflower.

Cumulonimbus

These are thunderstorm clouds. They form near Earth's surface and grow to nearly 15,000 m. Lightning, thunder, and strong winds are associated with cumulonimbus clouds.

Field GUIDE

Managing waste properly can reduce the use of resources and prevent pollution. To cut down on waste production and reduce harm to the environment, people can follow the three Rs of waste management: reduce, reuse, and recycle. For example, a 450-g, family-sized box of cereal uses a lot less cardboard than 18 25-g single-serving boxes use. Finding other functions for used items could include donating magazines that you already have read to local hospitals, nursing homes, or even your doctor's office. You can use many products every day that are recyclable. Also, you can help complete the cycle by purchasing items that are made from recycled materials.

Types of Recyclables

Recycled items are either pre- or post-consumer. Pre-consumer items are made of pieces that were left over during manufacturing. Pre-consumer items have never been used by a consumer. Post-consumer items are made of products that have been used, recycled, and made into something new. Cereal boxes, plastic bottles, writing paper, paper towels, and tissues often include post-consumer recycled materials.

Waste Management

Paper

White paper, newspaper, magazines, and telephone books are common paper goods that can be recycled. New products made from these items include newsprint, cardboard, egg cartons, and building materials. Not all glossy and colored papers are recyclable at all recycling centers. As with any recycled material, you should check with your local recycling center for specific instructions about properly sorting your paper goods.

Field Activity

Collect recyclable plastics in your school for one week. Using the Plastics Code System Table, organize the plastic products by code number so that they can be recycled. In your Science Journal, make a bar graph showing the amount and types of plastic you collected.

Plastics

Plastics are difficult products to recycle. Most plastics are composed of complex molecules that tend not to break down easily. Many types of plastics exist, and they often cannot be recycled together. **Table 1** lists the codes that are used to sort plastic items for recycling. Plastic is recycled into lumber, containers, carpet, and other products.

Table 1 The Plastics Code System		
Code	**Material**	
1	PETE	Polyethylene terephthalate
2	HDPE	High-density polyethylene
3	PVC	Polyvinyl chloride
4	LDPE	Low-density polyethylene
5	PP	Polypropylene
6	PS	Polystyrene
7	Other	All others and mixed

Glass

Glass often is separated by color before recycling. Most glass bottles are recyclable, but some glassware such as lightbulbs is too thin to be recycled. New products made from recycled glass include food and beverage containers.

Metals

A variety of metals are recyclable. Steel is the most common material recycled in the United States. Steel is recycled into any item new steel would be used for such as food cans, structural beams for buildings, and parts for automobiles. Aluminum is a commonly recycled material. It is recycled into new beverage cans, lawn chair frames, siding, and cookware. Even precious metals that are used in laboratories or in jewelry such as gold, silver, and platinum are recyclable. Recycling steel saves enough energy to supply one-fifth of the households in the United States (18 million homes) with electricity for one year. A recycled aluminum beverage can uses 95 percent less energy to produce than a can made from new aluminum, which is enough energy to burn a 100W light bulb for three and a half ($3\frac{1}{2}$) hours.

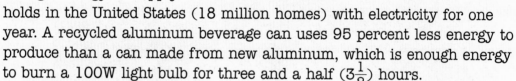

Organic Waste

Organic waste such as yard trimmings, food waste, paper, and wood make up a large percentage of solid waste in the United States. Composting is one way to recycle organic wastes. Food and yard wastes can be placed in compost bins and converted into nutrient rich soil. Mulching is another way to recycle yard waste and wood. In some communities grass clippings, leaves, sticks, and other yard waste can be shredded at local recycling centers to make mulch. Mulch is used to reduce water loss from soil and to keep weeds from growing. As the mulch decays it enriches the soil.

Household Hazardous Wastes

Household Hazardous Wastes (HHWs) include household and car batteries, bleach, household cleaners, paint, paint thinner, motor oil, gasoline, herbicides, pesticides, solvents, and automotive fluids. They contain chemicals that can cause injury or are harmful if used, stored, or discarded improperly. HHWs are marked with a skull and crossbones, caution words or special handling directions. HHWs should not be sent with other garbage to landfills or to incinerators. Water seeps into landfill areas and the toxins could end up in the water supply. Burning may send toxins into the air you breathe. Garbage companies and recycling agencies have free HHW drop-off days or locations.

Follow These Steps to Reduce HHWs:

1. Whenever possible, buy nontoxic alternatives to hazardous products.
2. If you buy a hazardous product, buy only what you need to do the job.
3. Before you store leftover products on the shelf, try to find someone who can use them.

Field GUIDE

For thousands of years, people have looked at the night sky and wondered about what they saw. To early Greek and Roman astronomers, groups of stars seemed to form pictures of animals, heroes, and other characters in myths. These star groups are called **constellations.** Constellations have guided travelers and have been an important part of some religions for a long time.

The Stars of the Zodiac

Are you a Leo, a Libra, a Capricorn, or a Virgo? The sign you were born under represents the constellation that the Sun was in at the moment of your first breath. What are the Zodiac signs, and why are they represented as constellations in the sky?

The Ancient Greeks noticed that the planets moved along a path in the sky. They also noticed star groups behind the revolving planets. They divided this path in the sky into 12 equal sectors, or divisions, and named them according to the star group that is found there. For example, Leo is a constellation resembling a lion; Aries is a ram. Ancient people referred to these star groups as the *Zodiac,* which means "carved figure" in Greek.

Backyard Astronomy

Big Dipper

Ursa Major

You've probably heard of the Big Dipper and even found it in the night sky. The Big Dipper is not a constellation. It's one part of a large constellation named Ursa Major, which is Latin for "Great Bear." The next time you look at the night sky, find the Big Dipper. Then try to find the rest of the stars that form the Great Bear.

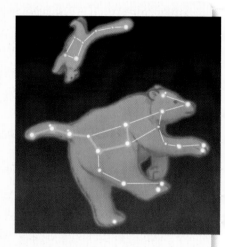

The Great Bear is just one of 88 identified constellations. You can't see all 88 constellations from the northern hemisphere, which is where you live. Some constellations are visible only in the northern hemisphere, and others are visible only in the southern hemisphere.

Field Activity

Observe the sky each night at the same time for a week. Use the appropriate star map and the instructions on the third page of this Field Guide to find the Big Dipper, Little Dipper, and Polaris. Then find at least three constellations. Draw the constellations in your Science Journal and label them with their names.

Constellations

The constellations always are in the
same part of the sky, but they appear to
move over time. As Earth rotates on its
polar axis, the constellations appear to
move from east to west across the sky
each night. As Earth revolves around the
Sun and the seasons change, some constellations come into view and others move out
of view. For example, in the northern hemisphere, you can see Orion in winter and
spring but not during summer and autumn.
Pegasus is visible in autumn and winter but
not in spring and summer. Other constellations, including Cassiopeia (ka see uh PEE
uh) and the Great Bear, are visible to you all year.

In the northern hemisphere, all of the constellations rotate in
a big circle around a bright star called Polaris, the North Star,
which lies at the tail end of a group of stars called the Little
Dipper. The Little Dipper is part of a constellation called Ursa Minor, which is Latin for
"Little Bear." Ancient sailors used the North
Star to navigate the oceans and find their
way home.

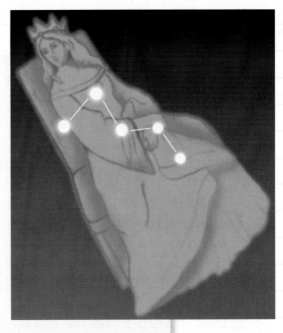

To find Cassiopeia,
look for five
bright stars that
form a large *W* or
a large *M*.

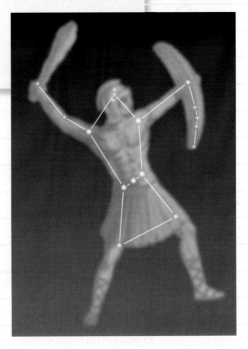

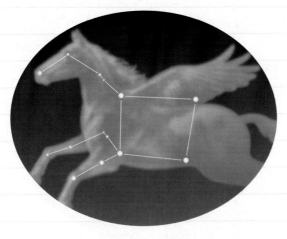

To find Pegasus, look for four bright
stars that form a large square.

To find Orion, look for three bright
stars close together in a straight line.
These stars form Orion's belt.

How to Find Constellations

Use these instructions to find constellations in the sky. The only time you'll be able to see the stars well is on a clear night without bright moonlight or light from buildings.

1. Choose the star map that shows the night sky for the season it is now. Take this book and a compass outside with you.

2. Use the compass to determine which direction is north, and stand facing that way. Turn the star map so the word *North* is right-side up. Compare the star patterns on the map with the stars you see in the northern sky. Find the Big Dipper. Use the picture on the first page of the field guide to help you.

3. Find the two stars that form the outer side of the Big Dipper's bowl. Imagine a line that connects those two stars and extends straight up from the dipper's bowl. Follow that line until you find a bright star. That star is Polaris, the North Star.

4. Starting from Polaris, find the stars that form the Little Bear. Find the stars that form the Great Bear.

5. Use the star map to find other constellations in the northern sky.

6. Turn around to face south. Turn the star map so the word *South* is right-side up. Use the map to find constellations in the southern sky.

7. Repeat step 6 facing west, then east.

8. As you search, try to find Cassiopeia. Look for Orion and/or Pegasus, depending on the season. Use the star map and the pictures on the second page of the field guide to help you.

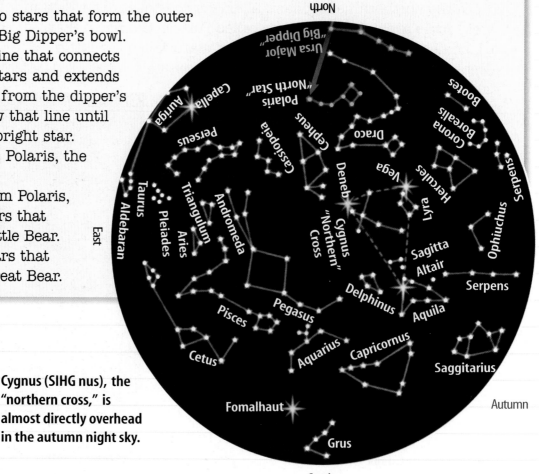

Cygnus (SIHG nus), the "northern cross," is almost directly overhead in the autumn night sky.

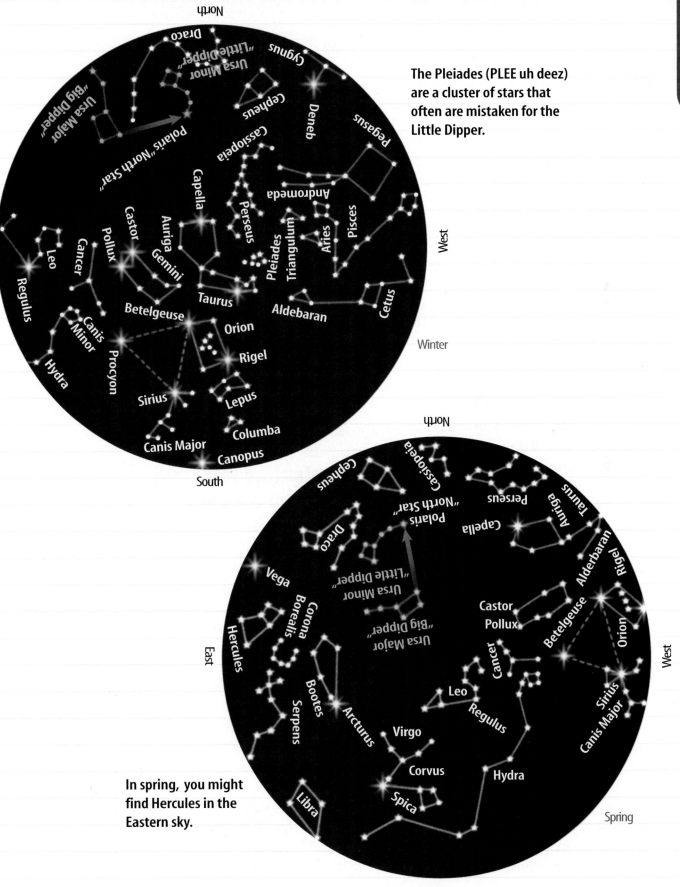

The Pleiades (PLEE uh deez) are a cluster of stars that often are mistaken for the Little Dipper.

Winter

In spring, you might find Hercules in the Eastern sky.

Spring

Skill
Handbooks

As you study science, you will make many observations and conduct investigations and experiments. You will also research information that is available from many sources. These activities will involve organizing and recording data. The quality of the data you collect and the way you organize it will determine how well others can understand and use it. In **Figure 1,** the student is obtaining and recording information using a thermometer.

Putting your observations in writing is an important way of communicating to others the information you have found and the results of your investigations and experiments.

Researching Information

Scientists work to build on and add to human knowledge of the world. Before moving in a new direction, it is important to gather the information that already is known about a subject. You will look for such information in various reference sources. Follow these steps to research information on a scientific subject:

Step 1 Determine exactly what you need to know about the subject. For instance, you might want to find out what happened when Mount St. Helens erupted in 1980.

Step 2 Make a list of questions, such as: When did the eruption begin? How long did it last? What kind of material was expelled and how much?

Step 3 Use multiple sources such as textbooks, encyclopedias, government documents, professional journals, science magazines, and the Internet.

Step 4 List where you found the sources. Make sure the sources you use are reliable and the most current available.

Figure 1
Collecting data is one way to gather information directly.

Evaluating Print and Nonprint Sources

Not all sources of information are reliable. Evaluate the sources you use for information, and use only those you know to be dependable. For example, suppose you live in an area where earthquakes are common and you want to know what to do to keep safe. You might find two Web sites on earthquake safety. One Web site contains "Earthquake Tips" written by a company that sells metal scrapings to help secure your hot-water tank to the wall. The other is a Web page on "Earthquake Safety" written by the U.S. Geological Survey. You would choose the second Web site as the more reliable source of information.

In science, information can change rapidly. Always consult the most current sources. A 1985 source about the Moon would not reflect the most recent research and findings.

Interpreting Scientific Illustrations

As you research a science topic, you will see drawings, diagrams, and photographs. Illustrations help you understand what you read. Some illustrations are included to help you understand an idea that you can't see easily by yourself. For instance, you can't see the layers of Earth, but you can look at a diagram of Earth's layers, as labeled in **Figure 2,** that helps you understand what the layers are and where they are located. Visualizing a drawing helps many people remember details more easily. Illustrations also provide examples that clarify difficult concepts or give additional information about the topic you are studying.

Most illustrations have a label or caption. A label or caption identifies the illustration or provides additional information to better explain it. Can you find the caption or labels in **Figure 2?**

Venn Diagram

Although it is not a concept map, a Venn diagram illustrates how two subjects compare and contrast. In other words, you can see the characteristics that the subjects have in common and those that they do not.

The Venn diagram in **Figure 3** shows the relationship between two types of rocks made from the same basic chemical. Both rocks share the chemical calcium carbonate. However, due to the way they are formed, one rock is the sedimentary rock limestone, and the other is the metamorphic rock marble.

Concept Mapping

If you were taking a car trip, you might take some sort of road map. By using a map, you begin to learn where you are in relation to other places on the map.

A concept map is similar to a road map, but a concept map shows relationships among ideas (or concepts) rather than places. It is a diagram that visually shows how concepts are related. Because a concept map shows relationships among ideas, it can make the meanings of ideas and terms clear and help you understand what you are studying.

Overall, concept maps are useful for breaking large concepts down into smaller parts, making learning easier.

Figure 2
This cross section shows a slice through Earth's interior and the positions of its layers.

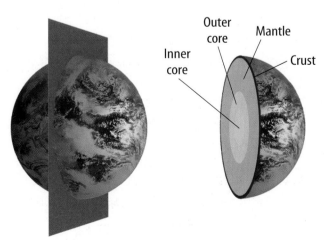

Figure 3
A Venn diagram shows how objects or concepts are alike and how they are different.

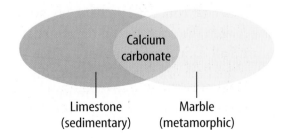

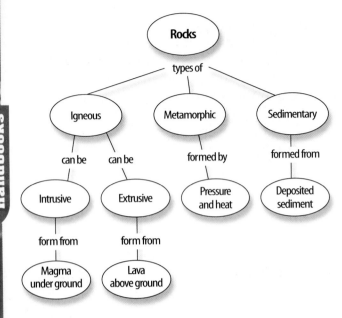

Figure 4
A network tree shows how concepts or objects are related.

Network Tree Look at the network tree in **Figure 4,** that shows the three main types of rock. A network tree is a type of concept map. Notice how some words are in ovals while others are written across connecting lines. The words inside the ovals are science terms or concepts. The words written on the connecting lines describe the relationships between the concepts.

When constructing a network tree, write the topic on a note card or piece of paper. Write the major concepts related to that topic on separate note cards or pieces of paper. Then arrange them in order from general to specific. Branch the related concepts from the major concept and describe the relationships on the connecting lines. Continue branching to more specific concepts. Write the relationships between the concepts on the connecting lines until all concepts are mapped. Then examine the network tree for relationships that cross branches, and add them to the network tree.

Events Chain An events chain is another type of concept map. It models the order of items or their sequence. In science, an events chain can be used to describe a sequence of events, the steps in a procedure, or the stages of a process.

When making an events chain, first find the one event that starts the chain. This event is called the *initiating event.* Then, find the next event in the chain and continue until you reach an outcome. Suppose you are asked to describe why and how a sound might make an echo. You might draw an events chain such as the one in **Figure 5.** Notice that connecting words are not necessary in an events chain.

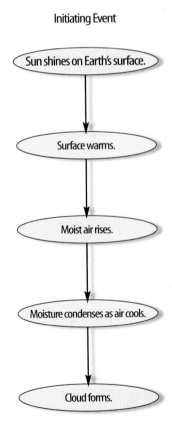

Figure 5
Events chains show the order of steps in a process or event.

Cycle Map A cycle concept map is a specific type of events chain map. In a cycle concept map, the series of events does not produce a final outcome. Instead, the last event in the chain relates back to the beginning event.

You first decide what event will be used as the beginning event. Once that is decided, you list events in order that occur after it. Words are written between events that describe what happens from one event to the next. The last event in a cycle concept map relates back to the beginning event. The number of events in a cycle concept varies but is usually three or more. Look at the cycle map shown in **Figure 6.**

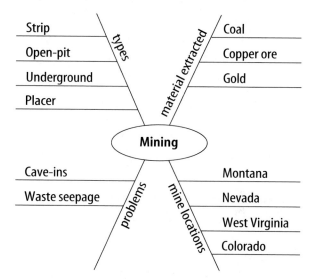

Figure 7

A spider map allows you to list ideas that relate to a central topic but not necessarily to one another.

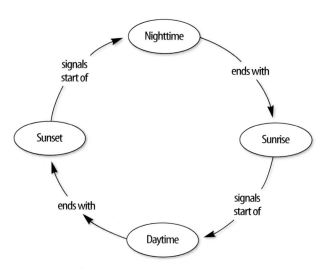

Figure 6

A cycle map shows events that occur in a cycle.

Spider Map A type of concept map that you can use for brainstorming is the spider map. When you have a central idea, you might find you have a jumble of ideas that relate to it but are not necessarily clearly related to each other. The spider map on mining in **Figure 7** shows that if you write these ideas outside the main concept, then you can begin to separate and group unrelated terms so they become more useful.

Writing a Paper

You will write papers often when researching science topics or reporting the results of investigations or experiments. Scientists frequently write papers to share their data and conclusions with other scientists and the public. When writing a paper, use these steps.

Step 1 Assemble your data by using graphs, tables, or a concept map. Create an outline.

Step 2 Start with an introduction that contains a clear statement of purpose and what you intend to discuss or prove.

Step 3 Organize the body into paragraphs. Each paragraph should start with a topic sentence, and the remaining sentences in that paragraph should support your point.

Step 4 Position data to help support your points.

Step 5 Summarize the main points and finish with a conclusion statement.

Step 6 Use tables, graphs, charts, and illustrations whenever possible.

Investigating and Experimenting

You might say the work of a scientist is to solve problems. When you decide to find out why one corner of your yard is always soggy, you are problem solving, too. You might observe that the corner is lower than the surrounding area and has less vegetation growing in it. You might decide to see whether planting some grass will keep the corner drier.

Scientists use orderly approaches to solve problems. The methods scientists use include identifying a question, making observations, forming a hypothesis, testing a hypothesis, analyzing results, and drawing conclusions.

Scientific investigations involve careful observation under controlled conditions. Such observation of an object or a process can suggest new and interesting questions about it. These questions sometimes lead to the formation of a hypothesis. Scientific investigations are designed to test a hypothesis.

Identifying a Question

The first step in a scientific investigation or experiment is to identify a question to be answered or a problem to be solved. You might be interested in knowing why a rock like the one in **Figure 8** looks the way it does.

Figure 8
When you find a rock, you might ask yourself, "How did this rock form?"

Forming Hypotheses

Hypotheses are based on observations that have been made. A hypothesis is a possible explanation based on previous knowledge and observations.

Perhaps a scientist has observed that thunderstorms happen more often on hot days than on cooler days. Based on these observations, the scientist can make a statement that he or she can test. The statement is a hypothesis. The hypothesis could be: *Warm temperatures cause thunderstorms.* A hypothesis has to be something you can test by using an investigation. A testable hypothesis is a valid hypothesis.

Predicting

When you apply a hypothesis, or general explanation, to a specific situation, you predict something about that situation. First, you must identify which hypothesis fits the situation you are considering. People use predictions to make everyday decisions. Based on previous observations and experiences, you might form a prediction that if warm temperatures cause thunderstorms, then more thunderstorms will occur in summer months than in spring months. Someone could use these predictions to plan when to take a camping trip or when to schedule an outdoor activity.

Testing a Hypothesis

To test a hypothesis, you need a procedure. A procedure is the plan you follow in your experiment. A procedure tells you what materials to use, as well as how and in what order to use them. When you follow a procedure, data are generated that support or do not support the original hypothesis statement.

For example, premium gasoline costs more than regular gasoline. Does premium gasoline increase the efficiency or fuel mileage of your family car? You decide to test the hypothesis: "If premium gasoline is more efficient, then it should increase the fuel mileage of my family's car." Then you write the procedure shown in **Figure 9** for your experiment and generate the data presented in the table below.

Figure 9
A procedure tells you what to do step by step.

Procedure

1. Use regular gasoline for two weeks.
2. Record the number of kilometers between fill-ups and the amount of gasoline used.
3. Switch to premium gasoline for two weeks.
4. Record the number of kilometers between fill-ups and the amount of gasoline used.

Gasoline Data			
Type of Gasoline	Kilometers Traveled	Liters Used	Liters per Kilometer
Regular	762	45.34	0.059
Premium	661	42.30	0.064

These data show that premium gasoline is less efficient than regular gasoline in one particular car. It took more gasoline to travel 1 km (0.064) using premium gasoline than it did to travel 1 km using regular gasoline (0.059). This conclusion does not support the hypothesis.

Are all investigations alike? Keep in mind as you perform investigations in science that a hypothesis can be tested in many ways. Not every investigation makes use of all the ways that are described on these pages, and not all hypotheses are tested by investigations. Scientists encounter many variations in the methods that are used when they perform experiments. The skills in this handbook are here for you to use and practice.

Identifying and Manipulating Variables and Controls

In any experiment, it is important to keep everything the same except for the item you are testing. The one factor you change is called the independent variable. The factor that changes as a result of the independent variable is called the dependent variable. Always make sure you have only one independent variable. If you allow more than one, you will not know what causes the changes you observe in the dependent variable. Many experiments also have controls—individual instances or experimental subjects for which the independent variable is not changed. You can then compare the test results to the control results.

For example, in the fuel-mileage experiment, you made everything the same except the type of gasoline that was used. The driver, the type of automobile, and the type of driving were the same throughout. In this way, you could be sure that any mileage differences were caused by the type of fuel—the independent variable. The fuel mileage was the dependent variable.

If you could repeat the experiment using several automobiles of the same type on a standard driving track with the same driver, you could make one automobile a control by using regular gasoline over the four-week period.

Collecting Data

Whether you are carrying out an investigation or a short observational experiment, you will collect data, or information. Scientists collect data accurately as numbers and descriptions and organize it in specific ways.

Observing Scientists observe items and events, then record what they see. When they use only words to describe an observation, it is called qualitative data. For example, a scientist might describe the color, texture, or odor of a substance produced in a chemical reaction. Scientists' observations also can describe how much there is of something. These observations use numbers, as well as words, in the description and are called quantitative data. For example, if a sample of the element gold is described as being "shiny and very dense," the data are clearly qualitative. Quantitative data on this sample of gold might include "a mass of 30 g and a density of 19.3 g/cm^3." Quantitative data often are organized into tables. Then, from information in the table, a graph can be drawn. Graphs can reveal relationships that exist in experimental data.

When you make observations in science, you should examine the entire object or situation first, then look carefully for details. If you're looking at a rock sample, for instance, check the general color and pattern of the rock before using a hand lens to examine the small mineral grains that make up its underlying structure. Remember to record accurately everything you see.

Scientists try to make careful and accurate observations. When possible, they use instruments such as microscopes, metric rulers, graduated cylinders, thermometers, and balances. Measurements provide numerical data that can be repeated and checked.

Sampling When working with large numbers of objects or a large population, scientists usually cannot observe or study every one of them. Instead, they use a sample or a portion of the total number. To *sample* is to take a small, representative portion of the objects or organisms of a population for research. By making careful observations or manipulating variables within a portion of a group, information is discovered and conclusions are drawn that might apply to the whole population.

Estimating Scientific work also involves estimating. To *estimate* is to make a judgment about the amount or the number of something without measuring every part of an object or counting every member of a population. Scientists first measure or count the amount or number in a small sample. A chemist, for example, might remove a 10-g piece of a large rock that is rich in copper ore. Then the chemist can determine the percentage of copper by mass and multiply that percentage by the mass of the rock to get an estimate of the total mass of copper in the rock. See **Figure 10** for another example.

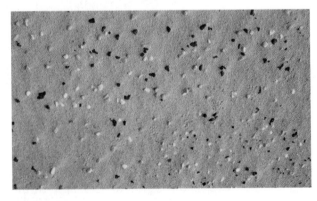

Figure 10
In a 1 m^2 frame positioned on a beach, count all the pebbles that you can see on the surface that are longer than 2.5 cm. Multiply this number by the area of the beach. This will give you an estimate for the total number of pebbles on the beach.

Measuring in SI

The metric system of measurement was developed in 1795. A modern form of the metric system, called the International System, or SI, was adopted in 1960. SI provides standard measurements that all scientists around the world can understand.

The metric system is convenient because unit sizes vary by multiples of 10. When changing from smaller units to larger units, divide by a multiple of 10. When changing from larger units to smaller, multiply by a multiple of 10. To convert millimeters to centimeters, divide the millimeters by 10. To convert 30 mm to centimeters, divide 30 by 10 (30 mm equal 3 cm).

Prefixes are used to name units. Look at the table below for some common metric prefixes and their meanings. Do you see how the prefix *kilo-* attached to the unit *gram* is *kilogram*, or 1,000 g?

Metric Prefixes			
Prefix	**Symbol**	**Meaning**	
kilo-	k	1,000	thousand
hecto-	h	100	hundred
deka-	da	10	ten
deci-	d	0.1	tenth
centi-	c	0.01	hundredth
milli-	m	0.001	thousandth

Now look at the metric ruler shown in **Figure 11.** The centimeter lines are the long, numbered lines, and the shorter lines are millimeter lines.

When using a metric ruler, line up the 0-cm mark with the end of the object being measured, and read the number of the unit where the object ends. In this instance, it would be 4.5 cm.

Figure 11
This metric ruler shows centimeters and millimeter divisions.

Liquid Volume In some science activities, you will measure liquids. The unit that is used to measure liquids is the liter. A liter has the volume of 1,000 cm³. The prefix *milli-* means "thousandth (0.001)." A milliliter is one thousandth of 1 L, and 1 L has the volume of 1,000 mL. One milliliter of liquid completely fills a cube measuring 1 cm on each side. Therefore, 1 mL equals 1 cm³.

You will use beakers and graduated cylinders to measure liquid volume. A graduated cylinder, as illustrated in **Figure 12,** is marked from bottom to top in milliliters. This one contains 79 mL of a liquid.

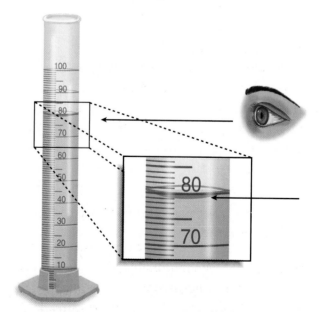

Figure 12
Graduated cylinders measure liquid volume.

Mass Scientists measure mass in grams. You might use a beam balance similar to the one shown in **Figure 13.** The balance has a pan on one side and a set of beams on the other side. Each beam has a rider that slides on the beam.

Before you find the mass of an object, slide all the riders back to the zero point. Check the pointer on the right to make sure it swings an equal distance above and below the zero point. If the swing is unequal, find and turn the adjusting screw until you have an equal swing.

Place an object on the pan. Slide the largest rider along its beam until the pointer drops below zero. Then move it back one notch. Repeat the process on each beam until the pointer swings an equal distance above and below the zero point. Sum the masses on each beam to find the mass of the object. Move all riders back to zero when finished.

Figure 13
A triple beam balance is used to determine the mass of an object.

You should never place a hot object on the pan or pour chemicals directly onto the pan. Instead, find the mass of a clean container. Remove the container from the pan, then place the chemicals in the container. Find the mass of the container with the chemicals in it. To find the mass of the chemicals, subtract the mass of the empty container from the mass of the filled container.

Making and Using Tables

Browse through your textbook and you will see tables in the text and in the activities. In a table, data, or information, are arranged so that they are easier to understand. Activity tables help organize the data you collect during an activity so results can be interpreted.

Making Tables To make a table, list the items to be compared in the first column and the characteristics to be compared in the first row. The title should clearly indicate the content of the table, and the column or row heads should tell the reader what information is found in there. The table below lists materials collected for recycling on three weekly pick-up days. The inclusion of kilograms in parentheses also identifies for the reader that the figures are mass units.

Recyclable Materials Collected During Week			
Day of Week	Paper (kg)	Aluminum (kg)	Glass (kg)
Monday	5.0	4.0	12.0
Wednesday	4.0	1.0	10.0
Friday	2.5	2.0	10.0

Using Tables How much paper, in kilograms, is being recycled on Wednesday? Locate the column labeled "Paper (kg)" and the row "Wednesday." The information in the box where the column and row intersect is the answer. Did you answer "4.0"? How much aluminum, in kilograms, is being recycled on Friday? If you answered "2.0," you understand how to read the table. How much glass is collected for recycling each week? Locate the column labeled "Glass (kg)" and add the figures for all three rows. If you answered "32.0," then you know how to locate and use the data provided in the table.

Recording Data

To be useful, the data you collect must be recorded carefully. Accuracy is key. A well-thought-out experiment includes a way to record procedures, observations, and results accurately. Data tables are one way to organize and record results. Set up the tables you will need ahead of time so you can record the data right away.

Record information properly and neatly. Never put unidentified data on scraps of paper. Instead, data should be written in a notebook like the one in **Figure 14.** Write in pencil so information isn't lost if your data gets wet. At each point in the experiment, record your data and label it. That way, your information will be accurate and you will not have to determine what the figures mean when you look at your notes later.

Figure 14
Record data neatly and clearly so it is easy to understand.

Recording Observations

It is important to record observations accurately and completely. That is why you always should record observations in your notes immediately as you make them. It is easy to miss details or make mistakes when recording results from memory. Do not include your personal thoughts when you record your data. Record only what you observe to eliminate bias. For example, when you record the time required for five students to climb the same set of stairs, you would note which student took the longest time. However, you would not refer to that student's time as "the worst time of all the students in the group."

Making Models

You can organize the observations and other data you collect and record in many ways. Making models is one way to help you better understand the parts of a structure you have been observing or the way a process for which you have been taking various measurements works.

Models often show things that are too large or too small for normal viewing. For example, you normally won't see the inside of an atom. However, you can understand the structure of the atom better by making a three-dimensional model of an atom. The relative sizes, the positions, and the movements of protons, neutrons, and electrons can be explained in words. An atomic model made of a plastic-ball nucleus and pipe-cleaner electron shells can help you visualize how the parts of the atom relate to each other.

Other models can be devised on a computer. Some models, such as those that illustrate the chemical combinations of different elements, are mathematical and are represented by equations.

Making and Using Graphs

After scientists organize data in tables, they might display the data in a graph that shows the relationship of one variable to another. A graph makes interpretation and analysis of data easier. Three types of graphs are the line graph, the bar graph, and the circle graph.

Line Graphs A line graph like in **Figure 15** is used to show the relationship between two variables. The variables being compared go on two axes of the graph. For data from an experiment, the independent variable always goes on the horizontal axis, called the x-axis. The dependent variable always goes on the vertical axis, called the y-axis. After drawing your axes, label each with a scale. Next, plot the data points.

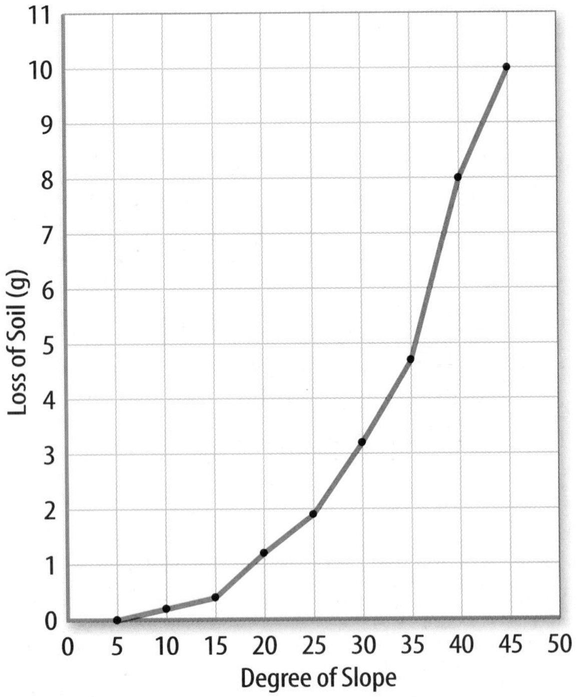

A data point is the intersection of the recorded value of the dependent variable for each tested value of the independent variable. After all the points are plotted, connect them.

Bar Graphs Bar graphs compare data that do not change continuously. Vertical bars show the relationships among data.

To make a bar graph, set up the y-axis as you did for the line graph. Draw vertical bars of equal size from the x-axis up to the point on the y-axis that represents the value of x.

Figure 16
The amount of aluminum collected for recycling during one week can be shown as a bar graph or circle graph.

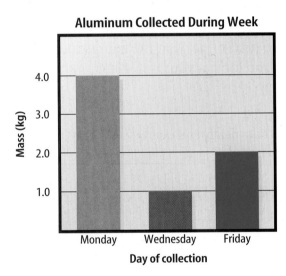

Circle Graphs A circle graph uses a circle divided into sections to display data as parts (fractions or percentages) of a whole. The size of each section corresponds to the fraction or percentage of the data that the section represents. So, the entire circle represents 100 percent, one-half represents 50 percent, one-fifth represents 20 percent, and so on.

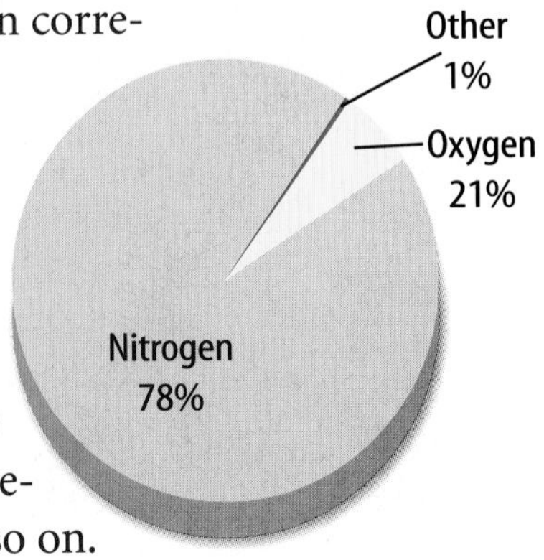

Figure 15
This line graph shows the relationship between degree of slope and the loss of soil in grams from a container during an experiment.

Analyzing Results

To determine the meaning of your observations and investigation results, you will need to look for patterns in the data. You can organize your information in several of the ways that are discussed in this handbook. Then you must think critically to determine what the data mean. Scientists use several approaches when they analyze the data they have collected and recorded. Each approach is useful for identifying specific patterns in the data.

Forming Operational Definitions

An operational definition defines an object by showing how it functions, works, or behaves. Such definitions are written in terms of how an object works or how it can be used; that is, they describe its job or purpose.

For example, a ruler can be defined as a tool that measures the length of an object (how it can be used). A ruler also can be defined as something that contains a series of marks that can be used as a standard when measuring (how it works).

Classifying

Classifying is the process of sorting objects or events into groups based on common features. When classifying, first observe the objects or events to be classified. Then select one feature that is shared by some members in the group but not by all. Place those members that share that feature into a subgroup. You can classify members into smaller and smaller subgroups based on characteristics.

How might you classify a group of rocks? You might first classify them by color, putting all of the black, white, and red rocks into separate groups. Within each group, you could then look for another common feature to classify further, such as size or whether the rocks have sharp or smooth edges.

Remember that when you classify, you are grouping objects or events for a purpose. For example, classifying rocks can be the first step in identifying them. You might know that obsidian is a black, shiny rock with sharp edges. To find it in a large group of rocks, you might start with the classification scheme mentioned. You'll locate obsidian within the group of black, sharp-edged rocks that you separate from the rest. Pumice could be located by its white color and by the fact that it contains many small holes called vesicles. Keep your purpose in mind as you select the features to form groups and subgroups.

Figure 17
Color is one of many characteristics that are used to classify rocks.

Skill Handbooks

Comparing and Contrasting

Observations can be analyzed by noting the similarities and differences between two or more objects or events that you observe. When you look at objects or events to see how they are similar, you are comparing them. Contrasting is looking for differences in objects or events. The table below compares and contrasts the characteristics of two minerals.

Mineral Characteristics		
Mineral	Graphite	Gold
Color	black	bright yellow
Hardness	1–2	2.5–3
Luster	metallic	metallic
Uses	pencil "lead"	jewelry, electronics

Recognizing Cause and Effect

Have you ever heard a loud pop right before the power went out and then suggested that an electric transformer probably blew out? If so, you have observed an effect and inferred a cause. The event is the effect, and the reason for the event is the cause.

When scientists are unsure of the cause of a certain event, they design controlled experiments to determine what caused it.

Interpreting Data

The word *interpret* means "to explain the meaning of something." Look at the problem originally being explored in an experiment and figure out what the data show. Identify the control group and the test group so you can see whether or not changes in the independent variable have had an effect. Look for differences in the dependent variable between the control and test groups.

These differences you observe can be qualitative or quantitative. You would be able to describe a qualitative difference using only words, whereas you would measure a quantitative difference and describe it using numbers. If there are differences, the independent variable that is being tested could have had an effect. If no differences are found between the control and test groups, the variable that is being tested apparently had no effect.

For example, suppose that three beakers each contain 100 mL of water. The beakers are placed on hot plates, and two of the hot plates are turned on, but the third is left off for a period of 5 min. Suppose you are then asked to describe any differences in the water in the three beakers. A qualitative difference might be the appearance of bubbles rising to the top in the water that is being heated but no rising bubbles in the unheated water. A quantitative difference might be a difference in the amount of water that is present in the beakers.

Inferring Scientists often make inferences based on their observations. An inference is an attempt to explain, or interpret, observations or to indicate what caused what you observed. An inference is a type of conclusion.

When making an inference, be certain to use accurate data and accurately described observations. Analyze all of the data that you've collected. Then, based on everything you know, explain or interpret what you've observed.

Drawing Conclusions

When scientists have analyzed the data they collected, they proceed to draw conclusions about what the data mean. These conclusions are sometimes stated using words similar to those found in the hypothesis formed earlier in the process.

Conclusions To analyze your data, you must review all of the observations and measurements that you made and recorded. Recheck all data for accuracy. After your data are rechecked and organized, you are almost ready to draw a conclusion such as "salt water boils at a higher temperature than freshwater."

Before you can draw a conclusion, however, you must determine whether the data allow you to come to a conclusion that supports a hypothesis. Sometimes that will be the case; other times it will not.

If your data do not support a hypothesis, it does not mean that the hypothesis is wrong. It means only that the results of the investigation did not support the hypothesis. Maybe the experiment needs to be redesigned, but very likely, some of the initial observations on which the hypothesis was based were incomplete or biased. Perhaps more observation or research is needed to refine the hypothesis.

Avoiding Bias Sometimes drawing a conclusion involves making judgments. When you make a judgment, you form an opinion about what your data mean. It is important to be honest and to avoid reaching a conclusion if there is no supporting evidence for it or if it is based on a small sample. It also is important not to allow any expectations of results to bias your judgments. If possible, it is a good idea to collect additional data. Scientists do this all the time.

For example, the *Hubble Space Telescope* was sent into space in April, 1990, to provide scientists with clearer views of the universe. *Hubble* is the size of a school bus and has a 2.4-m-diameter mirror. *Hubble* helped scientists answer questions about the planet Pluto.

For many years, scientists had only been able to hypothesize about the surface of the planet Pluto. *Hubble* has now provided pictures of Pluto's surface that show a rough texture with light and dark regions on it. This might be the best information about Pluto scientists will have until they are able to send a space probe to it.

Evaluating Others' Data and Conclusions

Sometimes scientists have to use data that they did not collect themselves, or they have to rely on observations and conclusions drawn by other researchers. In cases such as these, the data must be evaluated carefully.

How were the data obtained? How was the investigation done? Was it carried out properly? Has it been duplicated by other researchers? Were they able to follow the exact procedure? Did they come up with the same results? Look at the conclusion, as well. Would you reach the same conclusion from these results? Only when you have confidence in the data of others can you believe it is true and feel comfortable using it.

Communicating

The communication of ideas is an important part of the work of scientists. A discovery that is not reported will not advance the scientific community's understanding or knowledge. Communication among scientists also is important as a way of improving their investigations.

Scientists communicate in many ways, from writing articles in journals and magazines that explain their investigations and experiments, to announcing important discoveries on television and radio, to sharing ideas with colleagues on the Internet or presenting them as lectures.

Skill Handbooks

People who study science rely on computers to record and store data and to analyze results from investigations. Whether you work in a laboratory or just need to write a lab report with tables, good computer skills are a necessity.

Using a Word Processor

Suppose your teacher has assigned a written report. After you've completed your research and decided how you want to write the information, you need to put all that information on paper. The easiest way to do this is with a word processing application on a computer.

A computer application that allows you to type your information, change it as many times as you need to, and then print it out so that it looks neat and clean is called a word processing application. You also can use this type of application to create tables and columns, add bullets or cartoon art to your page, include page numbers, and check your spelling.

Helpful Hints

- If you aren't sure how to do something using your word processing program, look in the help menu. You will find a list of topics there to click on for help. After you locate the help topic you need, just follow the step-by-step instructions you see on your screen.
- Just because you've spell checked your report doesn't mean that the spelling is perfect. The spell check feature can't catch misspelled words that look like other words. If you've accidentally typed *mind* instead of *mine*, the spell checker won't know the difference. Always reread your report to make sure you didn't miss any mistakes.

Figure 18
You can use computer programs to make graphs and tables.

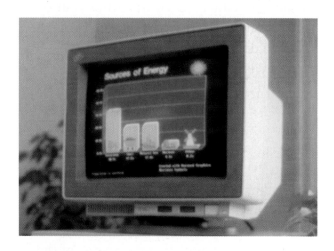

Using a Database

Imagine you're in the middle of a research project, busily gathering facts and information. You soon realize that it's becoming more difficult to organize and keep track of all the information. The tool to use to solve information overload is a database. Just as a file cabinet organizes paper records, a database organizes computer records. However, a database is more powerful than a simple file cabinet because at the click of a mouse, the contents can be reshuffled and reorganized. At computer-quick speeds, databases can sort information by any characteristics and filter data into multiple categories.

Helpful Hints

- Before setting up a database, take some time to learn the features of your database software by practicing with established database software.
- Periodically save your database as you enter data. That way, if something happens such as your computer malfunctions or the power goes off, you won't lose all of your work.

Doing a Database Search

When searching for information in a database, use the following search strategies to get the best results. These are the same search methods used for searching internet databases.

■ Place the word *and* between two words in your search if you want the database to look for any entries that have both the words. For example, "Earth *and* Mars" would give you information that mentions both Earth and Mars.

■ Place the word *or* between two words if you want the database to show entries that have at least one of the words. For example "Earth *or* Mars" would show you information that mentions either Earth or Mars.

■ Place the word *not* between two words if you want the database to look for entries that have the first word but do not have the second word. For example, "Moon *not* phases" would show you information that mentions the Moon but does not mention its phases.

In summary, databases can be used to store large amounts of information about a particular subject. Databases allow biologists, Earth scientists, and physical scientists to search for information quickly and accurately.

Using an Electronic Spreadsheet

Your science fair experiment has produced lots of numbers. How do you keep track of all the data, and how can you easily work out all the calculations needed? You can use a computer program called a spreadsheet to record data that involve numbers. A spreadsheet is an electronic mathematical worksheet.

Type your data in rows and columns, just as they would look in a data table on a sheet of paper. A spreadsheet uses simple math to do data calculations. For example, you could add, subtract, divide, or multiply any of the values in the spreadsheet by another number. You also could set up a series of math steps you want to apply to the data. If you want to add 12 to all the numbers and then multiply all the numbers by 10, the computer does all the calculations for you in the spreadsheet. Below is an example of a spreadsheet that records weather data.

Helpful Hints

■ Before you set up the spreadsheet, identify how you want to organize the data. Include any formulas you will need to use.

■ Make sure you have entered the correct data into the correct rows and columns.

■ You also can display your results in a graph. Pick the style of graph that best represents the data with which you are working.

Figure 19
A spreadsheet allows you to display large amounts of data and do calculations automatically.

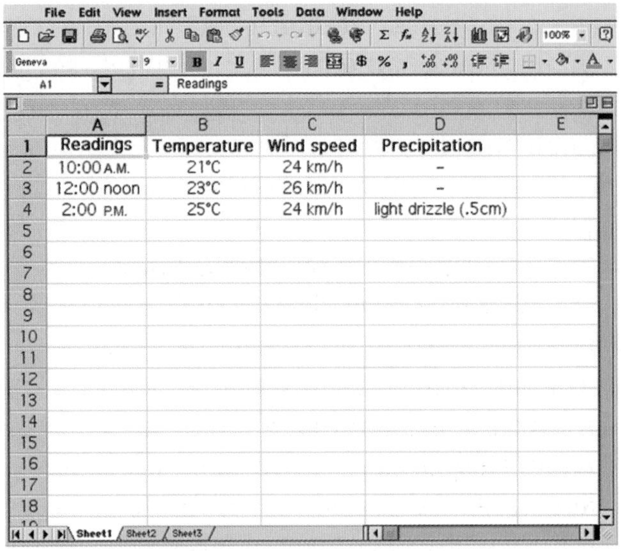

Using a Computerized Card Catalog

When you have a report or paper to research, you probably go to the library. To find the information you need in the library, you might have to use a computerized card catalog. This type of card catalog allows you to search for information by subject, by title, or by author. The computer then will display all the holdings the library has on the subject, title, or author requested.

A library's holdings can include books, magazines, databases, videos, and audio materials. When you have chosen something from this list, the computer will show whether an item is available and where in the library to find it.

Helpful Hints

■ Remember that you can use the computer to search by subject, author, or title. If you know a book's author but not the title, you can search for all the books the library has by that author.

■ When searching by subject, it's often most helpful to narrow your search by using specific search terms, such as *and, or,* and *not.* If you don't find enough sources, you can broaden your search.

■ Pay attention to the type of materials found in your search. If you need a book, you can eliminate any videos or other resources that come up in your search.

■ Knowing how your library is arranged can save you a lot of time. If you need help, the librarian will show you where certain types of materials are kept and how to find specific items.

Using Graphics Software

Are you having trouble finding that exact piece of art you're looking for? Do you have a picture in your mind of what you want but can't seem to find the right graphic to represent your ideas? To solve these problems, you can use graphics software. Graphics software allows you to create and change images and diagrams in almost unlimited ways. Typical uses for graphics software include arranging clip art, changing scanned images, and constructing pictures from scratch. Most graphics software applications work in similar ways. They use the same basic tools and functions. Once you master one graphics application, you can use other graphics applications.

Figure 20
Graphics software can use your data to draw bar graphs.

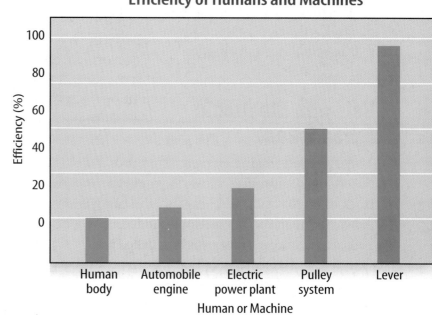

Efficiency of Humans and Machines

Figure 21
You can use this circle graph to find the names of the major gases that make up Earth's atmosphere.

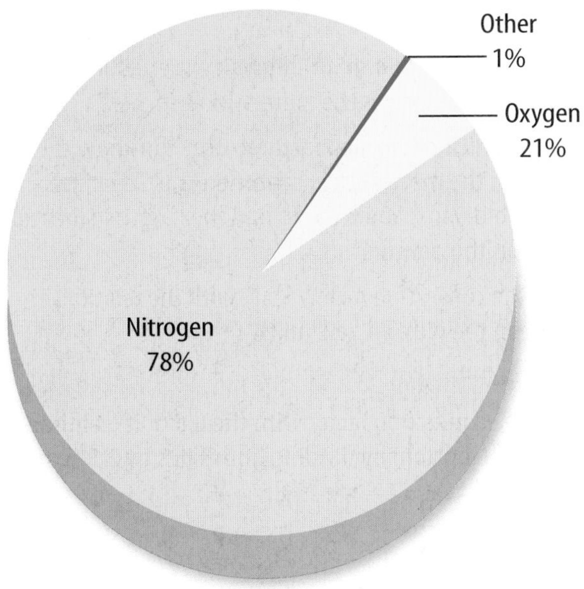

Other
1%

Oxygen
21%

Nitrogen
78%

Helpful Hints

- As with any method of drawing, the more you practice using the graphics software, the better your results will be.
- Start by using the software to manipulate existing drawings. Once you master this, making your own illustrations will be easier.
- Clip art is available on CD-ROMs and the Internet. With these resources, finding a piece of clip art to suit your purposes is simple.
- As you work on a drawing, save it often.

Developing Multimedia Presentations

It's your turn—you have to present your science report to the entire class. How do you do it? You can use many different sources of information to get the class excited about your presentation. Posters, videos, photographs, sound, computers, and the Internet can help show your ideas.

First, determine what important points you want to make in your presentation. Then, write an outline of what materials and types of media would best illustrate those points. Maybe you could start with an outline on an overhead projector, then show a video, followed by something from the Internet or a slide show accompanied by music or recorded voices. You might choose to use a presentation builder computer application that can combine all these elements into one presentation. Make sure the presentation is well constructed to make the most impact on the audience.

Figure 22
Multimedia presentations use many types of print and electronic materials.

Helpful Hints

- Carefully consider what media will best communicate the point you are trying to make.
- Make sure you know how to use any equipment you will be using in your presentation.
- Practice the presentation several times.
- If possible, set up all of the equipment ahead of time. Make sure everything is working correctly.

Math Skill Handbook

Use this Math Skill Handbook to help solve problems you are given in this text. You might find it useful to review topics in this Math Skill Handbook first.

Converting Units

In science, quantities such as length, mass, and time sometimes are measured using different units. Suppose you want to know how many miles are in 12.7 km.

Conversion factors are used to change from one unit of measure to another. A conversion factor is a ratio that is equal to one. For example, there are 1,000 mL in 1 L, so 1,000 mL equals 1 L, or:

$$1,000 \text{ mL} = 1 \text{ L}$$

If both sides are divided by 1 L, this equation becomes:

$$\frac{1,000 \text{ mL}}{1 \text{ L}} = 1$$

The **ratio** on the left side of this equation is equal to 1 and is a conversion factor. You can make another conversion factor by dividing both sides of the top equation by 1,000 mL:

$$1 = \frac{1 \text{ L}}{1,000 \text{ mL}}$$

To **convert units,** you multiply by the appropriate conversion factor. For example, how many milliliters are in 1.255 L? To convert 1.255 L to milliliters, multiply 1.255 L by a conversion factor.

Use the **conversion factor** with new units (mL) in the numerator and the old units (L) in the denominator.

$$1.255 \text{ L} \times \frac{1,000 \text{ mL}}{1 \text{ L}} = 1,255 \text{ mL}$$

The unit L divides in this equation, just as if it were a number.

Example 1 There are 2.54 cm in 1 inch. If a meterstick has a length of 100 cm, how long is the meterstick in inches?

Step 1 Decide which conversion factor to use. You know the length of the meterstick in centimeters, so centimeters are the old units. You want to find the length in inches, so inch is the new unit.

Step 2 Form the conversion factor. Start with the relationship between the old and new units.

$$2.54 \text{ cm} = 1 \text{ inch}$$

Step 3 Form the conversion factor with the old unit (centimeter) on the bottom by dividing both sides by 2.54 cm.

$$1 = \frac{2.54 \text{ cm}}{2.54 \text{ cm}} = \frac{1 \text{ inch}}{2.54 \text{ cm}}$$

Step 4 Multiply the old measurement by the conversion factor.

$$100 \text{ cm} \times \frac{1 \text{ inch}}{2.54 \text{ cm}} = 39.37 \text{ inches}$$

The meterstick is 39.37 inches long.

Example 2 There are 365 days in one year. If a person is 14 years old, what is his or her age in days? (Ignore leap years).

Step 1 Decide which conversion factor to use. You want to convert years to days.

Step 2 Form the conversion factor. Start with the relation between the old and new units.

$$1 \text{ year} = 365 \text{ days}$$

Step 3 Form the conversion factor with the old unit (year) on the bottom by dividing both sides by 1 year.

$$1 = \frac{1 \text{ year}}{1 \text{ year}} = \frac{365 \text{ days}}{1 \text{ year}}$$

Step 4 Multiply the old measurement by the conversion factor:

$$14 \text{ years} \times \frac{365 \text{ days}}{1 \text{ year}} = 5,110 \text{ days}$$

The person's age is 5,110 days.

Practice Problem A book has a mass of 2.31 kg. If there are 1,000 g in 1 kg, what is the mass of the book in grams?

Using Fractions

A **fraction** is a number that compares a part to the whole. For example, in the fraction $\frac{2}{3}$, the 2 represents the part and the 3 represents the whole. In the fraction $\frac{2}{3}$, the top number, 2, is called the numerator. The bottom number, 3, is called the denominator.

Sometimes fractions are not written in their simplest form. To determine a fraction's **simplest form,** you must find the greatest common factor (GCF) of the numerator and denominator. The greatest common factor is the largest common factor of all the factors the two numbers have in common.

For example, because the number 3 divides into 12 and 30 evenly, it is a common factor of 12 and 30. However, because the number 6 is the largest number that evenly divides into 12 and 30, it is the **greatest common factor.**

After you find the greatest common factor, you can write a fraction in its simplest form. Divide both the numerator and the denominator by the greatest common factor. The number that results is the fraction in its **simplest form.**

Example Twelve of the 20 peaks in a mountain range have elevations over 10,000 m. What fraction of the peaks in the mountain range are over 10,000 m? Write the fraction in simplest form.

Step 1 Write the fraction.

$$\frac{part}{whole} = \frac{12}{20}$$

Step 2 To find the GCF of the numerator and denominator, list all of the factors of each number.

Factors of 12: 1, 2, 3, 4, 6, 12 (the numbers that divide evenly into 12)

Factors of 20: 1, 2, 4, 5, 10, 20 (the numbers that divide evenly into 20)

Step 3 List the common factors.

1, 2, 4.

Step 4 Choose the greatest factor in the list of common factors.

The GCF of 12 and 20 is 4.

Step 5 Divide the numerator and denominator by the GCF.

$$\frac{12 \div 4}{20 \div 4} = \frac{3}{5}$$

In the mountain range, $\frac{3}{5}$ of the peaks are over 10,000 m.

Practice Problem There are 90 rides at an amusement park. Of those rides, 66 have a height restriction. What fraction of the rides has a height restriction? Write the fraction in simplest form.

Math Skill Handbook

Calculating Ratios

A **ratio** is a comparison of two numbers by division.

Ratios can be written 3 to 5 or 3:5. Ratios also can be written as fractions, such as $\frac{3}{5}$. Ratios, like fractions, can be written in simplest form. Recall that a fraction is in **simplest form** when the greatest common factor (GCF) of the numerator and denominator is 1.

Example A particular geologic sample contains 40 kg of shale and 64 kg of granite. What is the ratio of shale to granite as a fraction in simplest form?

Step 1 Write the ratio as a fraction. $\frac{\text{shale}}{\text{granite}} = \frac{40}{64}$

Step 2 Express the fraction in simplest form. The GCF of 40 and 64 is 8.

$$\frac{40}{64} = \frac{40 \div 8}{64 \div 8} = \frac{5}{8}$$

The ratio of shale to granite in the sample is $\frac{5}{8}$.

Practice Problem Two metal rods measure 100 cm and 144 cm in length. What is the ratio of their lengths in simplest fraction form?

Using Decimals

A **decimal** is a fraction with a denominator of 10, 100, 1,000, or another power of 10. For example, 0.854 is the same as the fraction $\frac{854}{1,000}$.

In a decimal, the decimal point separates the ones place and the tenths place. For example, 0.27 means twenty-seven hundredths, or $\frac{27}{100}$, where 27 is the **number of units** out of 100 units. Any fraction can be written as a decimal using division.

Example Write $\frac{5}{8}$ as a decimal.

Step 1 Write a division problem with the numerator, 5, as the dividend and the denominator, 8, as the divisor. Write 5 as 5.000.

Step 2 Solve the problem.

```
    0.625
8)5.000
  48
   20
   16
    40
    40
     0
```

Therefore, $\frac{5}{8} = 0.625$.

Practice Problem Write $\frac{19}{25}$ as a decimal.

Using Percentages

The word *percent* means "out of one hundred." A **percent** is a ratio that compares a number to 100. Suppose you read that 77 percent of Earth's surface is covered by water. That is the same as reading that the fraction of Earth's surface covered by water is $\frac{77}{100}$. To express a fraction as a percent, first find an equivalent decimal for the fraction. Then, multiply the decimal by 100 and add the percent symbol. For example, $\frac{1}{2} = 1 \div 2 = 0.5$. Then $0.5 \cdot 100 = 50 = 50\%$.

Example Express $\frac{13}{20}$ as a percent.

Step 1 Find the equivalent decimal for the fraction.

$$20)\overline{13.00} \quad \begin{array}{r} 0.65 \\ \hline \end{array}$$
$$\begin{array}{r} 120 \\ \hline 100 \\ 100 \\ \hline 0 \end{array}$$

Step 2 Rewrite the fraction $\frac{13}{20}$ as 0.65.

Step 3 Multiply 0.65 by 100 and add the % sign.

$$0.65 \cdot 100 = 65 = 65\%$$

So, $\frac{13}{20} = 65\%$.

Practice Problem In one year, 73 of 365 days were rainy in one city. What percent of the days in that city were rainy?

Using Precision and Significant Digits

When you make a **measurement,** the value you record depends on the precision of the measuring instrument. When adding or subtracting numbers with different precision, the answer is rounded to the smallest number of decimal places of any number in the sum or difference. When multiplying or dividing, the answer is rounded to the smallest number of significant figures of any number being multiplied or divided. When counting the number of **significant figures,** all digits are counted except zeros at the end of a number with no decimal such as 2,500, and zeros at the beginning of a decimal such as 0.03020.

Example The lengths 5.28 and 5.2 are measured in meters. Find the sum of these lengths and report the sum using the least precise measurement.

Step 1 Find the sum.

5.28 m	2 digits after the decimal
+ 5.2 m	1 digit after the decimal
10.48 m	

Step 2 Round to one digit after the decimal because the least number of digits after the decimal of the numbers being added is 1.

The sum is 10.5 m.

Practice Problem Multiply the numbers in the example using the rule for multiplying and dividing. Report the answer with the correct number of significant figures.

Solving One-Step Equations

An **equation** is a statement that two things are equal. For example, $A = B$ is an equation that states that A is equal to B.

Sometimes one side of the equation will contain a **variable** whose value is not known. In the equation $3x = 12$, the variable is x.

The equation is solved when the variable is replaced with a value that makes both sides of the equation equal to each other. For example, the solution of the equation $3x = 12$ is $x = 4$. If the x is replaced with 4, then the equation becomes $3 \cdot 4 = 12$, or $12 = 12$.

To solve an equation such as $8x = 40$, divide both sides of the equation by the number that multiplies the variable.

$$8x = 40$$
$$\frac{8x}{8} = \frac{40}{8}$$
$$x = 5$$

You can check your answer by replacing the variable with your solution and seeing if both sides of the equation are the same.

$$8x = 8 \cdot 5 = 40$$

The left and right sides of the equation are the same, so $x = 5$ is the solution.

Sometimes an equation is written in this way: $a = bc$. This also is called a **formula.** The letters can be replaced by numbers, but the numbers must still make both sides of the equation the same.

Example 1 Solve the equation $10x = 35$.

Step 1 Find the solution by dividing each side of the equation by 10.

$$10x = 35 \qquad \frac{10x}{10} = \frac{35}{10} \qquad x = 3.5$$

Step 2 Check the solution.

$$10x = 35 \qquad 10 \times 3.5 = 35 \qquad 35 = 35$$

Both sides of the equation are equal, so $x = 3.5$ is the solution to the equation.

Example 2 In the formula $a = bc$, find the value of c if $a = 20$ and $b = 2$.

Step 1 Rearrange the formula so the unknown value is by itself on one side of the equation by dividing both sides by b.

$$a = bc$$
$$\frac{a}{b} = \frac{bc}{b}$$
$$\frac{a}{b} = c$$

Step 2 Replace the variables a and b with the values that are given.

$$\frac{a}{b} = c$$
$$\frac{20}{2} = c$$
$$10 = c$$

Step 3 Check the solution.

$$a = bc$$
$$20 = 2 \times 10$$
$$20 = 20$$

Both sides of the equation are equal, so $c = 10$ is the solution when $a = 20$ and $b = 2$.

Practice Problem In the formula $h = gd$, find the value of d if $g = 12.3$ and $h = 17.4$.

Using Proportions

A **proportion** is an equation that shows that two ratios are equivalent. The ratios $\frac{2}{4}$ and $\frac{5}{10}$ are equivalent, so they can be written as $\frac{2}{4} = \frac{5}{10}$. This equation is an example of a proportion.

When two ratios form a proportion, the **cross products** are equal. To find the cross products in the proportion $\frac{2}{4} = \frac{5}{10}$, multiply the 2 and the 10, and the 4 and the 5. Therefore $2 \cdot 10 = 4 \cdot 5$, or $20 = 20$.

Because you know that both proportions are equal, you can use cross products to find a missing term in a proportion. This is known as **solving the proportion.** Solving a proportion is similar to solving an equation.

Example The heights of a tree and a pole are proportional to the lengths of their shadows. The tree casts a shadow of 24 m at the same time that a 6-m pole casts a shadow of 4 m. What is the height of the tree?

Step 1 Write a proportion.

$$\frac{\text{height of tree}}{\text{height of pole}} = \frac{\text{length of tree's shadow}}{\text{length of pole's shadow}}$$

Step 2 Substitute the known values into the proportion. Let h represent the unknown value, the height of the tree.

$$\frac{h}{6} = \frac{24}{4}$$

Step 3 Find the cross products.

$$h \cdot 4 = 6 \cdot 24$$

Step 4 Simplify the equation.

$$4h = 144$$

Step 5 Divide each side by 4.

$$\frac{4h}{4} = \frac{144}{4}$$

$$h = 36$$

The height of the tree is 36 m.

Practice Problem The ratios of the weights of two objects on the Moon and on Earth are in proportion. A rock weighing 3 N on the Moon weighs 18 N on Earth. How much would a rock that weighs 5 N on the Moon weigh on Earth?

Math Skill Handbook

Statistics is the branch of mathematics that deals with collecting, analyzing, and presenting data. In statistics, there are three common ways to summarize the data with a single number—the mean, the median, and the mode.

The **mean** of a set of data is the arithmetic average. It is found by adding the numbers in the data set and dividing by the number of items in the set.

The **median** is the middle number in a set of data when the data are arranged in numerical order. If there were an even number of data points, the median would be the mean of the two middle numbers.

The **mode** of a set of data is the number or item that appears most often.

Another number that often is used to describe a set of data is the range. The **range** is the difference between the largest number and the smallest number in a set of data.

A **frequency table** shows how many times each piece of data occurs, usually in a survey. The frequency table below shows the results of a student survey on favorite color.

Color	Tally	Frequency
red	||||	4
blue	⊬⊬⊬	5
black	||	2
green	|||	3
purple	⊬⊬⊬ ||	7
yellow	⊬⊬⊬ |	6

Based on the frequency table data, which color is the favorite?

Example The high temperatures (in °C) on five consecutive days at a desert observation station are 39°, 37°, 44°, 36°, and 44°. Find the mean, median, mode, and range of this set.

To find the mean:

Step 1 Find the sum of the numbers.

$$39 + 37 + 44 + 36 + 44 = 200$$

Step 2 Divide the sum by the number of items, which is 5.

$$200 \div 5 = 40$$

The mean high temperature is 40°C.

To find the median:

Step 1 Arrange the temperatures from least to greatest.

$$36, \ 37, \ \underline{39}, \ 44, \ 44$$

Step 2 Determine the middle temperature.

The median high temperature is 39°C.

To find the mode:

Step 1 Group the numbers that are the same together.

44, 44, 36, 37, 39

Step 2 Determine the number that occurs most in the set.

$\underline{44, 44}$, 36, 37, 39

The mode measure is 44°C.

To find the range:

Step 1 Arrange the temperatures from largest to smallest.

44, 44, 39, 37, 36

Step 2 Determine the largest and smallest temperature in the set.

$\underline{44}$, 44, 39, 37, $\underline{36}$

Step 3 Find the difference between the largest and smallest temperatures.

$$44 - 36 = 8$$

The range is 8°C.

Practice Problem Find the mean, median, mode, and range for the data set 8, 4, 12, 8, 11, 14, 16.

Safety in the Science Classroom

1. Always obtain your teacher's permission to begin an investigation.

2. Study the procedure. If you have questions, ask your teacher. Be sure you understand any safety symbols shown on the page.

3. Use the safety equipment provided for you. Goggles and a safety apron should be worn during most investigations.

4. Always slant test tubes away from yourself and others when heating them or adding substances to them.

5. Never eat or drink in the lab, and never use lab glassware as food or drink containers. Never inhale chemicals. Do not taste any substances or draw any material into a tube with your mouth.

6. Report any spill, accident, or injury, no matter how small, immediately to your teacher; then follow his or her instructions.

7. Know the location and proper use of the fire extinguisher, safety shower, fire blanket, first aid kit, and fire alarm.

8. Keep all materials away from open flames. Tie back long hair and tie down loose clothing.

9. If your clothing should catch fire, smother it with the fire blanket, or get under a safety shower. NEVER RUN.

10. If a fire should occur, turn off the gas; then leave the room according to established procedures.

Follow these procedures as you clean up your work area

1. Turn off the water and gas. Disconnect electrical devices.

2. Clean all pieces of equipment and return all materials to their proper places.

3. Dispose of chemicals and other materials as directed by your teacher. Place broken glass and solid substances in the proper containers. Make sure never to discard materials in the sink.

4. Clean your work area. Wash your hands thoroughly after working in the laboratory.

First Aid	
Injury	**Safe Response ALWAYS NOTIFY YOUR TEACHER IMMEDIATELY**
Burns	Apply cold water.
Cuts and Bruises	Stop any bleeding by applying direct pressure. Cover cuts with a clean dressing. Apply ice packs or cold compresses to bruises.
Fainting	Leave the person lying down. Loosen any tight clothing and keep crowds away.
Foreign Matter in Eye	Flush with plenty of water. Use eyewash bottle or fountain.
Poisoning	Note the suspected poisoning agent.
Any Spills on Skin	Flush with large amounts of water or use safety shower.

REFERENCE Handbook B

SI—Metric/English, English/Metric Conversions

	When you want to convert:	To:	Multiply by:
Length	inches	centimeters	2.54
	centimeters	inches	0.39
	yards	meters	0.91
	meters	yards	1.09
	miles	kilometers	1.61
	kilometers	miles	0.62
Mass and Weight*	ounces	grams	28.35
	grams	ounces	0.04
	pounds	kilograms	0.45
	kilograms	pounds	2.2
	tons (short)	tonnes (metric tons)	0.91
	tonnes (metric tons)	tons (short)	1.10
	pounds	newtons	4.45
	newtons	pounds	0.22
Volume	cubic inches	cubic centimeters	16.39
	cubic centimeters	cubic inches	0.06
	liters	quarts	1.06
	quarts	liters	0.95
	gallons	liters	3.78
Area	square inches	square centimeters	6.45
	square centimeters	square inches	0.16
	square yards	square meters	0.83
	square meters	square yards	1.19
	square miles	square kilometers	2.59
	square kilometers	square miles	0.39
	hectares	acres	2.47
	acres	hectares	0.40
Temperature	To convert °Celsius to °Fahrenheit	°C × 9/5 + 32	
	To convert °Fahrenheit to °Celsius	5/9 (°F − 32)	

*Weight is measured in standard Earth gravity.

Weather Map Symbols

Sample Station Model

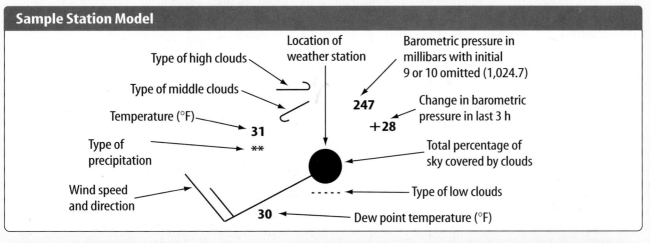

Type of high clouds

Type of middle clouds

Temperature (°F)

Type of precipitation

Wind speed and direction

Location of weather station

Barometric pressure in millibars with initial 9 or 10 omitted (1,024.7)

247

+28

Change in barometric pressure in last 3 h

Total percentage of sky covered by clouds

Type of low clouds

Dew point temperature (°F)

31

**

30

Sample Plotted Report at Each Station

Precipitation		Wind Speed and Direction		Sky Coverage		Some Types of High Clouds	
≡	Fog	○	0 calm	○	No cover	⌐⌐⌐⌐>	Scattered cirrus
★	Snow	╱	1-2 knots	◔	1/10 or less	⌐⌐⌐>	Dense cirrus in patches
●	Rain	╲	3-7 knots	◕	2/10 to 3/10	⌐⌐⌐⌐C	Veil of cirrus covering entire sky
⊺	Thunderstorm	╲	8-12 knots	◑	4/10	⌐⌐⌐C	Cirrus not covering entire sky
,	Drizzle	╲	13-17 knots	◐	–		
▽	Showers	╲	18-22 knots	◒	6/10		
		╲	23-27 knots	◓	7/10		
		╲	48-52 knots	◉	Overcast with openings		
		1 knot = 1.852 km/h		●	Completely overcast		

Some Types of Middle Clouds		Some Types of Low Clouds		Fronts and Pressure Systems	
∠	Thin altostratus layer	⌒	Cumulus of fair weather	(H) or High (L) or Low	Center of high- or low-pressure system
⫽	Thick altostratus layer	⌣	Stratocumulus	▲▲▲▲	Cold front
⟋	Thin altostratus in patches	-----	Fractocumulus of bad weather	●●●●	Warm front
⟋	Thin altostratus in bands	—	Stratus of fair weather	▲●▲▲	Occluded front
				●▲●▲	Stationary front

REFERENCE HANDBOOK D

Reference Handbook

Minerals

Mineral (formula)	Color	Streak	Hardness	Breakage Pattern	Uses and Other Properties
Graphite (C)	black to gray	black to gray	1–1.5	basal cleavage (scales)	pencil lead, lubricants for locks, rods to control some small nuclear reactions, battery poles
Galena (PbS)	gray	gray to black	2.5	cubic cleavage perfect	source of lead, used for pipes, shields for X rays, fishing equipment sinkers
Hematite (Fe_2O_3)	black or reddish-brown	reddish-brown	5.5–6.5	irregular fracture	source of iron; converted to pig iron, made into steel
Magnetite (Fe_3O_4)	black	black	6	conchoidal fracture	source of iron, attracts a magnet
Pyrite (FeS_2)	light, brassy, yellow	greenish-black	6–6.5	uneven fracture	fool's gold
Talc ($Mg_3 Si_4 O_{10} (OH)_2$)	white, greenish	white	1	cleavage in one direction	used for talcum powder, sculptures, paper, and tabletops
Gypsum ($CaSO_4 \cdot 2H_2O$)	colorless, gray, white, brown	white	2	basal cleavage	used in plaster of paris and dry wall for building construction
Sphalerite (ZnS)	brown, reddish-brown, greenish	light to dark brown	3.5–4	cleavage in six directions	main ore of zinc; used in paints, dyes, and medicine
Muscovite ($KAl_3Si_3 O_{10}(OH)_2$)	white, light gray, yellow, rose, green	colorless	2–2.5	basal cleavage	occurs in large, flexible plates; used as an insulator in electrical equipment, lubricant
Biotite ($K(Mg,Fe)_3 (AlSi_3O_{10}) (OH)_2$)	black to dark brown	colorless	2.5–3	basal cleavage	occurs in large, flexible plates
Halite (NaCl)	colorless, red, white, blue	colorless	2.5	cubic cleavage	salt; soluble in water; a preservative

Minerals

Mineral (formula)	Color	Streak	Hardness	Breakage Pattern	Uses and Other Properties
Calcite ($CaCO_3$)	colorless, white, pale blue	colorless, white	3	cleavage in three directions	fizzes when HCl is added; used in cements and other building materials
Dolomite ($CaMg(CO_3)_2$)	colorless, white, pink, green, gray, black	white	3.5–4	cleavage in three directions	concrete and cement; used as an ornamental building stone
Fluorite (CaF_2)	colorless, white, blue, green, red, yellow, purple	colorless	4	cleavage in four directions	used in the manufacture of optical equipment; glows under ultraviolet light
Hornblende ($(CaNa)_{2-3}$ $(Mg,Al,$ $Fe)_5-(Al,Si)_2$ Si_6O_{22} $(OH)_2)$	green to black	gray to white	5–6	cleavage in two directions	will transmit light on thin edges; 6-sided cross section
Feldspar ($KAlSi_3O_8$) $(NaAl$ $Si_3O_8),$ $(CaAl_2Si_2$ $O_8)$	colorless, white to gray, green	colorless	6	two cleavage planes meet at 90° angle	used in the manufacture of ceramics
Augite ((Ca,Na) (Mg,Fe,Al) $(Al,Si)_2 O_6)$	black	colorless	6	cleavage in two directions	square or 8-sided cross section
Olivine ($(Mg,Fe)_2$ $SiO_4)$	olive, green	none	6.5–7	conchoidal fracture	gemstones, refractory sand
Quartz (SiO_2)	colorless, various colors	none	7	conchoidal fracture	used in glass manufacture, electronic equipment, radios, computers, watches, gemstones

REFERENCE HANDBOOK E

Rocks		
Rock Type	**Rock Name**	**Characteristics**
Igneous (intrusive)	Granite	Large mineral grains of quartz, feldspar, hornblende, and mica. Usually light in color.
	Diorite	Large mineral grains of feldspar, hornblende, and mica. Less quartz than granite. Intermediate in color.
	Gabbro	Large mineral grains of feldspar, augite, and olivine. No quartz. Dark in color.
Igneous (extrusive)	Rhyolite	Small mineral grains of quartz, feldspar, hornblende, and mica, or no visible grains. Light in color.
	Andesite	Small mineral grains of feldspar, hornblende, and mica or no visible grains. Intermediate in color.
	Basalt	Small mineral grains of feldspar, augite, and olivine or no visible grains. No quartz. Dark in color.
	Obsidian	Glassy texture. No visible grains. Volcanic glass. Fracture looks like broken glass.
	Pumice	Frothy texture. Floats in water. Usually light in color.
Sedimentary (detrital)	Conglomerate	Coarse grained. Gravel or pebble size grains.
	Sandstone	Sand-sized grains 1/16 to 2 mm.
	Siltstone	Grains are smaller than sand but larger than clay.
	Shale	Smallest grains. Often dark in color. Usually platy.
Sedimentary (chemical or organic)	Limestone	Major mineral is calcite. Usually forms in oceans, lakes, and caves. Often contains fossils.
	Coal	Occurs in swampy areas. Compacted layers of organic material, mainly plant remains.
Sedimentary (chemical)	Rock Salt	Commonly forms by the evaporation of seawater.
Metamorphic (foliated)	Gneiss	Banding due to alternate layers of different minerals, of different colors. Parent rock often is granite.
	Schist	Parallel arrangement of sheetlike minerals, mainly micas. Forms from different parent rocks.
	Phyllite	Shiny or silky appearance. May look wrinkled. Common parent rocks are shale and slate.
	Slate	Harder, denser, and shinier than shale. Common parent rock is shale.
Metamorphic (non-foliated)	Marble	Calcite or dolomite. Common parent rock is limestone.
	Soapstone	Mainly of talc. Soft with greasy feel.
	Quartzite	Hard with interlocking quartz crystals. Common parent rock is sandstone.

Topographic Map Symbols

━━━━━━	Primary highway, hard surface	∿∿∿	Index contour
▬▬▬▬	Secondary highway, hard surface	··········	Supplementary contour
══════	Light-duty road, hard or improved surface	∿∿∿	Intermediate contour
=========	Unimproved road	🖾	Depression contours
┼──┼──┼──┼	Railroad: single track		
┼╪═╪═╪═	Railroad: multiple track	━ ━ ━ ━	Boundaries: national
┼╪┼╪┼╪┼	Railroads in juxtaposition	── ── ──	State
		── ── ── ──	County, parish, municipal
▫▪▄▆	Buildings	── ── ──	Civil township, precinct, town, barrio
▴▴⊞ cem	Schools, church, and cemetery	─·─·─·─	Incorporated city, village, town, hamlet
▫▭▨	Buildings (barn, warehouse, etc)	·─·─·─··	Reservation, national or state
○ ○	Wells other than water (labeled as to type)	──────────	Small park, cemetery, airport, etc.
●●● ⊘	Tanks: oil, water, etc. (labeled only if water)	─··─··─··	Land grant
⊙ ⚐	Located or landmark object; windmill	───────	Township or range line, U.S. land survey
⚹ ✕	Open pit, mine, or quarry; prospect	─ ─ ─ ─	Township or range line, approximate location
🗺	Marsh (swamp)		
🗺	Wooded marsh	∿∿	Perennial streams
▢	Woods or brushwood	→──←	Elevated aqueduct
⠿	Vineyard	○ ∿	Water well and spring
⠿	Land subject to controlled inundation	∿⊬	Small rapids
🗺	Submerged marsh	∿∿	Large rapids
🗺	Mangrove	🗺	Intermittent lake
⠿	Orchard	∿∿	Intermittent stream
⠿	Scrub	→=====←	Aqueduct tunnel
▢	Urban area	🗺	Glacier
		∿╳	Small falls
x7369	Spot elevation	🗺	Large falls
670	Water elevation	🗺	Dry lake bed

PERIODIC TABLE OF THE ELEMENTS

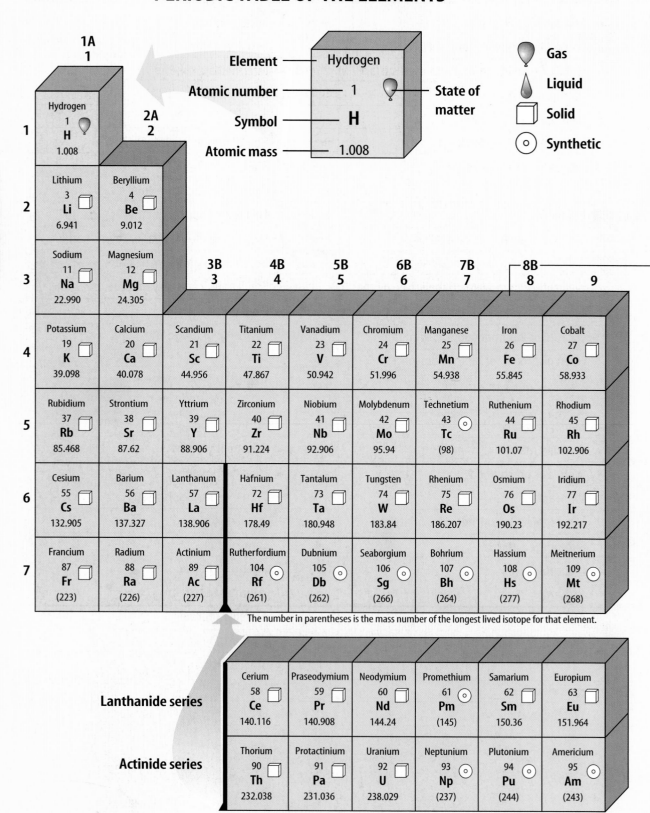

Element — Hydrogen
Atomic number — 1
Symbol — H
Atomic mass — 1.008

State of matter

Gas
Liquid
Solid
Synthetic

The number in parentheses is the mass number of the longest lived isotope for that element.

1A
1

2A
2

3B 3 **4B 4** **5B 5** **6B 6** **7B 7** **8B 8** **9**

1	Hydrogen 1 **H** 1.008								
2	Lithium 3 **Li** 6.941	Beryllium 4 **Be** 9.012							
3	Sodium 11 **Na** 22.990	Magnesium 12 **Mg** 24.305							
4	Potassium 19 **K** 39.098	Calcium 20 **Ca** 40.078	Scandium 21 **Sc** 44.956	Titanium 22 **Ti** 47.867	Vanadium 23 **V** 50.942	Chromium 24 **Cr** 51.996	Manganese 25 **Mn** 54.938	Iron 26 **Fe** 55.845	Cobalt 27 **Co** 58.933
5	Rubidium 37 **Rb** 85.468	Strontium 38 **Sr** 87.62	Yttrium 39 **Y** 88.906	Zirconium 40 **Zr** 91.224	Niobium 41 **Nb** 92.906	Molybdenum 42 **Mo** 95.94	Technetium 43 **Tc** (98)	Ruthenium 44 **Ru** 101.07	Rhodium 45 **Rh** 102.906
6	Cesium 55 **Cs** 132.905	Barium 56 **Ba** 137.327	Lanthanum 57 **La** 138.906	Hafnium 72 **Hf** 178.49	Tantalum 73 **Ta** 180.948	Tungsten 74 **W** 183.84	Rhenium 75 **Re** 186.207	Osmium 76 **Os** 190.23	Iridium 77 **Ir** 192.217
7	Francium 87 **Fr** (223)	Radium 88 **Ra** (226)	Actinium 89 **Ac** (227)	Rutherfordium 104 **Rf** (261)	Dubnium 105 **Db** (262)	Seaborgium 106 **Sg** (266)	Bohrium 107 **Bh** (264)	Hassium 108 **Hs** (277)	Meitnerium 109 **Mt** (268)

Lanthanide series

| Cerium 58 **Ce** 140.116 | Praseodymium 59 **Pr** 140.908 | Neodymium 60 **Nd** 144.24 | Promethium 61 **Pm** (145) | Samarium 62 **Sm** 150.36 | Europium 63 **Eu** 151.964 |

Actinide series

| Thorium 90 **Th** 232.038 | Protactinium 91 **Pa** 231.036 | Uranium 92 **U** 238.029 | Neptunium 93 **Np** (237) | Plutonium 94 **Pu** (244) | Americium 95 **Am** (243) |

Reference Handbook

REFERENCE HANDBOOK G

SCIENCE Online
Visit the Glencoe Science Web site at **science.glencoe.com** for updates to the periodic table.

Metal

Metalloid

Nonmetal

Recently discovered

| | | | | | | | | **8A** |
| | | | | | | | | **18** |

								Helium
								2
								He
								4.003

| | | **3A** | **4A** | **5A** | **6A** | **7A** | |
| | | **13** | **14** | **15** | **16** | **17** | |

		Boron	Carbon	Nitrogen	Oxygen	Fluorine	Neon
		5	6	7	8	9	10
		B	**C**	**N**	**O**	**F**	**Ne**
		10.811	12.011	14.007	15.999	18.998	20.180

1B	**2B**	Aluminum	Silicon	Phosphorus	Sulfur	Chlorine	Argon
11	**12**	13	14	15	16	17	18
		Al	**Si**	**P**	**S**	**Cl**	**Ar**
		26.982	28.086	30.974	32.065	35.453	39.948

10

Nickel	Copper	Zinc	Gallium	Germanium	Arsenic	Selenium	Bromine	Krypton
28	29	30	31	32	33	34	35	36
Ni	**Cu**	**Zn**	**Ga**	**Ge**	**As**	**Se**	**Br**	**Kr**
58.693	63.546	65.39	69.723	72.64	74.922	78.96	79.904	83.80
Palladium	Silver	Cadmium	Indium	Tin	Antimony	Tellurium	Iodine	Xenon
46	47	48	49	50	51	52	53	54
Pd	**Ag**	**Cd**	**In**	**Sn**	**Sb**	**Te**	**I**	**Xe**
106.42	107.868	112.411	114.818	118.710	121.760	127.60	126.904	131.293
Platinum	Gold	Mercury	Thallium	Lead	Bismuth	Polonium	Astatine	Radon
78	79	80	81	82	83	84	85	86
Pt	**Au**	**Hg**	**Tl**	**Pb**	**Bi**	**Po**	**At**	**Rn**
195.078	196.967	200.59	204.383	207.2	208.980	(209)	(210)	(222)
Ununnilium	Unununium	Ununbium		Ununquadium		Ununhexium		Ununoctium
* 110	* 111	* 112		* 114		* 116		* 118
Uun	**Uuu**	**Uub**		**Uuq**		**Uuh**		**Uuo**
(281)	(272)	(285)		(289)		(289)		(293)

* Names not officially assigned. Discovery of elements 114, 116, and 118 recently reported. Further information not yet available.

Gadolinium	Terbium	Dysprosium	Holmium	Erbium	Thulium	Ytterbium	Lutetium
64	65	66	67	68	69	70	71
Gd	**Tb**	**Dy**	**Ho**	**Er**	**Tm**	**Yb**	**Lu**
157.25	158.925	162.50	164.930	167.259	168.934	173.04	174.967
Curium	Berkelium	Californium	Einsteinium	Fermium	Mendelevium	Nobelium	Lawrencium
96	97	98	99	100	101	102	103
Cm	**Bk**	**Cf**	**Es**	**Fm**	**Md**	**No**	**Lr**
(247)	(247)	(251)	(252)	(257)	(258)	(259)	(262)

English Glossary

This glossary defines each key term that appears in bold type in the text. It also shows the chapter, section, and page number where you can find the word(s) used.

A

abrasion: a type of erosion that occurs when windblown sediments strike rocks and sediments, polishing and pitting their surface. (Chap. 8, Sec. 3, p. 224)

absolute age: age, in years, of a rock or other object; can be determined by using properties of the atoms that make up materials. (Chap. 13, Sec. 3, p. 383)

absolute magnitude: measure of the amount of light a star actually gives off. (Chap. 25, Sec. 1, p. 738)

abyssal (uh BIH sul) **plain:** flat seafloor area from 4,000 m to 6,000 m below the ocean surface, formed by the deposition of sediments. (Chap. 19, Sec. 1, p. 551)

acid: substance with a pH lower than 7. (Chap. 21, Sec. 2, p. 621)

acid rain: acidic moisture, with a pH below 5.6, that falls to Earth as rain or snow and can damage forests, harm organisms, and corrode structures. (Chap. 21, Sec. 2, p. 621)

adaptation: any structural or behavioral change that helps an organism survive in its particular environment. (Chap. 17, Sec. 2, p. 496)

air mass: large body of air that has the same characteristics of temperature and moisture content as the part of Earth's surface over which it formed. (Chap. 16, Sec. 2, p. 470)

apparent magnitude: measure of the amount of light from a star that is received on Earth. (Chap. 25, Sec. 1, p. 738)

aquifer (AK wuh fur): layer of permeable rock that allows water to flow through. (Chap. 9, Sec. 2, p. 252)

asteroid: a piece of rock made up of material similar to that which formed the planets; mostly found in the asteroid belt between the orbits of Mars and Jupiter. (Chap. 24, Sec. 4, p. 724)

asthenosphere (as THE nuh sfihr): plastic-like layer of Earth on which the lithospheric plates float and move around. (Chap. 10, Sec. 3, p. 284)

atmosphere: Earth's air, which is made up of a thin layer of gases, solids, and liquids; forms a protective layer around the planet and is divided into five distinct layers. (Chap. 15, Sec. 1, p. 434)

atom(s): tiny building blocks of matter, made up of protons, neutrons, and electrons. (Chap. 2, Sec. 1, p. 34)

atomic number: the number of protons in an atom. (Chap. 2, Sec. 1, p. 37)

axis: imaginary vertical line that cuts through the center of Earth and around which Earth spins. (Chap. 23, Sec. 1, p. 673)

B

basaltic: dense, dark-colored igneous rock formed from magma; rich in magnesium and iron and poor in silica. (Chap. 4, Sec. 2, p. 97)

base: substance with a pH above 7. (Chap. 21, Sec. 2, p. 621)

basin: low area on Earth in which an ocean formed when the area filled with water from torrential rains. (Chap. 18, Sec. 1, p. 523)

batholith: largest intrusive igneous rock body that forms when magma being forced upward toward Earth's crust cools slowly and solidifies underground. (Chap. 12, Sec. 3, p. 350)

beach: deposit of sediment whose materials vary in size, color, and composition and is most commonly found on a smooth, gently sloped shoreline. (Chap. 9, Sec. 3, p. 259)

benthos: marine plants and animals that live on or in the ocean floor. (Chap. 19, Sec. 2, p. 561)

bias: personal opinion. (Chap. 1, Sec. 2, p. 21)

big bang theory: states that about 12 billion to 15 billion years ago, the universe began with a huge, fiery explosion. (Chap. 25, Sec. 4, p. 757)

biomass energy: renewable energy derived from burning organic materials such as wood and alcohol. (Chap. 5, Sec. 2, p. 133)

black hole: final stage in the evolution of a supernova, where the core's mass collapses to a point that it has no volume and whose gravity is so strong that not even light can escape. (Chap. 25, Sec. 3, p. 750)

blizzard: winter storm that lasts at least three hours with temperatures of −12°C or below, poor visibility, and winds of at least 51 km/h. (Chap. 16, Sec. 2, p. 477)

breaker: collapsing ocean wave that forms in shallow water and breaks onto the shore. (Chap. 18, Sec. 3, p. 533)

C

caldera: large, circular-shaped opening formed when the top of a volcano collapses. (Chap. 12, Sec. 3, p. 352)

carbon film: thin film of carbon residue preserved as a fossil. (Chap. 13, Sec. 1, p. 370)

carbon monoxide: colorless, odorless gas that reduces the oxygen content in the blood, is found in car exhaust, and contributes to air pollution. (Chap. 21, Sec. 2, p. 622)

carrying capacity: maximum number of individuals of a given species that the environment will support. (Chap. 20, Sec. 1, p. 585)

cast: a type of body fossil that forms when crystals fill a mold or sediments wash into a mold and harden into rock. (Chap. 13, Sec. 1, p. 371)

cave: underground opening that can form when acidic groundwater dissolves limestone. (Chap. 9, Sec. 2, p. 255)

cementation: sedimentary rock-forming process in which large sediments are held together by natural cements that are produced when water soaks through rock and soil. (Chap. 4, Sec. 4, p. 105)

Cenozoic (sen uh ZOH ihk) **Era:** era of recent life that began about 66 million years ago and continues today; includes the first appearance of *Homo sapiens* about 400,000 years ago. (Chap. 14, Sec. 3, p. 418)

channel: groove created by water moving down the same path. (Chap. 9, Sec. 1, p. 242)

chemical weathering: occurs when chemical reactions dissolve the minerals in rocks or change them into different minerals. (Chap. 7, Sec. 1, p. 187)

chemosynthesis (kee moh SIHN thuh sihs): food-making process using sulfur or nitrogen compounds, rather than light energy from the Sun, that is used by bacteria living near thermal vents. (Chap. 19, Sec. 2, p. 559)

chlorofluorocarbons (CFCs): group of chemical compounds used in refrigerators, air conditioners, foam packaging, and aerosol sprays that may enter the atmosphere and destroy ozone. (Chap. 15, Sec. 1, p. 440)

chromosphere: layer of the Sun's atmosphere above the photosphere. (Chap. 25, Sec. 2, p. 741)

cinder cone volcano: steep-sided, loosely packed volcano formed when tephra falls to the ground. (Chap. 12, Sec. 2, p. 344)

cleavage: physical property of some minerals that causes them to break along smooth, flat surfaces. (Chap. 3, Sec. 2, p. 71)

climate: average weather pattern in an area over a long period of time; can be classified by temperature, humidity, precipitation, and vegetation. (Chap. 7, Sec. 1, p. 188) (Chap. 17, Sec. 1, p. 492)

coal: sedimentary rock formed from decayed plant material; the world's most abundant fossil fuel. (Chap. 5, Sec. 1, p. 121)

comet: space object formed from dust and rock particles mixed with frozen water, methane, and ammonia that forms a bright coma as it approaches the Sun. (Chap. 24, Sec. 4, p. 722)

compaction: process that forms sedimentary rocks when layers of small sediments are compressed by the weight of the layers above them. (Chap. 4, Sec. 4, p. 104)

composite volcano: volcano built by alternating explosive and quiet eruptions that produce layers of tephra and lava; found mostly where Earth's plates come together and one plate sinks below the other. (Chap. 12, Sec. 2, p. 345)

composting: conservation method in which yard wastes such as cut grass, pulled weeds, and raked leaves are piled and left to decompose gradually. (Chap. 20, Sec. 3, p. 597)

compound: matter that is made of two of more elements and has physical and chemical properties different from each of the elements that make it up. (Chap. 2, Sec. 2, p. 40)

condensation: process in which water vapor changes to a liquid. (Chap. 15, Sec. 2, p. 445)

conduction: transfer of energy that occurs when molecules bump into each other. (Chap. 15, Sec. 2, p. 444)

conic projection: map made by projecting points and lines from a globe onto a cone. (Chap. 6, Sec. 3, p. 167)

conservation: careful use of resources to reduce damage to the environment through such methods as composting and recycling materials. (Chap. 20, Sec. 3, p. 596)

constant: variable that does not change in an experiment. (Chap. 1, Sec. 1, p. 10)

constellation: group of stars that forms a pattern in the sky that looks like a familiar object (Big Dipper), animal (Pegasus), or character (Orion). (Chap. 25, Sec. 1, p. 736)

continental drift: Wegener's hypothesis that all continents were once connected in a single, large landmass that broke apart about 200 million years ago and drifted slowly to their current positions. (Chap. 10 Sec. 1, p. 276)

continental shelf: gradually sloping end of a continent that extends beneath the ocean and provides a home for most marine organisms. (Chap. 19, Sec. 1, p. 550)

continental slope: ocean basin feature that dips steeply down from the continental shelf. (Chap. 19, Sec. 1, p. 551)

contour line: line on a map that connects points of equal elevation. (Chap. 6, Sec. 3, p. 168)

control: standard for comparison in an experiment. (Chap. 1, Sec. 1, p. 10)

convection: transfer of heat by the flow of material. (Chap. 15, Sec. 2, p. 444)

convection current: current in Earth's mantle that transfers heat in Earth's interior and is the driving force for plate tectonics. (Chap. 10, Sec. 3, p. 289)

Coriolis (kor ee OH lus) **effect:** causes moving air and water to turn left in the southern hemisphere and turn right in the northern hemisphere due to Earth's rotation. (Chap. 15, Sec. 3, p. 448)

Coriolis effect: shifting of winds and surface currents caused by Earth's rotation that turns currents north of the equator clockwise and south of the equator counterclockwise. (Chap. 18, Sec. 2, p. 527)

corona: outermost, largest layer of the Sun's atmosphere; extends millions of kilometers into space and has temperatures up to 2 million K. (Chap. 25, Sec. 2, p. 741)

crater: steep-walled depression around a volcano's vent. (Chap. 12, Sec. 1, p. 336)

creep: a type of mass movement in which sediments move downslope very slowly; is common in areas of freezing and thawing, and can cause walls, trees, and fences to lean downhill. (Chap. 8, Sec. 1, p. 214)

crest: highest point of a wave. (Chap. 18, Sec. 3, p. 532)

crystal: solid in which the atoms are arranged in an orderly, repeating pattern. (Chap. 3, Sec. 1, p. 63)

cyanobacteria: chlorophyll-containing, photosynthetic bacteria thought to be one of Earth's earliest life-forms. (Chap. 14, Sec. 2, p. 407)

D

deflation: a type of erosion that occurs when wind blows over loose sediments, removes small particles, and leaves coarser sediments behind. (Chap. 8, Sec. 3, p. 224)

deforestation: destruction and cutting down of forests—often to clear land for mining, roads, and grazing of cattle—resulting in increased atmospheric CO_2 levels. (Chap. 17, Sec. 3, p. 509)

density: a physical property of matter that can be determined by dividing the mass of an object by its volume. (Chap. 2, Sec. 3, p. 46)

density current: circulation pattern in the ocean that forms when a mass of more dense seawater sinks beneath less dense seawater. (Chap. 18, Sec. 2, p. 529)

dependent variable: factor being measured in an experiment. (Chap. 1, Sec. 1, p. 10)

deposition: dropping of sediments that occurs when an agent of erosion, such as gravity, a glacier, wind, or water, loses its energy and can no longer carry its load. (Chap. 8, Sec. 1, p. 213)

dew point: temperature at which air is saturated and condensation forms. (Chap. 16, Sec. 1, p. 465)

dike: igneous rock feature formed when magma is squeezed into a vertical crack that cuts across rock layers and hardens underground. (Chap. 12, Sec. 3, p. 351)

drainage basin: land area from which a river or stream collects runoff. (Chap. 9, Sec. 1, p. 244)

dune (DOON): mound formed when wind-blown sediments pile up behind an obstacle; common landform in desert areas. (Chap. 8, Sec. 3, p. 227)

E

Earth: third planet from the Sun; has an atmosphere that protects life and surface temperatures that allow water to exist as a solid, liquid, and gas. (Chap. 24, Sec. 2, p. 710)

earthquake: vibrations produced when rocks break along a fault. (Chap. 11, Sec. 1, p. 305)

Earth science: study of Earth and space, including rocks, fossils, climate, volcanoes, land use, ocean water, earthquakes, and objects in space. (Chap. 1, Sec. 1, p. 9)

electromagnetic spectrum: arrangement of electromagnetic waves according to their wavelengths. (Chap. 22, Sec. 1, p. 641)

electrons: negatively charged particles that move around the nucleus of an atom and form an electron cloud. (Chap. 2, Sec. 1, p. 36)

element: substance that contains only one type of atom—for example, oxygen, aluminum, and iron. (Chap. 2, Sec. 1, p. 35)

English Glossary

ellipse (ee LIHPS): elongated, closed curve that describes Earth's yearlong orbit around the Sun. (Chap. 23, Sec. 1, p. 675)

El Niño (el NEEN yoh): climatic event that begins in the tropical Pacific Ocean; may occur when trade winds weaken or reverse, and can disrupt normal temperature and precipitation patterns around the world. (Chap. 17, Sec. 3, p. 501)

enzyme: substance that causes chemical reactions to happen more quickly. (Chap. 20, Sec. 2, p. 593)

eon: longest subdivision in the geologic time scale that is based on the abundance of certain types of fossils and is subdivided into eras, periods, and epochs. (Chap. 14, Sec. 1, p. 399)

epicenter (EP ih sent ur): point on Earth's surface directly above an earthquake's focus. (Chap. 11, Sec. 2, p. 309)

epoch: next-smaller division of geologic time after the period; is characterized by differences in life-forms that may vary regionally. (Chap. 14, Sec. 1, p. 399)

equator: imaginary line that wraps around Earth at 0° latitude, halfway between the north and south poles. (Chap. 6, Sec. 2, p. 162)

equinox (EE kwuh nahks): twice-yearly time—each spring and fall—when the Sun is directly over the equator and the number of daylight and nighttime hours is equal worldwide. (Chap. 23, Sec. 1, p. 677)

era: second-longest division of geologic time; is subdivided into periods and is based on major worldwide changes in types of fossils. (Chap. 14, Sec. 1, p. 399)

erosion: process in which surface materials are worn away and transported from one place to another by agents such as gravity, water, wind, and glaciers. (Chap. 8, Sec. 1, p. 212)

estuary: area where a river meets the ocean that contains a mixture of freshwater and ocean water and provides an important habitat to many marine organisms. (Chap. 19, Sec. 2, p. 564)

ethics: study of moral values about what is good or bad. (Chap. 1, Sec. 2, p. 20)

extrusive: fine-grained igneous rock that forms when magma cools quickly at or near Earth's surface. (Chap. 4, Sec. 2, p. 95)

F

fault: surface along which rocks move when they pass their elastic limit and break. (Chap. 11, Sec. 1, p. 304)

fault-block mountains: mountains formed from huge, tilted blocks of rock that are separated from surrounding rocks by faults. (Chap. 6, Sec. 1, p. 160)

fertilizer: chemical that helps plants and other organisms grow. (Chap. 21, Sec. 1, p. 611)

focus: in an earthquake, the point below Earth's surface where energy is released in the form of seismic waves. (Chap. 11, Sec. 2, p. 308)

fog: a stratus cloud that forms when air is cooled to its dew point near the ground. (Chap. 16, Sec. 1, p. 467)

folded mountains: mountains formed when horizontal rock layers are squeezed from opposite sides, causing them to buckle and fold. (Chap. 6, Sec. 1, p. 159)

foliated: metamorphic rock, such as slate and gneiss, whose mineral grains flatten and line up in parallel layers. (Chap. 4, Sec. 3, p. 101)

fossil fuel: nonrenewable energy resource, such as oil and coal, formed over millions of years from the remains of dead plants and other organisms. (Chap. 5, Sec. 1, p. 120)

fossils: remains, imprints, or traces of prehistoric organisms that can tell when and where organisms once lived and how they lived. (Chap. 13, Sec. 1, p. 369)

fracture: physical property of some minerals that causes them to break with uneven, rough, or jagged surfaces. (Chap. 3, Sec. 2, p. 71)

front: boundary between two air masses with different temperatures, density, or moisture; can be cold, warm, occluded, and stationary. (Chap. 16, Sec. 2, p. 471)

full moon: phase that occurs when all of the Moon's surface facing Earth reflects light. (Chap. 23, Sec. 2, p. 679)

G

galaxy: large group of stars, dust, and gas held together by gravity; can be elliptical, spiral, or irregular. (Chap. 25, Sec. 4, p. 752)

gem: beautiful, rare, highly prized mineral that can be worn in jewelry. (Chap. 3, Sec. 3, p. 73)

geologic time scale: division of Earth's history into time units based largely on the types of life-forms that lived only during certain periods. (Chap. 14, Sec. 1, p. 398)

geothermal energy: inexhaustible energy resource that uses hot magma or hot, dry rocks from below Earth's surface to generate electricity. (Chap. 5, Sec. 2, p. 132)

geyser: hot spring that erupts periodically and shoots water and steam into the air—for example, Old Faithful in Yellowstone National Park. (Chap. 9, Sec. 2, p. 255)

giant: late stage in the life of a main sequence star when hydrogen in the core is used up; the core contracts and temperatures inside the star increase, causing its outer layers to expand and cool. (Chap. 25, Sec. 3, p. 749)

glaciers: large, moving masses of ice and snow that change large areas of Earth's surface through erosion and deposition. (Chap. 8, Sec. 2, p. 217)

global warming: increase in the average global temperature of Earth. (Chap. 17, Sec. 3, p. 508)

granitic: light-colored, silica-rich igneous rock that is less dense than basaltic rock. (Chap. 4, Sec. 2, p. 97)

Great Red Spot: giant, high-pressure gas storm on Jupiter. (Chap. 24, Sec. 3, p. 714)

greenhouse effect: natural heating that occurs when certain gases in Earth's atmosphere, such as methane, CO_2, and water vapor, trap heat. (Chap. 17, Sec. 3, p. 507)

groundwater: water that soaks into the ground and collects in pores and empty spaces and is an important source of drinking water. (Chap. 9, Sec. 2, p. 251)

H

half-life: time it takes for half the atoms of an isotope to decay. (Chap. 13, Sec. 3, p. 384)

hardness: measure of how easily a mineral can be scratched; is determined by the arrangement of the mineral's atoms. (Chap. 3, Sec. 2, p. 69)

hazardous waste: poisonous, ignitable, or cancer-causing waste. (Chap. 20, Sec. 2, p. 592)

heterogeneous mixture: a mixture which is not mixed evenly and each component retains its own properties. (Chap. 2, Sec. 2, p. 43)

hibernation: behavioral adaptation for winter survival in which an animal's activity is greatly reduced, its body temperature drops, and body processes slow down. (Chap. 17, Sec. 2, p. 498)

homogeneous mixture: a mixture which is evenly mixed throughout. (Chap. 2, Sec. 2, p. 43)

English Glossary

horizon: each layer in a soil profile—horizon A (top layer of soil), horizon B (middle layer), and horizon C (bottom layer). (Chap. 7, Sec. 2, p. 192)

hot spot: unusually hot area at the boundary between Earth's mantle and core that forms volcanoes when melted rock is forced upward and breaks through the crust. (Chap. 12, Sec. 1, p. 338)

humidity: amount of water vapor held in the air. (Chap. 16, Sec. 1, p. 464)

humus (HYEW mus): dark-colored, decayed organic matter that supplies nutrients to plants and is found mainly in topsoil. (Chap. 7, Sec. 2, p. 192)

hurricane: large, severe storm that forms over tropical oceans, has winds of at least 120 km/h, and loses power when it reaches land. (Chap. 16, Sec. 2, p. 476)

hydroelectric energy: electricity produced by waterpower using large dams in a river. (Chap. 5, Sec. 2, p. 132)

hydrosphere: all the water on Earth's surface. (Chap. 15, Sec. 2, p. 445)

hypothesis: an educated guess. (Chap. 1, Sec. 1, p. 7)

I

ice wedging: mechanical weathering process that occurs when water freezes in the cracks of rocks and expands, causing the rock to break apart. (Chap. 7, Sec. 1, p. 186)

igneous rock: intrusive or extrusive rock formed when hot magma cools and hardens. (Chap. 4, Sec. 2, p. 94)

impact basin: a hollow left on the surface of the Moon caused by an object striking its surface. (Chap. 23, Sec. 3, p. 689)

impermeable: describes materials that water cannot pass through. (Chap. 9, Sec. 2, p. 252)

independent variable: factor that changes in an experiment. (Chap. 1, Sec. 1, p. 10)

index fossils: remains of species that existed on Earth for a relatively short period of time, were abundant and widespread geographically, and can be used by geologists to assign the ages of rock layers. (Chap. 13, Sec. 1, p. 373)

intrusive: a type of igneous rock that generally contains large crystals and forms when magma cools slowly beneath Earth's surface. (Chap. 4, Sec. 2, p. 95)

ion: electrically charged atom whose charge results from an atom losing or gaining electrons. (Chap. 2, Sec. 2, p. 41)

ionosphere: layer of electrically charged particles in the thermosphere that absorbs AM radio waves during the day and reflects them back at night. (Chap. 15, Sec. 1, p. 437)

isobars: lines drawn on a weather map that connect points having equal atmospheric pressure; also indicate the location of high- and low-pressure areas and can show wind speed. (Chap. 16, Sec. 3, p. 479)

isotherm (I suh thurm): line drawn on a weather map that connects points having equal temperature. (Chap. 16, Sec. 3, p. 479)

isotopes: atoms of the same element that have different numbers of neutrons. (Chap. 2, Sec. 1, p. 37)

J

jet stream: narrow belt of strong winds that blows near the top of the troposphere. (Chap. 15, Sec. 3, p. 450)

Jupiter: largest and fifth planet from the Sun; contains more mass than all the other planets combined; has continuous storms of high-pressure gas and an atmosphere mostly of hydrogen and helium. (Chap. 24, Sec. 3, p. 714)

L

land breeze: movement of air from land to sea at night, created when cooler, denser air from the land forces warmer air up over the sea. (Chap. 15, Sec. 3, p. 451)

latitude: distance in degrees north or south of the equator. (Chap. 6, Sec. 2, p. 162)

lava: thick, gooey, molten rock material flowing from volcanoes onto Earth's surface. (Chap. 4, Sec. 2, p. 94)

leaching: removal of minerals that have been dissolved in water. (Chap. 7, Sec. 2, p. 193)

light-year: distance light travels in one year—about 9.5 trillion km—which is used to record distances between stars and galaxies. (Chap. 25, Sec. 1, p. 739)

liquefaction: occurs when wet soil acts more like a liquid during an earthquake. (Chap. 11, Sec. 3, p. 319)

lithosphere (LIH thuh sfihr): rigid layer of Earth about 100 km thick, made of the crust and a part of the upper mantle. (Chap. 10, Sec. 3, p. 284)

litter: twigs, leaves, and other organic matter that help prevent erosion, hold water and eventually might be changed into humus by decomposing organisms. (Chap. 7, Sec. 2, p. 193)

loess (LOOS): windblown deposit of tightly packed, fine-grained sediments. (Chap. 8, Sec. 3, p. 227)

longitude: distance in degrees east or west of the prime meridian. (Chap. 6, Sec. 2, p. 163)

longshore current: current that runs parallel to the shoreline, is caused by waves colliding with the shore at slight angles, and moves tons of loose sediment. (Chap. 9, Sec. 3, p. 258)

lunar eclipse: occurs when Earth's shadow falls on the Moon. (Chap. 23, Sec. 2, p. 682)

luster: describes the way a mineral reflects light from its surface; can be either metallic or nonmetallic. (Chap. 3, Sec. 2, p. 70)

M

magma: hot, melted rock material beneath Earth's surface. (Chap. 3, Sec. 1, p. 65)

magnitude: measure of the energy released during an earthquake. (Chap. 11, Sec. 3, p. 318)

map legend: explains the meaning of symbols used on a map. (Chap. 6, Sec. 3, p. 170)

map scale: relationship between distances on a map and distances on Earth's surface that can be represented as a ratio or as a small bar divided into sections. (Chap. 6, Sec. 3, p. 170)

maria (MAHR ee uh): dark-colored, relatively flat regions of the Moon formed when ancient lava reached the surface and filled craters on the Moon's surface. (Chap. 23, Sec. 2, p. 683)

Mars: fourth planet from the Sun; has polar ice caps, a thin atmosphere, and a reddish appearance from iron oxide in weathered rocks. (Chap. 24, Sec. 2, p. 710)

mass movement: any type of erosion that occurs as gravity moves materials downslope. (Chap. 8, Sec. 1, p. 213)

mass number: the number of protons plus the number of neutrons in an atom. (Chap. 2, Sec. 1, p. 37)

matter: anything that has mass and takes up space; matter's properties are determined by the structure of its atoms and how they are joined. (Chap. 2, Sec. 1, p. 34)

meander (mee AN dur): broad, c-shaped curve in a river or stream, formed by erosion of its outer bank. (Chap. 9, Sec. 1, p. 245)

mechanical weathering: physical process that breaks rocks apart without changing their chemical makeup; can be caused by ice wedging, animals, and plant roots. (Chap. 7, Sec. 1, p. 185)

English Glossary

Mercury: smallest planet, closest to the Sun; has a thin atmosphere and a surface with many craters and high cliffs. (Chap. 24, Sec. 2, p. 708)

Mesozoic (mez uh ZOH ihk) **Era:** middle era of Earth's history, during which Pangaea broke apart, dinosaurs appeared, and reptiles and gymnosperms were the dominant land life-forms. (Chap. 14, Sec. 3, p. 414)

metamorphic rock: forms when heat, pressure, or fluids act on igneous, sedimentary, or other metamorphic rock and affect its form or composition, or both. (Chap. 4, Sec. 3, p. 99)

meteor: a meteoroid that burns up in Earth's atmosphere. (Chap. 24, Sec. 4, p. 723)

meteorite: a meteoroid that strikes the surface of a moon or planet. (Chap. 24, Sec. 4, p. 724)

meteorologist (meet ee uh RAHL uh just): studies weather and uses information from Doppler radar, weather satellites, computers and other instruments to make weather maps and provide forecasts. (Chap. 16, Sec. 3, p. 478)

mid-ocean ridge: area where new ocean floor is formed when lava erupts through cracks in Earth's crust. (Chap. 19, Sec. 1, p. 552)

mineral: naturally occurring inorganic solid that has a definite chemical composition and an orderly internal atomic structure. (Chap. 3, Sec. 1, p. 62)

mineral resources: resources from which metals are obtained. (Chap. 5, Sec. 3, p. 137)

mixture: describes two or more substances that retain their own properties even when they are combined and that can be separated from one another by physical means. (Chap. 2, Sec. 2, p. 43)

mold: a type of body fossil that forms in rock when an organism with hard parts is buried, decays or dissolves, and leaves a cavity in the rock. (Chap. 13, Sec. 1, p. 371)

moon phase: change in appearance of the Moon as viewed from Earth, due to the relative positions of the Moon, Earth, and the Sun. (Chap. 23, Sec. 2, p. 679)

moraine: large ridge of rocks and soil deposited by a glacier when it stops moving forward. (Chap. 8, Sec. 2, p. 219)

N

natural gas: fossil fuel formed from marine organisms that is often found in tilted or folded rock layers and is used for heating and cooking. (Chap. 5, Sec. 1, p. 123)

natural selection: process by which organisms that are suited to a particular environment are better able to survive and reproduce than organisms that are not. (Chap. 14, Sec. 1, p. 401)

nebula: large cloud of gas and dust that contracts under gravitational force and breaks apart into smaller pieces, each of which will collapse to form a star. (Chap. 25, Sec. 3, p. 748)

nekton: marine organisms that actively swim in the ocean. (Chap. 19, Sec. 2, p. 560)

Neptune: usually the eighth planet from the Sun; is large and gaseous, has rings that vary in thickness, and is bluish-green in color. (Chap. 24, Sec. 3, p. 718)

neutron: particle without an electric charge that is located in the nucleus of an atom. (Chap. 2, Sec. 1, p. 36)

neutron star: collapsed core of a supernova that can shrink to about 20 km in diameter and contains only neutrons in the dense core. (Chap. 25, Sec. 3, p. 750)

new moon: moon phase that occurs when the Moon is between Earth and the Sun, at which point the Moon cannot be seen because its lighted half is facing the Sun and its dark side faces Earth. (Chap. 23, Sec. 2, p. 679)

nonfoliated: metamorphic rock, such as quartzite and marble, whose mineral grains grow and rearrange but do not form layers. (Chap. 4, Sec. 3, p. 102)

nonpoint source pollution: pollution that enters water from a large area and cannot be traced to a single location. (Chap. 21, Sec. 1, p. 610)

normal fault: break in rock caused by tension forces, where rock above the fault surface moves down relative to the rock below the fault surface. (Chap. 11, Sec. 1, p. 306)

nuclear energy: alternative energy source that is based on atomic fission. (Chap. 5, Sec. 1, p. 127)

O

observatory: building that can house an optical telescope; often has a dome-shaped roof that can be opened for viewing. (Chap. 22, Sec. 1, p. 642)

oil: liquid fossil fuel formed from marine organisms that is burned to obtain energy and used in the manufacture of plastics. (Chap. 5, Sec. 1, p. 123)

orbit: curved path followed by a satellite as it revolves around an object. (Chap. 22, Sec. 2, p. 649)

ore: deposit in which a mineral exists in large enough amounts to be mined at a profit. (Chap. 5, Sec. 3, p. 137)

organic evolution: change of organisms over geologic time. (Chap. 14, Sec. 1, p. 400)

outwash: material deposited by meltwater from a glacier. (Chap. 8, Sec. 2, p. 219)

oxidation (ahk sih DAY shun): chemical weathering process that occurs when metallic material is exposed to oxygen and water over time. (Chap. 7, Sec. 1, p. 188)

ozone layer: layer of the stratosphere with a high concentration of ozone; absorbs most of the Sun's harmful ultraviolet radiation. (Chap. 15, Sec. 1, p. 440)

P

Paleozoic Era: era of ancient life, which began about 544 million years ago, when organisms developed hard parts, and ended with mass extinctions about 245 million years ago. (Chap. 14, Sec. 2, p. 408)

Pangaea (pan JEE uh): large, ancient land-mass that was composed of all the continents joined together. (Chap. 10, Sec. 1, p. 276) (Chap. 14, Sec. 1, p. 405)

particulate (par TIHK yuh layt) **matter:** fine solids such as pollen, dust, mold, ash and soot as well as liquid droplets in the air that can irritate and damage lungs when breathed in. (Chap. 21, Sec. 2, p. 623)

period: third-longest division of geologic time; is subdivided into epochs and is characterized by the types of life that existed worldwide. (Chap. 14, Sec. 1, p. 399)

permeable (PUR mee uh bul): describes soil and rock with connecting pores through which water can flow. (Chap. 9, Sec. 2, p. 252)

permineralized remains: fossils in which the spaces inside are filled with minerals from groundwater. (Chap. 13, Sec. 1, p. 370)

pesticide: substance used to keep insects and weeds from destroying crops and lawns. (Chap. 21, Sec. 1, p. 611)

photochemical smog: hazy, yellow-brown blanket of smog found over cities that is formed with the help of sunlight, contains ozone near Earth's surface, and can damage lungs and plants. (Chap. 21, Sec. 2, p. 620)

photosphere: lowest layer of the Sun's atmosphere; gives off light and has temperatures of about 6,000 K. (Chap. 25, Sec. 2, p. 741)

photosynthesis: food-making process using light energy from the Sun, carbon dioxide, and water. (Chap. 19, Sec. 2, p. 557)

pH scale: scale used to measure how acidic or basic something is. (Chap. 21, Sec. 2, p. 621)

plain: large, flat landform that often has thick, fertile soil and is usually found in the interior region of a continent. (Chap. 6, Sec. 1, p. 156)

plankton: marine plants and animals that drift in ocean currents. (Chap. 19, Sec. 2, p. 560)

plate: a large section of Earth's oceanic or continental crust and rigid upper mantle that moves around on the asthenosphere. (Chap. 10, Sec. 3, p. 284)

plateau (pla TOH): flat, raised landform made up of nearly horizontal rocks that have been uplifted. (Chap. 6, Sec. 1, p. 158)

plate tectonics: theory that Earth's crust and upper mantle are broken into plates that float and move around on a plastic-like layer of the mantle. (Chap. 10, Sec. 3, p. 284)

plucking: process that adds gravel, sand, and boulders to a glacier's bottom and sides as water freezes and thaws, breaking off pieces of surrounding rock. (Chap. 8, Sec. 2, p. 218)

Pluto: considered to be the ninth planet from the Sun; has a solid icy-rock surface and a single moon. (Chap. 24, Sec. 3, p. 719)

point source pollution: pollution that enters water from a specific location and can be controlled or treated before it enters a body of water. (Chap. 21, Sec. 1, p. 610)

polar zones: climate zones that receive solar radiation at a low angle, extend from 66°N and S latitude to the poles, and are never warm. (Chap. 17, Sec. 1, p. 492)

pollutant: any substance that contaminates the environment. (Chap. 20, Sec. 1, p. 586)

pollution: introduction of wastes to an environment, such as sewage and chemicals, that can damage organisms. (Chap. 19, Sec. 3, p. 565)

population: total number of individuals of one species occupying the same area. (Chap. 20, Sec. 1, p. 584)

Precambrian (pree KAM bree un) **time:** longest part of Earth's history, lasting from 4 billion to about 544 million years ago. (Chap. 14, Sec. 2, p. 406)

precipitation: water falling from clouds—including rain, snow, sleet, and hail—whose form is determined by air temperature. (Chap. 16, Sec. 1, p. 468)

primary wave: seismic wave that moves rock particles back and forth in the same direction that the wave travels. (Chap. 11, Sec. 2, p. 309)

prime meridian: imaginary line that represents 0° longitude and runs from the north pole through Greenwich, England, to the south pole. (Chap. 6, Sec. 2, p. 163)

principle of superposition: states that in undisturbed rock layers, the oldest rocks are on the bottom and the rocks become progressively younger toward the top. (Chap. 13, Sec. 2, p. 376)

Project Apollo: final stage in the U.S. program to reach the Moon in which Neil Armstrong was the first human to step onto the Moon's surface. (Chap. 22, Sec. 2, p. 654)

Project Gemini: second stage in the U.S. program to reach the Moon in which an astronaut team connected with another spacecraft in orbit. (Chap. 22, Sec. 2, p. 653)

Project Mercury: first step in the U.S. program to reach the Moon that orbited a piloted spacecraft around Earth and brought it back safely. (Chap. 22, Sec. 2, p. 653)

proton: positively charged particle that is located in the nucleus of an atom. (Chap. 2, Sec. 1, p. 36)

R

radiation: energy transferred by waves or rays. (Chap. 15, Sec. 2, p. 444)

radioactive decay: process in which some isotopes break down into other isotopes and particles. (Chap. 13, Sec. 3, p. 383)

radiometric dating: process used to calculate the absolute age of rock by measuring the ratio of parent isotope to daughter product in a mineral and knowing the half-life of the parent. (Chap. 13, Sec. 3, p. 385)

radio telescope: collects and records radio waves traveling through space; can be used day or night under most weather conditions. (Chap. 22, Sec. 1, p. 645)

recycling: conservation method in which old materials are processed to make new ones. (Chap. 5, Sec. 3, p. 141) (Chap. 20, Sec. 3, p. 597)

reef: rigid, wave-resistant, ocean margin habitat built by corals from skeletal materials and calcium. (Chap. 19, Sec. 2, p. 564)

reflecting telescope: optical telescope that uses a concave mirror to focus light and form an image at the focal point. (Chap. 22, Sec. 1, p. 642)

refracting telescope: optical telescope that uses a double convex lens to bend light and form an image at the focal point. (Chap. 22, Sec. 1, p. 642)

relative age: the age of something compared with other things. (Chap. 13, Sec. 2, p. 377)

relative humidity: measure of the amount of moisture held in the air compared with the amount it can hold at a given temperature; can range from 0 percent to 100 percent. (Chap. 16, Sec. 1, p. 464)

reserve: amount of a fossil fuel that can be extracted from Earth at a profit using current technology. (Chap. 5, Sec. 1, p. 125)

reverse fault: break in rock caused by compressive forces, where rock above the fault surface moves upward relative to the rock below the fault surface. (Chap. 11, Sec. 1, p. 306)

revolution: Earth's yearlong elliptical orbit around the Sun. (Chap. 23, Sec. 1, p. 675)

rock: mixture of one or more minerals, volcanic glass, organic matter, or other materials; can be igneous, metamorphic, or sedimentary. (Chap. 4, Sec. 1, p. 90)

rock cycle: model that describes how rocks slowly change from one form to another through time. (Chap. 4, Sec. 1, p. 91)

rocket: special motor that can work in space and burns liquid or solid fuel. (Chap. 22, Sec. 2, p. 647)

rotation: spinning of Earth on its imaginary axis, which takes about 24 hours to complete and causes day and night to occur. (Chap. 23, Sec. 1, p. 673)

runoff: any rainwater that does not soak into the ground or evaporate but flows over Earth's surface; generally flows into streams and has the ability to erode and carry sediments. (Chap. 9, Sec. 1, p. 240)

S

salinity (say LIHN ut ee): a measure of the amount of salts dissolved in seawater. (Chap. 18, Sec. 1, p. 524)

sanitary landfill: area where garbage is deposited and covered with soil and that is designed to prevent contamination of land and water. (Chap. 20, Sec. 2, p. 592)

satellite: any natural or artificial object that revolves around another object. (Chap. 22, Sec. 2, p. 649)

Saturn: second-largest and sixth planet from the Sun; has a complex ring system, at least 18 moons, and a thick atmosphere made mostly of hydrogen and helium. (Chap. 24, Sec. 3, p. 716)

English Glossary

science: process of looking at and studying things in the world in order to gain knowledge. (Chap. 1, Sec. 1, p. 8)

scientific law: rule that describes the behavior of something in nature; usually describes what will happen in a situation but not why it happens. (Chap. 1, Sec. 2, p. 19)

scientific methods: problem-solving procedures that can include identifying the problem or question, gathering information, developing a hypothesis, testing the hypothesis, analyzing the results, and drawing conclusions. (Chap. 1, Sec. 1, p. 8)

scientific theory: explanation that is supported by results from repeated experimentation or testing. (Chap. 1, Sec. 2, p. 18)

scrubber: device that lowers sulfur emissions from coal-burning power plants. (Chap. 21, Sec. 2, p. 624)

sea breeze: movement of air from sea to land during the day when cooler air from above the water moves over the land, forcing the heated, less dense air above the land to rise. (Chap. 15, Sec. 3, p. 451)

seafloor spreading: Hess's theory that new seafloor is formed when magma is forced upward toward the surface at a mid-ocean ridge. (Chap. 10, Sec. 2, p. 281)

season: short period of climate change in an area caused by the tilt of Earth's axis as Earth revolves around the Sun. (Chap. 17, Sec. 3, p. 500)

secondary wave: seismic wave that moves rock particles at right angles to the direction of the wave. (Chap. 11, Sec. 2, p. 309)

sedimentary rock: forms when sediments are compacted and cemented together or when minerals come out of solution or are left behind by evaporation. (Chap. 4, Sec. 4, p. 103)

sediments: loose materials, such as rock fragments, mineral grains, and the remains of once-living plants and animals, that have been moved by wind, water, ice, or gravity. (Chap. 4, Sec. 4, p. 103)

seismic (SIZE mihk) **wave:** wave generated by an earthquake. (Chap. 11, Sec. 2, p. 308)

seismograph: instrument used to register earthquake waves and record the time that each arrived. (Chap. 11, Sec. 2, p. 311)

sewage: water that goes into drains and contains human waste, household detergents, and soaps. (Chap. 21, Sec. 1, p. 612)

sheet erosion: a type of surface water erosion caused by runoff that occurs when water flowing as sheets picks up sediments and carries them away. (Chap. 9, Sec. 1, p. 243)

shield volcano: broad, gently sloping volcano formed by quiet eruptions of basaltic lava. (Chap. 12, Sec. 2, p. 344)

silicate: mineral that contains silicon and oxygen and usually one or more other elements. (Chap. 3, Sec. 1, p. 66)

sill: igneous rock feature formed when magma is squeezed into a horizontal crack between layers of rock and hardens underground. (Chap. 12, Sec. 3, p. 351)

slump: a type of mass movement that occurs when a mass of material moves down a curved slope. (Chap. 8, Sec. 1, p. 213)

soil: mixture of weathered rock, decayed organic matter, mineral fragments, water, and air that can take thousands of years to develop. (Chap. 7, Sec. 2, p. 190)

soil profile: vertical section of soil layers, each of which is a horizon. (Chap. 7, Sec. 2, p. 192)

solar eclipse: occurs when the Moon passes directly between the Sun and Earth and casts a shadow over part of Earth. (Chap. 23, Sec. 2, p. 681)

solar energy: energy from the Sun that is clean, inexhaustible, and can be transformed into electricity by solar cells. (Chap. 5, Sec. 2, p. 130)

solar system: system of nine planets, including Earth, and other objects that revolve around the Sun. (Chap. 24, Sec. 1, p. 703)

solstice: twice-yearly point at which the Sun reaches its greatest distance north or south of the equator. (Chap. 23, Sec. 1, p. 676)

solution: a kind of mixture in which one substance is completely and evenly mixed in another substance and is the same throughout. (Chap. 2, Sec. 2, p. 43)

space probe: instrument that travels far into the solar system, gathers data, and sends them back to Earth. (Chap. 22, Sec. 2, p. 650)

space shuttle: reusable spacecraft that can carry cargo, astronauts, and satellites to and from space. (Chap. 22, Sec. 3, p. 655)

space station: large facility with living quarters, work and exercise areas, and equipment and support systems for humans to live and work in space and conduct research not possible on Earth. (Chap. 22, Sec. 3, p. 656)

species: group of organisms that reproduces only with other members of their own group. (Chap. 14, Sec. 1, p. 400)

specific gravity: ratio of a mineral's weight compared with the weight of an equal volume of water. (Chap. 3, Sec. 2, p. 70)

sphere (SFIHR): a round, three-dimensional object whose surface is the same distance from its center at all points; Earth is a sphere that bulges somewhat at the equator and is slightly flattened at the poles. (Chap. 23, Sec. 1, p. 672)

spring: forms when the water table meets Earth's surface; often found on hillsides and used as a freshwater source. (Chap. 9, Sec. 2, p. 255)

station model: indicates weather conditions at a specific location, using a combination of symbols on a map. (Chap. 16, Sec. 3, p. 479)

streak: color of a mineral when it is in powdered form. (Chap. 3, Sec. 2, p. 71)

stream discharge: volume of water that flows past a specific point per unit of time. (Chap. 20, Sec. 2, p. 591)

strike-slip fault: break in rock caused by shear forces, where rocks move past each other without much vertical movement. (Chap. 11, Sec. 1, p. 307)

sunspots: areas of the Sun that are cooler and less bright than surrounding areas, are caused by the Sun's magnetic field, and occur in cycles. (Chap. 25, Sec. 2, p. 742)

supergiant: late stage in the life cycle of a massive star in which the core heats up, heavy elements form by fusion, and the star expands; can eventually explode to form a supernova. (Chap. 25, Sec. 3, p. 750)

surface current: wind-powered ocean current that moves water horizontally—parallel to Earth's surface—and moves only the upper few hundred meters of seawater. (Chap. 18, Sec. 2, p. 526)

surface wave: seismic wave that moves rock particles up and down in a backward rolling motion and side to side in a swaying motion. (Chap. 11, Sec. 2, p. 309)

T

technology: use of scientific discoveries for practical purposes, making people's lives easier and better. (Chap. 1, Sec. 1, p. 12)

temperate zones: climate zones with moderate temperatures that are located between the tropics and the polar zones. (Chap. 17, Sec. 1, p. 492)

tephra (TEFF ruh): bits of rock or solidified lava dropped from the air during an explosive volcanic eruption; ranges in size from volcanic ash to volcanic bombs and blocks. (Chap. 12, Sec. 2, p. 344)

terracing: farming method used to reduce erosion on steep slopes. (Chap. 7, Sec. 3, p. 201)

tidal range: the difference between the level of the ocean at high tide and the level at low tide. (Chap. 18, Sec. 3, p. 536)

English Glossary

tide: daily rise and fall in sea level caused, for the most part, by the interaction of gravity in the Earth-Moon system. (Chap. 18, Sec. 3, p. 535)

till: mixture of different-sized sediments that is dropped from the base of a retreating glacier and can cover huge areas of land. (Chap. 8, Sec. 2, p. 218)

topographic map: map that shows the changes in elevation of Earth's surface and indicates such features as roads and cities. (Chap. 6, Sec. 3, p. 168)

tornado: violent, whirling windstorm that crosses land in a narrow path and can result from wind shears inside a thunderhead. (Chap. 16, Sec. 2, p. 474)

trench: long, narrow, steep-sided depression in the seafloor formed where one crustal plate sinks beneath another. (Chap. 19, Sec. 1, p. 553)

trilobite (TRI luh bite): organism with a three-lobed exoskeleton that was abundant in Paleozoic oceans and is considered to be an index fossil. (Chap. 14, Sec. 1, p. 398)

tropics: climate zone that receives the most solar radiation, is located between latitudes 23° N and 23° S, and is always hot, except at high elevations. (Chap. 17, Sec. 1, p. 492)

troposphere: layer of Earth's atmosphere that is closest to the ground and contains 99 percent of the water vapor and 75 percent of the atmospheric gases; where clouds and weather occur. (Chap. 15, Sec. 1, p. 436)

trough: lowest point of a wave. (Chap. 18, Sec. 3, p. 532)

tsunami (soo NAHM ee): seismic sea wave that begins over an earthquake focus and can be highly destructive when it crashes on shore. (Chap. 11, Sec. 3, p. 320)

U

ultraviolet radiation: a type of energy that comes to Earth from the Sun, can damage skin and cause cancer, and is mostly absorbed by the ozone layer. (Chap. 15, Sec. 1, p. 440)

unconformity (un kun FOR mih tee): gap in the rock layer that is due to erosion or periods without any deposition. (Chap. 13, Sec. 2, p. 378)

uniformitarianism: principle stating that Earth processes occurring today are similar to those that occurred in the past. (Chap. 13, Sec. 3, p. 387)

upwarped mountains: mountains formed when blocks of Earth's crust are pushed up by forces inside Earth. (Chap. 6, Sec. 1, p. 160)

upwelling: circulation in the ocean that brings deep, cold water to the ocean surface. (Chap. 18, Sec. 2, p. 529)

Uranus (YOOR uh nus): seventh planet from the Sun, is large and gaseous, has a distinct bluish-green color, and rotates on an axis nearly parallel to the plane of its orbit. (Chap. 24, Sec. 3, p. 717)

V

variables: different factors that can be changed in an experiment. (Chap. 1, Sec. 1, p. 10)

vent: opening where magma is forced up and flows out onto Earth's surface as lava, forming a volcano. (Chap. 12, Sec. 1, p. 336)

Venus: second planet from the Sun; similar to Earth in mass and size; has a thick atmosphere and a surface with craters, faultlike cracks, and volcanoes. (Chap. 24, Sec. 2, p. 709)

volcanic mountains: mountains formed when molten material reaches Earth's surface through a weak crustal area and piles up into a cone-shaped structure. (Chap. 6, Sec. 1, p. 161)

volcanic neck: solid igneous core of a volcano left behind after the softer cone has been eroded. (Chap. 12, Sec. 3, p. 351)

volcano: opening in Earth's surface that erupts sulfurous gases, ash, and lava; can form at Earth's plate boundaries, where plates move apart or together, and at hot spots. (Chap. 12, Sec. 1, p. 334)

W

waning: describes phases that occur after a full moon, as the visible lighted side of the Moon grows smaller. (Chap. 23, Sec. 2, p. 679)

water table: upper surface of the zone of saturation; drops during a drought. (Chap. 9, Sec. 2, p. 252)

wave: rhythmic movement that carries energy through matter or space; can be described by its crest, trough, wavelength, and wave height. (Chap. 18, Sec. 3, p. 532)

waxing: describes phases following a new moon, as more of the Moon's lighted side becomes visible. (Chap. 23, Sec. 2, p. 679)

weather: state of the atmosphere at a specific time and place, determined by factors including air pressure, amount of moisture in the air, temperature, and wind. (Chap. 16, Sec. 1, p. 462)

weathering: mechanical or chemical surface processes that break rocks into smaller and smaller pieces. (Chap. 7, Sec. 1, p. 184)

white dwarf: late stage in the life cycle of a main sequence star in which its core uses up its helium and its outer layers escape into space, leaving behind a hot, dense core. (Chap. 25, Sec. 3, p. 749)

wind farm: area where many windmills use wind to generate electricity. (Chap. 5, Sec. 2, p. 131)

Spanish Glossary

Este glossario define cada término clave que aparece en negrillas en el texto. También muestra el capítulo, sección y el número de página en donde se usa dicho término.

A

abrasion / abrasión: tipo de erosión que ocurre cuando los sedimentos soplados por el viento golpean rocas y sedimentos, puliendo y dejando huecos en su superficie. (Cap. 8, Sec. 3, pág. 224)

absolute age / edad absoluta: edad, expresada en años, de una roca u otro material; se puede determinar usando las propiedades de los átomos que componen tales materiales. (Cap. 13, Sec. 3, pág. 383)

absolute magnitude / magnitud absoluta: medida de la cantidad de luz que en realidad emite una estrella. (Cap. 25, Sec. 1, pág. 738)

abyssal plain / planicie abisal: área plana de suelo marino que desciende de 4,000 a 6,000 m debajo de la superficie del océano y que se forma de la depositación de sedimentos. (Cap. 19, Sec. 1, pág. 551)

acid / ácido: sustancia cuyo pH es inferior a 7. (Cap. 21, Sec. 2, pág. 621)

acid rain / lluvia ácida: humedad ácida, cuyo pH es inferior a 5.6, que cae a la Tierra como lluvia o nieve y puede causar daños a bosques, perjudicar organismos y corroer las estructuras. (Cap. 21, Sec. 2, pág. 621)

adaptation / adaptación: todo cambio estructural o de comportamiento que le ayuda a un organismo a sobrevivir en un ambiente en particular. (Cap. 17, Sec. 2, pág. 496)

air mass / masa de aire: flujo enorme de aire que tiene las mismas características de temperatura y contenido de humedad que la superficie terrestre sobre la cual se formó. (Cap. 16, Sec. 2, pág. 470)

apparent magnitude / magnitud aparente: medida de la cantidad de luz proveniente de una estrella que recibe la Tierra. (Cap. 25, Sec. 1, pág. 738)

aquifer / acuífero: capa de roca permeable que permite la infiltración del agua. (Cap. 9, Sec. 2, pág. 252)

asteroid / asteroide: fragmento rocoso formado por material semejante a aquel que formó los planetas; se encuentra principalmente en el cinturón de asteroides entre las órbitas de Marte y Júpiter. (Cap. 24, Sec. 4, pág. 724)

asthenosphere / astenosfera: capa de la Tierra tipo plástico en la cual las placas litosféricas flotan y se mueven. (Cap. 10, Sec. 3, pág. 284)

atmosphere / atmósfera: el aire de la Tierra, el cual está compuesto por una capa tenue de gases, sólidos y líquidos; forma una capa protectora alrededor del planeta y está dividida en cinco capas distintivas. (Cap. 15, Sec. 1, pág. 434)

atom(s) / átomos: partículas diminutas de materia compuestas de protones, neutrones y electrones. (Cap. 2, Sec. 1, pág. 34)

atomic number / número atómico: el número de protones en un átomo. (Cap. 2, Sec. 1, pág. 37)

axis / eje: línea vertical imaginaria que pasa a través del centro de la Tierra y alrededor de la cual gira nuestro planeta. (Cap. 23, Sec. 1, pág. 673)

B

basaltic / basáltica: roca ígnea densa y oscura que se forma del magma rico en magnesio y hierro y deficiente en silicio. (Cap. 4, Sec. 2, pág. 97)

base / base: sustancia cuyo pH es superior a 7. (Cap. 21, Sec. 2, pág. 621)

basin / cuenca: depresión en la Tierra en la cual se formó un océano cuando el área se llenó de agua debido a lluvias torrenciales. (Cap. 18, Sec. 1, pág. 523)

batholith / batolito: cuerpo rocoso ígneo e intrusivo más grande que se forma cuando el magma que es forzado a ascender hacia la corteza terrestre se enfría lentamente y se solidifica bajo tierra. (Cap. 12, Sec. 3, pág. 350)

beach / playa: depósito de sedimentos cuyos materiales varían en tamaño, color y composición y que se halla comúnmente en un litoral liso y levemente inclinado. (Cap. 9, Sec. 3, pág. 259)

benthos / bentos: plantas y animales marinos que viven en el suelo oceánico o sobre él. (Cap. 19, Sec. 2, pág. 561)

bias / sesgo: opinión personal. (Cap. 1, Sec. 2, pág. 21)

big bang theory / teoría de la gran explosión: afirma que el universo comenzó hace unos 12 a 15 billones de años con una inmensa y ardiente explosión. (Cap. 25, Sec. 4, pág. 757)

biomass energy / energía de biomasa: energía renovable proveniente de la quema de materiales orgánicos, como la leña y el alcohol. (Cap. 5, Sec. 2, pág. 133)

black hole / agujero negro: etapa final en la evolución de una supernova, cuando la masa del núcleo colapsa hasta el punto de que no existe volumen y cuya gravedad es tan poderosa que ni siquiera la luz puede escapar. (Cap. 25, Sec. 3, pág. 750)

blizzard / ventisca: tormenta invernal que dura por lo menos tres horas, con temperaturas de −12°C o más bajas, poca visibilidad y vientos de por lo menos 51 km/h. (Cap. 16, Sec. 2, pág. 477)

breaker / cachón: ola oceánica que se forma en aguas poco profundas y que rompe en la playa. (Cap. 18, Sec. 3, pág. 533)

C

---

caldera / caldera: extensa abertura circular que se forma cuando colapsa la parte superior de un volcán. (Cap. 12, Sec. 3, pág. 352)

carbon film / película carbonácea: fina película de residuo carbonoso preservada como fósil. (Cap. 13, Sec. 1, pág. 370)

carbon monoxide / monóxido de carbono: gas incoloro e inodoro que reduce el contenido de oxígeno en la sangre, se encuentra en los gases de escape de los vehículos y contribuye a la contaminación del aire. (Cap. 21, Sec. 2, pág. 622)

carrying capacity / capacidad de carga: número máximo de individuos de una especie dada que el ambiente puede mantener. (Cap. 20, Sec. 1, pág. 585)

cast / impresión fósil: tipo de fósil corporal que se forma cuando los cristales llenan un molde o cuando los sedimentos se asientan en un molde y se endurecen convirtiéndose en roca. (Cap. 13, Sec. 1, pág. 371)

cave / caverna: abertura subterránea que se puede formar cuando las aguas subterráneas ácidas disuelven la piedra caliza. (Cap. 9, Sec. 2, pág. 255)

cementation / cementación: proceso formador de rocas sedimentarias en el cual los cementos naturales que se producen cuando el agua se filtra por la roca y el suelo mantienen unidos los sedimentos grandes. (Cap. 4, Sec. 4, pág. 105)

Cenozoic Era / Era Cenozoica: era de vida reciente que comenzó hace 66 millones de años aproximadamente y que continúa hoy en día; incluye la primera aparición del *Homo sapiens* hace unos 400,000 años. (Cap. 14, Sec. 3, pág. 418)

channel / cauce: surco creado por agua que corre hacia abajo del mismo sendero. (Cap. 9, Sec. 1, pág. 242)

chemical weathering / meteorización química: se presenta cuando las reacciones químicas disuelven los minerales de las rocas o los transforman en minerales diferentes. (Cap. 7, Sec. 1, pág. 187)

chemosynthesis / quimiosíntesis: proceso mediante el cual las bacterias que viven cerca de respiraderos térmicos elaboran el alimento a partir de compuestos de azufre o nitrógeno, en lugar de la energía luminosa proveniente del Sol. (Cap. 19, Sec. 2, pág. 559)

chlorofluorocarbons (CFCs) / clorofluorocarbonos: grupo de compuestos químicos que se utilizan en refrigeradores, acondicionadores de aire, empaques de espuma y rociadores de aerosol; estos compuestos químicos pueden penetrar en la atmósfera y destruir el ozono. (Cap. 15, Sec. 1, pág. 440)

chromosphere / cromosfera: capa de la atmósfera del Sol ubicada encima de la fotosfera. (Cap. 25, Sec. 2, pág. 741)

cinder cone volcano / volcán de cono de carbonilla: volcán de laderas abruptas y vagamente compreso que se forma cuando la tefrita cae al suelo. (Cap. 12, Sec. 2, pág. 344)

cleavage / crucero: propiedad física de algunos minerales de poder romperse a lo largo de superficies suaves y planas. (Cap. 3, Sec. 2, pág. 71)

climate / clima: patrón de tiempo promedio en una área a lo largo de un período largo de tiempo; puede clasificarse según la temperatura, la humedad, la precipitación y la vegetación. (Cap. 7, Sec. 1, pág. 188; Cap. 17, Sec. 1, pág. 492)

coal / carbón: roca sedimentaria formada de material vegetal en descomposición; el combustible fósil más abundante en el mundo. (Cap. 5, Sec. 1, pág. 121)

comet / cometa: astro formado por polvo y partículas rocosas mezcladas con agua congelada, metano y amoníaco, que forma una cola brillante a medida que se acerca al Sol. (Cap. 24, Sec. 4, pág. 722)

compaction / compactación: proceso que forma rocas sedimentarias cuando las capas de sedimentos pequeños se comprimen debido al peso de las capas superiores. (Cap. 4, Sec. 4, pág. 104)

composite volcano / volcán compuesto: volcán que se ha formado por la alternación de erupciones explosivas y silenciosas que producen capas de tefrita y lava; se halla principalmente donde se juntan las placas terrestres y una placa se hunde debajo de otra. (Cap. 12, Sec. 2, pág. 345)

composting / abono orgánico: método de conservación en que los desperdicios del jardín, como el pasto, la maleza y las hojas rastrilladas, se amontonan y se dejan descomponer gradualmente. (Cap. 20, Sec. 3, pág. 597)

compound / compuesto: materia que está hecha de dos o más elementos y que tiene propiedades físicas y químicas diferentes a las de los elementos que la formaron. (Cap. 2, Sec. 2, pág. 40)

condensation / condensación: proceso en el cual el vapor de agua se transforma en un líquido. (Cap. 15, Sec. 2, pág. 445)

conduction / conducción: transferencia de energía que ocurre cuando las moléculas chocan entre sí. (Cap. 15, Sec. 2, pág. 444)

conic projection / proyección cónica: mapa que se hace proyectando puntos y líneas de un globo terráqueo a un cono. (Cap. 6, Sec. 3, pág. 167)

conservation / conservación: uso cuidadoso de los recursos para disminuir el daño al ambiente a través de métodos como el abono orgánico y el reciclaje de los materiales. (Cap. 20, Sec. 3, pág. 596)

constant / constante: variable que no cambia en un experimento. (Cap. 1, Sec. 1, pág. 10)

constellation / constelación: grupo de estrellas que forma un patrón en el cielo que se ve como un objeto conocido (el Carro Mayor), un animal (Pegaso) o un personaje (Orión). (Cap. 25, Sec. 1, pág. 736)

continental drift / deriva continental: hipótesis de Wegener que afirmaba que todos los continentes estuvieron unidos en algún momento formando una sola masa continental, la cual se separó hace unos 200 millones de años, haciendo que los continentes derivaran lentamente a sus posiciones actuales. (Cap. 10, Sec. 1, pág. 276)

continental shelf / plataforma continental: extremo gradualmente inclinado de un continente que se extiende debajo del océano y que provee morada a la mayoría de los organismos marinos. (Cap. 19, Sec. 1, pág. 550)

continental slope / talud continental: relieve de la cuenca oceánica que desciende abruptamente a partir de la plataforma continental. (Cap. 19, Sec. 1, pág. 551)

contour line / curva de nivel: línea en un mapa que conecta puntos con igual elevación. (Cap. 6, Sec. 3, pág. 168)

control / control: estándar de comparación en un experimento. (Cap. 1, Sec. 1, pág. 10)

convection / convección: transferencia de calor a través del flujo de un material. (Cap. 15, Sec. 2, pág. 444)

convection current / corriente de convección: corriente en el manto terrestre que transfiere energía en el interior de la Tierra y que provee la potencia de la tectónica de placas. (Cap. 10, Sec. 3, pág. 289)

Coriolis effect / efecto de Coriolis: es la causa de que el aire y el agua en movimiento giren a la izquierda en el hemisferio sur y a la derecha en el hemisferio norte, debido a la rotación de la Tierra. (Cap. 15, Sec. 3, pág. 448)

Coriolis effect / efecto de Coriolis: cambio de vientos y corrientes superficiales provocados por la rotación de la Tierra; hace que las corrientes al norte del ecuador fluyan en dirección de las manecillas del reloj y las corrientes al sur del ecuador en dirección contraria. (Cap. 18, Sec. 2, pág. 527)

corona / corona: la capa más externa y grande de la atmósfera del Sol; se extiende millones de kilómetros y soporta temperaturas de hasta 2 millones K. (Cap. 25, Sec. 2, pág. 741)

crater / cráter: depresión de murallas escarpadas alrededor de la chimenea de un volcán. (Cap. 12, Sec. 1, pág. 336)

creep / corrimiento: tipo de movimiento de masas en el cual los sedimentos se mueven cuesta abajo paulatinamente; es común en áreas de congelamiento y derretimiento; puede hacer que las paredes, los árboles y las cercas se inclinen. (Cap. 8, Sec. 1, pág. 214)

crest / cresta: punto más alto de una onda. (Cap. 18, Sec. 3, pág. 532)

crystal / cristal: sólido cuyos átomos están arreglados en un patrón ordenado y repetitivo. (Cap. 3, Sec. 1, pág. 63)

cyanobacteria / cianobacterias: bacterias fotosintéticas que contienen clorofila y que se piensa son unas de las primeras formas de vida terrestre. (Cap. 14, Sec. 2, pág. 407)

D

deflation / deflacción: tipo de erosión que se presenta cuando el viento sopla sobre sedimentos sueltos, extrae micropartículas y deja atrás sedimentos más gruesos. (Cap. 8, Sec. 3, pág. 224)

deforestation / deforestación: destrucción y tala de bosques en que a menudo se despeja la tierra para la minería, la construcción de caminos y el pastoreo del ganado y la cual resulta en aumentos en los niveles atmosféricos de CO_2. (Cap. 17, Sec. 3, pág. 509)

density / densidad: cambio físico de la materia que puede calcularse dividiendo la masa de un cuerpo entre su volumen. (Cap. 2, Sec. 3, pág. 46)

density current / corriente de densidad: patrón de circulación en el océano que se forma cuando una masa de agua salada más densa se hunde debajo de agua salada menos densa. (Cap. 18, Sec. 2, pág. 529)

dependent variable / variable dependiente: factor que se mide en un experimento. (Cap. 1, Sec. 1, pág. 10)

Spanish Glossary

deposition / depositación: acción de dejar caer los sedimentos, la cual ocurre cuando un agente erosivo, como la gravedad, un glaciar, el viento o el agua, pierde su energía y no puede seguir transportando su carga. (Cap. 8, Sec. 1, pág. 213)

dew point / punto de condensación: temperatura a la cual el aire se satura y se forma la condensación. (Cap. 16, Sec. 1, pág. 465)

dike / dique: filón de roca ígnea que se forma cuando el magma es inyectado en una fisura vertical que atraviesa capas rocosas y se endurece bajo tierra. (Cap. 12, Sec. 3, pág. 351)

drainage basin / cuenca hidrográfica: terreno del cual un río u otra corriente de agua recoge las aguas de escorrentía. (Cap. 9, Sec. 1, pág. 244)

dune / duna: montículo que se forma cuando los sedimentos arrastrados por el viento se acumulan detrás de una barrera; relieve común en las regiones desérticas. (Cap. 8, Sec. 3, pág. 227)

E

Earth / la Tierra: tercer planeta más cercano al Sol; posee una atmósfera que protege la vida y las temperaturas de su superficie permiten la existencia de agua como sólido, líquido y gas. (Cap. 24, Sec. 2, pág. 710)

earthquake / terremoto: vibraciones producidas cuando las rocas se rompen a lo largo de una falla. (Cap. 11, Sec. 1, pág. 305)

Earth science / ciencias terrestres: estudio de la Tierra y del espacio, incluye el estudio de rocas, fósiles, clima, volcanes, uso del terreno, agua marina, terremotos y astros en el espacio. (Cap. 1, Sec. 1, pág. 9)

electromagnetic spectrum / espectro electromagnético: arreglo de ondas electromagnéticas según sus longitudes de onda. (Cap. 22, Sec. 1, pág. 641)

electrons / electrones: partículas con carga negativa que se mueven alrededor del núcleo de un átomo y forman la nube electrónica. (Cap. 2, Sec. 1, pág. 36)

element / elemento: sustancia que sólo contiene un tipo de átomo; por ejemplo, el oxígeno, el aluminio y el hierro. (Cap. 2, Sec. 1, pág. 35)

ellipse / elipse: trayectoria curva cerrada y alargada que describe la órbita anual alrededor del Sol que efectúa la Tierra. (Cap. 23, Sec. 1, pág. 675)

El Niño / El Niño: fenómeno climático que comienza en el océano Pacífico tropical; puede ocurrir debido al debilitamiento o inversión de los vientos alisios, y puede interrumpir los patrones normales de temperatura y precipitación por todo el mundo. (Cap. 17, Sec. 3, pág. 501)

enzyme / enzima: sustancia que acelera las reacciones químicas. (Cap. 20, Sec. 2, pág. 593)

eon / eón: la subdivisión más larga de la escala del tiempo geológico, basada en la abundancia de ciertos tipos de fósiles; se subdivide en eras, períodos y épocas. (Cap. 14, Sec. 1, pág. 399)

epicenter / epicentro: punto sobre la superficie terrestre directamente sobre el foco de un terremoto. (Cap. 11, Sec. 2, pág. 309)

epoch / época: la siguiente subdivisión de tiempo geológico que le sigue al período; se caracteriza por diferencias en formas de vida que pueden variar regionalmente. (Cap. 14, Sec. 1, pág. 399)

equator / ecuador: línea imaginaria que rodea la Tierra alrededor de 0° de latitud, equidistante del polo norte y el polo sur. (Cap. 6, Sec. 2, pág. 162)

equinox / equinoccio: ocurre dos veces al año, en la primavera y en el verano, cuando el Sol está directamente sobre el ecuador, ocasionando que la noche y el día tengan la misma duración en todo el mundo. (Cap. 23, Sec. 1, pág. 677)

era / era: segunda división mayor del tiempo geológico; se subdivide en períodos y se basa en cambios importantes a nivel mundial en los tipos de fósiles. (Cap. 14, Sec. 1, pág. 399)

erosion / erosión: proceso en el cual los materiales superficiales se desgastan y son transportados de un lugar a otro por agentes como la gravedad, el agua, el viento y los glaciares. (Cap. 8, Sec. 1, pág. 212)

estuary / estuario: área donde un río desemboca en el océano, la cual contiene una mezcla de agua dulce y agua marina y provee un hábitat importante para muchos organismos marinos. (Cap. 19, Sec. 2, pág. 564)

ethics / ética: estudio de los valores morales sobre el bien y el mal. (Cap. 1, Sec. 2, pág. 20)

extrusive / extrusiva: roca ígnea de grano fino que se forma cuando el magma se enfría rápidamente sobre o cerca de la superficie terrestre. (Cap. 4, Sec. 2, pág. 95)

F

fault / falla: superficie a lo largo de la cual se mueven y se rompen las rocas cuando exceden su límite de elasticidad. (Cap. 11, Sec. 1, pág. 304)

fault-block mountains / montañas de bloques de falla: montañas que se forman de enormes bloques rocosos inclinados, pero separados de las rocas circundantes por fallas. (Cap. 6, Sec. 1, pág. 160)

fertilizer / fertilizante: sustancia química que ayuda en el crecimiento de plantas y otros organismos. (Cap. 21, Sec. 1, pág. 611)

focus / foco: en un terremoto, es el punto sobre la superficie terrestre donde se libera la energía en forma de ondas sísmicas. (Cap. 11, Sec. 2, pág. 308)

fog / neblina: una nube estrato que se forma cuando el aire se enfría hasta su punto de rocío, cerca de la superficie terrestre. (Cap. 16, Sec. 1, pág. 467)

folded mountains / montañas plegadas: montañas que se forman cuando las capas rocosas horizontales son comprimidas desde lados opuestos, lo cual hace que se encorven y se doblen. (Cap. 6, Sec. 1, pág. 159)

foliated / foliada: roca metamórfica, como la pizarra y el gneiss, cuyos granos minerales se aplanan y se alinean en capas paralelas. (Cap. 4, Sec. 3, pág. 101)

fossil fuel / combustible fósil: recurso energético no renovable, como el petróleo y el carbón, que se formó hace millones de años a partir de los restos de plantas y otros organismos muertos. (Cap. 5, Sec. 1, pág. 120)

fossils / fósiles: restos, impresiones o trazas de organismos prehistóricos que pueden indicar cuándo y dónde los organismos vivieron y cómo vivieron. (Cap. 13, Sec. 1, pág. 369)

fracture / fractura: propiedad física de algunos minerales de romperse a lo largo de superficies disparejas, ásperas o dentadas. (Cap. 3, Sec. 2, pág. 71)

front / frente: límite entre dos masas de aire que poseen diferentes temperaturas, densidad o humedad; puede ser frío, cálido, ocluido y estacionario. (Cap. 16, Sec. 2, pág. 471)

full moon / luna llena: fase lunar que ocurre cuando toda la superficie lunar que da la cara a la Tierra refleja luz. (Cap. 23, Sec. 2, pág. 679)

G

galaxy / galaxia: extenso grupo de estrellas, polvo y gas que se mantiene unido gracias a la gravedad; puede ser elíptica, espiral o irregular. (Cap. 25, Sec. 4, pág. 752)

gem / gema: mineral precioso, muy valioso que se puede usar en joyería. (Cap. 3, Sec. 3, pág. 73)

Spanish Glossary

geologic time scale / escala del tiempo geológico: división de la historia de la Tierra en unidades cronológicas basadas en gran parte en los tipos de formas de vida que existieron solamente en ciertos períodos. (Cap. 14, Sec. 1, pág. 398)

geothermal energy / energía geotérmica: recurso energético interminable que hace uso del magma caliente o del calor de las rocas calientes y secas debajo de la superficie terrestre para generar electricidad. (Cap. 5, Sec. 2, pág. 132)

geyser / géiser: manantial de agua caliente que brota periódicamente y lanza agua y vapor al aire. Un ejemplo es el Old Faithful en el parque nacional de Yellowstone. (Cap. 9, Sec. 2, pág. 255)

giant / gigante: etapa avanzada en la vida de una estrella de la secuencia principal en la cual el hidrógeno del núcleo se agota, el núcleo se contrae y las temperaturas en el interior de la estrella aumentan, haciendo que sus capas externas se expandan y enfríen. (Cap. 25, Sec. 3, pág. 749)

glaciers / glaciares: extensas masas de hielo y nieve en movimiento que cambian grandes regiones de la superficie terrestre, a través de la erosión y la depositación. (Cap. 8, Sec. 2, pág. 217)

global warming / calentamiento global: aumento en el promedio de la temperatura global de la Tierra. (Cap. 17, Sec. 3, pág. 508)

granitic / granítica: roca ígnea de color claro y rica en sílice que es menos densa que la roca basáltica. (Cap. 4, Sec. 2, pág. 97)

Great Red Spot / Gran Mancha Roja: gigantesca tormenta de gases de altas presiones en Júpiter. (Cap. 24, Sec. 3, pág. 714)

greenhouse effect / efecto de invernadero: calentamiento natural que ocurre cuando ciertos gases en la atmósfera de la Tierra, por ejemplo, el metano, el CO_2 y el vapor de agua, atrapan el calor. (Cap. 17, Sec. 3, pág. 507)

groundwater / agua subterránea: agua que se filtra en el suelo y se junta en poros y espacios vacíos y la cual es una fuente importante de agua potable. (Cap. 9, Sec. 2, pág. 251)

H

half-life / media vida: tiempo que se demora en desintegrarse la mitad de los átomos de un isótopo. (Cap. 13, Sec. 3, pág. 384)

hardness / dureza: medida del grado de facilidad con que se puede rayar un mineral; se determina según el arreglo de los átomos del mineral. (Cap. 3, Sec. 2, pág. 69)

hazardous waste / desechos peligrosos: desechos venenosos, inflamables o carcinógenos. (Cap. 20, Sec. 2, pág. 592)

heterogeneous mixture / mezcla heterogénea: mezcla que no está distribuida uniformemente y en la cual cada componente retiene sus propiedades. (Cap. 2, Sec. 2, pág. 43)

hibernation / hibernación: adaptación del comportamiento para sobrevivir el invierno en que un animal disminuye considerablemente sus actividades corporales, la temperatura de su cuerpo baja y los procesos corporales se vuelven más lentos. (Cap. 17, Sec. 2, pág. 498)

homogeneous mixture / mezcla homogénea: mezcla que está distribuida uniformemente. (Cap. 2, Sec. 2, pág. 43)

horizon / horizonte: cada capa del perfil del suelo: horizonte A (capa superior del suelo), horizonte B (capa intermedia) y horizonte C (capa inferior). (Cap. 7, Sec. 2, pág. 192)

hot spot / foco caliente: zona de intenso calor ubicada en el límite entre el manto y el núcleo de la Tierra que forma volcanes cuando la roca derretida es forzada a ascender y atraviesa la corteza. (Cap. 12, Sec. 1, pág. 338)

humidity / humedad: cantidad de vapor de agua que sostiene el aire. (Cap. 16, Sec. 1, pág. 464)

humus / humus: materia orgánica negruzca y descompuesta que suministra nutrientes a las plantas y que se halla principalmente en la capa de suelo arable. (Cap. 7, Sec. 2, pág. 192)

hurricane / huracán: tormenta extensa y severa que se forma sobre los océanos tropicales, con vientos de por lo menos 120 km/h y que pierde fuerza al llegar a tierra firme. (Cap. 16, Sec. 2, pág. 476)

hydroelectric energy / energía hidroeléctrica: electricidad que produce la potencia del agua al hacer grandes represas en un río. (Cap. 5, Sec. 2, pág. 132)

hydrosphere / hidrosfera: toda el agua de la superficie terrestre. (Cap. 15, Sec. 2, pág. 445)

hypothesis / hipótesis: una conjetura informada. (Cap. 1, Sec. 1, pág. 7)

I

---

ice wedging / grietas debido al hielo: proceso de meteorización mecánica que ocurre cuando el agua se congela en las grietas de las rocas y se expande, haciendo que la roca se fragmente. (Cap. 7, Sec. 1, pág. 186)

igneous rock / roca ígnea: roca intrusiva o extrusiva que se forma cuando el magma caliente se enfría y se endurece. (Cap. 4, Sec. 2, pág. 94)

impact basin / cuenca de impacto: cavidad formada sobre la superficie de la Luna por el impacto de un objeto. (Cap. 23, Sec. 3, pág. 689)

impermeable / impermeable: describe los materiales por los cuales el agua no puede filtrarse. (Cap. 9, Sec. 2, pág. 252)

independent variable / variable independiente: factor que cambia en un experimento. (Cap. 1, Sec. 1, pág. 10)

index fossils / fósiles guía: restos de especies que existieron en la Tierra durante un período relativamente corto de tiempo, fueron abundantes y se extendieron geográficamente ; los geólogos pueden usar estos fósiles para determinar las edades de las capas rocosas. (Cap. 13, Sec. 1, pág. 373)

intrusive / intrusiva: tipo de roca ígnea que, por lo general, contiene cristales de gran tamaño y que se forma cuando el magma se enfría lentamente debajo de la superficie terrestre. (Cap. 4, Sec. 2, pág. 95)

ion / ion: átomo con carga eléctrica cuya carga resulta cuando un átomo pierde o gana electrones. (Cap. 2, Sec. 2, pág. 41)

ionosphere / ionosfera: capa de partículas cargadas eléctricamente en la termosfera que absorbe las ondas radiales AM durante el día y las vuelve a reflejar durante la noche. (Cap. 15, Sec. 1, pág. 437)

isobars / isobaras: líneas que se trazan en un mapa meteorológico conectando puntos que tienen la misma presión atmosférica; también indican la ubicación de las áreas de alta y de baja presión y pueden mostrar la velocidad del viento. (Cap. 16, Sec. 3, pág. 479)

isotherm / isoterma: línea que se traza en un mapa meteorológico conectando puntos que tienen la misma temperatura. (Cap. 16, Sec. 3, pág. 479)

isotopes / isótopos: átomos del mismo elemento con distintos números de neutrones. (Cap. 2, Sec. 1, pág. 37)

J

---

jet stream / corriente de chorro: franja estrecha de vientos fuertes que sopla cerca de la troposfera. (Cap. 15, Sec. 3, pág. 450)

Jupiter / Júpiter: el planeta más grande de nuestro sistema solar y el quinto más cercano al Sol; contiene más masa que todos los otros planetas juntos, tiene tormentas continuas de gases a altas presiones y una atmósfera compuesta en su mayor parte de hidrógeno y helio. (Cap. 24, Sec. 3, pág. 714)

L

land breeze / brisa terrestre: movimiento de aire nocturno desde la tierra hacia el mar y que se forma cuando el aire más frío y denso proveniente de la tierra fuerza el aire más cálido a ascender sobre el mar. (Cap. 15, Sec. 3, pág. 451)

latitude / latitud: distancia en grados al norte o al sur del ecuador. (Cap. 6, Sec. 2, pág. 162)

lava / lava: materia rocosa derretida, espesa y viscosa, que fluye de los volcanes hacia la superficie terrestre. (Cap. 4, Sec. 2, pág. 94)

leaching / lixiviación: extracción de minerales que se han disuelto en agua. (Cap. 7, Sec. 2, pág. 193)

light-year / año-luz: la distancia que viaja la luz en un año (cerca de 9.5 trillones de km), la cual se usa para registrar distancias entre estrellas y galaxias. (Cap. 25, Sec. 1, pág. 739)

liquefaction / liquefacción: ocurre cuando el suelo mojado actúa como un líquido durante un terremoto. (Cap. 11, Sec. 3, pág. 319)

lithosphere / litosfera: capa rígida de la Tierra de unos 100 km de grosor formada por la corteza y parte del manto superior. (Cap. 10, Sec. 3, pág. 284)

litter / desechos orgánicos: ramitas, hojas y otras materias orgánicas que ayudan a evitar la erosión y que retienen agua y pueden, a la larga, convertirse en humus gracias a la acción de organismos descomponedores. (Cap. 7, Sec. 2, pág. 193)

loess / loess: depósito de sedimentos firmemente compactados y de granos finos que es arrastrado por el viento. (Cap. 8, Sec. 3, pág. 227)

longitude / longitud: distancia en grados al este o al oeste del primer meridiano. (Cap. 6, Sec. 2, pág. 163)

longshore current / corriente costera: corriente que corre paralela a la costa, producto de olas que chocan contra la costa haciendo ángulos leves y que mueven toneladas de sedimento suelto. (Cap. 9, Sec. 3, pág. 258)

lunar eclipse / eclipse lunar: ocurre cuando la sombra de la Tierra cubre la Luna. (Cap. 23, Sec. 2, pág. 682)

luster / lustre: describe la manera en que un mineral refleja la luz desde su superficie; puede ser metálico o no metálico. (Cap. 3, Sec. 2, pág. 70)

M

magma / magma: material rocoso caliente y fundido que se halla debajo de la superficie terrestre. (Cap. 3, Sec. 1, pág. 65)

magnitude / magnitud: medida de la energía liberada durante un movimiento sísmico. (Cap. 11, Sec. 3, pág. 318)

map legend / leyenda de mapa: explica los símbolos que se usan en un mapa. (Cap. 6, Sec. 3, pág. 170)

map scale / escala de un mapa: relación entre la distancia en un mapa y la distancia real sobre la superficie terrestre, la cual se puede representar como una razón o como una pequeña barra dividida en secciones. (Cap. 6, Sec. 3, pág. 170)

maria / mares: regiones de la Luna relativamente planas y de color oscuro que se formaron cuando la lava alcanzó la superficie y llenó los cráteres en la superficie lunar. (Cap. 23, Sec. 2, pág. 683)

Mars / Marte: cuarto planeta más cercano al Sol; tiene casquetes polares de hielo, una atmósfera tenue y una apariencia rojiza que proviene del óxido de hierro de rocas desgastadas. (Cap. 24, Sec. 2, pág. 710)

mass movement / movimiento de masas: cualquier tipo de erosión que ocurre a medida que la gravedad mueve materiales cuesta abajo. (Cap. 8, Sec. 1, pág. 213)

mass number / número de masa: el número de protones más el número de electrones en un átomo. (Cap. 2, Sec. 1, pág. 37)

matter / materia: cualquier cosa que tiene masa y ocupa espacio; las propiedades de la materia están determinadas según la estructura y enlace de sus átomos. (Cap. 2, Sec. 1, pág. 34)

meander / meandro: curva ancha y en forma de c en un río u otra corriente de agua, formada por la erosión de su ribera externa. (Cap. 9, Sec. 1, pág. 245)

mechanical weathering / meteorización mecánica: proceso físico que fragmenta rocas sin alterar su composición química; puede ser el resultado de las grietas debido al hielo, a la acción de los animales y a las raíces vegetales. (Cap. 7, Sec. 1, pág. 185)

Mercury / Mercurio: el planeta más pequeño y el más cercano al Sol; posee una atmósfera tenue y una superficie con muchos cráteres y altos acantilados. (Cap. 24, Sec. 2, pág. 708)

Mesozoic Era / Era Mesozoica: era intermedia de la historia de la Tierra, durante la cual se separó Pangaea, aparecieron los dinosaurios y los reptiles y las gimnospermas eran las formas de vida dominantes en tierra. (Cap. 14, Sec. 3, pág. 414)

metamorphic rock / roca metamórfica: se forma cuando el calor, la presión o los líquidos actúan sobre rocas ígneas, sedimentarias u otras rocas metamórficas y les afectan la forma o composición o ambas. (Cap. 4, Sec. 3, pág. 99)

meteor / meteoro: meteoroide que se incendia en la atmósfera terrestre. (Cap. 24, Sec. 4, pág. 723)

meteorite / meteorito: meteoroide que choca contra la superficie de una luna o planeta. (Cap. 24, Sec. 4, pág. 724)

meteorologist / meteorólogo: persona que estudia el tiempo y usa información del radar Doppler, de los satélites meteorológicos, computadoras y otros instrumentos para hacer mapas meteorológicos y pronósticos del tiempo. (Cap. 16, Sec. 3, pág. 478)

mid-ocean ridge / dorsal medioocéanica: área donde se forma el nuevo suelo oceánico cuando la lava sale a través de las grietas en la corteza terrestre. (Cap. 19, Sec. 1, pág. 552)

mineral / mineral: sólido inorgánico que ocurre en forma natural y que posee una composición química definida y una estructura atómica interna ordenada. (Cap. 3, Sec. 1, pág. 62)

mineral resources / recursos minerales: recursos a partir de los cuales se obtienen metales. (Cap. 5, Sec. 3, pág. 137)

mixture / mezcla: describe dos o más sustancias que retienen sus propiedades a pesar de estar combinadas y la cual se puede separar mediante medios físicos. (Cap. 2, Sec. 2, pág. 43)

mold / molde: tipo de fósil corporal que se forma en la roca cuando un organismo con partes duras se entierra, se descompone o se disuelve y deja una cavidad en la roca. (Cap. 13, Sec. 1, pág. 371)

moon phase / fase lunar: cambio en la apariencia de la Luna, vista desde la Tierra, debido a las posiciones relativas de la Luna, la Tierra y el Sol. (Cap. 23, Sec. 2, pág. 679)

moraine / morrena: extenso banco de rocas y suelo que un glaciar deposita cuando se detiene su movimiento. (Cap. 8, Sec. 2, pág. 219)

N

natural gas / gas natural: combustible fósil que se formó de organismos marinos y que con frecuencia se encuentra en capas rocosas inclinadas o plegadas y que se usa en calefacción y cocción. (Cap. 5, Sec. 1, pág. 123)

natural selection / selección natural: proceso por el cual los organismos que están adaptados a un entorno en particular son más capaces de sobrevivir y reproducirse que organismos que no lo están. (Cap. 14, Sec. 1, pág. 401)

Spanish Glossary

nebula / nebulosa: extensa nube de gas y polvo que se contrae bajo la fuerza gravitatoria y se separa en fragmentos más pequeños, cada uno de los cuales colapsará para formar una estrella. (Cap. 25, Sec. 3, pág. 748)

nekton / necton: organismos marinos que nadan activamente en el océano. (Cap. 19, Sec. 2, pág. 560)

Neptune / Neptuno: por lo general el octavo planeta más alejado del Sol; es grande y gaseoso, posee anillos que varían en grosor y es de color verde-azulado. (Cap. 24, Sec. 3, pág. 718)

neutron / neutrón: partícula sin carga eléctrica ubicada en el núcleo del átomo. (Cap. 2, Sec. 1, pág. 36)

neutron star / estrella de neutrones: núcleo colapsado de una supernova que se puede compactar hasta alcanzar unos 20 km de diámetro y que contiene solamente neutrones en el núcleo denso. (Cap. 25, Sec. 3, pág. 750)

new moon / luna nueva: fase lunar que ocurre cuando la Luna está entre el Sol y la Tierra. En esta fase no se puede ver la Luna porque su mitad iluminada mira hacia el Sol mientras que su mitad oscura mira hacia la Tierra. (Cap. 23, Sec. 2, pág. 679)

nonfoliated / no foliada: roca metamórfica, como la cuarcita y el mármol, cuyos granos minerales crecen y se reordenan, pero sin formar capas. (Cap. 4, Sec. 3, pág. 102)

nonpoint source pollution / contaminación de fuente no puntual: contaminación proveniente de un área extensa que penetra en el agua y que no puede rastrearse a una sola localidad. (Cap. 21, Sec. 1, pág. 610)

normal fault / falla normal: ruptura en la roca causada por las fuerzas de tensión, en donde la roca sobre la superficie de la falla se mueve hacia abajo en relación con la roca debajo de la falla. (Cap. 11, Sec. 1, pág. 306)

nuclear energy / energía nuclear: fuente energética alterna que se basa en la fisión del átomo. (Cap. 5, Sec. 1, pág. 127)

O

observatory / observatorio: centro que puede albergar un telescopio óptico; tiene a menudo un techo en forma de domo que se puede abrir para observar el espacio. (Cap. 22, Sec. 1, pág. 642)

oil / petróleo: combustible fósil líquido formado a partir de organismos marinos que se quema para obtener energía y el cual se utiliza en la manufactura de plásticos. (Cap. 5, Sec. 1, pág. 123)

orbit / órbita: trayectoria curva que sigue un satélite a medida que gira alrededor de un objeto. (Cap. 22, Sec. 2, pág. 649)

ore / mena: depósito en que un mineral existe en cantidades lo suficientemente grandes para ser minado con fines de lucro. (Cap. 5, Sec. 3, pág. 137)

organic evolution / evolución orgánica: cambio de organismos a través del tiempo geológico. (Cap. 14, Sec. 1, pág. 400)

outwash / derrubio: material depositado por el agua derretida de un glaciar. (Cap. 8, Sec. 2, pág. 219)

oxidation / oxidación: proceso de meteorización química que ocurre cuando el material metálico se expone al oxígeno y al agua durante un período de tiempo. (Cap. 7, Sec. 1, pág. 188)

ozone layer / capa de ozono: capa de la estratosfera con una alta concentración de ozono; absorbe la mayor parte de la radiación ultravioleta dañina proveniente del Sol. (Cap. 15, Sec. 1, pág. 440)

P

Paleozoic Era / Era Paleozoica: era de vida antigua, la cual comenzó hace unos 544 millones de años, cuando los organismos desarrollaron partes duras y terminó con extinciones en masa hace unos 245 millones de años. (Cap. 14, Sec. 2, pág. 408)

Pangaea / Pangaea: masa de tierra extensa y antigua que una vez estuvo formada por el conjunto de todos los continentes. (Cap. 10, Sec. 1, pág. 276; Cap. 14, Sec. 1, pág. 405)

particulate matter / materia particulada: sólidos finos como el polen, el polvo, el moho, la ceniza y el hollín como también las gotitas líquidas del aire que pueden irritar y dañar los pulmones cuando se aspiran. (Cap. 21, Sec. 2, pág. 623)

period / período: tercera división mayor del tiempo geológico; se subdivide en épocas y se caracteriza por los tipos de vida que existieron por todo el planeta. (Cap. 14, Sec. 1, pág. 399)

permeable / permeable: describe el suelo y la roca con poros que se conectan y a través de los cuales puede correr el agua. (Cap. 9, Sec. 2, pág. 252)

permineralized remains / restos permineralizados: fósiles cuyos espacios interiores están llenos de minerales provenientes de aguas subterráneas. (Cap. 13, Sec. 1, pág. 370)

pesticide / pesticida: sustancia que se utiliza para evitar que los insectos y la maleza destruyan las cosechas y los céspedes. (Cap. 21, Sec. 1, pág. 611)

photochemical smog / smog fotoquímico: capa de smog brumoso y de color café amarillento que se presenta sobre las ciudades; se forma con la ayuda de la luz solar, contiene ozono cerca de la superficie terrestre y puede dañar los pulmones y las plantas. (Cap. 21, Sec. 2, pág. 620)

photosphere / fotosfera: la capa más baja de la atmósfera del Sol; emite luz y tiene temperaturas de hasta 6,000 K. (Cap. 25, Sec. 2, pág. 741)

photosynthesis / fotosíntesis: proceso de elaboración de alimentos que usa energía luminosa proveniente del Sol, dióxido de carbono y agua. (Cap. 19, Sec. 2, pág. 557)

pH scale / escala del pH: escala utilizada para medir el grado de acidez o basicidad de algo. (Cap. 21, Sec. 2, pág. 621)

plain / llanura: relieve enorme y plano que con frecuencia posee suelos gruesos y fértiles; por lo general se encuentra en las regiones interiores de un continente. (Cap. 6, Sec. 1, pág. 156)

plankton / plancton: plantas y animales marinos que nadan a la deriva en las corrientes oceánicas. (Cap. 19, Sec. 2, pág. 560)

plate / placa: región extensa del manto superior rígido y de la corteza oceánica o continental de la Tierra que se mueve sobre la astenosfera. (Cap. 10, Sec. 3, pág. 284)

plateau / meseta: relieve levantado formado principalmente de rocas casi horizontales que han sido levantadas. (Cap. 6, Sec. 1, pág. 158)

plate tectonics / tectónica de placas: teoría que afirma que la corteza y el manto superior terrestres se separan en placas que flotan y se mueven sobre una capa viscosa del manto. (Cap. 10, Sec. 3, pág. 284)

plucking / ablación: proceso mediante el cual se agrega grava, arena y piedras a las partes inferior y lateral de un glaciar conforme el agua se congela y derrite, rompiendo fragmentos de la roca circundante. (Cap. 8, Sec. 2, pág. 218)

Pluto / Plutón: se le considera el noveno planeta más alejado del Sol; posee una superficie sólida, rocosa y glacial, y una sola luna. (Cap. 24, Sec. 3, pág. 719)

point source pollution / contaminación de fuente puntual: contaminación proveniente de una fuente específica que penetra en el agua y que puede controlarse o tratarse antes de que entre a una masa de agua. (Cap. 21, Sec. 1, pág. 610)

polar zones / zonas polares: zona climática que recibe la radiación solar a un ángulo bajo; se extiende desde la latitud 66°N y S hasta los polos y en donde nunca hace calor. (Cap. 17, Sec. 1, pág. 492)

pollutant / contaminante: cualquier sustancia que contamina el ambiente. (Cap. 20, Sec. 1, pág. 586)

pollution / contaminación: introducción de desperdicios en el ambiente; por ejemplo, aguas negras y sustancias químicas, las cuales pueden causar daños a los organismos. (Cap. 19, Sec. 3, pág. 565)

population / población: número total de individuos de una especie que ocupan la misma área. (Cap. 20, Sec. 1, pág. 584)

Precambrian time / Era Precámbrica: la parte más larga de la historia de la Tierra, la cual duró de 4 billones de años hasta hace cerca de 544 millones de años. (Cap. 14, Sec. 2, pág. 406)

precipitation / precipitación: agua que cae de las nubes, incluye la lluvia, la nieve, la cellisca y el granizo, y cuya forma la determina la temperatura del aire. (Cap. 16, Sec. 1, pág. 468)

primary wave / onda primaria: onda sísmica que mueve las partículas rocosas en un movimiento oscilatorio en la misma dirección en que viaja la onda. (Cap. 11, Sec. 2, pág. 309)

prime meridian / primer meridiano: línea imaginaria que representa 0° de longitud y corre desde el polo norte, atravesando Greenwich, Inglaterra, hasta el polo sur. (Cap. 6, Sec. 2, pág. 163)

principle of superposition / principio de sobreposición: establece que en capas rocosas inalteradas, las rocas más antiguas se encuentran en el fondo y las rocas más recientes se hallan en la parte superior. (Cap. 13, Sec. 2, pág. 376)

Project Apollo / Proyecto Apolo: etapa final del programa espacial de EE.UU. para llegar a la Luna, en la cual el astronauta Neil Armstrong fue el primer ser humano en poner pie sobre la superficie lunar. (Cap. 22, Sec. 2, pág. 654)

Project Gemini / Proyecto Géminis: segunda etapa del programa espacial de EE.UU. para llegar a la Luna, en la cual un equipo de astronautas se conectó con otra astronave en órbita. (Cap. 22, Sec. 2, pág. 653)

Project Mercury / Proyecto Mercurio: primer paso en el programa espacial de EE.UU. para llegar a la Luna que orbitó una astronave piloteada alrededor de la Tierra, la cual regresó a salvo. (Cap. 22, Sec. 2, pág. 653)

proton / protón: partícula con carga positiva ubicada en el núcleo de un átomo. (Cap. 2, Sec. 1, pág. 36)

R

radiation / radiación: energía que transmiten las ondas o los rayos. (Cap. 15, Sec. 2, pág. 444)

radioactive decay / desintegración radiactiva: proceso en el cual algunos isótopos se desintegran en otros isótopos y partículas. (Cap. 13, Sec. 3, pág. 383)

radiometric dating / datación radiométrica: proceso que se usa para calcular la edad absoluta de las rocas midiendo la razón del isótopo original al producto descendiente en un mineral y conociendo la vida media del isótopo original. (Cap. 13, Sec. 3, pág. 385)

radio telescope / radiotelescopio: tipo de telescopio que recopila y registra ondas radiales que viajan por el espacio; se puede usar de día o de noche bajo casi cualquier condición meteorológica. (Cap. 22, Sec. 1, pág. 645)

recycling / reciclaje: método de conservación en que los materiales usados se procesan para hacer materiales nuevos. (Cap. 5, Sec. 3, pág. 141; Cap. 20, Sec. 3, pág. 597)

reef / arrecife: hábitat oceánico resistente a las olas y rígido construido por corales a partir de esqueletos y calcio. (Cap. 19, Sec. 2, pág. 564)

reflecting telescope / telescopio reflector: telescopio óptico que usa un espejo cóncavo para enfocar la luz y formar una imagen en el punto focal. (Cap. 22, Sec. 1, pág. 642)

Spanish Glossary

refracting telescope / telescopio refractor: telescopio óptico que usa una lente convexa doble para doblar la luz y formar una imagen en el punto focal. (Cap. 22, Sec. 1, pág. 642)

relative age / edad relativa: la edad de algo comparada con otras cosas. (Cap. 13, Sec. 2, pág. 377)

relative humidity / humedad relativa: medida de la cantidad de humedad que sostiene el aire, comparada con la cantidad de humedad que el aire puede sostener a una temperatura dada; puede variar de 0 por ciento a 100 por ciento. (Cap. 16, Sec. 1, pág. 464)

reserve / reserva: cantidad de combustible fósil que se puede extraer de la Tierra con fines de lucro, usando tecnología contemporánea. (Cap. 5, Sec. 1 pág. 125)

reverse fault / falla invertida: ruptura en la roca causada por las fuerzas de compresión, en que la roca sobre la superficie de la falla se mueve hacia arriba en relación con la roca debajo de la falla. (Cap. 11, Sec. 1, pág. 306)

revolution / traslación: órbita elíptica de la Tierra alrededor del Sol, la cual dura todo un año. (Cap. 23, Sec. 1, pág. 675)

rock / roca: mezcla de uno o más minerales, vidrio volcánico, materia orgánica u otros materiales; las rocas pueden ser ígneas, metamórficas o sedimentarias. (Cap. 4, Sec. 1, pág. 90)

rock cycle / ciclo de las rocas: modelo que describe el cambio lento de las rocas de una forma a otra a través del tiempo. (Cap. 4, Sec. 1, pág. 91)

rocket / cohete: motor especial que puede funcionar en el espacio y que quema combustible líquido o sólido. (Cap. 22, Sec. 2, pág. 647)

rotation / rotación: giro de la Tierra, alrededor de su eje imaginario, que dura 24 horas y que ocasiona el día y la noche. (Cap. 23, Sec. 1, pág. 673)

runoff / escorrentía: cualquier agua lluvia que no se filtra en el suelo o que no se evapora pero que corre por la superficie terrestre: por lo general, corre hacia corrientes de agua y tiene la capacidad de erosionar y transportar sedimentos. (Cap. 9, Sec. 1, pág. 240)

S

salinity / salinidad: medida de la cantidad de sales disueltas en el agua marina. (Cap. 18, Sec. 1, pág. 524)

sanitary landfill / vertedero controlado: área en donde la basura es depositada y cubierta con tierra; está diseñado para prevenir la contaminación del terreno y del agua. (Cap. 20, Sec. 2, pág. 592)

satellite / satélite: cualquier cuerpo natural o artificial que gira alrededor de otro objeto. (Cap. 22, Sec. 2, pág. 649)

Saturn / Saturno: es el segundo planeta más grande del sistema solar y el sexto más alejado del Sol; posee un complejo sistema de anillos, por lo menos 18 lunas y una atmósfera densa compuesta principalmente de hidrógeno y helio. (Cap. 24, Sec. 3, pág. 716)

science / ciencia: proceso de observación y estudio de la naturaleza con el propósito de adquirir conocimientos. (Cap. 1, Sec. 1, pág. 8)

scientific law / ley científica: regla que describe el comportamiento de algo en la naturaleza; por lo general describe lo que ocurrirá en cierta situación, pero no por qué ocurrirá ese fenómeno. (Cap. 1, Sec. 2, pág. 19)

scientific methods / métodos científicos: procedimientos para resolver problemas que pueden incluir: identificar e problema o pregunta, recoger información, desarrollar una hipótesis, probar la hipótesis, analizar los resultados y sacar conclusiones. (Cap. 1, Sec. 1 pág. 8)

Spanish Glossary

scientific theory / teoría científica: explicación que es apoyada por los resultados de experimentos o pruebas repetidas. (Cap. 1, Sec. 2, pág. 18)

scrubber / depurador: dispositivo que reduce las emisiones sulfúricas de las plantas que queman carbón. (Cap. 21, Sec. 2, pág. 624)

sea breeze / brisa marina: movimiento de aire diurno desde el mar hacia la tierra; se forma cuando el aire más frío sobre el agua se mueve hacia el interior forzando el ascenso del aire calentado y menos denso sobre la tierra. (Cap. 15, Sec. 3, pág. 451)

seafloor spreading / expansión del suelo marino: teoría de Hess que afirma que el nuevo suelo marino se forma cuando el magma es forzado a subir a la superficie en una dorsal medioceánica. (Cap. 10, Sec. 2, pág. 281)

season / estación: período corto de cambio climático en un área causado por la inclinación de eje terrestre a medida que la Tierra gira alrededor del Sol. (Cap. 17, Sec. 3, pág. 500)

secondary wave / onda secundaria: onda sísmica que al moverse hace que las partículas rocosas vibren formando un ángulo recto a la dirección de la onda. (Cap. 11, Sec. 2, pág. 309)

sedimentary rock / roca sedimentaria: se forma cuando los sedimentos se compactan y se cementan o cuando los minerales salen de una solución o cuando la evaporación los deja atrás. (Cap. 4, Sec. 4, pág. 103)

sediments / sedimentos: materiales sueltos, como fragmentos rocosos, granos minerales y restos de plantas y animales, dejados por el viento, el agua, el hielo o la gravedad. (Cap. 4, Sec. 4, pág. 103)

seismic wave / onda sísmica: onda generada por un movimiento sísmico. (Cap. 11, Sec. 2, pág. 308)

seismograph / sismógrafo: instrumento que registra las ondas sísmicas y anota el momento de llegada de cada onda. (Cap. 11, Sec. 2, pág. 311)

sewage / aguas negras: agua que se va por las alcantarillas y que contiene residuos humanos, detergentes caseros y jabones. (Cap. 21, Sec. 1, pág. 612)

sheet erosion / erosión laminar: tipo de erosión hídrica de superficie causada por el agua de escorrentía que ocurre cuando el agua que fluye sobre capas extensas recoge sedimentos y los transporta. (Cap. 9, Sec. 1, pág. 243)

shield volcano / volcán de escudo: volcán ancho y de laderas levemente inclinadas que se forma gracias a erupciones silenciosas de lava basáltica. (Cap. 12, Sec. 2, pág. 344)

silicate / silicato: mineral que contiene sílice y oxígeno y, por lo general, uno o más elementos. (Cap. 3, Sec. 1, pág. 66)

sill / intrusión: rasgo rocoso ígneo que se forma cuando el magma es inyectado en una grieta horizontal entre capas de roca y se endurece bajo tierra. (Cap. 12, Sec. 3, pág. 351)

slump / desprendimiento: tipo de movimiento de masas que se presenta cuando una masa de material se mueve cuesta abajo por una pendiente curva. (Cap. 8, Sec. 1, pág. 213)

soil / suelo: mezcla de roca meteorizada, materia orgánica descompuesta, fragmentos minerales y aire; su formación puede tomar miles de años. (Cap. 7, Sec. 2, pág. 190)

soil profile / perfil del suelo: corte vertical de las capas del suelo, cada una de las cuales constituye un horizonte. (Cap. 7, Sec. 2, pág. 192)

solar eclipse / eclipse solar: ocurre cuando la Luna se atraviesa entre el Sol y la Tierra y proyecta su sombra sobre una parte de la Tierra. (Cap. 23, Sec. 2, pág. 681)

solar energy / energía solar: energía proveniente del sol que es limpia, inagotable y que se puede transformar en electricidad por medio de células solares. (Cap. 5, Sec. 2, pág. 130)

solar system / sistema solar: sistema de nueve planetas, entre ellos, la Tierra, y otros cuerpos sólidos que giran alrededor del Sol. (Cap. 24, Sec. 1, pág. 703)

solstice / solsticio: ocurre dos veces al año y es el punto en que el Sol se aleja más del ecuador, hacia el norte o hacia el sur. (Cap. 23, Sec. 1, pág. 676)

solution / solución: tipo de mezcla en que una sustancia se mezcla completa y uniformemente en otra y que es idéntica en toda su extensión. (Cap. 2, Sec. 2, pág. 43)

space probe / sonda espacial: instrumento que viaja a gran distancia en el sistema solar, recopila datos y los envía a la Tierra. (Cap. 22, Sec. 2, pág. 650)

space shuttle / transbordador espacial: astronave reutilizable que puede transportar cargamento, astronautas y satélites hacia y desde el espacio. (Cap. 22, Sec. 3, pág. 655)

space station / estación espacial: instalaciones con zonas de habitación, de trabajo y de ejercicio y equipo y sistemas de apoyo para que los seres humanos vivan y trabajen en el espacio y efectúen investigación que no es posible llevar a cabo en la Tierra. (Cap. 22, Sec. 3, pág. 656)

species / especie: grupo de organismos que se reproduce sólo con otros miembros de su propio grupo. (Cap. 14, Sec. 1, pág. 400)

specific gravity / gravedad específica: razón del peso de un mineral comparada con el peso de un volumen igual de agua. (Cap. 3, Sec. 2, pág. 70)

sphere / esfera: objeto tridimensional redondo en que cualquier punto sobre su superficie está equidistante del centro. La Tierra es una esfera un poco alargada en el ecuador y ligeramente achatada en los polos. (Cap. 23, Sec. 1, pág. 672)

spring / manantial: se forma cuando el nivel hidrostático se junta con la superficie terrestre; a menudo se halla en las laderas de las colinas y se utiliza como una fuente de agua dulce. (Cap. 9, Sec. 2, pág. 255)

station model / código meteorológico: indica las condiciones del tiempo en un lugar específico, mediante el uso de símbolos en un mapa. (Cap. 16, Sec. 3, pág. 479)

streak / veta: color de un mineral cuando se encuentra en forma de polvo. (Cap. 3, Sec. 2, pág. 71)

stream discharge / descarga de corriente: volumen de agua que fluye por un punto específico por unidad de tiempo. (Cap. 20, Sec. 2, pág. 591)

strike-slip fault / falla transformante: lugar donde las fuerzas de cizalleamiento han ocasionado el rompimiento de las rocas, las cuales se deslizan una al lado de la otra en direcciones opuestas, pero sin mucho movimiento vertical. (Cap. 11, Sec. 1, pág. 307)

sunspots / manchas solares: áreas del Sol que son menos calientes y brillantes que las áreas circundantes, son causadas por el campo magnético del Sol y ocurren en ciclos. (Cap. 25, Sec. 2, pág. 742)

supergiant / supergigante: etapa avanzada en el ciclo de vida de una estrella masiva en la cual el núcleo se calienta, se forman elementos pesados por fusión y la estrella se expande; con el tiempo, puede explotar para formar una supernova. (Cap. 25, Sec. 3, pág. 750)

surface current / corriente de superficie: corriente oceánica accionada por el viento; se mueve horizontalmente, paralela a la superficie de la Tierra y solamente mueve unos pocos cientos de metros superiores de agua marina. (Cap. 18, Sec. 2, pág. 526)

surface wave / onda de superficie: onda sísmica que mueve las partículas rocosas de arriba hacia abajo en un movimiento rotatorio y de lado a lado en un movimiento de vaivén. (Cap. 11, Sec. 2, pág. 309)

T

technology / tecnología: usa los descubrimientos científicos para propósitos prácticos y para facilitar y mejorar la vida de las personas. (Cap. 1, Sec. 1, pág. 12)

temperate zones / zonas templadas: zonas climáticas con temperaturas moderadas, las cuales se encuentran entre los trópicos y las zonas polares. (Cap. 17, Sec. 1, pág. 492)

tephra / tefrita: fragmentos pequeños de roca o lava solidificada que caen del aire durante una erupción volcánica explosiva; varía en tamaño desde cenizas volcánicas a bombas y bloques volcánicos. (Cap. 12, Sec. 2, pág. 344)

terracing / cultivo en terrazas: método de cultivo que se utiliza para reducir la erosión en laderas inclinadas. (Cap. 7, Sec. 3, pág. 201)

tidal range / amplitud de la marea: la diferencia entre el nivel del océano en pleamar y su nivel en bajamar. (Cap. 18, Sec. 3, pág. 536)

tide / marea: ascenso y descenso del nivel del mar provocado, en su mayor parte, por la interacción de la gravedad en el sistema Tierra-Luna. (Cap. 18, Sec. 3, pág. 535)

till / tilita: mezcla de sedimentos de diferentes tamaños que un glaciar en retirada deja caer de la base y que puede cubrir inmensas áreas de terreno. (Cap. 8, Sec. 2, pág. 218)

topographic map / mapa topográfico: mapa que muestra los cambios en elevación en la superficie terrestre e indica rasgos como caminos y ciudades. (Cap. 6, Sec. 3, pág. 168)

tornado / tornado: tormenta de viento violento y arremolinado que se mueve sobre una estrecha trayectoria sobre la tierra y que puede ser resultado de los vientos laterales dentro de una tormenta eléctrica. (Cap. 16, Sec. 2, pág. 474)

trench / fosa: depresión de lados empinados angosta y larga en el suelo marino que se forma cuando una placa se hunde debajo de otra placa. (Cap. 19, Sec. 1, pág. 553)

trilobite / trilobites: organismo con un exoesqueleto dividido en tres partes; existía en abundancia en los océanos del Paleozoico; se le considera un fósil guía. (Cap. 14, Sec. 1, pág. 398)

tropics / trópicos: zona climática que recibe la mayor cantidad de radiación solar; está ubicada entre los 23°N y los 23°S y en donde siempre hace calor, con excepción de las altas elevaciones. (Cap. 17, Sec. 1, pág. 492)

troposphere / troposfera: capa de la atmósfera terrestre más próxima a la tierra, contiene un 99 por ciento de vapor de agua y un 75 por ciento de los gases atmosféricos; es la región donde se forman las nubes y ocurre el estado del tiempo. (Cap. 15, Sec. 1, pág. 436)

trough / seno: punto más bajo de una onda. (Cap. 18, Sec. 3, pág. 532)

tsunami / tsunami: onda marina sísmica que comienza sobre el foco de un terremoto y la cual puede ser muy destructiva cuando llega al litoral. (Cap. 11, Sec. 3, pág. 320)

U

ultraviolet radiation / radiación ultravioleta: tipo de energía que llega a la Tierra proveniente del Sol; puede causar daños a la piel y ocasionar cáncer; gran parte de esta radiación es absorbida por la capa de ozono. (Cap. 15, Sec. 1, pág. 440)

unconformity / discordancia: brecha en una capa rocosa provocada por la erosión o por períodos cuando no hubo deposaitación. (Cap. 13, Sec. 2, pág. 378)

uniformitarianism / uniformitarianismo: principio que establece que los procesos terrestres que suceden en la actualidad son semejantes a aquéllos que ocurrieron en el pasado. (Cap. 13, Sec. 3, pág. 387)

upwarped mountains / montañas plegadas anticlinales: montañas que se forman cuando los bloques de corteza terrestre son forzados a ascender por fuerzas internas de la Tierra. (Cap. 6, Sec. 1, pág. 160)

upwelling / corriente resurgente: circulación del océano que lleva agua fría y profunda a la superficie oceánica. (Cap. 18, Sec. 2, pág. 529)

Uranus / Urano: séptimo planeta más alejado del Sol; es grande y gaseoso, posee un marcado color verde-azulado y gira en un eje casi paralelo al plano de su órbita. (Cap. 24, Sec. 3, pág. 717)

V

variables / variables: diferentes factores que pueden cambiarse en un experimento. (Cap. 1, Sec. 1, pág. 10)

vent / chimenea: abertura por donde sube el magma y sale a la superficie terrestre como lava, formando un volcán. (Cap. 12, Sec. 1, pág. 336)

Venus / Venus: segundo planeta más cercano al Sol; semejante a la Tierra en masa y tamaño; posee una atmósfera densa y una superficie con cráteres, grietas que parecen fallas y volcanes. (Cap. 24, Sec. 2, pág. 709)

volcanic mountains / montañas volcánicas: montañas que se forman cuando el material fundido llega a la superficie terrestre a través de partes debilitadas de la corteza y se amontona en una estructura que tiene forma de cono. (Cap. 6, Sec. 1, pág. 161)

volcanic neck / chimenea volcánica: núcleo ígneo sólido de un volcán que queda atrás después de la erosión del cono más suave. (Cap. 12, Sec. 3, pág. 351)

volcano / volcán: abertura en la superficie terrestre que arroja gases sulfurosos, cenizas y lava; se puede formar en el límite de las placas terrestres, donde éstas se separan o se juntan y también en los focos calientes. (Cap. 12, Sec. 1, pág. 334)

W

waning / octante menguante: describe la fase que ocurre después de la luna nueva, cuando el lado visible de la Luna que vemos desde la Tierra empieza a decrecer. (Cap. 23, Sec. 2, pág. 679)

water table / nivel hidrostático o capa freática: nivel superior de la zona de saturación; disminuye durante una sequía. (Cap. 9, Sec. 2, pág. 252)

wave / onda: movimiento rítmico que transporta energía a través de la materia o el espacio; se puede describir por su cresta, seno, longitud y altura. (Cap. 18, Sec. 3, pág. 532)

waxing / octante creciente: describe la fase después de la luna nueva, cuando la parte iluminada de la Luna que vemos desde la Tierra empieza a ser más visible. (Cap. 23, Sec. 2, pág. 679)

weather / tiempo atmosférico: estado de la atmósfera en un momento y lugar específicos, determinado por factores que incluyen la presión atmosférica, la cantidad de humedad en el aire, la temperatura y el viento. (Cap. 16, Sec. 1, pág. 462)

weathering / meteorización: procesos superficiales mecánicos o químicos que rompen las rocas en fragmentos cada vez más pequeños. (Cap. 7, Sec. 1, pág. 184)

white dwarf / enana blanca: etapa avanzada en el ciclo de vida de una estrella de la secuencia principal en la cual su núcleo agota todo su helio y sus capas externas escapan hacia el espacio, dejando atrás un núcleo caliente y denso. (Cap. 25, Sec. 3, pág. 749)

wind farm / fincas de energía eólica: área en que muchos molinos de viento usan el viento para generar electricidad. (Cap. 5, Sec. 2, pág. 131)

Index

The index for *Glencoe Earth Science* will help you locate major topics in the book quickly and easily. Each entry in the index is followed by the number of the pages on which the entry is discussed. A page number given in bold-faced type indicates the page on which that entry is defined. A page number given in italic type indicates a page on which the entry is used in an illustration or photograph. The abbreviation *act.* indicates a page on which the entry is used in an activity.

A

Abrasion, 224, *224*
Abrasives, 139
Absolute ages, 383–387
Absolute magnitude, 738
Abyssal plain, 551, *551*
Acid(s), 621; weathering and, 187–188, *187*
Acid precipitation, 621, *621,* 623, 624
Acid rain, 255, **621,** *621,* 623, 624
Activities, 23, 24–25, 45, 52–53, 67, 80–81, 98, 110–111, 136, 142–143, 173, 174–175, 197, 202–203, 223, 230–231, 261, 262–263, 283, 294–295, 316, 324–325, 348, 354–355, 382, 388–389, 413, 420–421, 442, 452–453, 481, 482–483, 511, 512–513, 539, 540–541, 556, 570–571, 595, 600–601, 618, 626–627, 646, 662–663, 687, 692–693, 707, 726–727, 745, 758–759
Adaptations, 496–499; behavioral, 498–499, *498, 499;* structural, 497, *498*
Adirondack Mountains, 160
Age: absolute, **383**–387; relative, **377**–381, *act.* 382
Aggregate, 140
Agriculture, 588–589, *588, 589;* no-till farming, 200, *200;* soil erosion and, 199, 200–201, *200, 201;* till deposits and, 218; water pollution and, 611, *611*
Air: heated, 447, *447;* movement of, 447–451, *448, 449;* oxygen in, 434–435, *435*

Air mass, 470, *470*
Air pollution, 619–627; acid precipitation and, 621, *621,* 623, 624, 625; car emissions and, 619, *619,* 624, 625; causes of, 619–625, *619;* health and, 622–623, *622;* law on, 624, 625; ozone depletion, 440–441, *440, 441;* particulate matter and, 623, *act.* 626–627; reducing, 623–625, *624, 625;* smog, 435, 619–620, *619, 620;* in United States, 624–625, *624*
Air Quality Index, 622
Air temperature, 463, *463,* 479
Alcohol: energy from, 134, *134*
Aldrin, Edwin, 654
Algae, 524; blooms of, 566, 568, *568;* oxygen produced by, 440; water pollution and, 611, *611*
Alluvial fan, 250
Almadine, 75, *75*
Alpha decay, 384, *384*
Alternative resources: biomass energy, 133–135, *134, 135;* geothermal energy, 132, *133;* hydroelectric power, 132, *132;* nuclear energy, 127–129, *127, 128, 129;* solar energy, 130–131, *130, act.* 136; wind energy, 131, *131*
Altitude: and atmospheric pressure, 438, *438*
Altostratus clouds, 467
Aluminum, 77, *77;* recycling, 597
Amethyst, 73, 75, *75*
Amphibians, 410
Amplitude, 532
Andesite, *96*
Andesitic magma, 343

Andesitic rock, *96,* 97
Anemometer, 463
Angiosperms, 416
Angular unconformities, 378, *378, 379*
Animal(s): behavioral adaptations of, 498–499, *498, 499;* bottom-dwelling, 561; Ediacaran, 408, *409;* habitats of, 561–564, *562, 563, 564;* hibernation of, 498. *see* Vertebrate animals. *see also* Invertebrate animals
Antarctica: continental glaciers in, 220; ozone hole in, 441, *441*
Antares, 747, *747*
Anticyclone, 471
Apatite, 69
Appalachian Mountains, 159, 411, *411*
Apparent magnitude, 738
Appendices. *see* Reference Handbook
Aquifer, 252, *253*
Archean Eon, 406
Aristotle, 672
Armstrong, Neil, 654
Artesian well, 254, *254*
Articles, scientific, *act.* 23
Asteroid, 724–725, *724, 725*
Asteroid belt, 724, *724*
Asthenosphere, 284, 313, 315
Asthma, 622
Astrolabe, *663*
Astronauts, 638, 653, *653,* 654, *654,* 656
Astronomical unit (AU), 718
Astronomy Integration, 336
Atlantic Coastal Plain, 157, *157*
Atlantic Ocean, *act.* 418
Atmosphere, 432–453, **434,** *434;*

Index

Index

Index

Emerald, 74, *74*

Energy: biomass, **133**–135, *134, 135;* in food chain, 558, *558;* from fusion, 129, *129,* 747–748, *748;* geothermal, **132,** *133;* hydroelectric, **132,** *132;* nuclear, **127**–129, *127, 128, 129;* solar, **130**–131, *130, act.* 136, 443, *443,* 510, *510;* transfer of, 443–446, *444, act.* 452–453; from water, 132, *132;* waves and, 532, *532,* 533, *533,* 534; wind, 131, *131*

Energy sources: fossil fuels, 120–127; nonrenewable, 120–129; nuclear, **127**–129, *127, 128, 129;* percentage used in U.S., *123;* renewable, 130–136

Environment: effect of volcanic eruptions on, 335, *335;* fossils and, 374–375, *374, 375;* model of, *act.* 397; population and, 586–587

Environmental Science Integration, 476

Enzymes, 593; in breaking down organic pollutants, 593

Eon, 399, *399*

Epicenter, *305, 310,* **309**–312, *312, act.* 316

Epoch, 399, *399*

Equator, 162, *162*

Equinox, 676, 677

Era, 399, *399*

Erie, Lake, 615, *615*

Erosion, 210–231, **212,** *238, act.* 239, 242–243; agents of, 212; consequences of, 215–216; deposition and, 212–213, *212;* by glaciers, 212, 217–223, *218, 219, 220, 221;* gravity and, 212–216, *212, 213, 214, 215,* 241, *241;* gully, 242, *242;* mass movement and, 213–215, *213, 214, 215;* reducing, 216, *216,* 226, *226;* rill, 242; of rocks, 104;

of sand, 260, *260;* sheet, 242–**243,** *242;* of soil. *see* Soil erosion; stream, 243, *243;* vegetation and, *240,* 241; water speed and, *act.* 262–263; by wind, 212, 224–231, *224, 225, 226, 227, 228, 229, act.* 230–231

Eruptions: effects of, 335, *335;* factors in, 340–341; quiet, 341; violent, 343, *343,* 346–347, *347,* 435, *435*

Esker, 219, *219*

Estuaries, 564, *564*

Ethics, 20–22

Etna, Mount (Italy), *332,* 346

Europa (moon of Jupiter), 652, *652,* 715, *715*

Everest, Mt., *553*

Evolution: of mammals, 419, *419;* organic, **400**–402, *400, 401, 402;* of stars, 748–751, *749*

Exosphere, *436,* 437

Experiments, 9–11, *10, 11*

Explore Activity, 5, 33, 61, 89, 119, 155, 183, 211, 239, 275, 303, 333, 367, 397, 433, 461, 491, 521, 549, 583, 609, 639, 671, 701, 735

Extinction, 412, *412,* 414, 417

Extrusive rock, 95, *95, 96*

F

Farming. *see* Agriculture

Fault(s), 288, *288,* **304,** *310;* causes of, 305, *305;* formation of, 304, *304;* normal, 290; stress buildup along, *act.* 303; strike-slip, 292, *292;* types of, 306–307, *306, 307*

Fault-block mountains, 160, *160,* 290, *290*

Feldspar, 66, 69, 90, *90,* 101, *act.* 110–111

Fertilizer, *35,* 139, **611,** *611*

Field Guides: Building Stones Field Guide, 770–773; Shorelines Field Guide, 774–777; Faults and Folds Field Guide,

778–781; Fossils Field Guide, 782–785; Clouds Field Guide, 786–789; Waste Management Field Guide, 790–793; Backyard Astronomy Field Guide, 794–797

Fish: deep-sea, 560, *561;* early, 408, *408,* 410, *410*

Fish kill, 566, *566*

Fission: nuclear, 127–128, *128*

Fissure, 344

Flooding, 248–249, *248, 249,* 476

Fluorite, 69, *70*

Focus: of earthquake, **308**

Fog, 467, *467*

Foldables, 5, 33, 61, 89, 119, 155, 183, 211, 239, 275, 303, 333, 367, 397, 433, 461, 491, 521, 549, 583, 609, 639, 671, 701, 735

Folded mountains, 159, *159*

Foliated rocks, 101, *101*

Food chains, 558, *558*

Food webs, 558, 559

Fool's gold (pyrite), 68, *68,* 71

Force(s): compression, 306, *306*

Forecasting weather, 478–481, *478, 479, 480, act.* 481

Forests. *see also* Rain forests; deforestation and, 509, *509;* harvesting, 199, *199;* as resource, 590, *590;* temperate deciduous, 195

Fossil(s), 368–375, *368,* **369;** ancient environments and, 374–375, *374, 375;* changes shown by, 404; of dinosaurs, *396,* 415–416, *415, 416;* Ediacaran, 408, *409;* formation of, 369, *369;* index, **373,** *373;* making model of, *act.* 367; organic remains, 372, *372;* from Paleozoic Era, 408; *act.* 413; from Precambrian time, 406; preservation of, 369–373, *370, 371, 372;* in rocks, 408; trace, 372–373, *372*

Index (side tab)

Index

Index

Index

Index

Index

Index

Index

Art Credits

Glencoe would like to acknowledge the artists and agencies who participated in illustrating this program: Absolute Science Illustration; Andrew Evansen; Argosy; Articulate Graphics; Craig Attebery represented by Frank & Jeff Lavaty; CHK America; Gagliano Graphics; Pedro Julio Gonzalez represented by Melissa Turk & The Artist Network; Robert Hynes represented by Mendola Ltd.; Morgan Cain & Associates; JTH Illustration; Laurie O'Keefe; Matthew Pippin represented by Beranbaum Artist's Representative; Precision Graphics; Publisher's Art; Rolin Graphics, Inc.; Wendy Smith represented by Melissa Turk & The Artist Network; Kevin Torline represented by Berendsen and Associates, Inc.; WILDlife ART; Phil Wilson represented by Cliff Knecht Artist Representative; Zoo Botanica.

Photo Credits

Abbreviation key: AA=Animals Animals; AH=Aaron Haupt; AMP=Amanita Pictures; BC=Bruce Coleman, Inc.; CB=CORBIS; DM=Doug Martin; DRK=DRK Photo; ES=Earth Scenes; FP=Fundamental Photographs; GH=Grant Heilman Photography; IC=Icon Images; KS=KS Studios; LA=Liaison Agency; MB=Mark Burnett; MM=Matt Meadows; PE=PhotoEdit; PD=PhotoDisc; PQ=PictureQuest; PR=Photo Researchers; SB=Stock Boston; TSA=Tom Stack & Associates; TSM=The Stock Market; VU=Visuals Unlimited.

Cover PD; **viii** Mark A. Schneider/VU; **ix** Hans Strand/Stone; **x** Dr. Russ Utgard; **xi** Burhan Ozbilici/AP/Wide World Photos; **xii** Stone; **xiii** Peter Miller/Science Source/PR; **xv** SuperStock; **xvi** NASA/Science Source/PR; **xviii** Ken Whitmore/Stone; **xix** Wyman P. Meinzer; **xx** Emory Kristof/National Geographic; **xxi** (l)Jose Manuel Sanchis/CB, (r)DM; **xxii** Fred Bavendam/Minden Pictures; **xxiii** Sigurjon Sindrason; **xxiv** Paul Rocheleau/Index Stock; **xxv** Geoff Butler; **1** Roine Magnusson/Stone; **2-3** Mike Zens/CB; **3** Mark A. Schneider/VU; **4** Ken Lucas/VU, **4-5** Jim Page/North Carolina Museum of Natural Sciences; **5** Richard Hutchings; **7** (t)Michael Habicht/ES, (b)Michael Wilhelm/ENP Images; **8** Richard Cummins/CB; **9** John Heseltine/Science Photo Library/PR; **10 11** AH; **12** (l)Michael Dwyer/SB, (c)Carolas, (r)Mark Segal/SB; **13** (t)Science Museum/Science & Society Picture Library, (cl)reprinted by permission of Parks Canada and Newfoundland Museum, (c cr)Dorling Kindersley, (b)NASA; **14** Russ Underwood/Lockheed Martin Space Systems; **15** Todd Gustafson/Danita Delimont; **16** Smithsonian Institution; **18** NASA/MSFC; **19** European Space Agency/Science Photo Library/PR; **20** (l)Frans Lanting/Minden Pictures, (c)Ted Levin/AA, (r)Al, Linda Bristor/VU; **21** MM; **22** Bates Littlehales/National Geographic Image Collection; **23** AH; **24** MB; **25** Timothy Fuller; **26** AP Photo; **27** Dan Mauersberg/DMH Photography; **28** (l)Krafft/PR, (r)D.A. Calvert, Royal Greenwich Observatory/Science Photo Library/PR; **29** (l)CABISCO/VU, (r)Jan Hinsch/Science Photo Libray/PR; **32** IBMRL/VU; **32-33** Roine Magnusson/Stone; **33** MM; **35** (t, l to r)Mark Schneider/Peter Arnold, Inc, Dane S. Johnson/VU, Ken Lucas/VU, Mark A. Schneider/PR, (b, l to r) AH, AMP, Charles D. Winters/PR, AH; **36** John Evans; **39** (l)Herbert Kehrer/OKAPIA/PR, (c)DM, (r)Bruce Hands/Stone; **40** Kenji Kerins; **42** Ken Whitmore/Stone; **43** AMP; **44** Stuart Westmorland; **46** John S. Lough/VU; **48** CB; **49** (t)Storm Pirate Productions/Artville/PQ, (c)Breck P. Kent/ES, (b)CB/PQ; **A50** (t)Paul Chesley/Stone, (b)David Muench/CB; **51** NASA/JPL/Malin Space Science Systems; **52** (t)StudiOhio, (b)MM; **53** (t)Tim Courlas, (b)AH; **54** (t)Geoff Butler, (bl)KS, (br)StudiOhio; **56** (tl)MM, (tr)Keith Kent/Science Photo Library, (b)John Evans; **57** (l)DM, (r)MM; **58** MM; **60** Mark A. Schneider/PR; **60-61** D. Boone/CB; **61** (t)MM, (c)DM, (b)José Manuel Sanchis Calvete/CB; **62** MM; **63** (t)John R. Foster/PR, (b)Mark A. Schneider/VU; **64** (t)Mark A. Schneider/VU, (cl)Albert J. Copley/VU, (cr)Harry Taylor/DK Images, (bl)Harry Taylor/DK Images, (bc)First Image, (br)Mark A. Schneider/VU; **65** (l)Patricia K. Armstrong/VU, (r)Dennis Flaherty Photography/PR; **67** KS; **68** (l)MB/PR, (c)Dan Suzio/PR, (r)Breck P. Kent/ES; **69** (t)Bud Roberts/VU, (c)Charles D. Winters/PR, (b)IC; **70** (l)Andrew McClenaghan/Science Photo Library/PR, (r)Charles D. Winters/PR; **71** (t)Geoff Butler, (bl)DM, (br)PR; **72** MM; **73** Reuters NewMedia Inc./CB; **74** (tl)Biophoto Associates/PR, (tr)H. Stern/PR, (cl)Biophoto Associates/PR, (cr)A.J. Copley/VU, (b, l to r)Mark A. Schneider/VU, VU, A.J. Copley/VU, H. Stern/PR; **75** (tl)University of Houston, (tr)Charles D. Winters/PR; (cl)Arthur R. Hill/VU, (cr)David Lees/CB, (b, l to r)DM, DM, A.J. Copley/VU, Vaughan Fleming/Science Photo Library/PR; **76** (l)Francis G. Mayer/CB, (r)Smithsonian Institution; **77** (l)Randolph King/PR, (r)DM; **78** (t) DM, (bl)Paul Silverman/FP, (br)Biophoto Associates/PR; **79** Jim Cummins/FPG; **80** (t)DM, (c)Jose Manuel Sanchis/CB, (b)MM; **81** (l)Charles D. Winter/PR, (r)Andrew J. Martinez/PR; **82** SPL/Custom Medical Stock Photo; **83** Bettmann/CB; **84** (tl)Mark A. Schneider/ PR, (tr)Charles R. Belinky/PR, (c)Phillip Hayson/PR, (b)Mark A. Schneider/PR; **88** John D. Cunningham/VU; **88-89** Cliff Leight; **89** Geoff Butler; **90** (l)CB, (r)DM; **91** (tl)Steve Hoffman, (tr)Breck P. Kent/ES, (bl)Brent Turner/BLT Productions, (br)Breck P. Kent/ES; **92** (bkgd) CB/PQ, (t)CB, (bl)Martin Miller, (bc)Jeff Gnass, (br)Doug Sokell/TSA; **93** Russ Clark; **94** USGS/HVO; **95** (t)Breck P. Kent/ES, (b)DM; **96** (tl tc tr)Breck P. Kent/ES, (c, l to r)Mark Steinmetz, Breck P. Kent/ES, Breck P. Kent/ES, DM, (bl)DM, (br)Tim Courlas; **98** (l)Breck P. Kent/ES, (r)DM/PR; **99** (tc)courtesy Kent Ratajeski & Dr. Allen Glazner, University of NC, (b)Alfred Pasieka/PR, (others) Breck P. Kent/ES; **101** (l)John Evans, (r)Robert Estall/CB; **102** Paul Rocheleau/Index Stock; **103** (l)Timothy Fuller, (r)Steve McCutcheon/VU; **105** (l)IC, (lc)DM, (rc) Andrew Martinez/PR, (r)John R. Foster/PR; **106** (l) Breck P. Kent/ES, (r)AH; **107** (t)Georg Gerster/PR, (b)IC; **109** Beth Davidow/VU; **110** (t)Breck P. Kent/ES, (c)IC, (bl)Jack Sekowski, (br)Tim Courlas; **112-113** Y. Kawasaki/Photonica; **113** Matt Turner/LA; **114** (t)Mark Segal/Index Stock, (c)Albert J. Copley/VU, (b)DM; **116** Jeremy Woodhouse/DRK; **118** Lowell Georgia/CB; **118-119** Bill Ross/CB; **119** MM; **121** VU; **124** (l)George Lepp/CB, (r)Carson Baldwin Jr./ES; **125** Paul A. Souders/CB; **126** (tl)Emory Kristof, (tr)National Energy Technology Laboratory, (b)Ian R. MacDonald/Texas A&M

PhotoVenture/VU; 381 (l)Michael T. Sedam/CB, (r)Pat O'Hara/CB; 383 AH; 385 James King-Holmes/Science Photo Library/PR; 386 (t)Breck P. Kent/ES, (b)Kenneth Garrett; 387 WildCountry/CB; 388 (t)A.J. Copley/VU, (b)MM; 389 Lawson Wood/CB; 390-391 Jacques Bredy; 391 (t)CMN Archives/Dinofish, (c)CMN Archives/Dinofish, (b)Norbert Wu; 392 (t)Francois Gohier/PR, (c)From Journal of Ecology 2000, 88, p. 45-53. Photo courtesy Michigan Technological University, (b)Tom Bean/DRK; 393 (tl)Francois Gohier/PR, (tr)Mark E. Gibson/DRK, (b)Sinclair Stammers/PR; 396 CB; 396-397 Reuters New Media Inc./CB; 397 KS; 398 Tom & Therisa Stack/TSA; 400 (l)Gerald & Buff Corsi/VU, (r)John Gerlach/AA; 402 (t)Mark Boulron/PR, (others)Walter Chandoha; 403 Jeff Lepore/PR; 407 (l)Mitsuaki Iwago/Minden Pictures, (r)R. Calentine/VU; 409 J.G. Gehling/Discover Magazine; 410 Gerry Ellis/ENP Images; 413 MM; 416 (l)David Burnham/Fossilworks, Inc., (r)Francois Gohier/PR; 418 Michael Andrews/ES; 419 Tom J. Ulrich/VU; 420 David M. Dennis; 421 MB; 422 The Bridgeman Art Library; 424 (l)Adam Jones/PR, (r)Ken Lucas/VU; 425 (l)E. Webber/VU, (r)Len Rue, Jr./AA; 427 John Cancalosi/SB; 430-431 Angela Wyant/Stone; 431 (t)Henry Diltz/CB, (c b)Granger Collection, NY; 432 Lester V. Bergman/CB; 432-433 David Keaton/TSM; 433 First Image; 434 NASA; 435 (t)David S. Addison/VU, (bl)Frank Rossotto/TSM, (br)Larry Lee/CB; 438 Laurence Fordyce/CB; 440 DM; 441 NASA/GSFC; 442 Michael Newman/PE; 444 Larry Fisher/Masterfile; 447 (t)Dan Guravich/PR, (b)Bill Brooks/Masterfile; 449 (t)Gene Moore/PhotoTake NYC/PQ, (cl)Phil Schermeister/CB, (cr)Stephen R. Wagner, (bl)Joel W. Rogers, (br)Kevin Schafer/CB; 450 Bill Brooks/Masterfile; 452 453 David Young-Wolff/PE; 454 Bob Rowan/CB; 455 (l)J.A. Kraulis/Masterfile, (r)CB; 457 (l)Tom Bean/DRK, (r)Keith Kent/Science Photo Library/PR; 460 NASA; 460-461 Michael S. Yamashita/CB; 461 KS; 462 Kevin Horgan/Stone; 463 Fabio Colombini/ES; 467 (t)Charles O'Rear/CB, (b)Joyce Photographics/PR; 468 (l)Roy Morsch/TSM, (r)Mark McDermott/Stone; 469 (l)Mark E. Gibson/VU, (r)EPI Nancy Adams/TSA; 471 Van Bucher/Science Source/PR; 473 Jeffrey Howe/VU; 474 (t)Warren Faidley/Weatherstock, (b)Robert Hynes; 476 NASA/Science Photo Library/PR; 477 Fritz Pölking/Peter Arnold, Inc; 478 Howard Bluestein/Science Source/PR; 481 MB; 482 (t)Marc Epstein/DRK, (b)Timothy Fuller; 484 Erik Rank/Photonica; 485 Courtesy Weather Modification Inc.; 486 (l)Peter Miller/Science Source/PR, Inc, (r)Gary Williams/LA; 487 (l)George D. Lepp/PR, (r)Janet Foster/Masterfile; 488 Bob Daemmrich; 490 Pekka Parviainen/PR; 490-491 Rod Planck/PR; 491 AH; 492 CB; 495 (l)William Leonard/DRK, (r)Bob Rowan, Progressive Image/CB; 496 John Shaw/TSA; 497 (tl)David Hosking/CB, (tr)Yva Momatiuk & John Eastcott/PR, (b)Michael Melford/The Image Bank; 498 (t)S.R. Maglione/PR, (bl)Fritz Pölking/VU, (br)Jack Grove/TSA; 499 Zig Zeszczynski/AA; 500 SuperStock; 501 (l)Jonathan Head/AP/Wide World Photos, (r)Jim Corwin/Index Stock; 503 (t)A. Ramey/PE, (b)Peter Beck/Pictor; 504 Galen Rowell/Mountain Light; 508 John Bolzan; 509 Chip & Jill Isenhart/TSA; 510 (l)Jim Sugar Photography/CB, (r)AFP/CB; 511 MM; 513 DM; 514 (b)Gary Rosenquist; 514-515 Alberto Garcia/Saba; 516 (t)Paul Sakuma/Associated Press, (c)Galen Rowell/Mountain Light, (b)Michael Melford/The Image Bank;

517 (l)Spencer Grant/PE, (r)Steve Kaufman/DRK; 520 Judy Griesedieck/CB; 520-521 Warren Bolster/Stone; 521 Raven/Explorer/PR; 522 (l)Norbert Wu/Peter Arnold, Inc., (r)Darryl Torckler/Stone; 525 Cathy Church/Picturesque/PQ; 527 Bob Daemmrich; 528 (t)Darryl Torckler/Stone, (b)Raven/Explorer/PR; 532 Jack Fields/PR; 533 Tom & Therisa Stack; 534 (bkgd)Stephen R. Wagner, (l)Spike Mafford/PD, (r)Douglas Peebles/CB; 535 Arnulf Husmo/Stone; 536 (tl)Groenendyk/PR, (tr)Patrick Ingrand/Stone, (b)Kent Knudson/SB; 539 AH; 540 (t)Mark E. Gibson/VU, (b)Timothy Fuller; 541 Timothy Fuller; 543 Seth Resnick/SB/PQ; 544 (t)Worldsat Productions/NRSC/Science Photo Library, (cl)Phillippe Diederich/Contact Press Images/PQ, (b)S.J. Krasemann/Peter Arnold, Inc.; 545 (l)Carl R. Sams II/Peter Arnold, Inc., (r)Edna Douthat; 548 L.P. Madin/Woods Hole Oceanographic Institution, Woods Hole, MA; 548-549 Stuart Westmorland/Stone; 549 MB; 550-551 "The Floor of the Oceans" by Bruce C. Heezen and Marie Tharp, ©1980 by Marie Tharp. Reproduced by permission of Marie Tharp; 552 Woods Hole Oceanographic Institution; 553 Thomas J. Abercrombie/National Geographic Society; 554 (t)J. & L. Weber/Peter Arnold, Inc., (bl)Arthur Hill/VU, (br)John Cancalosi/Peter Arnold, Inc.; 555 (t)Instutute of Oceanographic Sciences/NERC/Science Photo Library/PR, (b)Biophoto Associates/PR; 557 Fred Bavendam/Minden Pictures; 559 Nanct Sefton/PR; 560 (l)Manfred Kage/Peter Arnold, Inc., (r)M.I. Walker/Science Source/PR; 561 (l)Nick Caloyianis/National Geographic Society, (c)Herb Segars/AA, (r)Norbert Wu; 562 Fred Bavendam/Peter Arnold, Inc.; 563 (clockwise from top)Lloyd K. Townsend, Michael Abbey/PR, Andrew J. Martinez/PR, Peter Skinner/PR, Gregory Ochocki/PR, Zig Leszczynski/AA, Gerald & Buff Corsi/VU, Anne W. Rosenfeld/AA, AA, Andrew J. Martinez/PR, Andrew J. Martinez/PR, Hal Beral/VU; 564 James H. Robinson/PR; 567 (tl)C.C. Lockwood/ES, (bl)C.C. Lockwood/DRK, (others)David Young-Wolff/Photo Edit; 568 NASA; 569 David Young-Wolff/Photo Edit; 570 (t)Jim Nilsen/Stone, (bl)Jeff Rotman/Peter Arnold, Inc., (br)Fred Bavendam/Minden Pictures; 571 Fred Bavendam/Minden Pictures; 572 Rick Price/CB; 572-573 Emory Kristof/National Geographic; 573 (t)Ralph White/CB, (b)Emory Kristof/National Geographic; 574 (t)Hal Beral/VU, (b)AP Photo/Dolores Ochoa; 575 (l)D.P. Wilson/Science Source/PR, (r)Fred Bavendam/Minden Pictures; 576 Peter Johnson/CB; 580-581 Joseph Sohm/ChromoSohm Inc./CB; 581 Andrew A. Wagner; 582 George Lepp/CB; 582-583 George D. Lepp/PR; 583 AH; 584 (l)Bob Daemmrich/SB, (r)TK/CB; 586 Timothy Fuller; 587 AH; 588 Paul Bousquet; 589 Andy Sacks/Stone; 591 Rich Iwasaki; 592 Simon Fraser/Northumbrian Environmental Management Ltd./Science Photo Library/Photo Researchers; 594 (t)Gloria H. Chomica/Masterfile, (c)Raymond Gehman/CB, (b)David Muench/Stone; 595 John Evans; 598 (tl)Philip James Corwin/CB, (cl)Bill Gallery/SB/PQ, (cr)Skiold/PE/PQ, (b)Aerials Only, (toothpaste, egg shell, shoe, tin can, baseball, cassette, sneaker, leaf, tin can lid)Image Ideas, (olive oil)Digital Stock, (others)PD; 599 David Young-Wolff/Photo Edit; 600 MM; 602-603 VCG/FPG; 603 Andy Levin/PR; 604 (t)Timothy Fuller, (c)KS, (b)Eric Neurath/SB; 605 (l)Stacy Pick/SB, (r)Tom Bean/Stone; 608 Michael Abbey/PR; 608-609 Art Wolfe/Stone; 609 MB; 610 (l)Michael J. Pettypool/Pictor, (r)VU; 611 (t)Bob

Credits

Child/AP/Wide World Photos, (b)David Hoffman/Stone; **613** Stephen R. Wagner; **614** Colin Raw/Stone; **615** (t)Cleveland Public Library, (b)Jim Baron/The Image Finders; **617** (l)C. Squared Studios/PD, (c)Larry Lefever from GH, (r)Dominic Oldershaw; **618** MB; **619** Alan Pitcairn from GH; **622** John Evans; **626** (t)Dr. Ryder/Jason Burns/PhotoTake NYC, (b)Dominic Oldershaw; **627** file photo; **628** Jeff Vanuga/CB; **629** Eric Hartmann/Magnum Photos; **630** (t)Alan Pitcairn from GH, (c)John D. Cunningham/VU; **631** (l)David Woodfall/Stone, (r)DM; **636-637** Steve Murray/PQ; **637** Davis Meltzer; **638** NASA; **638-639** NASA/Roger Ressmeyer/CB; **639** CB; **640** (l)Weinberg-Clark/The Image Bank, (r)Stephen Marks/The Image Bank; **641** (l)PE, (r)Wernher Krutein/LA; **642** Chuck Place/SB; **643** NASA; **644** (t)Roger Ressmeyer/CB, (b)Simon Fraser/Science Photo Library/PR; **645** Raphael Gaillarde/LA; **646** (t)IC, (b)Diane Graham-Henry & Kathleen Culbert-Aguilar; **647** NASA; **648** NASA/Science Photo Library/PR; **649** NASA; **650** (tl tr) NASA/Science Source/PR, (bl)Julian Baum/Science Photo Library/PR, (br)M. Salaber/LA; **651** (t, l to r)Dorling Kindersley Images, TASS from Sovfoto, NASA, NASA/JPL, (c, l to r)NASA/JPL/Caltech, NASA/JPL/Caltech, NASA, NASA/JPL, (bl br)NASA; **652** AFP/CB; **653** NASA; **654** NASA/Science Source/PR; **655** NASA/LA; **656** (t)NASA, (b)NASA/LA; **657** NASA/Science Source/PR; **658** NASA/JPL/Malin Space Science Systems; **659** NASA/JPL/LA; **660** (t)David Ducros/Science Photo Library/PR, (b)NASA; **661** HED Foundation/NASA; **662** Roger Ressmeyer/CB; **663** DM; **664-665** Robert McCall; **665** NASA/Science Photo Library/PR; **666** (tl)David Parker/Science Photo Library/PR, (tr)NASA/Science Photo Library/PR, (b)NASA; **667** (l)Novosti/Science Photo Library/PR, (c)Roger K. Burnard, (r)NASA; **670** NASA; **670-671** Chad Ehlers/Stone; **671** Bob Daemmrich; **680** (t)Lick Observatory, (b)Richard J. Wainscoat/Peter Arnold, Inc.; **682** Dr. Fred Espenak/Science Photo Library/PR; **683** Bettmann/CB; **684** NASA; **686** Roger Ressmeyer/CB; **689** BMDO/NRL/LLNL/Science Photo Library/PR; **690** (t)Zuber et al/Johns Hopkins University/NASA/PR, (b)NASA; **691** NASA; **692** MM; **694-695** Cosmo Condina/Stone; **695** Brown Brothers; **696** (t)NASA/Peter Arnold, Inc., (b)NASA; **700** PD; **700-701** Ted Thai/TimePix; **701** MM; **704** John R. Foster/PR; **705** Davis Meltzer; **706** Bettmann/CB; **708** USGS/Science Photo Library/PR; **709** (t)NASA/PR, (b)JPL/TSADO/TSA; **710** (t)Science Photo Library/PR, (bl)USGS/TSADO/TSA, (bc br) USGS/TSA; **711** NASA/JPL/Malin Space Science Systems; **713** Science Photo Library/PR; **714** (l)NASA/Science Photo Library/PR, (r)CB; **715** (t)USGS/TSADO/TSA, (tr)NASA/JPL/PR, (bl)JPL, (bc)TSADO/NASA/TSA, (br)NASA; **716** JPL; **717** Heidi Hammel/NASA; **718** (l)NASA/Science Source/PR, (r)JPL/NASA/TSADO/TSA; **719** CB; **720** (tl)NASA/JPL/TSADO/TSA, (tr)NASA/Science Source/PR, (bl)USGS/NASA/TSADO/TSA, (br)CB; **721** (tl)NASA/Science Photo Library/PR, (tr)NASA/Science Source/PR, (c)ASP/Science Source/PR, (bl)CB, (br)W. Kaufmann/JPL/Sscience Source/PR; **722 723** Pekka Parviainen/Science Photo Library/PR; **724** Georg Gerster/PR; **725** JPL/TSADO/TSA; **727** Bettmann/CB; **728 729** Smithsonian National Museum of Natural History; **730** (t)NASA, (c)John Thomas/Science Photo Library/PR, (b)NASA/JPL; **731** (l)NASA; **731** (lc)JPL/NASA; **731** (l)NASA, (lc)JPL/NASA, (rc)file photo; **731** (r)NASA; **734** Science Photo Library/PR; **734-735** Space Telescope Science Institute/NASA/Science Photo Library/PR; **735** First Image; **739** Bob Daemmrich; **742** (t)Carnegie Institution of Washington, (b)NSO/SEL/Roger Ressmeyer/CB; **743** (tl)NASA, (tr)Picture Press/CB, (bl)AFP/CB, (br)Bryan & Cherry Alexander/PR; **744** Celestial Image Co./Science Photo Library/PR; **745** Tim Courlas; **747** Luke Dodd/Science Photo Library/PR; **750** AFP/CB; **751** NASA; **753** (t)Kitt Peak National Observatory, (b)CB; **756** Stephen R. Wagner; **757** AFP/CB; **758** MM; **760** Dennis Di Cicco/Peter Arnold, Inc.; **761** Bill Ross/CB; **762** (t)CB, (b)file photo; **763** (l)CB, (r)AURA/STScl/NASA; **768-769** PD; **770** Garry D. McMichael/PR; **770-771** Harvey Wood/The Still Moving Picture Co.; **771** (t c)AH, (b)Daniel Chester French/CB; **772** (t)SuperStock, (bl)Lindsay Hebberd/CB, (br)PhotoTake NYC/PQ; **773** (t)Ric Ergenbright/CB, (bl)IC, (br)Raymond Gehman/CB; **775** (t)Fred Habegger from GH, (b)Larry Lefever from GH; **776** (t)Murray & Associates, Inc./Picturesque/PQ, (c)Digital Vision/PQ, (b)Frank M. Hanna/VU; **777** (t)NASA/GH, (b)Yann Arthus-Bertrand/CB; **778** CNES/PR; **779** (t)Neil E. Johnson, (b)Lloyd Cluff/CB; **780** (t)Tom Bean/DRK, (b)Buddy Mays/CB; **781** (t)Joe Cornish/Stone, (b)SuperStock; **782** Townsend P. Dickinson/The Image Works; **783** (tl)A.J. Copley/VU, (tr)Townsend P. Dickinson/The Image Works, (bl)Stephen J. Krasemann/DRK, (br)Ken Lucas/VU; **784** (tl)Townsend P. Dickinson/The ImageWorks, (tr)Sinclair Stammers/Science Photo Library/PR, (bl)John Elk III/SB, (br)James L. Amos/PR; **787** (t)Ruth Dixon, (c)Bryan Pickering: Eye Ubiquitous/CB, (b)Rod Currie/Stone; **788** (t)Warren Faidley/Weatherstock, (c)James N. Westwater, (b)Steve Austin: Papilio/CB; **789** (tl)MB, (tr)Chinch Gryniewicz: Ecoscene/CB, (bl)Annie Griffiths Belt/CB, (br)James N. Westwater; **790** Tony Freeman/PE/PQ; **791** (t)Dominic Oldershaw, (b)Mary Kate Denny/PE; **792** (t)file photo, (b)John Evans; **793** (t)KS, (b)Ken Lax; **798** Michell D. Bridwell/PE; **802** David Young-Wolff/PE; **804** Kaz Chiba/PD; **805** Dominic Oldershaw; **806** StudiOhio; **807** MM; **809** (bl)Elaine Shay, (br)Brent Turner/BLT Productions, (others)Mark Steinmetz; **812** Paul Barton/TSM; **815** Davis Barber/PE.

Acknowledgments

From "Landscape, History and the Pueblo Imagination" by Leslie Marmon Silko. Copyright © 1986 by Leslie Marmon Silko, reprinted by permission of The Wylie Agency, Inc. "Listening In" by Gordon Judge. Reprinted by permission of the author. "Song of the Sky Loom" from *Wearing the Morning Star*. Reprinted by permission Brian Swann. "The Microscope" by Maxine Kumin. Copyright © 1963 by The Atlantic Monthly Company, Boston, Mass. Reprinted by permission. "The Jungle of Ceylon" by Pablo Neruda, from PASSIONS AND IMPRESSIONS by Pablo Neruda, translated by Margaret Sayers Peden. Translation copyright © 1983 by Farrar, Straus & Giroux. Reprinted by permission of Farrar, Straus & Giroux.

PERIODIC TABLE OF THE ELEMENTS

Columns of elements are called groups. Elements in the same group have similar chemical properties.

Gas

Liquid

Solid

Synthetic

Element — Hydrogen
Atomic number — 1
Symbol — H
Atomic mass — 1.008
State of matter

The first three symbols tell you the state of matter of the element at room temperature. The fourth symbol identifies human-made, or synthetic, elements.

1

	1	**2**	**3**	**4**	**5**	**6**	**7**	**8**	**9**
1	Hydrogen 1 **H** 1.008								
2	Lithium 3 **Li** 6.941	Beryllium 4 **Be** 9.012							
3	Sodium 11 **Na** 22.990	Magnesium 12 **Mg** 24.305							
4	Potassium 19 **K** 39.098	Calcium 20 **Ca** 40.078	Scandium 21 **Sc** 44.956	Titanium 22 **Ti** 47.867	Vanadium 23 **V** 50.942	Chromium 24 **Cr** 51.996	Manganese 25 **Mn** 54.938	Iron 26 **Fe** 55.845	Cobalt 27 **Co** 58.933
5	Rubidium 37 **Rb** 85.468	Strontium 38 **Sr** 87.62	Yttrium 39 **Y** 88.906	Zirconium 40 **Zr** 91.224	Niobium 41 **Nb** 92.906	Molybdenum 42 **Mo** 95.94	Technetium 43 **Tc** (98)	Ruthenium 44 **Ru** 101.07	Rhodium 45 **Rh** 102.906
6	Cesium 55 **Cs** 132.905	Barium 56 **Ba** 137.327	Lanthanum 57 **La** 138.906	Hafnium 72 **Hf** 178.49	Tantalum 73 **Ta** 180.948	Tungsten 74 **W** 183.84	Rhenium 75 **Re** 186.207	Osmium 76 **Os** 190.23	Iridium 77 **Ir** 192.217
7	Francium 87 **Fr** (223)	Radium 88 **Ra** (226)	Actinium 89 **Ac** (227)	Rutherfordium 104 **Rf** (261)	Dubnium 105 **Db** (262)	Seaborgium 106 **Sg** (266)	Bohrium 107 **Bh** (264)	Hassium 108 **Hs** (277)	Meitnerium 109 **Mt** (268)

The number in parentheses is the mass number of the longest lived isotope for that element.

Rows of elements are called periods. Atomic number increases across a period.

The arrow shows where these elements would fit into the periodic table. They are moved to the bottom of the page to save space.

Lanthanide series

Cerium 58 **Ce** 140.116	Praseodymium 59 **Pr** 140.908	Neodymium 60 **Nd** 144.24	Promethium 61 **Pm** (145)	Samarium 62 **Sm** 150.36

Actinide series

Thorium 90 **Th** 232.038	Protactinium 91 **Pa** 231.036	Uranium 92 **U** 238.029	Neptunium 93 **Np** (237)	Plutonium 94 **Pu** (244)